Lecture Notes in Networks and Systems

Volume 653

The series "Lecture Notes in Networks and Systems" publishes the latest developments in Networks and Systems—quickly, informally and with high quality. Original research reported in proceedings and post-proceedings represents the core of LNNS.

Volumes published in LNNS embrace all aspects and subfields of, as well as new challenges in, Networks and Systems.

The series contains proceedings and edited volumes in systems and networks, spanning the areas of Cyber-Physical Systems, Autonomous Systems, Sensor Networks, Control Systems, Energy Systems, Automotive Systems, Biological Systems, Vehicular Networking and Connected Vehicles, Aerospace Systems, Automation, Manufacturing, Smart Grids, Nonlinear Systems, Power Systems, Robotics, Social Systems, Economic Systems and other. Of particular value to both the contributors and the readership are the short publication timeframe and the world-wide distribution and exposure which enable both a wide and rapid dissemination of research output.

The series covers the theory, applications, and perspectives on the state of the art and future developments relevant to systems and networks, decision making, control, complex processes and related areas, as embedded in the fields of interdisciplinary and applied sciences, engineering, computer science, physics, economics, social, and life sciences, as well as the paradigms and methodologies behind them.

Indexed by SCOPUS, INSPEC, WTI Frankfurt eG, zbMATH, SCImago.

All books published in the series are submitted for consideration in Web of Science.

For proposals from Asia please contact Aninda Bose (aninda.bose@springer.com).

Swagatam Das · Snehanshu Saha ·
Carlos A. Coello Coello · Jagdish Chand Bansal
Editors

Advances in Data-driven Computing and Intelligent Systems

Selected Papers from ADCIS 2022, Volume 2

Springer

Editors
Swagatam Das
Department of Electronics
and Communication Sciences
Indian Statistical Institute
Kolkata, West Bengal, India

Snehanshu Saha
Department of Computer Science
and Information Systems
Birla Institute of Technology and Science
Goa, India

Carlos A. Coello Coello
Department of Computer Science
CINVESTAV-IPN
Mexico, Mexico

Jagdish Chand Bansal
Department of Mathematics
South Asian University
New Delhi, Delhi, India

ISSN 2367-3370 ISSN 2367-3389 (electronic)
Lecture Notes in Networks and Systems
ISBN 978-981-99-0980-3 ISBN 978-981-99-0981-0 (eBook)
https://doi.org/10.1007/978-981-99-0981-0

This Springer imprint is published by the registered company Springer Nature Singapore Pte Ltd.
The registered company address is: 152 Beach Road, #21-01/04 Gateway East, Singapore 189721, Singapore

Preface

This book contains outstanding research papers as the proceedings of the International Conference on *Advances in Data-driven Computing and Intelligent Systems* (ADCIS 2022), at BITS Pilani, K K Birla Goa Campus, India, under the technical sponsorship of the Soft Computing Research Society, India. The conference is conceived as a platform for disseminating and exchanging ideas, concepts, and results of researchers from academia and industry to develop a comprehensive understanding of the challenges of the advancements of intelligence in computational viewpoints. This book will help in strengthening congenial networking between academia and industry. We have tried our best to enrich the quality of the ADCIS 2022 through the stringent and careful peer-reviewed process. This book presents novel contributions to intelligent systems and serves as reference material for data-driven computing.

We have tried our best to enrich the quality of the ADCIS 2022 through a stringent and careful peer-reviewed process. ADCIS 2022 received many technical contributed articles from distinguished participants from home and abroad. ADCIS 2022 received 687 research submissions from 26 different countries, viz., Bangladesh, Belgium, Brazil, Canada, Germany, India, Indonesia, Iran, Ireland, Italy, Japan, Mexico, Morocco, Nigeria, Oman, Poland, Romania, Russia, Saudi Arabia, Serbia, South Africa, South Korea, Sri Lanka, United Arab Emirates, USA, and Viet Nam. After a very stringent peer-reviewing process, only 132 high-quality papers were finally accepted for presentation and the final proceedings.

This book presents second volume of 66 research papers data science and applications and serves as a reference material for advanced research.

Mexico	Carlos A. Coello Coello
Kolkata, India	Swagatam Das
Goa, India	Snehanshu Saha
New Delhi, India	Jagdish Chand Bansal

About This Book

This book contains outstanding research papers as the proceedings of the International Conference on *Advances in Data-driven Computing and Intelligent Systems* (ADCIS 2022), at BITS Pilani, K K Birla Goa Campus, India, under the technical sponsorship of the Soft Computing Research Society, India. The conference is conceived as a platform for disseminating and exchanging ideas, concepts, and results of researchers from academia and industry to develop a comprehensive understanding of the challenges of the advancements of intelligence in computational viewpoints. This book will help in strengthening congenial networking between academia and industry. We have tried our best to enrich the quality of the ADCIS 2022 through the stringent and careful peer-reviewed process. This book presents novel contributions to intelligent systems and serves as reference material for data-driven computing.

The topics covered: artificial intelligence and machine learning, pattern recognition and analysis, parallel and distributed algorithms, natural language processing and machine translation, emotional intelligence, computational engineering, control and robotics, etc.

Mexico — Carlos A. Coello Coello
Kolkata, India — Swagatam Das
Goa, India — Snehanshu Saha
New Delhi, India — Jagdish Chand Bansal

Contents

Editors and Contributors

About the Editors

Swagatam Das received the B.E. Tel.E., M.E. Tel.E. (Control Engineering specialization) and Ph.D. degrees, all from Jadavpur University, India, in 2003, 2005, and 2009 respectively. Swagatam Das is currently serving as an associate professor and Head of the Electronics and Communication Sciences Unit of the Indian Statistical Institute, Kolkata, India. His research interests include evolutionary computing and machine learning. Dr. Das has published more than 300 research articles in peer-reviewed journals and international conferences. He is the founding co-editor-in-chief of *Swarm and Evolutionary Computation*, an international journal from Elsevier. He has also served as or is serving as the associate editors of the *IEEE Transactions on Cybernetics*, *Pattern Recognition* (Elsevier), *Neurocomputing* (Elsevier), *Information Sciences* (Elsevier), *IEEE Transactions on Systems, Man, and Cybernetics: Systems*, and so on. He is an editorial board member of *Information Fusion* (Elsevier), *Progress in Artificial Intelligence* (Springer), *Applied Soft Computing* (Elsevier), *Engineering Applications of Artificial Intelligence* (Elsevier), and *Artificial Intelligence Review* (Springer). Dr. Das has 25,000+ Google Scholar citations and an H-index of 76 till date. He has been associated with the international program committees and organizing committees of several reputed international conferences including NeurIPS, AAAI, AISTATS, ACM Multimedia, BMVC, IEEE CEC, GECCO, etc. He has acted as guest editors for special issues in journals like *IEEE Transactions on Evolutionary Computation* and *IEEE Transactions on SMC, Part C*. He is the recipient of the 2012 Young Engineer Award from the Indian National Academy of Engineering (INAE). He is also the recipient of the 2015 Thomson Reuters Research Excellence India Citation Award as the highest cited researcher from India in Engineering and Computer Science category between 2010 to 2014.

Snehanshu Saha holds Master's Degree in Mathematical and Computational Sciences at Clemson University, USA and Ph.D. from the Department of Applied

Mathematics at the University of Texas at Arlington in 2008. He was the recipient of the prestigious Dean's Fellowship during Ph.D. and Summa Cum Laude for being in the top of the class. After working briefly at his Alma matter, Snehanshu moved to the University of Texas El Paso as a regular full-time faculty in the Department of Mathematical Sciences, University of Texas El Paso. Currently, He is a professor of Computer Science and Engineering at PES University since 2011 and heads the Center for AstroInformatis, Modeling and Simulation. He is also a visiting Professor at the department of Statistics, University of Georgia, USA and BTS Pilani, India. He has published 90 peer-reviewed articles in top-tier international journals and conferences and authored three text books on *Differential Equations*, *Machine Learning* and *System Sciences* respectively. Dr. Saha is an IEEE Senior member, ACM Senior Member, Vice Chair-International Astrostatistics Association and Chair, IEEE Computer Society Bangalore Chapter and Fellow of IETE. He's Editor of the *Journal of Scientometric Research.* Dr Saha is the recipient of PEACE Award for his foundational contributions in Machine Learning and AstroInformatics. Dr. Saha's current and future research interests lie in Data Science, Astronomy and theory of Machine Learning.

Carlos A. Coello Coello (Fellow, IEEE) received the Ph.D. degree in computer science from Tulane University, New Orleans, LA, USA, in 1996. He is currently a Professor with Distinction (CINVESTAV-3F Researcher), Computer Science Department, CINVESTAV-IPN, Mexico City, Mexico. He has authored and coauthored over 500 technical papers and book chapters. He has also co-authored the book *Evolutionary Algorithms for Solving Multiobjective Problems* (2nd ed., Springer, 2007) and has edited 3 more books with publishers such as World Scientific and Springer. His publications currently report over 60,000 citations in Google Scholar (his H-index is 96). His major research interests are evolutionary multiobjective optimization and constraint-handling techniques for evolutionary algorithms. He has received several awards, including the National Research Award (in 2007) from the Mexican Academy of Science (in the area of exact sciences), the 2009 Medal to the Scientific Merit from Mexico City's congress, the Ciudad Capital: Heberto Castillo 2011 Award for scientists under the age of 45, in Basic Science, the 2012 Scopus Award (Mexico's edition) for being the most highly cited scientist in engineering in the 5 years previous to the award and the 2012 National Medal of Science in Physics, Mathematics and Natural Sciences from Mexico's presidency (this is the most important award that a scientist can receive in Mexico). He also received the Luis Elizondo Award from the Tecnológico de Monterrey in 2019. Additionally, he is the recipient of the 2013 IEEE Kiyo Tomiyasu Award, "for pioneering contributions to single- and multiobjective optimization techniques using bioinspired metaheuristics", of the 2016 The World Academy of Sciences (TWAS) Award in "Engineering Sciences", and of the 2021 IEEE Computational Intelligence Society Evolutionary Computation Pioneer Award. Since January 2011, he is an IEEE Fellow. He is currently the Editor-in-Chief of the *IEEE Transactions on Evolutionary Computation.*

Dr. Jagdish Chand Bansal is an Associate Professor at South Asian University New Delhi and Visiting Faculty at Maths and Computer Science, Liverpool Hope University UK. Dr. Bansal has obtained his Ph.D. in Mathematics from IIT Roorkee. Before joining SAU New Delhi he has worked as an Assistant Professor at ABV-Indian Institute of Information Technology and Management Gwalior and BITS Pilani. His Primary area of interest is Swarm Intelligence and Nature Inspired Optimization Techniques. Recently, he proposed a fission-fusion social structure-based optimization algorithm, Spider Monkey Optimization (SMO), which is being applied to various problems from engineering domain. He has published more than 70 research papers in various international journals/conferences. He is the section editor of the journal *MethodsX* published by Elsevier. He is the series editor of the book series Algorithms for Intelligent Systems (AIS) and Studies in Autonomic, Data-driven and Industrial Computing (SADIC) published by Springer. He is the editor in chief of *International Journal of Swarm Intelligence* (IJSI) published by Inderscience. He is also the Associate Editor of *Engineering Applications of Artificial Intelligence* (EAAI) and *ARRAY* published by Elsevier. He is the general secretary of Soft Computing Research Society (SCRS). He has also received Gold Medal at UG and PG level.

Contributors

Bitthal Acharya Department of Computer Science and Engineering, Rajasthan Technical University, Kota, Rajasthan, India

Abhiroop Agarwal Thapar Institute of Engineering and Technology, Patiala, Punjab, India

Preeti Aggarwal UIET, Panjab University, Chandigarh, India

Farruk Ahmed Independent University Bangladesh, Dhaka, Bangladesh

Mushtaq Ahmed Department of Computer Science and Engineering, Malaviya National Institute of Technology Jaipur, Jaipur, India

Neelu Jyothi Ahuja University of Petroleum and Energy Studies (UPES), Dehradun, India

Pratijnya Ajawan Department of ECE, KLS Gogte Institute of Technology, Belagavi, Karnataka, India

S. Akash Electronics and Communication Engineering, Velammal Institute of Technology, Thiruvallur, India

S. Amutha School of Computer Science and Engineering, SCOPE, Vellore Institute of Technology, Chennai, India

Anuj Arora Amity University Rajasthan, Jaipur, Rajasthan, India

Rashmi Arora Guru Tegh Bahdur Institute of Technology, New Delhi, India

Daniel Marcelo González Arriaga Facultad de Ciencias de la Computación, Benemérita Universidad Autónoma de Puebla, Puebla, Mexico

Ravali Attivilli BITS Pilani, Dubai, UAE

Chandrajit Bajaj Department of Computer Science and Oden Institute of Engineering and Sciences, University of Texas, Austin, TX, USA

Prashant Bartakke Department of Electronics and Telecommunication, College of Engineering, Pune, India

Swami Nisha Bhagirath Manipal University Jaipur, Jaipur, Rajasthan, India

Kaushik Bharadwaj NIT Kurukshetra, Haryana, India

Vaibhav Bhatnagar Manipal University Jaipur, Jaipur, Rajasthan, India

R. Bhuvana Department of Biomedical Engineering, VISTAS, Pallavaram, Chennai, Tamil Nadu, India

M. Bindhu ECE, Sri Venkateshwara College of Engineering, Sriperumbudur, Chennai, Tamil Nadu, India

Vivek Bongale Department of Computer Science and Engineering, Presidency University, Bengaluru, India

Sunita Chahar Electrical Engineering Department, RTU, Kota, India

Priyam Chakraborty Happymonk AI Labs, Bengaluru, Karnataka, India

Chhagan Charan NIT Kurukshetra, Haryana, India

Basavraj Chinagundi Thapar Institute of Engineering and Technology, Patiala, Punjab, India

Aditya Choudhary Department of Civil Engineering, Malaviya National Institute of Technology Jaipur, Jaipur, India

Sagar Choudhury National Institute of Technology Karnataka, Surathkal, India

Sandeep Dalal Department of Computer Science and Applications, MDU, Rohtak, India

Ayan Kumar Das Birla Institute of Technology, Patna Campus, Mesra, Patna, India

Deepanwita Das National Institute of Technology, Durgapur, Durgapur, West Bengal, India

Sudha Das Khan National Institute of Technology Durgapur, Durgapur, India; BIT Sindri, Dhanbad, Jharkhand, India

Aloke Kumar Datta National Institute of Technology Durgapur, Durgapur, West Bengal, India

Carlos Leopoldo Carreón Díaz de León Facultad de Ciencias de la Computación, Benemérita Universidad Autónoma de Puebla, Puebla, Mexico

A. R. Deepa Department of Computer Science and Engineering, Koneru Lakshmaiah Education Foundation, Andhra Pradesh, India

Veena Desai Department of ECE, KLS Gogte Institute of Technology, Belagavi, Karnataka, India

Sandipan Dhar National Institute of Technology Durgapur, Durgapur, India

Veena Dhayal Department of Chemistry, Manipal University Jaipur, Jaipur, India

Kaushik Doddamani Department of ECE, KLS Gogte Institute of Technology, Belagavi, Karnataka, India

Alric Duarte Department of Electrical and Electronics Engineering Birla Institute of Technology and Science Pilani, Dubai, UAE

Sarthika Dutt COER University, Roorkee, India;
University of Petroleum and Energy Studies (UPES), Dehradun, India

Tran Quang Duy School of Industrial Engineering and Management, International University. Vietnam National, University, HCM City, Vietnam

E. Elakkiya SRM University, Amaravati, Andhra Pradesh, India

Amit Ganatra Department of Computer Science and Engineering, Faculty of Technology and Engineering (FTE), Devang Patel Institute of Advance Technology and Research (DEPSTAR), Charotar University of Science and Technology (CHARUSAT), Anand, India

R. Geetha Tamil Nadu E-Governance Agency, Chennai, Tamil Nadu, India

Arjun Ghosh National Institute of Technology Durgapur, Durgapur, India

C. N. Gireesh Babu BMS Institute of Technology and Management, Bangalore Affiliated to Visvesvaraya Technological University, Belagavi, Karnataka, India

Parth Goel Department of Computer Science and Engineering, Faculty of Technology and Engineering (FTE), Devang Patel Institute of Advance Technology and Research (DEPSTAR), Charotar University of Science and Technology (CHARUSAT), Anand, India

Rajlakshmi Gogoi Jorhat Engineering College, Jorhat, India

Jesús López Gómez División Académica de Ingeniería y Arquitectura, Universidad Juárez Autónoma de Tabasco, Villahermosa, Mexico

Dinesh Gopalani Department of Computer Science and Engineering, Malaviya National Institute of Technology Jaipur, Jaipur, India

Jitendra Goyal Department of Computer Science and Engineering, Malaviya National Institute of Technology Jaipur, Jaipur, India

Pranshu Goyal Thapar Institute of Engineering and Technology, Patiala, Punjab, India

Aastha Gupta Department of Mathematics, Punjab Engineering College, Chandigarh, India

Cota Navin Gupta Neural Engineering Lab, Department of Bioscience and Bioengineering, IIT Guwahati, Assam, India

Lakshya Gupta School of Computer Science and Engineering, Vellore Institute of Technology, Vellore, Tamil Nadu, India

A. G. Guru Dutt BMS Institute of Technology and Management, Bangalore Affiliated to Visvesvaraya Technological University, Belagavi, Karnataka, India

K. S. Harishkumar Department of Computer Science and Engineering, Presidency University, Bengaluru, India

Kalpajyoti Hazarika Neural Engineering Lab, Department of Bioscience and Bioengineering, IIT Guwahati, Assam, India

R. J. Hemalatha Department of Biomedical Engineering, VISTAS, Pallavaram, Chennai, Tamil Nadu, India

Mare Jagapathi Department of Information Technology, Velagapudi Ramakrishna Siddhartha Engineering College, Vijaywada, Andhra Pradesh, India

Mushrafa Jahan Suha Independent University Bangladesh, Dhaka, Bangladesh

Shubham K. Jain Department of Civil Engineering, Malaviya National Institute of Technology Jaipur, Jaipur, India

Nanda Dulal Jana National Institute of Technology Durgapur, Durgapur, India

Lakshminarayana Janjanam Department of ECE, National Institute of Technology, Raipur, Chhattisgarh, India

M. Jayapriya College of Engineering, Anna University, Chennai, India

S. Joyal Electrical and Electronics Engineering, Saveetha Engineering College, Chennai, India

Nitika Kadam Computer Science Engineering, Oriental University, Indore, India

Kaushik Kampli Department of Computer Science and Engineering, Ramaiah Institute of Technology MSR Nagar, Bengaluru, India

Rajib Kar Department of ECE, National Institute of Technology, Durgapur, West Bengal, India

Aryan Karchi Department of ECE, KLS Gogte Institute of Technology, Belagavi, Karnataka, India

S. Karthika Sri Sivasubramaniya Nadar College of Engineering, Chennai, Tamil Nadu, India

Arvinder Kaur Guru Gobind Singh Indraprastha University, New Delhi, India

Jatinder Kaur Department of Mathematics, Chandigarh University, Gharuan, Mohali, Punjab, India

Navdeep Kaur MCM DAV College for Women, Chandigarh, India

Abhishek Kaushik Dundalk Institute of Technology, Dundalk, Ireland

D. S. Kaustubha Department of Computer Science and Engineering, Ramaiah Institute of Technology MSR Nagar, Bengaluru, India

Amit Kumar Department of Civil Engineering, Malaviya National Institute of Technology Jaipur, Jaipur, India

Ankit Kumar Department of Computer Engineering and Applications, GLA University, Mathura, India

Ashwani Kumar Department of Civil Engineering, Malaviya National Institute of Technology Jaipur, Jaipur, India

Krishan Kumar UIET, Panjab University, Chandigarh, India

Manish Kumar #4206, Data Analytics Lab, Computer Center-2, Indian Institute of Information Technology Allahabad, Allahabad, Uttar Pradesh, India;
Department of Mechanical Engineering, MNIT JAIPUR, Jaipur, India

Munish Kumar Maharaja Ranjit Singh Punjab Technological University, Bathinda, Punjab, India

Pradeep Kumar Department of Computer Applications, Manipal University Jaipur, Jaipur, India

Sudhir Kumar Department of Civil Engineering, Malaviya National Institute of Technology Jaipur, Jaipur, India

Jyoti Kumari Sardar Vallabhbhai National Institute of Technology, Surat, India

Raj Kumari University Institute of Engineering and Technology, Panjab University, Chandigarh, India

L. G. Lasithan APJ Abdul Kalam Technological University, CET Campus, Thiruvananthapuram, Kerala, India

Sergio Vergara Limon Facultad de Ciencias de la Electrónica, Benemérita Universidad Autónoma de Puebla, Puebla, Mexico

P. Uma Maheswari College of Engineering, Anna University, Chennai, India

Banashree Mandal National Institute of Technology, Durgapur, Durgapur, West Bengal, India

D. H. Manjaiah Department of Computer Science, Mangalore University, Mangalore, India

T. N. Manjunath BMS Institute of Technology and Management, Bangalore Affiliated to Visvesvaraya Technological University, Belagavi, Karnataka, India

Tanvir Singh Mann UIET, Panjab University, Chandigarh, India

Manish Manohar Department of Computer Science and Engineering, Ramaiah Institute of Technology MSR Nagar, Bengaluru, India

Junaid Maqbool UIET, Panjab University, Chandigarh, India

Avirup Mazumder National Institute of Technology Durgapur, Durgapur, India

Gowri Namratha Meedinti School of Computer Science and Engineering, Vellore Institute of Technology, Vellore, Tamil Nadu, India

Shyam Sundar Meena Department of Computer Science and Engineering, Shri Vaishnav Institute of Information Technology, Indore, India

Yash Mehta Department of Computer Science and Engineering, Faculty of Technology and Engineering (FTE), Devang Patel Institute of Advance Technology and Research (DEPSTAR), Charotar University of Science and Technology (CHARUSAT), Anand, India

Ajay Mittal UIET, Panjab University, Chandigarh, India

Murari Lal Mittal Department of Mechanical Engineering, MNIT JAIPUR, Jaipur, India

Biju R. Mohan National Institute of Technology Karnataka, Surathkal, India

Prabhavathy Mohanraj Department of Artificial Intelligence and Data Science, Coimbatore Institute of Technology, Coimbatore, India

Maitri Mohanty Department of Computer Science & Engineering, GIET University, Gunupur, Odisha, India

Ambarish G. Mohapatra Department of Electronics and Instrumentation Engineering, Silicon Institute of Technology, Bhubaneswar, Odisha, India

Sunitha Munappa S R M Institute of Science and Technology, Chennai, India

Gayathri Murugesan College of Engineering, Anna University, Chennai, India

Shruthi Muthukumar College of Engineering, Anna University, Chennai, India

Mohammad Nadeem Department of Computer Science, Aligarh Muslim University, Aligarh, India

Janhavi Namjoshi Department of Mechatronics Engineering, School of Automobile, Mechatronics and Mechanical Engineering (SAMM), Manipal University Jaipur (MUJ), Jaipur, Rajasthan, India

Pranay S. Nankani Department of Electrical and Electronics Engineering Birla Institute of Technology and Science Pilani, Dubai Campus, Dubai, UAE

Bharati Narute Department of Electronics and Telecommunication, M.E.S. College of Engineering, Pune, India

Nishika University Institute of Engineering and Technology, MDU, Rohtak, India

Afraa Noureen BITS Pilani, Dubai, UAE

Fariya Oyshi Independent University Bangladesh, Dhaka, Bangladesh

Manojkumar Pal Department of Mechanical Engineering, MNIT JAIPUR, Jaipur, India

S. Parasuraman ECE, Karpaga Vinakaga College of Engineering and Technology, Chengalpattu, Tamil Nadu, India

Rekha Pasupuleti Trent University, Peterborough, ON, Canada

C. S. Pavan Kumar Department of Information Technology, Velagapudi Ramakrishna Siddhartha Engineering College, Vijaywada, Andhra Pradesh, India

Banothu Pavankalyan Department of ECE, Vardhaman College of Engineering, Hyderabad, India

Sukonya Phukan Jorhat Engineering College, Jorhat, India

Anannya Popat School of Computer Science and Engineering, Vellore Institute of Technology, Vellore, Tamil Nadu, India

Rajan Prasad Artificial Intelligence Research Center, Department of Computer Science and Engineering, School of Engineering, Babu Banarasi Das University, Lucknow, India

M. K. Praveena Kumari Department of Computer Science, Mangalore University, Mangalore, India

A. Prawin Electronics and Communication Engineering, Velammal Institute of Technology, Thiruvallur, India

S. K. Pushpa BMS Institute of Technology and Management, Bangalore Affiliated to Visvesvaraya Technological University, Belagavi, Karnataka, India

Linesh Raja Department of Computer Applications, Manipal University Jaipur, Jaipur, Rajasthan, India

Gomathi Bhavani Rajagopalan Department of Electrical and Electronics Engineering Birla Institute of Technology and Science Pilani, Dubai Campus, Dubai, UAE

S. Rajarajeswari Department of Computer Science and Engineering, Ramaiah Institute of Technology MSR Nagar, Bengaluru, India

V. G. Rajesh Department of Mechanical Engineering, College of Engineering, Chengannur, Alappuzha, Kerala, India

Saveeth Ramanathan Department of Computer Science Engineering, Coimbatore Institute of Technology, Coimbatore, India

Dipak Ramoliya Department of Computer Science and Engineering, Faculty of Technology and Engineering (FTE), Devang Patel Institute of Advance Technology and Research (DEPSTAR), Charotar University of Science and Technology (CHARUSAT), Anand, India

Dipti P. Rana Sardar Vallabhbhai National Institute of Technology, Surat, India

Prashant Singh Rana Thapar Institute of Engineering and Technology, Patiala, Punjab, India

Pratik Ranjan Motihari College of Engineering, Motihari, India

Rajeev Ranjan Bakhtiyarpur College of Engineering, Bakhtiyarpur, India

Swami Ranjan Birla Institute of Technology, Patna Campus, Patna, India

Vaibhavi Sachin Rao BITS Pilani, Dubai, UAE

Jawaad Rashid Independent University Bangladesh, Dhaka, Bangladesh

Premansu Sekhara Rath Department of Computer Science & Engineering, GIET University, Gunupur, Odisha, India

Deeksha Ratnawat Department of Computer Science and Engineering, Malaviya National Institute of Technology Jaipur, Jaipur, India

Ashwin Raut #4206, Data Analytics Lab, Computer Center-2, Indian Institute of Information Technology Allahabad, Allahabad, Uttar Pradesh, India

Manish Rawat Department of Mechatronics Engineering, School of Automobile, Mechatronics and Mechanical Engineering (SAMM), Manipal University Jaipur (MUJ), Jaipur, Rajasthan, India

Avanghapuram Surya Prakash Reddy Department of ECE, Vardhaman College of Engineering, Hyderabad, India

Duddugunta Bharath Reddy Department of Information Technology, Velagapudi Ramakrishna Siddhartha Engineering College, Vijaywada, Andhra Pradesh, India

K. A. Rohith Kumar Electronics and Communication Engineering, Velammal Institute of Technology, Thiruvallur, India

Asis Kumar Rout National Institute of Technology Karnataka, Surathkal, India

Swarup Roy National Institute of Technology Durgapur, Durgapur, India

Suman Kumar Saha Department of ECE, National Institute of Technology, Raipur, Chhattisgarh, India

Ajay Saini Central Analytical Facilities, Manipal University Jaipur, Jaipur, India

Shiv Sajan Saini Post Graduate Institute of Medical Education and Research, Chandigarh, India

Mohammad Sajid Department of Computer Science, Aligarh Muslim University, Aligarh, India

Uma Maheswari Sankareswaran Department of Electronics and Communication Engineering, Coimbatore Institute of Technology, Coimbatore, India

Madhumita Sardar Haji Md. Serafat Mondal Government Polytechnic, Birbhum, West Bengal, India

V. N. Sastry IDRBT, Hyderabad, India

Manas P. Shankar Department of Computer Science and Engineering, Ramaiah Institute of Technology MSR Nagar, Bengaluru, India

Harish Sharma Department of Computer Science and Engineering, Rajasthan Technical University, Kota, Rajasthan, India

Nirmala Sharma Department of Computer Science and Engineering, Rajasthan Technical University, Kota, Rajasthan, India

Sanjeev Kumar Sharma Computer Science Engineering, Oriental University, Indore, India

Shweta Sharma Manipal University Jaipur, Jaipur, Rajasthan, India

Shathanaa Rajmohan Indian Institute of Information Technology, Sri City, India

Anubhav Shivhare #4206, Data Analytics Lab, Computer Center-2, Indian Institute of Information Technology Allahabad, Allahabad, Uttar Pradesh, India

Sakshi Shringi Department of Computer Science and Engineering, Manipal University Jaipur, Jaipur, Rajasthan, India

Vivek Shrivastava International Institute of Professional Studies, Devi Ahilya University, Indore, Madhya Pradesh, India

P. V. Shouri Department of Mechanical Engineering, Model Engineering College, Thrikkakara, Cochin, Kerala, India

Praveen Kumar Shukla Artificial Intelligence Research Center, Department of Computer Science and Engineering, School of Engineering, Babu Banarasi Das University, Lucknow, India

Gagandeep Kaur Sidhu Department of Mathematics, Chandigarh University, Gharuan, Punjab, India

Dalip Singh Department of Automobile Engineering, Manipal University Jaipur, Jaipur, India

Namrata Singh Birla Institute of Technology, Mesra, Patna, India

Parth Singh Department of Computer Science and Engineering, Faculty of Technology and Engineering (FTE), Devang Patel Institute of Advance Technology and Research (DEPSTAR), Charotar University of Science and Technology (CHARUSAT), Anand, India

Pavitdeep Singh Natwest Group, Gurgaon, India

Pramod Kumar Singh #4206, Data Analytics Lab, Computer Center-2, Indian Institute of Information Technology Allahabad, Allahabad, Uttar Pradesh, India

Kriti Singhal Thapar Institute of Engineering and Technology, Patiala, Punjab, India

Kamna Solanki University Institute of Engineering and Technology, MDU, Rohtak, India

Gunjan Soni Department of Mechanical Engineering, MNIT JAIPUR, Jaipur, India

S. R. Sreeja Indian Institute of Information Technology, Sri City, India

Kolasani Sai Sri Lekha Department of Information Technology, Velagapudi Ramakrishna Siddhartha Engineering College, Vijaywada, Andhra Pradesh, India

B. Sridevi Electronics and Communication Engineering, Velammal Institute of Technology, Thiruvallur, India

J. Subhashini S R M Institute of Science and Technology, Chennai, India

Pallikonda Sarah Suhasini Velagapudi Ramakrishna Siddhartha Engineering College, Vijayawada, India

Pragnesh Thaker National Institute of Technology Karnataka, Surathkal, India

Rohit Kumar Thakur University Institute of Engineering and Technology, Panjab University, Chandigarh, India

Vrinda Tokekar Department of Information Technology, Institute of Engineering and Technology, Indore, India

Pijush Topdar National Institute of Technology Durgapur, Durgapur, West Bengal, India

María Aurora D. Vargas Treviño Facultad de Ciencias de la Electrónica, Benemérita Universidad Autónoma de Puebla, Puebla, Mexico

Talha Umar Department of Computer Science, Aligarh Muslim University, Aligarh, India

Siddhaling Urolagin BITS Pilani, Dubai, UAE

Madhusudan Rao Veldanda Geethanjali College of Engineering and Technology, Hyderabad, India

Himanshukamal Verma School of Computer Science & IT, Devi Ahilya University, Indore, Madhya Pradesh, India

Tran Duc Vi School of Industrial Engineering and Management, International University. Vietnam National, University, HCM City, Vietnam

A. Vijaya Lakshmi Department of ECE, Vardhaman College of Engineering, Hyderabad, India

Yi Wang Oden Institute of Engineering and Sciences, University of Texas, Austin, TX, USA

Ashwani Kumar Yadav Amity University Rajasthan, Jaipur, Rajasthan, India

D. K. Yadav Electrical Engineering Department, RTU, Kota, India

Sargam Yadav Dublin Business School, Dublin, D02, Ireland

Vaishali Yadav Manipal University Jaipur, Jaipur, Rajasthan, India

Galla Yagnesh Department of Information Technology, Velagapudi Ramakrishna Siddhartha Engineering College, Vijaywada, Andhra Pradesh, India

Yunhao Yang Department of Computer Science and Oden Institute of Engineering and Sciences, University of Texas, Austin, TX, USA

S. Yogeeswran ECE, P.T. Leechengalvarya Naicker College of Engineering and Technology, Veliyur, Tamil Nadu, India

Scenario-Based Neural Network Model for Integrated Lighting Schemes in Residential Buildings

Pranay S. Nankani, Alric Duarte, and Gomathi Bhavani Rajagopalan

Abstract Integrated lighting schemes have become a popular choice for building designers and occupants alike since these schemes help achieve the dual objectives of maximizing visual comfort while minimizing energy consumption in building interiors. Neural network is an intelligent technique that helps capture the complex patterns of occupant behavior to accurately model integrated lighting schemes wherein the available daylight is adequately complemented with artificial lighting to ensure optimal visual comfort for the occupants. The location and scenarios of occupants are taken as important factors in the design of the lighting control system, and simulation and modeling of such a system is a fascinating field of research in the domain of building systems. This paper presents an intelligent simulation model, based on neural networks, for daylight-artificial light integrated schemes to demonstrate the optimization of light energy in a university residence building in Dubai. The models for integrated lighting schemes are to contend with multivariant, nonlinear, and dynamic processes impacted by occupancy, environmental and geographical variables. Hence, the model is simulated in DIALux, considering real-world scenarios taking place in the residential building with high accuracy. Further, the neural network model is trained and tested to predict the levels of the lighting system, and the results exhibited consistently good performance across different architectures with a mean square error value of only 0.00016. When compared with the governmental lighting regulation, the levels predicted by the neural network model satisfy the minimum illuminance levels required for each scenario.

Keywords Building · Integrated lighting · Intelligent lighting · Daylighting · Neural network

P. S. Nankani (✉) · A. Duarte · G. B. Rajagopalan
Department of Electrical and Electronics Engineering Birla Institute of Technology and Science Pilani, Dubai Campus, Dubai, UAE
e-mail: pranaynankani@gmail.com

S. Das et al. (eds.), *Advances in Data-driven Computing and Intelligent Systems*, Lecture Notes in Networks and Systems 653, https://doi.org/10.1007/978-981-99-0981-0_1

1 Introduction

Integrated lighting schemes have become a popular choice for building designers and occupants alike since these schemes help achieve the dual objectives of maximizing visual comfort while minimizing energy consumption in building interiors. These two objectives which were till a couple of decades ago thought to be on the conflicting ends of the spectrum are now seen as part of the same problem, thanks to sustainability thresholds that have been risen by governments as well as the impetus given to research on occupant comfort and behavior.

Given the restrictions that comfort conditions in the interior of the building are satisfied, it becomes obvious that the problem of energy consumption is a multi-dimensional one [1]. Scientists from a variety of fields have been working on this problem for a few decades now; however, essentially it remains an open issue [1]. Many models have been proposed to design systems that fulfill this difficult-to-achieve dual objective. One of the approaches has been to utilize various tools to accurately capture a building's characteristics so as to provide an accurate and reliable basis for the control of its behavior. By collecting information about the past behavior of the lighting subsystem, the control scenario can be enhanced. This is achieved by optimizing the various control options by exploring different scenarios using advanced computational performance simulation algorithms.

Neural network is an intelligent technique that helps capture the complex patterns of human behavior to accurately model integrated lighting schemes wherein the available daylight is adequately complemented with electric lighting to ensure optimal visual comfort for the occupants. Daylight sensors and occupancy sensors are the essential elements in integrated lighting, which can provide control scenarios such as on–off control, dimming, time-based, load shedding at peak times, etc. There are three areas that are significant when it comes to daylighting the buildings: simulation, daylighting control, occupant behavior, and preferences [2]. People have an instinctive tendency to natural light rather than artificially lit spaces in a building and studies confirmed exposure to daylight can have a significant effect on wellbeing by diminishing headaches, eye tension, or stress [3]. Daylight coupled with artificial light can also be a suitable stimulus for setting the circadian rhythm which is crucial for sleep and other functions of the body [4].

The location and activities of occupants are taken as important factors in the design of the lighting control system, and the simulation and modeling of such a system has become an integral part of analysis and control in the domain of building systems. Building Automation Systems (BAS) can have in-built intelligent routines to adjust lighting based on schedule, occupancy, and daylight. Modification, extension, hardware implementation, and control over a large area are the resultant advantages since the system is more distributed over the entire network. This paper presents an intelligent simulation model based on back propagation neural network (BPNN), for daylight-artificial light integrated schemes to demonstrate the optimization of light energy in a university residence building. The simulation has been done with reference to Dubai, United Arab Emirates (UAE), where sufficient daylight is available to illuminate the interiors for major parts of the year.

2 Related Work

There have been studies that have focused on integrated lighting schemes that have used learning models to great benefit. The following researchers have specifically paid attention to predicting the light output. Günaydin et al. [5] in their work implemented the feed-forward model of artificial neural network (ANN) to predict the daylight illuminance paired with a neuro-solution software to analyze the information that was predicted. In [6], Katsanou et al. presented an ANN model that takes into consideration factors such as office orientation, usage of blinds and work plane distance from the window in order to efficiently project the optimum light levels needed to work in an office. Khairul et al. [7] developed an intelligent lighting system for minimizing the energy consumption in an office environment, and they performed daylight simulation in DIALux from 8 am to 5 pm using the clear sky condition. The results of the simulation were used to develop the ANN using the radial basis function network (RBFN) model, and the intelligent lighting system using LEDs reduced the electricity consumption by a third, while also being brighter [7].

Hidayat et al. [8] developed a smart activity-based lighting system using ANN. The proposed model aimed to reduce the energy consumption of the rooms in the Kinanthi dormitory while maintaining the occupant's visual comfort and made use of sensors placed in the dormitory. After simulating, they observed that the existing lighting was lower than that recommended by the Indonesian National Standard (SNI) and hence added a desk lamp and replaced the LED light bulbs with brighter and more efficient ones. The data obtained from the sensors was fed as the input to the neural network, and the ANN was trained using the Levenberg–Marquardt algorithm [8]. Higuera et al. [9] developed an intelligent lighting and climate control system for offices using the wireless sensor and actuator network (WSAN) and MATLAB. The researchers used daylight and occupancy sensors with a fuzzy logic algorithm for optimizing the lighting in the office, and subsequently, used a smart thermostat with a neural network to prevent excess air conditioner usage. The proposed system reduced the energy consumption by 43% while maintaining the occupants' visual and thermal comforts [9].

The work by Seyedolhosseini et al. [10–12] concentrated on developing a zone-based dimmable lighting system. Seyedolhosseini et al. [10] compared two different models, first was a model where a single neural network would control the entire system, and the second one comprised of a neural network for each zone and a state machine. The work opined that the first model performed better as there were common luminaires; however, the second model performed better in the second test bed as the luminaires were separate and since the network size was smaller, the complexity of calculation reduced thereby increasing the energy efficiency [10]. In another work reported by the same researchers [11], a similar model was examined to see the effect of the changing sky conditions on the accuracy of the result. The work used daylight as the bias for the neural network, and the authors emphasized the need to place the photodetectors at the workplace rather than adjacent to the luminaires to get the most accurate output as the difference could be as much as 70%. They found that their proposed system resulted in an error of only 3% which was much lesser than the previous works done in this system [11]. Seyedolhosseini et al. [12] presented two case studies, first with different number of sensors and luminaires, and the second with different daylighting. In the first case, the work reported that the difference between the desired and actual lighting was less than 1% for both work zones, and in the second condition where the daylight varied, it was 5.6%.

Zizhen et al. [13] also developed a lighting control system for LED in an office environment; however, the model took the users' illumination preferences and also considered lighting beyond the regular office hours. The work considered the dimming of the 14 LEDs as input and the illuminance level of the 12 tables as the output. After obtaining 600 sets of input data, the dataset was normalized and split into training (70%), validation (15%), and testing (15%), and the work reported 22% less energy consumption and illuminance of 320 lx was obtained in most areas [13]. The model was also tested for certain specific tables during non-office hours, but some of the constraints of this model were that it did not take natural lighting into consideration, and it was developed, keeping an individual's preferences in mind and hence the layout or preference could not be changed later [13].

Madias et al. [14] proposed a lighting optimization system using genetic algorithm in MATLAB and developed two models, one prioritizing energy saving and another one emphasizing uniformity of lighting; however, the proposed system did not take into account daylight or occupancy in the room. Despite this, the model yielded a reduction in energy by 22% in the first case and an increase in uniformity by 26% in the second case [14]. Zhang et al. [15] developed a smart lamp using BH1750 light sensor and Arduino wherein the user first had to input the dimensions of the room, the color of the walls, etc., and the program developed would then calculate the minimum luminance and send it to the smart lamp where the difference between the current lighting level and the result from the software would be compared every 30 minutes during daytime. This system managed to save 1747 kWh/year and could also be remotely adjusted over the WiFi [15].

In Ji-Qing et al. [16], Guo and Zhang [17], Wu et al. [18], the authors developed bioinspired algorithms for decreasing the energy consumption due to lighting. In [16], Ji-Qing Qu et al. proposed an improved version of the particle swarm optimization algorithm for optimizing the indoor lighting layout. Using a big data analysis algorithm, the authors found that average illuminance, overall uniformity, unified glare rating, lighting power density, and cost were the main constraints for both users and lighting designers alike. The authors designed an application using MATLAB where the users could input the details of the room and the IES file for the luminaire. The application would then output the optimal number of luminaires row-wise and column-wise along with the spacing between each of them. The proposed model offered better performance and faster convergence than the particle swarm optimization algorithm and genetic algorithm, and upon simulation in DIALux a maximum deviation of only 2.2% was found. This model however did not take into consideration daylight and user preferences. In [17], the authors used a radial basis function neural network for calculating the illuminance. Subsequently, a hybrid between the genetic and simulated annealing algorithms was developed for optimization in order

Table 1 Software used in literature

Software	References
DIALux	[7, 8, 10–13, 16–18]
Daysim	[9, 19]
Relux	[14]
Arduino	[15]
MATLAB	[7–14, 16, 17]

to overcome the constraints of the traditional particle swarm optimization algorithm. In [18], Wu et al. modeled a hybrid of the particle swarm optimization algorithm and genetic algorithm for predicting the number of luminaires and illuminance in a residential facility. This combined approach obtained better performance than a traditional ANN and ANFIS model. However, in this paper the months from January to April were not considered, and very few samples were used in the model overall.

A summary of the software used in the above works is given in Table 1. From the table, we can see that DIALux is the most commonly used lighting design software for residential buildings.

From the literature reviewed, we also observe that simulation-assisted lighting control schemes offer unique advantages when combined with supervised learning models. However, there are many models which do not take daylight into account. Daylight is an important natural source of lighting that creates an environment which is not only comfortable to work with, but also plays a significant role in minimizing the energy consumption and developing a realistic model [19]. Using daylight as one of the key lighting elements, this work aims at constructing a model for the hostel rooms of BITS Pilani, Dubai Campus with a focus to minimize the energy consumption while maintaining adequate lighting levels. This model also takes into account the time of the day and the scenario of the occupant towards personal lighting.

3 Background

BITS Pilani, Dubai Campus, is situated in the Academic City and caters to providing undergraduate and graduate engineering education to a diverse population in Dubai. The management is keen to demonstrate energy saving in the academic and hostel blocks so that the buildings would not only be environmentally friendly but also would inspire the student community to adopt sustainable practices in their lives. The institute has six hostel blocks, four blocks for male students and staff, and two for female students and employees with exclusive gym facilities for each. Each block consists of ground plus four stories.

The university buildings have been adopting some energy and water saving practices, and this model looked at the lighting profile of hostel blocks. Most of the lighting fixtures in the hostel are fluorescent fixtures of 2 $\times$ 36W with magnetic ballasts in single occupancy rooms. Most of the fluorescent fixtures in corridors (4 $\times$ 18W) and restaurants have been replaced with LED fixtures. In this work, an intelligent lighting system based on occupant's activity was designed using ANN, whose objective is to optimize energy consumption while maintaining adequate visual comfort. The neural network-based lighting system takes into consideration various parameters and provides an optimized result with just a small fraction of error that is negligible as well as keeps getting better as the model is being used. The model discussed in the study makes use of data that is generated using DIALux, which was used to create the room model; and lighting scenarios for various activities were simulated taking into consideration the visual comfort of the occupant as well as integrating daylight and artificial light to provide a suitable scene for the occupant. The data obtained from DIALux contained lux levels at various locations in the room and along with other parameters such as the location, date, time of the day, and activities of the occupant were fed to the neural network to train it.

DIALux is a popular lighting simulation software used by industry professionals for simulating lighting designs [20]. It is an intuitive software that helps users create real life simulations that help in lighting design, and it has provisions for simulating both indoor and outdoor lighting scenarios with high accuracy. Designers can import an already existing architectural design and overlay walls and furniture as per their individual project requirements and then simulate and calculate the lux levels in the room or architectural structure. Features such as the ability to import the layout from other applications, placing furniture (and adding textures to them), viewing the effect of different sky conditions on the lighting, etc. make it very user-friendly software for researchers as well as designers.

ANN is based on a collection of connected nodes called artificial neurons, mimicking the actual neurons in our body that carries processed information. ANN is a branch of deep learning network that makes use of input data and then processes it in the consequent layers to give the optimized output. Each neuron in a layer that the data passes through has a specific weight associated to it and these weights can be updated after every epoch in order to reduce the error as compared to the target output. Specific parameters such as the learning rate and the momentum constant have to be specified prior to the simulation.

Back propagation neural network (BPNN) used in this work involves three major steps namely the feed-forward stage, back propagation (to check the error), and then the updating of weights and biases. This method is useful as it reduces the total error in the feedback system to a minimum in successive iterations. Figure 1 shows a visual representation of a three-layer BPNN model. Considering we have 'n' neurons in the input layer (and 'i' is any intermediate neuron in it), 'p' neurons in the hidden layer (and 'j' is an intermediate neuron in it), 'm' neurons in the output layer (and 'k' is an intermediate neuron in it), each input unit to hidden unit has a bias 'v', and each hidden unit to output unit has a bias 'w'. The equations of the model would be as follows:

$$Z_j = f(v_{0j} + \sum_{i=1}^{n} x_i v_{ij}) \tag{1}$$

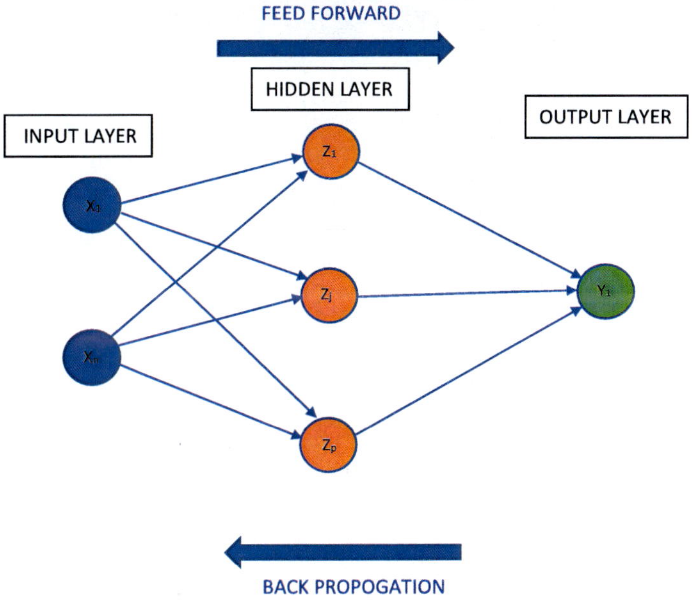

Fig. 1 ANN architecture

In Eq. 1, the hidden unit Z_j sums up its weights (and bias), applies the activation function, and sends the signal to the output layer. Once the output layer receives the signal, it will sum up its weights (and bias), apply the activation function, and compute the output signal which can be seen in Eq. 2. These stages are known as the feed-forward stage.

$$Y_k = f(w_{0k} + \sum_{j=1}^{p} z_j w_{jk}) \tag{2}$$

Now, comparing the output signal with the target vector (t_k) we find the weight adjustment for the output unit (δ_k) as well as for the hidden unit (δ_j) as seen in Eqs. 3 and 4 respectively (f' represents the derivative of the activation function).

$$\delta_k = (t_k - y_k) \cdot f'\left(w_{0k} + \sum_{j=1}^{p} z_j w_{jk}\right) \tag{3}$$

$$\delta_j = \sum_{k=1}^{m} \delta_k w_{jk} \cdot f'\left(v_{0j} + \sum_{i=1}^{n} x_i v_{ij}\right) \tag{4}$$

When η is considered as the learning rate, Eq. 5 represents the weight correction between the hidden layer and output layer, and Eq. 6 represents the weight correction between the input layer and hidden layer. Equations 3, 4, 5, and 6 form the backpropagation of error stage.

$$\Delta W_{jk} = \eta \cdot \delta_k \cdot z_j \tag{5}$$

$$\Delta V_{ij} = \eta \cdot \delta_j \cdot x_i \tag{6}$$

The old weights are updated using Eqs. 7 and 8. As the new weights are obtained, it is fed back into the system until the error is minimized and brought to the least possible value.

$$W_{jk}(\text{new}) = W_{jk}(\text{old}) + \Delta W_{jk} \tag{7}$$

$$V_{ij}(\text{new}) = V_{ij}(\text{old}) + \Delta V_{ij} \tag{8}$$

4 Methodology

4.1 Data Gathering and Analysis

In our work, information about the occupants' daily activities was gathered from the students residing in the hostel of BITS Pilani, Dubai Campus. The timing, location as well as the current lighting level corresponding to the activities in the room was noted. Upon analysis of the data gathered, it was found that ironing, eating, changing, resting, studying, working on the laptop, taking a break, and reading were the most commonly performed activities. These eight scenarios generally take place from 7:00 am to 10:00 pm on the weekdays throughout the entire semester. In addition, the AutoCAD layout, dimensions and placement of furniture, and existing lighting set up in the room were also obtained.

4.2 Simulation in DIALux

Using the above data, a model of the room was created in DIALux evo 10, and the room was divided into three work zones: the bed, the desk, and the rest of the room. The existing 72W fluorescent tubelight was replaced with a lower 46W dimmable LED tubelight, and an 8W desk lamp was added to optimize the light distribution and energy consumption. Setting Dubai as the location (latitude 25° 18′ N, longitude 54° 18′ E), 80 simulations for eight different scenarios were carried out from January 24, 2021 to January 21, 2022, as the study period. Figure 2 depicts the model of a room. The data obtained from each of these simulations were stored in an excel sheet and normalized to a range of 0–1.

Fig. 2 Room model

4.3 BPNN Model

In the next stage, a back propagation neural network model was developed in MATLAB 2016b. The model took in four inputs (date, time, activity, and sunlight in the room) and at the output of the neural network the following parameters were obtained: the dimming level of the tubelight and the desk lamp, the lux level at the bed, the desk, and the entire room. The dataset was divided using the dividerand function into 60% training, 20% validation, and 20% testing. The Levenberg–Marquardt (LM) back propagation algorithm was used for training the network with 1000 epochs. Using the hit and trial method, the best architecture for the neural network model was found to be 4:8:8:10:5 as shown in Fig. 3. The sigmoid activation function was used between the input layer and hidden layers as the normalized values lie between 0 and 1, and the linear activation function was used for the output layer. The performance of this neural network was assessed using the mean square error (MSE).

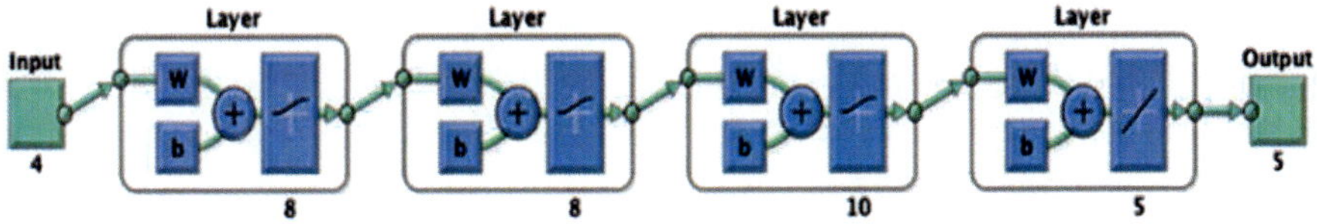

Fig. 3 BPNN architecture

Fig. 4 Scenario for ironing of clothes at 7 am

5 Results

5.1 DIALux Model

The first scenario considered was the ironing of clothes on the bed at 7:00 am, and the simulation for the scene with clear sky condition is shown in Fig. 4. We found that daylight is sufficient from April to September. The next scenario, changing clothes with the blinds completely shut also follows a similar trend. However, for the rest of the activities performed in the morning and afternoon, i.e., eating at the desk and resting on the bed, sunlight was found to meet the lighting requirement throughout the year.

The fifth scenario considered was studying at the desk in the evening at 5:15 pm. This scenario required the highest lux level, i.e., 400 lx, and the presence of sunlight helped in reducing the power required from the lighting system. Working on the laptop at 7:00 pm was the sixth scenario, and the simulation was obtained with no daylight. In both of these scenarios, the desk lamp was set to a higher intensity to provide adequate illumination at the desk (Fig. 5). The seventh scenario was the student using their phone or laptop while they were on their bed at 8:00 pm. From the simulation with the tubelight set to 38% on average, we obtained 116 lx at the bed. For the last scenario, reading before sleeping at 10:00 pm, an average of 210 lx was observed at the bed. In both these scenarios, the desk lamp was completely switched off in order to save electricity.

Figure 6 shows the variation of daylight at 12 noon in the room, and we can see that the amount of daylight in the room increases during the summer season and decreases during the winter season. This is consistent with the results obtained for ironing and changing clothes scenarios.

Fig. 5 Scenario for studying at the desk

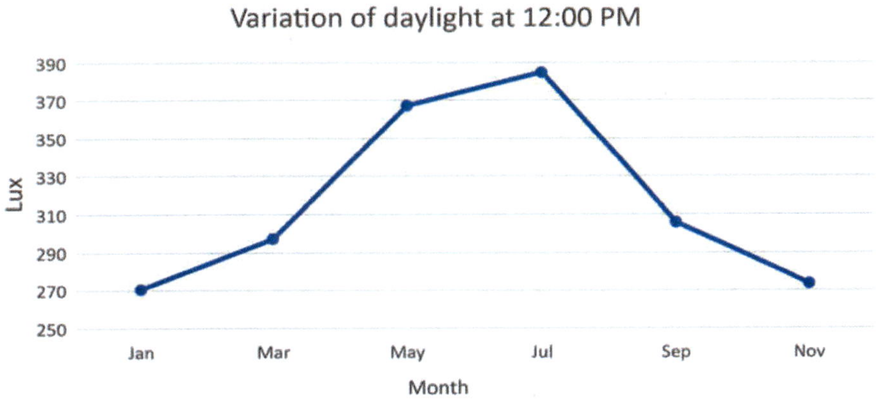

Fig. 6 Variation of daylight in the room at 12 noon

5.2 BPNN Model

The BPNN model was simulated, and the MSE metric was used to measure the performance of the model. MSE gives the average value of the sum of the deviation between the estimated value and the actual value. The model showed consistently good performance across different architectures as shown in Table 2.

Table 2 R value and MSE

Architecture	R value	Performance (MSE)
4:20:5	0.99803	2.8294×10^{-4}
4:10:10:5	0.99818	2.6188×10^{-4}
4:8:8:10:5	0.99887	1.6175×10^{-4}

The above models were run for 1000 epochs to avoid overfitting and using the hit-and-trial method the best performance was obtained for the 4:8:8:10:5 architecture. As shown in Fig. 7, the overall R value for this architecture was 0.99887 which is extremely close to unity and hence the model's regression performance on the test dataset specifically underscores an absence of overfitting.

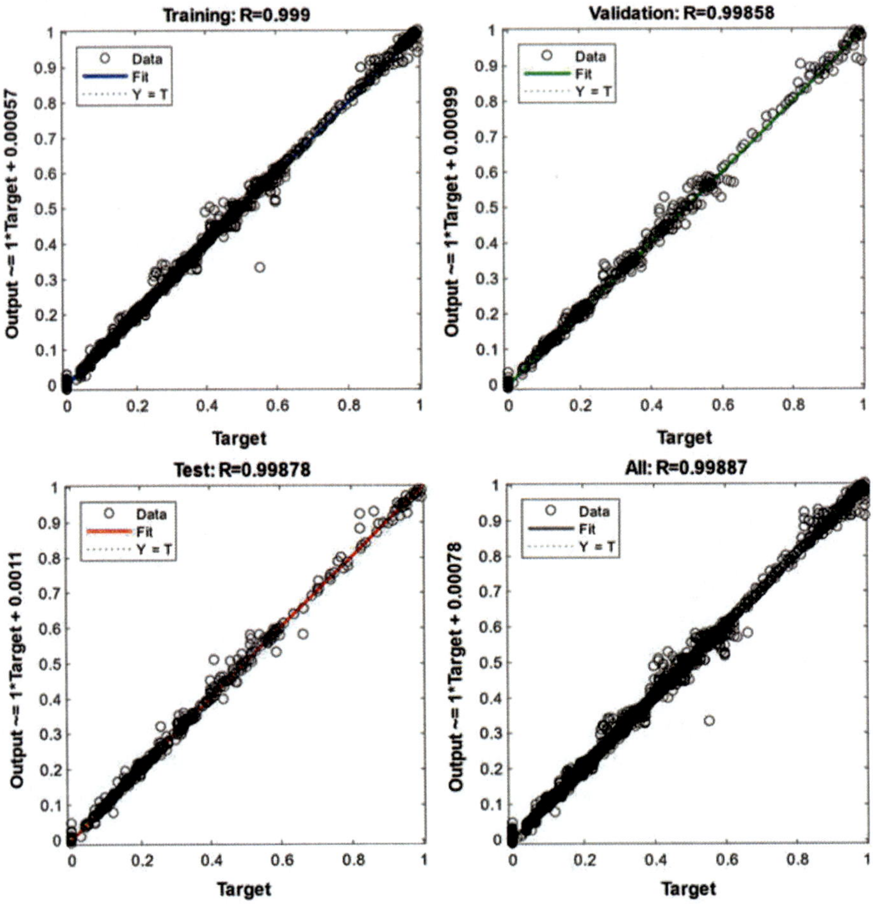

Fig. 7 Regression plots for training, validation, testing, and overall dataset

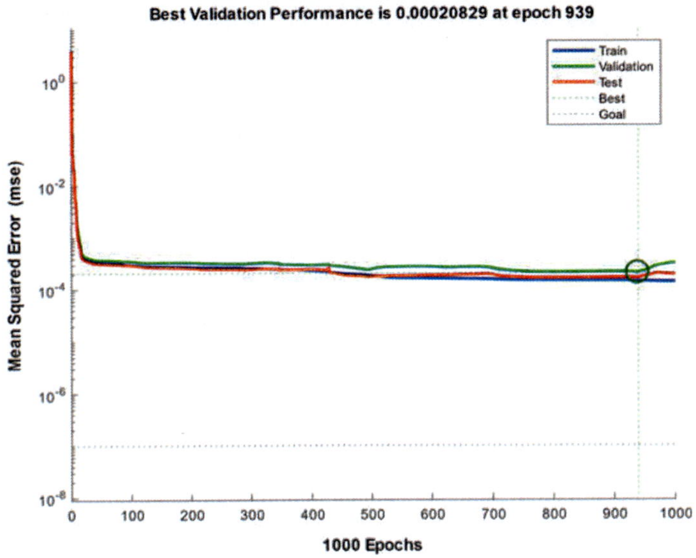

Fig. 8 Best validation performance for mean square error (versus) epochs plot

MSE for the model also steadily decreased as the number of epochs increased, and the best validation performance was at the 939th epoch as shown in Fig. 8. From the graph shown below, we can see that initially the value decreased rapidly with the increase in epochs as the network was being trained. At the 1000-epoch, the MSE obtained was 1.906×10^{-4} which is very close to 0.

As far as MSE values are concerned, the model showed better performance as evident from Table 3. It can be seen that most of the learning happened in the initial 100 epochs to limit the MSE to a negligibly small value, and hence, the model is able to capture the dynamics well. Because the training happened earlier in terms of epoch value, this model exhibits a better generalization capability without the need for too much training and has least tendency for overfitting.

Table 3 MSE of the model for different epoch values

Epoch	MSE
0	3.94
50	3.182×10^{-4}
100	2.960×10^{-4}
200	2.570×10^{-4}
500	1.77×10^{-4}
1000	1.906×10^{-4}

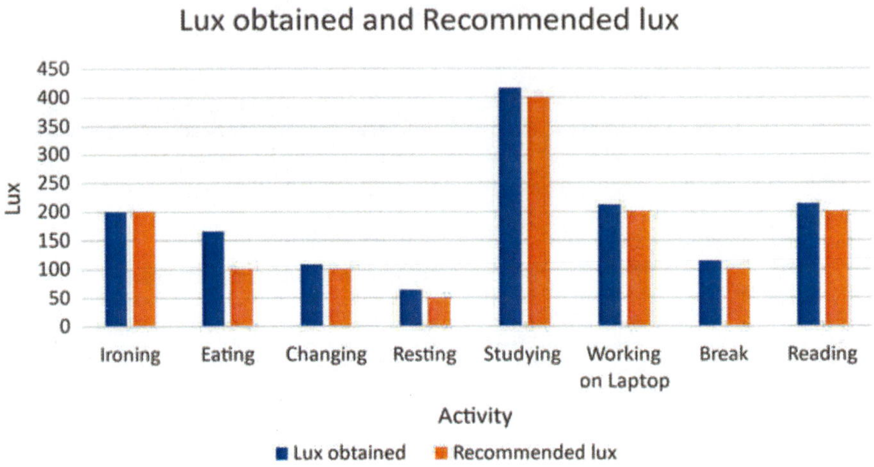

Fig. 9 Comparison of lux levels obtained by the model and that recommended by the regulation

To validate the model performance, the output obtained from the neural network model was denormalized and Fig. 9 illustrates the comparison between the simulated lux levels on December 14, 2021, with that recommended by the governmental lighting regulation in UAE [21]. We can see that the levels predicted by the BPNN model satisfy the minimum illuminance levels required for each scenario.

Table 4 gives the average annual dimming levels for the lighting system and the corresponding power saved for each activity during the academic year.

Hence by using daylight, this system was able to conserve power by 55% approximately and also maintain the occupants' visual comfort.

Table 4 Dimming levels and power saved for each scenario

Scenario	Tubelight (%)	Desk lamp (%)	Power saved (%)
Ironing	30	0	70
Eating	0	0	0
Changing	20	0	81
Resting	0	0	0
Studying	10	70	85
Working on laptop	20	40	82
Break	40	0	74
Reading	70	0	54

6 Conclusion

Lighting constitutes a major part of residential buildings energy consumption, but the energy demands due to lighting can be minimized without compromising on the visual comfort of the occupants. This paper presented an intelligent simulation model for daylight-artificial light integrated schemes which could be deployed in residential buildings to minimize their energy consumption due to lighting. The models for integrated lighting schemes are to contend with multivariant, nonlinear, and dynamic processes impacted by occupancy, environmental, and geographical variables, and hence, the model simulated in DIALux allowed for replicating real-world scenarios taking place in the residence of the BITS Pilani, Dubai Campus with high accuracy. Further, the BPNN model developed was able to predict the levels of the lighting system at the hostel with a MSE value of only 0.00016. By retrofitting the existing lighting system with a 46W dimmable LED tubelight and an 8W desk lamp system, we were able to achieve power savings of approximately 55% during the academic year.

This paper accentuates the usage of daylight in the modeling stage of lighting design, as incorporating daylight in the design feature not only helps maximize the occupants' visual comfort and wellbeing, but also minimizes the energy consumption. In addition to employing solutions such as LEDs, occupancy sensors, lighting control and building management systems; the implementation of intelligent systems like the one proposed in this paper can be used to further minimize the electricity consumption in residential buildings.

In the future, we would try to develop a hardware-based system using sensors which would be able to identify the activity being performed and automatically adjust the lighting in real time. For enhancing the accuracy and reliability, a larger dataset would be used and other parameters like glare and different sky conditions would also be considered.

Acknowledgements The authors would like to thank the management of Birla Institute of Technology and Science-Pilani, Dubai Campus, for offering their support and facilities in the completion of this work.

References

1. Dounis AI, Caraiscos C (2009) Advanced control systems engineering for energy and comfort management in a building environment. Renewable and Sustainable Energy Reviews 13(6–7):1246–1261. https://doi.org/10.1016/j.rser.2008.09.015
2. Maamari F, Fontoynont M, Adra N (2006) Application of the CIE test cases to assess the accuracy of lighting computer programs. Energy and Buildings 38(7):869–877. https://doi.org/10.1016/j.enbuild.2006.03.016
3. Rashid M, Zimring C (2008) A review of the empirical literature on the relationships between indoor environment and stress in health care and office settings. Environment and Behavior 40(2):151–190. https://doi.org/10.1177/0013916507311550

4. Aguilar-Carrasco M, Domínguez-Amarillo S, Acosta I, Sendra J (2021) Indoor lighting design for healthier workplaces: natural and electric light assessment for suitable circadian stimulus. Optics Express 29(19):29899–29917. https://doi.org/10.1364/OE.430747
5. Kazanasmaz T, Günaydin M, Binol S (2009) Artificial neural networks to predict daylight illuminance in office buildings. Building and Environment 44(8):1751–1757. https://doi.org/10.1016/j.buildenv.2008.11.012
6. Katsanou VN, Alexiadis MC, Labridis DP (2019) An ANN-based model for the prediction of internal lighting conditions and user actions in non-residential buildings. Journal of Building Performance Simulation 12(5):700–718. https://doi.org/10.1080/19401493.2019.1610067
7. Wagiman KR, Abdullah MN (2018) Intelligent lighting control system for energy savings in office building. Indonesian Journal of Electrical Engineering and Computer Science 11(1):195–202. https://doi.org/10.11591/ijeecs.v11.i1.pp%25p
8. Hidayat I, Faridah, Utami S (2018) Activity based smart lighting control for energy efficient building by neural network model. E3S Web of Conferences 43(9):1017–1025. https://doi.org/10.1051/e3sconf/20184301017
9. Higuera J, Hertog W, Perálvarez MJ, Carreras J (2015) Hybrid smart lighting and climate control system for buildings. In: IET conference on future intelligent cities. IEEE, UK, pp 1–5. https://doi.org/10.1049/ic.2014.0047
10. Seyedolhosseini A, Masoumi N, Modarressi M, Karimian N (2018) Zone based control methodology of smart indoor lighting systems using feedforward neural networks. In: 2018 9th international symposium on telecommunications (IST). IEEE, Tehran, pp 201–206. https://doi.org/10.1109/ISTEL.2018.8661118
11. Seyedolhosseini A, Masoumi N, Modarressi M, Karimian N (2018) Design and implementation of efficient smart lighting control system with learning capability for dynamic indoor applications. In: 2018 9th international symposium on telecommunications (IST). IEEE, Tehran, pp 241–246. https://doi.org/10.1109/ISTEL.2018.8661023
12. Seyedolhosseini A, Masoumi N, Modarressi M, Karimian N (2018) Illumination control of smart indoor lighting systems consists of multiple zones. In: 2018 smart grid conference (SGC). IEEE, Sanandaj, pp 1–4. https://doi.org/10.1109/SGC.2018.8777883
13. Wang Z, Tan YK (2013) Illumination control of LED systems based on neural network model and energy optimization algorithm. Energy and Buildings 62:514–521. https://doi.org/10.1016/j.enbuild.2013.03.029
14. Madias E, Kontaxis P, Topalis F (2016) Application of multi-objective genetic algorithms to interior lighting optimization. Energy and Buildings 125:66–74. https://doi.org/10.1016/j.enbuild.2016.04.078
15. Zhang X, Lu H, Li J, Peng X, Li Y, Liu L, Dai Z, Zhang W (2020) Design and implementation of intelligent light control system based on arduino. In: 2020 IEEE international conference on artificial intelligence and computer applications (ICAICA). IEEE, Dalian, pp 1369–1373. https://doi.org/10.1109/ICAICA50127.2020.9182657
16. Ji-Qing Q, Xu Q, Sun K (2022) Optimization of indoor luminaire layout for general lighting scheme using improved particle swarm optimization. Energies 15(4):1482. https://doi.org/10.3390/en15041482
17. Guo J, Zhang Y (2022) Research on control method of comfortable lighting and energy saving lighting. In: 2021 International conference on advanced technology of electrical engineering and energy (ATEEE). IEEE, Qingdao, pp 87–92. https://doi.org/10.1109/ATEEE54283.2021.00025
18. Wu Y, Zhang Y, Ilmin N, Sui J (2022) Residential energy-saving lighting based on bioinspired algorithms. Mathematical Problems in Engineering 2022:1–9. https://doi.org/10.1155/2022/7600021

19. Saraf R, Bhavani RG (2017) Assessment of daylight performance of a commercial office space in hot, arid climate for enhanced visual comfort conditions. In: 2017 International conference on technological advancements in power and energy (TAP Energy). IEEE, Kollam, pp1–6. https://doi.org/10.1109/TAPENERGY.2017.8397246
20. DIALux Homepage. https://www.dial.de/en/dialux/. Accessed 2022/06/18
21. Development of a lighting regulation in the UAE. https://www.emiratesnaturewwf.ae/sites/default/files/doc-2018-10/ews_wwf_rti_report_low_resolution.pdf. Accessed 2022/06/18

Electrical Muscle Stimulation Models Identification Based on Hammerstein Structure and Gravitational Search Algorithm

Lakshminarayana Janjanam, Suman Kumar Saha, and Rajib Kar

Abstract In this study, the electrical muscle stimulation (EMS) models are effectively estimated by representing them in Hammerstein structure form. Moreover, a highly computational efficient population-based evolutionary optimisation algorithm (EOA) named as gravitational search algorithm (GSA) is employed to get the optimal coefficients of the unknown EMS systems. Due to exploration and exploitation phases, GSA is able to avoid the local stagnation problem, unlike the genetic algorithm (GA) and fully informed particle swarm optimization (FIPSO). In this paper, three different EMS plants having different nonlinearities such as polynomial, sigmoid, and cubic spline, respectively, are successfully identified by using the real coded GA (RGA), FIPSO, and the GSA techniques. The simulation results confirm that the GSA exhibits more robust performance and accurate identification results as compared with the RGA and FIPSO methods which have been verified by using various quantitative metrics.

Keywords Electrical muscle stimulation models · Hammerstein structure · Parameters estimation · Evolutionary optimisation algorithms

1 Introduction

Unknown system modelling is a notable challenging task in distinct engineering areas, such as bio-medical, chemical, signal processing, and mechanical engineering [1]. To identify the unknown system parameters, various adaptive filtering techniques are developed by many researchers [2–5]. In an adaptive filtering technique, the adaptive algorithm adjusts the filter coefficients until the outputs of both unknown and known systems are equal. Unknown system identification can proceed in three

L. Janjanam (✉) · S. K. Saha
Department of ECE, National Institute of Technology, Raipur, Chhattisgarh 492010, India
e-mail: ljanjanam.phd2018.etc@nitrr.ac.in

R. Kar
Department of ECE, National Institute of Technology, Durgapur, West Bengal 713209, India

S. Das et al. (eds.), *Advances in Data-driven Computing and Intelligent Systems*, Lecture Notes in Networks and Systems 653, https://doi.org/10.1007/978-981-99-0981-0_2

ways, namely linear, nonlinear, and their combination. Linear plant (LP) estimation [2] is prevalent because of easy implementation, and it can be done by finite impulse response (FIR) [2] and infinite impulse response (IIR) filters [3–5]. However, the FIR filter requires large memory space, and hence, IIR models are preferred and widely used by the many researchers for solving the unknown plant estimation [3–5].

During recent days, EOAs are recurrently enforced for distinct signal processing applications because of their potential in extracting the global optimised parameters from the higher-dimensional problems. Hence, researchers are motivated to use the EOAs for unknown plant estimation [3–5]. Many recent works have identified both reduced and same order IIR models using different EOA techniques, such as variants of PSO [3], hybrid coral reefs optimisation [4], and teacher learner-based EOA approach [5] and also noticed more reasonable results in [5] than other competent algorithms. However, the linear models yield poor estimation results for industry-specific systems modelling because they are practically designed by nonlinear plants (NLPs), and hence, the parameter estimation of a NLP is also a primary challenging task [6, 7].

To model the NLPs, various prominent models called neural networks [8], bilinear [9], and Volterra [6, 10] have been used. In [7], Worden et al., have solved various nonlinear benchmark problems using heuristic approaches. The neural networks (NNs) are simple to design and are efficient to tackle the complex problems [8]. However, the NNs are easily locked at local solutions due its large computational cost (CC). Moreover, the bilinear models are not suitable for many applications. Furthermore, different Volterra systems are examined by using the Kalman filter (KF) which is optimised by the global gravitational search [6] and antlion [10] optimisation tools. Moreover, the benchmark plants [10] such as rational plant and heat exchanger are tracked by changing the memory sizes of the Volterra model, and the best results are obtained for the higher memory sizes which increase the CC.

In recent days, Hammerstein [11–13], Wiener [14], and their cascade modelling [15, 16] are frequently applied for the design of high complex systems. In general, the internal structure of these models comprises both LP and NLPs. In [11–16], various advanced optimisation techniques have been employed to achieve the better estimations. In [11], Pal et al., have proposed the colliding bodies optimisation (CBO) algorithm for estimation of open-loop and closed-loop Hammerstein type NLPs and also proved that CBO gives accurate parameters compared to GA, PSO, and DE approaches. Janjanam et al. [12] have derived optimal parameters of various Hammerstein benchmark plants by using social mimic optimiser-based KF and also approximated the real-time responses of cascaded tanks and heating systems. Inspired from [11, 12], in this work, EMS plants are modelled by the Hammerstein structure than other above presented models.

Identification of EMS is a very interesting and demanding task to the researchers. The EMS models are developed for rehabilitation of paralysed muscles [13, 17, 18]. The EMS is an efficient method that instigates muscle hypertrophy, i.e. an increase in mass of the muscle and area of the cross section, and increases the torque output and battles fatigue [13]. Moreover, the EMS also applied for restoration tasks like standing and reaching. In [13, 17, 18], the EMS modes of stroke patients are identified by using the Hammerstein structure. In [18], recursive least squares (RLS) technique is employed to exploit the coefficients of the Hammerstein-based EMS plants. However, the gradient (RLS) technique endured from slow convergence and yields sub-optimal coefficients. To bury this issue, Mehmood et al. [13] have proposed backtracking to extract the accurate estimates of the EMS model. Moreover, in [13], different types of EMS models are developed based on sigmoid, polynomial, and cubic spline nonlinear functions. Additionally, various standard metrics are assessed to investigate the parameter accuracy.

From the above literature [13, 17, 18], authors have motivated to apply the EOA (i.e., GSA) for identification of the EMS models and derived best estimates with faster convergent and small steady-state error results than other existing methods.

The highlights of this research work are presented below:

- The EMS plants are mathematically modelled by using the Hammerstein structure.
- A decent fitness function is proposed to achieve the global optimised EMS parameters.
- The GSA avoids the trapping in a local optimum due to exploration and exploitation phases. Hence, a new application of GSA technique is examined by representing the EMS plants with various nonlinear functions (cubic spline, polynomial, and sigmoid).
- This efficacious GSA technique for the modelling of EMS plants over other traditional algorithms like RGA and FIPSO is investigated.
- The achieved simulations show that the proposed GSA-based results yield better performance as compared with the rest of the approaches in terms of some standard metrics.

The rest of the paper is ordered as follows: Sect. 2 derives the Hammerstein structure and fitness function of the EMS model. Section 3 clearly describes the design methodology of the EMS model using the proposed GSA. Section 4 analyses the MATLAB software-based results of different kinds of EMS models, and finally, some conclusions derived in Sect. 5.

2 Problem Description

In general, the Hammerstein-controlled autoregressive (H-CAR) structure comprises NLP succeeded by LP. The output of NLP for the input $x(n)$ is represented by $f(x(n))$, where f represent the nonlinear function. The transfer function (TF) of LP block is $\frac{H(z)}{G(z)}$. In this work, the EMS model is mathematically represented based on the structure of H-CAR [13] and is given in (1), where the NLP block indicates the nonlinear relationship amongst the stimulus activation level and the output torque, whilst the LP block specifies the response of EMS dynamic moment [13, 17, 18].

$$y(n) = \frac{H(z)}{G(z)} f(x(n)) + \frac{1}{G(z)} \zeta(n) \tag{1}$$

where the noise $\zeta(n)$ is Gaussian (zero mean and 0.01variance) and its TF is $\frac{1}{G(z)}$. The Mth and Nth degree polynomials of $G(z)$ and $H(z)$, separately, defined as

$$G(z) = g_1 z^{-1} + g_2 z^{-2} + \ldots + g_M z^{-M} \tag{2}$$

$$H(z) = 1 + h_1 z^{-1} + h_2 z^{-2} + \ldots + h_N z^{-N} \tag{3}$$

where $g_1, g_2, \ldots, g_M$ and $h_1, h_2, \ldots, h_N$ are the real coefficients.

The nonlinear function $f(x(n))$ can be expressed by polynomial, sigmoid, and cubic spline functions [13], which are separately given in (4)–(6).

$$f(x(n)) = \lambda_1 x(n) + \lambda_2 x^2(n) + \ldots. + \lambda_p x^p(n) \tag{4}$$

$$f(x(n)) = \lambda_1 \frac{e^{\lambda_2 x(n)} - 1}{e^{\lambda_2 x(n)} + \lambda_3} \tag{5}$$

$$f(x(n)) = \sum_{i=1}^{p-2} \lambda_i |x(n) - x_{i+1}(n)|^3 + \lambda_{p-1} + \lambda_p x(n) + \lambda_{p+1} x^2(n) + \lambda_{p+2} x^3(n) \tag{6}$$

Let the spline function as given (6) having a knot value 150, then (6) becomes [13]

$$f(x(n)) = \lambda_1 |x(n) - 150|^3 + \lambda_2 + \lambda_3 x(n) + \lambda_4 x^2(n) + \lambda_5 x^3(n) \tag{7}$$

The parameter vector ϕ of the EMS model is $\phi = [\phi_L, \phi_Q]$, where ϕ_L and ϕ_Q denote the LP and NLP block parameters, respectively, which are defined in (8) and (9) based on the Eqs. (2)–(7).

$$\phi_L = [g_1, g_2, \ldots, g_M, h_1, h_2, \ldots, h_N] \tag{8}$$

$$\phi_Q = \begin{cases} [\lambda_1, \lambda_2, \ldots, \lambda_p] & \text{Polynomial} \\ [\lambda_1, \lambda_2, \lambda_3] & \text{Sigmoid} \\ [\lambda_1, \lambda_2, \lambda_3, \lambda_4, \lambda_5] & \text{Spline} \end{cases} \tag{9}$$

Substituting Eqs. (2)–(9) in Eq. (1), then the EMS plant output for the polynomial, sigmoid, and spline case, separately, expressed as [13]

$$y(n) = -\left(\sum_{i=1}^{M} g_i z^{-i}\right) y(n) + \left(\sum_{i=1}^{N} h_i z^{-i}\right) (\lambda_1 x(n) + \lambda_2 x^2(n) + \ldots + \lambda_p x^p(n)) + \zeta(n) \tag{10}$$

$$y(n) = -\left(\sum_{i=1}^{M} g_i z^{-i}\right) y(n) + \left(\sum_{i=1}^{N} h_i z^{-i}\right)\left(\lambda_1 \frac{e^{\lambda_2 x} - 1}{e^{\lambda_2 x} + \lambda_3}\right) + \zeta(n) \tag{11}$$

$$y(n) = -\left(\sum_{i=1}^{M} g_i z^{-i}\right) y(n) + \left(\sum_{i=1}^{N} h_i z^{-i}\right) \times \ldots$$
$$\left(\sum_{i=1}^{p-2} \lambda_i \left|x(n) - x_{i+1}(n)\right|^3 + \lambda_{p-1} + \lambda_p x(n) + \lambda_{p+1} x^2(n) + \lambda_{p+2} x^3(n)\right) + \zeta(n) \tag{12}$$

Consider, the output responses above models ((10)–(12)) are tracked with the same of the estimated outputs, which are given as follows:

$$\hat{y}(n) = -\left(\sum_{i=1}^{M} \hat{g}_i z^{-i}\right) \hat{y}(n) + \left(\sum_{i=1}^{N} \hat{h}_i z^{-i}\right)(\hat{\lambda}_1 x(n) + \hat{\lambda}_2 x^2(n) + \ldots + \hat{\lambda}_p x^p(n)) \tag{13}$$

$$\hat{y}(n) = -\left(\sum_{i=1}^{M} \hat{g}_i z^{-i}\right) \hat{y}(n) + \left(\sum_{i=1}^{N} \hat{h}_i z^{-i}\right)(\hat{\lambda}_1 \frac{e^{\hat{\lambda}_2 x(n)} - 1}{e^{\hat{\lambda}_2 x(n)} + \hat{\lambda}_3}) \tag{14}$$

$$\hat{y}(n) = -\left(\sum_{i=1}^{M} \hat{g}_i z^{-i}\right) \hat{y}(n) + \left(\sum_{i=1}^{N} \hat{h}_i z^{-i}\right) \times \ldots$$
$$\left(\sum_{i=1}^{p-2} \hat{\lambda}_i |x(n) - x_{i+1}(n)|^3 + \hat{\lambda}_{p-1} + \hat{\lambda}_p x(n) + \hat{\lambda}_{p+1} x^2(n) + \hat{\lambda}_{p+2} x^3(n)\right) \tag{15}$$

Based on awareness from [10, 13], the fitness function of EMS model is formulated which is normalised form of real and estimated outputs/parameters

$$\text{Fitness} = \sum_{k=1}^{K} \frac{(y(k) - \hat{y}(k))^2}{(y(k))^2} + \sum_{j=1}^{J} \frac{(\phi_j - \hat{\phi}_j)^2}{(\phi_j)^2} \tag{16}$$

where K and J denote the data samples length and the number of coefficients in ϕ, respectively; $y(k)$ is real output as defined in (10)–(12); $\hat{y}(k)$ is the estimated output as defined in (13)–(15); ϕ and $\hat{\phi}$ are the real and estimated parameters which are given below (polynomial case).

$$\phi = [\phi_L, \phi_Q] = [g_1, g_2, \ldots, g_M, h_1, h_2, \ldots, h_N, \lambda_1, \lambda_2, \ldots, \lambda_p] \tag{17}$$

$$\hat{\phi} = [\hat{\phi}_L, \hat{\phi}_Q] = [\hat{g}_1, \hat{g}_2, \ldots, \hat{g}_M, \hat{h}_1, \hat{h}_2, \ldots, \hat{h}_N, \hat{\lambda}_1, \hat{\lambda}_2, \ldots, \hat{\lambda}_p] \tag{18}$$

The primary requirement of the employed EOAs is to optimise (minimise) the objection function as given in (16) which results in the responses/parameters of both known and unknown plants are equal.

3 EMS System Design Methodology

In this paper, EMS systems are identified using different EOAs such as GSA [19], FIPSO [20], and RGA [21] techniques and also remarked that the GSA-based design approach offers reasonably improved results than others. Therefore, this work explains the procedure for the design of the EMS plant by using the proposed GSA technique only. GSA is highly computational efficient and able to attain the best global search accuracy in a higher-dimensional system. Hence, many researchers have used the GSA technique for solving various engineering optimization problems [19]. In GSA, the objects (search agents/masses) are interacted based on the principle of the law of gravity. In GSA, the position of each object is initialised with random numbers in the problem search space area. Due to the gravitational energy, the most massive mass object will attract the other objects, and it walks at a low speed. The position of the most massive mass object gives the best optimal solution.

In [19], Rashedi et al. have solved the various standard benchmark functions using GSA and achieved the best optimal solutions than other standard algorithms. In GSA, to update the position of each search agent, authors [19] have proposed various steps such as initialisation of objects, Euclidean distance measurement, calculation of the overall force, and updating the acceleration and velocity of the search agent. The same steps utilised in this work [19] for identification of the EMS systems. The step-by-step EMS plant design mechanism by using GSA is mentioned below:

Step 1: Randomise the location of all the *npop* search agents (SAs) within a specified search space boundary using (19).

$$X_i^l = (ub - lb) * \text{rand} + lb \tag{19}$$

where X_i^l is the location of the ith SA in the lth dimension search space for indexes $i = 1, 2, ...,$ npop and $l = 1, 2, ..., J$; *rand* generates the random numbers in the interval [0 1]; the upper (ub) and lower (lb) bounds of each problem variable and is selected as + 2 and − 2, separately; J is the number of variables of the problem, and *npop* is chosen as 50.

Step 2: Compute the fitness values of each search agent by using the objective function (16).

Step 3: Based on the achieved fitness values, select the lowest fitness SA as $\text{best}(t)(= h\text{best})$ and the highest fitness SA as $\text{worst}(t)(= h\text{worst})$.

Step 4: Evaluate masses of each SA using (20) and (21)

$$m_i(t) = \frac{\text{fit}_i(t) - \text{worst}(t)}{\text{best}(t) - \text{worst}(t)} \tag{20}$$

$$M_i(t) = \frac{m_i(t)}{\sum_{j=1}^{\text{npop}} m_j(t)} \tag{21}$$

where $\text{fit}_i(t)$ is the fitness value of ith SA in current iteration t.

Step 5: Determine Euclidean distance $(ED_{ij}(t))$ between the positions of the ith and jth search agents and gravitational constant $\lambda(t)$ using (22) and (23), respectively.

$$ED_{ij}(t) = \left\| X_i(t) - X_j(t) \right\|_2 \tag{22}$$

$$\gamma(t) = \gamma_0 * \exp(-\alpha \frac{t}{T}) \tag{23}$$

where γ_0 is the initial value; α is a decreasing constant parameter; T is the maximum epoch number. In this work, $\lambda_0, \alpha,$ and T are chosen as 500, 25, and 1000, respectively.

Step 6: Calculate the total force exerted on the ith SA using (24)

$$F_i^l(t) = \sum_{j=1, j \neq i}^{\text{npop}} \text{rand}_j F_{ij}^l(t) \tag{24}$$

where rand_j generates a uniform random number in the interval [0, 1], and $F_{ij}^l(t)$ is the force acting on ith SA at a time instant, t from the jth SA is given in (25).

$$F_{ij}^l(t) = \lambda(t) \frac{M_i(t) \times M_j(t)(X_j^l(t) - X_i^l(t))}{ED_{ij}(t) + \delta} \tag{25}$$

where δ is small positive constant and is selected as 0.0001.

Step 7: Evaluate the acceleration of ith SA at a time, t in lth dimension using (26)

$$a_i^l(t) = \frac{F_i^l(t)}{M_i(t)} \tag{26}$$

Step 8: The position and velocity of all the SAs updated using (27) and (28), respectively.

$$X_i^l(t+1) = X_i^l(t) + V_i^l(t+1) \tag{27}$$

$$V_i^l(t+1) = \text{rand}_i V_i^l(t) + a_i^l(t) \tag{28}$$

Step 9: Evaluate fitness/cost values using (16) for all the SAs and then update $h$best.

Step10: Repeat *Steps* 4–9 until the stopping criterion (T) is met.

Step11: Return $h$best which contains the globally optimised EMS model coefficients.

4 Simulation Results and Discussions

In this work, three distinct discrete-time EMS systems [13] are given in (29), (30), and (31), respectively, and are identified by using RGA, FIPSO, and GSA techniques. The control parameters such as mutation and crossover rate for RGA [21] are set at 0.3 and 0.6, respectively. Moreover, the constructive, acceleration, and group of neighbour particles parameters for FIPSO [20] are set at 0.9, 1.4, and [3, 5], respectively. Finally, the other parameters *npop* and T for all the three techniques mentioned above are chosen as 50 and 1000, respectively. To estimate and validate these plants, random input data sequences of length 1000 and 100 samples, respectively, are generated.

Experiment 1: *EMS plant having polynomial nonlinearity*

$$\begin{aligned} G(z) =& 1 - z^{-1} + 0.8z^{-2};\ H(z) = z^{-1} + 0.6z^{-2};\ f(x(n)) = 2.8x(n) \\ & - 4.8x^2(n) + 5.7x^3(n) \end{aligned} \tag{29}$$

Experiment 2: *EMS plant having sigmoid nonlinearity*

$$\begin{aligned} G(z) &= 1 - 1.9985z^{-1} + 0.9985z^{-2};\ H(z) = 0.0022z^{-1}; \\ f(x(n)) &= 6.8994\frac{e^{0.0410x(n)} - 1}{e^{0.0410x(n)} + 2389.70} \end{aligned} \tag{30}$$

Experiment 3: *EMS plant having cubic spline nonlinearity*

$$\begin{aligned} G(z) =& 1 - 1.094z^{-1} + 0.109z^{-2};\ H(z) = z^{-1} + 0.249z^{-2}; \\ f(x(n)) =& -0.028 + 1.90 \times 10^{-3}x(n) - 7.83 \times 10^{-6}x^2(n) + 1.78 \times 10^{-8}x^3(n)... \\ & + 2.36 \times 10^{-8}\left|x^3(n) - 150\right| \end{aligned} \tag{31}$$

The real and estimated responses of the Experiments 1–3 using RGA. FIPSO and GSA methods are shown in Fig. 1a–c, separately. It is perceived from Fig. 1 that the estimated responses due to all the methods appear very nearer to the real response. However, the zoomed plot as presented in the same figure confirms that the GSA-based output plot is in the vicinity of the real output than FIPSO- and RGA-based output plots.

In addition, various quantitative metrics [12, 16] are evaluated to measure the accuracy of the derived estimated coefficients. The evaluated metrics, namely variance norm (*VN*), $\left\| E(\hat{\phi} - E(\hat{\phi})) \right\|_2$, bias norm (*BN*), $\left\| \phi - E(\hat{\phi}) \right\|_2$, error variance account for ($E_{VAF} = 100 - VAF$), where $VAF = (1 - \text{var}(\phi_j - \hat{\phi}_j)/\text{var}(\phi_j)) \times 100$ and error Nash–Sutcliffe efficiency $E_{NSE} = 1 - NSE$, $NSE = 1 - (\sum_{j=1}^{J}(\phi_j - \hat{\phi}_j)^2 / \sum_{j=1}^{J}(\phi_j - \text{mean}(\hat{\phi}_j))^2)$ for all the experiments, are presented in Table 1. The obtained values for all the above mentioned metrics using GSA, FIPSO, and RGA methods lie in the order of 10^{-05} to 10^{-07}, 10^{-03} to10^{-05}, and 10^{-04} to10^{-02}, respectively.

Fig. 1 Real and estimated outputs comparison of various methods for EMS plants modelling

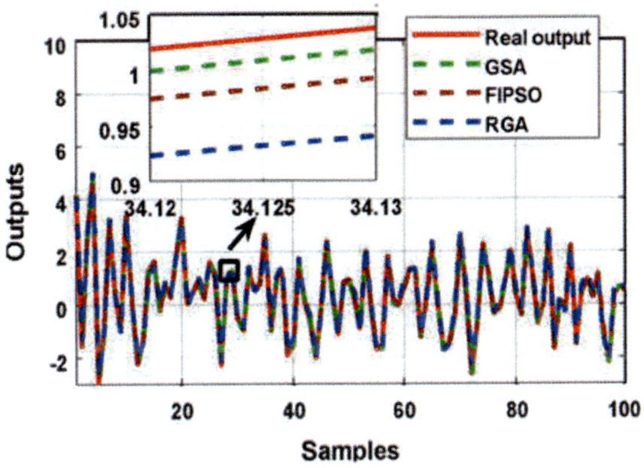

(a) Experiment 1

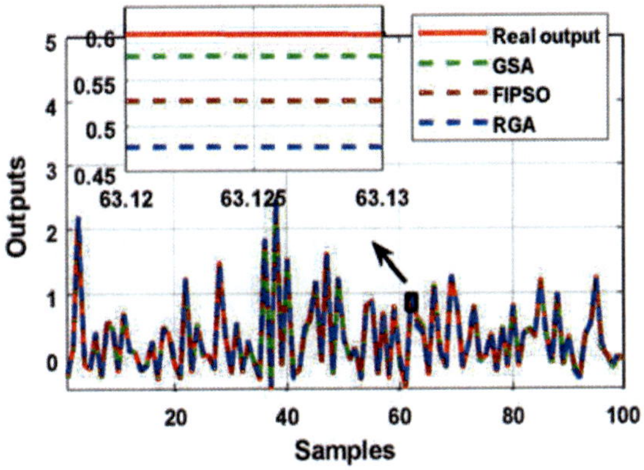

(b) Experiment 2

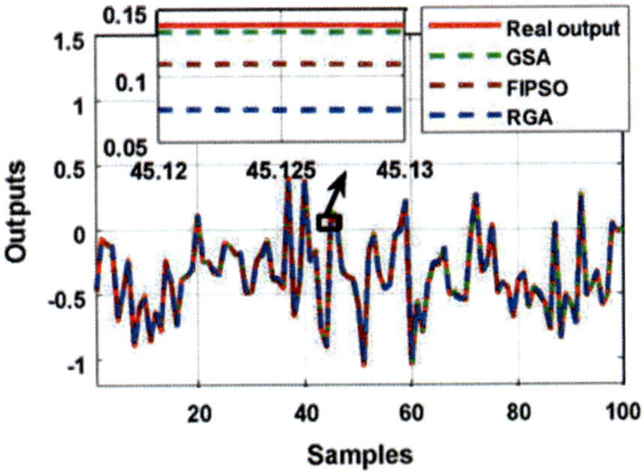

(c) Experiment 3

Table 1 Performance comparison of various methods for EMS plants modelling based on different metrics

Exp	Method	E_{VAF}	E_{NSE}	BN	VN	Fitness (dB)	QF (%)
1	GSA	2.44E−06	6.77E−07	2.23E−04	3.76E−05	− 65.72	99.34
	FIPSO	3.52E−04	5.46E−05	3.65E−03	5.44E−04	− 53.56	95.28
	RGA	5.35E−03	3.33E−04	1.68E−02	6.68E−03	− 45.87	92.66
2	GSA	7.22E−06	1.76E−06	7.32E−04	6.89E−05	− 58.52	98.92
	FIPSO	8.11E−04	4.44E−04	4.53E−03	9.21E−03	− 47.69	94.66
	RGA	9.22E−03	3.57E−-03	7.22E−02	2.35E−02	− 41.33	91.44
3	GSA	3.55E−05	5.67E−06	5.66E−04	6.74E−05	− 51.39	98.22
	FIPSO	5.21E−04	2.89E−05	7.87E−03	7.33E−04	− 43.44	94.37
	RGA	7.65E−03	6.63E−03	4.27E−02	3.18E−03	− 36.97	90.98

The convergence characteristics of lowest output fitness (dB) values for Experiments 1–3 using RGA, PSO, and the proposed GSA techniques are shown in Fig. 2a–c, respectively, and the corresponding optimum value achieved at 1000th epoch is reported in Table 1. From Fig. 2, GSA, FIPSO, and RGA methods converge to the lowest fitness values of − 65.72 dB, − 53.56 dB, and − 45.87 dB, separately, in 483, 433 and 490 epochs, for Experiment 1 modelling; in case of Experiment 2, 398, 459, and 584 epochs are required to yield best fitness of − 58.52 dB, − 47.69 dB, and − 41.33 dB, separately; finally in case of Experiment 3, 418, 476, and 560 epochs are necessary to get the lest fitness values of − 51.39 dB, − 43.44 dB, and − 36.97 dB, separately.

Moreover, based on real and estimated responses, the quality of fitness percentage ($QF(\%) = \left(1 - \|y(n) - \hat{y}(n)\| / \|y(n) - mean(\hat{y}(n))\|\right) \times 100$) [6] is measured, and obtained values are given in Table 1 and noticed that 99.34%, 98.92%, and 98.22%, respectively, for Experiments 1–3 by using GSA method.

5 Conclusions

In this paper, a new application of GSA is discussed for the effective modelling of EMS plants which is necessary for the rehabilitation of paralysed muscles. The EMS plant is mathematically modelled with the Hammerstein structure, where the nonlinear plant is realised with the sigmoid, polynomial, and cubic spline function. Moreover, greater results are obtained by using polynomial functions than others. For comparative analysis, the FIPSO and RGA methods are enforced to the EMS plant design. From the simulations, the proposed GSA-based EMS plant design exhibits highly satisfactory results than the FIPSO- and RGA-based design methods in terms of distinct quantitative metrics. The proposed GSA-based EMS design provides $QF(\%)$ of almost 99%, fitness of − 50 dB to − 65 dB, and 10^{-05} to $10^{-07}$for the other metrics (E_{VAF}, E_{NSE}, BN and VN). For future work, design the EMS models with

Fig. 2 Fitness (dB) values comparison of various methods for EMS plants modelling

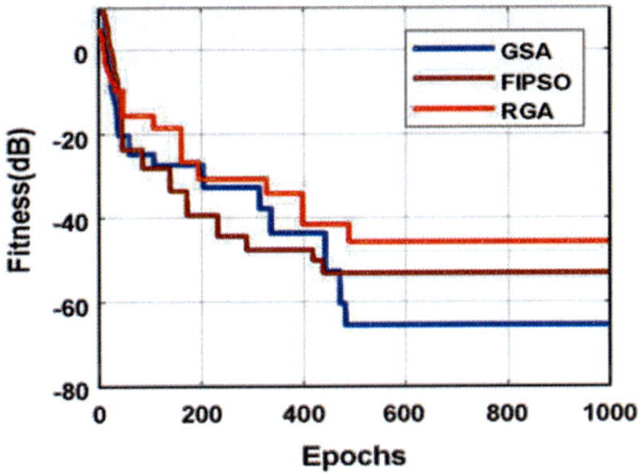

(a) Experiment 1

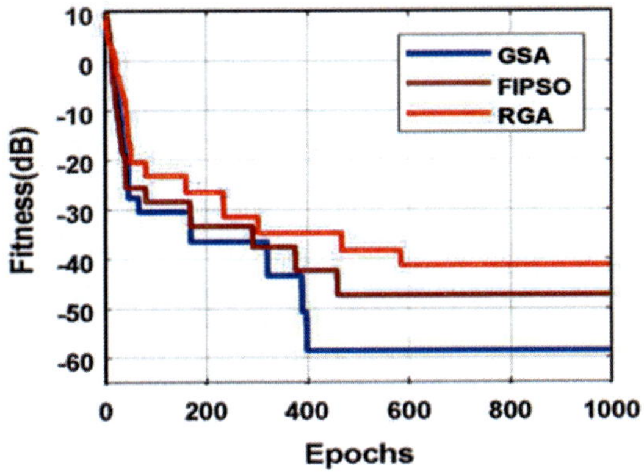

(b) Experiment 2

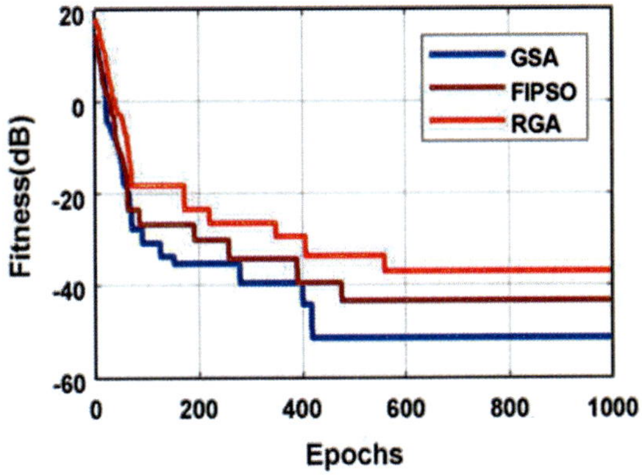

(c) Experiment 3

different nonlinear functions like dead zone, black lash, and hysteresis friction. In addition, optimise the EMS models with fractional order swarm and nature inspired algorithms.

References

1. Uncini A (2015) Fundamentals of adaptive signal processing. Springer International Publishing, Cham
2. Guo J, Wang LY, Yin G, Zhao Y, Zhang JF (2015) Identification of FIR systems with quantized inputs and observations. IFAC-Papers OnLine 48:674–679
3. Zou D-X, Deb S, Wang G-G (2018) Solving IIR system identification by a variant of particle swarm optimization. Neural Comput Appl 30:685–698
4. Yang Y, Yang B, Niu M (2018) Adaptive infinite impulse response system identification using opposition based hybrid coral reefs optimization algorithm. Appl Intell 48:1689–1706
5. Singh S, Ashok A, Kumar M, Rawat TK (2019) Adaptive infinite impulse response system identification using teacher learner based optimization algorithm. Appl Intell 49:1785–1802
6. Janjanam L, Saha SK, Kar R, Mandal D (2021) Global gravitational search algorithm-aided Kalman filter design for Volterra-based nonlinear system identification. Circuits Syst Signal Process 40:2302–2334
7. Worden K, Barthorpe RJ, Cross EJ, Dervilis N, Holmes GR, Manson G, Rogers TJ (2018) On evolutionary system identification with applications to nonlinear benchmarks. Mech Syst Signal Process 112:194–232
8. Perrusquía A, Yu W (2021) Identification and optimal control of nonlinear systems using recurrent neural networks and reinforcement learning: An overview. Neurocomputing 438:145–154
9. Hafezi Z, Arefi MM (2019) Recursive generalized extended least squares and RML algorithms for identification of bilinear systems with ARMA noise. ISA Trans 88:50–61
10. Janjanam L, Saha SK, Kar R, Mandal D (2021) An efficient identification approach for highly complex non-linear systems using the evolutionary computing method based Kalman filter. AEU—Int J Electron Commun 138:153890
11. Pal PS, Kar R, Mandal D, Ghoshal SP (2015) An efficient identification approach for stable and unstable nonlinear systems using colliding bodies optimization algorithm. ISA Trans 59:85–104
12. Janjanam L, Saha SK, Kar R, Mandal D (2022) Improving the modelling efficiency of Hammerstein system using Kalman filter and its parameters optimised using social mimic algorithm: application to heating and cascade water tanks. J Franklin Inst 359:1239–1273
13. Mehmood A, Zameer A, Chaudhary NI, Raja MAZ (2019) Backtracking search heuristics for identification of electrical muscle stimulation models using Hammerstein structure. Appl Soft Comput 84:105705
14. Janjanam L, Saha SK, Kar R, Mandal D (2022) Wiener model-based system identification using moth flame optimised Kalman filter algorithm. SIViP 16:1425–1433
15. Janjanam L, Kumar Saha S, Kar R, Mandal D (2022) Optimal design of cascaded Wiener-Hammerstein system using a heuristically supervised discrete Kalman filter with application on benchmark problems. Expert Syst Appl:117065
16. Janjanam L, Saha SK, Kar R, Mandal D (2022) Hammerstein-Wiener nonlinear system identification by using honey badger algorithm hybridized Sage-Husa adaptive Kalman filter with real-time applications. AEU—Int J Electron Commun 151:154218
17. Le F, Markovsky I, Freeman CT, Rogers E (2010) Identification of electrically stimulated muscle models of stroke patients. Control Eng Pract 18:396–407
18. Le F, Markovsky I, Freeman CT, Rogers E (2012) Recursive identification of Hammerstein systems with application to electrically stimulated muscle. Control Eng Pract 20:386–396

19. Rashedi E, Nezamabadi-pour H, Saryazdi S GSA: a gravitational search algorithm. Inf Sci 179:2232–2248
20. Mendes R, Kennedy J, Neves J (2004) The fully informed particle swarm: simpler, maybe better. IEEE Trans Evol. Comput 8:204–210
21. Valarmathi K, Devaraj D, Radhakrishnan TK (2009) Real-coded genetic algorithm for system identification and controller tuning. Appl Math Model 33:3392–3401

Experimental Analysis of "A Novel Swarm Intelligence Optimization Approach: Sparrow Search Algorithm"

Gagandeep Kaur Sidhu and Jatinder Kaur

Abstract The sparrow search algorithm (SSA) is the latest swarm optimization technique. SSA was invented not only to solve complex global optimization problems but also to obtain the best optimal solution with respect to convergence rate, preciseness, completeness, and reliability in comparison with existing Heuristic, Metaheuristic, and other algorithms. SSA was tested on 19 benchmark functions by Jiankai Xue and Bo Shen in their own paper and it was found that the output of SSA was much better in relation to convergence speed, accuracy, robustness, and stability as opposed to PSO, GWO, and GSA. After the invention of SSA, many researchers emphasized its steady convergence, deficiency of drop within local optimum together with poor accuracy, and to make its complete improvement, they modified it. In this paper, to check that the convergence rate of SSA is really slow, and it is implemented on 10 benchmark functions which are different from the 19 benchmark functions which were used in the sparrow search algorithm's paper. After the comparison of experimental results of SSA with GWO, BBA, and PSO, it is obtained that the convergence speed of SSA is slow in some benchmark functions.

Keywords Sparrow search algorithm (SSA) · Particle swarm optimization (PSO) · Gray Wolf Optimization (GWO) · Binary bat algorithm (BBA) · Revealer (producer) · Believer (scrounger) · Optimal solution · Convergence · Experimental analysis · Swarm optimization technique

1 Introduction

Optimization problems are faced almost everywhere in our day-to-day life. It can easily be understood by taking an example of a person who wanted to go on a trip but the place where he wanted to go was totally unknown to him. So, for that, he got 5–6 ways of traveling by talking with different people but he wanted to choose the best (optimal) way which takes less time to reach that particular location. So, by using

G. K. Sidhu (✉) · J. Kaur
Department of Mathematics, Chandigarh University, Gharuan, Punjab 140413, India
e-mail: gagan.kw777@gmail.com

S. Das et al. (eds.), *Advances in Data-driven Computing and Intelligent Systems*, Lecture Notes in Networks and Systems 653, https://doi.org/10.1007/978-981-99-0981-0_3

the GPS system he got the best way that means the optimal solution for his problem. This problem can also be referred as an optimization problem. Problems related to optimization can be found in every field such as engineering, transportation, medical line, supply chain, economics, share market. These optimization problems can be solved by using different algorithms such as exact, approximate, heuristic and meta-heuristic algorithms [1]. Metaheuristic algorithms are divided into two categories one is single-solution-based metaheuristics another is population-based metaheuristics. Further, population-based metaheuristics are divided into two parts called evolutionary algorithms and swarm-Intelligence algorithms.

So, the swarm-Intelligence (SI) algorithms are population-based metaheuristics algorithms. The collaborative behavior of animal herding, ant colonies, fish schooling, bird flocking etcetera is imitated by SI algorithms. SI algorithms plays a major role to solve problems related to global optimization. There is a significant number of SI algorithms, like ABC [2], GWO [3], ACO [4], PSO [5, 6], BBA [7], etc.

With the passage of time, researchers, for the best optimal solution with regard to stability, convergence and robustness, kept on modifying existing algorithms and researching new findings. In the same way, Xue and Shen discovered SSA [8] in 2020 to find a complete result of complex global optimization problems. SSA is a kind of SI algorithms which works on not only foraging approach but also anti-predation habit of sparrows. Properties of SSA are faster convergence, fine stability, powerful robustness. Although SSA gave better performance in comparison with GWO, PSO, and GSA, it had to face some limitations also, such as lack of population diversity, starting population quality, and search capability was seen in it [9], due to SSA's less global search ability it was modified to enhance its global search ability and implemented in sensor network [10], it had limitation of fall into local optimum, inadequate result accuracy, and steady convergence speed [11], while the problem was complex, global optimum was given up for the present best value as well as it provided random results [12], two issues were observed in SSA related to depending on starting population phase and s fall in the optimum which is local [13, 14], falling into local optima was observed many time and it was also seen that its run time was longer [15], it was noticed that SSA was deprived of mechanism which was related to global improvement [16], deficiency of global optimization proficiency was witnessed [17], etc.

No free lunch (NFL) theorem [18] states that a specific algorithm can display very auspicious results on a series of problems whereas on the different series of problems, the same algorithm can play the poor role. So, we can say nature of the function effect the optimality and convergence rate also. In this paper for checking the limitation of SSA, which were given by researchers, it is tested on 10 different benchmark functions. Results revealed that although SSA gives better results as compared to other state art algorithms, yet it needs to be modified because there are some drawbacks in the result which are less optimality, low convergence, etc.

2 Explanation of Sparrow Search Algorithm (SSA)

According to the law of nature in this whole world for the living creatures to stay alive, they need food. From these creatures, some creatures make their own food for example plants and other some creatures look for food made by others and eat it such as animals, sparrows. Family of sparrows plays a vital role in foraging process. In this whole process sparrows are distributed into two categories. The very first one is revealer (producer) and the second one is believer (scrounger). From the whole population, 20% of sparrows which contain a huge level of power (having the best fitness value) acts as revealer (producer) and rest 80% as a believer (scrounger). It can be seen from the given Fig. 1.

The believer walks on the path indicated by the revealer to obtain food. During this food searching process, sometimes believers play the role of revealers and vice versa, but the ratio of revealer and believer remains constant. Through this process, when some sparrows feel the presence of danger around them, then these sparrows lead the whole population to a safe area.

The position of revealer (producer) is updated by using the following Eq. no. (1).

$$X_{i,j}^{t+1} = \begin{cases} X_{i,j}^{t} \cdot \exp\left(\frac{-i}{\alpha \cdot \text{iter}_{\max}}\right) & \text{if } R_2 < \text{S T} \\ X_{i,j}^{t} + Q \cdot L & \text{if } R_2 \geq \text{ST} \end{cases} \tag{1}$$

In the above Eq. (1), j varies from 1 to d and the current iteration is represented by t. $\alpha \in (0, 1]$ considers as a random number. $X_{i,j}$ depicts the place of the present ith sparrow in jth dimension. A constant number having maximum number of iterations

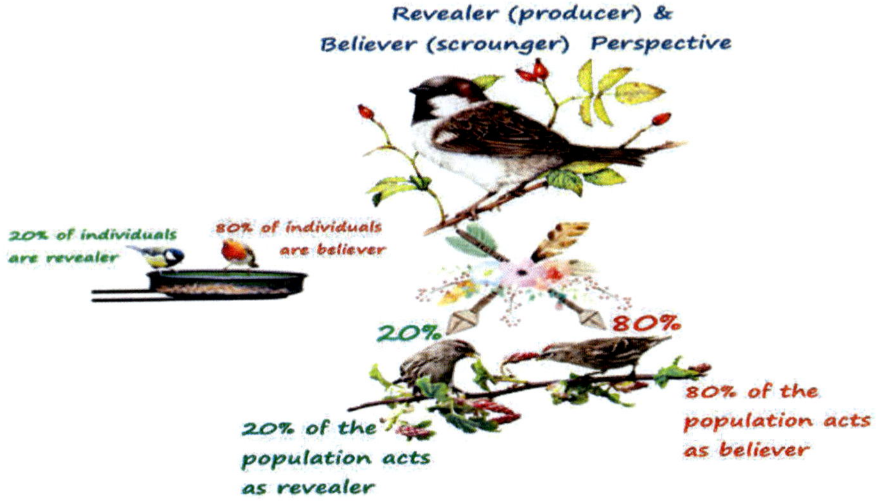

Fig. 1 Distribution of revealers (producers) and believers (scroungers)

denoted by " iter $_{max}$". "ST (ST $\in$ [0.5, 1.0])" perform the role of safety threshold whereas "$R_2(R_2 \in [0, 1])$" as alarm value. L stands for $1 \times d$ matrix in which each element is 1. Q appears as a random number and that follows normal distribution.

When "$R_2 <$ ST", that means there is no threat around the sparrows and revealer can search on the large-scale mode whereas "$R_2 \geq$ ST" shows the threat around the sparrows and then all sparrows immediately move toward the safe area.

The position of believer (scrounger) is updated by using the following Eq. no. (2).

$$X_{i,j}^{t+1} = \begin{cases} Q \cdot \exp\left(\frac{X_{\text{worst}}^t - X_{i,j}^t}{i^2}\right) & \text{if } i > n/2 \\ X_p^{t+1} + \left|X_{i,j}^t - X_p^{t+1}\right| \cdot A^+.L & \text{if } i \leq n/2 \end{cases} \tag{2}$$

In the above Eq. (2), X_{worst} is the present global worst position while X_p act as optimal location taken by revealer. Total number of sparrows is equal to n. A is a matrix of order $1 \times d$ in which each element is either 1 or -1, and also "$A^+ = A^T(A \cdot A^T)^{-1}$". When "$i > n/2$", it tells that the ith believer has bad fitness value does not get any food and can die because of hunger.

Sparrows, which are aware of the danger lead the whole population in the safe area by using the Eq. no. (3).

$$X_{i,j}^{t+1} = \begin{cases} X_{\text{best}}^t + \beta \cdot \left|X_{i,j}^t - X_{\text{best}}^t\right| & \text{if } f_i > f_g \\ X_{i,j}^t + K \cdot \left(\frac{\left|X_{i,j}^t - X_{\text{worst}}^t\right|}{(f_i - f_w) + \varepsilon}\right) & \text{if } f_i = f_g \end{cases} \tag{3}$$

In the above Eq. (3), X_{best} gives the present global optimal location. "β", being the step high control parameter follows the standard normal distribution of random numbers "$N(0,1)$" even if "$K \in [-1, 1]$" is a random number. The fitness value of the current sparrow is described by "f_i" moreover "f_w" and "f_g" personify the present worst and global best fitness values, subsequently. ε is the littlest constant to ignore the error of zero division. When "$f_i > f_g$", represents the sparrows on the border of the group however "$f_i = f_g$" personifies that the sparrows in the middle of the group are completely alert from danger and should have to shift their place near toward others.

3 Flow Chart of Sparrow Search Algorithm (SSA)

From the Fig. 2, we can see working of SSA for optimization by considering step-wise procedure.

Step 1: In order to start the working of the algorithm, firstly define the optimization problem and then initialize the population.The population (position of sparrows) can be expressed by the subsequent matrix:

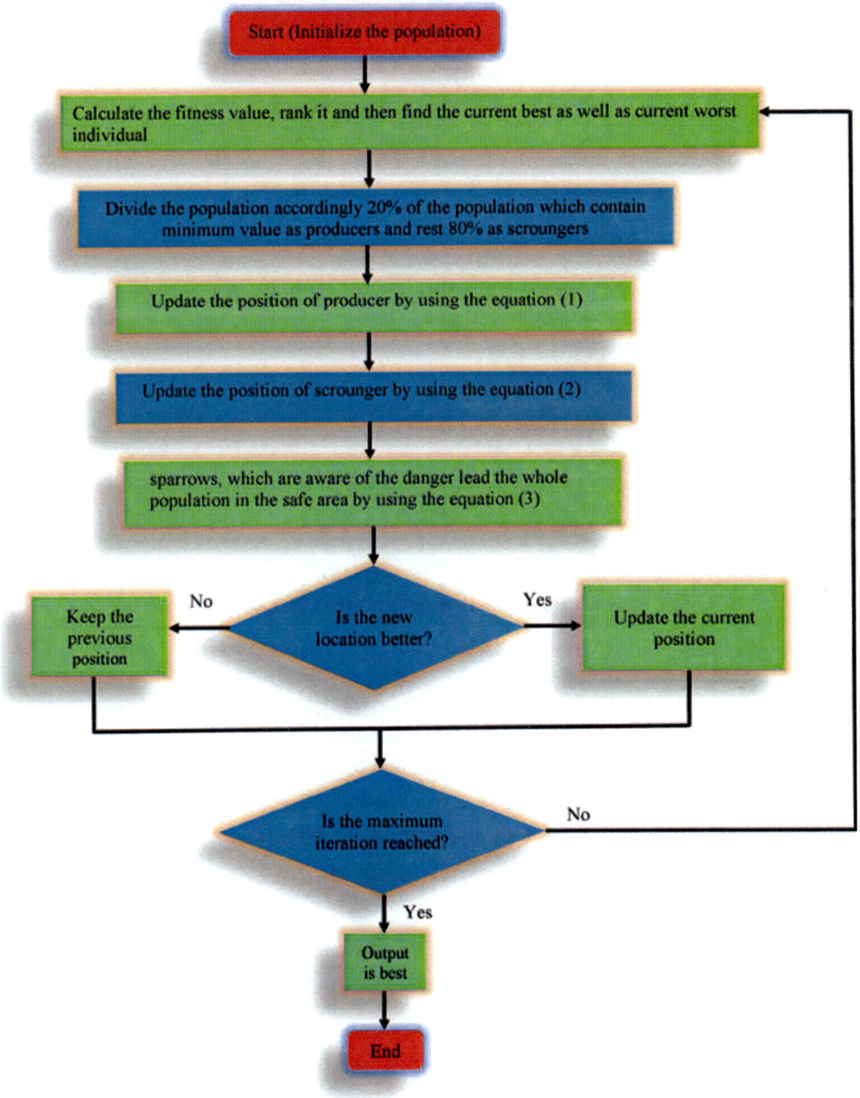

Fig. 2 Flow chart of SSA

$$X = \begin{bmatrix} x_{1,1} & x_{1,2} & \cdots\cdots & x_{1,d} \\ x_{2,1} & x_{2,2} & \cdots\cdots & x_{2,d} \\ \vdots & \vdots & \vdots \quad \vdots & \vdots \\ x_{n,1} & x_{n,2} & \cdots\cdots & x_{n,d} \end{bmatrix} \tag{4}$$

In the above Eq. (4), n is the number of sparrows and d depicts the dimension of the variables to be optimized.

Step 2: After the initialization step, calculate the fitness value from the objective function of optimization problem and then rank it and find the current best as well as the current worst individual. The fitness value of all sparrows can be represented by the given Eq. (5)

$$F_X = \begin{bmatrix} f\left(\begin{bmatrix} x_{1,1} & x_{1,2} & \cdots\cdots & x_{1,d} \end{bmatrix}\right) \\ f\left(\begin{bmatrix} x_{2,1} & x_{2,2} & \cdots\cdots & x_{2,d} \end{bmatrix}\right) \\ \vdots \quad \vdots \quad \vdots \quad \vdots \quad \vdots \\ f\left(\begin{bmatrix} x_{n,1} & x_{n,2} & \cdots\cdots & x_{n,d} \end{bmatrix}\right) \end{bmatrix} \tag{5}$$

Step 3: Divide the whole population into 20% of that sparrows which contain the best fitness value and are named as revealer and the rest 80% as a believer.

Step 4: Update the position of the revealer (producer) by using Eq. (1).

Step 5: Update the position of the believer (scrounger) by using Eq. (2).

Step 6: After the updation in the positions of revealer and believer, update the position of the sparrows who are aware of the danger and will lead the whole population in the safe area by using Eq. (3).

Step 7: In this step, get the current new location and if the new location is better then, update the current position otherwise keep the previous position.

Step 8: Termination criterion.

Stop if maximum iteration is reached that means output is best; otherwise, repeat it from Step 2.

4 Experimental Outcomes as Well as Analysis

4.1 Tuning of the Parameters and Environment of Experiment

Every algorithm runs on Window10, the 64-bit operating system having 16.0 GB RAM and the processor is Intel(R) Core(TM) i5-7300U CPU @ 2.60, 2.71 GHz. Coding of all algorithms implemented by MATLAB R2014a. Ten standard test functions are taken to check the effectiveness, reliability, optimality, and convergence also of the "sparrow search algorithm (SSA)", it is compared with other state art

Table 1 Standard benchmark test functions

Function	Function's name	DIM	Range	Min
$F_1(z) = 0.1\left\{\sin^2(3\pi z_1) + \sum_{i=1}^{n}(z_i-1)^2\left[1+\sin^2(3\pi z_1+1)\right] + (z_n-1)^2\left[1+\sin^2(2\pi z_n)\right]\right\} + \sum_{i=1}^{n} h(z_i, 5, 100, 4)$ $h(z_i,\alpha,l,k) = \begin{cases} l(z_i-\alpha)^k & z_i > \alpha \\ 0 & -\alpha < z_i < \alpha \\ l(-z_i-\alpha)^k & z_i < -\alpha \end{cases}$	Generalized Penalized function	30	[− 50, 50]	0
$F_2(z) = \left(\frac{1}{500} + \sum_{j=1}^{25}\frac{1}{j+\sum_{i=1}^{2}(z_i-\alpha_{ij})^6}\right)^{-1}$	Shekel's Foxholes function	02	[− 65.536, 65.536]	1
$F_3(z) = \sum_{i=1}^{11}\left[\alpha_i - \frac{z_1(\beta_i^2+\beta_i z_2)}{\beta_i^2+\beta_i z_3+z_4}\right]^2$	Kowalik's function	04	[− 5, 5]	0.00030
$F_4(z) = \left(z_2 - \frac{5.1}{4\pi^2}z_1^2 + \frac{5}{\pi}z_1 - 6\right)^2 + 10\left(1-\frac{1}{8\pi}\right)\cos z_1 + 10$	Branin function	02	[− 5, 5]	0.398
$F_5(z) = \left[1+(z_1+z_2+1)^2\left(19-14z_1+3z_1^2-14z_2+6z_1z_2+3z_2^2\right)\right] \times \left[30+(2z_1-3z_2)^2 \times \left(18-32z_1+12z_1^2+48z_2-36z_1z_2+27z_2^2\right)\right]$	Goldstein-price function	02	[− 2, 2]	3
$F_6(z) = -\sum_{i=1}^{4} c_i \exp\left(-\sum_{j=1}^{6}\alpha_{ij}\left(z-p_{ij}\right)^2\right)$	Hartman's family	06	[0, 1]	− 3.32
$F_7(z) = -\sum_{i=1}^{10}\left[(z-\alpha_i)(z-\alpha_i)^T + c_i\right]^{-1}$	Shekel's family	04	[0, 10]	− 10.5363
$F_8(z) = \sum_{i=1}^{D}\left(10^6\right)^{\frac{i-1}{D-1}} z_i^2$	High conditioned elliptic function	30	[− 10.0, 10.0]	0
$F_9(z) = \frac{1}{2}\sum_{i=1}^{n}\left(z_i^4 - 16z_i^2 + 5z_i\right)$	Styblinski-tang function	30	[− 5, 5]	− 78.332
$F_{10}(z) = -\exp\left(-0.5\sum_{i=1}^{D} z_i^2\right)$	Exponential Function	30	[− 1, 1]	1

algorithms such as "grey wolf optimization (GWO)", "particle swarm optimization (PSO)", and "binary bat algorithm (BBA)". Maximum search agents' number that means the size of population (n) in every algorithm is 100 and maximum number of iterations are 1000. The parameter of SSA is ST = 0.8. Parameters of GWO are: $r1$ and $r2$ are random numbers in [0, 1], whereas **a** decrease linearly from 2 to 0. Parameters of BBA are: $r = 0.1$ (pulse rate), $A = 0.25$ (Loudness), minimum frequency is 0, and on the other hand, the maximum frequency is 2. Parameters of PSO are: inertia weight is taken as according to the given formula which is "$\omega = 0.5 - (0.2 \times (1 \div \text{iter}_{\max}))$". To check the algorithm appropriately, in every case then we run each algorithm 30 times separately and calculate the optimal value, average value, and also standard deviation of every algorithm.

Table 2 Output of test functions

Fun	Algorithm	Best	Ave	Std
*F*1	SSA	4.5765e − 18	5.044207583333e − 15	1.3463723082180721e − 14
	GWO	3.8482e − 06	1.25673e − 01	1.059824013e − 01
	PSO	6.9898e − 26	1.464933e − 03	3.798710243e − 03
	BBA	1.3498e − 32	1.3498e − 32	0.0
*F*2	SSA	9.98e − 01	2.62815e + 00	3.5152357548e + 00
	GWO	9.98e − 01	2.4374767e + 00	2.9447274364e + 00
	PSO	9.98e − 01	9.98e − 01	0.0
	BBA	1.26705e + 01	1.26705e + 01	0.0
*F*3	SSA	3.0749e − 04	3.0749e − 04	0.0
	GWO	3.0749e − 04	2.37408867e − 03	6.1032003821e − 03
	PSO	3.0749e − 04	4.6486833e − 04	3.47955463e − 04
	BBA	1.4841e − 01	1.4841e − 01	0.0
*F*4	SSA	3.9789e − 01	3.9789e − 01	0.0
	GWO	3.9789e − 01	3.9789e − 01	0.0
	PSO	3.9789e − 01	3.9789e − 01	0.0
	BBA	2.77029e + 01	2.77029e + 01	0.0
*F*5	SSA	3.00e + 00	3.00e + 00	0.0
	GWO	3.00e + 00	3.00e + 00	0.0
	PSO	3.00e + 00	3.00e + 00	0.0
	BBA	6.00e + 02	6.00e + 02	0.0
*F*6	SSA	− 3.322e + 00	− 3.26255e + 00	6.046631294e − 02
	GWO	− 3.322e + 00	-3.2435767e + 00	7.929005975e − 02
	PSO	− 3.322e + 00	-3.25244e + 00	6.2983553409e − 02
	BBA	− 1.6572e − 01	− 1.6572e − 01	0.0
*F*7	SSA	− 1.05364e + 01	− 1.05364e + 01	0.0
	GWO	− 1.05364e + 01	− 1.053619e + 01	1.213430589e − 04
	PSO	− 1.05364e + 01	− 7.9694033e + 00	3.303911101e + 00
	BBA	− 5.12850e + 00	− 5.12850e + 00	0.0
*F*8	SSA	0.0	0.0	0.0
	GWO	2.1762e − 86	1.977419033333e − 84	3.2551578800987726e − 84
	PSO	2.2645e − 25	2.862509033333e − 23	3.7259065030217487e − 23
	BBA	0.0	0.0	0.0
*F*9	SSA	− 1.17499e + 03	− 1.17499e + 03	0.0
	GWO	− 1.03698e + 03	− 9.519211467e + 02	5.598879607e + 01
	PSO	− 1.14671e + 03	− 1.090164663e + 03	3.0613928372e + 01
	BBA	− 1.50e + 02	− 1.50e + 02	0.0
*F*10	SSA	− 1.00e + 00	− 1.00e + 00	0.0
	GWO	− 1.00e + 00	− 1.00e + 00	0.0
	PSO	− 1.00e + 00	− 1.00e + 00	0.0
	BBA	− 1.00e + 00	− 1.00e + 00	0.0

Ten benchmark test functions are given in Table 1, out of which one function is unimodal test function, three multimodal test functions, and six fixed-dimension test functions are to test the sparrow search algorithm. The experimental output of all algorithms is given in the following Table 2.

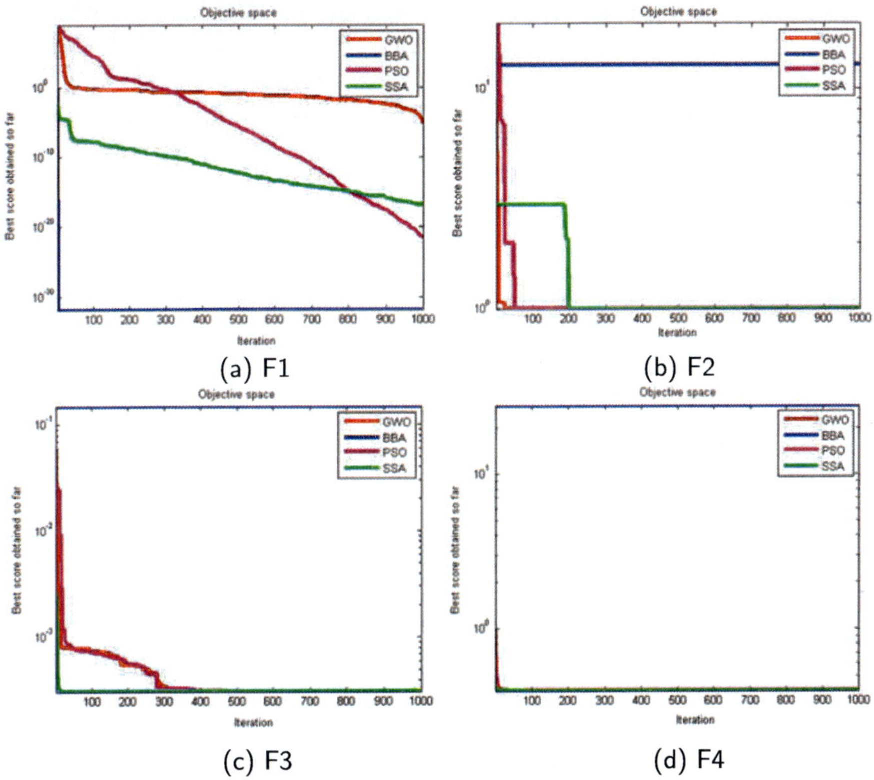

Fig. 3 Convergence curves of four algorithms tested on test functions ($F1$–$F4$)

4.2 Analysis From the Optimal Point of View

From the Table 2 it is very clear that sparrow search algorithm (SSA) is not superior on other three algorithms with respect to all ten test functions. In $F1$, BBA gives best optimal solution in relation to SSA, PSO, GWO. From the function $F2$ to $F7$ same optimal solution is taken by SSA, GWO, and PSO while BBA doesn't give optimal solution. It can be observed that in function $F8$ and $F10$ approximately all algorithms deliver the optimal solution although in $F9$ only BBA shows the best optimal solution as compare to SSA, PSO and GWO.

4.3 Analysis From Convergence Point of View

In this paper, we would like to study the convergence rate of SSA with standard evolutionary algorithms such as PSO, GWO, and BBA.

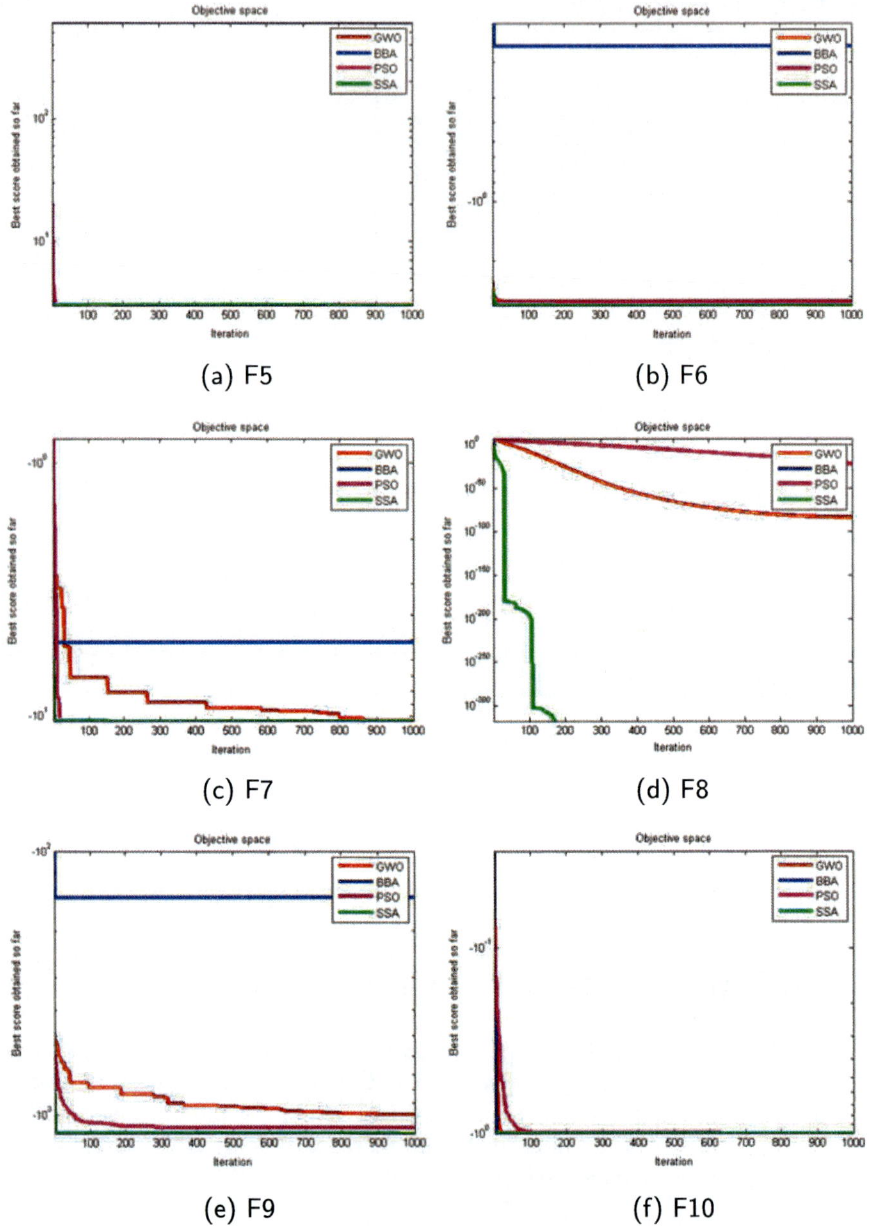

Fig. 4 Convergence curves of four algorithms tested on test functions ($F5$–$F10$)

From the Figs. 3 and 4, it is crystal clear that convergence rate of SSA cannot be faster than the other algorithms in all cases but in some cases it can be. Sparrow search algorithm (SSA) displays the better convergence rate in $F3$ and $F7$ as compare to BBA, PSO, and GWO but SSA, BBA, and GWO approximately converge on the same rate in the function $F10$. Convergence speed of BBA in F8 is faster. From the functions $F4$, $F5$, $F6$ it can be seen that SSA, GWO, and PSO shows the same good convergence rate but BBA not while in $F1$ and $F9$, BBA's convergence rate is more effective in comparison with others. Convergence of GWO in $F2$ function is excellent as compare to rest algorithms.

5 Conclusion

SSA is a metaheuristic swarm optimization technique which is originate to obtain best optimal solution in all aspects. But still some researchers find out few drawbacks in it such as slow converge speed and less optimality etc. In this paper, we have done experiment through MATLAB by using 10 different benchmark test function on SSA, GWO, PSO, and BBA. Then, it is recognized that the statements of researchers are correct that means yet certain number of limitations are found in SSA such as slow convergence, fall into the local optima etc. So, modification of SSA algorithm is suggested for further experiment.

References

1. Desale S, Rasool A, Andhale S, Rane P (2015) Heuristic and meta-heuristic algorithms and their relevance to the real world: a survey. Int J Comput Eng Res Trends 351(5):2349–7084
2. Karaboga D (2005) An idea based on honey bee swarm for numerical optimization. vol 200. Technical report-tr06, Erciyes university, engineering faculty, computer engineering department, pp 1–10
3. Mirjalili S, Mirjalili SM, Lewis A (2014) Grey wolf optimizer. Adv Eng Soft 69:46–61
4. Akhtar A (2019) Evolution of ant colony optimization algorithm–a brief literature review. arXiv preprint arXiv:1908.08007
5. Boeringer DW, Werner DH (2005) Efficiency-constrained particle swarm optimization of a modified Bernstein polynomial for conformal array excitation amplitude synthesis. IEEE Trans Antennas Propag 53(8):2662–2673
6. Sun W, Tang M, Zhang L, Huo Z, Shu L (2020) A survey of using swarm intelligence algorithms in IoT. Sensors 20(5):1420
7. Nakamura RYM, Pereira LAM, Rodrigues D, Costa KAP, Papa JP, Yang XS (2013) Binary bat algorithm for feature selection. In: Swarm intelligence and bio-inspired computation. Elsevier, pp 225–237
8. Xue J, Shen B (2020) A novel swarm intelligence optimization approach: sparrow search algorithm. Syst Sci Control Eng 8(1):22–34
9. Song W, Liu S, Wang X, Wu W (2020) An improved sparrow search algorithm. In: 2020 IEEE Intl conf on parallel & distributed processing with applications, big data & cloud computing, sustainable computing & communications, social computing & networking (ISPA/BDCloud/SocialCom/SustainCom). IEEE, pp 537–543

10. Peng Y, Liu Y, Li Q (2020) The application of improved sparrow search algorithm in sensor networks coverage optimization of bridge monitoring. In: MLIS, pp 416–423
11. Chengtian O, Yujia L, Donglin Z (2021) An adaptive chaotic sparrow search optimization algorithm. In: 2021 IEEE 2nd International conference on big data, artificial intelligence and internet of things engineering (ICBAIE). IEEE, pp 76–82
12. Li J (2021) Robot path planning based on improved sparrow algorithm. J Phys Conf Ser 1861(1):012017. IOP Publishing
13. Chen X, Huang X, Zhu D, Qiu Y (2021) Research on chaotic flying sparrow search algorithm. J Phys Conf Ser 1848(1):012044. IOP Publishing
14. Ouyang C, Qiu Y, Zhu D (2021) A multi-strategy improved sparrow search algorithm. J Phys Conf Ser 1848(1):012042. IOP Publishing
15. Lv X, Mu X, Zhang J (2021) Multi-threshold image segmentation based on improved sparrow search algorithm. Syst Eng Electr 43(2):318–327
16. Ouyang C, Zhu D, Qiu Y (2021) Lens learning sparrow search algorithm. Math Prob Eng
17. Zhang C, Ding S (2021) A stochastic configuration network based on chaotic sparrow search algorithm. Knowl-Based Syst 220:106924
18. Wolpert DH, Macready WG (1997) No free lunch theorems for optimization. IEEE Trans Evol Comput 1(1):67–82

Solving FJSP Using Multi-agent System with GA

Manojkumar Pal, Murari Lal Mittal, Gunjan Soni, and Manish kumar

Abstract Industry 4.0 is a new paradigm of manufacturing which is now being researched and implemented in academic and industrial domains. This new paradigm supports agility, automation, responsiveness, and distributed processes in manufacturing systems. In manufacturing, scheduling is a perennial problem and considered to be one of the most difficult problems to be solved. Flexible job shop scheduling problem (FJSP) got attention of researchers due to its closeness to current advanced manufacturing systems. In this paper, the problem of FJSP is considered to minimize the makespan. In terms of agility, flexibility, and adaptability, decentralized approaches like multi-agent systems are suited for addressing such complex problems. This paper proposes a multi-agent system (MAS) approach for solving the FJSP. To optimize the results, genetic algorithm (GA) is also implemented with a multi-agent system. The proposed approach is evaluated by solving standard problem instances, and the results are promising.

Keywords Flexible job shop problem · Scheduling · Decentralized approach, · Multi-agent system · Genetic algorithm

1 Introduction

Complexity in manufacturing processes has increased with time due to the need of high production flexibility which enforces machines to be more flexible and highly productive. The job shop scheduling problem (JSP), where machine selection is predefined for each operation, needs to find optimal sequencing of operations [1].

M. Pal · M. L. Mittal (✉) · G. Soni · M. kumar
Department of Mechanical Engineering, MNIT JAIPUR, Jaipur 302017, India
e-mail: mlmittal.mnit@mnit.ac.in

M. Pal
e-mail: 2017rme9086@mnit.ac.in

G. Soni
e-mail: gsoni.mech@mnit.ac.in

S. Das et al. (eds.), *Advances in Data-driven Computing and Intelligent Systems*, Lecture Notes in Networks and Systems 653, https://doi.org/10.1007/978-981-99-0981-0_4

An extension of JSP, popularly known as Flexible job shop scheduling problem (FJSP), considers flexible (multi-purpose) machines which can process all or some operations of the jobs. If a machine can do all the operations on all the jobs, then it is considered to be a total flexible FJSP and a partial flexible FJSP. Solution to FJSP involves solving two problems: machine selection (allocation of an operation on a machine from the set of available machines) and operation sequencing (sequencing of the operations on the selected machines).

In the last decade, FJSP has been a popular research topic in scheduling problems. JSP is considered an NP hard problem [2, 3]. So, the FJSP is also NP-hard problem for being its extension. Various approaches have been proposed for FJSPs, but exact algorithms are not suitable for large size problems. So, approximate algorithms, e.g., GA [1, 2, 4, 5], particle swarm optimization (PSO) [6], discrete artificial bee colony optimization (DABC) [3], Jaya algorithm [7], Tabu Search (TS) [8, 9], biogeography-based optimization (BBO) [10], simulated annealing (SA) [11] have been proposed to solve FJSPs. These approximate algorithms have considered a centralized decision-based approach. Though very popular to solve FJSPs, these have limitations in terms of high computation workload and lower convergence with optimal scheduling for large problem size. Therefore, decentralized approaches were suggested to solve FJSP [12]. Multi-agent system (MAS) is one of the decentralized approaches, but slight attention has been paid to use it for solving FJSP. This paper proposes a multi-agent approach for solving the FJSP.

2 Literature Review

The FJSP is probably first introduced by Brucker and Schlie [13], which later on attracted the attention of the researcher community. Several approximate algorithms using meta-heuristics have been proposed to solve FJSP such as TS by Brandimarte [8] and Kacem et al. [14]. Hybrid DABC is developed by Li et al. [3]. In the research by Pan et al. [4], effective GA is proposed to solve FJSP to minimize the makespan. The BBO algorithm is proposed by Rahmati et al. [10] while Li et al. [2] and Mokhtari et al. [15] have applied GA + TS and GA + SA to solve FJSP, respectively. Similarly, in the research article by Huang et al. [16], discrete PSO is applied to solve FJSP. Improved GA and improved PSO have also been applied to solve FJSP by Zhang et al. [1] and Ding et al. [6], respectively. Most of these existing approaches have considered, however, centralized decision-based approaches where a single decision maker decides all decisions to optimize the objective(s). These centralized approaches have disadvantages of high computation load and lower convergence for scheduling for large size problems [17, 18].

This limitation can be overcome by decentralized approaches such as MAS. In MAS, the problem is divided into small sub-problems, and these sub-problems are cooperatively handled by two or more agents to fulfill the objectives. In literature, few MAS have been applied to solve the FJSP. For optimization of the objective(s), these MAS is embedded with meta-heuristics, e.g., MAS developed by Ennigrou and

Ghedira [9] considered three types of agents (interface, job, and machine) with TS for minimizing makespan. This MAS is simplified by Henchiri and Ennigrou [19] in which two agents (interface and machine) are considered with PSO to solve FJSP. In the research work by Nouri et al. [20], MAS is proposed with two classes of agents (scheduler agent and cluster agents) embedded with GA and TS. In recent work by Xiong and Fu [18], humoral immunity-based new immune multi-agent systems are proposed using three agents. This work proposes a multi-agent system using three classes of agents with novel auction mechanisms to solve FJSP. GA is one of the most popular meta-heuristics for global search of the solution space. So, the proposed MAS is employed with GA for optimization.

3 The Problem Definition

The formulation of FJSP is described as follows: A set of n jobs (1... n) is considered. Job i need a number of operations. The operations are to be performed on m machines (1... k). OP_{ih} is the hth operation of ith job; PT_{ihk} is the processing time for hth operation of ith job on the machine k. $C_{\max}$ is the makespan (maximum completion time of all the jobs). C_{ih} is the completion time for hth operation of ith job. Followings are assumptions for the problem:

- Jobs and machines are independent.
- All the jobs and machines are considered to be available at time zero.
- Pre-emption is not permitted.
- Only one operation of a job can only be executed at a time on a machine selected from the set of available machines.
- At a time, machines can perform an operation only.

The objective is to minimize the makespan.

Decision Variables

$$X_{ihk} = \begin{cases} 1, & \text{if } OP_{ih} \text{ performed by } M_k \\ 0, & \text{otherwise} \end{cases}$$

Objective Function

$$\text{Min } Z = C_{\max}$$

Subject to:

$$C_{ih} - C_{i(h-1)} \geq PT_{ihk} \cdot X_{ihk} \tag{1}$$

$$\sum_k X_{ihk} = 1 \tag{2}$$

$$[C_{ihk} - PT_{ihk}, C_{ihk}] \cap [C_{dek} - PT_{dek}, C_{dek}] = \emptyset \quad (3)$$

Constraint (1) ensures the operation precedence constraints. Constraint (2) ensures that an operation can only be performed by a machine at a time. Constraint (3) enforces non-overlap of two consecutive operations (C_{ihk}, C_{dek}) where i ≠ d, on the same machine.

4 The Architecture of Multi-agent System

This section describes the proposed multi-agent system (MAS) for solving FJSP. Production reservation (PR) approach is developed by Saad et al. [21], where an auction mechanism is implemented between two types of agents to select the machine for a job operation, modified for the proposed approach. Figure 1 illustrates the architecture of the MAS. It has considered three agent classes: scheduler agent, job agent, and machine agent. The scheduler agent is a class which represents an individual agent. Job agent class has multiple agents, and each job agent represents one job. Similarly, machine agent classes have multiple agents, and each machine agent represents each resource/machine.

Scheduling agent creates job agents and machine agents and also generates the initial solutions in the form of operation sequence, e.g., {2, 1, 1, 2, 3, 3} where each

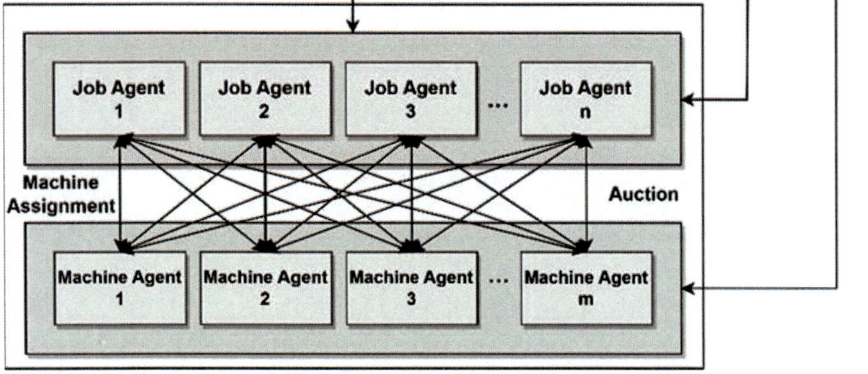

Fig. 1 Architecture of the multi-agent system

number represents a job and its multiple appearance shows the operation number to be carried out. As per job order in OS sequence, the respective job agent gets activated and sends calls for proposal (CFP) for processing of an operation to all machine agents. On receiving CFP, the machine agent submits a bid, i.e., expected completion time of the operation to the auctioneer (JA). Then, the job agent selects the minimum bid. In case of ties, a bid is selected randomly. The selection offer is sent to the winner MA for the operation. After assigning all job operations to the machines, the generated OS schedules are sent to the scheduler agent for optimization (global search). Here, GA, a popular meta-heuristic, is applied to optimize the makespan results. For application of GA, two-point crossover is used with swap mutation operator. This leads to generating new off-springs (solutions) which further become the part of the pool for the machine selection process as discussed above. The steps of the algorithm are detailed in the flowchart as shown in Fig. 2

5 Experimental Findings

This section discusses the evaluation of results obtained by the proposed multi-agent approach and its comparison of the solutions obtained by the state-of-the-art algorithms. The proposed algorithm is executed in Python programming (3.7.7). The algorithm was run on Intel Core i5-8250U CPU, 3.6 GHz, and 8 GB RAM machine. From the literature and experimental observation, the value of population size as well as number of iterations are considered 50. The crossover probability is considered as 0.8, and the mutation probability is taken as 0.1. Each problem instance has been solved independently for 10 times, and the best result is reported.

To illustrate a simple 3×3 problem with processing time, data are given in Table 1. The makespan obtained for an OS solution $\{1, 3, 2, 2, 1, 3, 1, 2, 2\}$ after using the proposed algorithm is 16, and the Gantt chart is shown in Fig. 3. For comparison, three problem instances of sizes 4×5, 8×8, and 10×7 taken from [14] and one problem instance (8×5) from [22] are solved, and the result is compared with the algorithms: GATS-HM proposed by Nouri [23], hGA by Li and Gao [2], improved hybrid-PSO (IH-PSO) by Zhang et al. [24], hybrid discrete PSO (HD-PSO) by Li and Kui [22]. Table 2 shows the minimum makespan value and average makespan by the proposed approach and benchmark algorithms. In order to test the efficacy of the problem, complex problems of size 10×10 and 15×10 taken from [25] are also solved. The results can be seen in Table 2. It can be seen that the proposed multi-agent approach obtains the best results in comparison with existing algorithms to be considered in this work. The Gantt chart for problem 8×5 is shown in Fig. 4. Different color rectangle bars show operation times (Op21 means second job's first operation).

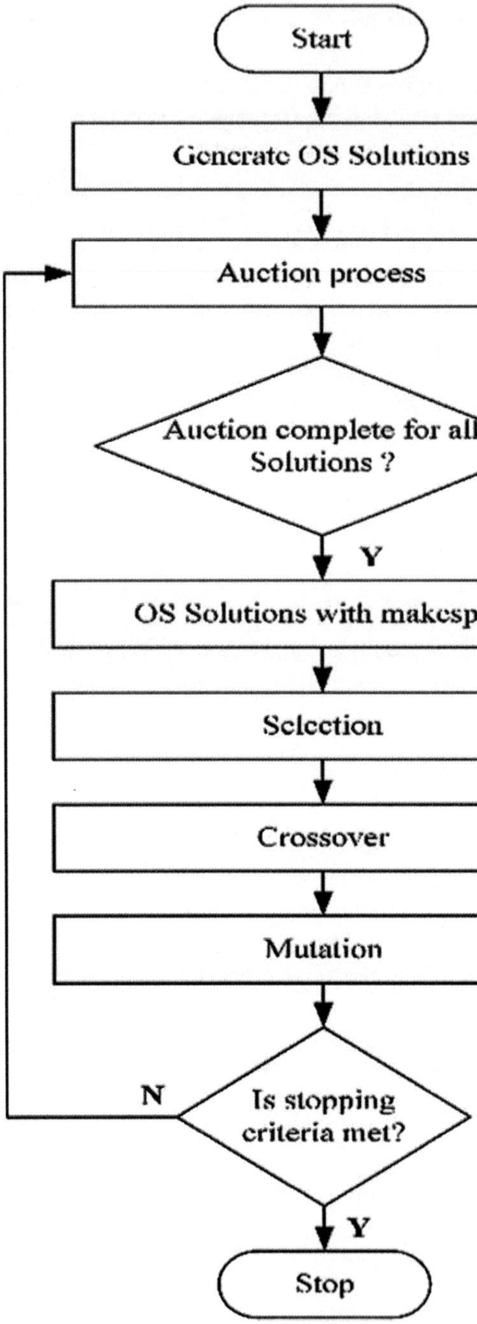

Fig. 2 Flowchart of the proposed approach

Table 1 Processing time data for the problem instance 3 × 3

Job No.	Operation	M_1	M_2	M_3
J_1	OP_{11}	3	4	5
	OP_{12}	4	1	3
	OP_{13}	7	5	2
J_2	OP_{21}	5	3	2
	OP_{22}	4	6	–
	OP_{23}	8	4	2
	OP_{24}	6	5	–
J_3	OP_{31}	2	4	4
	OP_{32}	4	5	3

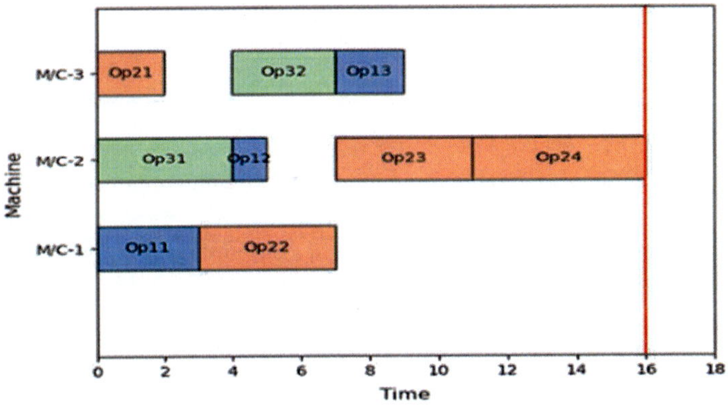

Fig. 3 Gantt chart for the problem 3 × 3

Table 2 Results of makespan for the three Kacem instances

Problem instance	GATS + HM	hGA	IH-PSO	HD-PSO		Our algorithm	
	Min.	Min.	Min.	Min.	Avg.	Min.	Avg.
4 × 5	11	11	11	11	11	11	11
8 × 8	14	14	14	14	14.2	14	14.2
10 × 7	11	11	11	11	11	11	11
10 × 10	7	7	7	7	7	7	7
15 × 10	11	11	11	11	12.7	11	11.1
8 × 5	NA	NA	NA	27	27.9	27	28.0

NA not available

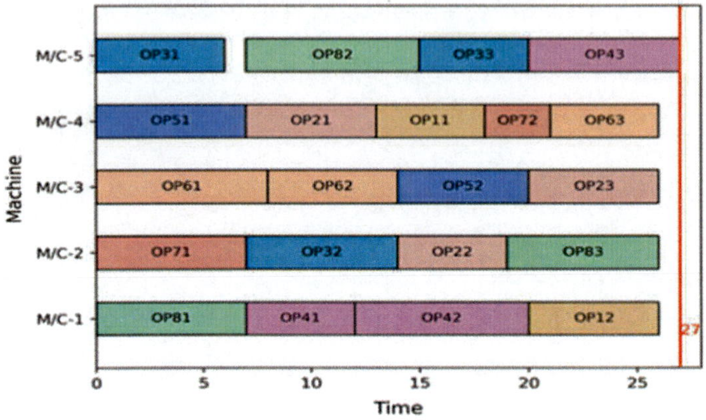

Fig. 4 Gantt chart for the problem instance 8 × 5

6 Conclusion

This paper has considered a multi-agent-based approach for solving FJSP. Decentralized system is considered very useful to solve such complex systems due to their inherent nature of decomposing the main complex problem into multiple and simple sub-problems. The multi-agent approach has been proposed with three agents: scheduler agent, job agents, and machine agents. These agents cooperatively generate initial solutions in form of operation sequences and select machines for each operation using auction-based approach. The various problem instances have been solved using the proposed approach. The results are compared with respect to makespan with the benchmarked algorithms. The promising results have shown the effectiveness of the proposed approach. In future work, we would like to perform more study on large-scale instances of the FJSP. It can be improved for inclusion of other objectives along with makespan which can also prove effectiveness of the approach.

References

1. Zhang G, Hu Y, Sun J, Zhang W (2020) An improved genetic algorithm for the flexible job shop scheduling problem with multiple time constraints. Swarm Evol Comput 54(2):100664
2. Li X, Gao L (2016) An effective hybrid genetic algorithm and tabu search for flexible job shop scheduling problem. Int J Prod Econ 174:93–110
3. Li JQ, Pan QK, Tasgetiren MF (2013) A discrete artificial bee colony algorithm for the multi-objective flexible job-shop scheduling problem with maintenance activities. Appl Math Model 38(3):1111–1132
4. Pan Y, Zhang WX, Gao TY, Ma QY, Xue DJ (2011) An adaptive genetic algorithm for the flexible job-shop scheduling problem. Proceedings–2011 IEEE Int Conf Comput Sci Autom Eng CSAE 2011 4(4):405–409

5. Meziane ME, Taghezout N (2018) A hybrid genetic algorithm with a neighborhood function for flexible job shop scheduling 14:161–175
6. Ding H, Gu X (2020) Improved particle swarm optimization algorithm based novel encoding and decoding schemes for flexible job shop scheduling problem. Comput Oper Res 121
7. Caldeira RH, Gnanavelbabu A (2019) Solving the flexible job shop scheduling problem using an improved Jaya algorithm. Comput Ind Eng 137(August):106064
8. Brandimarte P (1993) Routing and scheduling in a flexible job shop by tabu search. Ann Oper Res 41(3):157–183
9. Ennigrou M, Ghédira K (2008) New local diversification techniques for flexible job shop scheduling problem with a multi-agent approach. Auton Agent Multi-Agent Syst 17(2):270–287
10. Rahmati SHA, Zandieh M (2012) A new biogeography-based optimization (BBO) algorithm for the flexible job shop scheduling problem:1115–1129
11. Tamssaouet K, Dauzère-Pérès S, Knopp S, Bitar A, Yugma C (2022) Multiobjective optimization for complex flexible job-shop scheduling problems. Eur J Oper Res 296(1):87–100
12. Zhang J, Ding G, Zou Y, Qin S, Fu J (2019) Review of job shop scheduling research and its new perspectives under Industry 4.0. J Intell Manuf 30(4):1809–1830
13. Brucker P, Schlie R (1990) Job-shop scheduling with multi-purpose machines. Comput 45(4):369–375. https://doi.org/10.1007/BF02238804
14. Kacem I, Hammadi S, Borne P (2002) Approach by localization and multiobjective evolutionary optimization for flexible job-shop scheduling problems. IEEE Trans Syst Man Cybern Part C Appl Rev 32(1):1–13
15. Mokhtari H, Hasani A (2017) An energy-efficient multi-objective optimization for flexible job-shop scheduling problem. Comput Chem Eng 104:339–352
16. Huang S, Tian N, Wang Y, Ji Z (2016) Solving flexible job shop scheduling problem using a discrete particle swarm optimization with iterated local search. Commun Comput Inf Sci 643:603–612
17. Barbati M, Bruno G, Genovese A (2012) Applications of agent-based models for optimization problems: a literature review. Expert Syst Appl 39(5):6020–6028
18. Xiong W, Fu D (2018) A new immune multi-agent system for the flexible job shop scheduling problem. J Intell Manuf 29(4):857–873
19. Henchiri A, Ennigrou M (2013) Particle swarm optimization combined with tabu search in a multi-agent model for flexible job shop problem. In: Lecture notes in computer science (including subseries lecture notes in artificial intelligence and lecture notes in bioinformatics), 7929 LNCS(PART 2), pp 385–394
20. Nouri HE, Belkahla Driss O, Ghédira K (2018) Solving the flexible job shop problem by hybrid metaheuristics-based multiagent model. J Ind Eng Int 14(1):1–14
21. Saad A, Kawamura K, Biswas G (1997) Performance evaluation of contract net-based heterarchical scheduling for flexible manufacturing systems. Intell Autom Soft Comput 3(3):229–247
22. Li B, Kui C (2020) Research on FJSP with transportation time constraint based on improved particle swarm optimization. In: Proceedings—IEEE 18th international conference on dependable, autonomic and secure computing, IEEE 18th international conference on pervasive intelligence and computing, IEEE 6th international conference on cloud and big data computing and IEEE 5th Cybe, vol 63, pp 130–137
23. Nouri HE (2018) Solving the flexible job shop problem by hybrid metaheuristics-based multiagent model, pp 1–14
24. Zhang Y, Zhu H, Tang D (2020) An improved hybrid particle swarm optimization for multi-objective flexible job-shop scheduling problem. Kybernetes 49(12):2873–2892
25. Kacem I, Hammadi S, Borne P (2002) Pareto-optimality approach for flexible job-shop scheduling problems: hybridization of evolutionary algorithms and fuzzy logic. Math Comput Simul 60(3–5):245–276

A Comparative Analysis on Optimal Power Allocation and Pairing of User in Downlink NOMA System

Kaushik Bharadwaj and Chhagan Charan

Abstract Non-orthogonal multiple access (NOMA) is an emerging multiple access technology in obtaining higher capacity for the fifth-generation mobile communication network. To maximize the total data rate of the NOMA network, both user pairing and power allocation play a vital role. This paper compares the total system sum rate with two different user pairing techniques, Hungarian algorithm base user pairing and far-far (F–F) or near-near (N–N) user pairing and near far (N–F) user pairing technique. After the user pairing, two different power allocation techniques, optimal power allocation utilizing Karush–Kuhn–Tucker (KKT) constraints and dynamic power allocation (DYN), are applied to compare the total data rate of the system. The comparative results indicate that the system data rate of Hungarian algorithm base user pairing technique and optimum power allocation using KKT constraints is higher than that of the other user pairing technique with dynamic power allocation technique.

Keywords NOMA · Pairing of user and power allocation

1 Introduction

The 5G networks will play an eternal role in the era of smart generation [1–3] by serving mobile users with high data rates. However, this leads to a more challenging problem of spectrum scarcity. A wide variety of techniques such as, machine-to-machine (M2M) communication technique [5] and non-orthogonal multiple access techniques [4] were used to solve this problem. Yet, NOMA is the only successful technique that has arisen as a savior. Different power levels are used in non-orthogonal multiple access techniques to serve different users [6–8]. In NOMA, at the receiver

K. Bharadwaj (✉) · C. Charan
NIT Kurukshetra, Haryana 136119, India
e-mail: kaushiktez97@gmail.com

C. Charan
e-mail: chhagan.charan@nitkkr.ac.in

S. Das et al. (eds.), *Advances in Data-driven Computing and Intelligent Systems*, Lecture Notes in Networks and Systems 653, https://doi.org/10.1007/978-981-99-0981-0_5

side, the successive interference cancelation (SIC) process is used to remove the user's signal interference problem [9]. So, each user needs to perform SIC technique to identify its own signal. Superposition coding (SC) process is used to superimpose all individual signals into a single waveform in the transmitting end and the composite signal is transmitted by the transmitter [10]. In [11], the design of uplink transmission in the NOMA system has been addressed. The performance of fairness among the users in NOMA has been characterized in [12].

As we know, NOMA serves multiple users simultaneously at the same time but in reality, if many users are deployed in one frequency carrier, then its performance will begin to drop. So, we cannot deploy a large number of users per carrier permanently. As a solution to this problem, hybrid NOMA concept is brought into picture. Obviously, the performance of NOMA mostly depends on how we grouped the user within each available orthogonal resource and how we optimally allocate power for each of the multiplexed users. In recent era, most of the researchers mainly focus on the above mentioned problem so that they can increase the throughput in NOMA. To increase the throughput of the system, a fixed power allocation technique has been studied in [15]. However, for practical circumstances, this method is not considered suitable as it could not give an optimal solution. In [16], the power allotment problem in NOMA is also studied with the visible light communication system. In order to maximize the throughput of the system, an optimal power allocation technique considering fairness constraint is studied in [17]. A user pairing technique is applied in [18] for capacity maximization in NOMA, in which the pairing of the user is done based on their channel condition. The authors in [19] developed a genetic algorithm (GA)-based power allotment to increase the sum rate of the system. Its performance is then compared with full-search power allocation (FP) which has higher complexity than GA. Thus, it is necessary to recognize the balance between the system complexity and system performance. In [20], a greedy asynchronous distributed interference avoidance (GADIA) algorithm, which is a distributed algorithm, is applied to find the sub-optimal frequency band assignment to the users of the system. Device-to-device users applied GADIA for dynamic frequency allocation scheme.

Since both user pairing and optimum allocation of power have a significant consequence on the performance of downlink NOMA, this paper mainly focuses on the study and analysis of two optimum user pairing techniques and two optimum power allocation techniques. The optimum power allocation technique that we are using with the Hungarian algorithm is a non-convex problem. So, KKT constraints are used to find optimum solution of power allocation. Hungarian algorithm approach is adopted for optimally pairing two users in each of the subchannels. After pairing two users at each subchannel, KKT constraints are applied to find optimum power allocation. Another user pairing strategy that we discussed in this paper is N-F pairing and F–F or N–N user pairing, in which users are paired based on the distance gap of channel gain coefficient. After the user pairing, a dynamic power allocation scheme discussed in [21] is applied rather than considering fixed power allocation. We consider that the base station knows about the channel state information (CSI). At the receiving end, SIC technique is used for removing the user's signal interference problem. After all, the system data rate of the downlink NOMA network is compared among the user pairing approach by Hungarian algorithm with optimal power allocation and the conventional N–F pairing and F–F or N–N pairing with dynamic power allocation.

The remaining paper is organized as: Sect. 2 illustrates the system model. In Sect. 3, the user pairing strategy and power allotment schemes is described. In Sect. 4, simulation results are analyzed. At last, Sect. 5 concludes the work.

2 System Model

We consider a scenario of downlink NOMA system, where there is a single base station (BS) that servers K users $u = \{1, 2, \ldots, \mathrm{U}\}$ within a cell radius of R with C subchannels, respectively. We denote the group of subchannels as $c = \{1, 2, \ldots, \mathrm{C}\}$ and $\mathrm{UE}_{u,c}$ to be uth user in cth subchannel. The total available bandwidth we considered to be B and the bandwidth in each subchannel is $B_S = \frac{B}{C}$. The total BS transmit power is given by P_M and the power allotment at each subchannel is given by P_C. So, power budget over the subchannel is restricted to $\sum_{u=1}^{K} P_{u,c} = P_c$ and $\sum_{c=1}^{C} P_c = P_M$. . Let us assume that the base station has good information about the user's channel state information (CSI). The number of users in cth subchannel is given by K_c. The BS transmits the superposition of the user's modulated symbol on each subchannel c. The transmitted signal is given by:

$$x_c = \sum_{u=1}^{K} \sqrt{P_{u,c} Q_u} \tag{1}$$

where Q_u is the modulated symbol of uth user and $P_{u,c}$ is the allotment of power to the uth user in cth subchannel. The signal received at $\mathrm{UE}_{u,c}$ is given by:

$$y_{u,c} = h_{u,c} x_c + w_{u,c} \tag{2}$$

where $h_{u,c} = g_{u,c} d_u^{-\gamma}$ is the Rayleigh fading channel coefficient from base station (BS) to $\mathrm{UE}_{u,c}$, $g_{u,c}$ is the Rayleigh fading parameter. The distance between u^{th} user and the BS is denoted by d_u, γ represents pass loss exponent and $w_{u,c}$ is the Additive White Gaussian Noise (AWGN), i.e., $w_{u,c} \sim CN\left(0, \sigma_C^2\right)$. Let $H_{u,c} = |h_{u,c}|^2/N_O$ represents the channel response normalized by noise considering that uth user is allotted to cth subchannel. N_O is the noise power spectral density. We assume the user's channel are ordered as $H_{1,c} \geq H_{2,c} \geq H_{3,c} \geq \ldots H_{U,c}$. Therefore, $UE_{1,c}$ is the nearest user which has good channel condition and $UE_{U,c}$ is the farthest user that has bad channel condition on subchannel c. Basically, to increase the total system rate, we always allocate a higher power level to those users whose channel gain is low and low power level is allocated to those users whose channel condition is good. So, the power allocation in each user at each subchannel is given by:

$$P_{1,c} \leq P_{2,c} \leq P_{3,c} \leq \ldots \leq P_{U,c} \tag{3}$$

At the receiver end, when we deploy successive interference cancelation (SIC), the nearest user which has good channel condition will remove the interference from other users having bad channel condition and the farthest user which has bad channel condition will notice interference from all other users when decoding his own signal. Therefore, the SIC method is deployed in the receiver end to remove inter-user interference. Thus, the achievable data rate of the uth user in cth subchannel is [12]

$$R_{u,c} = B_S \log 2\left(1 + \frac{P_{u,c}H_{u,c}}{1 + \sum_{l=1}^{u-1} P_{l,c}H_{u,c}}\right) bps \tag{4}$$

So, the total rate of the system is given by:

$$R = \sum_{c=1}^{C}\sum_{u=1}^{U} B_S \log 2\left(1 + \frac{P_{u,c}H_{u,c}}{1 + \sum_{l=1}^{u-1} P_{l,c}H_{u,c}}\right) \tag{5}$$

3 User Pairing Strategy and Optimal Power Allocation Technique

The performance of NOMA mostly depends on power allocation and user pairing. We first analyze the user pairing approach based on channel condition and after that we apply the optimum power allocation technique to increase the overall system rate.

3.1 User Pairing Strategy

We apply the Hungarian algorithm to find the optimum matching between the users which we need to deploy in a subchannel. This user pairing approach is clearly defined in [22]. The aforementioned steps defined in [22] are used to find the optimum matching between the users which are (U_1, U_5), (U_2, U_6) and (U_3, U_4). To see the effectiveness of this algorithm, the performance of it is compared with the N-F user pairing and F-F user pairing method defined in [14]. In N–F pairing, the user which is near the BS is paired with the user which is far from the BS. In this paper, we are considering U_1 as the nearest user and U_6 as the farthest user. So, N-F pairing will pair U_1 with U_6, U_2 with U_5 and U_3 with U_4. Similarly, in N–N or F–F pairing case, the user which is near is paired with the next nearest user, the user which is far is paired with the next farthest user and so on. So, U_1 is paired with U_2 in one subchannel, U_3 is paired along U_4 and U_5 is paired along U_6 in one subchannel.

3.2 Power Allocation Technique

In this section, our aim is to maximize the total rate of the system by obtaining the optimal power allocation. To achieve this, we considered two constraints which are power and minimum achievable rate constraints. The total sum rate of the system is demonstrated as maximization problem which is given by:

$$\max B_S \log 2\left(1+\frac{P_{u,c}H_{u,c}}{1+\sum_{l=1}^{u-1}P_{l,c}H_{u,c}}\right) \tag{6}$$

$$s.t. \sum_{u=1}^{K} P_{u,c} \leq P_M,\ P_{u,c} \geq 0,\ \forall u \in K \tag{6a}$$

$$B_S \log 2\left(1+\frac{P_{u,c}H_{u,c}}{1+\sum_{l=1}^{u-1}P_{l,c}H_{u,c}}\right) \geq R_u,\ \forall u \in K \tag{6b}$$

where constraint (6a) denotes maximum transmit power constraint which means maximum transmit power to each user in a subchannel should not exceed the total transmit power by the BS and constraint (6b) denotes the minimum achievable data rate requirements of the users in the system. Let us assume the minimum achievable data rate requirements as R_u. However, this optimization problem is a convex optimization problem defined in [13]. So, to overcome this problem, KKT constraints are used. The correspondent Lagrange multiplier is stated as (7).

$$\begin{aligned} L(P_u,\varphi,\vartheta) =& B_S\sum_{u=1}^{K}\log 2\left(1+\frac{P_{u,c}H_{u,c}}{1+\sum_{l=1}^{u-1}P_{l,c}H_{u,c}}\right)-\varphi\left(\sum_{u=1}^{K}P_{u,c}-P_M\right)\\ &-\sum_{u=1}^{K}\vartheta\left(R_u-B_S\log 2\left(1+\frac{P_{u,c}H_{u,c}}{1+\sum_{l=1}^{u-1}P_{l,c}H_{u,c}}\right)\right) \end{aligned} \tag{7}$$

where φ and ϑ indicate Lagrange multiplier and R_u represents minimum target data rate. Let us suppose,

$$\Omega_u = \frac{H_{u,c}}{1+\sum_{l=1}^{u-1}P_{l,c}H_{u,c}} \tag{8}$$

$$L(P_u,\varphi,\vartheta) = B_S(1+\vartheta)\sum_{u=1}^{K}\log 2(1+P_{u,c}\Omega_u)-\varphi\left(\sum_{u=1}^{K}P_{u,c}-P_M\right)-\sum_{u=1}^{K}\vartheta R_u \tag{9}$$

Taking derivative of Eq. (9) with respect to P_u P_u

$$\frac{\partial L}{\partial P_u} = \frac{B_S(1+\vartheta^*)\Omega_u}{lnln2\left(1+P_{u,c}^*\Omega_u\right)} - \varphi^* = 0, \forall u \in K \tag{10}$$

$$\varphi^*\left(\sum_{u=1}^{K} P_{u,c}^* - P_M\right) = 0, \ \forall u \in K \tag{11}$$

$$\vartheta^*\left(R_u - B_S \log 2\left(1+P_{u,c}^*\Omega_u\right)\right) = 0, \ \forall u \in K \tag{12}$$

$$\varphi^* \geq 0 \tag{13}$$

$$\vartheta^* \geq 0 \tag{14}$$

To find the optimal solution, both φ^* and ϑ^* should be greater than zero.

$$\sum_{u=1}^{K} P_{u,c}^* = P_M, \ \forall u \in K \tag{15}$$

$$R_u = B_S \log 2(1+P_{u,c}^*\Omega_u) \tag{16}$$

Therefore, from Eq. (12), the optimum power allocation is achieved which is illustrated in Eq. (17).

$$P_{u,c}^* = \frac{1}{\Omega_u}\left(2^{R_u} - 1\right) \tag{17}$$

Thus, this implies that more power should be allocated to weak users first and then the rest of the power is allocated to users whose channel conditions are good. By taking into consideration those two constraints, we can achieve maximum system data rate. To investigate the efficiency of this power allocation technique, its system rate is compared with the dynamic power allocation technique described in [21]. In order to meet the far user minimum achievable rate, the power allocation coefficients are calculated. The rest of the power is allotted to the near user, after meeting the far user requirements. The derivation of power allocation coefficient of dynamic power allocation (DYN PA) is described in [21] where it considers only two user scenarios (far and near user case). The power allocation coefficient of far user for dynamic power allocation scheme is represented by:

$$\alpha_f = \min\left(1, \frac{\mu\left|h_f\right|^2 P_M + \sigma^2}{\left|h_f\right|^2 P_{M,}(1+\mu)}\right) \tag{18}$$

$$\alpha_n = 1 - \alpha_f \tag{19}$$

where $\mu = 2^{R^*-1}$. Let R^* denotes the minimum achievable rate of far users. The channel coefficient of far user is denoted by h_f, σ^2 is the noise power and total BS transmit power is denoted by P_M. The power allocation coefficient for far user is represented by α_f and power allocation coefficient for near user is expressed by α_n. By considering this power allocation scheme, when $\frac{\mu|h_f|^2 P_M+\sigma^2}{|h_f|^2 P_{M,}(1+\mu)}$ exceeds 1, even if the entire power is allotted to the far user, then the far user is not able to meet its minimum achievable data rate and as we are not allocating power to the near user so it will also be in outage. So, when $\frac{\mu|h_f|^2 P_M+\sigma^2}{|h_f|^2 P_{M,}(1+\mu)}$ exceeds 1, then instead of setting α_f to 1, let us limit $\alpha_f = 0$, which undoubtedly set α_n to 1. Furthermore, this dynamic power allocation scheme is used in N–F user pairing and F–F or N–N user pairing technique and its system sum rate is compared with Hungarian algorithm base user pairing scheme along with optimum power allocation technique obtained using KKT constraints in this paper.

4 Results and Discussions

In this simulation, we consider six users which are uniformly distributed in the cell radius of 500 m with random location and a single BS which is at the center of the cell. The distance between the BS and users is considered as 30 m and distance between the users is considered as 60 m. Total available bandwidth of the system is considered as 5 MHz, power consumption by the circuit is of 1 W and path loss exponent (γ) is considered as 2. We consider two users at each subchannel. The power of noise is considered as − 174 dBm. To obtain the results, simulation is being performed in MATLAB software.

Figure 1 depicts the system sum rate versus transmit power by BS, where transmission power of the BS is from 1 to 12 W. From the results, it can be observed that, the sum rate of NOMA Hungarian-based user pairing algorithm and optimal power allocation obtained using KKT condition is higher than those of N–F pairing and F–F or N–N pairing along with dynamic power allocation scheme applied in it. The performance of SC-NOMA is also compared which is poor compared to all other techniques discussed in this paper.

In Fig. 2, the effect of the sum rate of the system versus SNR in dB is explored. From Fig. 2, it can be analyzed that, the performance of Hungarian-based user pairing algorithm and optimal power allocation using KKT condition is much higher compared to other two user pairing techniques discussed in [14] along with dynamic power allocation scheme applied in it. Furthermore, the performance of SC-NOMA is also compared which is worse than all other techniques discussed in this paper.

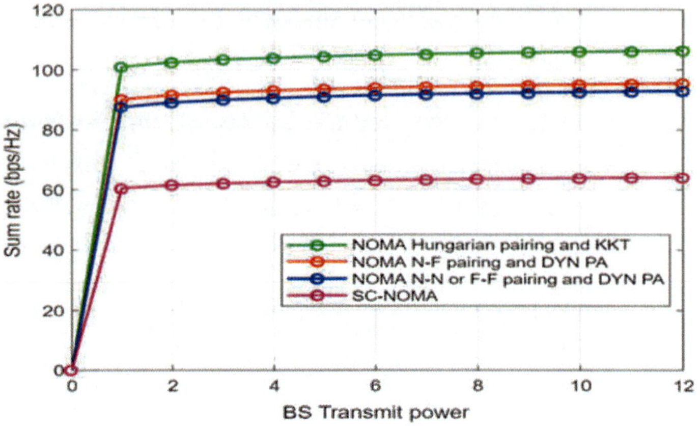

Fig. 1 System rate versus transmit power

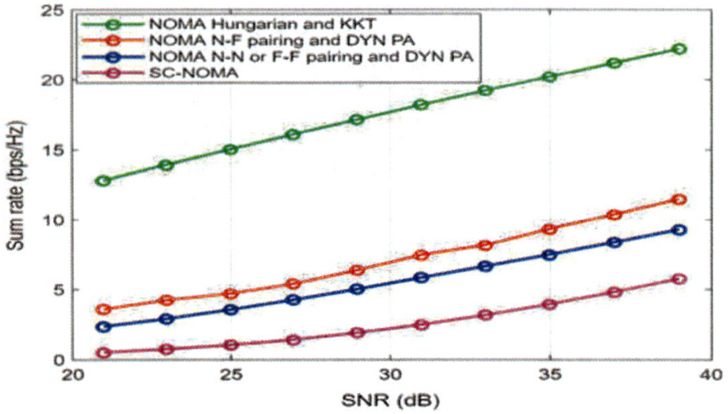

Fig. 2 System rate versus SNR

5 Conclusion

In this paper, a comparative study of two optimal power allocation techniques and two different user pairing approaches has been discussed. The performance of optimum power allocation obtained by applying KKT condition after the user pairing at each subchannel by Hungarian algorithm is being compared with dynamic power allocation applied in N–F pairing and F–F or N–N pairing. The simulation results show that the sum rate of optimal power allocation and Hungarian algorithm for user pairing is better, compared to dynamic power allocation applied in N–F pairing and F–F or N–N pairing techniques. Thus, the optimum power allocation obtained by employing KKT constraints and Hungarian algorithm for user pairing outperforms the other user pairing techniques and dynamic power allocation scheme discussed in this paper.

In the future, we can compare these two user pairing techniques and optimal power allocation techniques with other existing methods. We can also analyze these user pairing techniques and optimal power allocation schemes to improve the energy efficiency performance.

References

1. Chui KT, Lytras MD, Visvizi AJE (2018) Energy sustainability in smart cities: artificial intelligence, smart monitoring, and optimization of energy consumption. Energies 11:2869
2. Lytras MD, Chui KT, Visvizi A (2019) Data analytics in smart healthcare: the recent developments and beyond. Appl Sci 9:2812
3. Visvizi A, Lytras MD (2018) It's not a fad: smart cities and smart villages research in European and global contexts. Sustainability 10:2727
4. Al-Falahy N, Alani OY (2017) Technologies for 5G networks: challenges and opportunities. IT Prof 19:12–20
5. Wu Q, Li GY, Chen W, Ng DWK, Schober R (2017) An overview of sustainable green 5G networks. IEEE Wirel Commun 24:72–80
6. Saito Y, Benjebbour A, Kishiyama Y, Nakamura T (2013) System level performance evaluation of downlink non-orthogonal multiple access (NOMA). Proceedings of IEEE Annual Symposium PIMRC. London, U.K., Sept 2013, pp 611–615
7. Al-Imari M, Xiao P, Imran MA, Tafazolli R (2014) Uplink nonorthogonal multiple access for 5G wireless networks. In: Proceedings 11th ISWCS. Barcelona, Spain, Aug 2014, pp 781–785
8. Ding Z, Yang Z, Fan P, Poor HV (2014) On the performance of nonorthogonal multiple access in 5G systems with randomly deployed users. IEEE Signal Process Lett 21(12):1501–1505
9. Higuchi K, Benjebbour A (2015) Non-orthogonal multiple access (NOMA) with successive interference cancellation for future radio access. IEICE Trans Commun 98(3):403–414
10. Dai L, Wang B, Yuan Y, Han S, Chih Lin I, Wang Z (2015) Non orthogonal multiple access for 5G: solutions, challenges, opportunities, and future research trends. IEEE Commun Mag 53(9):74–81
11. Al-Imari M, Xiao P, Imran MA, Tafazolli R (2014)Uplink non-orthogonal multiple access for 5G wireless networks. In: Proceedings of 11th ISWCS. Barcelona, Spain, Aug 2014, pp 781–785
12. Timotheou S, Krikidis I (2015) Fairness for non-orthogonal multiple access in 5G systems. IEEE Signal Process Lett 22(10):1647–1651
13. Fang F, Zhang H, Cheng J, Leung VCM (2016) Energy-efficient resource allocation for downlink non-orthogonal multiple access network. IEEE Trans Commun 64:3722–3732
14. Ding Z, Fan P, Poor V (2016) Impact of user pairing on 5G non-orthogonal multiple access downlink transmissions. IEEE Trans Veh Technol 65(8):6010–6023
15. He J, Tang Z (2017) Low-complexity user pairing and power allocation algorithm for 5G cellular network non-orthogonal multiple access. Electron Lett 53(9):626–627
16. Fu Y, Hong Y, Chen L, Sung CW (2018) Enhanced power allocation for sum rate maximization in OFDM-NOMA VLC systems. IEEE Phot Techn Lett 30:1218–1221
17. Manglayev T, Kizilirmak RC, Kho YH (2017) Optimum power allocation for non-orthogonal multiple access (NOMA). Appl Inf Commun Technol AICT 2016—Conf Proc:5–8
18. Shahab MB, Irfan M, Kader F, Shin (2016) User pairing schemes for capacity maximization in non-orthogonal multiple access systems. Wirel Commun Mob Comput 16(September):2884–2894
19. Alghasmari W, Nassef L (2021) Optimal power allocation in downlink non-orthogonal multiple access (NOMA). Int J Adv Comput Sci Appl 12(2)
20. Rajab H, Benkhelifa F, Cinkler T (2021) Analysis of power allocation for NOMA-Based D2D communications using GADIA. Information 12:510. https://doi.org/10.3390/info12120510

21. Yang Z, Ding Z, Fan P, Al-Dhahir N (2017) The impact of power allocation on cooperative non-orthogonal multiple access with SWIPT. IEEE Trans Wirel Commun 6(7)
22. Ali ZJ, Noordin NK, Sali A, Hashim F, Balfaqih M (2019) An efficient method for resource allocation and user pairing in downlink non-orthogonal multiple access system. In: Proceedings of the 2019 IEEE 14th Malaysia international conference on communication (MICC). Selangor, Malaysia, 2–4 Dec

Sentiment-Based Community Detection Using Graph Transformation

Shyam Sundar Meena and Vrinda Tokekar

Abstract There are a variety of social media platforms where people can review products and services depending on personal expertise. The user's friends and followers may see each other's ideas and sentiments, which may spread to more users in the future. Therefore, this research proposes a concept named "sentiment community". The purpose is to explore the feelings and interactions of users on social networking sites. We have used graphs for the modelling of these networks. The suggested sentiment community discovery approach alters the original graph based on sentiment scores of users. Experiments have been conducted on networks generated using the LFR model and Twitter's data. In the experiments, the modularity score is used to determine the quality of the detected community structure. The communities were detected in each network with and without the sentiment concept. The results reveal that the inclusion of the sentiment score concept for community detection significantly improves the modularity score and quality of the resultant community structure.

Keywords Social network analysis · Community detection · Sentiment analysis · Modularity

1 Introduction

In the recent decade, technological advancements have opened many new opportunities. Social media Websites such as Twitter, Facebook, and LinkedIn have developed

---

S. S. Meena (✉)
Department of Computer Science and Engineering, Shri Vaishnav Institute of Information Technology, Shri Vaishnav Vidyapeeth Vishwavidyalaya, Indore, India
e-mail: sundar.shyam@svvv.edu.in

V. Tokekar
Department of Information Technology, Institute of Engineering and Technology, Devi Ahilya Vishwavidyalaya, Indore, India
e-mail: vtokekar@ietdavv.edu.in

S. Das et al. (eds.), *Advances in Data-driven Computing and Intelligent Systems*, Lecture Notes in Networks and Systems 653, https://doi.org/10.1007/978-981-99-0981-0_6

crucial social interaction platforms. People can interact in various ways through social networking platforms, such as by liking and following one another's posts and by placing their trust in the opinions of others. Innovators can utilise such social ties to try out means of communication and collaboration. There are a variety of social media platforms where people can review products and services depending on personal expertise.

Customer reviews on Amazon.com are informative and of good quality, which allows us to make more informed buying selections. These reviews may be written by reliable or unreliable persons. But, when someone you trust posts a review on social media platforms, the feelings and thoughts stated in the comment seem to be more able to influence the buying decisions [1]. People who have close personal relationships are more trusting of each other than those with whom they have only professional ties. Since social media sites allow users to interact with each other, the opinions and attitudes they share can spread faster.

Social networking sites (SNSs) have developed as crucial in Web marketing. A robust methodology is required for in-depth analysis of SNSs to get the most out of this data. Firms are using SNSs for consumer segmentation and target marketing. It requires firms to track customers' feedback on their products and services.

The currently available techniques for identifying communities in networks [18] divide individuals according to either their profiles or patterns of connections in the network. These algorithms can identify "cliques" of online members who are intimately connected. Unfortunately, most of them neglect the diversity of semantics and sentiments in social interactions. Users of the same community may follow similar or different sentiments. Utilising both network topological and sentimental differences between users is necessary to detect the communities in social networks. Therefore, in this research, an idea is proposed named "sentiment community". The sentiment community is a group of densely tied users that also share similar opinions, i.e. members of the sentiment community have tight connections and nearly the same feelings. Community detection based on sentiments can be used for a wide range of products purchasing, such as cloths, foods, movies, and books, etc.

Positive sentiment communities represent groups of consumers who frequently order goods. Manufacturers may suggest comparable items to all these consumers. Redesigning products, making improvements, and then recommending to communities with negative opinions about the product are all options available to entrepreneurs.

Finding out what others feel about things can have several benefits such as (1) sellers may suggest comparable products to positive groups whilst avoiding negative communities [5], and (2) manufacturers can build and launch separate new items for different groups based on their sentiments. To perform the above and similar types of activity, sentiment-based community discovery techniques can be helpful. This idea enables us to dig into social networks more deeply.

The study proposed in research transforms the original graph based on user sentiment scores before applying community detection. The experimental results suggest that the approach is successful and efficient. The paper is structured as follows. The relevant work is discussed in Sect. 2. Section 3 describes the sentiment community and the ways to discover it. Section 4 depicts the experimental results, whilst Sect. 5 summarises the study and discusses future research directions.

2 Related Work

This section focuses on works done in the area of sentiment analysis and community detection techniques.

2.1 Sentiment Analysis in Social Networks

Since social networking sites have become a prominent way to share news, reviews, and comments across the World Wide Web, such rankings and evaluations provided by consumers are influential factors that could affect customers' buying decisions about products or services. A significant amount of research has been done to investigate the connections between "word-of-mouth" and selling. Academicians and researchers are interested in user reviews because they provide details of items or brands and personal thoughts and sentiments expressed by consumers. On numerous online review sites, customer comments have been frequently presented in the form of a "smile versus sad face" scale or a five-star scoring system. Extracting customers' feelings from multiple formats requires more data pre-processing.

The user-generated content has already been widely studied for sentiment analysis [4]. These approaches perform analysis of the data at the attribute, sentence, or document level. The document level sentiment analysis approach categorises reviews into positive, negative, or neutral polarity [1]. The method used for sentence level analysis recognises and determines subjective sentences' polarities. In most of these investigations, machine learning was used [27]. Attribute level analysis aims to identify thoughts on items' particular features from review [13].

The examination of feelings can make use of a wide variety of characteristics and indicators. When determining the polarity of documents, it has been demonstrated that the presence of terms is more effective than the occurrence of terms. It has also been observed that the order in which words are presented substantially influences sentiment analysis [21]. Words with part-of-speech (POS) tags (e.g. adjectives and adverbs) are useful indications for determining sentiment polarity. Sentiment polarity categorization can occasionally surpass the bag-of-words technique [15]. Relationships between subject and sentiment are critical in attribute-level sentiment analysis [12].

Unsupervised and supervised methods are excellent alternatives for doing sentiment analysis. These categories have their own set of advantages and disadvantages. Unsupervised approaches commonly count the number of positive and negative words and phrases in a piece of text to assess its polarity. The supervised techniques used labelled data to train models, which subsequently use this model to make predictions about unlabelled data [21].

By utilizing these sentiment analysis tools, scholars can investigate the role of online reviews in electronic markets. The studies suggest a link between consumers' positive product evaluations and subsequent sales of the same product on a Website [6]. According to Duan et al. [9], the number of reviews available for a product may positively impact online sales. Forman et al. [10] concluded that reviews that incorporated identity-descriptive information had a better possibility of earning positive ratings, which resulted in the future sales increases.

2.2 Community Detection

Discovering the structure of a community in a social network is known as community monitoring or community discovery. There are three basic sorts of algorithms for community finding: topology-based community discovery, semantic-based community discovery, and community discovery that incorporate both. Community detection algorithms are classified as non-overlapping or overlapping community detection based on network topology. The spectral clustering methods (such as the minimum cut algorithm [23]), modularity optimization methods (such as the GN algorithm [11]), and label transfer-based methods strategy (such as LPA) [22] are examples of non-overlapping techniques. The techniques used to detect the overlapping communities include the CPM algorithm based on group penetration improvement [20] and the LMF algorithm based on seed set transfer [16]. Such approaches only consider the topological relationship between users.

The sentiment-based community detection approaches group users based on emotional traits. There have been a lot of studies on community detection, but very few has utilised the sentiments of users. For the first time, the term "emotional community" is defined by Xu et al. [26]. The algorithm aims to achieve the maximum possible degree of emotional consistency amongst community members. The suggested algorithm divided comments into three polarities: "positive," "neutral," and "negative". The authors in [25] suggested strategies to maximise modularity and minimise emotional heterogeneity. The approach can differentiate between communities that have different emotional polarities.

Deitrick et al. [8] performed several levels of sentiment analysis and verified the efficacy using data from the @technet social networking network. The authors of a paper [7] employed sentiment analysis to improve community identification. The Tweets features such as comments, retweets, and responses are considered.

3 Sentiment-Based Community Detection

The flowchart in Fig. 1 represents the proposed methodology. The steps followed in the proposed work are described in this section.

3.1 Sentiment Analysis

First, we analyse the users' sentiment from text descriptions written on a topic or about a product. The VADER sentiment analysis tool [14], a component of the Natural Language Toolkit (NLTK) library in Python, was utilised for sentiment analysis. The VADER can analyse text and assign a score. The score is calculated on a scale ranging

Fig. 1 Proposed approach for sentiment community detection

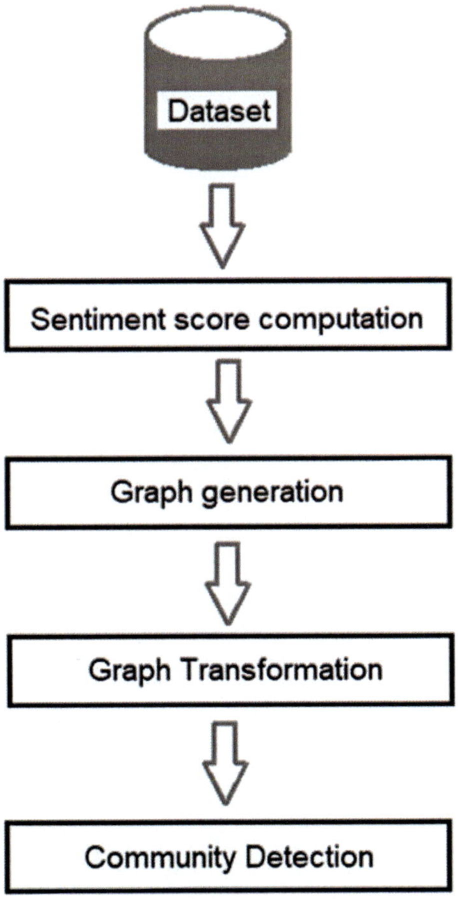

Table 1 Sentiment score categorization

Sentiment category	Score range
Very positive	0.5–1
Positive	0.1–0.5
Neutral	−0.1 to 0.1
Negative	−0.1 to −0.5
Very negative	−0.5 to −1

from minus one to plus one, as shown in Table 1, where minus one represents the most negative sentiment, and plus one indicates the most positive sentiment. These sentiment scores are used as an attribute of nodes in the graph.

3.2 Graph Generation

A graph such as shown in Fig. 2a was generated where nodes are connected based on their interactions in social network. The sentiment score was used as an attribute of nodes.

3.3 Graph Transformation

In the graph transformation step, weight of the edges of graphs was modified (see Fig. 2b). Edges are removed if the new calculated weight is zero (see Fig. 2c). For weight calculation, the sentiment scores of nodes that lie in the edge's corners have been added.

3.4 Community Detection

Finally, the community detection approach was applied on the transformed weighed graph to detect the communities of nodes that share similar sentiments.

4 Experimental Setup and Results

In the experiments, detection and visualisation of networks and community structures were performed with the help of the Gephi tool [2]. The Louvain algorithm [3] implemented in the Gephi was utilised for community detection. The modularity [19]

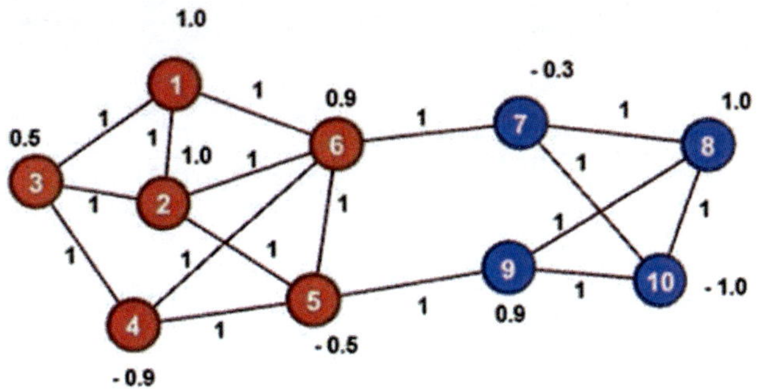

(a) A self-generated example graph with sentiment scores

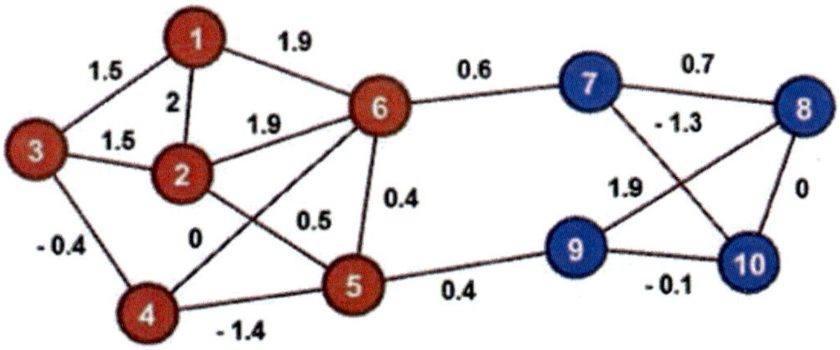

(b) Intermediate representation with updated weight of edges

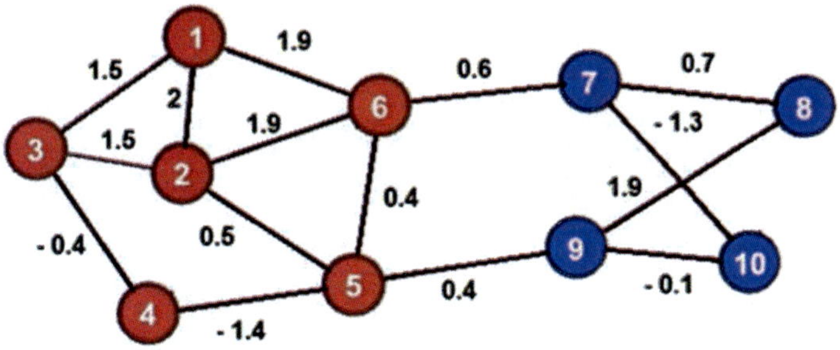

(c) Transformed graph

Fig. 2 Graph transformation process

Table 2 Modularity scores and number of communities detected in various networks

Network	$\|V\|$	$\|E\|$	Sentiment score detection method	Evaluation criteria			
				Without-sentiment score		With-sentiment score	
				# Communities	Modularity	# Communities	Modularity
Self-generated example network (Fig. 2a)	10	15	Manual	1	0.9	3	1
LFR generated network (Fig. 4)	1000	7657	Random	29	0.8	55	0.8
Twitter-network [24]	1103	3066	NLTK library	28	0.2	356	0.4

score was computed to determine the quality of the community discovery results. Modularity score takes value from − 1 to + 1. The high value of modularity indicates the goodness of network partitioning. The modularity scores achieved, and the number of communities detected with or without sentiment concept in each network are presented in Table 2.

For the validation of the proposed approach, we generated a network of 10 nodes and 15 edges (Fig. 2a). The sentiment score of each node was decided randomly. The community detection result is presented in Fig. 3. The Louvain algorithm detected a 1 community without the sentiment concept and 3 communities after incorporating the sentiment concept. Both 4th and 10th nodes were assigned to the different communities due to the high negative sentiment scores and weak connectivity. The result reveals that large number of communities detected due to the similarity and dissimilarity in the sentiment of nodes.

The scalability test of the proposed approach was performed with the help artificial network of 1000 nodes and 7657 edges (Fig. 4). This network was generated by a well-known LFR model [17]. The sentiment score of each node was decided randomly. The Louvain algorithm divided the network into 29 and 55 groups (i.e. communities) without and with the sentiment concept. The resultant structure of communities is shown in Fig. 5. This experimental study confirms that the proposed approach can process complex networks for sentiment-based community detection.

Finally, we have used Twitter social network site data from Kaggle [24]. This dataset includes tweets concerning the conflict between Russia and Ukraine that started at the end of February 2022. This data was pre-processed for the initial graph generation, and then, network transformation and community detection were accomplished. The proposed approach (see Sect. 3) detected 28 and 356 communities without and with the sentiment concept. From the results presented in Table 2, it was clear that incorporating a sentiment score concept for community detection can significantly improve the community structure and modularity score.

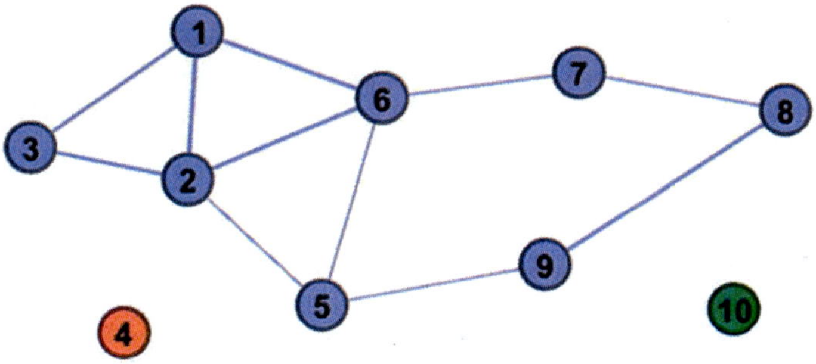

Fig. 3 Community structure detected by Gephi tool in the self-generated example graph (Fig. 2a)

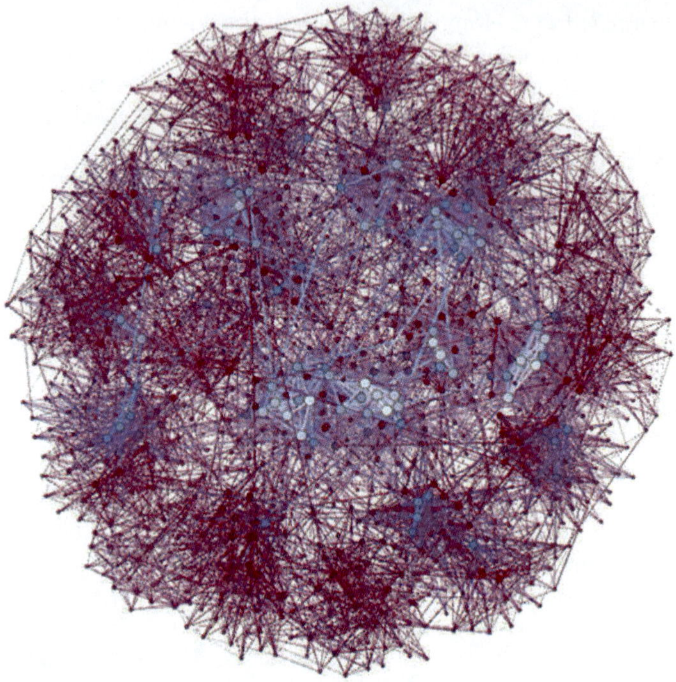

Fig. 4 Network of 1000 nodes generated by LFR model [17]

5 Conclusion

This study tries to identify communities of users who follow similar sentiments. The proposed work transforms the original graph based on user sentiment scores before applying community detection. It begins by analysing the users' sentiment from text descriptions. This score was used as an attribute of nodes. In graphs, nodes were linked together based on social interactions. The graph transformation changes the weights of edges and removes the zero-weight edges. Communities of nodes that share the same attitude were identified using a community detection approach. For the validation and scalability test of the proposed approach, we have generated networks by utilising the LFR model and Twitter's data. We compared the modularity scores and the number of communities found in networks. The results demonstrate that integrating sentiment concept improves the modularity score and quality of the resultant community structure. For community detection, it is evident that sentiment scores play an essential role.

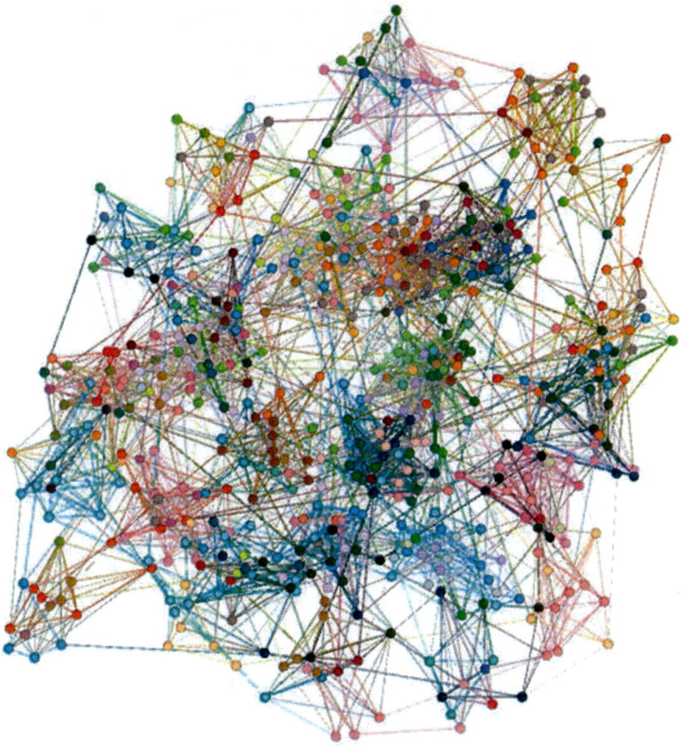

Fig. 5 Structure of communities detected in the network generated by LFR model

Acknowledgements I would like to thank my Ph.D. advisor, Dr. Vrinda Tokekar, Professor, Department of Information Technology, Institute of Engineering & Technology, Devi Ahilya Vishwavidyalaya, Indore, India, for all her guidance and support throughout this study. I'd also like to thank my family for their unwavering support, without which none of this would have been possible.

References

1. Abbasi A, Chen H, Salem A (2008) Sentiment analysis in multiple languages: feature selection for opinion classification in web forums. ACM Trans Inf Syst (TOIS) 26(3):1–34
2. Bastian M, Heymann S, Jacomy M (2009) Gephi: an open source software for exploring and manipulating networks. http://www.aaai.org/ocs/index.php/ICWSM/09/paper/view/154
3. Blondel VD, Guillaume JL, Lambiotte R, Lefebvre E (2008) Fast unfolding of communities in large networks. J statis Mech: Theory Exper 2008(10):P10008
4. Chau M, Xu J (2007) Mining communities and their relationships in blogs: a study of online hate groups. Int J Hum-Comput Studies 65(1):57–70
5. Chen L, Wang F (2014) Sentiment-enhanced explanation of product recommendations. In: Proceedings of the 23rd international conference on World Wide Web, pp 239–240

6. Chevalier JA, Mayzlin D (2006) The effect of word of mouth on sales: online book reviews. J Marketing Res 43(3):345–354
7. Deitrick W, Hu W (2013) Mutually enhancing community detection and sentiment analysis on twitter networks
8. Deitrick W, Valyou B, Jones W, Timian J, Hu W (2013) Enhancing sentiment analysis on twitter using community detection
9. Duan W, Gu B, Whinston AB (2008) Do online reviews matter?-an empirical investigation of panel data. Decision Supp Syst 45(4):1007–1016
10. Forman C, Ghose A, Wiesenfeld B (2008) Examining the relationship between reviews and sales: the role of reviewer identity disclosure in electronic markets. Inf Syst Res 19(3):291–313
11. Girvan M, Newman ME (2002) Community structure in social and biological networks. Proc Nat Acad Sci 99(12):7821–7826
12. Hagedorn BA, Ciaramita M, Atserias J (2007) World knowledge in broad-coverage information filtering. In: Proceedings of the 30th annual international ACM SIGIR conference on Research and development in information retrieval, pp 801–802
13. Hu M, Liu B (2004) Mining and summarizing customer reviews. In: Proceedings of the tenth ACM SIGKDD international conference on Knowledge discovery and data mining, pp 168–177
14. Hutto C, Gilbert E (2014) Vader: a parsimonious rule-based model for sentiment analysis of social media text. In: Proceedings of the international AAAI conference on web and social media, vol 8, pp 216–225
15. Kudo T, Matsumoto Y (2004) A boosting algorithm for classification of semi-structured text. In: Proceedings of the 2004 conference on empirical methods in natural language processing, pp 301–308
16. Lancichinetti A, Fortunato S, Kertész J (2009) Detecting the overlapping and hierarchical community structure in complex networks. New J Phys 11(3):033015
17. Lancichinetti A, Fortunato S, Radicchi F (2008) Benchmark graphs for testing community detection algorithms. Phys Rev E 78(4):046110
18. Leskovec J, Lang KJ, Mahoney M (2010) Empirical comparison of algorithms for network community detection. In: Proceedings of the 19th international conference on World wide web, pp 631–640
19. Newman ME (2006) Modularity and community structure in networks. Proceedings of the national academy of sciences 103(23):8577–8582
20. Palla G, Derényi I, Farkas I, Vicsek T (2005) Uncovering the overlapping community structure of complex networks in nature and society. Nature 435(7043):814–818
21. Pang B, Lee L, Vaithyanathan S (2002) Thumbs up? sentiment classification using machine learning techniques. arXiv preprint cs/0205070
22. Raghavan UN, Albert R, Kumara S (2007) Near linear time algorithm to detect community structures in large-scale networks. Phys Rev E 76(3):036106
23. Stoer M, Wagner F (1994) A simple minimum cut. Algorithms–ESA'94 141–147
24. Ukraine conflict twitter dataset – kaggle. https://www.kaggle.com/datasets/bwandowando/ukraine-russian-crisis-twitter-dataset-1-2-m-rows/discussion. Accessed on 17 Jun 2022
25. Wang D, Li J, Xu K, Wu Y (2017) Sentiment community detection: exploring sentiments and relationships in social networks. Electr Commer Res 17(1):103–132
26. Xu K, Li J, Liao SS (2011) Sentiment community detection in social networks. In: Proceedings of the 2011 iConference, pp 804–805
27. Yu H, Hatzivassiloglou V (2003) Towards answering opinion questions: Separating facts from opinions and identifying the polarity of opinion sentences. In: Proceedings of the 2003 conference on Empirical methods in natural language processing, pp 129–136

Multivariate Data-Driven Approach to Identify Reliable Neural Components and Latency in a P300 Dataset Using Correlated Component Analysis

Kalpajyoti Hazarika and Cota Navin Gupta

Abstract This paper demonstrates the use of correlated component analysis to identify reliable and highly correlated P300 components and its associated latency. Electroencephalogram (EEG) trials from eight channels collected during an oddball experiment were used in this analysis. All the trials were divided into non-target (non-P300) and target (P300) trial cohorts. Data-driven correlated component analysis (CorrCA) was applied to both cohorts separately. We observed that first CorrCA components from target trials showed more coherent structure of p300 latency than single electrode (C_z) trials when those were visualized in event-related potential (ERP) image plot. The averaged first CorrCA component across P300 trials cohort showed higher amplitude at P300 latency when compared to non-P300 trials. P300 amplitude of averaged first CorrCA component across all able bodied subjects was $0.37\mu V$ higher than P300 amplitude of averaged target trials at C_z electrode. Forward model considering eight channels showed lower neural activation in disabled subjects than abled-bodied subjects. Through this work, we show the utility of the CorrCA method for traditional P300 experiments. The presented application-level multivariate framework can be used to obtain critical parameters like latency from event-related potential datasets recorded for various psychiatric conditions.

Keywords P300 latency · Correlated component analysis · Signal-to-noise ratio · Forward model

1 Introduction

P300 has been used widely as a useful marker for understanding cognitive processes such as working memory load and attention [1]. It is elicited during an oddball paradigm when the subject attends to target stimuli (image/sound) in a sequence of stimuli presented in a random order. P300 amplitude is considered as the largest

K. Hazarika · C. N. Gupta (✉)
Neural Engineering Lab, Department of Bioscience and Bioengineering, IIT Guwahati, Assam, India
e-mail: cngupta@iitg.ac.in

S. Das et al. (eds.), *Advances in Data-driven Computing and Intelligent Systems*, Lecture Notes in Networks and Systems 653, https://doi.org/10.1007/978-981-99-0981-0_7

positive peak within 300–800 ms, and peak latency may vary depending on task conditions, stimulus type, and subject's age. Information of P300 amplitudes and peak latency has played a major role in understanding psychiatric disorders [2]. P300 wave has been used as a feature in the field of the brain-computer interface (BCI). BCI translates brain wave activity into some specific commands and later these commands are fed to a computer to interpret subject's behavior. BCI has also been used to replace or restore useful functions for disabled people suffering from amyotrophic lateral sclerosis, cerebral palsy, or stroke. Numerous applications using P300 potentials have been realized for real-time attention [3], cognitive biometrics [4], gauging joint attention in virtual reality applications [5], and for spelling devices [6]. A detailed review of recent P300-related research is discussed in [7–9]. More recently, it was reported yoga sessions seem to increase P300 peak amplitude recorded at P_z significantly [10]. Age and gender affected P300 amplitudes but not latency [11]. A meta-analysis showed prolonged P300 latency at the C_z electrode in Parkinson's disease dementia patients (PDD) compared to Parkinson's disease non-dementia patients (PDND), and reduced P300 amplitude was observed at F_z electrode in Parkinson's disease non-dementia patients compared to healthy control [12].

Correlated component analysis (CorrCA) is a data-driven statistical method that extracts the most correlated components from multi-trial or multi-subject electroencephalogram (EEG) data. The fundamental assumption is that it has at least one shared dimension between the trials or subjects of EEG data. The extracted CorrCA component captures the reliable part of the signal across trials or subjects. CorrCA performed significantly better than averaged PCA and canonical component analysis for dimension reduction [13]. CorrCA can also be applied to calculate inter-subject correlation (ISC) and ISC measures how different individuals' neural responses vary with shared stimulus. The application of CorrCA was first observed in a two-stage frequency recognition method for SSVEP-based BCI [14, 15]. In recent studies, CorrCA was used to measure the attention level of the subjects in different audio and visual platforms [16, 17]. In one study, the researchers applied CorrCA to find the strength of the relationship between explicit (judgment)-implicit (EEG) measures during sensory and hedonic processing of beers. It was based on the degree of tasting expertise, and they reported that beer experts showed a more efficient pattern of gustatory processing compared to general tasters and consumers [18]. Jens Madsen et al. applied CorrCA to identify robust within-subject coupling between the brain, heart, and eyes, and they reported that ISC reflected the strength of brain–body connection during an attentional task [19]. Marc Rosenkranz et al. also showed how brain activity changes with selective attentional engagement to speech using CorrCA [20].

In this work, we first show an increased signal-to-noise ratio (SNR) of CorrCA components when compared to EEG channel data. Secondly, we showed how to extract P300 amplitude and its latency from first CorrCA component. Finally, we showed utility of forward model in the source localization of neural activities by considering the first CorrCA component. Spatial distribution of the electrical activities from the topoplot indicated differences between healthy and disabled subject's neural activities.

2 Materials and Method

In this work, P300 datasets were collected from a previous BCI-related work. The data were recorded during an oddball paradigm experiment. Please refer to [21] to get more details about the experimental setup and the datasets. Refer to the work of [13] for more information about the CorrCA method. EEG data were recorded from four able-bodied subjects (S6–S9) and four disabled subjects (S1–S4). First, we separated the data into target and non-target cohorts based on given labels. MATLAB Scripts used in this work are available at the weblink: https://github.com/NeuralLabIITGuwahati/CorrCA_for_P300.git

2.1 Preprocessing

We followed the preprocessing steps from the paper [21]. Then we used a covariance trial rejection method to remove the artifact EEG trials [22]. We used the electrode configuration (F_z, C_z, P_z, O_z, P_7, P_3, P_4, P_8) reported in [21]. Then we formed three-dimensional data (channel by sample points by trials) by concatenating all the trials.

2.2 Application of Correlated Component Analysis

Correlated component analysis (CorrCA) was developed by Dmochowski et al. [23] and we used this algorithm to extract reliable brain responses across multiple trials. The function of CorrCA is to obtain the linear combination of EEG channels such that correlation between trials is maximum. If the target cohort has T sample points, D electrodes, and, N trials, then we have N data volume .$x_i^l \in R^D$ ($d = 1, 2, 3 \ldots d$ represent channels, $i = 1, 2, 3 \ldots T$ are time samples, and $l = 1, 2, 3 \ldots N$ represents EEG trials) is the neural response of the ith sample of lth trial. These two matrices are calculated as follows:

$$R_{\mathrm{B}} = \sum_{i=1}^{T}\sum_{l=1}^{N}\sum_{k=1,k\neq l}^{N}\left(x_i^l - \overline{x_*}^l\right)\left(x_i^k - \overline{x_*}^k\right)^T \tag{1}$$

$$R_{\mathrm{W}} = \sum_{i=1}^{T}\sum_{l=1}^{N}\left(x_i^l - \overline{x_*}^l\right)\left(x_i^l - \overline{x_*}^l\right)^T \tag{2}$$

where $\overline{x_*}^i = \frac{1}{T}\sum_{i=1}^{T} x_i^l$ is the sample-mean vector for trial l. R_{W} and R_{B} were calculated for both non-target and target cohorts. The projection vectors are derived from eigenvectors of matrix $R_{\mathrm{W}}^{-1} R_{\mathrm{B}}$ and are arranged in the projection matrix from left to right columns based on descending order of eigenvalues. For both cohorts, correlated component of ith trial, component $y_i^l \in R^D$ is expressed in terms of projection vector v as follows:

$$y_i^l = v^T x_i^l \tag{3}$$

Inter-trial correlation of the trial m and component k is computed by the correlation between the given trial and all the other trials that experience the same stimuli:

$$\rho = \frac{1}{N-1} \frac{v^T R_B v}{v^T R_W v} \tag{4}$$

ITC values are the sorted eigenvalues of $R_W^{-1} R_B$ (descending order). In this work, we used electroencephalogram (EEG) data from four disabled and four able-bodied subjects [21]. Each subject has four sessions. For each subject, we segregated the target and non-target trials based on the given labels. The size of the data is three-dimensional. Suppose T denotes sample points, D indicates electrodes, and N represents trials, then our data will be $T \times D \times N$. N has two classes: target and non-target. If EEG data are represented by X and the CorrCA component represented by Y, then:

$$Y = X W_i^T \tag{5}$$

where W_i (where, $i = 1, 2, 3 \ldots D$) are projection vectors. The number of projection vectors may be less than or equal to D. After applying CorrCA to both classes of data, we obtained projection vectors, inter-trial correlation (ρ), and CorrCA components from both classes. We validated the CorrCA components using leave one out of the sample trials. In this method, we first calculated the projection matrix from the $N - 1$ trials and then multiplied the projection matrix by the remaining trial.

To check statistical significance of CorrCA components, we used non-parametric test statistic ($\alpha = 0.05$ level). To perform this test, at first, we generated circularly shifted surrogate data. Each shuffle generated a set of inter-trial correlation (ITC) values corresponding to each component. Correlation between the trials of each component is expressed in terms of inter-trial correlation. The p-value is computed from each shuffle of the extracted trials. The p-value is defined as the number of times at which ITC computed from the randomly shuffled data is greater than the maximum ITC of the most correlated component in the original data [13].

2.3 Signal-to-Noise Ratio

In this work, signal-to-noise ratio (SNR) was used to measure repeat reliability across the trials. Theoretically, maximizing the correlation between the trials is equivalent to maximizing repeat reliability [13]. After application of CorrCA on both target and non-target cohorts, we evaluated the SNR of channel and CorrCA components. SNR can be important for two reasons. First, it determines the quality of the data during the evaluation of the robustness of the findings. Secondly, SNR can be an important metric, when we expect more cross-trial variability in one condition than the other [22].

2.4 Neural Activity Using the Forward Model Approach

The CorrCA projection matrix can be considered as a backward model. The limitation of backward model is that it cannot interpret the neurophysiological process at the sensor level [13, 24]. Forward model solves this problem by establishing a relationship between channels and component activity. It helps to interpret neural activities at the sensor level. The weights of the forward model matrix indicate the contribution from an individual channel contributed to an individual component. For example, if the first CorrCA component represents P300, then the weights from the first column of the forward model matrix represent the neural activities related to the P300 component.

3 Result

Figure 1 showed comparison between SNR in target trials and the non-target trials with SNR of the channels data after the application of CorrCA. Due to low signal-to-noise ratio of channels, CorrCA component can be used to detect the P300. Noise present in both channels decrease SNR value. Figure 1 shows that after the third CorrCA component, other components became non-significant. So, we can reject those non-significant components from further analysis. This function is similar to a dimensional reduction technique. We also observed that SNR present in non-target trials is lower than SNR present in target trials in both channel and component space. In the case of target trials, SNR in components is enhanced compared to SNR present in channels. However, the number of significant components may be reduced. On the other hand, SNR in non-target trials showed a slight increase in components compared to channels. Figure 1 shows that for target trials, the first three components are significant. This may remove the problems that arise in the case of feature (channels) selection for machine learning applications.

We compared the ERP image plot of channel C_z (target and non-target cohorts) with the extracted CorrCA components for the target cohort.

Bottom panel of Figs. 2 and 3 indicates amplitude for target trials which is higher than amplitude in non-target trials, and the upper panels of ERP image plot indicated less inter-trial variability in the target trials (Fig. 2) than in non-target trials (Fig. 3). After application of CorrCA on the target cohort ($T \times D \times N$), we observed that P300 peak latency was well captured by the first CorrCA component across target

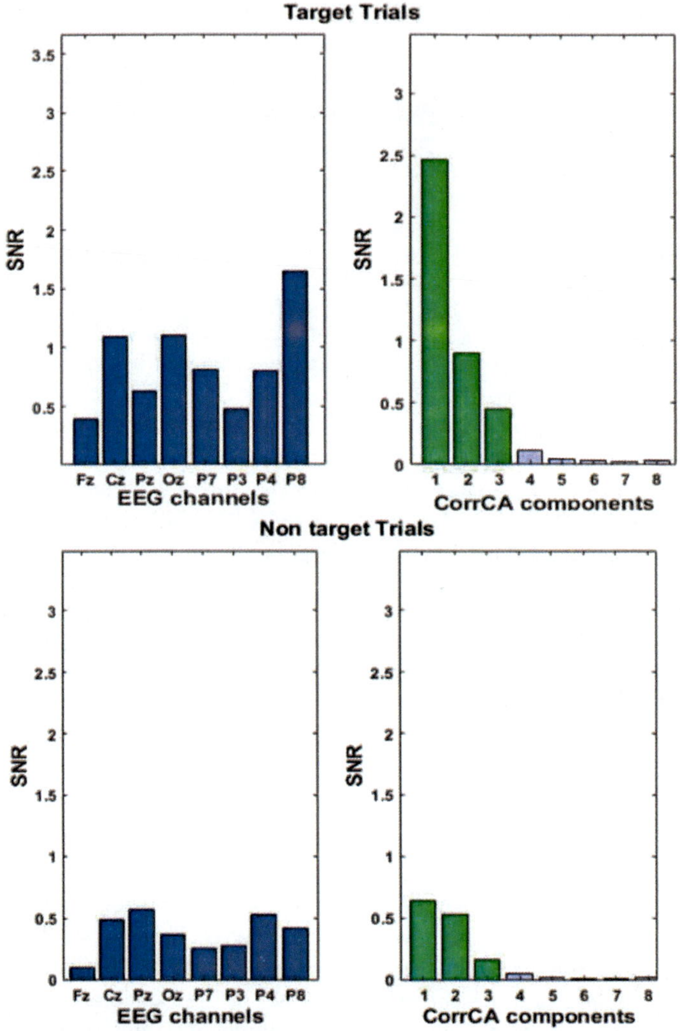

Fig. 1 Top figure corresponds to target trials, and the bottom figure corresponds to non-target trials. The green color bar denotes the significant ($p < 0.5$) CorrCA component. Components 4–8 in target trials and in non-target trials are non-significant

trials (indicated as red area within two black lines in Fig. 4). So, the first component can be selected instead of eight channels for feature extraction. Figure 4 (bottom) also shows a strong peak around 400 ms, indicating the P300 amplitude.

We used eight channels [15], mainly from the parietal, frontal, and central regions to illustrate topoplots. Figure 5 shows the spatial distribution of neural activities captured by the first CorrCA component in all the subjects. In Fig. 5, the disabled subjects showed lower neural activities. Only subject 1 and subject 3 showed similar

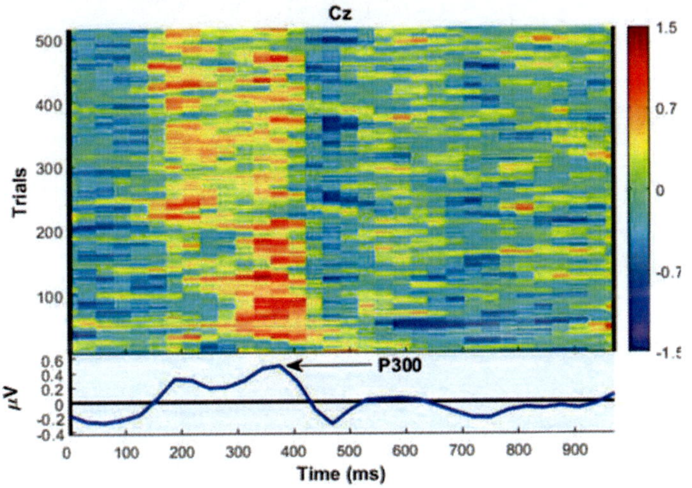

Fig. 2 ERP image plot of *Cz* electrode across target trials (upper); averaged target trial (bottom)

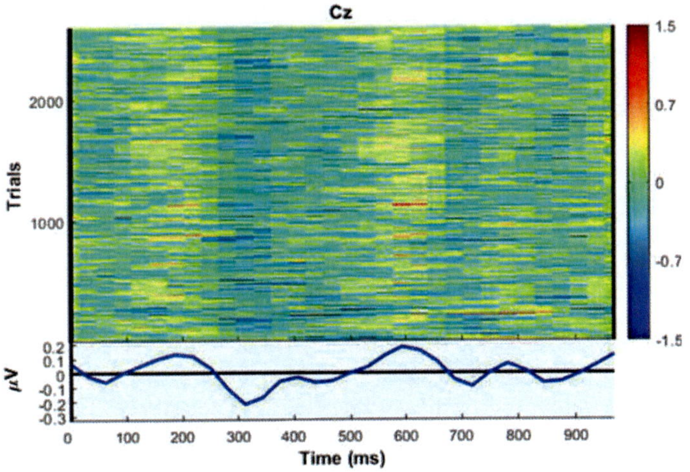

Fig. 3 ERP image plot of C_z electrode across non-target trials (upper); averaged non-target trial (bottom)

neural activities in O_z and P_4 electrodes. Subject 2 and subject 4 showed the most insufficient neural activation. In the case of subject 4, we observed a small amount of neural activity around the P_4 electrode. In Fig. 5, all the able bodied subjects showed a distinct pattern of neural activities across central and parietal regions. Subject 9 showed the lowest neural activation among all the able bodied subjects, and the most neural activities were observed at O_z electrode. Subject 6 and subject 7 had similar regions of neural activation. Similar neural activities may reflect the brain coupling

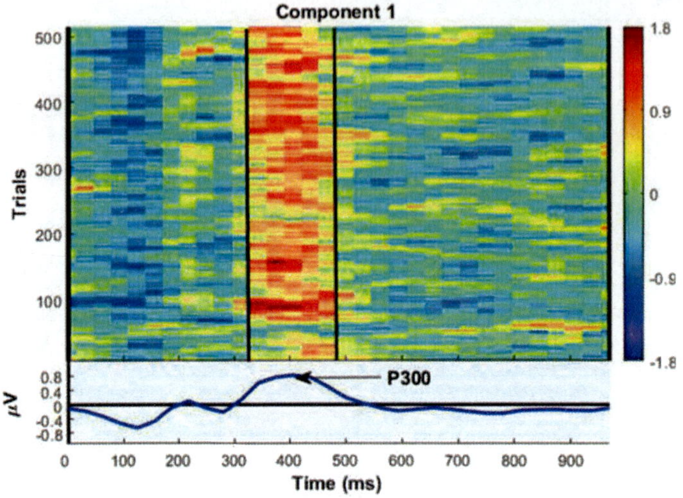

Fig. 4 First CorrCA components from target trials form ERP image plot (top). After taking average across CorrCA component of target trials (bottom)

within subjects with the shared stimulus. Neural activities of subject 8 were highly concentrated at the C_z electrode.

4 Discussion

These works applied correlated component analysis (CorrCA) to identify reliable correlated neural components from P300-based EEG data. CorrCA component gives information about peak latency by generating a coherent structure. This information of peak latency is important in some cases; for example, a shorter latency indicates superior cognitive performance than longer latency [1]. This work showed mean peak latency of P300 component of all disabled subjects which was 31.25 ms longer than able-bodied subjects. The reliable components were arranged according to the descending order of ITC values. We found that the first CorrCA component (Fig. 4) captured better peak latency structure than a single electrode across trials. The forward model approach helped to perform source localization in finding the spatial origins of current sources. The forward model approach in Fig. 5. explained how components captured distinct patterns of neural activities associated with P300. We observed that the topoplots between disabled and able bodied subjects were not similar. Disabled subjects showed higher dissimilarity of neural activation among themselves. The coherence in neural activation among able bodied subjects may be due to structural and functional similarity among individual brains. For disabled subjects, it is unclear what might be the reason for topoplot variations, which warrants further research.

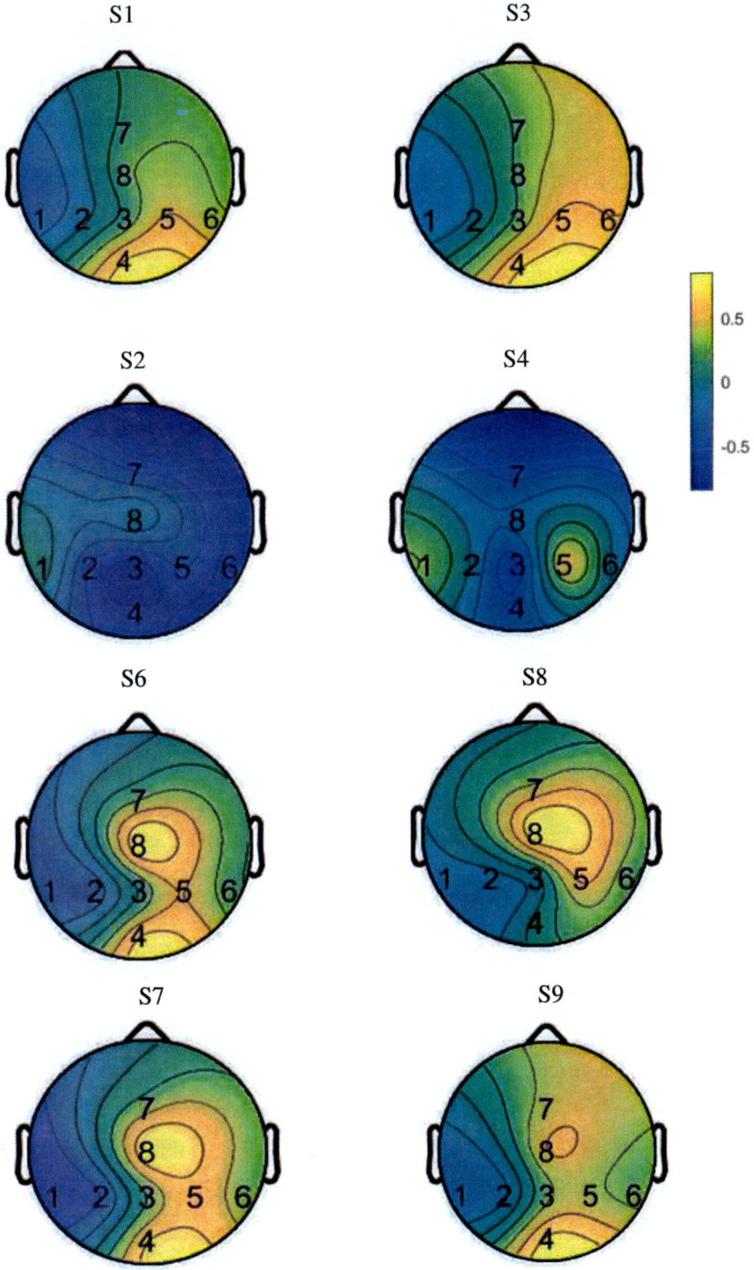

Fig. 5 First CorrCA component of target trials. (S1:S4-disabled subjects and S6:S9-healthy subjects). Here, electrodes, $1 = P_7$, $2 = P_3$, $3 = P_z$, $4 = O_z$, $5 = P_4$, $6 = P_8$, $7 = F_z$, $8 = C_z$

5 Conclusion

In this work, we extracted the reliable source component associated with the P300 dataset, reducing the limitations of channel-based P300 analysis. ERP plot of first CorrCA component from target trials showed a more coherent structure for P300 latency than ERP plot of single electrode (C_z) trials. First CorrCA components from the target trial cohort captured a P300 peak latency (around 400 ms). The forward model considering eight channels were used to identify the source of reliable neural activities for both able and disabled subjects and topoplots indicated comparatively lower neural activation among disabled subjects.

References

1. Sutton S, Braren M, Zubin J, John ER (1965) Evoked-potential correlates of stimulus uncertainty. Science 150(3700):1187–1188
2. Linden DEJ (2005) The P300: where in the brain is it produced and what does it tell us? Neuroscientist 11(6):563–576
3. Mijovic P et al (2017) Towards continuous and real-time attention monitoring at work: reaction time versus brain response. Ergonomics 60(2):241–254. https://doi.org/10.1080/00140139.2016.1142121
4. Palaniappan R, Paramesran R, Gupta CN (2012) Exploiting the P300 paradigm for cognitive. Int J Cogn Biometrics 1(1):26–38
5. Chatterjee B, Palaniappan R, Gupta CN (2020) Performance evaluation of manifold algorithms on a P300 paradigm based online BCI dataset. IFMBE Proc 76:1894–1898. https://doi.org/10.1007/978-3-030-31635-8_231
6. Farwell LA, Donchin E (1988) Talking off the top of your head: toward a mental prosthesis utilizing event-related brain potentials. Electroencephalogr Clin Neurophysiol 70(6):510–523. https://doi.org/10.1016/0013-4694(88)90149-6
7. Masood F, Hayat M, Murtaza T, Irfan A (2020) A review of brain computer interface spellers. In: International conference on emerging trends in smart technologies (ICETST), pp 1–6. https://doi.org/10.1109/ICETST49965.2020.9080743
8. Rezeika A, Benda M, Stawicki P et al (2018) Brain–computer interface spellers: a review. Brain Sci 8:57. https://doi.org/10.3390/BRAINSCI8040057
9. Simoes M et al (2020) BCIAUT-P300: a multi-session and multi-subject benchmark dataset on autism for P300-based brain-computer-interfaces. Front Neurosci 14:568104. https://doi.org/10.3389/fnins.2020.568104
10. Kala N, Telles S et al (2022) P300 following four voluntarily regulated yoga breathing practices and breath awareness. Clin EEG Neurosci. https://doi.org/10.1177/15500594221089369
11. Yerlikaya D et al (2022) The reliability of P300 and the influence of age, gender and education variables in a 50 years and older normative sample. Int J Psychophysiol 181:1–13. ISSN: 0167-8760. https://doi.org/10.1016/j.ijpsycho.2022.08.002
12. Xu H et al (2022) N200 and P300 component changes in Parkinson's disease: a meta-analysis. Neurol Sci 1:1–12
13. Lucas CP et al (2018) *Correlated components analysis—extracting reliable dimensions in multivariate data.* ArXiv abs/1801.08881 (2018)
14. Zhang Y et al (2018) Two-stage frequency recognition method based on correlated component analysis for SSVEP-based BCI. IEEE Trans Neural Syst Rehabil Eng 26(7):1314–1323. https://doi.org/10.1109/TNSRE.2018.2848222

15. Zhang Y et al (2018) Correlated component analysis for enhancing the performance of SSVEP-based brain-computer interface. IEEE Trans Neural Syst Rehabil Eng 26(5):948–956. https://doi.org/10.1109/TNSRE.2018.2826541
16. Ki JJ et al (2016) Attention strongly modulates reliability of neural responses to naturalistic narrative stimuli. J Neurosci 36(10):3092–3101. https://doi.org/10.1523/JNEUROSCI.2942-15.2016
17. Cohen SS, Parra LC (2016) Memorable audiovisual narratives synchronize sensory and supramodal neural responses. eNeuro. https://doi.org/10.1523/ENEURO.0203-16.2016
18. Aguayo IH et al (2022) Implicit and explicit measures of the sensory and hedonic analysis of beer: the role of tasting expertise. Food Res Int 152:110873. https://doi.org/10.1016/J.FOODRES.2021.110873
19. Madsen J, Parra LC (2022) Cognitive processing of a common stimulus synchronizes brains, hearts, and eyes. PNAS Nexus 1(1). https://doi.org/10.1093/PNASNEXUS/PGAC020
20. Rosenkranz M, Holtze B, Jaeger M, Debener S (2021) EEG-based inter subject correlations reflect selective attention in a competing speaker scenario. Front Neurosci 15. https://doi.org/10.3389/FNINS.2021.685774/PDF
21. Hoffmann U, Vesin JM, Ebrahimi T, Diserens K (2008) An efficient P300-based brain-computer interface for disabled subjects. J Neurosci Methods 167(1):115–125. https://doi.org/10.1016/j.jneumeth.2007.03.005
22. Cohen M (2014) Analyzing neural time series data. The MIT Press
23. Dmochowski JP et al (2012) Correlated components of ongoing EEG point to emotionally laden attention—a possible marker of engagement? Front Hum Neurosci 6(112). https://doi.org/10.3389/fnhum.2012.00112
24. Haufe S et al (2014) On the interpretation of weight vectors of linear models in multivariate neuroimaging. Neuroimage 87:96–110

A Framework for an Intelligent Voice-Assisted Language Translation System for Agriculture-Related Queries

Pratijnya Ajawan, Kaushik Doddamani, Aryan Karchi, and Veena Desai

Abstract Language translation plays a major role in responding to farmers queries in technology-assisted response systems for agriculture. It is imperative that there should exist a system that can handle the queries raised by a farmer in regional language and respond to queries asked with minimal human involvement. The proposed framework accepts the farmers' queries spoken in Kannada language and translates the Kannada query into English query. The translated query obtained in English from the proposed framework is utilized by the response system to retrieve an appropriate query from a sample set chosen from the Kisan Call Centre dataset for Belagavi district which consists of queries and answer pairs asked by the farmers in English language. Further, the corresponding English text answer obtained from the response model is converted to Kannada text answer, which is translated as Kannada voice answer. The Bilingual Evaluation Understudy score is utilized as a metric for analyzing the efficiency of the framework. The BLEU is calculated by comparing a candidate text translation against reference translations. BLEU 1-gram, 2-gram, and 3-gram scores have been analyzed for evaluating the performance of the translation model. Google translator is incorporated in the framework to translate the query from Kannada speech to English text and vice versa. For experimentation, 9 queries corresponding to the most frequently asked categories from around 50,000 KCC queries are considered. The response generated by the system has been compared with 4 reference translations. 83% translation accuracy is obtained by considering the BLEU score for 1-gram.

Keywords Agriculture · Querying system · Response system · Kannada · Kisan Call Centre · Belagavi · TTS · BLEU

P. Ajawan (✉) · K. Doddamani · A. Karchi · V. Desai
Department of ECE, KLS Gogte Institute of Technology, Belagavi 590008, Karnataka, India
e-mail: psajawan@git.edu

K. Doddamani
e-mail: kaushikdoddamani@gmail.com

S. Das et al. (eds.), *Advances in Data-driven Computing and Intelligent Systems*, Lecture Notes in Networks and Systems 653, https://doi.org/10.1007/978-981-99-0981-0_8

1 Introduction

There are several online sites that provide useful agriculture information; however, these sites may not always provide sufficient information in regional languages that address farmers' queries. From a broader perspective, many farmers are unfamiliar with the English language, making it difficult for them to comprehend the information offered on these websites.

Kisan Call Centre's (KCC) or agricultural helplines are set up throughout India. The main aim of these centre's is to respond to the farmer's queries and to guide them during times of uncertainty and distress. The predominant drawback of the KCC centers is the requirement of human intervention. The domain experts or operators have to be always available in person to answer the calls. Another drawback is that the response is dependent on the perception and expertise of the expert/operator answering the calls. During times of uncertainty like floods, pandemic, etc., the calls go unanswered due to unavailability of the experts/operator, and farmers face difficulty in these situations with no point of contact to get solutions to their queries. Making call to the Kisan Call Centre with toll-free number 1800-180-1551 is one of the ways for farmers in the Belagavi district to get assistance for their queries, which are related to agriculture [1].

At KCC, the framing of the questions for the KCC dataset is limited by the language proficiency of the operator and may be prone to error. The translation model assists the operators/experts to frame appropriate questions required for a querying system so as to answer the farmer's questions in the Kannada language. The framework proposed processes the voice typed Kannada query and translates the Kannada query into corresponding English query without changing the meaning. This English query further can be used by Natural language response generating systems to process the questions in English text and generate an appropriate answer. The answers generated by the response system in English text have to be narrated to farmer in Kannada language, i.e., text to speech [2–5]. This paper establishes a framework to translate the queries asked by farmers, with the language under consideration being Kannada.

The framework is developed using Python programming language with Google trans Pygame, a cross-platform set of Python modules designed for writing video games, and gTTS an application for converting the given text into audio format. The efficiency of the translation system is verified using BLEU score [6–10]. 1-gram, 2-gram, 3-gram, and 4-gram. BLEU scores are analyzed, and the accuracy of the translation framework is obtained by comparing with reference sentences.

2 Background

The core module in building the human computer interactive systems is to generate natural speech from the input text, i.e., text to speech (TTS), in particular, end-to-end framework-based TTS models [2]. Google Cloud Speech API makes it simple for developers to include Google's speech recognition technology into their applications. It uses the Cloud Speech API service to send audio and receive a text transcription. It supports a global user base by recognizing over 80 languages and variants. The words of users speaking into an app's microphone can be transcribed, allowing command-and-control by speech or transcription of audio files, among other possibilities. Before uploading audio to Speech API, users do not need to use advanced signal processing or noise cancelation [3]. Google translate is a useful tool for comparative research when using bag-of-words text models [4] is stated by comparing the term document matrices (TDMs) as well as topic model results from gold standard translated text and machine-translated text. The Myanmar script's different units, such as characters and syllables, are contrasted. The results obtained on syllable unit for Myanmar (Myan) to English (Eng) transliteration direction with 89.3% BLEU score and character unit for English (Eng) to Myanmar (Myan) transliteration direction with 82.0% BLEU score, according to the evaluations [6]. The BLEU statistic was employed to assess the translation quality. A high BLEU score implies that the translation is more accurate. The BLEU score for the Hiligaynon to English translation is 21.74, indicating that the translation is comprehensible but still needs a lot of work. Furthermore, the result of English to Hiligaynon is 24.4 [7]. People's perceptions of the challenge [8] are influenced by improvements in neural machine translation approaches. To address this issue, various online translation and mobile application technologies like Google Translate, DeepL, SYSTRAN, and others are developed. However, there are no convenient translation tools for the English-Mizo parallel languages. An attempt to improve the English to Mizo conversion in the Bilingual Evaluation Understudy metric by leveraging the superiority power of neural machine translation is made. The model is evaluated with hidden units 128, 256, and 512 in neural network layers 2 and 4. With the highest BLEU score of 12.3, GRU cells in encoder and decoder with attention, 2-layer neural network, and 512 hidden units appear to be better at translating English sentences into Nepali sentences [9]. For conversion of text from one language to the other language, six different Indian languages such as Hindi, Bengali, Gujarati, Malayalam, Tamil, and Telugu are worked on. BLEU varies with the usage of word embedding techniques which have been clearly shown with BLEU score of 21.97 [10].

3 Methodology

For the work carried out in this paper, the dataset from data.gov.in is considered a platform for supporting open data initiatives by the Government of India. This portal provides a single point of access to datasets, documents, services, tools, and applications made available by Indian government ministries, departments, and organizations. The dataset is a CSV file that comprises of the questions asked by farmers in the Belagavi region in English during 2017–2020. For testing and analyzing the proposed framework, 9 sample English language text queries are chosen from the most frequently asked categories, namely weather, plant protection, nutrient management, market information, and fertilizer use.

The implementation methodology of the framework is as shown in Fig. 1.

The description of Fig. 1 is as follows:

1. Mic: Integrated mic of the laptop/computer system is used as an input device. The farmer can make use of a cell phone mic for asking questions in Kannada language.
2. Speech to text converter: The captured Kannada speech is converted to Kannada text using speecktyping.com, a speech to text converter website.
3. Text translator: Google Translate API is used to translate Kannada language query in corresponding English text query without changing the meaning of the sentence.

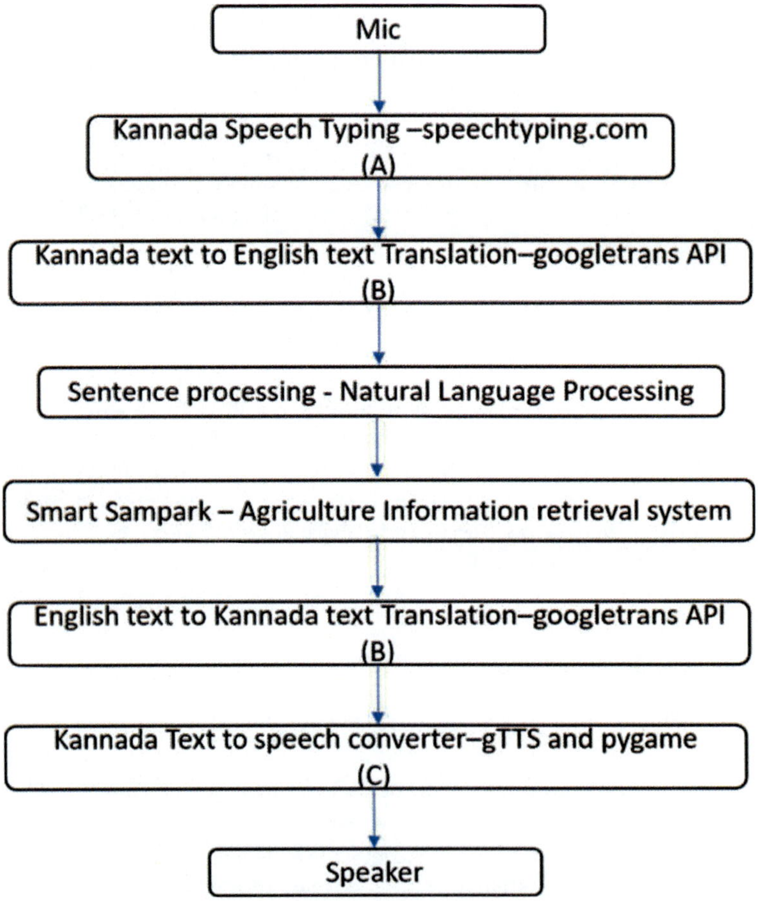

Fig. 1 Block diagram of the framework proposed

4. Smart Sampark an Intelligent Response System: The system generates an appropriate response to farmer's queries in English. The English answers are to be converted to the local language Kannada for the farmers to understand the responses generated by the Smart Sampark system.
5. Text translator: The text translator used is googletrans API. The English language response to the farmers queries obtained from Smart Sampark, an information retrieval system, is translated to Kannada language by the googletrans translator API.
6. Text-to-speech converter: The gTTS and pygame module for converting translated Kannada text to speech. The text is converted into mp3 format by the Google Text-to-Speech API. Pygame is employed to play the mp3 generated file back to the farmer as a Kannada voice response.
7. Speaker: The integrated speaker of the farmer's mobile device is used to play the mp3 voice response file.

The proposed framework comprises the following main modules.

A. Kannada Speech typing
B. Kannada text to English text Translation and vice versa
C. Kannada Text-to-Speech Conversion.

3.1 Kannada Speech Typing

The speechtyping.com, a web application, selenium package with chrome driver is used for Kannada speech typing. The framework built displays a window in which the user can speak the query in Kannada, the local language considered for experimentation.

In Fig. 2, the query asked by farmer is:
ಅಕ್ಕಿ ಬೆಲೆ ಎಷ್ಟು(Akki belle yeshtu)

Fig. 2 Pop up window for farmer to ask query

```
Please say your Query...
Start Speaking...
ಅಕ್ಕಿ ಬೆಲೆ ಎಷ್ಟು
How much is the price of rice
```

Fig. 3 Example of Kannada to English translation

```
The price of RIce is 30 per kg
ಅಕ್ಕಿಯ ಬೆಲೆ ಕೆ.ಜಿ.ಗೆ 30 ಆಗಿದೆ
```

Fig. 4 Example of English to Kannada translation

3.2 Kannada Text to English Text Translation and Vice Versa

Translation is a very complicated task that needs specialized and experienced translators. Mistakes as tiny as a comma missing can change the complete meaning of a sentence. Google Translator API is thus employed in the framework, and the performance is validated using BLEU score.

In Fig. 3, the Kannada query text is translated to corresponding English language query.

The English language query obtained from the previous step is utilized by Smart Sampark system [1] to process and generate a response in English language. For experimentation consider the query translated.

Q: How much is the price of rice.

The response for the above query obtained by Smart Sampark system [1] is:

A: "The price of Rice is 30/kg".

The framework translates the English language response in Kannada language as shown in Fig. 4.

3.3 Kannada Text to Speech Conversion

For the farmer who does not know how to read or write Kannada language, the answer is delivered to him in the form of Kannada speech. The Kannada text response obtained is converted into an audio format using the gTTS library in Python. The gTTS library produces a file in mp3 format played by Pygame library through the speakers.

4 Results

The framework is tested for 9 sample queries asked by the farmers, and BLEU is the metric considered for testing the framework proposed.

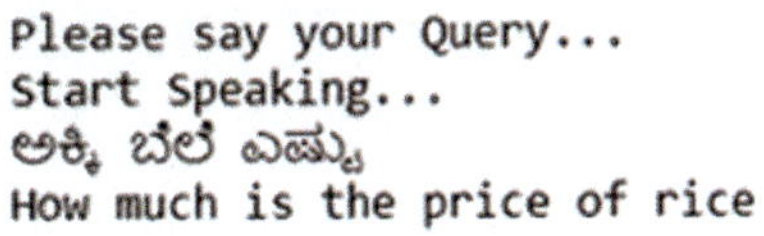

Fig. 5 Sample English language query considered for BLEU calculation

BLEU Score Calculation

The BLEU score for n-grams is defined as

$$\text{BLEU} = \text{BP}.\exp\left(\sum_{n=1}^{N} w_n \log p_n\right)$$

where

P_n is the modified precision score defined as,

$$P_n = \frac{\sum_{C\ \in\{\text{Candidates}\}} \sum_{n\text{-gram}\ \in\ C} \text{Count}_{\text{clip}}(n\text{-gram})}{\sum_{C'\in\{\text{Candidates}\}} \sum_{n\text{-gram}'\in\ C'} \text{Count}\ (n\text{-gram}')}$$

$w_n = 1/N$, N is the length of n-grams.

BP (Brevity penalty) which is defined as,

$$\text{BP} = 1 \ \text{ if } \ c > r; \ \ \text{e}^{(1-\frac{r}{c})} \ \text{ if } \ c \le r$$

where

r is the effective length of the reference sentence.

c is the total length of the candidate translation corpus.

Candidate with a BLEU score of 1 corresponds to a perfect machine-translated sentence with respect to the reference sentence.

Sample query asked by the farmer in IDE is depicted in Fig. 5,

Candidate: "How much is the price of rice"

Reference 1: "What is the price of rice"

Reference 2: "What is the cost of rice"

Reference 3: "How much does rice cost"

Reference 4: "How much is the price of rice"

The calculated BLEU score of 1 is obtained for 1-gram, 2-gram, 3-gram, and 4-gram. This implies that the candidate, i.e., the translation carried out by the proposed framework is a perfect match.

Table 1 depicts the result of the English language translations of the 9 sample queries in Kannada language. Each query translated is referred with 4 human generated reference queries. Table 2 depicts the BLEU score for different n-grams for each query in Table 1. Table 3 showcases the result of the Kannada language translation of 9 responses in English language. Table 4 depicts the BLEU score for different n-gram for each response in Table 3.

Table 1 Kannada to English translation

S. No.	Input text (Kannada)	Output text (English)	Reference 1	Reference 2	Reference 3	Reference 4
1	ಕಡಲೆಕಾಯಿಯ ಸಸ್ಯ ರಕ್ಷಣೆ	Plant protection of peanuts	Plant protection of peanuts	Peanut plant protection	Plant protection for peanuts	NA
2	ಗೋಧಿ ನಾ ಹೇಗೆ ಕಾಪಾಡೋದು	How to protect wheat NA	How to protect wheat	How can wheat be preserved	How is wheat preserved	How can wheat be protected
3	ಜೋಳದ ಬೆಲೆ ಎಷ್ಟು	How much is the price of corn?	How much is the price of corn	How much does corn cost	What is the price of corn	What is the cost of corn
4	ಅಕ್ಕಿಗೆ ಯಾವುದು ಗೊಬ್ಬರ ಹಾಕಬೇಕು	What to fertilize for rice	Which fertilizer should be used for rice	Which fertilizer can be used for rice	What fertilizer should be used for rice	Which fertilizer has to be added for rice
5	ಬೆಂಗಳೂರು ವೆದರ್ ರಿಪೋರ್ಟ್	Bangalore weather report	Bangalore weather report	Bengaluru weather report	Weather report of Bengaluru	Weather report of Bangalore
6	ಇವತ್ತಿನ ಮಾರುಕಟ್ಟೆ ಮಾಹಿತಿ	Market information today	Today's market information	Market information of today	What is market information of today	NA
7	ಬೆಳಗಾವಿಯ ಹವಾಮಾನ ವರದಿ	Belgaum weather report	Belgaum weather report	Weather report of Belgaum	Belagavi weather report	Weather report of Belagavi
8	ಮೇಜಿನ ಮಾರುಕಟ್ಟೆ ಮಾಹಿತಿಗಾಗಿ ಕೇಳಲಾಗಿದೆ	Asked for table market information	Asked for market information of maize	Asked for maize market information	Maize market information is asked	Market information of maize is asked
9	ಬೆಂಗಳೂರು ಗ್ರಾಮದ ನೀರಿನ ನಿರ್ವಹಣೆ ಕೇಳಲಾಗಿದೆ	Water management of Bangalore village asked	Asked for Bangalore gram water management	Asked for Bengaluru village water management	Water management of Bangalore village is asked	Water management of Bangalore gram is asked

Table 2 BLEU score for Kannada to English translation

S. No.	BLEU 1-gram score	BLEU 2-gram score	BLEU 3-gram score	BLEU 4-gram score	BLEU score	Time taken (s)
1	1	1	1	1	1	0.499964
2	0.4	0.25	0	0	0	0.641949
3	0.857143	0.833333	0.8	0.75	0.809	0.617650
4	0.536256	0.167580	0	0	0	0.776603
5	1	1	1	0	0	0.693631
6	0.666667	0.5	0	0	0	0.674297
7	1	1	1	0	0	0.631135
8	0.4	0.25	0	0	0	0.880728
9	0.833333	0.6	0.5	0.333333	0.5372	0.836585

5 Analysis

The BLEU 1-gram, 2-gram, 3-gram, 4-gram, and overall BLEU score calculated for the Kannada to English translation as depicted in Table 2 are plotted in Figs. 6, 7, 8, 9, 10, respectively.

Figure 6 shows the 1-gram BLEU score for all 9 queries considered in Table 1. It is observed that the BLEU score for 1-gram is comparatively higher as compared to other grams due to the fact that 1-gram compares the candidate sentence with other reference sentences word by word. When the calculation is done word by word, the chances of getting an exact match for the words of the candidate sentence with the reference sentence are comparatively higher as compared to any other n-gram.

The 2-gram score for the queries in Table 1 is shown in Fig. 7. It is observed that the 2-gram score has lesser values when compared with the 1-gram score. The reason is that the chances of exact match for composition of 2 words (2-gram) from candidate sentences with reference sentences are less when compared to the single word (1-gram). The sentences which have the BLEU score as 1 for 1-gram still retain the same score for BLEU 2-gram score as well.

Figure 8 consists of the 3-gram BLEU score for all the queries in Table 1. The 3-gram scores for most of the sentence are decreased except for those with BLEU score 1. The reduction in BLEU score is the same as that mentioned for reduction of BLEU score for 2-gram. However, the sentences which have BLEU 1-gram and 2-gram score as 1 have BLEU 3-gram score as also 1.

Figure 9 shows the BLEU 4-gram score for the queries mentioned in Table 1. The BLEU 4-gram score for queries numbered 4, 5, 6, 7, and 8 is 0, because the queries 5, 6, 7 consist of only 3 words in the translated text (refer Table 1). For query numbered 4, the output translated text is

"What to fertilize for rice", whereas the references are as follows:

1. "Which fertilizer should be used for rice",

Table 3 English to Kannada translation

S. No	Input Text (English)	Output Text (Kannada)	Reference 1	Reference 2	Reference 3	Reference 4
1	RECOMMENDED:SPRAY EMAMECTIN BENZOATE 80-100GM/ACRE	ಶಿಫಾರಸು ಮಾಡಲಾಗಿದೆ: ಸ್ಪ್ರೇ ಎಮಾಮೆಕ್ಟಿನ್ ಬೆಂಜೊಯೇಟ್ 80-100 ಜಿಎಂ / ಆಕ್ರೆ	ಶಿಫಾರಸು ಮಾಡಲಾಗಿದೆ: ಸ್ಪ್ರೇ ಎಮಾಮೆಕ್ಟಿನ್ ಬೆಂಜೊಯೇಟ್ 80-100 ಜಿಎಂ / ಆಕ್ರೆ	ಸ್ಪ್ರೇ ಎಮಾಮೆಕ್ಟಿನ್ ಬೆಂಜೊಯೇಟ್ 80 - 100 ಜಿಎಂ / ಆಕ್ರೆ ಶಿಫಾರಸು ಮಾಡಲಾಗಿದೆ	ಅನುಮೋದಿಸಲಾಗಿದೆ: ಸ್ಪ್ರೇ ಎಮಾಮೆಕ್ಟಿನ್ ಬೆಂಜೊಯೇಟ್ 80-100 ಜಿಎಂ / ಆಕ್ಲ	ಸ್ಪ್ರೇ ಎಮಾಮೆಕ್ಟಿನ್ ಬೆಂಜೊಯೇಟ್ 80-100 ಜಿಎಂ / ಆಕ್ ಅನುಮೋದಿಸಲಾಗಿದೆಲ
2	SUGGESTED TO CALL CROP INSURANCE HELPLINE NUMBER 1800-425-0505	ಕ್ರಾಪ್ ಇನ್ಶ್ಯುರೆನ್ಸ್ ಸಹಾಯ ಸಂಖ್ಯೆ 1800-425-0505 ಅನ್ನು ಕರೆಯಲು ಸೂಚಿಸಲಾಗಿದೆ	ಬೆಳೆ ವಿಮೆಯ ಸಹಾಯ ಸಂಕೆ 1800-425-0505 ನ್ನು ಕರೆಯಲು ಸೂಚಿಸಲಾಗಿದೆರೆ	ಕ್ರಾಪ್ ಇನ್ಶ್ಯುರೆನ್ಸ್ ಸಹಾಯ ಸಂಖ್ಯೆ 1800-425-0505 ನ್ನು ಕರೆಯಲು ಸೂಚಿಸಲಾಗಿದೆ	ಕ್ರಾಪ್ ಇನ್ಶ್ಯುರೆನ್ಸ್ ಸಹಾಯ ಸಂಖ್ಯೆ 1800-425-0505 ನ್ನು ಕರೆಯಲು ಅನುಮೋದಿಸಲಾಗಿದೆ	ಬೆಳೆ ವಿಮೆಯ ಸಹಾಯ ಸಂಕೆ 1800-425-0505 ನ್ನು ಕರೆಯಲು ಅನುಮೋದಿಸಲಾಗಿದೆ
3	NUTRIENT MANAGEMENT information given	ನ್ಯೂಟ್ರಿಯಂಟ್ ಮ್ಯಾನೇಜ್ಮೆಂಟ್ ಮಾಹಿತಿಯನ್ನು ನೀಡಲಾಗಿದೆ	ಪೋಷಕಾಂಶಗಳ ನಿರ್ವಹಣೆಯ ಮಾಹಿತಿಯನ್ನು ನೀಡಲಾಗಿದೆ	ನ್ಯೂಟ್ರಿಯಂಟ್ ಮ್ಯಾನೇಜ್ಮೆಂಟ್ ಮಾಹಿತಿಯನ್ನು ನೀಡಲಾಗಿದೆ	ಪೋಷಕಾಂಶಗಳ ನಿರ್ವಹಣೆಯ ಮಾಹಿತಿಯನ್ನು ತಿಳಿಸಲಾಗಿದೆ	ನ್ಯೂಟ್ರಿಯಂಟ್ ಮ್ಯಾನೇಜ್ಮೆಂಟ್ ಮಾಹಿತಿಯನ್ನು ತಿಳಿಸಲಾಗಿದೆ
4	Pleasant with plenty of sunshine	ಸಾಕಷ್ಟು ಬಿಸಿಲಿನೊಂದಿಗೆ ಆಹ್ಲಾದಕರ	ಸಾಕಷ್ಟು ಬಿಸಿಲಿನೊಂದಿಗೆ ಆಹ್ಲಾದಕರವಾಗಿರುತ್ತದೆ	ಸಾಕಷ್ಟು ಬಿಸಿಲಿನೊಂದಿಗೆ ಉಲ್ಲಾಸಕರವಾಗಿರುತ್ತದೆ	ಹೆಚ್ಚು ಬಿಸಿಲಿನೊಂದಿಗೆ ಉಲ್ಲಾಸಕರವಾಗಿರುತ್ತದೆ	ಗರಿಷ್ಠ ಮಟ್ಟದ ಬಿಸಿಲಿನೊಂದಿಗೆ ಆಹ್ಲಾದಕರವಾಗಿರುತ್ತದೆ

(continued)

Table 3 (continued)

5	Market rate information given	ಮಾರುಕಟ್ಟೆ ದರದ ಮಾಹಿತಿಯನ್ನು ನೀಡಲಾಗಿದೆ	ಮಾರುಕಟ್ಟೆ ದರದ ಮಾಹಿತಿಯನ್ನು ನೀಡಲಾಗಿದೆ	ಮಾರುಕಟ್ಟೆ ಬೆಲೆಯ ಮಾಹಿತಿ ನೀಡಲಾಗಿದೆ	ಮಾರುಕಟ್ಟೆ ದರದ ಮಾಹಿತಿಯನ್ನು ಕೊಡಲಾಗಿದೆ	ಮಾರುಕಟ್ಟೆ ಬೆಲೆಯ ಮಾಹಿತಿ ನೀಡಲಾಗಿದೆ
6	Dont irrigate the field during flowering stage	ಹೂಬಿಡುವ ಹಂತದಲ್ಲಿ ಕ್ಷೇತ್ರಕ್ಕೆ ನೀರಾವರಿ ಮಾಡಬೇಡಿ	ಹೂಬಿಡುವ ಹಂತದಲ್ಲಿ ಕ್ಷೇತ್ರಕ್ಕೆ ನೀರಾವರಿ ಮಾಡಬೇಡಿ	ಹೂಬಿಡುವ ಹಂತದಲ್ಲಿ ಭೂಮಿಗೆ ನೀರಾವರಿ ಮಾಡಬೇಡಿ	ಹೂಬಿಡುವ ಹಂತದಲ್ಲಿ ಸ್ಥಳಕ್ಕೆ ನೀರಾವರಿ ಮಾಡಬೇಡಿ	ಮಹೂಬಿಡುವ ಹಂತದಲ್ಲಿ ನೆಲಕ್ಕೆ ನೀರಾವರಿ ಮಾಡಬೇಡಿ
7	RECOMMENDED TO SPRAY GLYPHOSATE 41 WSC	ಗ್ಲೈಫೋಸೇಟ್ 41 ಡಬ್ಲ್ಯೂಎಸ್ಸಿಯನ್ನು ಸಿಂಪಡಿಸಲು ಶಿಫಾರಸು ಮಾಡಲಾಗಿದೆ	ಗ್ಲೈಫೋಸೇಟ್ 41 ಡಬ್ಲ್ಯೂಎಸ್ಸಿಯನ್ನು ಸಿಂಪಡಿಸಲು ಶಿಫಾರಸು ಮಾಡಲಾಗಿದೆ	ಗ್ಲೈಫೋಸೇಟ್ 41 ಡಬ್ಲ್ಯೂಎಸ್ಸಿಯನ್ನು ಸಿಂಪಡಿಸಲು ಅನುಮೋದಿಸಲಾಗಿದೆ	ಗ್ಲೈಫೋಸೇಟ್ 41 ಡಬ್ಲ್ಯೂಎಸ್ಸಿಯನ್ನು ಸಿಂಪಡಿಸಲು ಹೇಳಲಾಗಿದೆ	ಗ್ಲೈಫೋಸೇಟ್ 41 ಡಬ್ಲ್ಯೂಎಸ್ಸಿಯನ್ನು ಸಿಂಪಡಿಸಲು ತಿಳಿಸಲಾಗಿದೆ
8	Give water and complex fertilizer	ನೀರು ಮತ್ತು ಸಂಕೀರ್ಣ ಗೊಬ್ಬರವನ್ನು ನೀಡಿ	ನೀರು ಮತ್ತು ಸಂಯುಕ್ತ ಗೊಬ್ಬರವನ್ನು ನೀಡಿ	ನೀರು ಮತ್ತು ಮಿಶ್ರ ಗೊಬ್ಬರವನ್ನು ನೀಡಿ	ನೀರು ಮತ್ತು ಮಿಶ್ರ ಗೊಬ್ಬರವನ್ನು ಹಾಕಿ	ನೀರು ಮತ್ತು ಸಂಯುಕ್ತ ಗೊಬ್ಬರವನ್ನು ಕೊಡಿ
9	RECOMMENDED:- CHEMICAL SPRAY IS NOT REQUIRED AT THIS STAGE OF PLANT GROWTH	ಶಿಫಾರಸು ಮಾಡಲಾಗಿದೆ: - ಸಸ್ಯ ಬೆಳವಣಿಗೆಯ ಈ ಹಂತದಲ್ಲಿ ರಾಸಾಯನಿಕ ಸ್ಪ್ರೇ ಅಗತ್ಯವಿಲ್ಲ	ಶಿಫಾರಸು ಮಾಡಲಾಗಿದೆ: - ಸಸ್ಯ ಬೆಳವಣಿಗೆಯ ಈ ಹಂತದಲ್ಲಿ ರಾಸಾಯನಿಕ ಸಿಂಪರಣೆ ಅಗತ್ಯವಿಲ್ಲ	ಶಿಫಾರಸು : - ಸಸ್ಯ ಬೆಳವಣಿಗೆಯ ಈ ಹಂತದಲ್ಲಿ ರಾಸಾಯನಿಕ ಸಿಂಪರಣೆ ನಿರಾಕರಿಸಿರಿ	ಶಿಫಾರಸು : - ಸಸ್ಯ ಬೆಳವಣಿಗೆಯ ಈ ಹಂತದಲ್ಲಿ ರಾಸಾಯನಿಕ ಸಿಂಪರಣೆ ಅವಶ್ಯಕತೆ ಇಲ್ಲ	ಶಿಫಾರಸು : - ಸಸ್ಯ ಬೆಳವಣಿಗೆಯ ಈ ಹಂತದಲ್ಲಿ ರಾಸಾಯನಿಕ ಸಿಂಪಡಿಸುವುದು ಅವಶ್ಯಕತೆ ಇಲ್ಲ

Table 4 BLEU score for English to Kannada translation

S. No.	BLEU 1-gram score	BLEU 2-gram score	BLEU 3-gram score	BLEU 4-gram score	BLEU Score	Time taken (s)
1	1	1	1	1	1	0.571827
2	0.875	0.714286	0.5	0.4	0.5946036	0.481388
3	1	1	1	1	1	0.541568
4	0.666667	0.5	0	0	0	0.483490
5	1	1	1	1	1	0.494539
6	1	1	1	1	1	0.536633
7	1	1	1	1	1	0.532423
8	0.8	0.5	0	0	0	0.525391
9	0.9	0.777778	0.75	0.7142857	0.7825423	0.495032

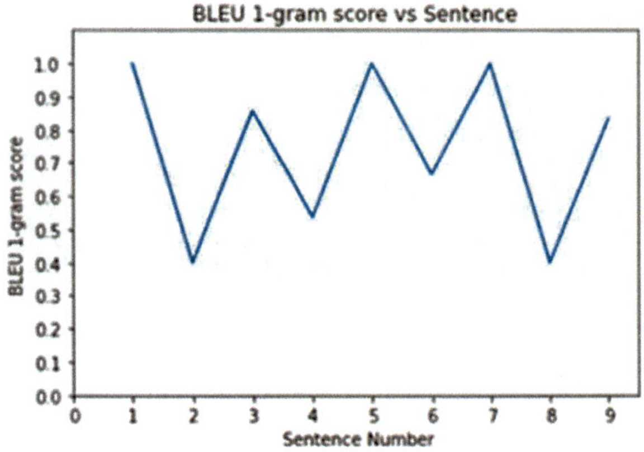

Fig. 6 BLEU 1-gram score of Kannada to English translation of 9 sample queries

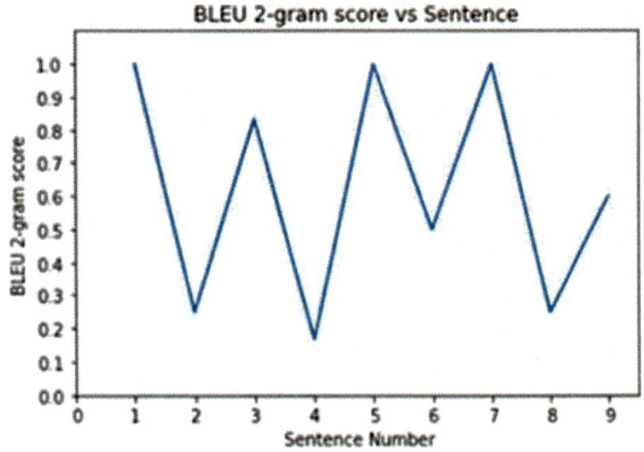

Fig. 7 BLEU 2-gram score of Kannada to English translation of 9 sample queries

Fig. 8 BLEU 3-gram score of Kannada to English translation of 9 sample queries

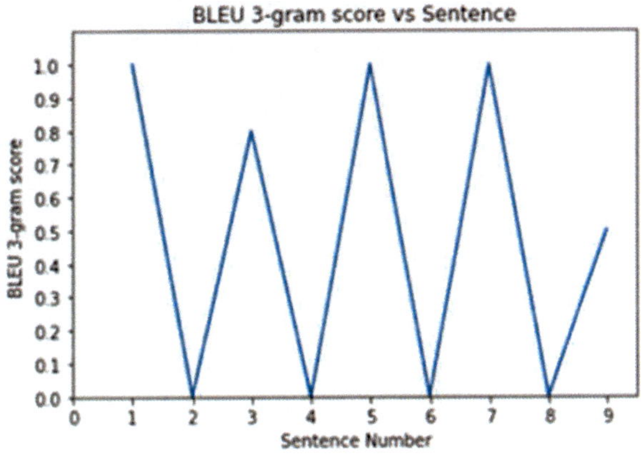

Fig. 9 BLEU 4-gram score of Kannada to English translation of 9 sample queries

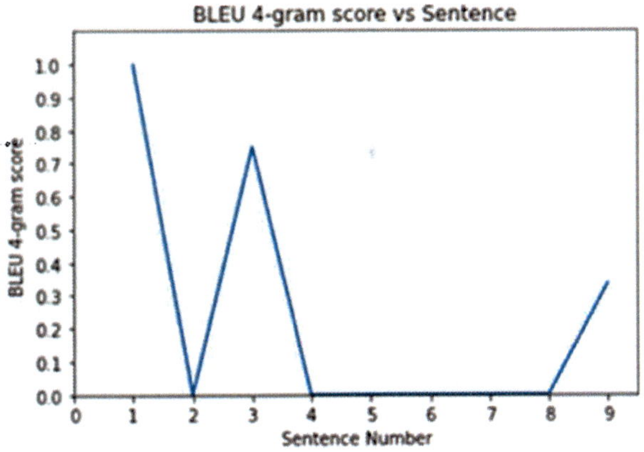

Fig. 10 Overall BLEU score of Kannada to English translation of 9 sample queries

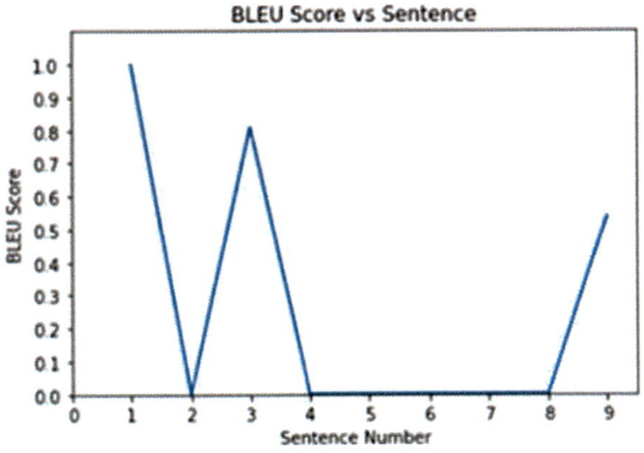

2. “Which fertilizer can be used for rice”,
3. “What fertilizer should be used for rice”,
4. “Which fertilizer has to be added for rice”.

When 4 words of sentence 4 are considered together it does not match with any of the 4 words from the reference sentence when taken together. Hence, the BLEU score for query 4 is 0 for 4-gram. Similar reason is valid for query 8.

Considering query 9, the translated output sentence is

“Water Management of Bangalore Village asked”, whereas the reference sentences are:

1. “Asked for Bangalore gram water management”,
2. “Asked for Bangalore gram water management”,
3. “Water Management of Bangalore village is asked”
4. “Water Management of Bangalore gram is asked”.

On further analysis, it is observed that candidate sentences match with reference 3 except for the missing word “is” in the candidate sentence and capitalization of letter “V” in the candidate sentence. This results in reduction of BLEU score for 3-gram to 0.5.

The overall BLEU score is depicted in Fig. 10. If the model has a good BLEU score for 1-gram, it is not necessary that the overall BLEU score would reflect the same. In fact, it is observed that BLEU penalizes heavily for all the n-grams which do not give satisfactory translation scores. As an example, consider query 5. According to Fig. 10, the overall BLEU score is 0. On the contrary, according to Fig. 6, 7, and 8 the BLEU score for 1-gram, 2-gram, and 3-gram is 1, which indicates that the model has given perfect translation. This contradictory result is due to the fact that overall BLEU score has penalized heavily for BLEU 4-gram score which is 0. The reason for the 4-gram score being 0 is because the total sequence of words for query 5 is 3 which is less than 4. This heavy penalization results in overall BLEU score giving the score as 0.

Consider query 3, the BLEU 1-gram, 2-gram, 3-gram, and 4-gram scores are 0.8571, 0.8333, 0.8, and 0.75, respectively. This is due to the extra question mark “?” given by the model. Hence, on comparing with higher n-gram scores, the BLEU score decreases.

The BLEU 1-gram, 2-gram, 3-gram, 4-gram, and overall BLEU score calculated for the English to Kannada translation of the responses as depicted in Table 4 are plotted in Figs. 11, 12, 13, 14, and 15, respectively.

Figure 11, consists of the 1-gram BLEU score for all the responses mentioned in Table 3 for English to Kannada translation.

The comparatively higher BLEU 1-gram score is the same as that mentioned in the analysis of 1-gram score for Kannada to English translation. It is also observed that the majority of the translations have a BLEU score of 1, indicating good translation results by the translation model.

The BLEU 2-gram score for the responses in Table 3 is shown in Fig. 12. Considering response 2, the output translated texted is:

Fig. 11 BLEU 1-gram score of English to Kannada translation of 9 sample responses

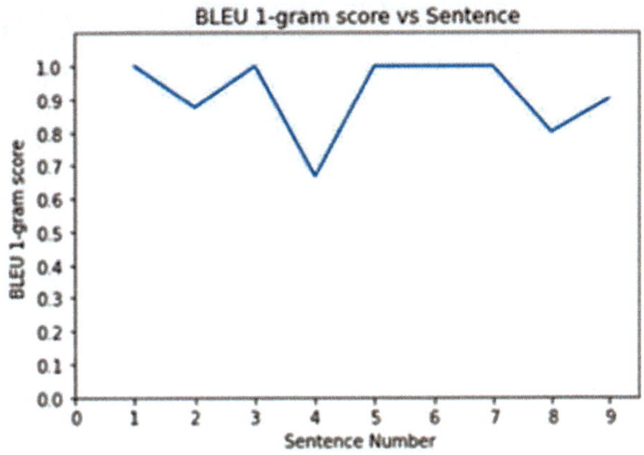

Fig. 12 BLEU 2-gram score of English to Kannada translation of 9 sample responses

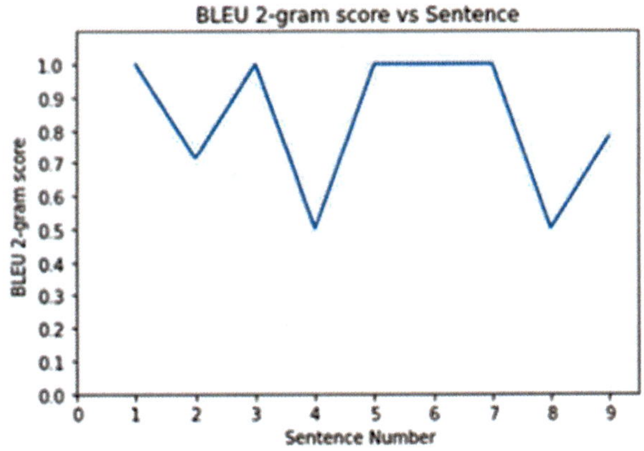

Fig. 13 BLEU 3-gram score of English to Kannada translation of 9 sample responses

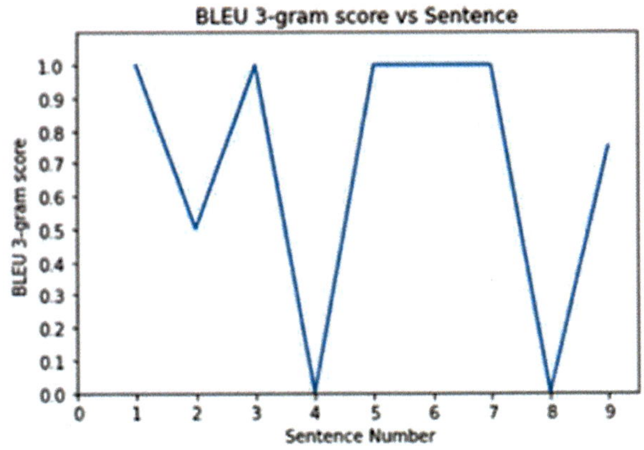

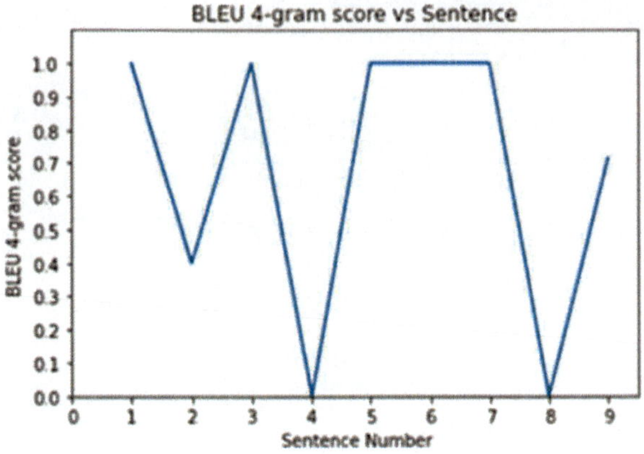

Fig. 14 BLEU 4-gram score of English to Kannada translation of 9 sample responses

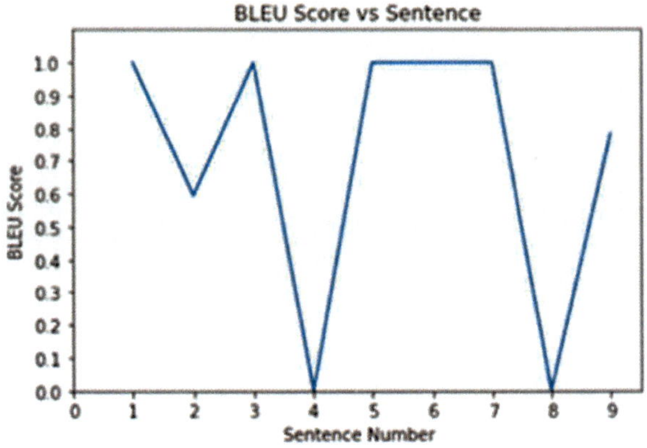

Fig. 15 Overall BLEU score of English to Kannada translation of sample responses

"ಸಾಕಷ್ಟು ಬಿಸಿಲಿನೊಂದಿಗೆ ಆಹ್ಲಾದಕರ"

Whereas the reference sentences are:

1. "ಸಾಕಷ್ಟು ಬಿಸಿಲಿನೊಂದಿಗೆ ಆಹ್ಲಾದಕರವಾಗಿರುತ್ತದೆ",
2. "ಸಾಕಷ್ಟು ಬಿಸಿಲಿನೊಂದಿಗೆ ಉಲ್ಲಾಸಕರ ವಾಗಿರುತ್ತದೆ",
3. "ಹೆಚ್ಚು ಬಿಸಿಲಿನೊಂದಿಗೆ ಉಲ್ಲಾಸಕರ ವಾಗಿರುತ್ತದೆ",
4. "ಗರಿಷ್ಠ ಮಟ್ಟದ ಬಿಸಿಲಿನೊಂದಿಗೆ ಆಹ್ಲಾದಕರವಾಗಿರುತ್ತದೆ".

It is found that candidate text (translated query) is matching with reference sentences when consider word by word. When considering a group of 2 words together, the total number of matches reduces which leads to the reduction in the BLEU score for 2-gram. Same applies for responses 4, 8, and 9.

According to Fig. 13, responses numbered 4 and 8 have a BLEU score as 0. This is due to the fact that there is no exact match for the candidate sentence with the referenced sentences when considered 3 words together. This results in a BLEU score for responses numbered 4 and 8 resulting in 0 value, whereas responses numbered 2 and 9 have some matches with the reference when 3 words together are considered. Hence, the value of BLEU 3-gram score for references numbered 2 and 9 is not 0 in fact it is about 0.5 and 0.75, respectively.

In Fig. 14. the reason for reduction in BLEU 4-gram score as compared to BLEU 3-gram is the total number of exact matches of the output sentence with the reference sentence decreases when considered 4 words simultaneously as compared to 3 words. The responses which have total length beyond 4 words and have exact match with the reference sentence have a BLEU 4-gram as 1.

The overall BLEU score for all the responses in Table 3 is shown in Fig. 15. The overall BLEU score can be considered as the average effect of the BLEU score for n-grams where n is a natural number.

It is observed that English to Kannada translation performs better as compared to that of Kannada to English translation. The number of queries with BLEU score as 1 is higher in case of English to Kannada translation as compared to that of Kannada to English translation.

5.1 Comparison with Other Translation Systems

The translation implementation in [11] is based on deep neural network (DNN) uses sequence to sequence (Seq2Seq) modeled dataset. With application of RNN unit comprising of long short-term memory (LSTM) performing the encoding and decoding mechanism, the neural machine translation (NMT) has accuracy of 86.32%. The translation is unidirectional, whereas this paper has ability to perform both bidirectional translations.

The comparison BLEU score of the deep neural network on the test data [11] with the translation system used in this paper is shown in Table 5.

In [12], the neural machine translation performed on the following five pairs of languages: Kannada to Tamil, Kannada to Telugu, Kannada to Malayalam, Kannada

Table 5 Comparison of Bilingual evaluation study (BLEU) scores

BLEU (weights)	Deep neural network translation system	Intelligent voice-assisted language translation system-proposed framework
BLEU—1	0.472143	0.743711
BLEU—2	0.360877	0.622323
BLEU—3	0.302902	0.477778
BLEU—4	0.173815	0.194444

Table 6 BLEU scores for various translation system

System	kn–ml	kn–ta	kn–te	kn–tu	kn–sn
LSTM	0.3521	0.3537	0.4292	0.5535	0.8085
BiLSTM	0.3352	0.3636	0.4477	0.4200	0.8059
Conv2Seq	0.0233	0.0303	0.0701	0.3975	0.4400
Transformer from scratch	0.3431	0.3496	0.4272	0.8123	0.5551
Pretrained model	0.3241	0.3778	0.4068	NR	NR
Fine tuned + back translation	0.2963	0.3536	0.3687	NR	NR

to Sanskrit, and Kannada to Tulu. Various translation models including Seq2Seq model such as LSTM, bidirectional LSTM, Conv2Seq, fine-tuning of already pre-trained models, and many more were trained using the datasets for each of the five language pairs. The BLEU scores on test data passed for each system are shown in Table 6.

In Table 6, NR represents "Not Recorded" due to no support for translation by the pre-trained models for those translations. Also, kn, ml, ta, te, tu, sn stand for Kannada, Malyalam, Tamil, Telugu, Tulu, and Sanskrit, respectively.

6 Conclusion

With the help of the proposed framework, farmers would be able to communicate and receive answers to their questions about agriculture in Kannada. The translations of simple sentences that are used in everyday speech, for example: "What is your name?" have a good BLEU score of 0.7–1.0. Translations for sentences longer than four words have a decent overall BLEU score for agriculture-related queries, ranging from 0.3 to 1.0. It is also observed that 1-gram BLEU score is higher when compared to 2-gram, 3-gram, 4-gram BLEU for both bidirectional translations. When comparing the BLEU score of the translation system with other implementations, it is observed that the translation system used in this paper performs better than that of deep neural network translation system for all the n-gram BLEU scores as shown in Table 5. It is observed that the English to Kannada translation performs better with higher BLEU score as compared to Kannada to English translation. The framework's overall accuracy, based on the 1-gram score for both the Kannada to English and English to Kannada translations, is 83%.

7 Future Scope

The framework could be integrated at KCC centers. The operator's/experts in KCC can make use of the framework to frame appropriate questions and responses. The framework could also be utilized to minimize the human intervention in responding to farmers questions and to build a standard agriculture query and response dataset. The translation system can be extended to other local languages such as Marathi, Tamil, Telugu, Malayalam in order to address the issue of farmers across different states of India.

Authors Declaration

Funding

No funding was received to assist with the preparation of this manuscript.

Conflicts of Interest/Competing Interests

The authors have no conflicts of interest to declare that are relevant to the content of this article.

Ethics Approval/Declarations

Not applicable

Consent to Participate

Not Applicable

Consent for Publication

Not applicable

Availability of data and material/Data availability (data transparency, if link please provide the link to access. For further information, go to

The datasets generated during and/or analyzed during the current study are available in https://data.gov.in

All data generated or analyzed during this study are included in this published article. The datasets generated during and/or analyzed during the current study are available from the corresponding author on reasonable request.

Code Availability

Code is available from the corresponding author on reasonable request.

Authors' Contributions

All authors contributed to the study conception and design. Material preparation, data collection and analysis were performed by [Pratijnya Ajawan], [Kaushik Doddamani] and [Dr. Veena Desai]. The first draft of the manuscript was written by [Pratijnya Ajawan] and all authors commented on previous versions of the manuscript. All authors read and approved the final manuscript.

References

1. Ajawan P, Desai P, Desai V (2020) Smart Sampark—an approach towards building a responsive system for Kisan Call Center. In: 2020 IEEE Bangalore humanitarian technology conference (B-HTC), pp 1–5. https://doi.org/10.1109/B-HTC50970.2020.9297854
2. Joo Y-S, Bae H, Kim Y-I, Cho H-Y, Kang H-G (2020) Effective emotion transplantation in an end-to-end text-to-speech system. IEEE Access 8:161713–161719. https://doi.org/10.1109/ACCESS.2020.3021758
3. Dimauro G, Di Nicola V, Bevilacqua V, Caivano D, Girardi F (2017) Assessment of speech intelligibility in Parkinson's disease using a speech-to-text system. IEEE Access 5:22199–22208. https://doi.org/10.1109/ACCESS.2017.2762475
4. Bano S, Jithendra P, Niharika GL, Sikhi Y (2020) Speech to text translation enabling multilingualism. IEEE Int Conf Innov Technol (INOCON) 2020:1–4. https://doi.org/10.1109/INOCON50539.2020.9298280
5. Kano T, Sakti S, Nakamura S (2020) End-to-end speech translation with transcoding by multi-task learning for distant language pairs. IEEE/ACM Trans Audio Speech Lang Proces 28:1342–1355. https://doi.org/10.1109/TASLP.2020.2986886
6. Mon AM, Soe KM (2020) Phrase-based named entity transliteration on Myanmar-English terminology dictionary. In: 2020 23rd conference of the oriental COCOSDA international committee for the co-ordination and standardisation of speech databases and assessment techniques (O-COCOSDA) 5 Nov 2020, pp 38–43. IEEE
7. Macabante DG et al (2017) Bi-directional English-Hiligaynon statistical machine translation. In: TENCON 2017—2017 IEEE region 10 conference, pp 2852–2853. https://doi.org/10.1109/TENCON.2017.8228347
8. Thihlum Z, Khenglawt V, Debnath S (2020) Machine translation of English language to Mizo language. IEEE Int Conf Cloud Comput Emerg Markets (CCEM) 2020:92–97. https://doi.org/10.1109/CCEM50674.2020.00028
9. Nemkul K, Shakya S (2021) English to Nepali sentence translation using recurrent neural network with attention. In: 2021 international conference on computing, communication, and intelligent systems (ICCCIS), pp 607–611. https://doi.org/10.1109/ICCCIS51004.2021.9397185
10. Gogineni S, Suryanarayana G, Surendran SK (2020) An effective neural machine translation for English to Hindi language. Int Conf Smart Electron Commun (ICOSEC) 2020:209–214. https://doi.org/10.1109/ICOSEC49089.2020.9215347
11. Nagaraj PK, Ravikumar KS, Kasyap MS, Murthy MHS, Paul J (2021) Kannada to English machine translation using deep neural network. Ingénierie des Systèmes d Inf 26:123–127. https://doi.org/10.18280/isi.260113
12. Vyawahare A, Tangsali R, Mandke A, Litake O, Kadam D (2022) PICT@ DravidianLangTech-ACL2022: neural machine translation on dravidian languages. arXiv preprint arXiv:2204.09098
13. Laskar SR, Dutta A, Pakray P, Bandyopadhyay S (2019) Neural machine translation: English to Hindi. IEEE Conf Inform Commun Technol 2019:1–6. https://doi.org/10.1109/CICT48419.2019.9066238

Speaker Identification Using Ensemble Learning With Deep Convolutional Features

Sandipan Dhar, Sukonya Phukan, Rajlakshmi Gogoi, and Nanda Dulal Jana

Abstract Speaker identification (SI) is an emerging area of research in the domain of digital speech processing. SI is the process of classification of speakers based on their speech features extracted from the speech utterances. After the recent developments of deep learning (DL) models, deep convolutional neural networks (DCNNs) have been widely used for solving the SI tasks. A CNN model consists of mainly two parts, deep convolutional feature extraction and classification. However, the training process of DCNN models is computationally expensive and time consuming. Therefore, in terms of reducing the computational cost of training a DCNN model an ensemble of machine learning (ML) models is proposed for the speaker identification task. The ensemble model classifies the speakers based on the deep convolutional features extracted from the input speech features. In this work, the deep convolutional features are extracted from mel-spectrograms in terms of flatten vectors (FVs) by utilizing a pre-trained DCNN model (VGG-16 model). The machine learning models that are considered for the hard voting ensemble approach are random forest (RF), extreme gradient boosting (XGBoost) and support vector machine (SVM). The models are trained and tested with voice conversion challenge (VCC) 2016 mono-lingual speech data, VCC 2020 multi-lingual speech data and multi-lingual emotional speech data (ESD). Moreover, three data augmentation techniques are used for increasing the samples of the speech data namely pitch-scaling, amplitude scaling and polarity inversion. The accuracy obtained from the proposed approach is significantly higher than the individual ML models.

S. Dhar (✉) · N. D. Jana
National Institute of Technology Durgapur, Durgapur 713209, India
e-mail: sd.19cs1101@phd.nitdgp.ac.in

N. D. Jana
e-mail: nandadulal@cse.nitdgp.ac.in

S. Phukan · R. Gogoi
Jorhat Engineering College, Jorhat 785007, India

S. Das et al. (eds.), *Advances in Data-driven Computing and Intelligent Systems*, Lecture Notes in Networks and Systems 653, https://doi.org/10.1007/978-981-99-0981-0_9

Keywords Speaker identification · Ensemble learning · Random forest · XGBoost · SVM · Convolutional neural network

1 Introduction

Speaker recognition (SR) is an application domain of digital speech processing which deals with the identification of speakers based on their vocal features [1]. The vocal features are considered as one of the important biological traits of humans by which people can be identified uniquely. Therefore, speech uttered by a speaker is considered as an important source of information for recognizing the identity of the speaker. Because speech carries the information of several vocal features of the speaker such as the unique vocal tract shape, larynx size, accent, rhythm. Hence, speech is just not an important form of verbal communication instead it is also considered as an important identification mark of speakers' individuality. This indicates the importance of SR domain in this era of advanced technology where identity of each individuals plays an important role in varieties of applications such as real-time authentication systems, surveillance systems, forensic investigations.

The domain of SR is mainly sub-divided into two categories: speaker identification (SI) and speaker verification (SV) [9]. The difference between SI and SV is that SI deals with the process of recognizing an unknown speakers' identity, whereas SV deals with the process of verifying the claimed identity of a speaker. This implies that in SI given the input speech features to the SI system the speakers' class is unknown and needs to be identified (i.e., 1:N matching), whereas in SV given the input features to the SI system, the speakers' class is known and needs to be verified (i.e., 1:1 matching). Although both of these domains are equally important based on their need in different areas of applications, SI is more challenging than SV as it requires correct classification of the input into one of the multiple target classes.

In the early developments of SI-based applications, mostly hidden Markov models (HMMs), recurrent neural networks (RNNs) and long short-term memory (LSTM) models are utilized. However, due to the progressive developments of convolutional neural network (CNN) models in computer vision (CV), deep CNN (DCNN) models become a better alternative for the SI task as it can extract features automatically given the input data. However, the training process of DCNN models is computationally expensive and thus very time consuming [13]. Therefore, in order to reduce the computational cost of training a DCNN model, a lightweight model is proposed in this work which is based on the ensemble of three machine learning (ML) models. Here, deep convolutional features are extracted in terms of flatten vectors (FVs) from the input speech feature by utilizing a pre-trained DCNN model. Thereafter, the FVs are used as input features to the ML models which are incorporated to the ensemble model.

In this work, a pre-trained VGG-16 DCNN [15] model is utilized to extract deep convolutional features from the input speech features. The input speech features considered here are mel-spectrograms. The deep convolutional features are extracted in terms of the FVs which are also called the latent representation (or latent feature) of the corresponding mel-spectrograms. The FVs are used as the input to the ensemble model which consists of random forest (RF) [12], extreme gradient boosting (XGBoost) [2] and support vector machine (SVM) [3]. The ensemble mechanism used here is hard voting ensemble. The models are trained and tested with three speech datasets such as voice conversion challenge (VCC) 2016 mono-lingual speech data [10], VCC 2020 multi-lingual speech data [14] and multi-lingual emotional speech data (ESD) [5]. Moreover, three data augmentation techniques are used for increasing the samples of the speech data namely pitch-scaling, amplitude scaling and polarity inversion. The accuracy obtained from the proposed approach is significantly higher than the considered state-of-the-art models which implies the superiority of the proposed approach.

The rest of the paper is organized as follows: Sect. 2 includes a brief background study of the related works on speaker identification. In Sect. 3, the details of the proposed approach are explained. In Sect. 4, the experimental setups, details of the datasets and training conditions are provided. Results and discussions with evaluation metrics are presented in Sect. 5. Finally, Sect. 6 concludes the paper unearthing some important future avenues of speaker identification research.

2 Related works

In the past few years, lots of research works have been done in this area of SI based on deep learning algorithms. The state-of-the-art research works related to SI are discussed in this section.

In [17], Ye et al. proposed an improvised 2D-CNN model where the authors stacked gated recurrent unit (GRU) layers to the proposed CNN model. Although their proposed model achieved sufficiently high accuracy for the SI task using mel-spectrograms but the total number of learnable parameters present in the model is 18.42 million (M), which is also sufficiently large in number. In [6], authors have utilized a hybrid model which consists of recurrent neural network (RNN) and long short-term memory (LSTM) model for the SI task, where three types of speech features are considered. The speech features considered are mel-frequency cepstral coefficients (MFCCs), spectrum and log-spectrum. In [16], Wu et al. proposed an improved ResNet architecture termed as ResNet34 for text independent speaker verification. Here, instead of taking fixed length speech samples, the authors have proposed a novel approach of considering variable length training samples for learning short-term and long-term contexts. Their proposed model has shown significant performance; however, the number of learnable parameters considered in the model is 10.1 M. In [18], Zhong et al. proposed a deep residual network that composed of a residual network (ResNet) and a convolutional attention statistics pooling (CASP) layer. Moreover, they also incorporated margin-based loss function to the proposed model for enhancing the accuracy of text independent speaker identification considering log filter bank feature vectors as speech features. Apart from single DL model for the SI task, hybrid DL model is incorporated in [8] for CNN-based speaker identification. Here, the authors proposed a two-stage feature extraction framework coupled with a CNN network. The authors used wavelet scattering transform (WST) for feature extraction. The model obtained better accuracy as compared to the considered state-of-the-art model; however, a large number of trainable parameters are used in the proposed model, i.e., 18.1 M. In [7], the authors proposed a VGG-13 model, and an offline feature extraction method is also incorporated by using a short segment of audio samples and online data augmentation technique. The total numbers of learnable parameters considered in the model are 7.3 M, and the accuracy obtained is sufficiently high using log-mel-spectrogram speech features. In [11], the authors have utilized genetic algorithm (GA)-based feature selection approach for microsleep prediction and speech emotion recognition. A similar approach is considered in [4] for the identification of Indian spoken languages using meta-heuristic feature selection algorithm.

The above-mentioned studies are mainly based on the improvement of CNN architectures for accuracy enhancement of SI task. Furthermore, most of the CNN models considered for SI task are consists of millions of learnable parameters, which takes huge time and resources for model training and performance evaluation. Therefore, there is a scope to explore the domain of SI with lightweight ML models which takes less training time without hampering the overall performance of the model. Moreover, there is also a scope to evaluate the performance of the SI models for multi-lingual speech data as well as speech data consists of different emotional speech conditions (e.g., happy, angry, sad etc).

3 Proposed Approach

In this section the details about the proposed approach is discussed.

Initially in the preprocessing phase, the audible speech samples are converted into mel-spectrograms by following mainly four steps. The first step is called framing or windowing where the time domain (amplitude versus time) speech samples are divided into multiple blocks or frames based on some fixed window size. The window size considered here is 512. Secondly, fast Fourier transform (FFT) is applied to each blocks for obtaining the time domain to frequency domain transformation of the corresponding blocks. Thirdly, mel-scaling is done using log-mel filter banks for each frequency components, and the mel-spectrogram is generated considering each blocks by decomposing the magnitude of the speech signal into its components corresponding to the mel-scale frequencies. Finally, the amplitudes are converted from power to decibel (dB). In this work, mel-spectrogram is used as inputs because it provides time-frequency representation along with the associated loudness of the speech signal, which includes the complete information of the speech signal in both spectral and temporal domain.

The three datasets used in this work (i.e., VCC 2016, VCC 2020 and ESD) have lesser numbers of speech samples. Therefore, three data augmentation techniques

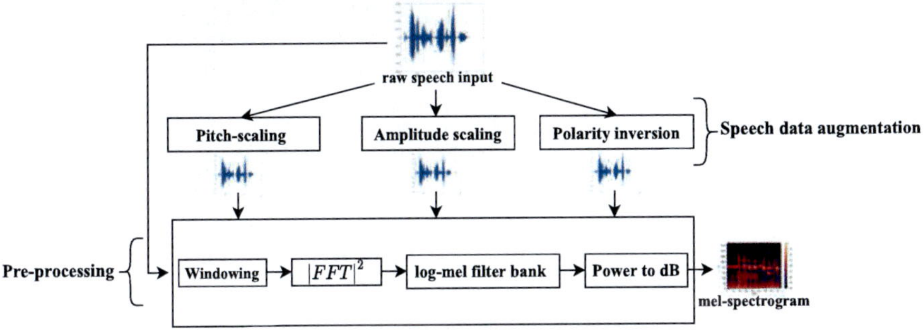

Fig. 1 Block diagram of the preprocessing and the data preparation process

are applied on the raw speech samples for generating augmented speech data corresponding to each datasets. The three data augmentation techniques used in this work are pitch-scaling, amplitude scaling and polarity inversion. Pitch-scaling or pitch shifting is the process of scaling the frequency of the speech without affecting the speed or the duration of an speech signal, similarly amplitude scaling in the process of scaling the amplitudes of a speech signal with some constant factor (i.e., gain rate) for amplification or attenuation of the speech signal, whereas polarity inversion is the process of inverting the complete speech signal in terms of its amplitude by multiplying each amplitude values by -1. Pitch and loudness of a speech uttered by a speaker carry the information of vocal characteristics of the speaker which specify the unique identity of that speaker. Therefore, in the augmentation process, these two features are specifically considered for generating new speech samples belonging to each speakers' identity. Thereafter, same preprocessing technique is applied to each augmented speech samples for extracting the mel-spectrograms as the speech features. The block diagram of the preprocessing and the data preparation process is depicted in Fig. 1.

The complete dataset is split into training and testing set. In this work, a pre-trained DCNN model is used as a deep convolutional feature extractor. In Fig. 2, the basic architectural overview of the CNN model considered in the proposed work is depicted. Here, in Fig. 2, each unique color code represents the identity of the corresponding operations shown with arrow symbol.

As shown in Fig. 2, five 2D convolutional blocks are used to transform the mel-spectrograms of the training and testing set into corresponding flatten vectors (FVs) which represent the deep convolutional features. The CNN architecture used in this work is adopted form VGG-16 DCNN model. Here, the first two convolutional blocks (Conv-Block-number) are consist of two convolution operations, batch normalization, max-pooling and dropout, whereas the consecutive three Conv-Blocks are consist of three convolution operations, batch normalization, max-pooling and dropout. The FVs are used as inputs to the ML models considered in this work. There are total three ML models considered in this work namely random forest (RF), extreme gradient boosting (XGBoost) and support vector machine (SVM) (considering radial

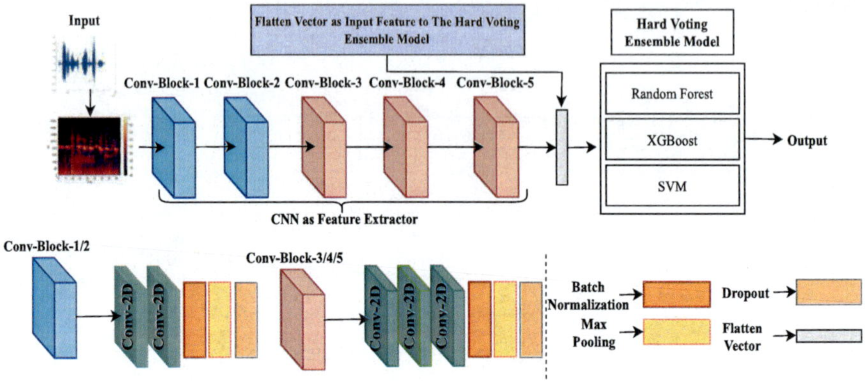

Fig. 2 Schematic overview of the proposed deep convolutional feature-based ensemble learning approach

basis function as the kernel). Each models are trained and tested with the FVs corresponding to each datasets respectively for evaluating their individual performances. Thereafter, an hard voting or majority voting ensemble model is formed by considering the three ML models for enhancing the performance of the SI task as depicted in in Fig. 2. The ensemble model is also trained and tested with FVs obtained from CNN-based feature extractor considering VCC 2016, VCC 2020 and ESD datasets. The proposed ensemble-based approach sufficiently increased the accuracy of the overall model. The objective of considering VGG-16-based deep-feature extraction process over the others is that the recent state-of-the-art deep CNN models are comprised of high number of learnable parameters than the VGG-16 model. This implies the requirement of high computational time. Therefore, in this work, VGG-16 model is considered for extracting the deep convolutional features from mel-spectrograms.

The algorithmic overview of the hard voting ensemble model is shown in Algorithm 1. As shown in Algorithm 1, given the test samples x_p to each of the models M, the corresponding class labels are predicted. The class labels y_p^d (where d implies one of the models out of N models considered in this work) corresponding to each x_p^d is stored in the output matrix O. The size of the matrix is $m \times N$, where m indicates total number of test samples. For each test samples, the final output (i.e., y_f^p) is obtained based on the class labels predicted by majority of the models.

4 Experimental Setups

4.1 Dataset Description

Algorithm 1: Algorithm for the hard voting or majority voting ensemble model

Input: Dataset (D) : Training dataset consisting of n samples belonging to one of the C classes, i.e., $\{x_i\}_{i=1}^{n} \in \{y_j\}_{j=1}^{C}$ (Here, $x_i \rightarrow i^{th}$ input data (i.e., FV) and $y_j \rightarrow j^{th}$ class label)

x_p: p^{th} data as the test sample from the test dataset T consisting of m samples (such that $m<n$)

M: ML models considered in the ensemble algorithm

N: Numbers of M used

Output: class label of each test samples

for p= 1 to m **do**

  for d= 1 to N **do**

    call M_d and train with a subset of D

    take x_p as the input sample and provide the output class label y_p^d using M_d

    store the output class label y_p^d corresponding to each data sample x_p^d in a matrix O $(O \in \mathbb{R}^{m \times N})$

    y_p^f=majority class belonging to p^{th} row of O

  end for

end for

There are total three datasets used in this work. These are VCC 2016, VCC 2020 and ESD. The VCC 2016 dataset is a uni-lingual or mono-lingual speech dataset recorded in English consists of parallel speech samples (i.e., similar speech content which are time aligned) of the speakers. However, both VCC 2020 and ESD datasets are recorded with multi-lingual speech contents. Each datasets contains both male and female speakers. The details of each sample (including augmented samples) are discussed as below[1]. Here, 30% of the speech samples corresponding to each datasets are obtained by augmentation techniques.

In this work, four speakers of VCC 2016 are considered, namely SF1, SF3, SM1 and SM2. Total speech samples considered are 162 for each speakers (including augmented samples). In Table 1, the details of the VCC 2016 dataset are given.

The VCC 2020 dataset used in this work is basically a multi-lingual non-parallel (i.e., dissimilar speech contents) speech dataset. Here, total four languages are considered such as English, Finnish, German and Mandarin. The details of the VCC 2020 dataset are also provided in Table 1.

The ESD dataset used in this work is basically a multi-lingual non-parallel speech dataset recorded in different emotional conditions. Here, total four emotional speech conditions are considered in two languages as given in Table 2. As presented in Table 2, the emotional speech conditions considered are surprise, angry, sad and happy. And the languages considered are Mandarin and English.

The length or time duration of each speech samples considered here varies between 3 and 5 s. In this work, each of the datasets split into 70% for training and 30% for testing.

[1] Links of all the datasets are available at GitHub https://github.com/SandyPanda-MLDL/Speaker-Identification-Using-Ensemble-Learning-Approach-Considering-CNN-As-Feature-Extractor

Table 1 Details of the VCC 2016 and VCC 2020 dataset

VCC 2016			
Speakers	Language	Gender	No. of samples
SF1	English	Female	162
SF3	English	Female	162
SM1	English	Male	162
SM2	English	Male	162
VCC 2020			
TEF1	English	Female	109
TFF1	Finnish	Female	109
TGM1	German	Male	109
TMM1	Mandarin	Male	109

Table 2 Details of the ESD dataset

Speakers	Language	Gender	Emotion	No. of samples
CFS	Mandarin	Female	Surprise	150
CMA	Mandarin	Male	Angry	150
EFS	English	Female	Sad	150
EMH	English	Male	Happy	150

Table 3 Details of the corresponding labels of each speakers

VCC 2016	VCC 2020	ESD	Labels
SF1	TEF1	CFS	0
SF3	TFF1	CMA	1
SM1	TGM1	EFS	2
SM2	TMM1	EMH	3

The identity of each speakers belonging to each datasets is converted into numerical labels based on one-hot encoding mechanism. The details of the corresponding labels of each speakers are given in Table 3.

4.2 Preprocessing and Training Details

In the preprocessing phase for obtaining the mel-spectrograms, window length considered is 512, number of frequency bins considered is 2048 and the number of mel filter banks considered is 128. The obtained mel-spectrograms are resized into the dimension of [3] to make the inputs compatible with VGG-16-based DCNN model used for feature extraction. The FVs obtained from the DCNN model are converted into the feature vectors of size [420, 32768] and used as inputs for the considered ML

Table 4 Performance evaluation of the models for VCC 2016 dataset

Models	Accuracy	Precision	Recall	$F1$ Score
RF	84.69	0.8469	0.8548	0.8478
XGBoost	89.28	0.8978	0.8989	0.8965
SVM	91.98	0.9192	0.9195	0.9192
Proposed	92.85	0.9336	0.9339	0.9331

models. The VGG-16 model deployed for feature extraction is a pre-trained model where 14 M non-trainable parameters are used up-to the convolution part. In pitch scaling, the sampling rate used is 22.05 kHz and number of semitones used is 2. In amplitude scaling, the amplitude values are scaled with a scaling factor between 2 and 4. The polarity inversion is done by multiplying the amplitude values by -1.

The proposed approach is implemented in Python 3.6.9. For preprocessing, libraries used are Librosa 0.7.2 and SoundFile 0.10.3. For CNN-based feature extraction, Python libraries used are TensorFlow 1.15.0 and Keras 3.6.9. And for ML-based model implementations, scikit-learn version 0.23.2 is used[2].

5 Results

In this section, the results obtained from each individual ML models and the hard voting ensemble model are discussed in details.

Firstly, the performance evaluation for each individual models (i.e., RF, XGBoost and SVM) is conducted using VCC 2016 dataset. Thereafter, the performance evaluation is also carried out for the proposed hard voting ensemble model based on VCC 2016 dataset. The accuracy of the models along with precision, recall and F1 score is provided in Table 4.

As shown in Table 4, the accuracy of the proposed hard voting ensemble model is 92.85%, whereas the accuracy of RF, XGBoost and SVM are 84.69%, 89.28% and 91.98%, respectively. This implies that the proposed approach has obtained the highest accuracy as compared to other individual ML models considered for the VCC 2016 dataset. The confusion matrix obtained for the proposed approach is also shown in Fig. 3.

As given in Table 4, the accuracy of SVM is also sufficiently high as SVM is a strong classifier. However, the proposed ensemble approach has enhanced the performance of the overall model by combining the individual performances of the models.

In Table 5, the accuracy, precision, recall and $F1$ score of the proposed hard voting ensemble model along with RF, XGBoost and SVM are provided for VCC 2020 dataset. As shown in Table 5, the proposed model obtained the highest accuracy, i.e.,

[2] Code implementation of the proposed work along with preprocessing part is available at GitHub https://github.com/SandyPanda-MLDL/Speaker-Identification-Using-Ensemble-Learning-Approach-Considering-CNN-As-Feature-Extractor

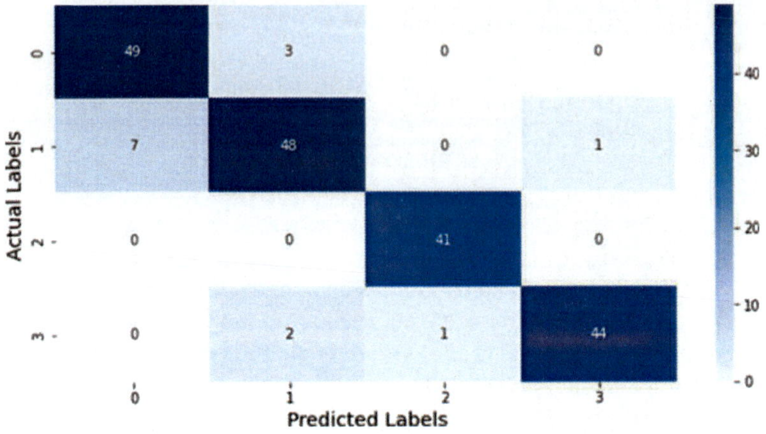

Fig. 3 Confusion matrix obtained for the proposed approach considering VCC 2016 dataset

Table 5 Performance evaluation of the models for VCC 2020 dataset

Models	Accuracy	Precision	Recall	$F1$ Score
RF	94.69	0.9450	0.9469	0.9470
XGBoost	94.69	0.9450	0.9469	0.9470
SVM	96.96	0.9700	0.9729	0.9704
Proposed	98.48	0.9848	0.9848	0.9843

98.48% for VCC 2020 dataset. Moreover, in Table 5, it can be seen that the precision, recall and $F1$ score of proposed model is 0.9336, 0.9339 and 0.9331, which is highest than the individual models. This indicates the superiority of the proposed ensemble approach over the other individual models. The confusion matrix obtained for the proposed approach is shown in Fig. 4.

In Table 6, the performance evaluation of the individual models and the proposed model are shown for the ESD dataset. As provided in Table 6, the accuracy of the proposed hard voting ensemble model is highest, i.e., 99.44% as compared to other individual models considered for the ESD dataset. The confusion matrix obtained using the proposed approach is shown in Fig. 5. Along with that, it can be seen in Table 6 that the precision, recall and $F1$ score of the proposed model is highest among the other individual models. This implies the superiority of the proposed ensemble approach.

As provided in Tables 4, 5 and 6, it can be seen that among the individual ML models, the performance of SVM and XGBoost is comparatively better over RF. However, it is well observed from the obtained results that ensemble learning approach has the ability to enhanced overall accuracy for the deep convolutional feature-based SI task.

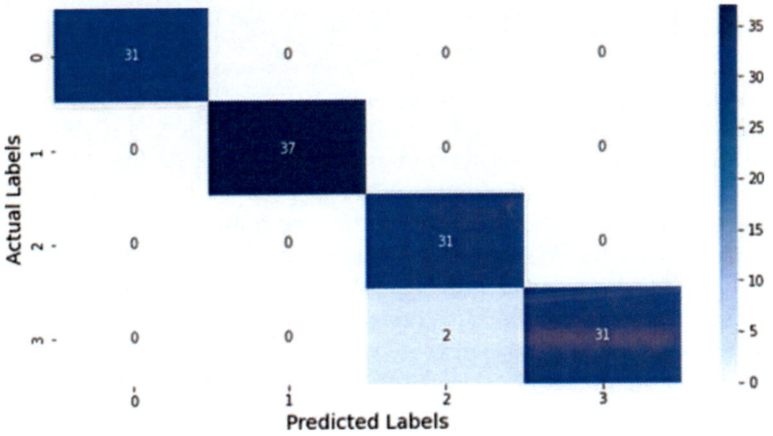

Fig. 4 Confusion matrix obtained for the proposed approach considering VCC 2020 dataset

Table 6 Performance evaluation of the models for ESD dataset

Models	Accuracy	Precision	Recall	$F1$ Score
RF	93.33	0.9346	0.9325	0.9334
XGBoost	97.77	0.9775	0.9782	0.9777
SVM	98.98	0.9892	0.9895	0.9892
Proposed	99.44	0.9943	0.9948	0.9939

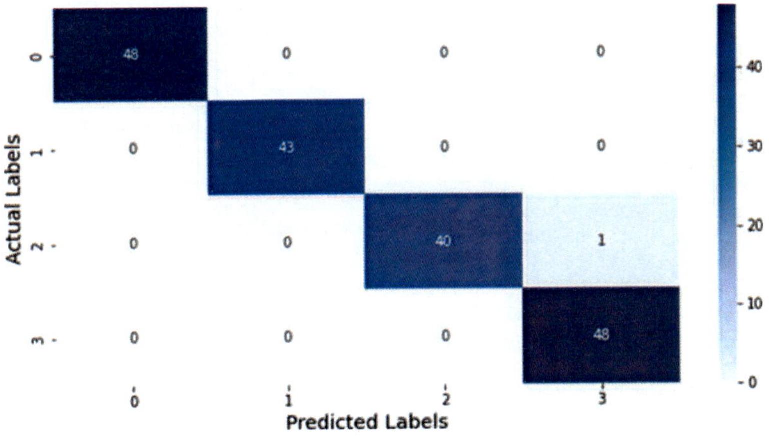

Fig. 5 Confusion matrix obtained for the proposed approach considering ESD dataset

6 Conclusion

In this paper, a majority voting or hard voting ensemble learning approach is proposed for the multi-lingual SI task. The ML models considered in this work are RF, XGBoost and SVM. Moreover, in this work, a unique feature extraction process is incorporated where a pre-trained VGG-16 DCNN model is used as a feature extractor. The deep convolutional features extracted from the mel-spectrograms are obtained in terms of flatten vectors. The flatten vectors are the latent representation of the mel-spectrograms and considered as the input features for the individual ML models as well as the proposed hard voting ensemble model. The proposed ensemble learning approach has sufficiently increased the accuracy of the SI task. In this work, the objective of using ML model is to incorporate lightweight classification models for the real-time applications which can make the training process less time consuming. In future studies, there is a scope to utilize meta-heuristic approaches for both speech features selection and the selection of classification models in terms of enhancing the performance of the classification task.

References

1. Bai Z, Zhang XL (2021) Speaker recognition based on deep learning: an overview. Neural Netw: Official J Int Neural Netw Soc 140:65–99
2. Chen T, Guestrin C (2016) Xgboost: a scalable tree boosting system. In: Proceedings of the 22nd ACM SIGKDD international conference on knowledge discovery and data mining
3. Cristianini N, Ricci E (2008) Support vector machines. Springer US, Boston, MA, pp 928–932
4. Das A, Guha S, Singh PK, Ahmadian A, Senu N, Sarkar R (2020) A hybrid meta-heuristic feature selection method for identification of Indian spoken languages from audio signals. IEEE Access 8:181432–181449. https://doi.org/10.1109/ACCESS.2020.3028241
5. Dhar S, Jana ND, Das S (2022) An adaptive learning based generative adversarial network for one-to-one voice conversion. In: IEEE Transactions on artificial intelligence, pp 1–1
6. El-Moneim SA, Nassar MA, Dessouky MI, Ismail NA, El-Fishawy AS, Abd El-Samie FE (2020) Text-independent speaker recognition using lstm-rnn and speech enhancement. Multimedia Tools Appl 79(33):24013–24028
7. Farsiani S, Izadkhah H, Lotfi S (2022) An optimum end-to-end text-independent speaker identification system using convolutional neural network. Comput Electr Eng 100:107882
8. Ghezaiel W, Brun L, Lézoray O (2021) Hybrid network for end-to-end text-independent speaker identification. In: 2020 25th International conference on pattern recognition (ICPR), pp 2352–2359
9. Kabir MM, Mridha MF, Shin J, Jahan I, Ohi AQ (2021) A survey of speaker recognition: Fundamental theories, recognition methods and opportunities. IEEE Access 9:79236–79263
10. Kobayashi K, Takamichi S, Nakamura S, Toda T (2016) The nu-naist voice conversion system for the voice conversion challenge 2016. In: INTERSPEECH
11. Krajewski J, Golz M, Sommer D, Wieland R (2009) Genetic algorithm based feature selection applied on predicting microsleep from speech. In: Vander Sloten J, Verdonck P, Nyssen M, Haueisen J (eds) 4th European conference of the international federation for medical and biological engineering. Springer, Berlin Heidelberg, Berlin, Heidelberg, pp 184–187
12. Liu Y, Wang Y, Zhang J (2012) New machine learning algorithm: random forest. In: Liu B, Ma M, Chang J (eds) Inf Comput Appl. Springer, Berlin Heidelberg, Berlin, Heidelberg, pp 246–252

13. Mazumder A, Ghosh S, Roy S, Dhar S, Jana ND (2022) Rectified adam optimizer-based cnn model for speaker identification. In: Advances in intelligent computing and communication. Springer Nature Singapore, Singapore, pp 155–162
14. Peng YH, Hu CH, Kang ACF, Lee HS, Chen PY, Tsao Y, Wang HM (2020) The academia sinica systems of voice conversion for vcc2020. ArXiv abs/2010.02669
15. Simonyan K, Zisserman A (2015) Very deep convolutional networks for large-scale image recognition. CoRR abs/1409.1556
16. Wu Y, Zhao J, Guo C, Xu J (2021) Improving Deep CNN Architectures with variable-length training samples for text-independent speaker verification. In: Proceedings interspeech 2021, pp 81–85
17. Ye F, Yang J (2021) A deep neural network model for speaker identification. Appl Sci 11(8)
18. Zhong Q, Dai R, Zhang H, Zhu Y, Zhou G (2021) Text-independent speaker recognition based on adaptive course learning loss and deep residual network. EURASIP J Adv Signal Process 2021(1):45

Hyperparameter Optimization of CNN Using Genetic Algorithm for Speech Command Recognition

Sandipan Dhar, Arjun Ghosh, Swarup Roy, Avirup Mazumder, and Nanda Dulal Jana

Abstract Identification or recognition of speaker uttered short speech commands, known as speech command recognition (SCR), is a challenging problem in digital speech processing. It plays a vital role in various real-life applications such as Internet of things (IoT) devices and assistive technology to name a few. The deep learning models like convolutional neural networks (CNNs) have shown the potential ability to solve SCR tasks. However, these models' performance heavily depends on the appropriate choice of associated hyperparameters values, which are very labor-intensive and time-consuming. This paper proposes a genetic algorithm (GA)-based hyperparameters optimization approach to fine-tune CNN hyperparameters for performing SCR tasks. The proposed method is trained and tested with the google speech command (GSC) dataset by considering eight speech commands. The experimental results demonstrate the efficiency and effectiveness of the proposed method for solving speech features-based classification problems like SCR.

Keywords Speech command recognition · Hyperparameters selection · Convolutional neural networks · Genetic algorithm

1 Introduction

Automatic speech recognition (ASR) is the domain of digital speech processing that helps the machines to recognize human spoken words by classifying unique phonetic patterns uttered by a speaker belonging to its corresponding textual transcripts [11, 12]. Speech commands recognition (SCR) [3] is a sub-domain of ASR in which the machine recognizes short spoken commands. SCR systems are extensively deployed in a wide variety of real-life applications such as IoT-based smart home assistants,

S. Dhar · A. Ghosh · S. Roy · A. Mazumder · N. D. Jana (✉)
National Institute of Technology Durgapur, Durgapur 713209, India
e-mail: nandadulal@cse.nitdgp.ac.in

S. Dhar
e-mail: sd.19cs1101@phd.nitdgp.ac.in

S. Das et al. (eds.), *Advances in Data-driven Computing and Intelligent Systems*, Lecture Notes in Networks and Systems 653, https://doi.org/10.1007/978-981-99-0981-0_10

assistive technology for blind or handicapped people, and speech command control-based wheel chairs for patients to name a few. So, a speech command-based control system plays a valuable role as speech is the most important form of verbal communication.

Earlier, SCR systems are developed by using hidden Markov model (HMM), Gaussian mixture model (GMM), multi-layered perceptron model, recurrent neural networks (RNNs), and long short-term memory (LSTM) [8]. Recently, deep learning models specifically convolutional neural networks (CNNs) exhibited a promising alternative methods for solving SCR task [5, 10] due to its ability to deal with massive datasets and automatic learning of significant features from the datasets.

A CNN model consists of several hyperparameters such as number of convolution layers, convolution filter numbers, convolution filter size, pooling type, and activation function. The values of these hyperparameters are very sensitive to the performance of any CNN model for solving a particular task [6, 7, 19]. Several CNN models have been introduced for SCR task which are designed based on manual choice of associated hyperparameters value. In general, the manual configurations of hyperparameter values for a CNN are trial-and-error task and very time-consuming. Therefore, it is very difficult to find optimal configuration of hyperparameters for a CNN model within a reasonable cost. To the best of our knowledge, this is the first attempt to optimize hyperparameters of CNN model for carried out the SCR task.

In this work, genetic algorithm (GA)-based CNN hyperparameters selection approach is proposed for the SCR task. Here, each individual of the population is represented by six hyperparameters of a CNN model. All the operators of GA such as selection, crossover, and mutation are applied in a convectional way to keep GA framework unchanged. Moreover, spectrograms are considered as the input speech feature which is fed to the CNN model. The performance of the proposed method in solving speech command recognition task is evaluated on the google speech command (GSC) dataset by considering eight speech commands. The experimental results demonstrate the effectiveness and efficiency of the proposed method in terms of achieved accuracy (89.1%) compared to other models. The results of this study reveal that employing evolutionary algorithms such as GA contributes to solve the SCR task.

In Sect. 2, the background study and the related works on SCR are discussed. In Sect. 3, the proposed approach is explained in details. In Sect. 4, the experimental details along with the details of the datasets and information about the training conditions are provided. In addition to that the results have been discussed with the evaluation metrics in Sect. 5. Finally, Sect. 6 concludes the paper and discussed about some important future avenues of SCR research.

2 Background and Related Works

2.1 *Genetic Algorithm (GA)*

GA is a meta-heuristic approach which is based on the natural selection process [9]. The algorithm relies on mainly five operations: population initialization, fitness evaluation, selection, crossover, and mutation [13]. The first step of GA is generating an initial population that incorporates a diverse range of possible solutions. A possible solution is represented by an individual which is also called as chromosome that is comprised of a collection of genes. Genes are variables that define the characteristics of each individuals. The individuals are then evaluated using the fitness function. Individuals with higher fitness values have a greater chance of being selected to reproduce in subsequent generations. Several fitness functions are used depending on the applications and the search problems. In selection, individuals are joined as parents to produce new individuals for the next generation. Usually, individuals that score higher on the fitness calculation are selected as parents and pass their genes to the next generation. Then comes the crossover operation which is one of the essential components of the GA. This genetic procedure generates two new offspring using the exchange of information of genes based on the crossover probability. Some of the genes in the newly generated offspring are mutated through mutation operations to increase the diversity in the population. The processes involved in GA ends when the stopping criteria are met or the maximum generation is reached. Finally, the best individual from all generations reported as the optimal solution.

2.2 *Related Works*

Speech command recognition (SCR) is the task of detecting specific speech commands by machines using traditional speech processing and AI-based algorithms. However, in recent years, the flourishing developments of DL models like the variants of CNN in computer vision have also shown a better alternative for carrying out SCR task using CNN models [11].

In [2], andrade et al. have used convolution operation for extracting local relationship from mel-scale spectrograms that are considered as input to the model. And then, a bidirectional LSTM is used along with attention mechanism for capturing two-way (forward and backward) long-term dependencies. Finally, a weighted average of LSTM is incorporated to multiple fully connected layers for obtaining the final outcome. However, compared to deep CNN (DCNN) models, it performed almost similar in terms of accuracy. In [17], authors have used a depth-wise separable CNN (DS-CNN) model for KWS, and the input feature considered was mel-frequency spectral coefficients (MFSC). Although, the incorporation of depth-wise separable convolution operation enhanced the accuracy of the task but also increased the architectural complexity of the model. In [16], soltanian et al. have proposed quadratic self-organized operational neural network for SCR considering mel-frequency cepstral coefficients (MFCCs) as the input features. They have compared the model with LeNet-1 like CNN architecture and observed better performance. In [4], bakouri et al. have deployed a CNN model to mobile application for operating wheelchair using speech command. In their work, MFCC features are considered for the light weighted multi-layered CNN model.

The above discussed literatures are mainly focused on manual selection of CNN hyperparameters for the accuracy enhancement of SCR or KWS task. However, finding the optimal set of hyperparameters based on manual selection approach that provides the best feasible solution in terms of accuracy is a time-consuming task. This indicates that there is a scope to explore the automatic selection of CNN hyperparameters using optimization techniques for the SCR task.

3 Proposed Approach

In this section, the preprocessing of GSC speech dataset along with the proposed methodology of speech command recognition (SCR) task using GA-based CNN hyperparameters selection is explained.

In this work, eight speech commands are considered from the GSC dataset such as "down", "go", "left", "no", "right", "stop", "up", and "yes". Here, 1000 speech samples belonging to each speech commands (i.e., total 8000 speech samples) are considered which are one-second to two-second long audio files (.wav extension). Each speech samples are converted into spectrograms of shape 124×129 for making the input data compatible to work with $2D$ CNN. To obtained the spectrograms, short-time fourier transform (STFT) is applied on the wave-forms by considering frame length as 255 and frame step as 128. In this work, spectrograms are used as inputs because it provides information of variations of frequency and loudness over the time in speech data, which helps in distinguishing unique phonemes uttered by speakers.

The 2D CNN considered in this work is obtained by selecting 6 hyperparameters based on GA-based selection approach, by fixing the number of convolution layers and dense layers as 13 and 2 (including input and hidden layer), respectively. The range of the hyperparameters in this work is set on the basis of the literature review. The 6 hyperparameters considered for GA-based selection approach are the number of convolution filters, convolution filter size, activation function, dropout value, number of neurons in dense layers, and optimizer. For each convolution layers, the number of convolution filters are selected among certain choices which are 16, 32, 64, and 128. For selecting the size of 2D filters, the choices are made between 3 and 5. The choices for activation functions are considered as "relu", "elu", and "selu" for the hyperparameters selection approach [14]. The dropout value is selected between the range of 0.1–1 uniformly. The choices for the number of neurons in the dense layers are considered among three values which are 128, 256, and 512. Finally, for the

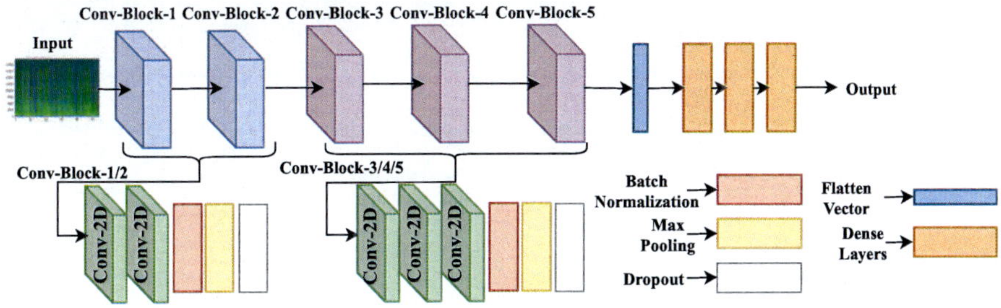

Fig. 1 Layer architecture of the CNN model used in the proposed method

selection of optimizer, four choices are considered, namely "adamax", "adadelta", "adam", and "adagrad" [18]. In this study, the stopping criteria of training each generation are set as early stopping.

In the proposed method, chromosomal structure representation and fitness function evaluation are started first. After that, three major components such as selection, crossover, and mutation are conducted sequentially. The VGG-16 CNN model [15] architecture is considered as layout framework of CNN model in our study. The basic architectural overview of the CNN model is presented in Fig. 1.

The layer architecture of CNN model used in the study consists of five 2D convolution blocks denoted as Conv-Block-Number as shown in Fig. 1. The first two convolution blocks are comprised of two 2D convolution (Conv-2D) operations, batch normalization, max pooling, and dropout operation, respectively. The rest of three convolution blocks are consists of three Conv-2D operations, batch normalization, max pooling, and dropout operation, respectively. Therefore, the total 13 number of Conv-2D operations are performed in the CNN model which is incorporated in our proposed method. Thereafter, the flatten vector passes through three dense layers, namely input layer, hidden layer, and the output layer. Batch normalization operation is also incorporated in the dense layers. Finally, the output is obtained in terms of probability of class belongingness of the input speech commands using softmax activation function.

3.1 Chromosome Structure

In this work, each chromosome is encoded in terms of associated hyperparameters of CNN model. Each position of the chromosomes is known as gene. The associated genes represent the possible hyperparameter settings which are incorporated in a complete structure of the CNN belonging to each generation. In the proposed method, six number of hyperparameters such as number of convolution filters, number of neurons in dense layers, convolution filter size, activation function, dropout value, and optimizer are considered to be optimized for CNN model. These six hyperparameters are encoded into ten genes in each chromosome structure, i.e., each chromosome

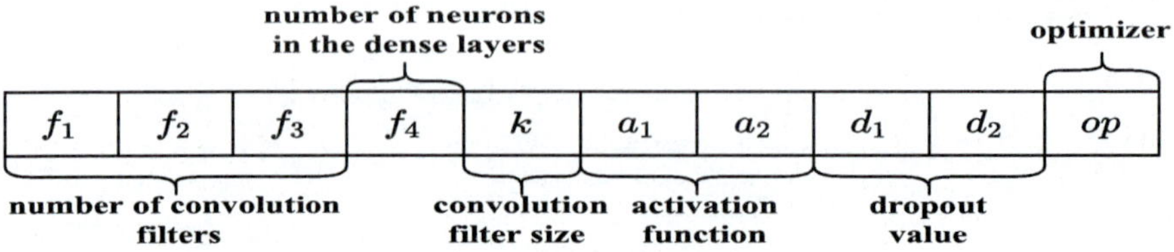

Fig. 2 Encoded chromosome of the proposed GA-based hyperparameters selection approach

length is ten. An example of a chromosome structure is presented in Fig. 2. Here, three choices are considered for the number of convolution filters denoted as f_1, f_2, and f_3 to perform 13 Conv-2D operations. One gene represents the number of neurons in dense layers and denoted as f_4. The convolution filter size is denoted as k. Finally, two choices for the activation functions denoted as a_1 and a_2, and also two choices for the dropout values denoted as d_1 and d_2. And one choice for the optimizer is represented as op.

3.2 *Fitness Evaluation*

Fitness evaluation is one of the important components of GA which formulate the fitness function. Every chromosome in the population has a fitness function used to determine how well such chromosome does at a particular task. In case of classification problem, accuracy of any deep learning (DL) model evaluated on the validation datasets after being trained on training datasets. Therefore, fitness is associated with prediction classification accuracy of the corresponding model on the validation dataset. In this paper, categorical cross-entropy (CE) [20] is used as the prediction accuracy of the model. The mathematical representation of the categorical cross-entropy function is presented in Eq. 1,

$$\text{categorical}CE = -\frac{1}{K}\sum_{i=1}^{K}[y_i \log(p_i) + (1 - y_i)\log(1 - p_i)]. \tag{1}$$

Where K represents the number of classes (i.e., class of each speech commands). y_i and p_i indicate the actual and predicted class of the ith sample.

3.3 *Selection*

Selection mechanism is also an important component of GA. The selection mechanism is used to choose individuals for the initial population. In GA, there are several selection mechanisms present. However, mostly, tournament selection mechanism is used as the selection operator in GA [13] with noisy fitness. The objective behind the use of noisy fitness function is that there is no guarantee that training is taking place for the same neural architecture for multiple times. Therefore, as a result, tournament selection mechanism gives a unified fitness value. Individuals with high fitness (accuracy) scores or values are selected for the next population.

3.4 Crossover

The crossover operation of GA creates an offspring by combining two parents $parent_1$ and $parent_2$ selected through selection operation. A random number r is chosen between 0 and 1 initially. If r is less than the crossover probability p_c, then crossover is performed on $parent_1$ and $parent_2$. Otherwise, they are incorporated into a new generation following a mutation process. In this work, the value of p_c is set to as 0.4. In this case, a crossing point was chosen at random from both parents. Then, with respect to its crossing point, each genome is divided in half. To produce the offspring, the left portion of $parent_1$ is concatenated with the right part of $parent_2$, and similarly, the left part of $parent_2$ is concatenated with the right part of $parent_1$.

3.5 Mutation

GA allows to modify one of the gene values for each individuals through the mutation operation. Here, some of the hyperparameter settings are randomly mutated in each individual. The mutation operation incorporated in this work will take place for each individuals with the mutation probability $p_m = 0.2$.

The overall framework of the proposed method is depicted in Algorithm 12.

4 Experimental Setups

4.1 Dataset Description

The proposed GA-based hyperparameters selection approach is trained and tested on google speech command (GSC) dataset [1]. The GSC dataset is consists of 65000 samples of 30 short words. However, in this work, 8 speech commands are considered, namely "down", "go", "left", "no", "right", "stop", "up", and "yes". In this work, total 8000 speech samples are considered by taking 1000 samples belonging to each speech commands. The dataset is split into training, validation, and test set. In this work, the model is trained with 6400 training samples and 1000 validation samples. After the completion of training, the trained model is tested with 600 test samples.

Algorithm 1: Pseudo-code of the proposed method

Input: maximum number of generation: g_max, size of the population: N
Output: the best chromosome: Y_{best}

```
begin
generate initial population of N chromosome Y_i (i = 1, 2, 3, ....N)
set the generation g ← 0
Evaluate the fitness of each chromosome
while g ≤ g_max do
    Select a pair of chromosome as parents from initial population randomly using
    tournament selection approach
    Apply crossover operation on selected pair with crossover probability p_c and generate
    the offspring
    Apply mutation on the offspring with mutation probability p_m
    Replace old population with newly generated population
    g ← g + 1
end
return the best chromosome Y_best
```

4.2 *Parameters Settings*

In the proposed study, all parameters are set according to the conventional GA and deep learning communities along with available computational resources in hand. All the associated parameters associated with proposed method used in our paper are presented in Table 1.

For the corresponding GA parameters setting, we set the population and maximum generation number both as 10 according to our available computational resources. We set the corresponding ρ_c and ρ_m as 0.4 and 0.2. Regarding CNN hyperparameters settings, we randomly define the number of filters in convolution layers among f_1, f_2, f_3 and stride size k. Furthermore, we also set the activation function a_1 and a_2 among the specified options. The dropout values change according to the range specified corresponding to d_1, d_2 in Table 1. Finally, optimizer op is selected from the options given in specified Table 1. For CNN training, the value of learning rate is fixed to 0.001. The number of epochs is set between 50 and 100 for the evaluation of each CNN model and also incorporated the early stopping mechanism. Finally, batch size is fixed to 64 for whole the proposed method.

5 Results

The proposed method is executed with fixed population size and generation number, which is set to 10 in both cases. After every generation, the top 5 best networks are kept as it is, and 3 loser networks are randomly kept for the next generation. In Fig. 3, the generation wise best networks are shown which are selected in terms of their performance accuracy (training and validation accuracy).

Table 1 Parameter settings in the proposed method

GA Parameters	Value
Population size	10
Maximum generation	10
Crossover probability ρ_c	0.4
Mutation probability ρ_m	0.2
CNN Hyperparameters	*Value*
f_1 (no. of Conv filters)	{16, 32, 64}
f_2 (no. of Conv filters)	{32, 64, 128}
f_3 (no. of Conv filters)	{32, 64, 128}
f_4 (no. of neurons in dense layers)	{128, 256, 512}
k (Conv filter size)	{3, 5}
a_1 (activation function)	{ReLU, SELU, ELU}
a_2 (activation function)	{ReLU, SELU, ELU}
d_1 (dropout value)	{0.1, 0.2, 0.3, 0.4, 0.5}
d_2 (dropout value)	{0.1, 0.2, 0.3, 0.4, 0.5}
op (optimizer)	{Adam, Adamax, Adadelta, Adagrad}
Training of CNN	*Value*
Learning-rate	0.001
Epochs number	50–100 (using early stopping)
Batch size	64

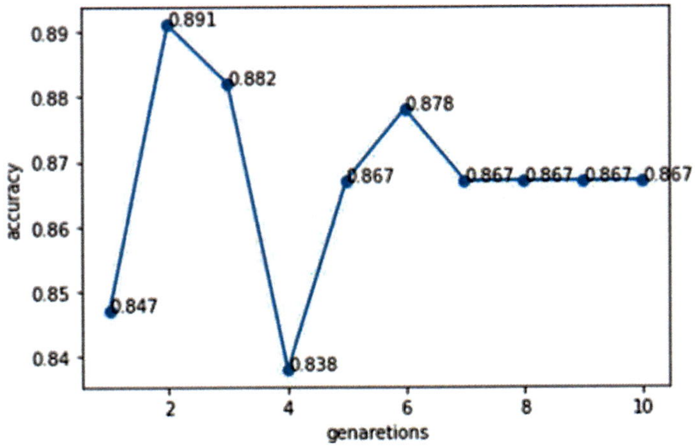

Fig. 3 Generation wise accuracy by the proposed approach

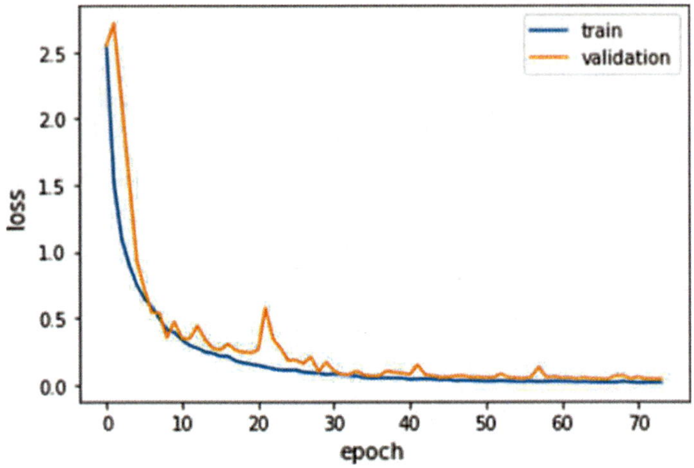

Fig. 4 Training and validation loss plot for the proposed GA-based hyperparameters selection approach

Table 2 Precision, recall, and F-1 score of each speech commands belonging to GSC dataset

Command (Label)	Precision	Recall	F-1 score	Accuracy
UP (0)	0.86	0.90	0.88	0.86
NO (1)	0.97	0.97	0.97	0.98
YES (2)	0.82	0.85	0.84	0.83
DOWN (3)	0.91	0.86	0.88	0.91
RIGHT (4)	0.83	0.73	0.78	0.84
GO (5)	0.93	0.95	0.94	0.93
LEFT (6)	0.98	0.94	0.96	0.99
STOP (7)	0.78	0.90	0.84	0.79

As shown in Fig. 3, the highest accuracy obtained is 89.1% in generation number 2. The training and validation loss of the highest accuracy model (i.e., the best model) are shown in Fig. 4. As shown in Fig. 4, both training and validation losses are decreasing with the number of epochs. This indicates the satisfactory performance of the CNN model obtained using GA-based hyperparameters selection approach. The corresponding confusion matrix of the obtained best model is shown in Fig. 5. Here, 0–7 is the corresponding labels of each of the 8 speech commands.

The precision, recall, and F-1 score of the corresponding 8 speech commands (along with their corresponding labels) are depicted in Table 2. As shown in Table 2, the accuracy of the speech commands (i.e., UP, YES, RIGHT, STOP) is around $80 \pm 2\%$; however, the average accuracy is 89.1% (i.e., almost 90%). The corresponding hyperparameters of the best model are shown in Table 3.

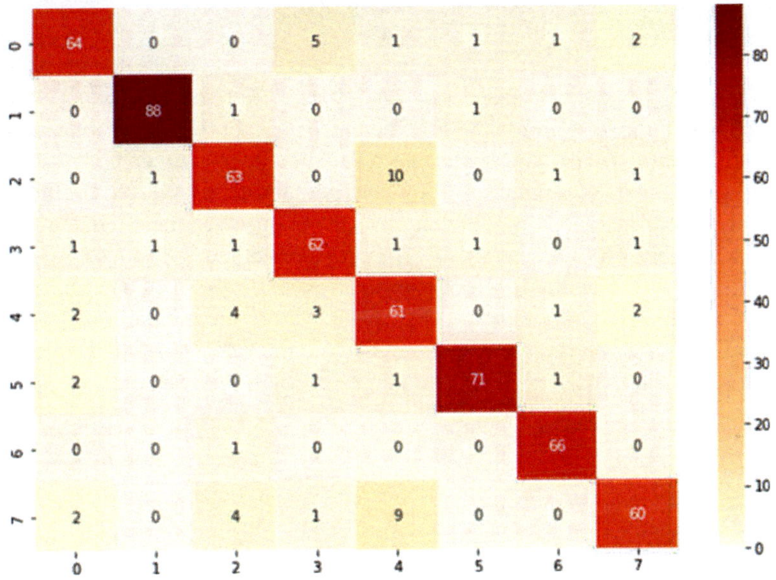

Fig. 5 Confusion matrix of the best model selected using the proposed approach

Table 3 Hyperparameters obtained from the best model

Hyperparameters	Value
f_1 (no. of Conv filters)	{32}
f_2 (no. of Conv filters)	{32}
f_3 (no. of Conv filters)	{32}
f_4 (no. of neurons in dense layers)	{128}
k (Conv filter size)	{3}
a_1 (activation function)	{ELU}
a_2 (activation function)	{ELU}
d_1 (dropout value)	{0.2}
d_2 (dropout value)	{0.4}
op (optimizer)	{Adamax}

The performance of the model is compared with grid search-based hyperparameters selection used in this work [17]. From the literature review, it is observed that the highest accuracy obtained in the grid search-based hyperparameters work is 84.7%. Whereas in our proposed GA-based hyperparameters selection approach, the highest accuracy obtained is 89.1%. This implies the superiority of the proposed work over the preceding grid search-based approach. However, there is a scope to explore the problem with other optimization algorithms to analyze the performance and the impact of the algorithms on the SCR task.

6 Conclusion

In this paper, an efficient GA-based hyperparameters selection approach is proposed for speech command recognition (SCR). The traditional approach of manual selection of hyperparameters is a time-consuming and labor-intensive task. Therefore, the proposed approach of automatic selection of hyperparameters reduced the time taken for finding the optimal CNN model that gives the highest accuracy. This implies the significance of the proposed work in determining the problem specific neural network architecture. In future work, different optimization algorithms can be utilized for the selection of speech features as well as for the selection of hyperparameters of CNN for SCR task. Moreover, there is also a scope to incorporate the mechanism of neural architecture search (NAS) in the area of speech recognition research.

References

1. Launching the speech commands dataset (Aug 24 2017), [Online; accessed 04 July 2022]
2. de Andrade DC, Leo S, Viana MLDS, Bernkopf C (2018) A neural attention model for speech command recognition. CoRR **abs/1808.08929**, http://dblp.uni-trier.de/db/journals/corr/corr1808.html#abs-1808-08929
3. Ayache M, Kanaan H, Kassir K, Kassir Y (2021) Speech command recognition using deep learning. In: 2021 sixth international conference on advances in biomedical engineering (ICABME), pp 24–29. https://doi.org/10.1109/ICABME53305.2021.9604862
4. Bakouri M, Alsehaimi M, Ismail HF, Alshareef K, Ganoun A, Alqahtani A, Alharbi Y (2022) Steering a robotic wheelchair based on voice recognition system using convolutional neural networks. Electronics 11(1). https://www.mdpi.com/2079-9292/11/1/168
5. Dhar S, Jana ND, Das S (2022) An adaptive learning based generative adversarial network for one-to-one voice conversion. IEEE Trans Artif Intell 1–1. https://doi.org/10.1109/TAI.2022.3149858
6. Ghosh A, Jana ND (2020) Neural architecture search with improved genetic algorithm for image classification. In: 2020 International conference on computational performance evaluation (ComPE), pp 344–349. https://doi.org/10.1109/ComPE49325.2020.9200164
7. Junior FEF, Yen GG (2019) Particle swarm optimization of deep neural networks architectures for image classification. Swarm Evol Comput 49:62–74
8. Kabir MM, Mridha MF, Shin J, Jahan I, Ohi AQ (2021) A survey of speaker recognition: Fundamental theories, recognition methods and opportunities. IEEE Access 9:79236–79263. https://doi.org/10.1109/ACCESS.2021.3084299
9. Katoch S, Chauhan SS, Kumar V (2021) A review on genetic algorithm: past, present, and future. Multimedia Tools Appl 80(5):8091–8126
10. Lee KA, Vestman V, Kinnunen T (2021) Asvtorch toolkit: speaker verification with deep neural networks. SoftwareX 14:100697
11. Malik M, Malik M, Mehmood K, Makhdoom I (2021) Automatic speech recognition: a survey. Multimedia Tools Appl 80(03):1–47. https://doi.org/10.1007/s11042-020-10073-7
12. Mazumder A, Ghosh S, Roy S, Dhar S, Jana ND (2022) Rectified adam optimizer-based cnn model for speaker identification. In: Advances in intelligent computing and communication. Springer Nature Singapore, pp 155–162
13. Mirjalili S (2019) Genetic algorithm. Evol Algorithms Neural Netw 43–55
14. Nwankpa C, Ijomah WL, Gachagan A, Marshall S (2018) Activation functions: comparison of trends in practice and research for deep learning. ArXiv **abs/1811.03378**

15. Simonyan K, Zisserman A (2015) Very deep convolutional networks for large-scale image recognition. CoRR abs/1409.1556
16. Soltanian M, Malik J, Raitoharju J, Iosifidis A, Kiranyaz S, Gabbouj M (2021) Speech command recognition in computationally constrained environments with a quadratic self-organized operational layer. In: 2021 International joint conference on neural networks (IJCNN) pp 1–6
17. Sørensen PM, Epp B, May T (2020) A depthwise separable convolutional neural network for keyword spotting on an embedded system. EURASIP J Audio Speech Music Process 2020(1):10. https://doi.org/10.1186/s13636-020-00176-2
18. Sun S, Cao Z, Zhu H, Zhao J (2020) A survey of optimization methods from a machine learning perspective. IEEE Trans Cyber 50:3668–3681
19. Xie L, Yuille AL (2017) Genetic CNN. In: IEEE International conference on computer vision (ICCV), pp 1388–1397
20. Zhou Y, Wang X, Zhang M, Zhu J, Zheng R, Wu Q (2019) Mpce: a maximum probability based cross entropy loss function for neural network classification. IEEE Access 7:146331–146341. https://doi.org/10.1109/ACCESS.2019.2946264

IoT-Based Plant Disease Detection and Classification: A Study on Learning Models

Pramod Kumar Singh, Anubhav Shivhare, Ashwin Raut, and Manish Kumar

Abstract The agriculture sector faces severe challenges due to variable and unpredictable agro-climatic conditions throughout the globe. Early disease detection and prevention measures help in improving the farm productivity. Internet of Things (IoT)-based monitoring techniques are required to reduce the manual efforts and improve the precision in decision making. Sensor-based Internet of Things (SBIoT) is capable of providing a framework for remote monitoring, and alarm generation in case any ailment is detected in crop. This is helpful in better decision making, information sharing, etc. The present work proposes a SBIoT framework to analyze the farm data for detecting the early stage of plant disease. Further, a comparative study of the various machine and deep learning models is provided over the plant village dataset to evaluate the algorithms on different parameters.

Keywords Sensor-based Internet of Things (SBIoT) · Deep learning (DL) · Plant disease · Classification

1 Introduction

Agricultural productivity has a significant impact on improving the efficiency of the economy and has the potential to impact the entire supply chain, food security concerns, etc. [1]. Plant disease detection is necessary to reduce crop failure risks and improve the cattle feed quality to boost livestock productivity. Further, early diagnosis of plant disease can significantly improve crop yield and quality.

Internet of Things (IoT) is a emerging technology to remotely sense and analyze a natural phenomena [2]. Multiple applications have been proposed for sensor-

P. K. Singh · A. Shivhare · A. Raut (✉) · M. Kumar
#4206, Data Analytics Lab, Computer Center-2, Indian Institute of Information Technology Allahabad, Allahabad, Uttar Pradesh, India
e-mail: rsi2018004@iiita.ac.in

M. Kumar
e-mail: manish@iiita.ac.in

S. Das et al. (eds.), *Advances in Data-driven Computing and Intelligent Systems*, Lecture Notes in Networks and Systems 653, https://doi.org/10.1007/978-981-99-0981-0_11

based IoT (SBIoT) among which agricultural-IoT (AIoT) is a prominent domain that helps in enhancing the agricultural productivity. Traditionally, human intelligence and experience were used for diagnosing plants diseases. Additionally, the manual plant disease detection methodology included laboratory testing on plant leaves and visual identification [3]. Initially, plant growth is observed visually at early stage (in situ observation), then diseased leaves are compared with healthy leaves. Thus, features of the leaf are observed like crop height, color, leaf body, leaf density on the arms, differences in root system, etc. Further, high-precision microscopes are used for accurate disease detection. Such traditional methods for plant disease detection require experts as well as continuous manual monitoring of farm. Hence, such methods are expensive, time consuming, laborious, less accurate and inefficient. Further, manual monitoring lacks reliability and it is not scalable to newer plant varieties and agricultural locations. Also, rising population, along with declining natural resources, availability of limited cultivable field, unpredictable weather, etc.,. are newer challenges that require early disease detection to minimize loss of crop productivity. Hence, technological interventions are required that can formulate plant disease as an event and detect it at the earliest in spatio-temporal scale. As a result of the challenges encountered with traditional methodologies, there is a shift from toward Smart Agriculture Practices (SAPs).

There is a correlation between the adoption of wireless sensor networks (WSNs) as a major driver of smart agriculture and the utilization of IoT which offers increased bandwidth, lower latency and increased data analysis capabilities. As a result, the AIoT integrates several existing technologies, such as SBIoT, middleware systems, WSN, radio frequency identification (RFID), end-user applications, cloud computing and user-specific hardware and software resources [4, 5]. The application domain of SBIoT is expanded further through the use of robotics, drones, remote sensors and computer vision with continually improving classical learning methods. This is useful for crop monitoring, surveying and analyzing fields, and giving proof to farmers for suitable farm management techniques which save money and time. The major works related to plant disease detection are surveyed and categorized into 'classical learning', 'deep learning' and 'IoT-based methodologies' shown in Tables 1, 2 and 3.

2 Research Challenges

Plant diseases are major factor responsible declining crop productivity. The manual methods for plant disease detection are not scalable and lack reliability in spite of being costly in terms of human resources required. Moreover, plant disease detection as an 'event' has slow initiation in temporal scale but may spread to vast areas at a faster pace in spatial domain. This makes the plant disease detection a real-time problem because it requires decision making at a very early stage to minimize the losses.

Table 1 CL approaches for plant disease detection

CL approaches for plant disease detection					
Paper	Method	Dataset	Accuracy (%)	Advantages	Limitation
[6]	SVM and CNN for feature extraction	Rice dataset	91.37	Different types of disease are classified	Limited dataset is used
[7]	SVM, PNN, BPNN random forest and GLCM	Grapes dataset	86	Highest disease detection as compared to other models.	Dimensionality reduction could improve the accuracy
[3]	*K*-NN, decision tree, Naive Bayes and logistic regression	Rice dataset	97.90	Decision tree perform better than the other learning models	Low-quality dataset is used
[8]	SVM, *k*-means, PHOG, KSNNC, SIFT and IRT	Apple and cucumber dataset	90.43	Various types of disease are considered	Proposed model is unable to extract suitable features

Further, using the classical learning and deep learning models requires prior training which is heavily dependent upon the dataset. Unfortunately, the agricultural-specific sensor-based data is not readily available [16]. Hence, SBIoT intervention is required for remote monitoring and early disease detection. This will not only help in precise data collection but also in analyzing its spatio-temporal variations while using the SBIoT motes and wireless networks few newer challenges are encountered. For example, multiple types of cameras are available for data collection which is necessary to suit user-specific needs. Yet they provide heterogeneity to data that hinders the performance on the data processing and analysis side. The available SBIoT cameras are categorized into five different types along with their advantages and limitations given in Table 4, [17]. Further, the collected image data has background noise in the form of colors other than black and white, images of other plants, leaves, soil, grass, etc. As a result, the segmentation becomes difficult. Therefore, crop-specific and quality dataset collection is required using SBIoT devices for faster and better detection.

The next challenge is diseased area segmentation and disease identification that is further aggravated when multiple diseases show similar symptoms. Additionally, these superimposed symptoms may further vary due to environmental factors and create challenges in real-time disease identification. Further, It is also challenging to quantify an infection because the same diseases exhibit different symptoms at various stages. The alteration in tissue hue, as well as external conditions such as humidity, exposure to sunlight, temperature, wind and other climatological phenomena, all have a considerable influence on the characteristics of disease symptoms. Hence, it is imperative to capture the images at each stage of disease symptoms and store in

Table 2 DL approaches for plant disease detection

DL approaches for plant disease detection					
Paper	Method	Dataset	Accuracy (%)	Advantages	Disadvantages
[9]	CNN, data augmentation (image flipping, gamma correction, noise injection, PCA, rotation, scaling and color augmentation)	Plant village dataset	96	Better performance than traditional and deep learning approaches in this paper like AlexNet, ResNet, VGG-16, and Inception-v3, SVM, decision tree, logistic regression, and K-NN	Expensive to implement in real time. Continuous monitoring of plant diseases is not happen
[10]	VGG-16, Inception V4, ResNet and DenseNets	Plant village dataset	99.75	Better performance than shallow network	Network complexity and depth do not always imply better precision
[11]	AlexNet and GoogleNet	Plant village dataset	99.35	Identification of crop diseases through neural networks was also considered	Small dataset is used for disease detection and identification
[12]	VGG-16, ResNet, ResNet, ResNet InceptionRes-Netv2 and MLP-CNNs	Apple disease and pests	99.2	Higher accuracy than single model classifier	Training of large dimension feature vector is very high

a database for accurately training the prediction models and increase the accuracy. Also the deployment of SBIoT network for data collection and image data processing at edge and fog level suffers from a low resource constraint like computational power, data storage, heterogeneity of data sources.

3 SBIoT Architecture for Plant Disease Detection

Traditional technological intervention for early disease detection in plants included knowledge extraction from manually collected plant images. The authors propose an end-to-end data collection and plant disease detection framework consisting of agriculture-specific SBIoT network (see Fig. 1).

Table 3 IoT-based approaches for plant disease detection

IoT-based approaches for plant disease detection					
Paper	Method	Dataset	Accuracy (%)	Advantages	Limitation
[1]	SVM, GLCM, CS	Pomegranate, brinjal and tomato	98.4	Various crops are taken for disease detection	Proposed method failed to follow weather conditions
[13]	Decision tree, random forest, Naive Bayes, SVM, K-NN, logistic regression color histogram	Rice dataset	97.5	Since, classification is done at the edge devices, so, it reduces the latency and connectivity over the network	Small dataset is used for classification
[14]	MCFN-CNN	Plant village dataset	97.5	On the basis of weather conditions, an excessive classical learning technique was used to forecast tomato powdery mildew disease	Noisy data is used for disease detection
[15]	CNN, SVM, decision tree and random forest	Rice dataset	89.4	Perform better than SVM, decision tree and random forest	High computation resources are used

The network consists of deployed CCD camera (Waveshare IMX477-160 12.3MP Camera 160° FOV) [18]. The deployed area is fully accessible to the network administrator. The deployment strategy is targeted at covering the deployment area or region of interest (ROI). Each of the camera is capable of independently communicating with sink or Base Station (BS) using communication protocol (say zigbee) at a sampling rate of 1hr. The sink is equipped with a higher storage capacity and pushes the data to dedicated cloud on daily basis.

At the cloud system, a trained model is available for classification of the collected data into categories of 'diseased' or 'non-diseased'. Additionally, the application user (say farmer) can also use an Internet-enabled mobile unit to take the images manually and push it to cloud for real-time disease detection and identification. Furthermore, before sending the image to the cloud, the captured data is preprocessed, segmented and features are extracted to increase the disease detection accuracy while using the pre-trained model at cloud. The other major steps are enumerated as following:

1. **Image acquisition and data transmission** phase consists of capturing the plant images by using high-resolution SBIoT cameras, drone, smartphone, etc. [17].

Table 4 Types of SBIoT cameras used for plant disease detection in agriculture

Types of SBIoT cameras used for IoT plant disease detection in agriculture			
S. No.	Types of camera	Advantages	Limitation
1.	RGB camera	Capture high-resolution images as compared to multispectral camera. It provides finer spatial features of plants for disease detection and monitoring	Less accurate multispectral and hyperspectral camera for spectral resolution of camera
2.	Multispectral camera	It is best suited camera for plant disease detection and pests detection	It is very expansive and required more calibration efforts for a specific task like disease detection and image processing
3.	Hyperspectral camera	It can identify and discriminate object details even in fewer spectral difference	It produces large amount of data generated when not properly calibrated which causes difficulties in extracting relevant information of the plant
4.	Thermal camera	Low cost as compared to spectral cameras. Easily converted into RGB camera with certain modifications	Low-resolution images are captured
5.	Depth sensor	It provides extra depth information of RGB pixels	The object may not be detected if the distance between the sensor and the object is more

The captured images can be heterogeneous in terms of storage formats, capturing spectrum etc based on the data requirements. After the successful acquisition of required images the data are sent through communication protocol (IPv-6 RPL) [19] which are widely used for IoT applications for preprocessing, based upon the requirement. Applied image enhancement techniques are used on the captured images if they are not suitable for processing.

2. **Image preprocessing** phase is a data preparation step that includes image collection through SBIoT-based camera with field experts, image labeling and data augmentation. This step is required to enhance higher detection accuracy that can be hampered due to presence of water drops, dust, small insect, etc., on plants. To overcome these problems, there are many image preprocessing techniques that are available to enhance the image quality [20].
3. **Image segmentation** phase simply divides the image into several pieces of objects or regions. Further, the image classification label finding method is used, and each image pixels and gradients is calculated accordingly. A detailed study of different types of image segmentation methods is presented in [21].
4. **Feature extraction** is next steps where important features reflect the intrinsic content of the image [22]. In general, the extracted features are divided into three categories, i.e., shape, color and texture. The shape and appearance of an

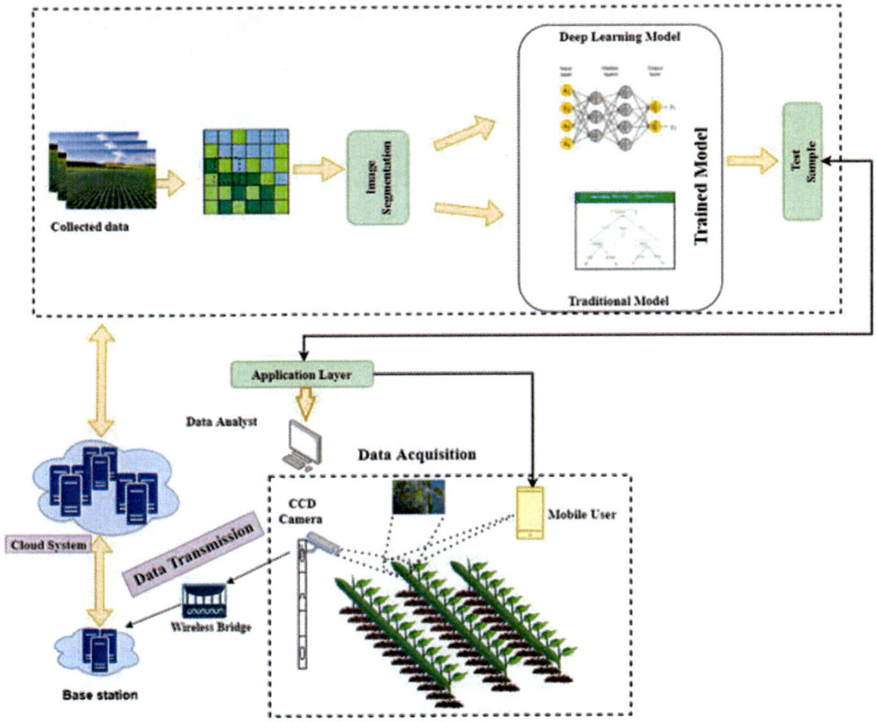

Fig. 1 AIoT architecture for plant disease detection

image due to diseases help classical learning models to identify diseases from the formation of characters. Further, these feature structures vary in their axis, locations and angles. The color parameter is helpful in distinguishing diseases from each other. The third parameter, i.e., the texture, describes how the color patterns are sprayed on the images.

5. **Disease classification** is the final phase that includes identification of input data into different known classes. Classification of plants leaves and identification of plant diseases happens in this phase. Normally, leaves are classified into healthy and unhealthy leaves. Disease classification is much difficult than disease detection, so it requires appropriate classifiers based upon the user-specific application.

4 Comparative Analysis of Models

The comparative analysis of the classical learning and deep learning models gives a clear picture of the best suited model for a particular situation. The situations depend on the location (where the model will be deployed), the machine's processing capacity, the time complexity of deploying the model and the model's accuracy.

Deep learning algorithms do not require any separate feature extraction phase. Additionally, the feature extraction process is embedded within the deep learning algorithms where features are extracted in fully automated way without any external interventions. After the data preparation, the pre-training of deep learning algorithms like convolutional neural network [23] (CNN) is carried out using back-propagation algorithms [24]. The advantage of using back-propagation algorithms is that it reduces the cost function that measure the total error of the learning model during training time. Further, the error is reduced with respect to all weights and the optimum of the cost function is calculated using gradient descent [25]. It is notable that classical learning in methodologies use expert help that is time consuming process. But in deep learning methods, the case is handled by back-propagation [26]. Further, authors in [27] have shown that feature extraction process in deep learning methods produces better result than classical learning methods.

For empirical comparison of both the methodologies, multiple datasets were used which are summarized in Table 5. Firstly, the dataset was divided into 20% for evaluation and 80% for training. All the experiment was done on the plant village dataset [28]. The performance of the deep learning models (AlexNet, DenseNet, Inception, ResNet, SqueezeNet and VGG-13/16, DCNN, BPNN, ANN, RBFNN, FR-CNN, SDD) which are analyzed for the plant diseases detection task. Furthermore, the performance of the classical learning models support vector machine (SVM), logistic regression (LR), K-NN, decision tree (DT), linear discriminant analysis (LDA), random forest (RF) and Naive Bayes (NB), multi-class SVM, SVM-kernel, Minimum Distance Classifer (MDC), Probabilistic Neural Network (PNN), maximum-minimum color difference and euclidean distance (MMCDE) is also discussed in comparison with deep learning algorithms.

Note that, deep learning algorithms uses back-propagation optimization algorithm for the pre-training of the neurons. It fine tunes the parameters at all the network layers for the feature extraction. Using these two approaches, the deep learning models try to learn more specific features for plant diseases detection starting from pre-trained networks.

Further, the detection accuracy of classical and deep learning models will depend on factors such as the types of datasets used which are best suited to the models, image preprocessing, segmentation of diseased leaf part and feature extraction (color, texture and shape). But in classical learning, models perform limited outcomes. This is because in classical learning models, the feature extraction process is done manually which depends upon researcher's selection. While the deep learning model, extract features automatically at the learning phase. Deep learning models extract suitable features from image with providing enough datasets during training phase. Further, RF gives better performance when it is trained for high-dimensional data features as compared to other models. Further, if the segmented data features of leaves are linearly separable, then SVM performs better than other classical learning models. For deep learning models, the deep network performs better than the shallow network because deep network models learn high-dimensional data features. The accuracy of

Table 5 Comparative analysis of classical and deep learning models

Dataset	Deep learning models	Classical learning models
Plant village dataset	AlexNet, DenseNet, Inception, ResNet, SqueezeNet, VGG-13	SVM, K-NN, RF, NB, LR, DT, LDA
Rice leaf dataset	DCNN, BPNN, DNN, CNN, RBFNN, ANN	MDC, NB, SVM, *K*-NN
Apple leaf dataset	DCNN, CNN, ANN	SVM, RF, *M*-SVM, SVM-*K*
Tomato leaf dataset	ResNet, Inception, CNN, FR-CNN, SDD, VGG-16	RF, PNN, *K*-NN, SVM
Potato leaf dataset	CNN, AlexNet, DCNN, ANN, BPNN	MMCDE, MC-SVM, SVM, RF

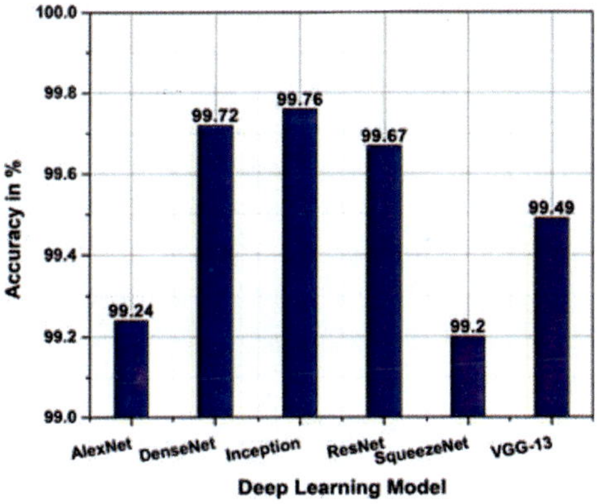

Fig. 2 Deep learning models

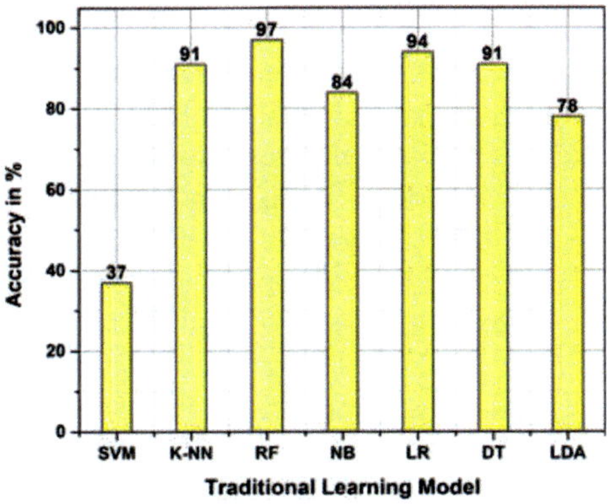

Fig. 3 Classical learning models

deep learning models also depends on the architecture. Additionally, deep learning gives poor result when it exceeds hidden layers threshold without hyperparameter tuning.

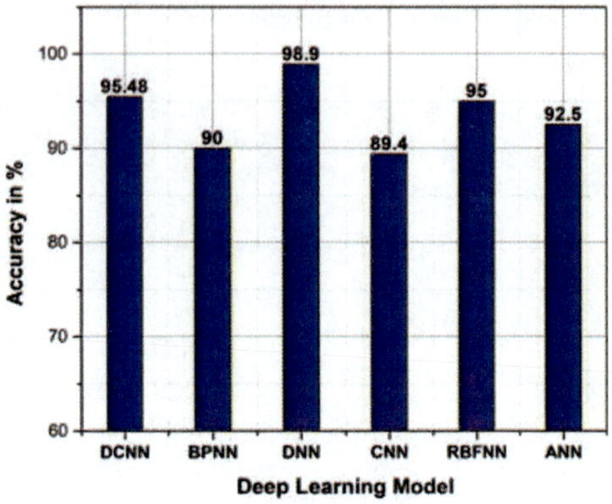

Fig. 4 Deep learning models for rice disease detection

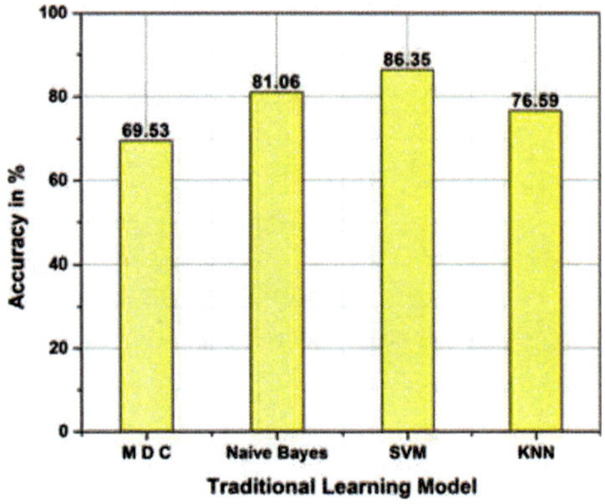

Fig. 5 Classical learning models for rice disease detection

The accuracy of seven classical and six deep learning models which are experimented on the same dataset (Plant village dataset). For deep learning algorithms, inception gives highest disease detection accuracy of given in (see Fig. 2), and similarly for classical learning algorithms, random forest tree shows highest disease detection accuracy of 99.2% with low error rate given in (see Fig. 3). AlexNet and SqueezeNet have almost same least detection accuracy of 99.24 and 99.2% (see Fig. 2). Overall for classical learning models, SVM gives least disease detection accuracy of 37% given in (see Fig. 3).

Disease detection performance was also analyzed for deep learning methodologies and classical learning methods over four different types of plants, i.e., rice, apple, tomato and potato with respective datasets. (see Fig. 4) shows that the DNN gives overall highest disease detection accuracy of 98.9% and CNN gives highest accuracy compliment with 89.4%. For classical learning models shown in (see Fig. 5), SVM gives highest detection accuracy with accuracy 86.35% and MDC performed least with accuracy of 69.53%.

Further, for apple leaves disease detection, the deep learning-based CNN gives overall highest accuracy of 98.9% shown in (see Fig. 6) and ANN gives highest

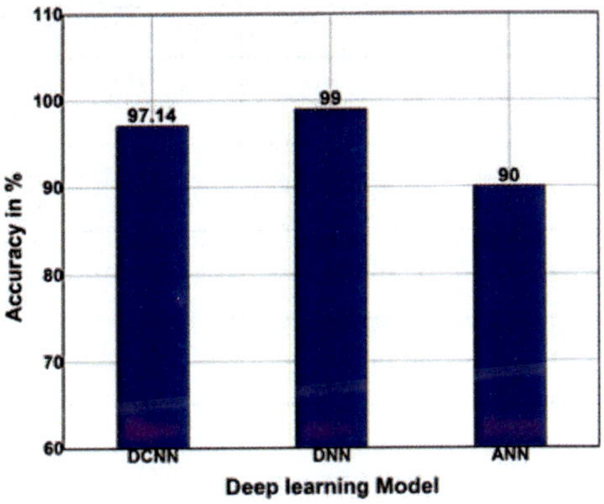

Fig. 6 Deep learning models for apple leaf disease detection

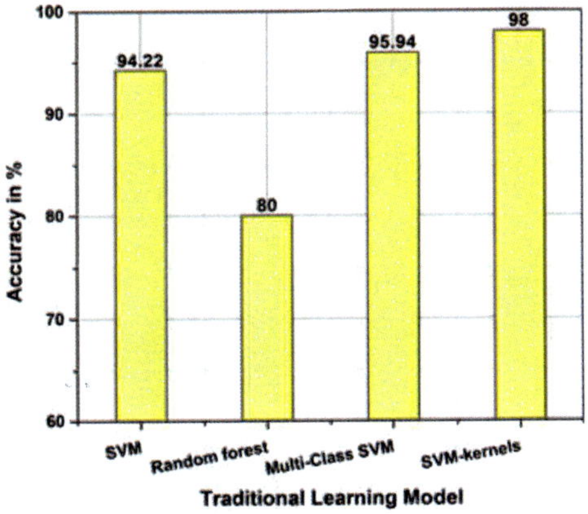

Fig. 7 Classical learning models for apple leaf disease detection

accuracy compliment with 89.4%. For classical learning models, SVM kernels gives highest detection with accuracy of 98% and RF performed least with accuracy 80% shown in (see Fig. 7).

Furthermore, for tomato leaves disease, in deep learning models, CNN gives overall highest disease detection accuracy of 99.95% and CNN gives highest accuracy compliment with 83.06% shown in (see Fig. 8). For classical learning models, M-C-SVM gives highest detection accuracy of 95.5% and SVM performed least with accuracy 89% shown in (see Fig. 9).

At last, for potato leaves disease detection, among the deep learning models, CNN gives overall highest disease detection accuracy of 99% and CNN give highest accuracy of 90.85% shown in (see Fig. 10). For classical learning models, MC-SVM gives highest detection accuracy of 95% and RF performed least with accuracy 79% shown in (see Fig. 11).

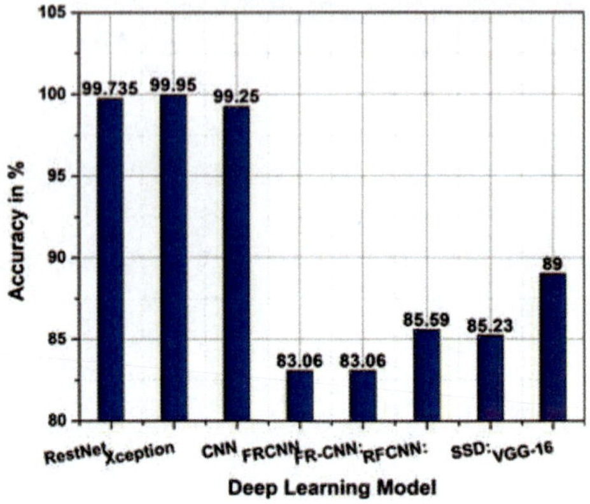

Fig. 8 Deep learning models for tomato leaf disease detection

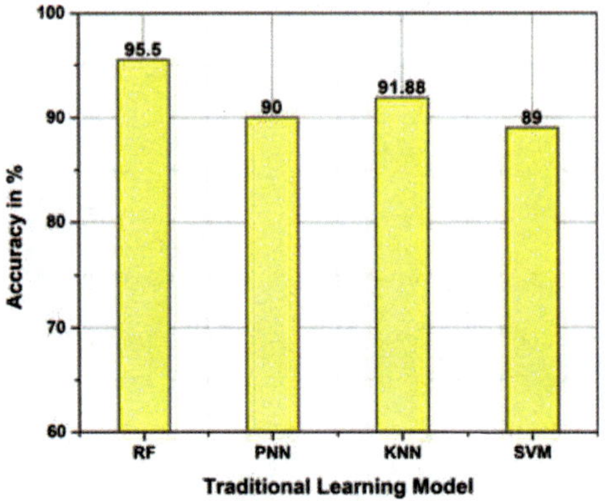

Fig. 9 Classical learning models for tomato leaf disease detection

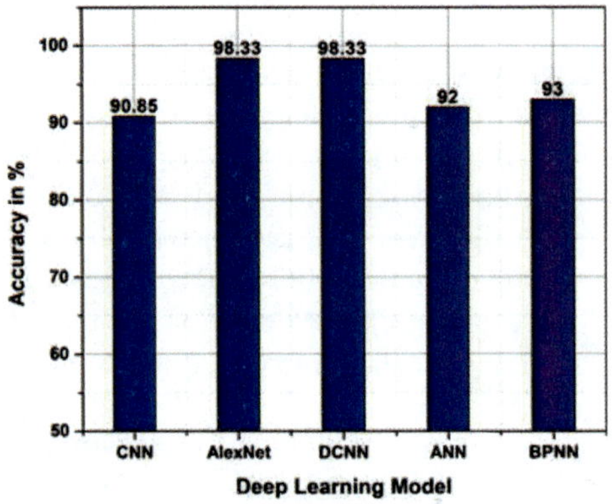

Fig. 10 Deep learning models for potato leaf disease detection

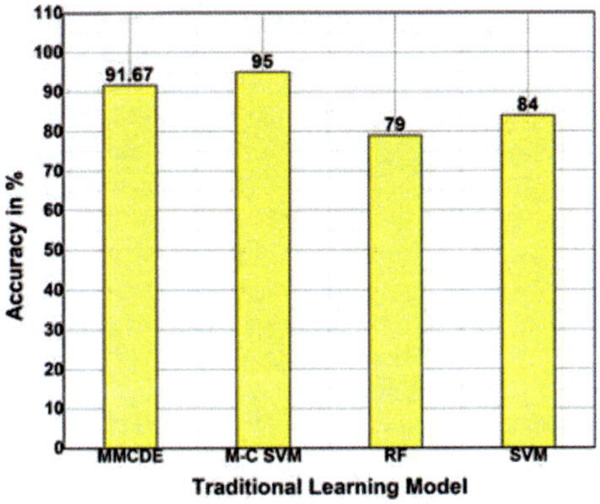

Fig. 11 Classical learning models for potato leaf disease detection

5 Conclusion and Future Work

The problem of plant disease detection has been explored. The specific need for automatic data collection, agricultural-specific dataset generation and need for agricultural domain specific methodologies have been emphasized upon. It has been maintained that new methods and techniques are required to be developed to address various existing problems like image background noise, low-cost camera, appropriate deep and classical learning models for disease identification and classification. Additionally, SBIoT-driven monitoring is required for early disease detection. Hence, a SBIoT architecture has also been discussed for the same. Lastly, multiple methods from deep learning and classical learning have been experimented with over available datasets like plant village dataset, potato dataset, tomato and rice dataset. The results have been shown to compare various algorithms on each of the cases. The disease detection accuracy of different learning models depends on the following steps like data preprocessing, segmentation algorithms, feature extraction algorithms (shape, texture and color) and learning models properties. Additionally, classical and deep learning model give better result when these steps are processed very well. Given, SBIoT architecture can reduce the processing time for disease detection in real environment.

References

1. Aasha Nandhini S, Hemalatha R, Radha S, Indumathi K (2018) Web enabled plant disease detection system for agricultural applications using WMSN. Wirel Pers Commun 102(2):725–740
2. Raut A, Shivhare A, Purohit N, Chaurasiya VK, Kumar M (2021) Improving classification accuracy using data fusion technique in IoT
3. Ahmed K, Shahidi TR, Irfanul Alam SM, Momen S (2019) Rice leaf disease detection using machine learning techniques. In: 2019 International conference on sustainable technologies for industry 4.0 (STI), pp 1–5

4. Ray PP (2017) Internet of things for smart agriculture: technologies, practices and future direction. J Amb Intell Smart Environ 9(4):395–420
5. Sinha BB, Dhanalakshmi R (2022) Recent advancements and challenges of Internet of Things in smart agriculture: a survey. Future Gener Comput Syst 126:169–184
6. Shrivastava VK, Pradhan MK, Minz S, Thakur MP (2019) Rice plant disease classification using transfer learning of deep convolution neural network. Int Arch Photogram Remote Sens Spatial Inf Sci 42(3):W6
7. Sandika B, Avil S, Sanat S, Srinivasu P (2016) Random forest based classification of diseases in grapes from images captured in uncontrolled environments. In: 2016 IEEE 13th international conference on signal processing (ICSP), pp 1775–1780
8. Zhang S, Wang H, Huang W, You Z (2018) Plant diseased leaf segmentation and recognition by fusion of superpixel. K-means PHOG Optik 157:866–872
9. Geetharamani G, Pandian A (2019) Identification of plant leaf diseases using a nine-layer deep convolutional neural network. Comput Electr Eng 76:323–338
10. Too EC, Yujian L, Njuki S, Yingchun L (2019) A comparative study of fine-tuning deep learning models for plant disease identification. Comput Electr Agricult 161:272–279
11. Agarwal M, Gupta SK, Biswas K (2020) Development of efficient CNN model for tomato crop disease identification. Sustain Comput Inf Syst 28:100407
12. Turkoglu M, Hanbay D, Sengur A (2019) Multi-model LSTM-based convolutional neural networks for detection of apple diseases and pests. J Amb Intell Hum Comput 1–11
13. Rumy SSH, Hossain MIA, Jahan F, Tanvin T (2021) An IoT based system with edge intelligence for rice leaf disease detection using machine learning. In: IEEE International IOT, electronics and mechatronics conference (IEMTRONICS). IEEE, pp 1–6
14. Verma S, Chug A, Singh AP (2018) Prediction models for identification and diagnosis of tomato plant diseases. In: 2018 International conference on advances in computing, communications and informatics (ICACCI). IEEE, pp 1557–1563
15. Chen WL, Lin YB, Ng FL, Liu CY, Lin YW (2019) RiceTalk: rice blast detection using Internet of Things and artificial intelligence technologies. IEEE Internet of Things J 7(2):1001–1010
16. Ngugi LC, Abelwahab M, Abo-Zahhad M (2021) Recent advances in image processing techniques for automated leaf pest and disease recognition-a review. Inf Process Agric 8(1):27–51
17. Singh V, Sharma N, Singh S (2020) A review of imaging techniques for plant disease detection. Artif Intell Agric 4:229–242
18. IMX477-160 12.3MP Camera, 160o Applicable for jetson nano/compute module. https://www.waveshare.com/imx477-160-12.3mp-camera.htm. [Online; accessed 04 July 2022]
19. Iova O, Picco P, Istomin T, Kiraly C (2016) Rpl: the routing standard for the internet of things... or is it? IEEE Commun Magaz 54(12):16–22
20. Kodagali JA, Balaji S (2012) Computer vision and image analysis based techniques for automatic characterization of fruits-a review. Int J Comput Appl 50(6)
21. Minaee S, Boykov YY, Porikli F, Plaza AJ, Kehtarnavaz N, Terzopoulos D (2021) Image segmentation using deep learning: a survey. IEEE Trans Patt Anal Mach Intell
22. Ping Tian D et al. (2013) A review on image feature extraction and representation techniques. Int J Multimedia Ubiquitous Eng 8(4):385–396
23. Singh T, Bhisikar S, Satakshi, Kumar M Fingerprint identification using modified capsule Network. In: 2021 12th International conference on computing communication and networking technologies (ICCCNT), pp 1–6
24. Shahabi H, Shirzadi A, Ronoud S, Asadi S, Pham BT, Mansouripour F et al (2021) Flash flood susceptibility mapping using a novel deep learning model based on deep belief network, back propagation and genetic algorithm. Geosci Front 12(3):101100
25. Wu H (2021) A stochastic gradient descent theorem and the back-propagation algorithm. arXiv preprint arXiv:2104.00539
26. LeCun Y, Bengio Y, Hinton G (2015) Deep learning. Nature 521(7553):436–444
27. Brahimi M, Boukhalfa K, Moussaoui A (2017) Deep learning for tomato diseases: classification and symptoms visualization. Appl Artif Intell 31(4):299–315
28. Mohanty SP, Hughes DP, Salathé M (2016) Using deep learning for image-based plant disease detection. Front Plant Sci 7:1419

Analysis of Online Health-Related Private Data using Named Entity Recognition by Deep Correlation Techniques

R. Geetha, Rekha Pasupuleti, and S. Karthika

Abstract Online social network analysis has become a valuable source for health research, as it recognizes convictions and behaviors hidden in traditional surveys. Twitter is a highly dynamic social media platform that enables people to interact and is growing as a valuable source of personal health data. The presence of voluminous health data on the Internet will lead to privacy leaks on user-specific private or sensitive information. Therefore, it is necessary to understand and predict the health data disclosed by social media users. The study focused on private content classification on real-time tweets extracted based on 22 health domain keywords. The tweets were categorized by the proposed Private Content Detection (PCD) algorithm constructed based on personal experience and established regulations/standards to identify the presence of sensitive private health data. The disclosed sensitive private health data are generally in the form of real-time entities like disease information, health and physical attributes like height and weight of a person, and so on. Therefore, the sensitive health entities in social media were identified by the novel Principal Sensitive Health Entity Extraction process which applies spearman's correlation coefficient as the feature extraction phase. This research work built Deep Privacy Models with deep neural networks namely CNN, RNN, C-LSTM, CRF, Bi-LSTM, and Bi-LSTM-CRF with correlation features to classify the private and non-private health data. Highly correlated private health entities identified by Spearman Correlation Coefficient factor in tweets were found to be PRODUCT, GPE, PERSON, and TIME. The Bi-LSTM-CRF model outperformed the other two models with an accuracy of 95% on account of the effective utilization of deep feature-based learning.

R. Geetha (✉)
Tamil Nadu E-Governance Agency, Chennai, Tamil Nadu, India
e-mail: geethkaajal@gmail.com

R. Pasupuleti
Trent University, Peterborough, ON K9L0G2, Canada

S. Karthika
Sri Sivasubramaniya Nadar College of Engineering, Kalavakkam, Chennai, Tamil Nadu, India
e-mail: skarthika@ssn.edu.in

S. Das et al. (eds.), *Advances in Data-driven Computing and Intelligent Systems*, Lecture Notes in Networks and Systems 653, https://doi.org/10.1007/978-981-99-0981-0_12

Keywords Online social networks · Deep neural network · Sensitive named entity recognition · User privacy · User regrets

1 Introduction

The Internet is a huge source of information that most people utilize to look up medical information. It is now widely utilized not only as a source of information but also for sharing and commenting on personal health information and experiences [1–3]. Social media platforms like Twitter and Facebook are used to disseminate a lot of health-related information to the general population. These large amounts of User-Generated Content (UGC) can be used in a variety of ways, opening new possibilities for data and text mining methodologies in a variety of sectors. Mining the personal data of the monitored users offers an instantaneous picture of the public's opinions, while longitudinal surveillance identifies changes in attitudes [4]. As seen by several types of research that use tweets and other UGC to analyze and categorize the information sought by individuals to infer health status or measure the spread of a disease in a population will have health-related information. The greatest advantage of these methods over traditional ones is instant feedback from UGCs such as tweets and search engine query logs could be obtained and processed instantly. Both experimental and anecdotal evidence suggests that Computer-Mediated-Communication (CMC) and general Internet-based behavior can be characterized as containing high levels of self-disclosure [5, 6]. It is a place where people often end up revealing themselves far more intimately than they would be inclined to do without the intermediation of screens [7]. Indeed, medical patients tend to report more symptoms and undesirable behaviors when interviewed by computer rather than face-to-face [8]. Various government authorities, including the National Institute of Standards and Technology (NIST), the European Union's (EU) General Data Protection Regulation (GDPR) [9], global norms for sharing data during Public Health Emergencies [10], and various social media in their privacy policies, have designated such health-related personally identifiable data as sensitive. The popularity of social networking sites is linked to the adoption of more healthy lifestyles and coping abilities for day-to-day health difficulties, according to the Pew Research Centre. Many recent studies have found that patients enjoy using social media to connect with one another and share their healthcare experiences [8]. These data can be mined for an instantaneous snapshot of public sentiment in terms of trust and context [11]. It can be used to assess and categorize the kind of information that people want, to infer health status, or to track the spread of a disease in a population. The four main contributions of this work are as follows:

1. A corpus of 100 K tweets is created for health-specific cyber-keywords.
2. A keyword set of most common words used for self-disclosures by Twitter users extracted from vulnerable health-related posts.

3. A novel ruleset for annotating tweets as private and non-private tweets was framed based on standards recommended by WHO, GDPR, and NIST—Privacy Content Detection (PCD) Algorithm.
4. A deep privacy model was built by experimenting with the optimal performance of six neural network models to classify tweets with self-disclosure on health data.
5. Highly correlated private health entities are identified by Spearman Correlation.

2 Related Works

2.1 Neural Networks for Short Text Classification

Machine learning techniques that use linear models, such as support vector machines or logistic regression, have dominated fundamental NLP techniques [12], which are learned over high-dimensional but sparse feature vectors. Neural networks are powerful learners, providing opportunities ranging from nonlinear classification to non-Markovian modeling of sequences and trees [13].

A short text classification using Bi-directional Recurrent Neural Network (Bi-RNN) with LSTM is proposed in combination with the topic model [14]. This combination helped in capturing contextual and semantic information. A sentiment analysis and emotion identification system were proposed by Stojanovski et al. [15] using a deep learning system that applied CNN and various classification algorithms. The aspect sentiment classification in tweets experimented with CNN in combination with Target-Dependent LSTM, Target-Connection LSTM, AE LSTM, AT LSTM, and AEAT-LSTM [16]. The CNN had attention-based input layers for imposing context words into features in which the aspect information is injected into the neural network. The networks with convolution and pooling layers are optimal for the short text classification process [17].

Self-taught learning for identifying drug abuse in tweets was modeled using deep nets (CNN, LSTM) and boosting algorithms (SVM, Naïve Bayes, and Random Forest). The deep learning models were proved to be efficient in handling the challenges of noise and sparsity in Twitter data [18]. Binary classification of tweets for identifying Personal Experience Tweets (PET) was performed by constructing a vector space model (VSM) and feeding it to an LSTM neural network [19]. It was once again evident that deep networks outperformed other statistical models and conventional algorithms in identifying PETs. It also insisted that scaling up the features of social media text has a direct significance on deep learning models.

2.2 Self-Disclosures in Health Domain

The authors of [20] experimented with instances of Private Sensitive Information detection tool alerts and evaluated the participant's responses by both quantitatively and qualitatively. The findings revealed that participants thought the PSI detection tool was useful, had a positive interest in it, and had a low level of anxiety about it, despite having a neutral intention to use it. Humans have an innate desire to reveal themselves, which is frequent on social networking sites with the features of modern social media. People are usually good at using self-disclosure to boost enjoyment while minimizing downsides—for example, by avoiding traps like disclosing too much personal information to a big audience. The users are not always able to make reasonable decisions when balancing self-disclosure gratifications with privacy risks on social media platforms, as recommended by the 'privacy calculus' approach [21].

To master self-disclosure decisions, users must be assisted. A chatbot was designed, developed, and tested by Lee et al. [22] that incorporates self-disclosure features when making small talk. They conducted a three-week study with 47 participants, dividing them into three groups to employ distinct chatbot conversing techniques. They discovered that chatbot self-disclosure had a reciprocal effect on generating deeper participant self-disclosure that lasted throughout the study period, whereas other conversation styles lacking self-disclosure characteristics did not.

Research by Oghazi et al. [23] investigated a cross-cultural comparison of users' self-disclosure of personal information on Facebook in Germany and Norway by considering three antecedents of privacy, such as concerns, attitudes, and intentions, and assessed their potential implications on self-disclosure. The findings demonstrate that users' privacy intentions are the sole antecedent that has a direct impact on their self-disclosure. The effects of stress on the relationship between privacy concerns and self-disclosure on social media sites are investigated in Zhang and Fu [24], as well as gender differences in the effects of stress. According to the findings of a survey conducted with 556 Hong Kong university students, privacy concerns are negatively related to the amount, intimacy, and honesty of self-disclosure on social media sites. This research sheds light on the emotional aspect of privacy management.

The presence of potential information for revealing people's opinions was identified in Twitter posts. Therefore, various natural language processing and text mining tools like AYLIEN API were applied to tweets to extract sentiment subjectivities and polarities of users [25]. A set of regular expressions was used to extract tweets for analyzing health conditions prevailing in a society [26]. The authors considered four health conditions, namely flu, depression, eating disorders, and pregnancy in locations of Portugal and Spain. The features were selected using an n-gram and correlation-based feature selection (CFS) approach that resulted in better accuracy. The research work of [27] investigated some critical self-disclosures and responses on Instagram using visual and textual features based on statistical methods.

2.3 *Named Entity Recognition in Social Media Text*

The named entity recognition is the most important feature extraction method used to discover and categorize entities in social media text into meaningful categories such as personal names, organization, locations, and quantities like percent, money, etc. A stacked neural architecture [28] is composed of CRF on top of a Bi-LSTM with an embedding vector fused with Glove, and BERT was trained with different embeddings say, a strong class classifier with word embeddings and a weak class classifier with glove feature extraction method which are finally converging to a majority voting system to arrive at the neural model's prediction results. The classification task was optimized by locating the presence of weak classes in a sentence as a significant enhancer condition. A combination of linguistic and statistical approaches was identified as the most effective way to accomplish a named entity recognition task [29]. The approaches like MaxEnt, HMM, Finite Automata Acquisition, SVM, ANN, C4.5 decision tree, Dynamic Bayesian Network, Morphological and contextual clues, Editex, Soundex, CRF, hybrid models, handcrafted rules, domain rules, and condition-based lookups are proven to outperform for NER identification task.

A custom NER model using LSTM trained with a dataset labeled by IOB (inside, outside, beginning) tagging [8] to quickly recognize patterns in a stream of text data. The custom NER models had proved to outperform the OpenNLP and SpaCy in terms of time taken for prediction, ease of training, and size of the model but lacked understanding of the relation between entities and tags. In Portuguese, [30] extract and classify sensitive data like personal identification number, socio-economic information, and primary communication sources from unstructured text. It brings together techniques such as rule-based approaches, statistical models, machine learning models, and neural nets. The authors realized that rule-based and lexical-based techniques were only effective for a few classes and that statistical models such as conditional random fields, random forest, and Bi-LSTM needed to be adapted.

3 Proposed Work

A brief process of building a Deep Privacy Model (DPM) to classify tweets into private and non-private health tweets is proposed as shown in Fig. 1. The real-time tweets for analyzing the presence of sensitive health information are collected using Docteur Tweety [31], the most efficient tweet extraction tool. The authors choose 22 health domain keywords based on two factors [32]. Firstly, the keywords were the most frequent search items by users in social media and the most discussed health-related topics in online social networks. Secondly, it was also determined manually based on the influential factors in the health domain. The Privacy Content Detection (PCD) algorithm guides the efficient utilization of the 22 health domain keywords in the discovery of self-disclosures in tweets which will be adapted in the data annotation task.

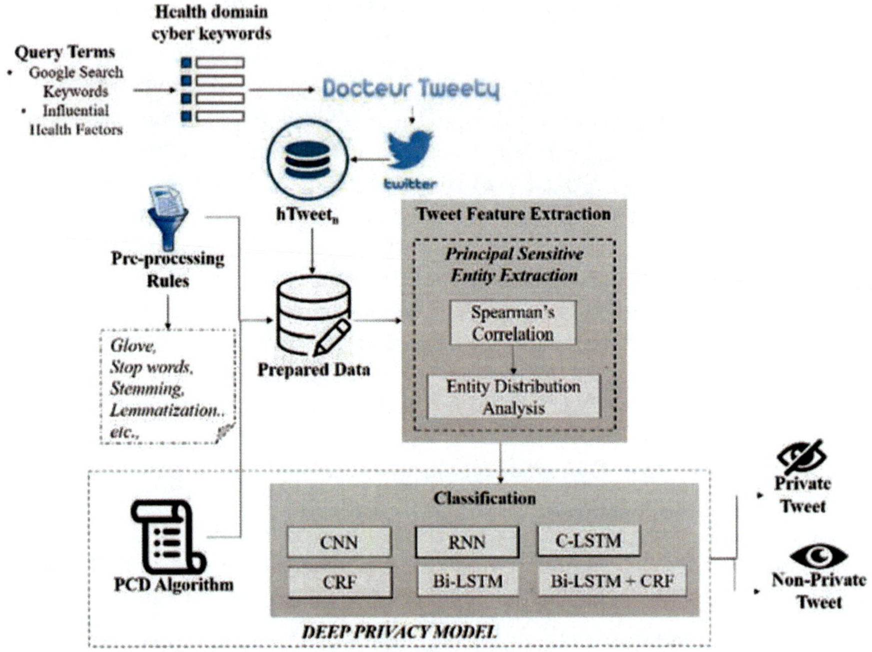

Fig. 1 Proposed workflow for identifying private health data

The cyber-keywords mentioned in the PCD algorithm contributed about 100 k tweets out of which 1250 tweets were found suitable for the classification. The collected tweets are pre-processed with a set of standard pre-processing rules that are essential for short-cleaning texts on social media. The data fetched after pre-processing is taken for feature extraction where the principal named entities in the health domain are extracted using Spearman's correlation. The textual features are now analyzed based on the identified entities and their distribution in the dataset which will be further classified by various DPM for sensitive private health data prediction in tweets. The following section describes the tweet annotation, feature extraction, and classification in detail.

3.1 Data Preparation

The DPM was built by formulating the baseline of the PCD algorithm that classified tweets into private and non-private health tweets. Private health tweets refer to the personal medical information being shared in a user's tweet. Private tweets convey a person's personal health information, personal experience pertains to a person's

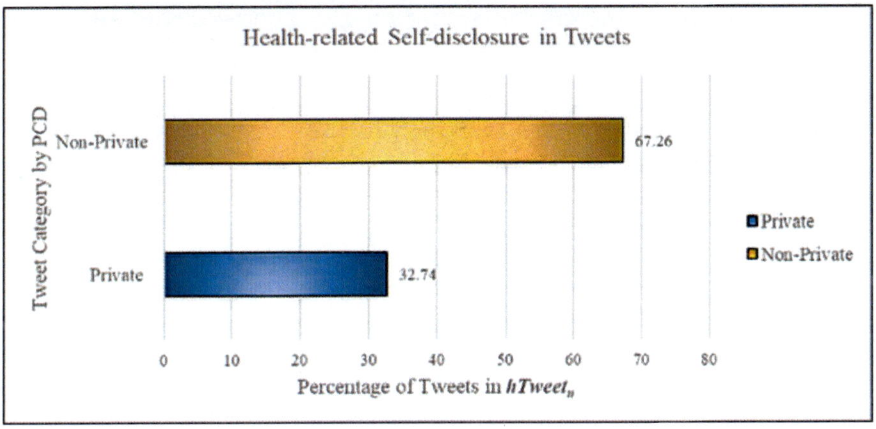

Fig. 2 Percentage of Private Tweets and Non-Private Tweets of health information present in the dataset taken after analysis

encounters or observations related to his or her life, and it can be the changes experienced by an individual in his or her health, which can be an illness, a disease, or a medical treatment. Such experience shared by users on social media is widely utilized as data for public health surveillance. The non-private health tweets, on the other hand, refer to any health information without sensitive health or medical data being disclosed.

The collected data based on 22 cyber-keywords *(birth, baby, blood, breast, medical, diagnosis, accident, vaccine, therapy, death, mental, hospital, treatment, pain, life, cancer, injury, weight, influenza, natural, report, alcohol)* related to health ailments is annotated based on the presence of the intensity of private content. Out of the selected 1250 tweets, 67.26% of tweet contents were found to have private or personal health-related information and the remaining 32.74% are common health-related tweets [33]. Figure 2 depicts the percentage of self-disclosure in health tweets.

3.2 *Textual Feature Extraction*

The data pre-processing and relevant feature extraction plays a very crucial role in generating accurate results. To analyze the dataset, each tweet is transformed into a feature vector based on weighted TF-IDF. Let $\{f_1, f_2, \ldots, f_p\}$ be a well-defined set of p keywords (features) that can appear in the health dataset hTweet$_n$ and n_i (hTweet$_n$) be the number of occurrences of f_i in dataset hTweet$_n$. Then, each dataset d is represented by a p-dimensional vector as shown in Eq. (1).

$$\text{hTweet}^{\rightarrow} = (n_1(\text{hTweet}), n_2(\text{hTweet}), \ldots, n_p(\text{hTweet})) \quad (1)$$

3.3 Correlation Analysis for Social Health Data—Principal Sensitive Health Entity Extraction

The identification of sensitive entities through the NER approach can be considered a sequence-labeling problem, where correlations are represented by the joint probability between adjacent labels and the entire sequence of labels. Therefore, the words were jointly decoded as labeled sequences using different deep learning models to learn the sensitive data presence instead of deciding by independent modeling [34–37]. The counterintuitive behavior of sensitive private data disclosures [38] can be predicted by applying elementary univariate statistical approaches like correlation coefficient derivatives.

Therefore, to leverage both the significance of correlation values and identified entities in health data, a novel approach is proposed by applying correlation to the identified entities in each tweet instance instead of unigram features. Furthermore, to handle this binary classification problem against cosine similarity, the authors choose rank-based correlation coefficients, such as Spearman's ρ in (2) as similarity measures between entities and embed them in textual features. This can be utilized to discover the influence of one entity on another in personal health disclosure.

$$\rho = 1 - \frac{6\sum_{i=1}^{n}(u_i - v_i)^2}{n(n^2 - 1)} \tag{2}$$

For n raw scores P_i, Q_i for $i = 1, \ldots, n$ the entity representation is based on TF-IDF values for two vectors P and Q where p_i and $\mathrm{q_i}$ are the respective ranks of P_i and Q_i for $i = 0, .., n\text{-}1$. The ranks for entities are allotted based on the correlation values; thus, the largest is assigned the rank value of n-1 and the least is 0. Here, the analysis was performed by considering the entities (ORG, GPE, PERSON, DATE, TIME, NORP, LOC, PRODUCT, EVENT, PERCENT) in the user-presented health data in tweets.

Entity Distribution Analysis. The distribution of entities is analyzed for various combinations like one-to-one, one-to-two, and two-to-two entity pairing scenarios, and feature vectors are generated accordingly. The pairing of entities is performed by considering each entity as an independent and dependent entity feature based on the entity distribution in each cyber-keyword. The principal entity derived is a normalized absolute value which is a product of correlation value and the ratio of cyber-keyword-based entity distribution and global entity distribution under different pairing scenarios. The results of entity-based correlation analysis and principal entities identified are presented in Sect. 4.

3.4 *Deep-Net Models for Privacy Detection through NER*

The primary aim of generating DPM is to handle the uncertainty of short texts and classify health-related private tweets with sensitive health entities. The authors chose CNN, RNN, CNN-LSTM, CRF, Bi-LSTM, and Bi-LSTM + CRF models to experiment with the health tweets and find the best operational deep net for private health data classification. The following subsections will showcase how the deep neural networks were modeled for health-related tweets in DPM.

Convolutional Neural Networks (CNN). The proposed DPM used the basic working architecture of CNN to classify the tweet with three layers namely the convolution filter (input) layer, max-pooling (dense) layer, and the SoftMax (flatten) layer. The convolution filters for each tweet were built in 1-dimension to act as unigram detectors to predict the closely related health keywords of n-grams. The max-pooling layer is added to CNN to extract the most relevant health unigrams for deciding on self-disclosure. The max-pooling layer induces a thresholding behavior that enables the neural nets to ignore values of less significance during the learning process of classification. The final layer has SoftMax to classify tweets based on the textual information and weights provided throughout the learning process. When a specific pattern is discovered in CNN, the output of each convolution layer is fired. It can recognize patterns of several sizes (a sliding window that focuses on 2, 3, or 5 consecutive words) by altering the size of the kernels and concatenating their outputs. Ultimately, CNN was chosen as it can recognize private health content in the sentence regardless of their position.

Recurrent Neural Networks (RNN). The DPM adopted a RNN for classifying health tweets to recursively learn and classify the health-related self-disclosures by applying its knowledge (input layer) over a transition function. The traditional strategy for modeling any text sequence in RNN includes mapping the input sequence to a fine-tuned-sized vector representing the context (hidden layer) and then propagating the vector over a softmax layer (output layer) for classification. Therefore, the recursive property of traditional RNN will be gradient dependent that might exponentially grow or decay over long feature sequences in DPM to efficiently learn the existing long-distance correlations in a sequence of features in a tweet. This network learns by applying weights ranging in small values from 0.1 and computes various gradient descent values to identify the optimal epoch and log-likelihood values for an accurate DPM. The training process includes the calculation of errors according to the cross-entropy criteria of the back-propagation algorithm to update weights.

CNN-LSTM. The architecture of the C-LSTM in DPM consists of two main components: CNN and LSTM which were trained to classify the health tweets, given a sequence of words already seen in the training set. This classification is performed by monitoring the output of the hidden layers at the last time step of LSTM as the feature vectors of tweet followed by adding a softmax layer.

Conditional Random Fields. The CRF model is built in terms of unigram features where the global features are represented sequentially. The global features are extracted by aggregating the local features for an input sequence and labels which are then fed to the conditional probabilistic model to derive the predictions with normalized refactored weights.

Bi-LSTM. A Bi-LSTM model [39] is employed to process every sentence token-by-token and produce an intermediate representation by exploring sequence in forward and reverse (bi-directional) directions as two separate layers. Accordingly, the two hidden states, h $^{\rightarrow}$ and h $^{\leftarrow}$, are concatenated to represent the final output. For an input tweet x, containing n-words like (x_1,x_2,...,x_n), an LSTM computes a left representation h $^{\rightarrow}$t of the given sentence at every word t. Similarly, a representation of the right context h$^{\leftarrow}$t can be achieved from the same sequence in reverse. As a result, Bi-LSTM yields the representation of a keyword by concatenating the outputs of its left and right contexts, $h_t = [h^{\rightarrow}{}_t, h^{\leftarrow}{}_t]$. The output of LSTM is processed by the classifier during every stage, one among 3 classes is predicted, corresponding to IOB tags common in NER: B(eginning), I(nside entity), and O(ut of entity).

Bi-LSTM + CFR. The Bi-LSTM-CRF is a recurrent neural network obtained from the mix of a long short-term memory (LSTM) and a conditional random field (CRF). The model with Bi-LSTM uses intermediate representation for the CRF to give the prediction of all the token labels. The input data to Bi-LSTM will be a sequence of word embedding generated from the correlation feature set, which returns a representation of the left and right context for each entity disclosed related to health data. These representations are concatenated and linearly projected onto a layer whose size is equal to the number of distinct tags. The CRF is responsible for tracking neighboring/co-occurring tags, yielding the final predictions for every cyber-keyword and target class.

Therefore, the generated features with identified entities as feature vectors are fed to each of the DPM discussed above and its operational performance is monitored. The best-suited model is determined by tuning the DPM with suitable hyper-parameters discussed in the following Sect. 4.

4 Results and Discussions

This research work has been carried out to identify health-related self-disclosures in social media texts. An extensive research-oriented exploration was performed to find the optimal approaches that suit the fuzziness and inconsistency in user health data on social media platforms. Figure 3a presents the entity distribution in each health domain cyber-keywords. This entity distribution analysis revealed the fact that the presence of entities differs with cyber-keyword and user preferences. According to the observations in the correlation between entities shown in Fig. 3b, there is no negative correlation in health-based NER for the current set of data. Table 1 contains the top sensitive health entities in one-to-two and two-to-two combinations based on correlation value. Entities like 'EVENT' and 'PERCENT' are ignored for combinational entity analysis as they contributed zero influential data. The combinations are considered based on the highest correlation values, and other entities are ignored which are less than the nominal correlation value.

Therefore, the authors conclude that the presence of any health-related entities will increase the risk of privacy leaks in OSN. The entity distribution analyzer eliminated the less significant entities (EVENT and PERCENT) during ranking and derives the entities with high coupling effects in disclosing private information. The most correlated entities observed from the research dataset are ORG-PRODUCT, ORG-GPE, ORG-PERSON, and ORG-TIME, ORG-TIME: PERSON-NORP.

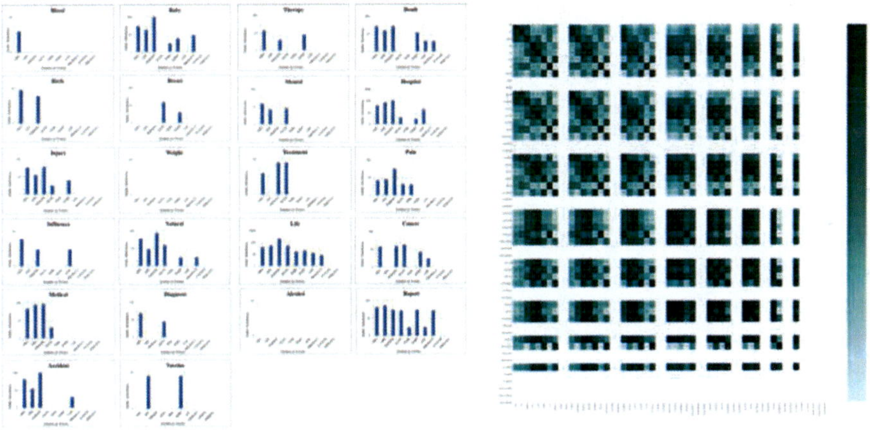

(a) (b)

Fig. 3 **a** Sensitive health-related entity distribution across various health-related cyber-keywords. **b** Correlation between sensitive health-related entities identified from the dataset

Table 1 Entity pairing scenarios and sensitive entities for health-related tweets

No	Dependent entity (Ignore 'Event', 'Percent')	Independent entity	Correlation value
One-to-One Entity (No Entity with significant + ve Correlation Value) (all values are < 0.5)			
One-to-two entity			
1	Product	ORG-Product	1
2	GPE	ORG-GPE	0.99
3	Person	ORG-Person	0.99
4	Time	ORG-Time	0.98
Two-to-two entity			
1	ORG-Time	Person-NORP	0.99
		Date-Time	
		Time-NORP	
2	GPE-Product	ORG-Product	0.99

Table 2 Hyperparameters for deep privacy models

Models	Hyperparameters				
	Dropout	Activation function	Loss	Optimizer	Epoch
CNN	–	Relu, softmax	Sparse_categorial_crossentropy	Adam	5, *10*, 15
RNN	0.5	Relu, sigmoid	Binary_crossentrpy	Adam	5, *10*, 15
CNN-LSTM	0.2	Relu, softmax	Sparse_categorial_crossentropy	Adam	5, *10*, 15
CRF	0.1	LBFGS	Elastic net regularization	–	*5*, 10, 15
Bi-LSTM	0.1	Softmax	Categorical_crossentropy	Adam	*5*, 10, 15
Bi-LSTM-CRF	0.1	ReLU	CRF loss_function	rmsprop	*5*, 10, 15

There is no negative correlation observed for health-based NER in the dataset. Therefore, any presence of health-related entities will increase the risk of privacy leaks in OSN. Table 1 presents the most correlated entities namely ORG-PRODUCT, ORG-GPE, ORG-PERSON, ORG-TIME, PERSON-NORP, TIME-NORP, and GPE-PRODUCT. This research derived a hypothesis stating as the number of entities disclosed by users increases, the sensitivity of a tweet increases and there is a high possibility to lose privacy in the health prospectus of a social media user. From the observed principal entities, the textual feature vectors are re-structured with entity information and fed to the DPM for the classification process. The DPM had a different set of hyper-parameters for each model as listed in Table 2.

Figure 4 illustrates the operational performance of various DPM with metrics like Accuracy, ROC, and Loss values. The optimal model that efficiently classified health-based sensitive data with private entities was Bi-LSTM-CRF with an accuracy of 95.5%, roc curve with 0.9412, and a loss value of 0.17. A deep analytical study inferred that the Bi-LSTM-CRF models provide an advantage from complimentary modeling features: as a complex, nonlinear model, the LSTM effectively captured the consecutive relationships among the input features. Whereas the CRF permits optimum, joint prediction of all the labels within the sentence, capturing the relationships at the label level.

Table 3 presents the comparative study on the performance of the proposed model and previous research works based on the state-of-the-art approaches. This will help researchers to further explore various possibilities and perspectives of health data disclosures in social media and prevent privacy leaks by creating some data detectors or filters with the generated models.

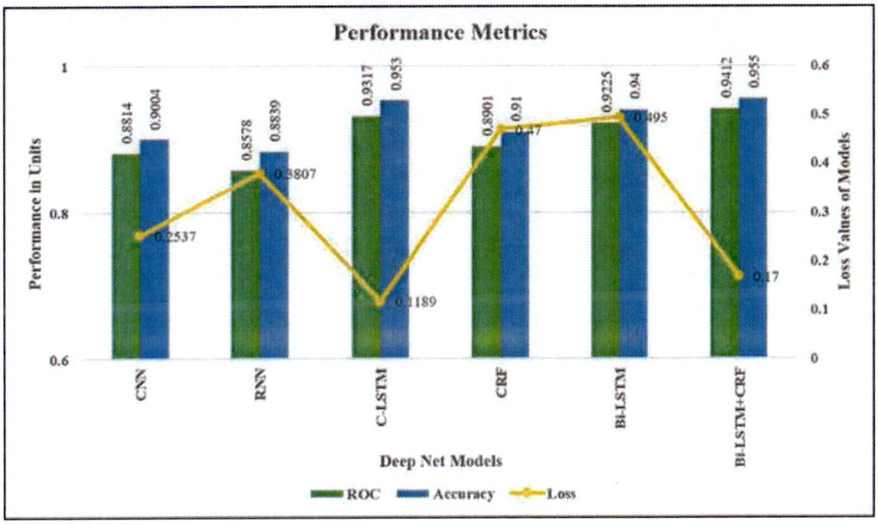

Fig. 4 Observed performance of deep-net classifiers in identifying health-related sensitive personal information

Table 3 Comparison of DPM classifier performance against SOTA

Models	Data source	Accuracy (%)
Bi-RNN + Topic [14]	Google snippets dataset	86.36
Word2vec + Softmax [15]	Twitter	58.84
Word embedding + LSTM [19]	Twitter	81.5
RNN-CNN [28]	Generic sentences	84
Bi-LSTM [30]	SIGARRA news corpus	78.27
Bi-LSTM-CRF [DPM]	Twitter	95.5

5 Conclusion and Future Works

In this study, a novel Private Content Detection (PCD) algorithm was proposed for identifying self-disclosures of health-related content in online social networks. A hypothesis derived from the research demonstrates that the number of entities disclosed by the user has a direct relationship with sensitivity and effects of self-disclosure in social media. The research also discovered principal sensitive entities like ORG, PRODUCT, GPE, PERSON TIME, and NORP that contributed to sensitive health data self-disclosure in OSN. The presence of health-related private data classification was ideal when the learning and prediction were through Bi-LSTM +

CRF which associates the Long Short-Term Memory network (LSTM) with Conditional Random Field (CRF). The future work of this research work would be extended to content analysis of Twitter messages on one health-related issue. Hence, a closer look into the content of self-disclosure would provide a better understanding of such benefit-exchanging activities. Other factors like affecting one's choice to uncover him/herself on the web can also be one of the reasons for disclosing false information. Researchers may consider exploring the impact of different components, particularly mental inclinations, in the process. Lastly, the relationship between online self-disclosure and offline self-disclosure warrants more investigation in future studies.

References

1. Lin W-Y, Zhang X, Song H, Omori K (2016) Health information seeking in the Web2.0 age: trust in social media, uncertainty reduction, and self-disclosure. Comput Hum Behav 56:289–294
2. Paul MJ, Dredze M (2011) You are what you tweet: Analyzing twitter for public health. In: Fifth international AAAI conference on weblogs and social media
3. Scanfeld D, Scanfeld V, Larson EL (2010) Dissemination of health information through social networks: Twitter and antibiotics. Am J Infect Control 38(3):182–188
4. Chew C, Eysenbach G (2010) Pandemics in the age of Twitter: content analysis of Tweets during the 2009 H1N1 outbreak. PLoS ONE 5(11):e14118
5. Joinson AN (2001) Self-disclosure in computer- mediated communication: the role of self-awareness and visual anonymity. Eur J Soc Psychol 31(2):177–192
6. Thompson N, Brindley J (2020) Who are you talking about? Contrasting determinants of online disclosure about self or others. Inform Technol People
7. Reingold H (1993) The virtual community Addison. Wesley, New York
8. Thomas J, Prabhu AV, Heron DE, Beriwal S (2018) Twitter and brachytherapy: an analysis of "tweets" over six years by patients and health care professionals. Brachytherapy 17(6):1004–1010
9. General Data Protection Regulation, http://gdpr-info.eu
10. Modjarrad K, Moorthy VS, Millett P, Gsell P-S, Roth C, Kieny M-P (2016) Developing global norms for sharing data and results during public health emergencies. PLoS Med 13(1):e1001935
11. Singh MM, Chin TY (2016) Hybrid multi-faceted computational trust model for online social network (OSN). Int J Adv Comput Sci Appl 7:1–11
12. Young T, Hazarika D, Poria S, Cambria E (2018) Recent trends in deep learning based natural language processing. IEEE Comput Intell Mag 13(3):55–75
13. Schmidhuber J (2015) Deep learning in neural networks: an overview. Neural Netw 61:85–117
14. Gu Y, Gu M, Long Y, Xu G, Yang Z, Zhou J, Qu W (2018) An enhanced short text categorization model with deep abundant representation. World Wide Web 21(6):1705–1719
15. Stojanovski D, Strezoski G, Madjarov G, Dimitrovski I, Chorbev I (2018) Deep neural network architecture for sentiment analysis and emotion identification of Twitter messages. Multimedia Tools Appl 77(24):32213–32242
16. Xing Y, Xiao C, Wu Y, Ding Z (2018) A convolutional neural network for aspect sentiment classification. arXiv preprint arXiv:1807.01704
17. Goldberg Y (2016) A primer on neural network models for natural language processing. J Artif Intell Res 57:345–420
18. Hu H, Phan N, Geller J, Vo H, Manasi B, Huang X, Lorio SD, Dinh T, Chun SA (2018) Deep Self-taught learning for detecting drug abuse risk behavior in tweets. In: International conference on computational social networks. Springer, Cham, pp 330–342

19. Jiang K, Feng S, Song Q, Calix RA, Gupta M, Bernard GR (2018) Identifying tweets of personal health experience through word embedding and LSTM neural network. BMC Bioinf 19(8):67–74
20. Bracamonte V, Tesfay WB, Kiyomoto S (2021) Towards exploring user perception of a privacy sensitive information detection tool. In: ICISSP, pp 628–634
21. Krämer NC, Schäwel J (2020) Mastering the challenge of balancing self-disclosure and privacy in social media. Curr Opin Psychol 31:67–71
22. Lee Y-C, Yamashita N, Huang Y, Fu W (2020) I hear you, i feel you: encouraging deep self-disclosure through a Chatbot. In: Proceedings of the 2020 CHI conference on human factors in computing systems, pp 1–12
23. Oghazi P, Schultheiss R, Chirumalla K, Kalmer NP, Rad FF (2020) User self-disclosure on social network sites: a cross-cultural study on Facebook's privacy concepts. J Bus Res 112:531–540
24. Zhang R, Fu JS (2020) Privacy management and self-disclosure on social network sites: the moderating effects of stress and gender. J Comput-Mediated Commun 25(3):236–251
25. Kolajo T, Kolajo JO (2018) Sentiment analysis on twitter health news. Fudma J Sci 2(2):14–20, ISSN: 2616–1370
26. Prieto VM, Matos S, Alvarez M, Cacheda F, Oliveira JL (2014) Twitter: a good place to detect health conditions. PloS one 9(1):e86191
27. Andalibi N, Ozturk P, Forte A (2017) Sensitive self-disclosures, responses, and social support on Instagram: the case of# depression. In: Proceedings of the 2017 ACM conference on computer supported cooperative work and social computing. ACM, pp 1485–1500
28. Nguyen T, Nguyen D, Rao P (2020) Adaptive name entity recognition under highly unbalanced data. arXiv preprint arXiv:2003.10296
29. Patil N, Patil AS, Pawar BV (2016) Survey of named entity recognition systems with respect to Indian and foreign languages. Int J Comput Appl 134(16)
30. Dias M, Boné J, Ferreira JC, Ribeiro R, Maia R (2020) Named entity recognition for sensitive data discovery in Portuguese. Appl Sci 10(7):2303
31. Docteur Tweety, https://docter-tweety.com/
32. Lu X, Qu Z, Li Q, Hui P (2015) Privacy information security classification for internet of things based on internet data. Int J Distrib Sens Netw 11(8):932–941
33. Health_Tweet_Repo_for_PrivateData, https://github.com/GeethR/Health_Tweet_Repo_for_PrivateData
34. Gridach M (2017) Character-level neural network for biomedical named entity recognition. J Biomed Inform 70:85–91
35. Ma X, Hovy E (2016) End-to-end sequence labeling via bi-directional lstm-cnns-crf. arXiv preprint arXiv:1603.01354
36. Wang X, Yang C, Guan R (2018) A comparative study for biomedical named entity recognition. Int J Mach Learn Cybern 9(3):373–382
37. Wang X, Zhang Y, Ren X, Zhang Y, Zitnik M, Shang J, Langlotz C, Han J (2018) Cross-type biomedical named entity recognition with deep multi-task learning. Bioinformatics 35(10):1745–1752
38. Botti-Cebriá V, del Val E, García-Fornes A (2019) Automatic detection of sensitive information in educative social networks. In: Computational intelligence in security for information systems conference. Springer, Cham, pp 184–194
39. Graves A, Schmidhuber J (2005) Framewise phoneme classification with bidirectional LSTM and other neural network architectures. Neural Netw 18(5–6):602–610

Trust-Based DSR Protocol for Secure Communication in Mobile Ad-hoc Networks

Pratik Ranjan and Rajeev Ranjan

Abstract Identification of a malicious node in the ad-hoc network is a very challenging issue due to some limitations, such as no base station or centralized server, dynamic topology, low computational capacity and limited battery power. In ad-hoc network each node having a limited range of communication to communicate with remaining nodes. Because of the limited range of communication, source node sends the packet/message with the help of intermediate neighbor nodes to the destination node. Dependency of other nodes poses security threats in ad-hoc networks. In this paper we are using Boneh-Franklin's IBC authentication scheme before deployments of nodes in MANETs and after setup of ad-hoc network, we are using DSR protocol with trust value of each node to find out a secure route between source and destination nodes. Trust-based DSR protocol will find a route that is free from black hole nodes. From security analysis, we have shown that the proposed scheme is providing the secure communication in MANETs.

Keywords Ad-hoc networks · Black hole attack · IBC · MANETs · Trust based DSR

1 Introduction

Security in Mobile ad-hoc networks (MANETs) is a very challenging issue because ad-hoc networks are infrastructure less multi-hop networks without any centralized server or base station [23]. Ad-hoc network is a group of mobile nodes which are self configurable and follow dynamic network topologies. In MANETs, node may be mobile phones, computer, laptop, personal digital assistant (PDA) and hand held digital devices etc. The structure of MANETs shown in Fig. 1. Nodes can change

P. Ranjan (✉)
Motihari College of Engineering, Motihari, India
e-mail: dr.pratik.ranjan@gmail.com

R. Ranjan
Bakhtiyarpur College of Engineering, Bakhtiyarpur, India
e-mail: rajeev.macet@gmail.com

S. Das et al. (eds.), *Advances in Data-driven Computing and Intelligent Systems*, Lecture Notes in Networks and Systems 653, https://doi.org/10.1007/978-981-99-0981-0_13

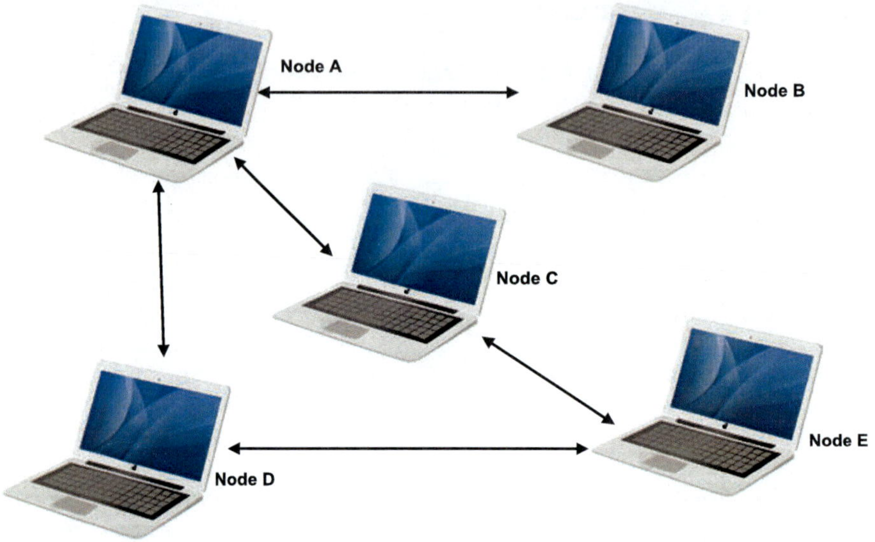

Fig. 1 Mobile Ad-hoc network structure

their position arbitrarily and they can communicate directly with neighbor nodes. Those nodes which are not in the range of direct communication will communicate through intermediate nodes [4].

The use of ad-hoc networks in various fields like military exercise, battlefields, earth quake, flood, rescue operation and mine site operation. There are some limitations of MANETs such as dynamic topology, limited battery power, lack of infrastructure, low computational complexity and vulnerability of nodes that creates more challenges to achieve security goals. There are two types of threat possible in ad-hoc networks such as Internal threat and External threat. Internal threats are more difficult to detect than external threats. Due to these threats, to achieve security goals are more difficult in MANETs. The security goals are Confidentiality, Integrity, Auhentication and Availability [13].

Confidentiality: It allows only authorized node to access the sensitive and protected data.

Integrity : There should not be any modification in message during transmission of messages.

Authentication: It is a process by which nodes verify the identity of others nodes. This is essential for effective security.

Availability: Authorized nodes are able to access the information when needed to nodes.

There are various attacks in MANETs such as Black hole, Denial of Services, Sink hole, Worm hole. Black hole attack is one of the well known attack in ad-hoc networks [6]. In this attack, malicious node advertise that it is the shortest path from source to destination immediately after receiving the route request message from the source node, without forwarding the route request packet to the neighbor nodes [7]. Hence, the source node sends the message through the malicious node to the destination. In this attack message dropped by malicious node and it does not reach the destination.

This paper is organized further as follows: in Sect. 2, related work on secure routing protocols in MANETs is presented. Black hole attack is discussed in Sect. 3. We have presented the proposed scheme in Sect. 4 and its security analysis is discussed in Sect. 5. Section 6 conclude this paper.

2 Related Works

There are so many security issues on routing protocols in MANETs. Secure route selection, malicious node detection and resists from various attacks in MANETs is a challenging task. Though in recent years, these challenges are addressed and the probable solutions have been proposed. But, still it is an open challenge to provide a secure communication in MANETs.

Authenticated Routing (ARAN) protocol [19] was proposed in 2002 which works in MANETs that uses trusted certificate server for secure routing. In 2005, Aridane [8] was proposed which is derived from Dynamic Source Routing (DSR) protocol that prevents denial of service attack. Modified AODV routing protocol [25] uses digital signature and hash chain for securing the protocol. Digital signature makes secure non-mutable fields while hash chain makes secure mutable fields of AODV protocol. Luo et al. [12] have proposed a black hole attack prevention mechanism in authenticated AODV routing protocol for secure communication, through combining of hash function, pseudo random function and message authentication code. Bhalaji et al. [2] have proposed an association-based DSR protocol that selects the path from the source to destination node on the basis of trust ratings of nodes. Raja et al. [18] have proposed a guard node-based technique to find out a black hole node in the ad-hoc networks. This scheme suffers from heavy energy loss because of continuous listening of nodes in the network. Kurosawa et al. [11] have proposed to detect a black hole attack by dynamic learning method. Deng et al. [5] have proposed routing security protocol to detect a black hole attack for on-demand distance vector routing protocol in MANETs. Ahmed et al. [1] have presented a survey on trust-based security in MANETs and identify some of the unresolved issues that need to be addressed in further research. Similar research published in recent years by authors in [10, 14, 17, 21, 22, 24]

Mohanapriya et al. [15] proposed a trust-based DSR protocol. This scheme find out a secure route free from black hole nodes with co-operation from the neighbors. We found that the nodes are transmitting the data packets to each other. This scheme have not given much attention on node authentication technique for formation of network. Authentication is a very important phase of a secure network. We have proposed a scheme for MANETs by introducing the authentication mechanism as per the method discussed in [16].

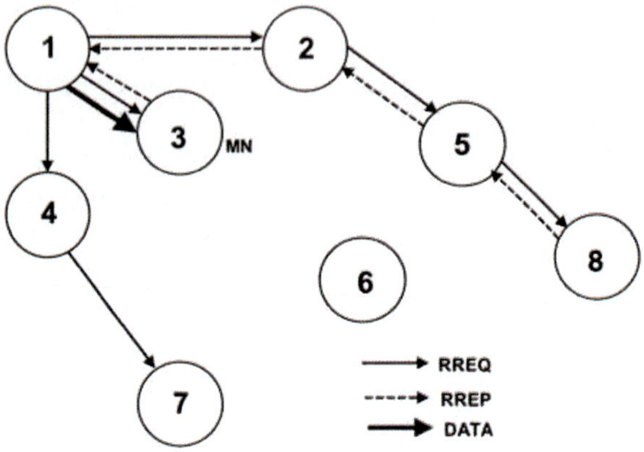

Fig. 2 Black hole attack

3 Black Hole Attack

This is an attack found on Network Layer. In this attack, malicious node sends the route reply message to the source node and claims that it has a shortest, fresh and valid route to the destination node. This results that a malicious node takes all the data traffic toward itself. But rather forwarding any data packets to the neighbor nodes, malicious node dropped these data packets. Hence, with this attack, attacker restrict the nodes to forward the data packets to the destination. Black hole attacks are of two types [20]:

(i) **Single Black Hole Attack** which involve a single malicious node claim about the shortest route to the destination and drop the data packets.
(ii) **Cooperative Black Hole Attack** in which more than one malicious node collaborate together and perform multiple black hole attacks in a large network.

In Fig. 2, source node 1 wants to communicate with node 8. Source node broadcast the route request (RREQ) message to its immediate neighbor nodes 2, 3 and 4. Node 3 is a malicious node (MN) which sends a route reply (RREP) message immediately to source node 1 and claims that it has a shortest, fresh and valid route to the destination node. MN does not even checks the route for destination node 8. Hence source node 1, recognize that the node MN has a shortest route to reach the destination node and transmit all the data packets to node MN. After receiving the data packets from source node, MN drops all the data packets which results that the data packets are not forwarded to further neighbor nodes. This attack is known as black hole attack and the malicious node (node 3) is called as black hole node. In this attack, an attacker can drop all the data packets and restrict any packet to reach to the destination.

Table 1 List of notations used

Symbol	Description
Z	Finite integer set
Z_n	Set of integer after modulo n
F_q	q elements with finite fields
Z_q^*	The multiplicative group of integers modulo prime number q
E/F_p	Elliptic curve over finite field (F_p)
d_{id}	Private key corresponding to node identity
Q_{id}	Public key corresponding to node identity
S	Secret key (Master)
P_{pub}	Public key (System)
H_i	An i, integer usage as subscript, in the system multiple hash function are used
G_1	The additive group, sub group of points on E/F_{p}
G_2	The multiplicative group, subgroup of the finite field
P, Q	Two points on elliptic curve

4 Proposed Scheme

There are two phases in our proposed scheme. First is the Authentication Phase and second is the Secure Route Selection Phase. In Authentication Phase, nodes are authenticated before deployment to establish an ad-hoc network. The Private Key Generator (PKG) distributes the system parameters to the authenticated nodes only. We have used the system parameters same as in Boneh-Fraklin's IBC crypto-system [3]. The next phase of the proposed scheme is Secure Route Selection Phase in which authenticated nodes are selecting the secure route for transmission of data packets to the destination node. Notations used in the scheme are presented in Table 1.

4.1 Authentication Phase

The PKG selects a random number $s_0 \in Z_q^*$ and $P_0, Q \in E/F_{\text{P}}$ are arbitrary points order q. The system parameters are para $=< p, q, P_0, Q, P_{\text{pub0}}, H, H_0, H_1 >$, ($P_{\text{pub0}} = s_0 P_0$) distributed through public channels. The private keys and real system parameters are applied on nodes. A node A selects a random number $r_0 \in Z_q^*$, encrypts its identity ID with $K = e(P_{\text{pub0}}, r_0 Q)$ and sends the encrypted identity C_{ID} with $r_0 P_0$ to the PKG: $< C_{\text{ID}}, r_0 P_0 >$

$$K = e(P_{\text{pub0}}, r_0 Q) = e(s_0 P_0, r_0 Q) = e(P_0, Q)^{r_0 s_0} \tag{1}$$

$$C_{\text{ID}} = \text{AES}(K, \text{ID}_{\text{A}}) \tag{2}$$

The PKG decrypts ID with $K_0 = e(s_0.Q, (r_0 P_0)) = K$, authenticate identity of node A. The parameter P, P_{pub} and $A's$ private key with K_0 encrypts updated system if that gets through it and through public channels, it is sent back to the respective node. Then updated system parameters and its private key are decrypted by node A.

$$K_0 = e(r_0 P_0, s_0 Q) = e(P_0, Q)^{r_0 s_0} \tag{3}$$

If $K = K_0$, verifies identity of node A after that PKG updated system parameter P, P_{pub} in place of P_0 and P_{pub0}. It is sent back to the respective nodes through public channels. The updated new system parameter is $< p, q, P, Q, P_{\text{pub}}, H, H_0, H_1 >$. Now nodes are authenticated for secure communication.

4.2 *Secure Route Selection Phase*

After node authentication phase, the next phase has to select secure route for data forwarding. Dynamic Source Routing (DSR) [9] is a reactive routing protocol. In our proposed scheme, route discovery and route maintenance works same as discussed in DSR protocol [9]. We have added one extra attribute, i.e., trust value. Trust value is initialized to 1 for every node in the network at the time of deployment of node. Then source node starts route discovery process to find a route from source to destination node. When the source node sends a route request (RREQ) packet to immediate neighbor nodes, then the trust value of the source node will increase by one. For every n number of RREQ messages, the trust value of the source node will become $1 + n$. If the trust value of node becomes 2 or more then it indicates that the nodes are forwarding the packets to the neighbor nodes. In black hole attack, the malicious nodes are not forwarding the data packets to the neighbor nodes and hence their trust values remain unchanged. After getting the route reply (RREP) message from nodes, source node selects the shortest path to reach to the destination among various routes found through RREP messages. Source node will consider the shortest path only from those routes where trust value of nodes are more than 1. Nodes having trust value 1 have chance to be malicious node (MN) because it does not forward any data packet. In this way, our proposed scheme will find the malicious node present in the network then provide the secure path for forwarding the data. Hence, our scheme is able to resists from black hole attack.

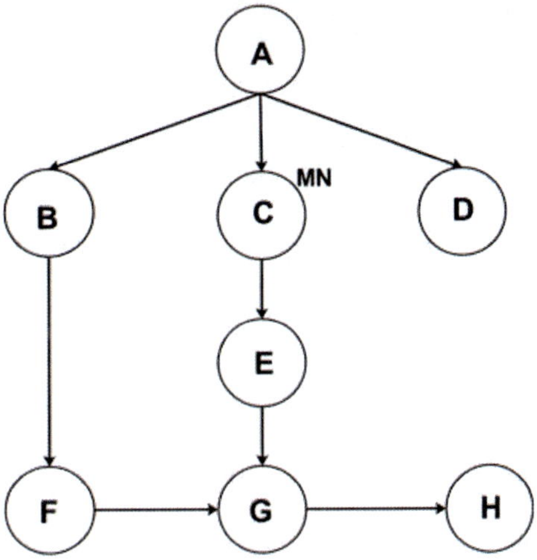

Fig. 3 Network for secure route selection phase

Table 2 Trust value of nodes after step 1

Node	A	B	C	D	E	F	G	H
Trust value	4	1	1	1	1	1	1	1

$$\text{Trust Value (Node } N1 \longrightarrow N2) = \text{High if Trust Value} >= 2$$
$$\text{Trust Value (Node } N1 \longrightarrow N2) = \text{Low if Trust Value} = 1$$

5 Security Analysis

We present the security analysis of our proposed scheme in this section. We have taken a network as shown in Fig. 3. Node A, is the source node and node H is the destination node. All other nodes (node B, C, D, E, F, G) are intermediate nodes that helps to forward the data packets. The initial trust value of each node is 1.

Step 1. To send the data packets to destination node (H), Node A first sends the RREQ message to immediate neighbor nodes (B, C, D) as shown in Fig. 4. The trust value of nodes after this step are presented in Table 2.

Step 2. After receiving the RREQ message, node B will forward this message to node F, which will be further forwarded to node G and H. Destination node H, will send the RREP message to source node A, that will reach through intermediate nodes G, F and B. Hence the new trust value of nodes B, F and G are 2.

Node C which is a malicious node (MN) does not forward the RREQ message message to any other node and simply send the RREP message immediately to source

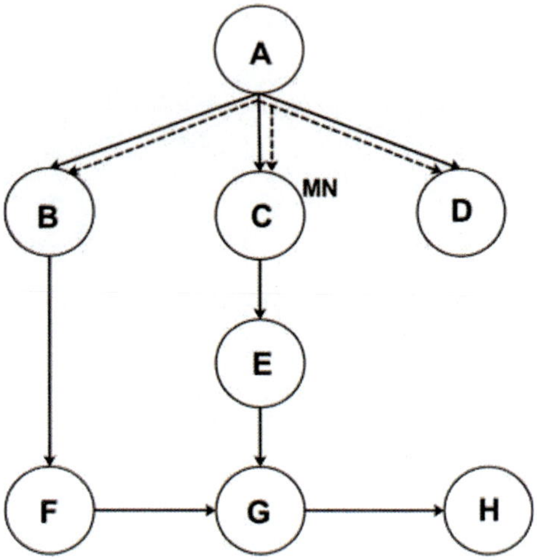

Fig. 4 Secure route selection phase—step 1

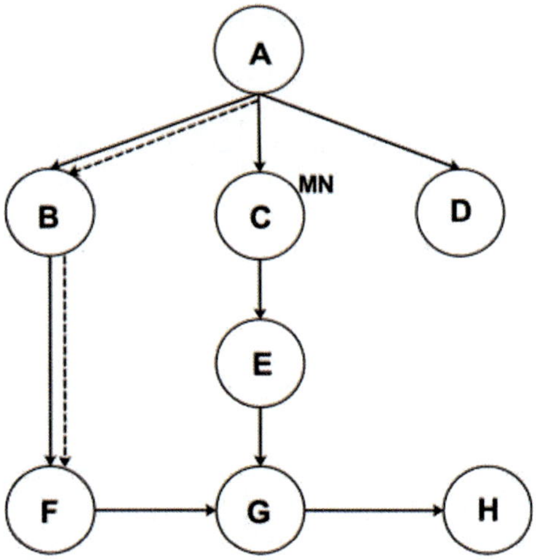

Fig. 5 Secure route selection phase—RREQ from B to F

node A claiming that it has a shortest route to the destination node. So, the update trust value of node C remain unchanged.

Node D cannot able to forward the RREQ message to any other node because of unavailability of neighbor nodes. Node D sends the RREP message to source node with the information that destination node is unreachable from this node. hence the trust value of this node is 1.

The updated trust value of each node is presented in Tables 3, 4 and 5.

Finally as discussed in Sect. 4.2, based on reachability to the destination node and the trust value of nodes, source node A have decided to send the data packets through node B, F and G to destination as shown in Figs. 5, 6 and 7.

Table 3 Trust value of nodes after RREQ from B to F

Node	A	B	C	D	E	F	G	H
Trust Value	4	2	1	1	1	1	1	1

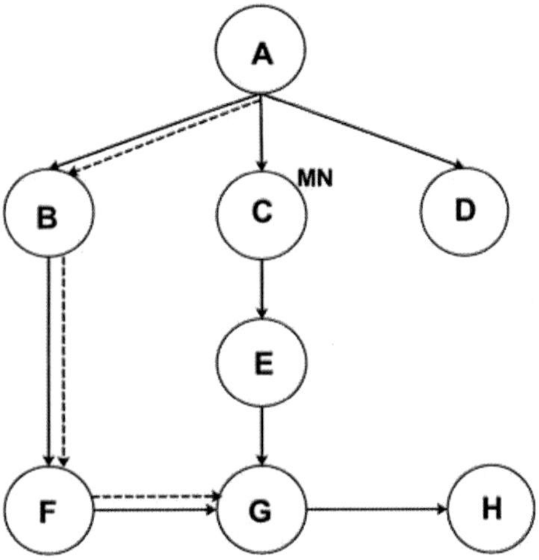

Fig. 6 Secure route selection phase—RREQ from F to G

Table 4 Trust value of nodes after RREQ from F to G

Node	A	B	C	D	E	F	G	H
Trust value	4	2	1	1	1	2	1	1

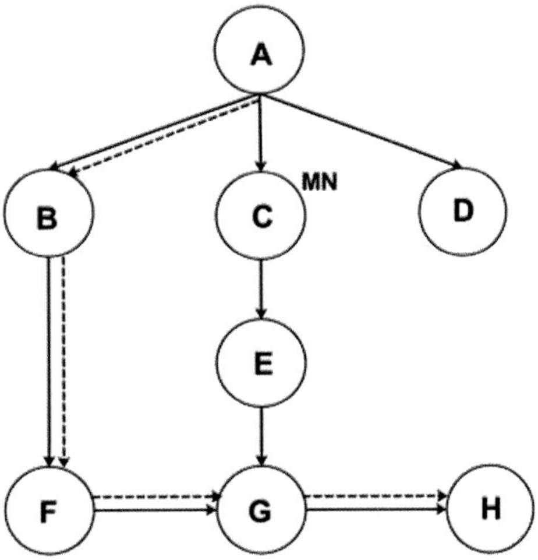

Fig. 7 Secure route selection phase—RREQ from G to H

Table 5 Trust value of nodes after RREQ from G to H

Node	A	B	C	D	E	F	G	H
Trust value	4	2	1	1	1	2	2	1

From the above analysis, it is clear that based on trust value of the nodes, the source node selects a secure route among the authenticated nodes of the network for data forwarding. Hence our scheme resists against black hole attack.

6 Conclusion

In this paper, we have used Boneh-Franklin's IBC authentication scheme before deployments of nodes in MANETs and after setup of ad-hoc network, we have used DSR protocol with trust value of every node that finds a secure route between source and destination nodes. Trust-based DSR protocol will find a route that is free from black hole nodes. Black hole attack found on network layer in which malicious node sends the route reply message to the source node and claims that it has a shortest, fresh and valid route to the destination node and hence the malicious node takes all the data traffic toward itself. But rather forwarding any data packets to the neighbor nodes, malicious node dropped these data packets. From security analysis, it is clear that based on trust value of the nodes, the source node selects a secure route among the authenticated nodes of the network for data forwarding. Hence our scheme resists against black hole attack. we have shown that the proposed scheme is providing the secure communication in MANETs.

References

1. Ahmed A, Bakar KA, Channa MI, Haseeb K, Khan AW (2015) A survey on trust based detection and isolation of malicious nodes in ad-hoc and sensor networks. Front Comput Sci 9(2):280–296
2. Bhalaji N, Kanakeri AV, Chaitanya KP, Shanmugam A (2010) Trust based strategy to resist collaborative blackhole attack in manet. In: Information processing and management. Springer, pp 468–474
3. Boneh D, Franklin M (2001) Identity-based encryption from the weil pairing. In: Annual international cryptology conference. Springer, pp 213–229
4. Bruzgiene R, Narbutaite L, Adomkus T (2017) Manet network in internet of things system. Ad Hoc Netw 66:89–114
5. Deng H, Li W, Agrawal DP (2002) Routing security in wireless ad hoc networks. IEEE Commun Mag 40(10):70–75
6. Dhanaraj RK, Krishnasamy L, Geman O, Izdrui DR (2021) Black hole and sink hole attack detection in wireless body area networks. Comput Mater Continua 68(2):1949–1965

7. Farahani G (2021) Black hole attack detection using k-nearest neighbor algorithm and reputation calculation in mobile ad hoc networks. Secur Commun Netw 2021
8. Hu YC, Perrig A, Johnson DB (2005) Ariadne: a secure on-demand routing protocol for ad hoc networks. Wirel Netw 11(1–2):21–38
9. Johnson DB, Maltz DA, Broch J (2006) DSR: the dynamic source routing protocol for multi-hop wireless ad hoc networks. Ad Hoc Netw 5(1):139–172
10. Juneja K (2022) Design of a novel micro-zone adaptive energy-trust evaluation active on-demand vector protocol to optimize communication in challenging and attacked mobile network. Int J Commun Syst e5328
11. Kurosawa S, Nakayama H, Kato N, Jamalipour A, Nemoto Y (2007) Detecting blackhole attack on AODV-based mobile ad hoc networks by dynamic learning method. IJ Netw Secur 5(3):338–346
12. Luo J, Fan M, Ye D (2008) Black hole attack prevention based on authentication mechanism. In: 11th IEEE Singapore international conference on communication systems, 2008. ICCS 2008. IEEE, pp 173–177
13. Mallarmé (2018)S The impressionists and edouard manet. In: Modern art and modernism: a critical anthology. Routledge, pp 39–44
14. Mandhare A, Kadam S (2019) Performance analysis of trust-based routing protocol for manet. In: Computing, communication and signal processing. Springer, pp 389–397
15. Mohanapriya M, Krishnamurthi I (2014) Trust based DSR routing protocol for mitigating cooperative black hole attacks in ad hoc networks. Arab J Sci Eng 39(3)
16. Ranjan R (2016) Cryptanalysis of secure routing among authenticated nodes in manets. In: Proceedings of the 5th international conference on frontiers in intelligent computing: theory and applications (FICTA) 2016. Springer, pp 1–6
17. Rao GBN, Veeraiah D, Rao DS (2020) Power and trust based routing for manet using rrrp algorithm. In: 2020 2nd international conference on innovative mechanisms for industry applications (ICIMIA). IEEE, pp 160–164
18. Raza I, Hussain SA (2008) Identification of malicious nodes in an AODV pure ad hoc network through guard nodes. Comput Commun 31(9):1796–1802
19. Sanzgiri K, Dahill B, Levine BN, Shields C, Belding-Royer EM (2002) A secure routing protocol for ad hoc networks. In: Proceedings of 10th IEEE international conference on network protocols 2002. IEEE, pp 78–87
20. Sathish M, Arumugam K, Pari SN, Harikrishnan V (2016) Detection of single and collaborative black hole attack in manet. In: International conference on wireless communications, signal processing and networking (WiSPNET). IEEE, pp 2040–2044
21. Shafi S, Ratnam DV (2022) A trust based energy and mobility aware routing protocol to improve infotainment services in VANETs. Peer-to-Peer Netw Appl 15(1):576–591
22. Sharma N, Sharma M, Sharma DP (2020) A trust based scheme for spotting malicious node of wormhole in dynamic source routing protocol. In: 2020 4th international conference on I-SMAC (IoT in social, mobile, analytics and cloud)(I-SMAC). IEEE, pp 1232–1237
23. Srinivasan V (2021) Detection of black hole attack using honeypot agent-based scheme with deep learning technique on manet. Ingénierie des Systèmes d'Information 26(6)
24. Syed SA, Shahzad A (2022) Enhanced dynamic source routing for verifying trust in mobile ad hoc network for secure routing. Int J Electr Comput Eng 12(1):425
25. Zapata MG, Asokan N (2002) Securing ad hoc routing protocols. In: Proceedings of the 1st ACM workshop on wireless security. ACM, pp 1–10

The Modified Binary Sparrow Search Algorithm (mbSSA) and Its Implementation

Gagandeep Kaur Sidhu and Jatinder Kaur

Abstract In our large world, there are various optimization problems that cannot be solved exactly through heuristic and meta-heuristic algorithms due to their large randomness in their initial population. In this paper, we opted the SSA which is known for its best optimal capability as well as speedy convergence including many more advantages. But still, SSA is facing some disadvantages such as massive randomness in the initial population and fall into the local optima. To overcome this deprivation, we have modified SSA by converting the random initial population into the binary initial population. And this modified SSA is named modified binary SSA (mbSSA). In this paper, the newly invented modified binary SSA (mbSSA) is implemented on 10 benchmark test functions as well as the outcomes of mbSSA are contrasted with the original SSA, PSO, and GWO. In this work, we analyzed mbSSA from various aspects like optimal value, mean for convergence accuracy, the standard deviation for stability, and convergence curves to check the convergence rate, and also we applied the Wilcoxon signed-rank test on mbSSA. In all aspects, the output of the experiment demonstrates that the modified binary SSA (mbSSA) is best in relation to SSA, GWO, and PSO.

Keywords Sparrow search algorithm (SSA) · Modified binary SSA (mbSSA) · GWO · PSO · Optimization problems · Meta-heuristic algorithms

1 Introduction

Optimization problems occur in our day-to-day life. So, almost in every field, we can see optimization problems, e.g., engineering design problems, feature selection in data classification, software defects prediction, traveling salesman problems, etc. So, with respect to different optimization problems, various algorithms exist in the literature for their solution such as exact, approximate, heuristic, and meta-heuristic

G. Kaur Sidhu (✉) · J. Kaur
Department of Mathematics, Chandigarh University, Gharuan, Punjab 140413, India
e-mail: gagan.kw777@gmail.com

S. Das et al. (eds.), *Advances in Data-driven Computing and Intelligent Systems*, Lecture Notes in Networks and Systems 653, https://doi.org/10.1007/978-981-99-0981-0_14

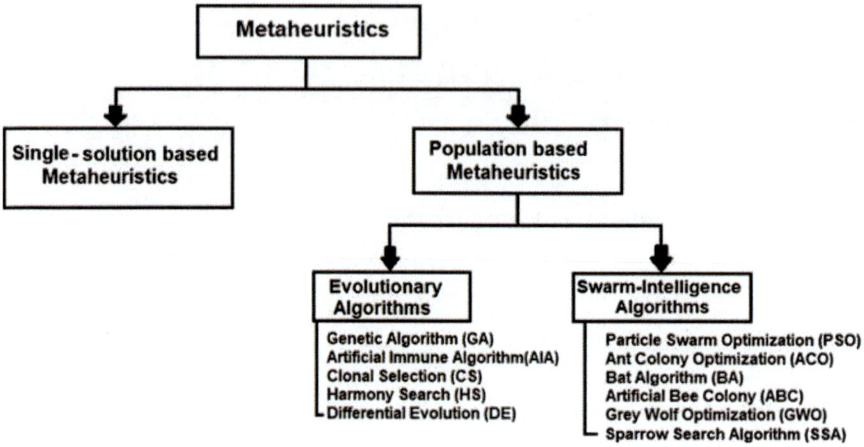

Fig. 1 Classification of meta-heuristic algorithms

algorithms [1]. Exact algorithms at all times provide the best solution for the given problem of optimization, while approximate algorithms can determine a close optimum solution. Heuristic algorithms are problem-dependent and specific. On the other hand, a meta-heuristic is an algorithmic structure that is high-level problem independent that gives a group of strategies or guidelines.

Meta-heuristic algorithms can be divided as follows:

Although there are a lot of algorithms in the literature as mentioned in Fig. 1, still researchers keep on researching to invent new algorithms and modify or hybrid the existing algorithms in order to search for the optimal solution with regards to stability, convergence accuracy, and optimality. If we take our sight on the new SI algorithms, then we found sparrow search algorithm (SSA) [2] which is recently proposed by Xue and Shen [2]. SSA plays a very good role in both the exploitation and exploration of the search space with regard to optimization. Still, some limitations are found in it. One of them is massive randomness in the initial population of SSA which leads to slow convergence and low optimality of SSA. So, to overcome this disadvantage, we proposed a modified binary sparrow search algorithm (mbSSA). In this mbSSA, we generate the initial population into binary form. Then, the rest of the procedure kept as same as it is. The remaining paper is systematized as follows:

Section 2 depicts a literature review of meta-heuristic algorithms. Section 3 shows the interpretation of SSA, whereas Sect. 4 presents modified binary SSA. Settling of parameters and experimental environment, experimental outcomes are represented in Sects. 5 and 6, respectively. Ultimately, Sect. 7 concludes the whole work, and the results are done in this paper.

2 Literature Review

There are various examples of heuristic and meta-heuristic swarm intelligence (SI) optimization techniques that modified or hybridized by researchers to obtain the desired solution. So, for this, we can take our sight on the following literature.

Wang et al. [3] modified traditional PSO into BPSO to solve the problems related to discrete binary optimization because traditional PSO was unable to solve optimization problems related to the binary combination as it works only in continuous space. Mohana Roopa [4] has done the investigation to recognize the component cluster in parallel by employing a multiagent adaptive algorithm known as the SPARROW algorithm. The outcomes of the experiment were crucial as these will be utilized to create proficient component-based software architecture. Doreswamy et al. [5] consider the problem in their research that the number of breast cancer patients is increasing day by day over the entire globe. Its treatment is greatly impacted if it is diagnosed earlier. So, the hybrid model feed forward neural network (FNN) based on binary bat algorithm was suggested due to the efficiency of FNN and the pros of BBA. Classification using FNNBBA provided 92.61 and 89.95% accuracy for training and testing data, respectively. Uros Mlakar [6] presented the hybrid cuckoo search algorithm which was used to solve the optimization problems of engineering design including constraints by stabilizing the exploration strategies in CS. Çelik et al. [7] proposed an improved firefly algorithm to solve the tension/compression spring design problem by adapting the neighborhood method in FA because the modification was needed in FA. Stephen et al. [8] used a new non-traditional optimization algorithm to calculate the global optimum solution for minimizing the weight of the speed reducer design problem because conventional optimization techniques were occasionally unable to provide global optima. Lei [9] improved SSA to deal with the situation of low positioning accuracy depending upon the DV-Hop (positioning technology) in the wireless sensor network (WSN). Peng et al. [10] improved SSA to increase its global search capability and implemented it to the bridge monitoring's sensor network converge optimal problem and got the good results. Kishor Yadav et al. [11] presented a novel chaotic Henry gas solubility optimization technique that is dependent on Henry's law and implemented on 47 benchmark functions. Ouyang et al. [12] suggested learning SSA for solving the problem of robot path planning and the CEC 2017 test function. Yang et al. [13] applied T-distribution mutation along with chaotic mapping in adaptive SSA to examine the optimization problems in order to overcome the problem of dropping in the local minima in the original SSA. Wu et al. [14] used a novel sparrow search algorithm for the problem of a traveling salesman (TSP). SSA does not have only the problem of falling into local optima but also has inadequate stagnation while applying to TSP. To overcome this problem, a cosine and sign search strategy based on a new greedy genetic sparrow search algorithm was used. Quan et al. [15] did the hybridization of SSA with the model of DELM known as ESSA-DELM for predicting the end-point phosphorus content of BOF. Cauchy mutation and trigonometric substitution mechanism were added in order to neglect to fall in the local optima as well as enhance the capacity of global exploration in SSA.

Peraza-Vazquez et al. [16] represented how the optimization problem of engineering design was solved by the bio-inspired method which was motivated by the strategies of dingoes hunting. Tang et al. [17] proposed a chaotic SSA for controlling the problems like dropping in the local minima as well as a steady convergence rate. So, strategies of adaptive step and logarithmic spiral were introduced in chaotic SSA to solve the engineering problems. Yang et al. [18] presented the hybrid PSO and SSA to predict software defects. PSO and SSA were hybridized because the convergence of PSO is very fast, whereas its solution accuracy is low, and SSA contains strong robustness, good stability, speedy convergence, and high search accuracy. Verma et al. [19] introduced an advanced hybrid meta-heuristic algorithm (haDEPSO) to figure out the solution to the optimization problems related to large- and small-scale engineering design. Kumar et al. [20] told how on the wind farm, the wind turbine's best location was found by SSA. Li [21] explained how robot path planning problems become the hotspot in the research area. For the path planning in raster, SSA was improved in order to cure the disadvantage of the classical raster method. Wang et al. [22] upgraded SSA into the IHSSA algorithm by using the strategy of hybrid reverse learning and also iterative chaotic mapping which is infinitely folded to resolve the optimization problems of engineering containing constraints. Wang [23] has done the hybridization of ESSA and PSO to deal with unmanned aerial vehicle path planning problems. Xie et al. [24] presented an improved SSA in concern with local and global optimization problems by applying a random walk strategy to it.

From the above literature, we can say that still, researchers are revealing limitations in the sparrow search algorithm (SSA). This motivated us to modify SSA. So, to overcome the disadvantages found in SSA, we proposed a modified binary sparrow search algorithm (mbSSA). In this mbSSA, we generate the initial population into binary form. Then, the rest of the procedure kept as same as it is, then investigated it on 10 benchmark test functions.

3 Interpretation of SSA

SSA algorithm basically works on group intelligence, searching for food, and the anti-predation nature of the sparrows. There are many species of the bird sparrow but in the experiment virtual sparrows were considered and sparrows were classified into two roles namely locater and asker. Of the entire population, 20% of the sparrows that have a high level of strength worked as locators and the remaining 80% as an asker. The asker walks on the path indicated by the locater to obtain food. During this food searching process, sometimes locators play the role of asker and vice versa, but the ratio of asker and locators remains constant. Through this process, when some sparrows feel the presence of danger around them, they lead the whole population to a safe area.

The position of locators is updated by using Eq. (1).

$$X_{i,j}^{t+1} = \begin{cases} X_{i,j}^{t} \cdot \exp\left(\frac{-i}{\alpha \cdot \text{iter}_{\max}}\right) & \text{if} R_2 < ST \\ X_{i,j}^{t} + Q \cdot L & \text{if } R_2 \geq ST \end{cases} \tag{1}$$

In Eq. (1), j varies from 1 to d and the current iteration is represented by t. $\alpha \in (0, 1]$ considers as a random number. $X_{i,j}$ depicts the place of the present ith sparrow in the jth dimension. A constant number having the max iterations is denoted by "$\text{iter}_{\max}$". "ST (ST $\in$ [0.5, 1.0])" performs the role of safety threshold whereas "$R_2(R_2 \in [0, 1])$" as alarm value. L stands for $1 \times d$ matrix in which each element is 1. Q appears as a random number and that follows the normal distribution.

When "$R_2 < ST$", that claim there is no threat around the sparrows, and the locators can search on the large-scale mode, whereas "$R_2 \geq ST$" shows the threat around the sparrows, and then, all sparrows immediately move toward the safe area.

The position of the asker is updated by using Eq. (2).

$$X_{i,j}^{t+1} = \begin{cases} Q \cdot \exp\left(\frac{X_{\text{worst}}^{t} - X_{i,j}^{t}}{i^2}\right) & \text{if } i > n/2 \\ X_p^{t+1} + \left|X_{i,j}^{t} - X_p^{t+1}\right| \cdot A^{+}.L & \text{if } i \leq n/2 \end{cases} \tag{2}$$

In Eq. 2, X_{worst} is the present global worst position while X_p acts as the optimal location taken by the locator. The total number of sparrows is equal to n. A is a matrix of order $1 \times d$ in which each element is either 1 or -1, and also "$A^{+} = A^T (A \cdot A^T)^{-1}$". When "$i > n/2$", it tells that the ith asker has a bad fitness value does not get any food and can die because of hunger.

Sparrows, who are aware of the danger lead the whole population in the safe area by using Eq. (3).

$$X_{i,j}^{t+1} = \begin{cases} X_{\text{best}}^{t} + \beta \cdot \left|X_{i,j}^{t} - X_{\text{best}}^{t}\right| & \text{if } f_i > f_g \\ X_{i,j}^{t} + K \cdot \left(\frac{\left|X_{i,j}^{t} - X_{\text{worst}}^{t}\right|}{(f_i - f_w) + \varepsilon}\right) & \text{if } f_i = f_g \end{cases} \tag{3}$$

In Eq. (3), X_{best} gives the present global optimal location. "β", being the step high control parameter follows the standard normal distribution of random numbers "N(0,1)" even if "$K \in [-1, 1]$" is a random number. The fitness value of the current sparrow is described by "f_i" moreover "f_w" and "f_g" personify the present worst and global best fitness values, subsequently. ε is the littlest constant to ignore the error of zero division. When "$f_i > f_g$" represents the sparrows on the border of the group, however, "$f_i = f_g$" personifies that the sparrows in the center of the population are completely alert from danger and should have to shift their place near toward others.

The working of the algorithm is shown by the following flow chart of SSA:

4 Modified Binary Sparrow Search Algorithm (mbSSA)

Modified binary SSA (mbSSA) is conceptually identical to the original SSA, only the difference between them can be found in the initialization stage. The real SSA implements in the continuous search space, although, mbSSA implements in the binary space. Initialization of the population of mbSSA can be done by using Eq. (4).

$$X = \begin{bmatrix} x_{1,1} & \cdots & x_{1,d} \\ \vdots & \ddots & \vdots \\ x_{n,1} & \cdots & x_{n,d} \end{bmatrix} \tag{4}$$

In Eq. (4), each $x_{i,j}$ is defined by the given Eq. (5)

$$x_{i,j} = \begin{cases} 0 & \text{if rand } () \leq 0.5 \\ 1 & \text{otherwise} \end{cases} \tag{5}$$

After the initialization, we have to follow all the steps from the step no. 2 that are given in Fig. 2 flow chart of SSA.

5 Settling of Parameters and Experimental Environment

All the four algorithms run on Window10, the "64-bit operating system" has "16.0 GB RAM" and the processor is "Intel(R) Core(TM) i5-7300U CPU @ 2.60GHz 2.71 GHz". Coding of all algorithms implemented by MATLAB R2014a. Ten standard benchmark functions are taken in this paper for analyzing the optimality, liability, and convergence of the modified binary sparrow search algorithm (mbSSA). The execution of the mbSSA is contrasted with different meta-heuristic techniques like SSA, PSO, as well as GWO. The number of search agents for every technique is the same that is 100 and the max iterations are 1000 for all techniques. Parameters of all algorithms are referred to Table 1.

To check the algorithm appropriately, in every case then we run each algorithm 30 times separately and calculate the best solution, mean, and also standard deviation of all the techniques. Ten benchmark test functions are given in Table 2, out of which seven are unimodal test functions, and three multimodal test functions are taken to test the modified binary SSA (mbSSA).

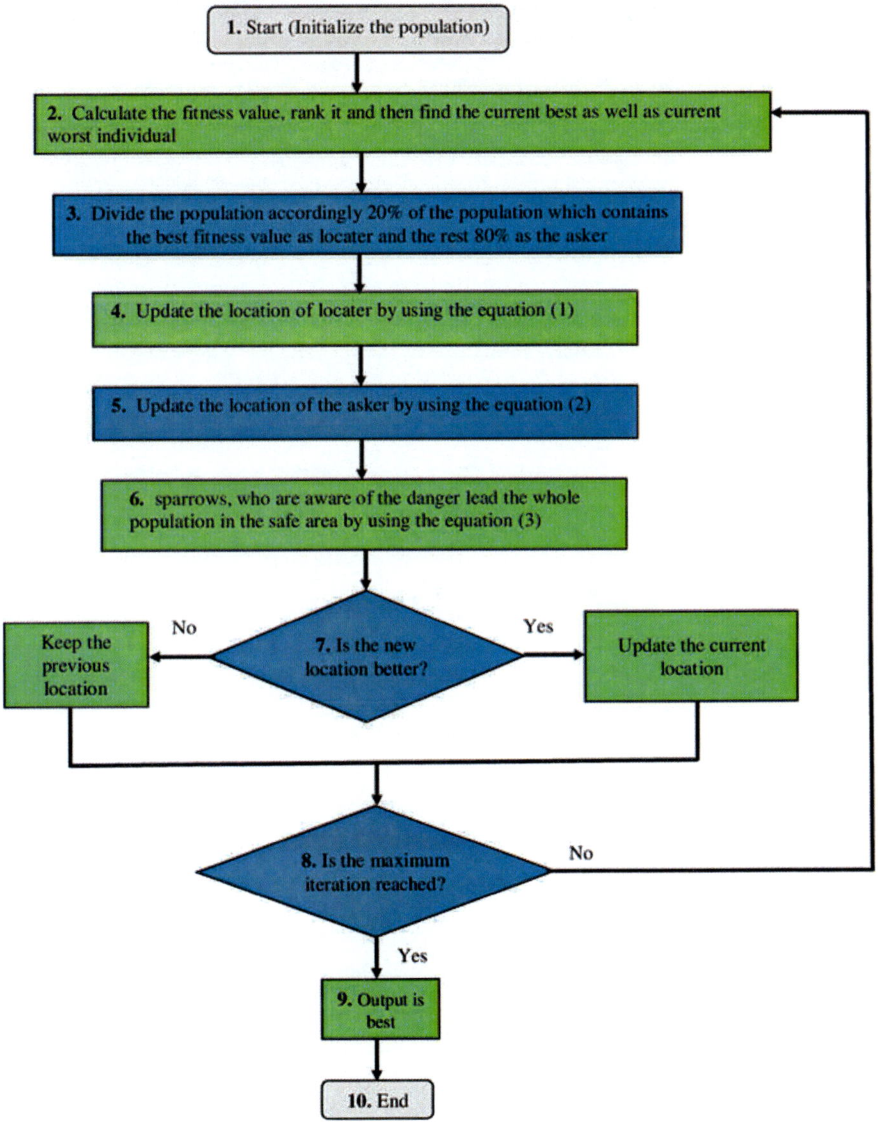

Fig. 2 Flow chart of SSA

Table 1 Parameters of all four algorithms

Algorithm	Parameter	Value
PSO	Inertial weight: ω	$\omega =$ $0.5 - (0.2 \times (1 \div \text{iter}_{\max}))$
GWO	Random numbers: r_1, r_2	[0, 1], [0, 1]
	$\vec{a}$	Decrease linearly from 2 to 0
mbSSA	Safety threshold: ST	0.8
SSA	Safety threshold: ST	0.8

Table 2 Standard benchmark test functions

Function	DIM	Range	Min
$F_1(z) = \sum_{i=1}^{n} z_i^2$	30	[−100, 100]	0
$F_2(z) = \sum_{i=1}^{n} \lvert z_i \rvert + \prod_{i=1}^{n} \lvert z_i \rvert$	30	[−10,10]	0
$F_3(z) = \sum_{i=1}^{n} \left(\sum_{j=1}^{i} z_j\right)^2$	30	[−100, 100]	0
$F_4(z) = \max\{\lvert z_i \rvert, 1 \le i \le n\}$	30	[−100, 100]	0
$F_5(z) = \sum_{i=1}^{n-1} \left[100\left(z_{i+1} - z_i^2\right)^2 + (z_i - 1)^2\right]$	30	[−30, 30]	0
$F_6(z) = \sum_{i=1}^{n} i z_i^4 + \text{random}[0, 1)$	30	[−1.28, 1.28]	0
$F_7(z) = \sum_{i=1}^{n} \left[z_i^2 - 10\cos(2\pi z_i) + 10\right]$	30	[−5.12, 5.12]	0
$F_8(z) = -20\exp\left(-0.2\sqrt{\frac{1}{n}\sum_{i=1}^{n} z_i^2}\right) - \exp\left(\frac{1}{n}\sum_{i=1}^{n}\cos(2\pi z_i)\right) + 20 + e$	30	[−32, 32]	0
$F_9(z) = 0.1\{\sin^2(3\pi z_1) + \sum_{i=1}^{n} (z_i - 1)^2\left[1 + \sin^2(3\pi z_1 + 1)\right] + (z_n - 1)^2[1 + \sin^2(2\pi z_n)]\} + \sum_{i=1}^{n} h(z_i, 5, 100, 4)$ $h(z_i, \alpha, l, k) = \begin{cases} l(z_i - \alpha)^k & z_i > \alpha \\ 0 & -\alpha < z_i < \alpha \\ l(-z_i - \alpha)^k & z_i < -\alpha \end{cases}$	30	[−50, 50]	0
$F_{10}(z) = \sum_{i=1}^{n} \left(10^6\right)^{\frac{i-1}{n-1}} z_i^2$	30	[−10.0, 10.0]	0

6 Experimental Output and Scrutinization

6.1 Analysis of Optimality, Convergence Accuracy, and Stability

In this paper, we performed 30 independent runs of all the four algorithms namely mbSSA, SSA, PSO, and GWO on ten standard benchmark test functions and calculated optimal value, mean, and also standard deviation of every algorithm. Imple-

Table 3 Output of test functions

Fun	Algorithm	Best	Ave	Std
F1	mbSSA	0.0	0.0	0.0
	SSA	0.0	0.0	0.0
	GWO	2.4259e-88	1.85001503e-85	3.548332678788218e-85
	PSO	2.1143e-26	1.4251928e-23	2.879720205526266e-23
F2	mbSSA	0.0	0.0	0.0
	SSA	0.0	0.0	0.0
	GWO	4.283e-50	4.6982113333333334e-49	7.39484339518295e-49
	PSO	1.8588e-16	1.3313575776666666e-12	5.925871126955444e-12
F3	mbSSA	0.0	0.0	0.0
	SSA	0.0	3.829667e-317	0.0
	GWO	2.7774e-31	2.1118924413333333e-27	6.423915723120329e-27
	PSO	21.3621	70.75083666666667	40.86328751441629
F4	mbSSA	0.0	0.0	0.0
	SSA	0.0	5.38266666666665e-310	0.0
	GWO	2.9915e-23	8.239680666666666e-22	1.5243875080511771e-21
	PSO	0.62629	1.318996	0.5502869406083865
F5	mbSSA	0.0	0.0	0.0
	SSA	2.8332e-09	2.11601265e-06	4.8779919739955815e-06
	GWO	24.8794	26.119313333333334	0.6461401362538632
	PSO	6.9511	52.37619333333333	39.83015856307823
F6	mbSSA	4.5498e-06	5.003998e-05	3.6590256091934904e-05
	SSA	1.5464e-05	7.521863333333334e-05	5.179177269604556e-05
	GWO	7.3798e-05	0.0002916551	0.00018260433385391224
	PSO	0.0062281	0.013931773333333333	0.004527625874023431
F7	mbSSA	0.0	0.0	0.0
	SSA	0.0	0.0	0.0
	GWO	0.0	1.8947666666666667e-15	1.0378064445422053e-14
	PSO	9.9496	32.03809666666667	10.148146270862284
F8	mbSSA	8.8818e-16	8.8818e-16	0.0
	SSA	8.8818e-16	8.8818e-16	0.0
	GWO	7.9936e-15	1.0598893333333332e-14	3.35528190160892e-15
	PSO	2.5757e-14	1.2107025666666666e-12	1.3172682886402264e-12
F9	mbSSA	1.3498e-32	1.3498e-32	0.0
	SSA	5.0929e-18	1.0112520666666667e-15	2.2659037286750873e-15
	GWO	6.0543e-06	0.18804640708666667	0.1110321955135683
	PSO	1.3139e-24	0.0032961	0.005120948405897298
F10	mbSSA	0.0	0.0	0.0
	SSA	0.0	0.0	0.0
	GWO	2.1653e-86	1.4113164066666667e-83	6.592465413922989e-83
	PSO	1.9068e-25	1.7745910933333334e-22	5.25964545418008e-22

mentation of all algorithms on the benchmark problems has done through MATLAB, to test the optimality, convergence accuracy, and stability of mbSSA. From Table 3, it could be visible that the finest value and mean of mbSSA is the better than that of other algorithms such as SSA, PSO, and GWO. That means from the optimal point of view and also from convergence accuracy, the performance of mbSSA is best. It can clearly see from Table 3 that standard deviation of mbSSA is zero which means stability of mbSSA is also excellent. So, from convergence accuracy, stability, and optimal point of view, it can be said that performance of mbSSA is better than the original SSA. That indicates SSA is superior with respect to convergence accuracy, stability, and optimality.

6.2 *Wilcoxon Signed-Rank Test*

The Wilcoxon signed-rank test is a purely statistical technique based on the order of observations within the sample [25]. The one with the smallest ranking will be regarded as the best out of everyone, and vice versa. This statistical rank test's findings performed on all of the algorithms are shown in Table 4, whereas Table 5 shows their rank summary. The findings show that when compared to the other optimization algorithms, mbSSA had the lowest rank for the majority of the benchmark functions. This demonstrates mbSSA's superior performance in comparison with others. Although, SSA and GWO competed in the company of mbSSA nearly and rated second and third, respectively. The prevalent execution of mbSSA does not imply that it is more predominant than all other optimization techniques that exist within the literature which will also lead to the infringement of the "free lunch theorem" [26]. Its execution essentially implies that it is way finer than other algorithms consider in this work only.

Table 4 Pair-wise Wilcoxon signed-rank test results

Function	Wilcoxon signed-rank test order
F1	**mbSSA** = SSA < GWO < PSO
F2	**mbSSA** = SSA < GWO < PSO
F3	**mbSSA** = SSA < GWO < PSO
F4	**mbSSA** = SSA < GWO < PSO
F5	**mbSSA** < SSA < PSO < GWO
F6	**mbSSA** < SSA < GWO < PSO
F7	**mbSSA** = SSA = GWO < PSO
F8	**mbSSA** = SSA < GWO < PSO
F9	**mbSSA** < PSO < SSA < GWO
F10	**mbSSA** = SSA < GWO < PSO

Table 5 Rank summary of statistical assessment results

Function	mbSSA	SSA	GWO	PSO
F1	1.5	1.5	3	4
F2	1.5	1.5	3	4
F3	1.5	1.5	3	4
F4	1.5	1.5	3	4
F5	1	2	4	3
F6	1	2	3	4
F7	2	2	2	4
F8	1.5	1.5	3	4
F9	1	3	4	2
F10	1.5	1.5	3	4
Total	**14**	18	31	37

6.3 Analysis of Convergence Speed

In this paper, we would like to study the convergence rate of mbSSA with standard evolutionary techniques like real SSA, PSO, and GWO.

From Figs. 3 and 4, it is crystal clear that the rapid convergence of mbSSA can be seen in all cases. So, the modified binary sparrow search algorithm (mbSSA) displays a better convergence rate in all benchmark functions in comparison with SSA, PSO, and GWO.

7 Conclusion

SSA is a meta-heuristic swarm optimization technique that originated to obtain the best optimal solution in all aspects. But still, some researchers find out a few drawbacks in it such as slow converge speed and less optimality. In this paper, we have proposed a modified binary sparrow search algorithm (mbSSA) by generating the initial population into binary form instead of random initialization. And we have done experiments through MATLAB by using 10 different benchmark test functions on mbSSA, SSA, GWO, and PSO. Results depict that mbSSA plays an extremely good role in all aspects such as optimality, convergence accuracy, stability, and convergence rate. We have applied the Wilcoxon signed-rank test on mbSSA, SSA, GWO, and also on PSO where mbSSA got the smallest ranking which means mbSSA's performance is superior in comparison with others. This concludes that mbSSA is the best from each point of view.

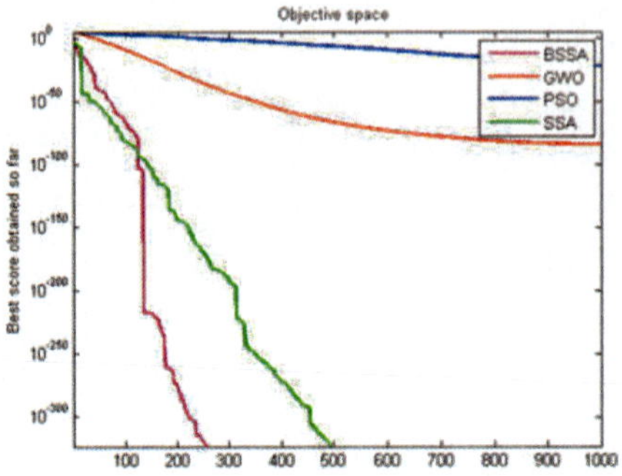

(a) "F1"

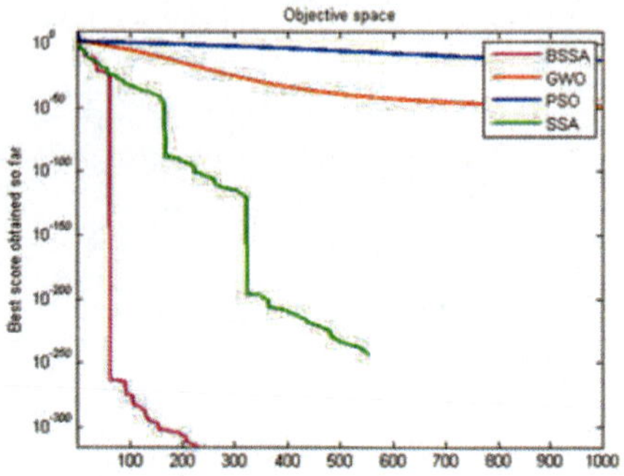

(b) "F2"

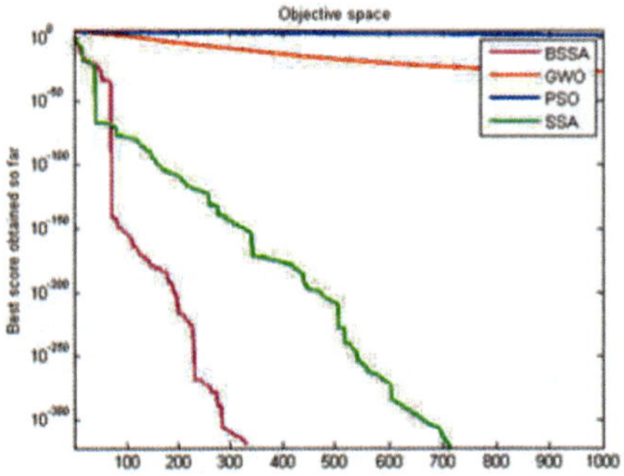

(c) "F3"

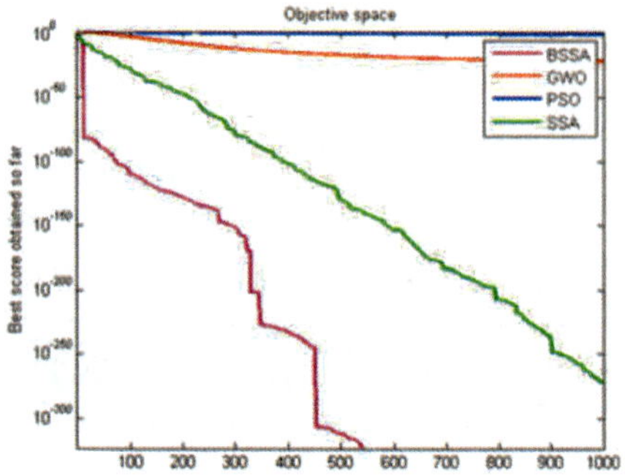

(d) "F4"

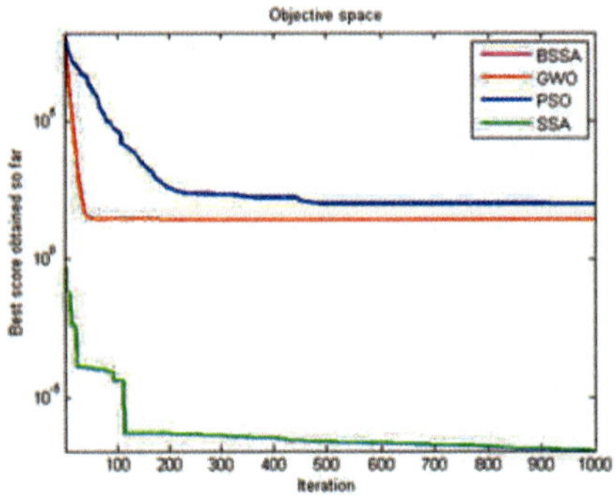

(e) "F5"

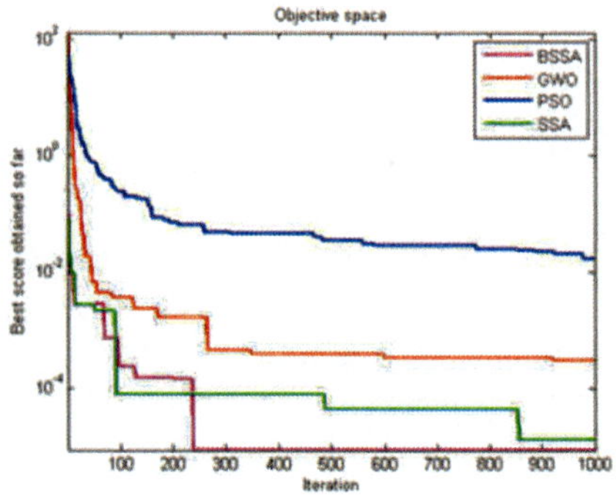

(f) "F6"

Fig. 3 Convergence curves of four algorithms tested on test functions (F1-F6)

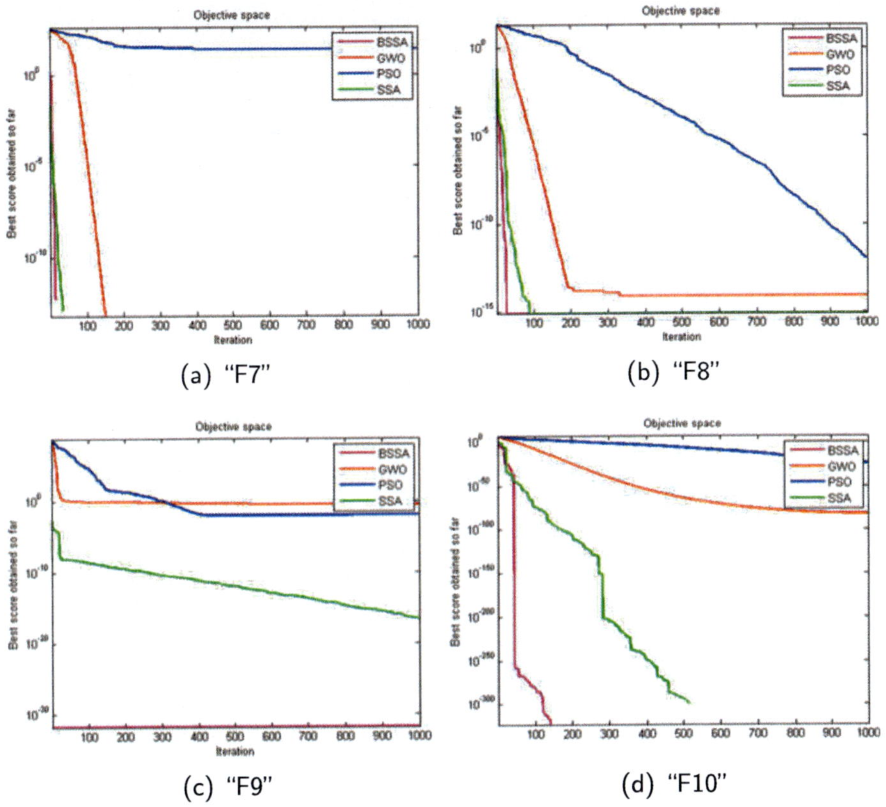

Fig. 4 Convergence curves of four algorithms tested on test functions (F7-F10)

References

1. Desale S, Rasool A, Andhale S, Rane P (2015) Heuristic and meta-heuristic algorithms and their relevance to the real world: a survey. Int J Comput Eng Res Trends 351(5):2349–7084
2. Xue J, Shen B (2020) A novel swarm intelligence optimization approach: sparrow search algorithm. Syst Sci Control Eng 8(1):22–34
3. Wang L, Wang X, Fu J, Zhen L (2008) A novel probability binary particle swarm optimization algorithm and its application. J Softw 3(9):28–35
4. Roopa YM (2014) SPARROW algorithm for clustering software components
5. Doreswamy H, Salma UM (2016) A binary bat inspired algorithm for the classification of breast cancer data. Int J Soft Comput Artifi Intell Appl (IJSCAI) 5(2/3):1–21
6. Mlakar U (2016) Hybrid cuckoo search for constraint engineering design optimization problems. In: Proceedings of StuCoSReC, pp 57–60
7. Çelik Y, Kutucu H (2018) Solving the tension/compression spring design problem by an improved firefly algorithm. IDDM 1(2255):1–7
8. Stephen S, Christu D, Dalvi AE (2018) Design optimization of weight of speed reducer problem through matlab and simulation using ansys. Int J Mech Eng Technol (IJMET) 9:339–349
9. Lei Y, De G, Fei L (2020) Improved sparrow search algorithm based DV-Hop localization in WSN. In: 2020 Chinese automation congress (CAC). IEEE, pp 2240–2244

10. Peng Y, Liu, Y, Li Q (2020) The application of improved sparrow search algorithm in sensor networks coverage optimization of bridge monitoring. In: MLIS, pp 416–423
11. Kishor Yadav N, Saraswat M (2020) Chaotic Henry gas solubility optimization algorithm. Congress on intelligent systems. Springer, Singapore, pp 247–257
12. Ouyang C, Zhu D, Wang F (2021) A learning sparrow search algorithm. Comput Intell Neurosci 2021
13. Yang X, Liu J, Liu Y, Xu P, Yu L, Zhu L, Chen H, Deng W (2021) A novel adaptive sparrow search algorithm based on chaotic mapping and t-distribution mutation. Appl Sci 11(23):11192
14. Wu C, Fu X, Pei J, Dong Z (2021) A novel sparrow search algorithm for the traveling salesman problem. IEEE Access 9:153456–153471
15. Quan L, Li A, Cui G, Xie S (2021) Using enhanced sparrow search algorithm-deep extreme learning machine model to forecast end-point phosphorus content of BOF
16. Peraza-Vazquez H, Pena-Delgado AF, Echavarria-Castillo G, Morales-Cepeda AB, Velasco-Alvarez J, Ruiz-Perez F (2021) A bio-inspired method for engineering design optimization inspired by dingoes hunting strategies. Math Probl Eng 2021
17. Tang A, Zhou H, Han T, Xie L (2021) A chaos sparrow search algorithm with logarithmic spiral and adaptive step for engineering problems. CMES Comput Model Eng Sci
18. Yang L, Li Z, Wang D, Miao H, Wang Z (2021) Software defects prediction based on hybrid particle swarm optimization and sparrow search algorithm. IEEE Access 9:60865–60879
19. Verma P, Parouha RP (2021) An advanced hybrid meta-heuristic algorithm for solving small- and large-scale engineering design optimization problems. J Electr Syst Inf Technol 8(1):1–43
20. Kumar KK, Reddy GN (2021) The sparrow search algorithm for optimum position of wind turbine on a wind farm. Int J Renew Energy Res (IJRER) 11(4):1939–1946
21. Li J (2021) Robot path planning based on improved sparrow algorithm. J Phys Conf Ser 1861(1):012017
22. Wang Z, Huang X, Zhu D (2022) A multistrategy-integrated learning sparrow search algorithm and optimization of engineering problems. Comput Intell Neurosci 2022
23. Wang Z (2022) A parallel particle swarm optimization and enhanced sparrow search algorithm for unmanned aerial vehicle path planning
24. Xie S, He S, Cheng J (2022) Research on improved sparrow algorithm based on random walk. J Phys Conf Ser 2254(1):012051
25. Wilcoxon F, Katti SK, Wilcox RA (1963) Critical values and probability levels for the Wilcoxon rank sum test and the Wilcoxon signed rank test, vol 1. American Cyanamid, Pearl River
26. Wolpert DH, Macready WG (1997) No free lunch theorems for optimization. IEEE Trans Evol Comput 1(1):67–82

Stacked Ensemble Architecture to Predict the Metastasis in Breast Cancer Patients

Sunitha Munappa, J. Subhashini, and Pallikonda Sarah Suhasini

Abstract Prediction of metastasis in breast cancer (BC) patients plays a vital role in treatment and to prolong lifespan of BC patients. The critical part is to locate the tumor cells in lymph node. Whole slide image (WSI) of BC patients requires the expertise in the domain, and it will take more time for pathologists. During this observation, a single tumor cell may be skipped from observation, resulting in false label. Automation of identifying the tumor cells in WSI using deep learning architectures will reduce the false labeling and make the classification process faster than the conventional diagnosis system. In this paper, ResNet-50, Efficient Net B3, and MobileNetV2 deep learning models are implemented on the CAMELYON 17 challenge dataset containing lymph node WSI of BC patients. Stacking Ensemble of ResNet-50, Efficient Net B3, and MobileNetV2 deep learning models is done and implemented to improve the accuracy metrics. A comparison of ResNet-50, Efficient Net B3, MobileNetV2 individual, and Stacking ensemble of these architectures was investigated in terms of AUC, accuracy, specificity, and sensitivity. Result shows that ResNet-50 has better AUC, accuracy, sensitivity, and specificity (0.899, 0.949, 0.921, and 0.9539) than other two networks. Stacking ensemble of these networks has better AUC, accuracy, sensitivity, and specificity (0.934, 0.955, 0.9612, and 0.954) than individual networks.

Keywords Metastasis · WSI · Stacking ensemble · Deep learning · Classification

S. Munappa · J. Subhashini (✉)
S R M Institute of Science and Technology, Chennai, India
e-mail: subhashj@srmist.edu.in

S. Munappa
e-mail: sm4461@srmist.edu.in

P. S. Suhasini
Velagapudi Ramakrishna Siddhartha Engineering College, Vijayawada, India
e-mail: sarahsuhasini@vrsiddhartha.ac.in

S. Das et al. (eds.), *Advances in Data-driven Computing and Intelligent Systems*, Lecture Notes in Networks and Systems 653, https://doi.org/10.1007/978-981-99-0981-0_15

1 Introduction

Breast cancer (BC) metastasis is the leading cause of increased mortality in BC patients [1]. Detecting BC at an early stage is critical for patients' survival, but detecting lumps in the breast at an early stage is difficult because women can detect the tumor by palpation only after it has grown, and few women will undergo regular mammograms, so most women are unable to detect the tumor by palpation at an early stage. As a result, most BC patients are diagnosed after the initial stage [2]. Metastasis is the recurrence of disease, i.e., spreading of cancer cells from first place to other body parts like brain, liver, bone, and lungs [3]. In the event that metastasis is anticipated, by legitimate prescription patients' life expectancy can be extended. Distinguishing cancer cells in lymph nodes from other cells can anticipate the event of metastasis. For examination of lymph nodes, small pieces of lymph nodes are taken and the cancer cells on WSI through a magnifying instrument are observed. Distinguishing the cancer cells from ordinary tissues is extremely challenging, it requires greater investment for pathologists to dissect the slide and once in a while false labeling will occur due to missing single cancer cells. Automation of distinguishing the cancer cells in WSI utilizing profound learning strategies will lessen the false labeling and make the classification quicker.

2 Related Work

Munappa et al. [4] has reviewed the methods to predict the metastasis in breast cancer (BC) patients by using image processing techniques. By detecting the CTC cells in blood and by detecting the cancer cells in lymph nodes of BC, a patient's metastasis can be predicted. To detect cancer cells manually or with conventional methods, it requires the expertise in the domain and it will take more time for pathologists. To make faster and automated classification, deep learning algorithms will be used. For deep learning algorithms, a large data set should be required to train the model for accurate classification for that Wang et al. [5] in their work proposed a novel data augmentation method called Random Centre Cropping (RCC). And used different architectures, viz. ResNet-50, Efficient Net B3, Dense Net 121, and Boosted Efficient Net B3 to predict and classify the lymph node metastasis of breast cancer patients [5]. With Boosted Efficient Net B3, high accuracy and AUC are obtained on CAMELYON 16 and 17 challenges. To make sure that the assistance of deep learning algorithms to pathologists for classification of WSI helps to improve accuracy and reduce the classification time, for that David et al. [6] proposed Lymph Node Assistant (LYNA) with inception V3 deep learning structure to detect the tumor outline and was classified into macro- and micro-metastases based on tumor outline diameter. Results show that with assistance of DL pathologist's classification achieved more accuracy than classification with pathologists alone. WSI are very large in size so those images are divided into patches and then given as inputs to

DL models, for that Wang et al. [7] done patient-level lymph node classification with CAMELYON 17 data set. Deep segmentation network (DSNet) was designed to detect metastases in patch level. Density-based spatial clustering of applications with noise (DBSCAN) was designed to detect metastases in slide level, and finally, Deep regional metastases segmentation (DRMS) was designed to detect metastases in patient level. Lin et al. [8] used anchor layers for model conversion for metastasis detection on CAMELYON 2016 data set with basic VGG16 network architecture. Wollmann et al. [9] classification of WSI images is done based on tumor cells by combining the Cycle-Consistent Generative Adversarial Network (Cycle-GAN) with a densely connected deep neural network (DenseNet) on CAMELYON 17 challenge dataset. In metastasis prediction, accurate classification is more important, i.e., false negatives and false positives must be less in classification. For that, Liang et al. [10] presented an attention mechanism incorporated in CNN. ResNet-50 using Convolution Block Attention Module (CBAM) is implemented on histopathology images of breast cancer patients. CAMELYON 17 data is used and CBAM attained 0.976 as AUC score.

3 Section-I

Model Training

Initially, the number of test patches is increased by data augmentation and then the inputs are preprocessed according to the model prerequisites. Then, preprocessed inputs are applied to the base model followed by applying the global average pooling. As a last stage, the prediction layer with sigmoid activation function whose value exists between the range of (0, 1) is used. ResNet-50, Efficient Net B3, and MobileNetV2 models are investigated in this work. Global average pooling (GAP) layer performs the average of features from the previous layers. Addition of GAP layer reduces or avoids the overfitting problem.

Prediction and Classification

Sliding window technique is used for prediction, with window size of 64×64 over the entire image/slide. For each window over the slide, prediction of the probability of having cancer is done. To decide whether the patch is normal or cancerous, the prediction value is compared with a pre-selected threshold value. Here, the threshold value is taken as 0.5. If the prediction value is greater than threshold, then it will be considered that the slide has metastasis or else not. Based on the classified prediction values, heatmap is obtained which provides the information regarding the location/presence of the tumor in the slide. Based on the heatmap, classification was done. CAMELYON 17 challenge dataset provides nearly 400 whole slide images (WSI) collected from different medical centers [11]. The dataset consists of tumor slides which have metastases, normal slides, and test slides which may or not have metastases.

ResNet-50

The ResNet-50 model consists of 5 stages, each with a convolution and Identity block. Each convolution block has 3 convolution layers, and each identity block also has 3 convolution layers. In the last convolution layer, output is added to the main input.

Figure 1 is the input WSI to architecture; due to large size and resolution, each slide is divided into number of patches each of size 256 × 256, and model was trained with patch-level annotations; finally, model will give slide-level classification, and any one of the patches in slide is 1 then that slide classified as metastasis. Figure 2 shows the heat map generated by ResNet-50 architecture [12].

Figure 3 shows the validation results, loss with respect to Epochs, AUC with respect to Epochs, and accuracy with respect to Epochs. As the number of Epochs increases, loss decreases and AUC increases. Different evaluation metrics obtained are AUC 0.899, accuracy 0.949, sensitivity 0.925, and specificity 0.953. Figure 4

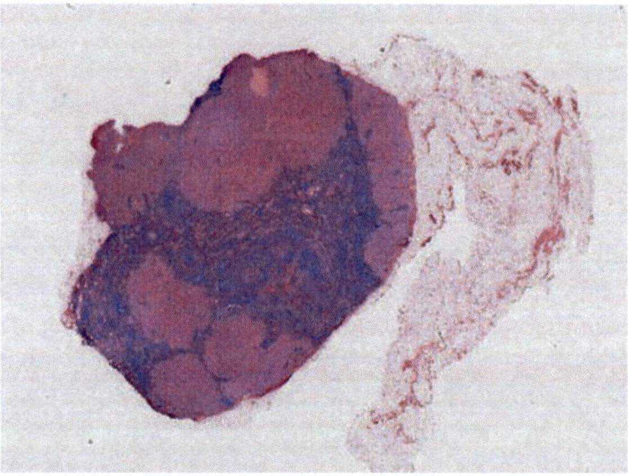

Fig. 1 Input image from CAMELYON data set [10]

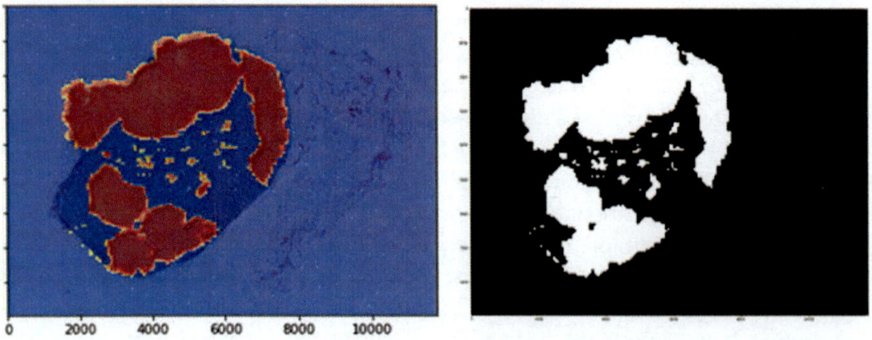

Fig. 2 Heat map by ResNet-50 architecture

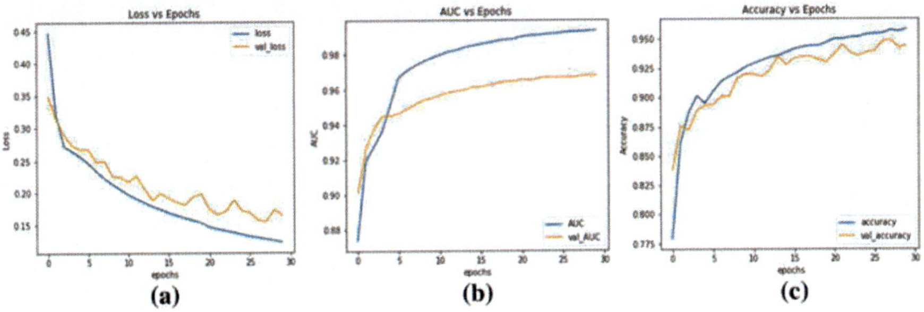

Fig. 3 ResNet-50 outputs, training, and validation **a** loss with respect to Epochs **b** AUC with respect to Epochs, and **c** accuracy with respect to Epochs

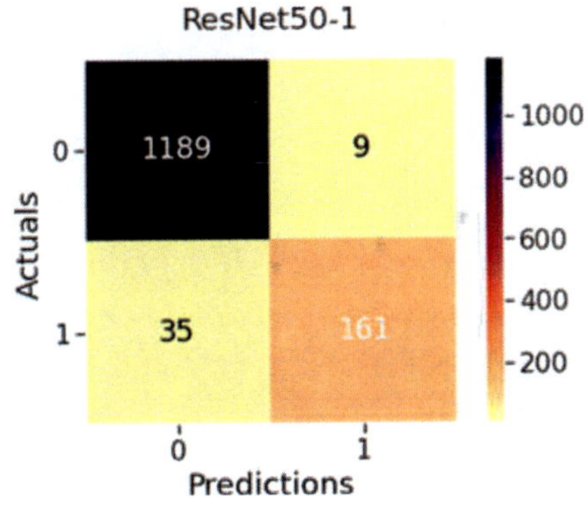

Fig. 4 Confusion matrix

shows the Confusion matrix of ResNet-50, from which it is evident that the false positives are less.

Efficient network B3:

Efficient Net B3 is convolutional neural network architecture and basically is a scaling method that uniformly scales all dimensions of depth/width/resolution using a compound coefficient [13]. Here, each slide is divided into a number of patches each of size 256×256 as done in the previous model trained with patch-level annotations. Finally, the model will give slide-level classification, if any one of the patches in the slide is 1 then that slide is classified as metastasis. Figure 5 shows the heat map generated.

Figure 6 shows the validation results, loss with respect to Epochs, AUC with respect to Epochs, and accuracy with respect to Epochs, as observed in graphs number of Epochs increases, loss decreases, and AUC increases. Different evaluation metrics obtained are AUC 0.825, accuracy 0.901, sensitivity 0.881, and specificity 0.903. Figure 7 shows the Confusion matrix of Efficient Net B3.

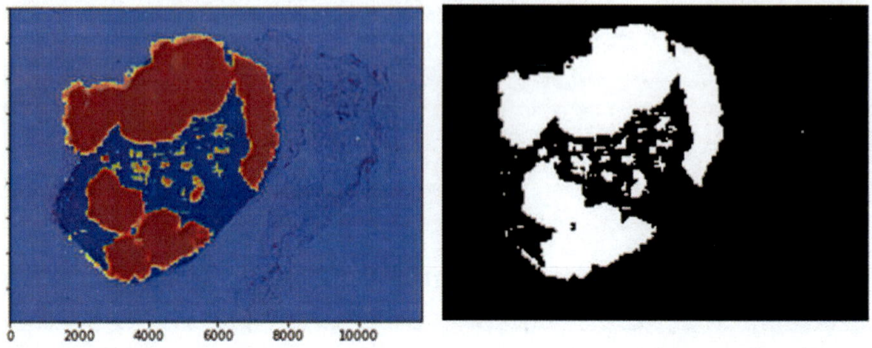

Fig. 5 Heat map by Efficient Net B3 architecture

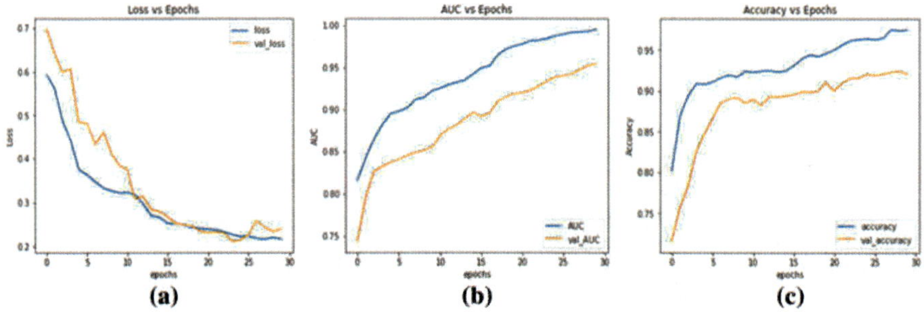

Fig.6 Efficient Net B3 outputs, training, and validation **a** loss with respect to Epochs **b** AUC with respect to Epochs, and **c** accuracy with respect to Epochs

Fig. 7 Confusion matrix

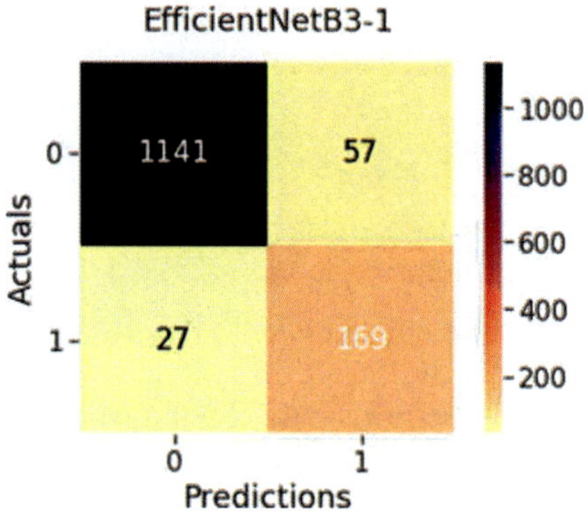

MobileNetV2

MobileNetV2 is a convolution neural network architecture based on an inverted residual structure where the residual connections are between the bottleneck layers. The intermediate expansion layer uses lightweight depth-wise convolutions to filter

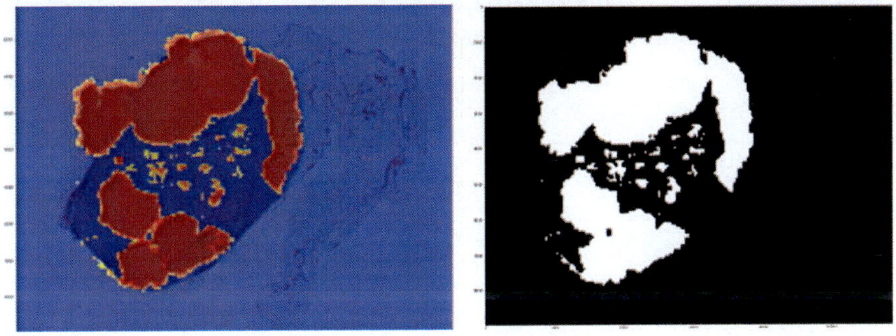

Fig. 8 Heat map by MobileNetV2 architecture

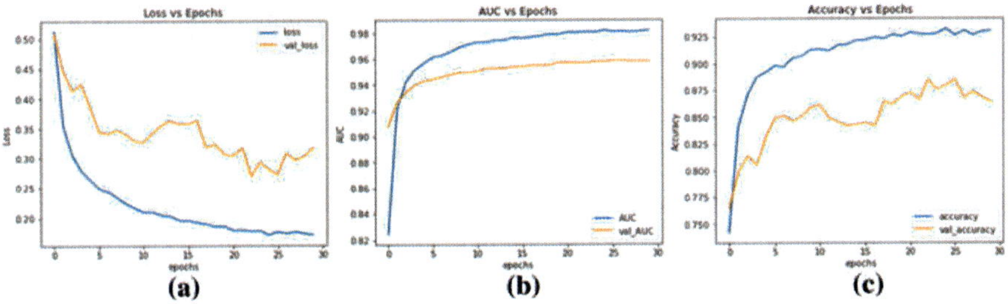

Fig. 9 MobileNetV2 outputs, training, and validation, **a** loss with respect to Epochs **b** AUC with respect to Epochs, and **c** accuracy with respect to Epochs

features as a source of non-linearity. The architecture of MobileNetV2 contains the initial fully convolution layer with 32 filters, followed by 19 residual bottleneck layers [14, 15]. Each slide is divided into a number of patches each of size 256 × 256, the model was trained with patch-level annotations, and finally, the model will give slide-level classification, if any one of the patches in the slide is 1 then that slide is classified as metastasis. Figure 8 shows the heat map generated.

Figure 9 shows the validation results, loss with respect to Epochs, AUC with respect to Epochs, and accuracy with respect to Epochs, as observed in graphs number of Epochs increases, loss decreases, and AUC increases. Different evaluation metrics obtained are AUC 0.869, accuracy 0.929, sensitivity 0.919, and specificity 0.930. Figure 10 shows the Confusion matrix of MobileNetV2, and it has more false positives and false negatives.

4 Section-II

Ensemble learning is a machine learning approach that combines different base models to produce an optimal model [16, 17]. There are different approaches to

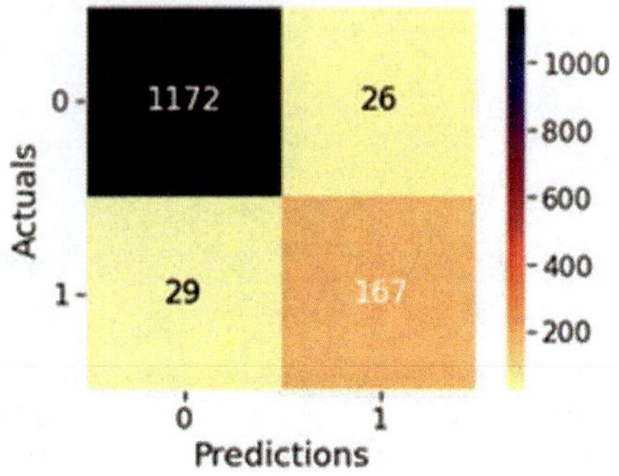

Fig. 10 Confusion matrix

implement the ensemble learning, viz. Bagging, Boosting, and stacking. Here, a stacked ensemble was implemented. To do this, all the trained models are loaded and freeze the layers so that their weights are not modified further [18]. The outputs of all the loaded models are concatenated and require dense layers to it [19, 20] and finally generated the ensemble model which yielded more accurate results and better prediction values than the base models. Stacking Ensemble mode is shown in Fig. 12. Figure 11 shows the heat map generated. Here, ResNet-50, Efficient Net B3, and MobileNetV2 are stacking.

Total 8364 patches are taken from 400 images from CAMELYON 17 data set, in that 4062 positive patches, i.e., having the metastasis, and 4302 negative patches, i.e., no metastasis. From this, 5576 patches were used to train the model, 1394 patches for validating the model, and 1394 patches for testing the model. For efficient training, Deep Learning models need large datasets. Size of the data set can be increased with the data augmentation techniques, and used techniques in this paper were given below.

1. Random Rotation: By choosing angle, randomly the image was rotated.
2. Random Horizontal Flip: With a given probability, the image was flipped horizontally.

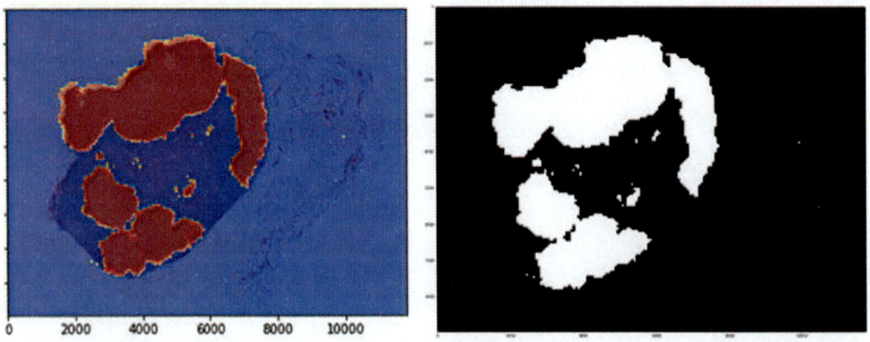

Fig. 11 Heat map for Ensemble procedure

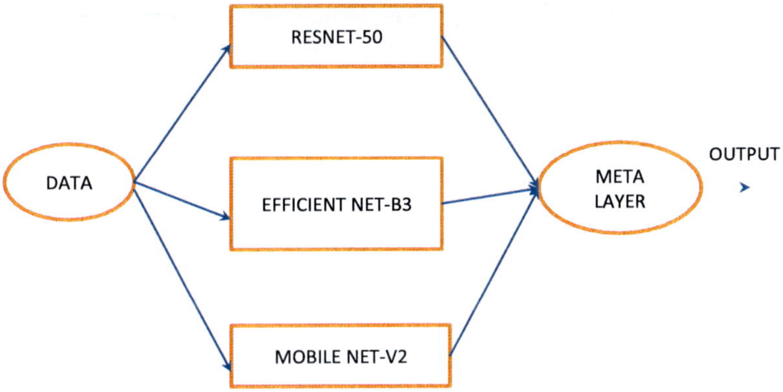

Fig. 12 Stacking ensemble

- Random Rotation angle range: [−20°, 20°]
- Probability of Random Horizontal Flip: 0.2.

Evaluation metrics for proposed architecture

$$\text{TN} = \text{True Negatives} \quad \text{TP} = \text{True Positives}$$
$$\text{FP} = \text{False Positives} \quad \text{FN} = \text{False Negatives}$$

- Precision: It is the fraction of positive predictions that are positive.

$$\text{Precision} = \text{TP}/\text{TP} + \text{FP}$$

- Recall: It is the fraction of positive samples that are predicted as positive.

$$\text{Recall} = \text{TP}/\text{TP} + \text{FN}$$

- F1 Score: F1 Score is the harmonic mean of precision and recall.

$$\text{F1 Score} = 2 \times \text{precision} \times \text{recall}/\text{precision} + \text{recall}$$

- Accuracy: Accuracy is the fraction of the total correct predictions.

$$\text{Accuracy} = \text{TP} + \text{TN}/\text{TP} + \text{FP} + \text{FN} + \text{TN}$$

Figure 13 shows the validation results, loss with respect to Epochs, AUC with respect to Epochs, and accuracy with respect to Epochs, as observed in graphs number

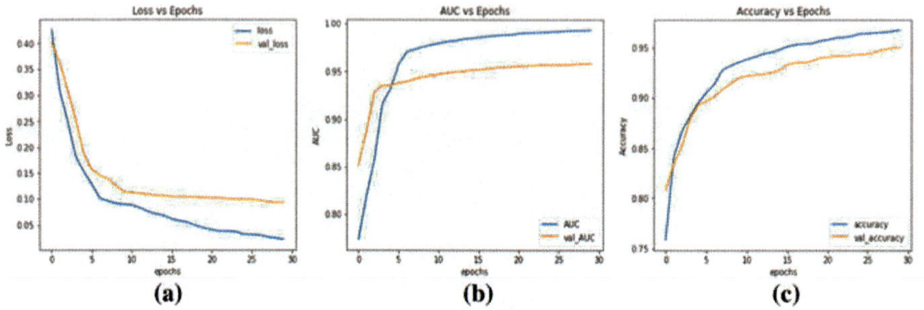

Fig. 13 Ensemble outputs, training, and validation **a** loss with respect to Epochs, **b** AUC with respect to Epochs, and **c** accuracy with respect to Epochs

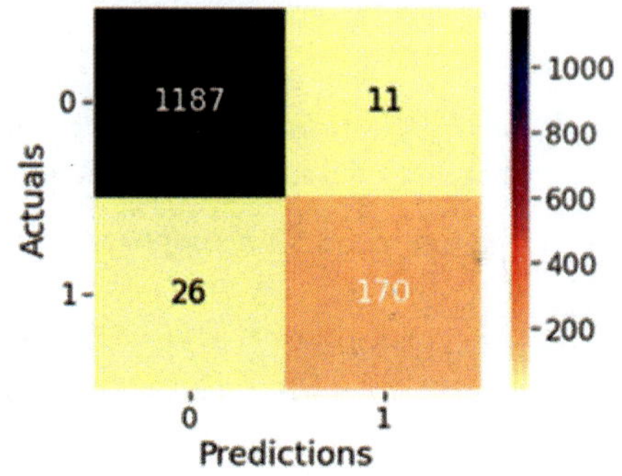

Fig. 14 Ensemble evaluation metrics

of Epochs increases, loss decreases, and AUC and accuracy increase. Different evaluation metrics obtained are AUC 0.934, accuracy 0.955, sensitivity 0.961, and specificity 0.954. Figure 14 shows the Confusion matrix and has less false positives and false negatives.

5 Section-III

Table 1 shows comparisons of ResNet-50, Efficient Net B3, MobileNetV2, and Ensemble of these three architectures with respect to different evaluation metrics. Figure 15 shows the comparisons through the graph.

To classify the images accurately, there should be less false positives and false negatives, which means the model should have high sensitivity and specificity. Result shows that ResNet-50 has better AUC, accuracy, sensitivity, and specificity than other two base models. It has 9 false positives and 35 false negatives at 0.5 threshold value out of 1394 patches, whereas Efficient Net B3 has 57 false positives and 27 false negatives and MobileNetV2 has 26 false positives and 29 false negatives. The stacking ensemble of ResNet-50, Efficient Net B3, and MobileNetV2 networks

Table 1 Comparisons of evaluation metrics for ResNet-50, Efficient Net B3, MobileNetV2 architectures, and ensemble of these three architectures

Model/Metric	AUC	Accuracy	Sensitivity	Specificity
Efficient net B3	0.8254	0.9013	0.8810	0.9037
MobilenetV2	0.8691	0.9292	0.9199	0.9306
Resnet-50	0.8993	0.9495	0.9251	0.9539
Ensemble	0.9341	0.9557	0.9612	0.9546

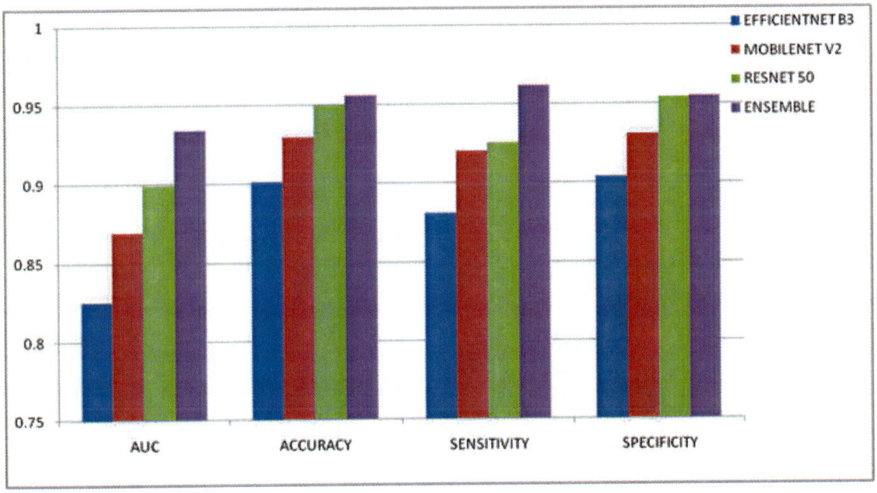

Fig. 15 Comparison of results

performed better than individual networks in terms of AUC, accuracy, sensitivity, and specificity. It has 37 false predictions out of 1394 patches at 0.5 threshold value.

6 Conclusion

ResNet-50, Efficient Net B3, and MobileNetV2 deep learning models are implemented on the CAMELYON 17 challenge dataset containing lymph node WSI of BC patients. Stacking ensemble of ResNet-50, Efficient Net B3, and MobileNetV2 is done. Total patches taken are 8364, modeled 5576 for training, 1394 patches for validation, and 1394 patches for testing. ResNet-50 demonstrated better performance than the other two base networks. Stacking Ensemble of architectures improved the performance than individual networks. Further, accuracy may be increased by increasing training patches and including more dense layers in the architectures.

References

1. Allemani C, Matsuda T, Di Carlo V, Harewood R, Matz M, Niksic M (2018) Global surveillance of trends in cancer survival 2000–14 (CONCORD-3). Analysis of individual records for 37 513 025 patients diagnosed with one of 18 cancers from 322 population-based registries in 71 countries. Lancet 391:1023–1075
2. Boyle P, Levin B (2008) World cancer report, World Health Organization, Geneva, Switzerland
3. Liang Y, Zhang H, Song X, Yang Q (2020) Metastatic heterogeneity of breast cancer. Molecular mechanism and potential therapeutic targets, Elsevier. Semin Cancer Biol 60:14–27
4. Sunitha M, Subhashini J, Suhasini PS (2022) Review on methods to predict metastasis of breasr cancer using Artificial Intelligence. ICECMSN, Springer Lecture notes on Data Engineering and Communication Technologies. http://doi.org/10.1007/978-981-16-9605-3_32
5. Wang J, Liu Q, Xie H, Yang Z (2021) Boosted efficient net: detection of lymph node metastases in breast cancer using convolutional neural networks. MDPI J Cancers 13661
6. Steiner DF, MacDonald R, Liu Y (2018) Impact of deep learning assistance on the histopathology review of lymph nodes for metastatic breast cancer, impact of deep learning assistance. Am J Surg Pathol (Am J Surg Pathol)
7. Wang L, Song T, Katayama T, Jiang X (2021) Deep regional metastases segmentation for patient level lymph node status classification. IEEE Access Multi Open Access J 9:129293–129302
8. Lin H, Chen H, Graham S, Dou Q (2018) Fast scan net: fast and dense analysis of multi-gigapixel whole slide images for cancer metastasis detection. IEEE Trans Med Imaging
9. Wollmann T, Eijkman CS, Rohr K (2018) Adversarial domain adaptation to improve automatic breast cancer grading in lymph nodes. In: IEEE 15th international symposium on biomedical imaging. Washington, USA, April, 978-1-5386-3636-7/18
10. Liang Y, Jinglong, Quan X (2019) Metastatic breast cancer recognition in histopathology images using convolutional neural network with attention mechanism. IEEE Conf 01, 978-1-7281-4094
11. CAMELYON17 contest home page. https://camelyon17.grand-challenge.org
12. Hung G, Liu, VanDer Maaten Z (2017) Densely connected convolutional networks. In: IEEE conference on computer vision and pattern recognition. Honolulu, HI, USA
13. Tan M, Le QV (2019) Efficient Rethinkinh model scaling for convolutional neural networks. Online: https://arxiv.org/abs/1905.11946
14. Chiu Y-C, Tsai C-Y, Ruan M-D, Shen G-Y, Lee T-T (2020) Mobilenet-SSDv2: an improved object detection model for embedded systems. In: International conference on system science and engineering (ICSSE), Oct. https://doi.org/10.1109/ICSSE50014.2020.9219319
15. ShenWang YZ, Qin X (2020) Label-free detection of rare circulating tumor cells by image analysis and machine learning. Sci Rep 10:12226. https://doi.org/10.1038/s41598-020-69056-1
16. Huang F, Xie G, Xiao R (2010) Research on ensemble learning. In: International conference on artificial intelligence and computational intelligence. https://doi.org/10.1109/AICI.2009.235
17. Wang K, Liu X, Zhao J, Gao H, Zhang Z (2020) Application research of ensemble learning frameworks. Chin Autom Congr (CAC), Nov, Electronic ISSN: 2688-0938
18. Jangam E, Chandra Sekhar Rao A (2021) A stacked ensemble for the detection of COVID-19 with high recall and accuracy. Comput Biol Med 135:104608. https://doi.org/10.1016/j.compbiomed.2021.104608
19. Loddo A, Buttau S (2022) Deep learning based pipelines for Alzheimer's disease diagnosis: a comparative study and a novel deep-ensemble method. Comput Biol Med 141. https://doi.org/10.1016/j.compbiomed.2021.105032
20. Shakeel PM, Tolba A, Al-Makhadmeh Z, Jaber MM (2020) Automatic detection of lung cancer from biomedical data set using discrete AdaBoost optimized ensemble learning generalized neural networks. Neural Comput Appl 32(3):777–790

A New Sensor Neuro-controller-based MPPT Technique for Solar PV Module

Sunita Chahar and D. K. Yadav

Abstract The integration of solar PV generation systems in energy generation is growing very fast. The MPPT technique for solar PV systems has become an emerging research area to draw maximum power output to fulfill increased energy needs and to make a more promising solar PV generation system. In recent years, much more work in the field of soft computing-based MPPT control schemes has been done. Nevertheless, all these different types of MPPT tactics have their pros and cons. In this work, a new sensor neuro-controller-based MPPT technique is designed for a solar PV module to extract maximum power output. The main purpose of implementing this new single-sensor neuro-controller is to reduce the input variables from two or more to only one input. The variations in current are found large in comparison to voltage therefore current is used as the input variable. A comparison of the proposed new sensor neuro-controller-based MPPT technique with the perturb and observe (P&O) methodology is presented. The performance of the proposed technique for the different case studies has been demonstrated. The better performance of the proposed technique in comparison to P&O has been validated under different environmental conditions in terms of tracking speed and conversion efficiency. It has been found that the tracking speed is fast, and efficiency is 99.71 and 99.78% for the proposed MPPT technique.

Keywords P&O · MPPT · Neuro-controller · Solar PV · Single sensor

1 Introduction

Over the past few years, soft-computing-based tactics in the field of renewable energy sources have drawn attention worldwide. In the field of energy production, the development of different types of renewable energy sources and research in the field of

S. Chahar (✉) · D. K. Yadav
Electrical Engineering Department, RTU, Kota, India
e-mail: schahar@rtu.ac.in

D. K. Yadav
e-mail: dkyadav@rtu.ac.in

S. Das et al. (eds.), *Advances in Data-driven Computing and Intelligent Systems*, Lecture Notes in Networks and Systems 653, https://doi.org/10.1007/978-981-99-0981-0_16

output power control using intelligent control methods to make them more popular have increased. Solar PV is one of the highly famous sustainable energy sources among different types of clean sources like wind, solar, biomass, biogas, and many more. The Sun's irradiation is freely available everywhere and every day and this is a great source of energy to generate power. The main reason for capturing the market as a big player by solar PV system is various attractive government schemes and its low prices year by year [1].

On the contrary, electricity generation by the solar PV system is afflicted by uncertain atmospheric conditions, seasons, and surroundings. Everyday electricity generation from the PV panel may vary due to weather. Its power production is unpredictable in all seasons. Therefore, inputs to the solar system in terms of irradiation and temperature affect the current, voltage, power output, and efficiency of the solar PV system. The conversion efficiency has been found to be 15–22%. Therefore, to improve efficiency and to get maximum output from solar panels MPPT technique is used [2].

Researchers have investigated different types of soft-computing-based MPPT techniques to find the point of maximum power. The main objective of the MPPT technique is to make the solar PV system capable of supplying the maximum possible electricity generation under unpredictable variations in temperature and solar insolation. Many researchers have investigated and tried to compare different MPPT techniques. Various external uncertain environmental variables and their impact on solar PV modules using intelligent computing methodology have been researched by scholars. From the survey, various MPPT techniques can be classified as the traditional, modified traditional, novel, modified novel, the combination of traditional and novel, and modified combination of traditional and novel MPPT schemes. Each class has its pros and cons. Conventional techniques are simple but not able to identify global peak values on power and voltage characteristics. Novel-type computations are complicated and not able to trace global peak value on the power and voltage characteristics [3, 4]. In hybrid, combination of the traditional and advanced methods is fast but not able to trace GPMP during the search [5]. Therefore, It is required to choose the correct MPPT technique to improve the efficiency and performance of solar PV systems, components, converters, and inverters under unknown weather conditions and shedding conditions [6–8]. In recent times, out of different intelligent tactics, ANN has gained acceptance in controlling and harvesting maximum power for renewable energy sources [9]. It can learn, identify patterns, and respond quickly to changes. Another main reason for its popularity is the no need for computational effort. ANN has proven its better performance for load management, and power management [10, 11] ANN played an important role in battery energy management for continuous and stable power supply [12].

In this paper, a new sensor neuro-controller-based MPPT technique and P&O method has been used. The conventional type P&O MPPT technique has been used for comparison. The capability and fulfillment of the proposed tactic have been investigated for tracking speed, oscillations, efficiency, and behavior of the controller under fast and slow-changing atmospheric conditions in terms of efficiency. A comparative analysis has been presented.

This work is categorized into the following sections: Part 2 is about the organization of the proposed configuration and its implementation of the neuro-controller-based MPPT for the solar PV module. This section is further divided into subsections. This describes the proposed MPPT methodology and the basics of the solar PV module. The second subsection explains the parameters for the proposed controller and the procedure of working. Part 3 is about analysis, comparison, and results. The final section is about the conclusion.

2 Design and Implementation of Proposed New Neuro-controller-based MPPT for Solar PV Module

2.1 Description of Proposed MPPT Methodology

In this section, a neuro-controller is used to make a solar PV module efficient to supply maximum power to the connected load. A key feature of the neuro-controller is to remove the nonlinearities of the system [13] Different samples of weather conditions are used as input variables to the solar PV model. Switching pulses are stimulated in such a manner to obtain maximum power conditions. These pulses are generated by a neuro-controller and given to the switches of a DC-DC boost converter. This controller is performed efficiently to learn from given patterns and follow them.

The following are the steps to implement the proposed scheme:

1. Recognition of the solar plant.
2. Configured the neuro-controller to develop the controller.

Recognition of the Solar Plant

In this process, a plant model is used as a test model. Here, solar PV module 1Soltech 1STH-215-P has been used. A current source of I_p current in parallel with the diode is connected with the DC-DC converter and load to supply generated I_SolarPV current as shown in Fig. 1. The solar PV configuration is supplied maximum power supply to the load. The characteristics of the solar PV module are presented in Fig. 2 [14, 15]. For the required power rating of the solar PV module, the combinations of series and parallel PV modules are used. Each cell's power rating is 1–2 W.

Configured the Neuro-controller to Develop the Controller

In this section, a neuro-controller is taught and instructed to operate for the desired value. This is accomplished by selecting a neural network and parameters. These parameters for the neuro-controller are brought up in Table 1. This process makes the controller acquainted with the condition.

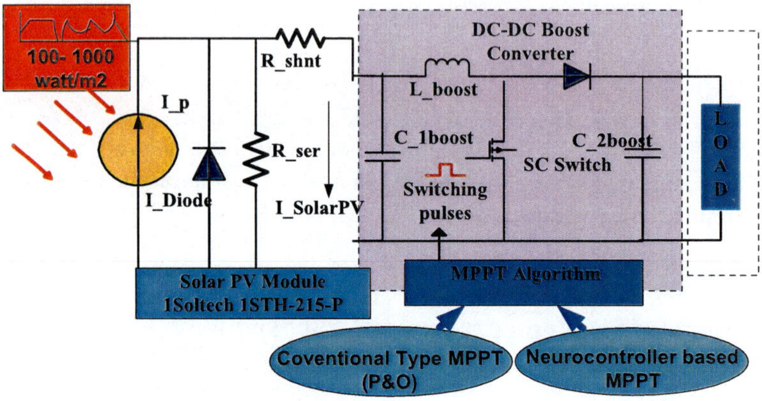

Fig. 1 Solar PV module

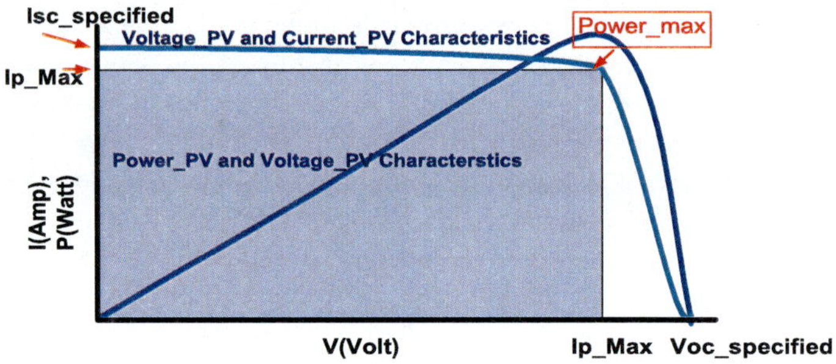

Fig. 2 Solar PV module characteristics

Table 1 Specifications of neuro-controller for two different sets of weather conditions

Case-study variables	No. of hidden layer	No. of variables	Epochs	Variables for validation	Variables for testing
Case-I	3	132	300	20% of the total variable	20% of the total variable
Case-II	5	50,001	300	20% of the total variable	20% of the total variable

$$\begin{aligned}&\text{Approxinmation model equation}\\&= f\_{\text{mnn}}\left[O_{(T)}, O_{(T-1)}, \ldots O_{(T-n+1)}, I_{\text{ref}(T-1)}, \ldots I_{\text{ref}(T-\text{pre}+1)}\right]\\&\quad + g\_{\text{mnn}}\left[O_{(T)}, O_{(T-1)}, \ldots O_{(T-n+1)}, I_{\text{ref}(T-1)}, \ldots I_{\text{ref}(T-tn+1)}\right]\\&\quad \times I_{\text{ref}(T)}\end{aligned} \tag{1}$$

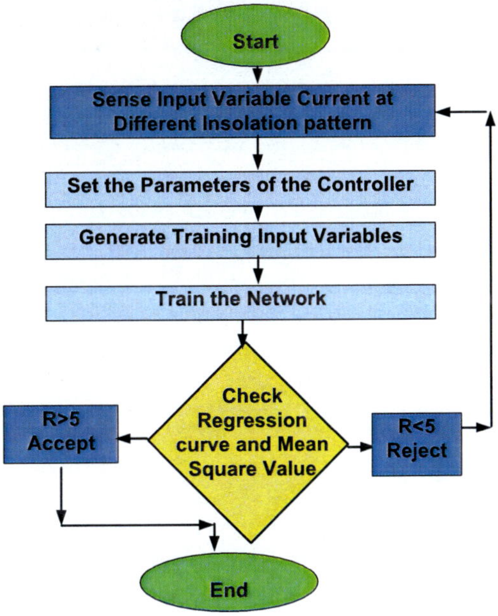

Fig. 3 Flowchart of proposed MPPT

Equation 1 is representing an approximate model of the standard nonlinear structure such that output (O) is following the reference value (I_{ref}) with reduced possible mean square error.

In Fig. 3, the working of the proposed MPPT methodology is presented through the flow chart.

The control signal produced by the controller is represented by the mathematical equation Eqs. (2)–(3) [16].

$$= \frac{I_{(T+d)} - f\_mnn\begin{bmatrix} I_{(T)}, I_{(T-1)}, \ldots I_{(T-\text{pre}+1)}, \\ I_{(inT)} I_{(inT-1)}, \ldots I_{(inT-tn+1)} \end{bmatrix}}{g\_mnn\begin{bmatrix} I_{(T)}, I_{(T-1)}, \ldots I_{(T-\text{pre}+1)}, \\ I_{(inT)} I_{(inT-1)}, \ldots I_{(inT-tn+1)} \end{bmatrix}} \tag{2}$$

When plant output is following the reference path, the control signal is in the form of switching pulses for the DC–DC converter's switches. The control signal-

$$\text{Control signal} = \frac{I_{(T+d)ref} - f\_mnn\begin{bmatrix} I_{(T)}, I_{(T-1)}, \ldots I_{(T-\text{pre}+1)}, \\ I_{(inT)} I_{(inT-1)}, \ldots I_{(inT-tn+1)} \end{bmatrix}}{g\_mnn\begin{bmatrix} I_{(T)}, I_{(T-1)}, \ldots I_{(T-\text{pre}+1)}, \\ I_{(inT)} I_{(inT-1)}, \ldots I_{(inT-tn+1)} \end{bmatrix}} \tag{3}$$

Here, I and I_{in} are the plant input and output. Pre and tn represent the previous value and the total number of inputs. d is the bias signal, and f_mnn and g_mnn are the multi-neural networks.

In the proposed module, the author has compared the output current value to the reference value of current. Therefore, by comparing the output value to the reference value, an error signal is generated. The controller has induced a signal to reduce this error. The proposed MPPT has been compared with the P&O MPPT technique. This is a simple and conventional method. This MPPT technique need not bother about previous data. It swings about the maximum power point. Its applicability is affected due to its speed and oscillations.

2.2 Design Experimental Set-Up

The proposed solar PV module with a neuro-controller-based MPPT scheme has been designed in a MATLAB 2018b domain, and the performance of the proposed scheme has been tested under different, unpredictable, and variable weather conditions. Component values have been calculated by mathematical equations. These values have been used to simulate the test system.

Figure 4 is showing the rated Power_pv and Voltage_pv performance curve. From the characteristics, the rated maximum value of power_pv is 213.1 W at 29 V. Rated values are mentioned in Table 2.

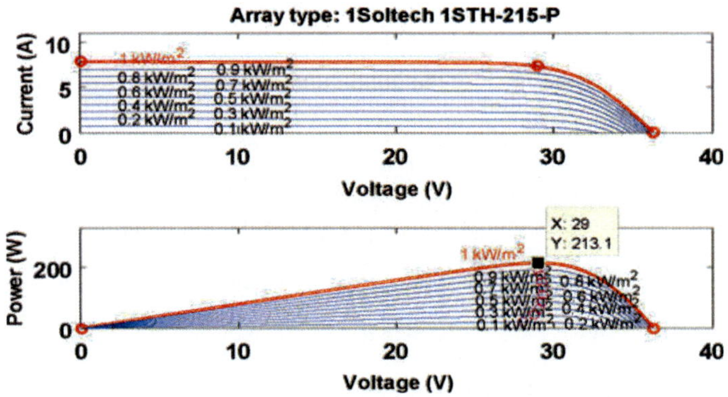

Fig. 4 Rated Power_pv and Voltage_pv performance characteristics of solar PV System

Table 2 Design and operating characteristics values used in the test model

Parameters	Rated Power_max	Voc_specified	No. of cells in each module (N_cell)	Isc_specified
Rated values	213.1 W	36.3 V	60	7.84 A

3 Results

This section is divided into two parts. The first part consists of specimens of two different cases of uncertain insolation conditions. In this section, two cases of variable weather conditions are simulated in a MATLAB environment. In the second part, the simulation results are executed under variable slow and fast-changing insolation patterns to obtain maximum power. In this section, a comparison has been made with the popular conventional MPPT tactic to prove the proposed controller's efficiency.

3.1 Case Study

In this work, two cases have been incorporated to validate the performance of the proposed neuro-controller-based MPPT methodology. In the first case, 132 insolation patterns have been included in the study. The first condition is shown in Fig. 5. This case has been considered here to check the working of the proposed controller's tracking time and reliability.

In the second case, a fast-changing insolation pattern has been considered to check the neuro-controller's operational capability and sensitivity for large no of insolation weather inputs and at each small change in insolation. The second case is demonstrated in Fig. 5. Variations in insolation samples have been considered from 1000 to 100 W/m^2 at a 25 °C temperature for evaluation as mentioned in Table 3.

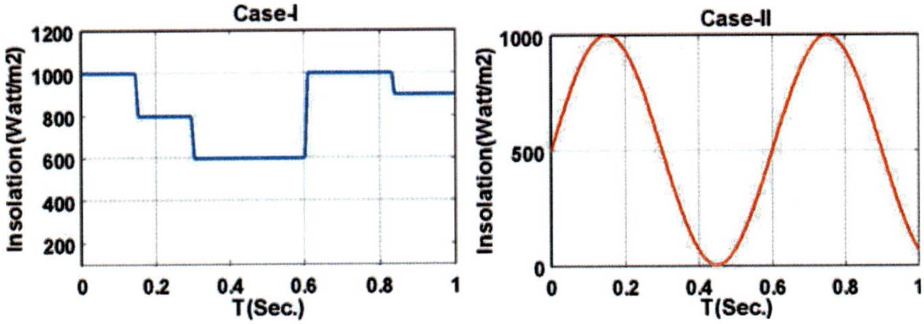

Fig. 5 Case study

Table 3 Input data sample for evaluation

S. No.	1	2	3	4	5	6	7	8	9	10
Insolation (W/m^2)	1000	900	800	700	600	500	400	300	200	100

3.2 *Modeling of Proposed System, Comparison, and Validation of Results*

In this section,complete modeling of the solar PV module to supply the load with DC to DC boost converter has been done. The proposed neuro-controller-based MPPT scheme has been implemented. Two cases have been applied as input to the proposed solar module. And the observation has been made to validate the suggested scheme's performance in contrast to the traditional MPPT tactic to get the maximum possible electricity generation. The MATLAB model of the proposed scheme for the solar PV system is demonstrated in Fig. 6.

Comparative Analysis for Insolation Condition Case-I

In Fig. 7, simulation results have been demonstrated at the slow-changing variable environmental condition for P&O and proposed MPPT techniques. This is the daily case of the uncertain conditions of different levels and the transition phase of the insolation. In Fig. 7, the response is obtained for 132 data of weather conditions varying from 1000 to 600 W/m^2. A comparative investigation has been done for the proposed scheme with the P&O scheme. From the results, it is observed that the output in the case of the neuro-controller-based MPPT tactic is following the exact pattern of insolation variation without oscillation. In this case, the value of maximum power obtained is 212.5 W at 1000 W/m^2 and 25 °C. In this context for the P&O MPPT technique, maximum power is 212 W. The rated maximum power for the test system is 213.1 W as shown in Fig. 4. Therefore, the efficiency of the proposed controller is 99.71%, and the efficiency of the traditional tactic is 99.48%. It proves the effective and efficient performance of the presented controller.

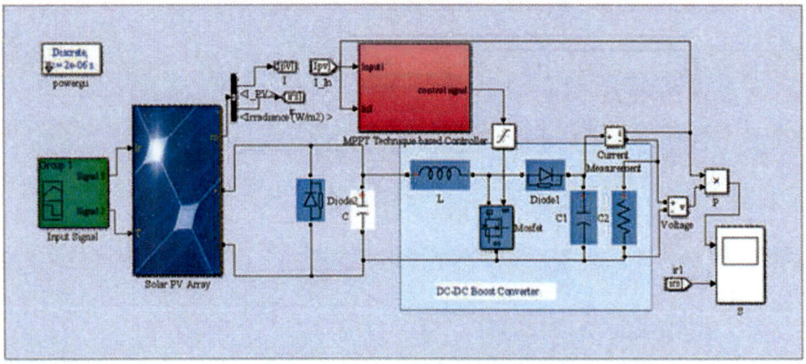

Fig. 6 MATLAB model of proposed MPPT technique for solar PV

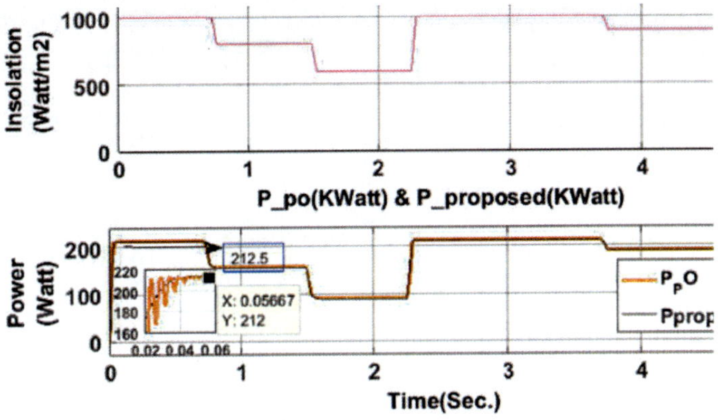

Fig. 7 Comparative evaluation of P_po and P_proposed for Case-I

This maximum value of power is achieved at 0.04 s. In comparison to the proposed scheme, this is achieved at 0.05667 s. Therefore, the proposed controller has demonstrated rapid response to get the maximum value of power in contrast to the P&O scheme.

The power output corresponds to insolation variations obtained without oscillations for the proposed concept. The power output corresponds to insolation variations in the case of the conventional technique obtained with oscillations as shown in the window in Fig. 7.

Comparative Analysis for Insolation Condition Case-II

In Fig. 8, simulation results have been demonstrated for the fast-changing variable environmental condition and variation in very small steps for P&O and proposed MPPT techniques. In this figure, the response of the controller is obtained for 50,001 data samples varying from 1000 to 100 W/m^2.

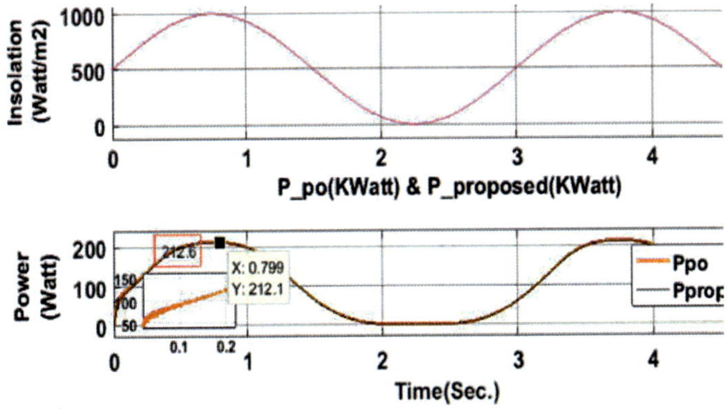

Fig. 8 Comparative evaluation of P_po and P_proposed for Case-II

Table 4 Summary of comparative analysis

S. No.	Case	No. of weather data	Comparative analysis	P&O MPPT technique	Proposed neuro-controller-based MPPT tactic for PV system
1.	Case-I	132	Efficiency (%)	99.48	99.71
2.			Tracking speed	Medium	Fast
3.			Time to reach at the point of maximum power	0.05667	0.04
1.	Case-II	50,001	Efficiency (%)	99.53	99.76
2.			Tracking speed	Medium	Fast
3.			Time to reach at the point of maximum power	0.79	0.7497

A comparative evaluation has been done for the proposed scheme with the P&O scheme. From the results, it is realized that the power is following the exact pattern of insolation variation without oscillation for the case of the neuro-controller-based MPPT technique. In this case, the value of maximum power obtained is 212.6 W at 1000 W/m^2 and 25 °C. In contrast to the proposed tactic, maximum power is 212.1 W. The rated maximum power for the test system is 213.1 W as shown in Fig. 4. Therefore, the efficiency of the proposed controller is 99.76%, and the efficiency of the traditional tactic is 99.53%.

The maximum value of power has been achieved at 0.7497 s for the proposed methodology. But in the case of the P&O MPPT scheme, the maximum value has been achieved at 0.79 s. Therefore, the proposed controller has demonstrated a speedy response to fast-changing conditions with improved efficiency to get the maximum power in contrast to the P&O scheme.

In the case of the proposed tactic, the output power that has been attained without oscillations corresponds to fast insolation variations for the proposed tactic. In contrast to the proposed method, the power output corresponds to fast insolation variations in the case of the conventional technique obtained with oscillations as shown in the window in Fig. 8. A concise comparative analysis is presented in Table 4.

4 Conclusion

In this work, a single sensor neuro-controller-based MPPT to track maximum power for solar PV module has proved its better performance in comparison to the traditional P&O MPPT tactic. The maximum power has been traced only by using one sensing variable current. Therefore, in this proposed methodology, the requirement of the input variable is reduced from two or more to only one input variable. The proposed controller has demonstrated a faster speed to reach a maximum power condition with no oscillations in comparison to the P&O MPPT method. Simultaneously, it can respond to every small variation in weather conditions in comparison to conventional tactics. The conversion efficiency of the new single sensor neuro-controller-based MPPT is higher, i.e., 99.71 and 99.78% for slow and fast-changing atmospheric conditions in comparison to the P&O MPPT tactic. Therefore, improved efficiency, faster response to the variation, maximum power output without oscillation, and requirement of less number of variables are the important achievement of this proposed controller. The future work will be implementing the proposed methodology for the wind energy system and other renewable energy sources.

References

1. Moorthy JG, Manual S, Moorthi S, Raja P (2020) Performance analysis of solar PV based DC optimizer distributed system with simplified MPPT method. SN Appl Sci 2:1–8
2. El-Khozondar HJ, El-Khozondar RJ, Matter K, Suntio T (2016) A review study of photovoltaic array maximum power tracking algorithms. Renew Wind Water Sol 3:1–8
3. Belay Kebede A, Biru Worku G (2020) Comprehensive review and performance evaluation of maximum power point tracking algorithms for photovoltaic system. Glob Energy Interconnect 3:398–412
4. Podder AK, Roy NK, Pota HR (2019) MPPT methods for solar PV systems: a critical review based on tracking nature. IET Renew Power Gener 13:1615–1632
5. Baba AO, Liu G, Chen X (2020) Classification and evaluation review of maximum power point tracking methods. Sustain Futur 2:100020
6. Singh BP, Goyal SK, Siddiqui SA, Kumar P (2020) A study and comprehensive overview of inverter topologies for grid-connected photovoltaic systems (PVS). 607:1009–1017
7. Sridhar R, Jeevananthan S, Dash SS, Selvan NT (2014) Unified MPPT controller for partially shaded panels in a photovoltaic array. Int J Autom Comput 11:536–542
8. Killi M, Samanta S (2019) Voltage-sensor-based MPPT for stand-alone PV systems through voltage reference control. IEEE J Emerg Sel Top Power Electron 7:1399–1407
9. Haseeb I, Armghan A, Khan W, Alenezi F, Alnaim N, Ali F, Muhammad F, Albogamy FR, Ullah N (2021) Solar power system assessments using ann and hybrid boost converter based mppt algorithm. Appl Sci 11
10. Zhang H, Cheng S (2011) A new MPPT algorithm based on ANN in solar PV systems. Lecture notes in electronics and engineering 121 LNEE, pp 77–84
11. Charrouf O, Betka A, Abdeddaim S, Ghamri A (2020) Artificial Neural Network power manager for hybrid PV-wind desalination system. Math Comput Simul 167:443–460
12. Santhoshi BK, Mohanasundaram K, Kumar LA (2021) ANN-based dynamic control and energy management of inverter and battery in a grid-tied hybrid renewable power system fed through switched Z-source converter. Electr Eng 103:2285–2301
13. Jang JO, Jeon GJ (2000) A parallel neuro-controller for DC motors containing nonlinear friction. Neurocomputing 30:233–248
14. Salman S, Ai X, Wu Z (2018) Design of a P-&-O algorithm based MPPT charge controller for a stand-alone 200W PV system. Prot Control Mod Power Syst 3:25
15. Mohamed SA, Abd El Sattar M (2019) A comparative study of P&O and INC maximum power point tracking techniques for grid-connected PV systems. SN Appl Sci 1:1–13
16. Jalil NA, Darus IZM (2013) NARMA-L2 vibration controller for flexible structure with non-collocated. n y n. 17–22

Data-Driven Prediction of Effluent BOD_5 from an Institutional Wastewater Treatment Plant

Shubham K. Jain, Ashwani Kumar, Sudhir Kumar, Amit Kumar, and Aditya Choudhary

Abstract The study develops a data-driven based model to predict biological oxygen demand of the treated wastewater from an institutional wastewater treatment plant. First, the data was collected for six parameters (i.e. pH, conductivity, temperature, total suspended solids, total organic carbon and biological oxygen demand) from the selected wastewater treatment plant by collecting and analysing samples over a period of more than two months. The quality parameters were then combined in seven different scenarios for training of machine learning models. These scenarios were tested using four machine learning models, i.e. multiple linear regression, artificial neural network, support vector machine and random forest. Finally, these models were compared to select the best model for prediction of biological oxygen demand using a set of indicators, i.e. coefficient of determination and root mean square error.

Keywords Machine learning · Wastewater treatment · BOD_5 · Regulatory compliance

1 Introduction

The treatment of wastewater is the most significant step in reducing environmental pollution and improving the water quality. Modern wastewater treatment facilities must comply with increasingly stringent pollution requirements, as well as new energy efficiency and resource recycling guidelines given by the government bodies. Many uncertainties exist in wastewater treatment systems due to the complexity of natural events, anthropogenic activities and the wastewater treatment process [1].

S. K. Jain (✉) · A. Kumar · S. Kumar · A. Kumar · A. Choudhary
Department of Civil Engineering, Malaviya National Institute of Technology Jaipur, Jaipur, India
e-mail: 2020PCE5581@mnit.ac.in

S. Kumar
e-mail: skumar.ce@mnit.ac.in

A. Kumar
e-mail: amitrathi.ucf@gmail.com

S. Das et al. (eds.), *Advances in Data-driven Computing and Intelligent Systems*, Lecture Notes in Networks and Systems 653, https://doi.org/10.1007/978-981-99-0981-0_17

The composition of wastewater is quite diverse, with influent quality, i.e. pollutant concentrations, and treated wastewater varying among wastewater treatment plants. Furthermore, wastewater treatment is a multi-step process that may be influenced by a multitude of variables.

Among various parameters, biological oxygen demand (BOD_5) is widely used for assessing organic pollutant contamination in wastewater engineering, as well as for monitoring, designing and modelling wastewater systems [2, 3]. Analysing BOD_5 in the laboratory is an arduous and time-consuming task. A feedback control system based on BOD_5 cannot adjust for influent wastewater perturbations in the interim since it takes five days for BOD_5 analysis to complete. Rapid and simple BOD_5 measurement is needed for the control of wastewater treatment [4]. Moreover, the analysis of BOD_5 is very sensitive as it may be affected by the presence of chlorides and other chemicals. To monitor the treatment efficiency of wastewater treatment plants (WWTPs) and compliance with existing regulations, it would be better to know the value of wastewater quality (in terms of BOD_5) in real time.

In view of the difficulties discussed above, machine learning may be helpful in advance prediction of effluent quality from a wastewater treatment plant. Machine learning models have the ability for high-speed operation, handling multiple complex variables and the ease of use. Hence, this study has been undertaken to develop a ML model for data-driven prediction of the BOD_5 of wastewater. The objectives of the study are as follows: (1) to analyse the various quality parameters for treated effluent from the selected WWTP; (2) selection, construction, training and validation of multiple ML models; (3) to identify a suitable ML model for BOD_5 prediction from other easily measurable parameters.

2 Literature Review

The review indicated that the frequently used AI systems for the modelling and optimisation of WWT processes included artificial neural network (ANN), support vector machine (SVM), decision trees (DT), long short-term memory (LSTM), convolutional neural network (CNN) and recurrent neural networks (RNNs) (Fig. 1). The literature review focused on assessing the various AI models used in WWT applications, extracted from the latest 21 studies published between 2012 and 2021. ANN was the most recurrent ML model. The indicators used for the analysis of model efficiency are root mean square error (RMSE), coefficient of determination R^2, correlation coefficient R and mean absolute percentage error (MAPE) [5].

The parameter selection has a significant role in prediction of any quality parameter because of the interrelationship between the input parameters and the predicted parameter. The level of accuracy of prediction may change by changing the selection of various input parameters. For the same, earlier studies were reviewed and the frequency of various parameters used for BOD_5 prediction was recorded (Table 1).

For the application of machine learning models, the amount of data available is one of the most important criteria. A review of the literature was also performed for

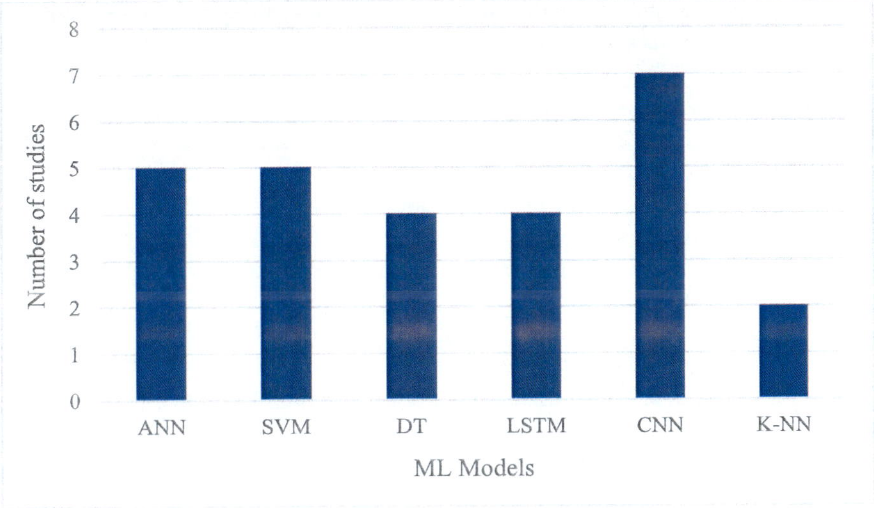

Fig. 1 Distribution of publications by ML model type ($n = 21$)

Table 1 Frequency of input parameter used for prediction of BOD5

Input parameter	Frequency	References
pH	3	[6–8]
COD	5	[6–10]
TSS	5	[6, 7, 9, 10]
Conductivity	3	[6–8]
Temperature	1	[7]
TN	4	[7–10]
TP	2	[7, 10]
DO	1	[7]
Inflow	2	[6, 11]
Nitrate	1	[7]
Sulphide	1	[9]

determining the number of data points for various applications. The minimum data set points used were 50 for maximising the water flux through membrane [12], and the maximum data was used 1800 for the prediction of total nitrogen [13].

3 Methodology

This section covers the sampling site, treatment scheme at the site, sampling plan, wastewater quality parameters for laboratory analysis and the selection of machine learning models for training and prediction. The wastewater samples were collected from the WWTP located in the campus of MNIT Jaipur, Jaipur. The WWTP employs moving bed biofilm reactor (MBBR) technology for wastewater treatment. The treated wastewater is being reused for the purpose of landscaping in the campus itself, which makes it more important for compliance with the regulations. The samples were collected twice a day from the effluent of the secondary sedimentation tank prior to chlorination (between 9:00 a.m. and 10:00 a.m. in the morning and 3:00 p.m. and 4:00 p.m. in the afternoon). Total six parameters, i.e. pH, conductivity, temperature, TSS, TOC and BOD_5, were analysed for the samples collected. All the parameters were analysed in the laboratory as per the standard methods provided in [14].

Initially collected data set was first evaluated for descriptive statistics to determine the following values: minimum, maximum, mean, standard deviation and the range. Afterwards, the data is first analysed by multiple machine-learning models. For the machine learning model, the models which have been frequently used in previous studies have been selected for this study. The selected ML models are multiple linear regression (MLR), artificial neural network (ANN), support vector machine (SVM) and random forest (RF). Google Colab was the simulation platform used for training the machine learning models. The operational parameters used for the analysis are RMSE and R^2.

4 Results

The following section describes the results from the study. The descriptive statistic for various parameters is presented in Table 2.

For determining the optimum combination of input parameters, total seven scenarios were developed as follows: Scenario-1: pH, temperature, conductivity, TSS, TOC versus BOD_5; Scenario-2: temperature, conductivity, TSS, TOC versus BOD_5; Scenario-3: conductivity, TSS, TOC versus BOD_5; Scenario-4: temperature, TSS, TOC versus BOD_5; Scenario-5: TSS, TOC versus BOD_5; Scenario-6: temperature, TOC versus BOD_5; Scenario-7: TSS versus BOD_5. Afterwards, the selected machine learning models (MLR, SVM, ANN and RF) were applied to these scenarios for training and validation (Fig. 2).

Table 2 Descriptive statistics for the water quality parameters ($n = 186$)

Variables	Minimum	Maximum	Mean	Standard deviation
pH	6.6	7.6	7.1	0.21
Temperature (°C)	15.4	31	24.5	3.71
Conductivity (ms/cm)	0.72	1.32	1.1	0.15
TSS (mg/L)	32.8	338	122.6	49.5
TOC (mg/L)	7.5	64.28	29.2	13
BOD_5 (mg/L)	22	230	124	45.3

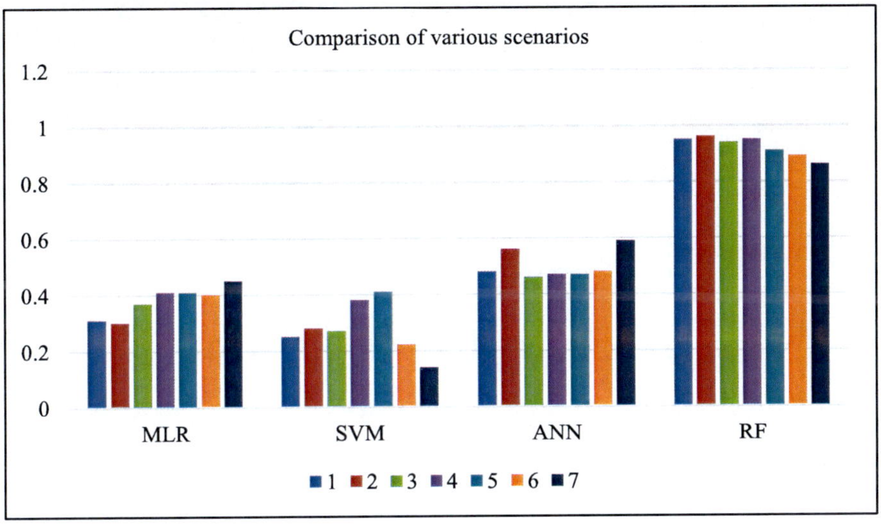

Fig. 2 Determination coefficient (R^2) values for various scenarios with ML models

For MLR, Scenario-7 gives the highest values of R^2 (i.e. 0.45) which includes only TSS values. For SVM, the highest coefficient value is observed for Scenario-5 (i.e. 0.41) which includes TSS and TOC only. For ANN, Scenario-7 shows the best value of 0.59 followed by Scenario-2 (having R^2 value of 0.56). For RF, Scenario-2 gives the optimum result. Furthermore, Scenario-2 includes four parameters, i.e. temperature, conductivity, TSS and TOC. All these factors have been shown to be affecting BOD_5 of a wastewater sample.

Measured values of BOD_5 concentration of wastewater at the effluent from WWTP in MNIT campus were then compared to the predicted values by the machine learning models (MLR, SVM, ANN and RF) to check the model's performance. Two performance indicators were determined to identify the best suitable model: coefficient of determination (R^2) and root mean square error (RMSE).

The values of both the performance indicators have been calculated and provided in Table 3. The observed correlation coefficient R^2 of MLR is 0.30 which is higher than SVM's 0.28 but lower than ANN's 0.56 and RF's 0.95 (Fig. 3). Furthermore, RMSE values are the lowest for RF models only.

Table 3 Performance indicators for Scenario-2

Model	R^2	RSME
MLR	0.30	34.21
SVM	0.28	34.51
ANN	0.56	27.18
RF	0.95	8.87

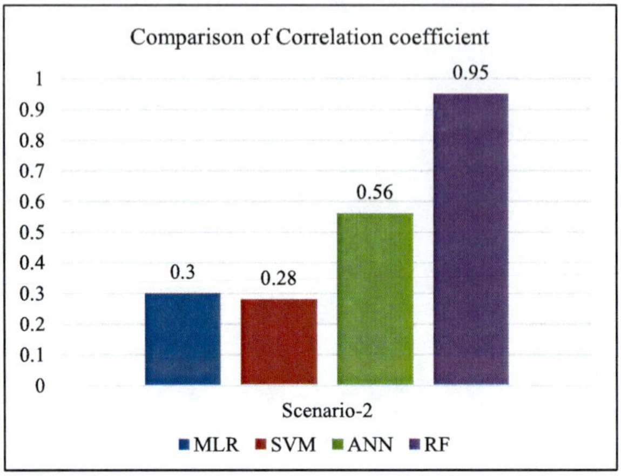

Fig. 3 Correlation coefficients for various ML models in case of Scenario-2

Finally, all the trained ML models (i.e. ANN, RF, SVM and MLR) are used to predict the BOD_5 values. The plots between predicted and observed values of BOD_5 have been provided in Fig. 4. The scatter plots also represent the suitability of the RF model for BOD_5 prediction in the present study. As evident from the figure, the predictions from the RF model fall almost on a straight line.

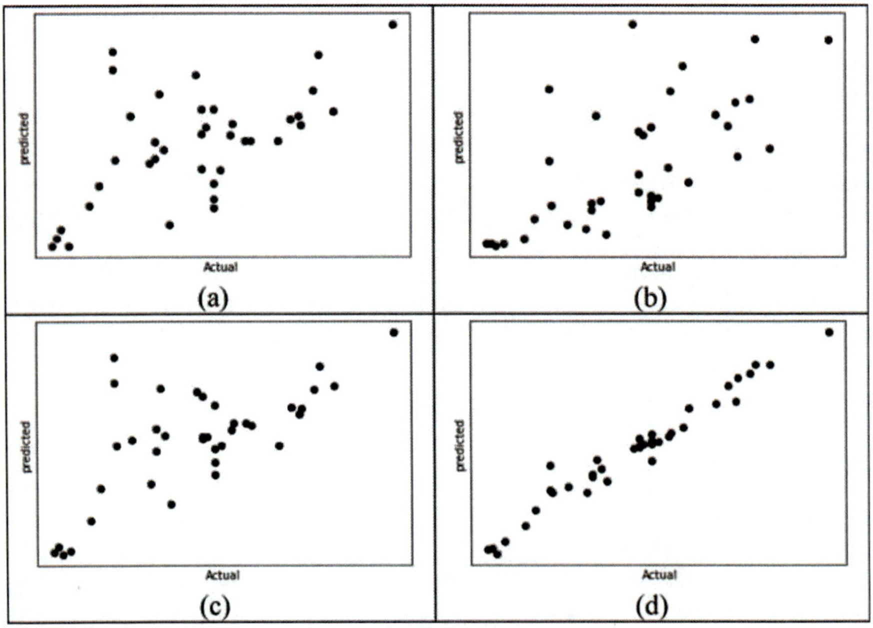

Fig. 4 Scattered plot of **a** MLR, **b** SVM, **c** ANN, **d** RF

5 Conclusions

This study was conducted to develop a machine learning model for prediction of effluent quality from an institutional WWTP, located in the campus of MNIT Jaipur, Jaipur. The samples for lab analysis were collected from the selected WWTP and analysed for six parameters in the laboratory: pH, conductivity, temperature, TSS, TOC and BOD5. For prediction of effluent BOD_5, seven scenarios were developed taking into consideration various quality parameters analysed in the laboratory. Each of the four machine learning models, i.e. ANN, MLR, SVM and RF, was applied onto these scenarios. Scenario-2 involving temperature, conductivity, TSS and TOC was found to be performing best in terms of R^2 values. For Scenario-2, the values of R^2 were found to be the highest (i.e. 0.96) for RF model among all the machine learning models. Similarly, the RF model performed better than other models in terms of the values of RMSE having lowest value (i.e. 8.87). Based on the above findings, the RF model is suitable for prediction of BOD_5 of treated wastewater from the selected WWTP. The machine learning model developed in the study can be used to predict BOD_5 of treated wastewater on the same day by measuring temperature, conductivity, TSS and TOC in the laboratory.

Limitation for the study mainly lies in terms of the number of data points collected and specificity of the developed model for the selected WWTP. Moreover, other parameters, e.g. dissolved oxygen and total nitrogen, may also be included in the sampling for achieving better degree of predication accuracy.

References

1. Zhao L, Dai T, Qiao Z, Sun P, Hao J, Yang Y (2020) Application of artificial intelligence to wastewater treatment: a bibliometric analysis and systematic review of technology, economy, management, and wastewater reuse. Process Saf Environ Prot 133(92):169–182. https://doi.org/10.1016/j.psep.2019.11.014
2. Granata F, Papirio S, Esposito G, Gargano R, de Marinis G (2017) Machine learning algorithms for the forecasting of wastewater quality indicators. Water (Switzerland) 9(2):1–12. https://doi.org/10.3390/w9020105
3. Lee J, Lee S, Yu S, Rhew D (2016) Relationships between water quality parameters in rivers and lakes: BOD5, COD, NBOPs, and TOC. Environ Monit Assess 188(4):1–8. https://doi.org/10.1007/s10661-016-5251-1
4. Luo F, Qiao X, Liao W (2017) Soft-sensing modeling based on GK-LSSVM method for online predictions of BOD in activated sludge process. In: 2017 International conference on robotics and automation science, ICRAS 2017, pp 134–138. https://doi.org/10.1109/ICRAS.2017.8071931
5. Butterfield A, Ngondi GE, Kerr A (2016) A dictionary of computer science. Oxford University Press
6. Fan W, Zhang Z (2020) A CNN-SVR hybrid prediction model for wastewater index measurement. In: Proceedings—2020 2nd international conference on advanced computing and technology information science and communications, CTISC 2020, pp 90–94. https://doi.org/10.1109/CTISC49998.2020.00022

7. Chen Y, Song L, Liu Y, Yang L, Li D (2020) A review of the artificial neural network models for water quality prediction. Appl Sci 10(17). https://doi.org/10.3390/app10175776
8. Nourani V, Asghari P, Sharghi E (2021) Artificial intelligence based ensemble modeling of wastewater treatment plant using jittered data. J Clean Prod 291:125772. https://doi.org/10.1016/j.jclepro.2020.125772
9. El-Rawy M, Abd-Ellah MK, Fathi H, Ahmed AKA (2021) Forecasting effluent and performance of wastewater treatment plant using different machine learning techniques. J. Water Process Eng 44(October):102380. https://doi.org/10.1016/j.jwpe.2021.102380
10. Mundi G, Zytner RG, Warriner K, Bonakdari H, Gharabaghi B (2021) Machine learning models for predicting water quality of treated fruit and vegetable wastewater. Water (Switzerland) 13(18):1–17. https://doi.org/10.3390/w13182485
11. Nourani V, Elkiran G, Abba SI (2018) Wastewater treatment plant performance analysis using artificial intelligence—An ensemble approach. Water Sci Technol 78(10):2064–2076. https://doi.org/10.2166/wst.2018.477
12. Mungray AKA, Agarwal S, Ali J, Chandra Garg M (2021) Performance optimisation of forward-osmosis membrane system using machine learning for the treatment of textile industry wastewater. J Clean Prod 289:125690. https://doi.org/10.1016/j.jclepro.2020.125690
13. Guo H et al (2015) Prediction of effluent concentration in a wastewater treatment plant using machine learning models. J Environ Sci (China) 32:90–101. https://doi.org/10.1016/j.jes.2015.01.007
14. APHA (1998) Standard methods for the examination of water and wastewater. Washington, DC

Predictive Models for Equipment Fault Detection: Application in Semiconductor Industry

Tran Quang Duy and Tran Duc Vi

Abstract Predictive maintenance (PdM) in manufacturing using machine learning (ML) techniques has received much attention from many researchers and practitioners in recent years, particularly in the capital-intensive industries such as semiconductor manufacturing. Developing a decision tool that detects problems in equipment or processes early to maintain high process efficiencies is a critical step in process control for cost reduction. In this study, we conducted experiments on real semiconductor manufacturing data (SECOM). Our experimental results show that performance of predictive models using all the data balancing techniques that we studied including both the data-driven and algorithmic-based techniques outperformed models using unbalanced data. The total of 30 combinations of experiments are conducted for this study (six scenarios of data balancing techniques × 5 ML classification algorithms), we find the best model providing the optimal performance with the highest area under the ROC curve (AUC) is support vector machine (SVM) in combination with cost sensitive learning.

Keywords Predictive models · Imbalanced dataset · Machine learning

1 Introduction

PdM has attracted many interests for both academic and practitioners recently due to one of the most important steps in the production process and a huge benefit on cost reduction [1]. The outcome of this function has a significant influence on the control and decision-making in manufacturing. This process allows for the exploration of future possible flaws and defect patterns, which may be regarded as a firm basis for effective production. PdM is used to keep industrial assets like gadgets and production quality and quantity in good working order.

Semiconductor manufacturing is a subset of microelectronics manufacturing, a bigger industrial discipline. It is a highly complex and multi-step manufacturing

T. Q. Duy · T. D. Vi (✉)
School of Industrial Engineering and Management, International University. Vietnam National, University, Block 6, Linh Trung Ward, Thu Duc District, HCM City, Vietnam
e-mail: tdvi@hcmiu.edu.vn

S. Das et al. (eds.), *Advances in Data-driven Computing and Intelligent Systems*, Lecture Notes in Networks and Systems 653, https://doi.org/10.1007/978-981-99-0981-0_18

process with hundreds of processes [2]. So, the prediction of any detection becomes important and necessary work to keep the process running smoothly as well as maintain the high quality for output products. Typically, maintenance planning for fault prediction before an actual defect by creating a predicting model can help reduce downtime, improve work efficiency, productivity, as well as optimize maintenance costs and reduce failure to a minimum rate.

In this study, the problem of accurate detection of equipment fault states in the wafer fabrication of semiconductor manufacturing processes is investigated by using real data SECOM [3]. The dataset is intrinsically highly imbalanced in distribution. Therefore, we not only focus on developing a good predictive model of detecting failures that can foresee future occurrences and prevent but also propose the investigation about the efficiency of several approaches used for handling this imbalanced dataset. The remainder of the paper is structured as follows: Sect. 2 briefly reviews related studies. Proposed general framework for fault detection is presented in Sect. 3. Experiments and results are shown in Sect. 4 with the focus on application of several data balancing techniques for classification and prediction, including different metrics for evaluation. Conclusion is given in Sect. 5.

2 Related Works

2.1 Predictive Maintenance—Fault Detection

PdM is a new maintenance strategy. Growing interest in this research area has been started since the 2010s. The lack of study in this field has led to difficulty and complexity of applying effective PdM strategy in production [4]. ML has been proven as a great potential approach on predictive maintenance in recent years [5]. Due to interactive behaviors among industrial equipment, unexpected breakdown might result in terrible accidents and losses for manufacturers [6]. Three types of maintaining strategies and terminology are defined [7, 8]: (i) run-to-failure (R2F)—the simplest maintenance plan for repairing of the broken parts, (ii) preventive maintenance (PvM)—a time-based maintenance or scheduled maintenance which is carried out on a regular basis based on a predetermined schedule in time in order to predict breakdowns, and (iii) PdM—which is the main concern on this study.

2.2 Fault Detection in Semiconductor Industry

Semiconductor manufacturing is a multi-step process and one of the most sophisticated industries [9], in which several ML and deep learning (DL) approaches have been documented applied to defect detection. In terms of research in semiconductor manufacturing, K-nearest neighbor (KNN)-based approach applied in defect detection system was proposed in [10], where the squared closest neighbor distance threshold was calculated using a non-central chi-square distribution with a constrained normal zone. Susto et al. [11] demonstrated that supervised aggregative feature extraction (SAFE) outperforms other techniques in a time series dataset of semiconductor manufacturing by using the root mean square error (RMSE) for evaluation. In another research, evaluation of confusion matrix was first examined in [12] with generalized linear models, random forest (RF), gradient boosting (GB) and deep learning approach on a dataset collected from an equipment. Then Mourabit et al. [13] examined by researching four ML algorithms: Naive Bayes, C4.5 decision tree, multilayer perceptron (MLP), SVM on semiconductor data with several pre-processing techniques to evaluate the accuracy of algorithms on each data's variant. As a result, MLP was ranked the highest score (91.95% accuracy), and principal component analysis (PCA) was the optimal method in data pre-processing phase. Different techniques, data pre-processing process as well as several classifiers studied on SECOM were documented in [14].

Although there are many studies which researched fault detection in semiconductors by using ML approaches, few studies focused on techniques in terms of handling imbalanced data, which was demonstrated to significantly impact the performance of classifiers [15]. Moreover, there is a lack of research which compares the efficiency of different data balancing methods. In literature, different techniques are studied in novel and classified into two techniques: data-driven and algorithmic-based approaches. Within data-driven approaches, increasing minority class and decreasing majority class are two main ways. Random over sampling (ROS), adaptive synthetic sampling (ADASYN) [16], synthetic minority oversampling technique (SMOTE) [17], SMOTE extensions including Borderline-SMOTE [18], Safe-Level-SMOTE [19] are the case of the first way. With regard to the latter, random under sampling (RUS) [15] was studied. Methods for combining over sampling techniques with under sampling techniques were also mentioned in the literature: SMOTEENN [20], SMOTETomek [20]. Regarding algorithmic-based approaches, threshold method [21], one-class learning [22], and cost sensitive learning (CS) [23] were researched.

In this paper, the focus will be on the performance of several data balancing techniques in the following session on the SECOM dataset for fault prediction. Results of classification for five different ML models are then presented with different evaluation metrics following model recommendation.

3 Methodology

3.1 SECOM Dataset

SECOM was first presented in 2010 and contributed by Michael McCann for an investigation for selection of process related data taken from a semiconductor production line [3]. It was laid out as a table with 1567 instances and 591 attributes. There are two values in the label column: "0" and "1", "0" represented as "PASS", while "1" was "FAIL".

Observation time was also recorded corresponding to a selected functionality test. Due to a highly imbalanced dataset, missing values, and noisy features, SECOM has been considered a good example of semiconductor data as well as a sample of real production line data for studying purposes Therefore, it is a challenge for handling this kind of data in order to get good predicting results.

The overview about the dataset can be described as follows:

- Binary classification data
- Only 104 of 1567 instances were labeled as failure, approximately 1:14 proportion between FAIL/PASS.
- Not all attributes are useful for classification. SECOM contains features that present little to no data, which are noisy and do not affect the result.
- Nearly 7% missing data. These data points were either never recorded or were lost.

Overall, three problems and hence challenges were determined in SECOM data: missing values, unimportant features, and imbalanced classes distribution.

3.2 Proposed Framework for Classification

Figure 1 proposed a framework for the study. To begin with, the dataset needs to be cleaned by checking missing as well as unique data values, then others are filled by data imputation technique. The entire dataset is transformed to the range from 0 to 1 in the feature scaling phase. After the dataset is made to be cleaner, principal component analysis (PCA) is applied to identify most relevant and linked features with the outcome in classification. Approaches used for handling imbalanced data are then divided into two types as follows: (i) data-driven approaches, including: ROS, SMOTE, ADASYN, SMOTE + RUS, and (ii) algorithmic-based approach, including CS, then the data is split into train set 65% and test set 35%. Five ML algorithms, including logistic regression (LR), SVM, RF, MLP, and GB are applied. Finally, the results will be evaluated by several evaluation metrics: precision, recall, F1-score, and AUC.

1. Feature Engineering

The entire dataset is checked under the condition of 50% missing data. A column will be removed if it contains more than 50% missing values. The reason for this is that if a column contains an excessive number of missing values, it may not contribute to the predicting model. Moreover, if the column has the same value in all instances, it is also removed. Following these steps, 144 features are removed, leaving 448 features, while others are filled by KNN imputation, which replaces missing values with genuine ones that are both sensible and maximize the data point's expected value.

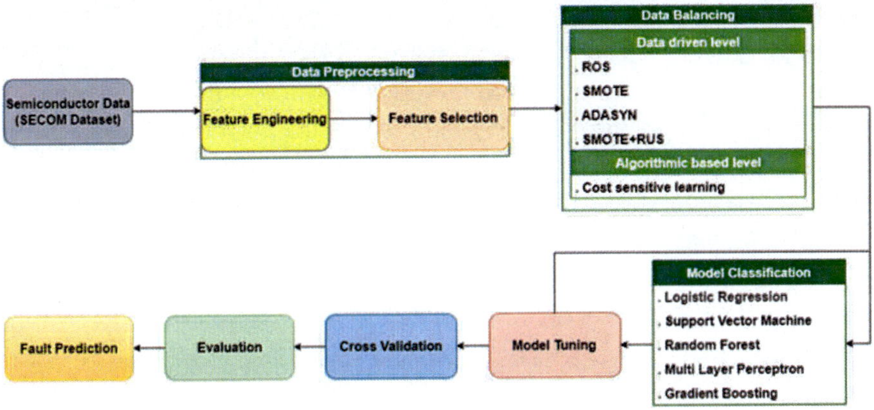

Fig. 1 Proposed of predictive model architecture

The correlation analysis is then used; it is a statistical measure that expresses how closely two variables are related linearly. Following the application of the approach, 272 features remain in the data. At that stage, feature scaling method is used to normalize the data into an appropriate range. The normalization procedure is defined as transforming continuous variables into values ranging from 0 to 1 or −1 to 1.

2. Features Selection

In order to keep only relevant data, this phase is required to consider which features are suitable and important for the classification. Unimportant and distracting features are removed. PCA, a dimensionality-reduction method, is applied. The basic idea is to reduce large datasets by transforming a set of variables into a smaller one that only contains the most important information [24]. There are 129 features that capture 95% of the variance.

3. Data Balancing

This phase is the focus of this research. Predictive models need data balancing methods that can solve the issue. For CS, the unbalanced dataset is used and no need to be balanced before.

4. Model Classification

This phase is carried out by dividing the data into two distinct sets: training and testing. The most effective model is then obtained by implementing a series of five ML algorithms based on different data balancing techniques.

5. Model Tuning

Parameter tuning is used to select the best parameters for each ML algorithm. To find the best parameters, the technique is tested in the form of loops inside each fold of the train set. The tuning's outcome is applied to the test set for result evaluation.

Table 1 Optimal parameters' value of each algorithm after tuning process

Classifier	Parameter	NB	ROS	SMOTE	ADASYN	SMOTE + RUS	CS
LR	**C**	1000	10,000	10,000	1000	1000	1
SVM	**C**	10	1	10,000	10,000	10,000	100
RF	**# of trees**	396	396	446	166	321	363
MLP	**# of neurons**	251	258	228	228	228	251
GB	**# of trees**	138	269	269	269	269	138

Bold represents parameters used for each classifier

The following is a summary of parameter values after the tuning process: (see Table 1).

Table 1 shows the optimal parameter values after tuning. The inverse of regularization strength is the key parameter of LR and SVM (C). The number of trees is the most important parameter in RF and GB, whereas the number of nodes in a hidden layer is the most influential parameter in MLP.

6. Cross-Validation

The models are validated using cross-validation. To perform cross-validation, the dataset is divided into ten sections. One portion is assumed as the test set, while the other nine are designated as training ones. This process is repeated until all of the sections are chosen as the test set. As a result, all of the cases in the dataset have been tested, and the data has been thoroughly examined. This study will use tenfold cross-validation.

7. Model Evaluation

There are several ways to evaluate a predictive model. In this study, five metrics precision, recall, F1-score, ROC-AUC, and false positive rate (FPR) are used. For an imbalanced dataset, accuracy often does not express the actual effect of the model [25], while precision, recall (TPR), F-measure, and AUC will be alternatively used.

Moreover, all elements of confusion matrix are defined as follows:

- True positive (TP): number of fail cases that are correctly identified as fail.
- True negative (TN): number of pass cases that are correctly identified as pass.
- False positive (FP): number of pass cases that are incorrectly identified as fail.
- False negative (FN): number of fail cases that are misclassified as pass.

The formula of each metric is presented as follows:

- Precision = TP/(TP + FP): ratio of true positives to all the positives predicted by the model.
- Recall (TPR) = TP/(TP + FN): ratio of true positives to all the positives in dataset.
- $F1$-Score = 2*(Precision * Recall)/(Precision + Recall): harmonic mean of precision and recall.
- FPR = 1 − TN/(TN + FP)
- Accuracy = (TP + TN)/(TP + TN + FP + FN)

4 Experiments and Results

4.1 Experimental Setup

The implementation procedure is divided into different scenarios with different data balancing approaches reflected in six scenarios: Scenario 1: Not Balanced (NB), Scenario 2: ROS, Scenario 3: SMOTE, Scenario 4: ADASYN, Scenario 5: SMOTE + RUS, Scenario 6: CS. Five different ML algorithms are then studied for each scenario, and five metrics are obtained.

4.2 Results

Table 2 shows the results of each ML algorithm based on different evaluation metrics. AUC is used as the primary metric of evaluation in this study since it is a fair indicator when categorizing an unbalanced dataset [14]. Recall, which measures how many errors were found, is another crucial performance indicator. The true negative ratio (TNR), on the other hand, shows the proportion of successes that were correctly foreseen and shows how few false alarms were generated by the classifier. Improved classification was reflected by higher AUC, recall, and TNR [14].

Data balancing results

The average and best performance with unbalanced data, balanced data consisting of single sampling methods, combination approaches, and CS are compared to evaluate the outcome of implementing data balancing strategies on the dataset.

According to Fig. 2, data balancing techniques outperform the results applied in unbalanced data. This is consistent with the study in the literature [14]. Although there is not much difference in average TNR value, the scores of the other three metrics are considerably higher than the NB scenario by increasing the AUC, TPR, and precision. Moreover, the results show the importance of balancing the data distribution compared to the performance on the unbalanced dataset through the large difference in TPR of the two ways. TNR, part of the trade-off between finding more defects and having fewer false alarms, is also noticeable. Moreover, the high TNR implies that this classifier gives lower false alarms in general. As a result, the models trained with the balanced dataset are able to detect more defects and provide higher performance with a change in the frequency of false alarms that is acceptable. While comparing the best results, will give insight into which combination of approaches can deliver the best model, comparing the average results helps the fairness of generality in comparison by making each model compete in the same combination of data balancing technique.

Table 2 Evaluation results of ML algorithms

Model	Metric	NB	ROS	SMOTE	ADASYN	SMOTE + RUS	CS
LR	Accuracy (%)	89.62	75.96	76.32	76.14	75.77	88.16
	Precision	0.18	0.16	0.14	0.15	0.14	0.19
	TPR	0.13	0.56	0.46	0.49	0.46	0.21
	F1	0.15	0.25	0.22	0.22	0.21	0.20
	AUC	0.54	0.67	0.62	0.64	0.62	0.57
	FPR	0.05	0.23	0.21	0.22	0.22	0.07
SVM	Accuracy (%)	91.26	93.08	92.53	92.53	92.53	77.41
	Precision	0.15	1.00	0.25	0.25	0.25	0.18
	TPR	0.05	0.03	0.03	0.03	0.03	0.59
	F1	0.08	0.05	0.05	0.05	0.05	0.27
	AUC	0.52	0.51	0.51	0.51	0.51	0.69
	FPR	0.02	0.00	0.01	0.01	0.01	0.21
RF	Accuracy (%)	92.90	92.90	90.53	90.89	90.89	90.89
	Precision	0.00	0.00	0.07	0.08	0.13	0.08
	TPR	0.00	0.00	0.03	0.03	0.05	0.03
	F1	0.00	0.00	0.04	0.04	0.07	0.04
	AUC	0.50	0.50	0.50	0.50	0.51	0.51
	FPR	0.00	0.00	0.03	0.02	0.03	0.02
MLP	Accuracy (%)	84.52	87.98	87.07	87.25	85.61	84.52
	Precision	0.13	0.17	0.18	0.20	0.17	0.129
	TPR	0.21	0.18	0.23	0.26	0.26	0.205
	F1	0.16	0.18	0.20	0.22	0.20	0.158
	AUC	0.55	0.56	0.58	0.59	0.58	0.550
	FPR	0.11	0.07	0.08	0.08	0.10	0.106
GB	Accuracy (%)	92.53	92.90	90.53	90.16	90.53	92.53
	Precision	0.33	0.00	0.12	0.06	0.12	0.333
	TPR	0.05	0.00	0.05	0.03	0.05	0.051
	F1	0.09	0.00	0.07	0.04	0.07	0.089
	AUC	0.52	0.50	0.51	0.50	0.51	0.520
	FPR	0.01	0.00	0.03	0.03	0.03	0.008

Classification results

As seen in Table 1, parameters of algorithms chosen by the tuning process are considered as the optimal parameters for implementation of the classification in this dataset. In order to discover which algorithm performs better, the detailed results on all algorithms are shown in Figs. 3, 4 and 5 by AUC, TPR, F1-score:

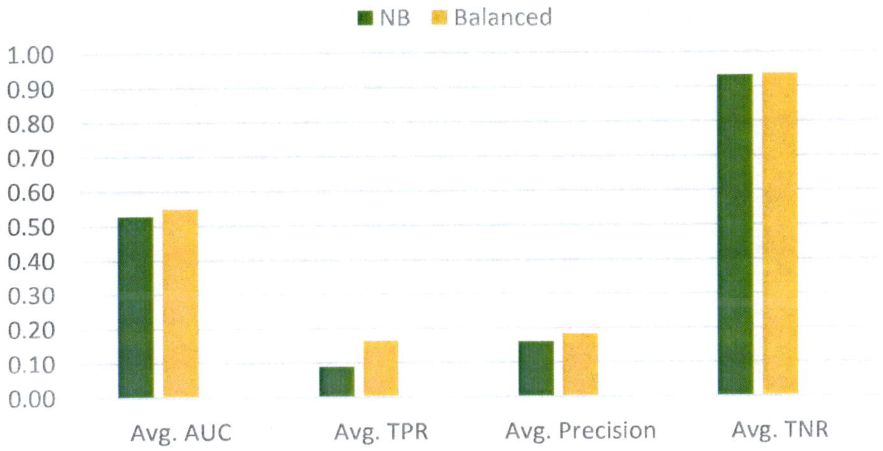

Fig. 2 Average evaluation results on different data balancing techniques

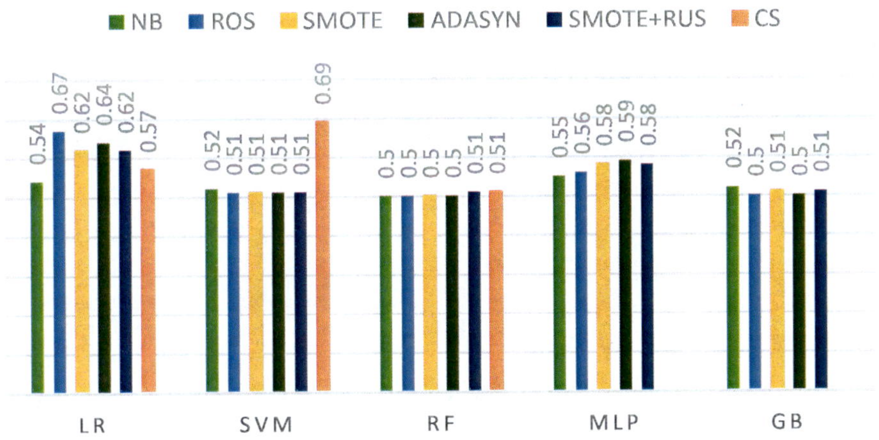

Fig. 3 AUC result of each classifier on data balancing techniques

Based on the results and comparing evaluation metrics in general from Figs. 3, 4 and 5, LR, SVM, MLP would be preferred models due to good evaluation performance. However, to find the best algorithm in the best scenario, SVM presents significant performance in scenarios of cost sensitive learning.

Table 3 presents the results in details of ML algorithms on each data balancing technique. According to Table 3, three final results of algorithms with highest AUC are presented. All of the values reflect the results of the optimal parameter by tuning process only for this dataset.

The table shows that training the SVM model on the SECOM data may produce optimal results in terms of AUC and recall with the scenario of CS. The number of

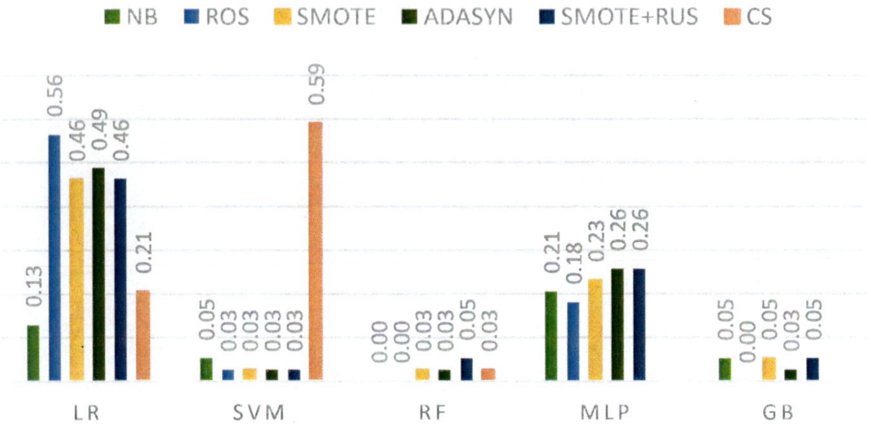

Fig. 4 TPR result of each classifier on data balancing techniques

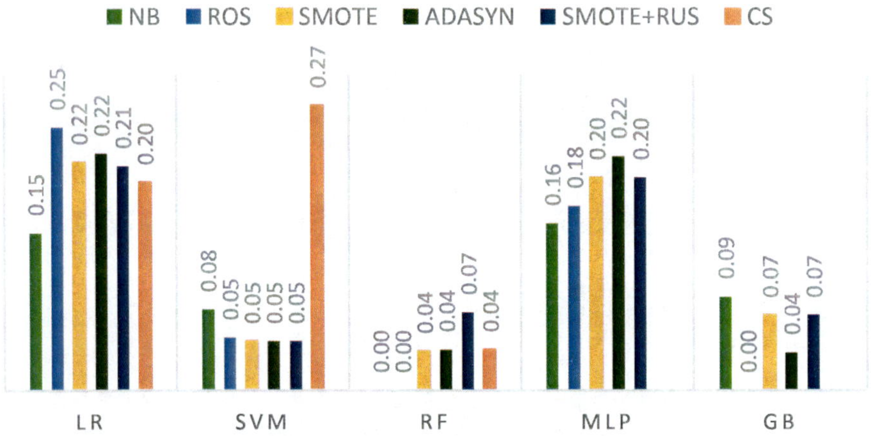

Fig. 5 F1 result of each classifier on data balancing techniques

true positive labels which SVM predicted is also the highest—23 labels of 39 labels in the test set. There is a huge difference between the most effective models and the worst by comparing AUC scores. Three models on the top of the table perform well and are considered as good classifiers: SVM algorithm with CS, LR algorithm trained on data balanced by ROS or ADASYN.

Table 3 Classification results with highest AUC

Classifier	Data balancing	True negative	False positive	False negative	True positive	Precision	Recall	AUC	TNR
SVM	CS	402	108	16	23	0.18	0.59	0.69	0.96
LR	ROS	395	115	17	22	0.16	0.56	0.67	0.96
LR	ADASYN	399	111	20	19	0.15	0.49	0.64	0.95
–	–	–	–	–	–	–	–	–	–
GB	ROS	510	0	39	0	0.00	0.00	0.50	0.93
RF	NB	510	0	39	0	0.00	0.00	0.50	0.93

5 Conclusion

In this study, an investigation of the application of ML for fault prediction on the real SECOM data was studied. The experiments are conducted using 30 combinations including different data balancing techniques using five ML algorithms. Our experimental results show that performance of predictive models using all the data balancing techniques that we studied including both the data-driven and algorithmic-based techniques outperformed models using unbalanced data. The best model providing the optimal performance with highest AUC is SVM combined with cost sensitive learning. The result of this finding could help the decision maker to choose the best model for predicting failures in semiconductor manufacturing with imbalanced data and hence provide tools for cost reduction in semiconductor manufacturing.

For future work, a deep comparative analysis will be discussed and examined for example using generative adversarial networks (GAN) [26]. Its variants will be investigated along with popular existing methods. Second, an analysis of efficiency changing of algorithms in different parameters values should be implemented in the future. Also, a larger, more complex, and various labels dataset should be used for implementation in order to study other relevant features that impact the classification result of the dataset.

References

1. Hung Y-H (2021) Improved ensemble-learning algorithm for predictive maintenance in the manufacturing process. Appl Sci 11(15):6832
2. Moyne J, Iskandar J (2017) Big data analytics for smart manufacturing: case studies in semiconductor manufacturing. Processes 5(3):39
3. McCann M, Li Y, Maguire L, Johnston A (2010) Causality challenge: benchmarking relevant signal components for effective monitoring and process control. In Causality: objectives and assessment. PMLR, pp 277–288
4. Hashemian HM (2010) State-of-the-art predictive maintenance techniques. IEEE Trans Instrum Meas 60(1):226–236
5. Carvalho TP, Soares FA, Vita R, Francisco RDP, Basto JP, Alcalá SG (2019) A systematic literature review of machine learning methods applied to predictive maintenance. Comput Ind Eng 137:106024
6. Henriquez P, Alonso JB, Ferrer MA, Travieso CM (2013) Review of automatic fault diagnosis systems using audio and vibration signals. IEEE Trans Syst Man Cybern Syst 44(5):642–652
7. Susto GA, Beghi A, De Luca C (2012) A predictive maintenance system for epitaxy processes based on filtering and prediction techniques. IEEE Trans Semicond Manuf 25(4):638–649
8. Susto GA, Schirru A, Pampuri S, McLoone S, Beghi A (2014) Machine learning for predictive maintenance: a multiple classifier approach. IEEE Trans Industr Inf 11(3):812–820
9. May GS, Spanos CJ (2006) Fundamentals of semiconductor manufacturing and process control. Wiley
10. He QP, Wang J (2007) Fault detection using the k-nearest neighbor rule for semiconductor manufacturing processes. IEEE Trans Semicond Manuf 20(4):345–354
11. Susto GA, Beghi A (2016) Dealing with time-series data in predictive maintenance problems. In: 2016 IEEE 21st international conference on emerging technologies and factory automation (ETFA). IEEE, pp 1–4

12. Butte S, Prashanth AR, Patil S (2018) Machine learning based predictive maintenance strategy: a super learning approach with deep neural networks. In: 2018 IEEE workshop on microelectronics and electron devices (WMED). IEEE, pp 1–5
13. El Mourabit Y, El Habouz Y, Zougagh H, Wadiai Y (2020) Predictive system of semiconductor failures based on machine learning approach. Int J Adv Comput Sci Appl (IJACSA) 11(12):199–203
14. Salem M, Taheri S, Yuan JS (2018) An experimental evaluation of fault diagnosis from imbalanced and incomplete data for smart semiconductor manufacturing. Big Data Cogn Comput 2(4):30
15. Kotsiantis S, Pintelas P (2003) Mixture of expert agents for handling imbalanced data sets. Annal Math Comput TeleInf 1(1):46–55
16. He H, Bai Y, Garcia EA, Li S (2008) ADASYN: adaptive synthetic sampling approach for imbalanced learning. In: 2008 IEEE international joint conference on neural networks (IEEE world congress on computational intelligence). IEEE, pp 1322–1328
17. Chawla NV, Bowyer KW, Hall LO, Kegelmeyer WP (2002) SMOTE: synthetic minority over-sampling technique. J Artif Intell Res 16:321–357
18. Han H, Wang W-Y, Mao B-H (2005) Borderline-SMOTE: a new over-sampling method in imbalanced data sets learning. In: Proceedings of international conferences on intelligent computing. Springer, Berlin, pp 878–887
19. Bunkhumpornpat C, Sinapiromsaran K, Lursinsap C (2009) Safe-level-smote: safe-level-synthetic minority over-sampling technique for handling the class imbalanced problem. In Proceedings of Pacific–Asia conferences knowledge. Discovery data mining. Springer, Berlin, pp 475–482
20. Batista GE, Prati RC, Monard MC (2004) A study of the behavior of several methods for balancing machine learning training data. ACM SIGKDD Explorations Newsl 6(1):20–29
21. Weiss GM (2004) Mining with rarity: a unifying framework. ACM SIGKDD Explorations Newsl 6(1):7–19
22. Japkowicz N (2001) Concept-learning in the presence of between-class and within-class imbalances. In: Conference of the Canadian society for computational studies of intelligence. Springer, Berlin, pp 67–77
23. Domingos P (1999) Metacost: a general method for making classifiers cost-sensitive. In: Proceedings of the 5th ACM SIGKDD international conference on knowledge discovery and data mining, pp 155–164
24. Jolliffe IT, Cadima J (2016) Principal component analysis: a review and recent developments. Philos Trans Roy Soc A Math Phys Eng Sci 374(2065):20150202
25. Hossin M, Sulaiman MN (2015) A review on evaluation metrics for data classification evaluations. Int J Data Mining Knowl Manage Process 5(2):1
26. Tanaka FHKDS, Aranha C (2019) Data augmentation using GANs. arXiv preprint arXiv:1904.09135

Simulation Tools for Cloud Computing: A Comparative Study

Talha Umar, Mohammad Nadeem, and Mohammad Sajid

Abstract Computing is evolving toward a paradigm that consists of commoditized services supplied in a way similar to conventional utilities. A variety of computing paradigms have been proposed to realize this goal, the most recent of which is known as cloud computing. It is capable of performing complex and massive scale computing, which has changed roles and responsibilities in IT services. It relieves the consumers from the burden of maintaining massive computing hardware, applications, and data storage systems. As many people have moved to cloud computing because of its elastic resource allocations and pay-for-per-usage features, and also because the cloud environment is very complex and sensitive, it is essential to study, analyze, and evaluate the performance, security, and other features that might be confronted in cloud computing. It is expensive to set up and perform these tests in actual cloud systems. However, modeling and simulation tools are often viable alternatives to cloud computing environment emulation. There are numerous simulation programs available, and they all have different features and applications. As a contribution to knowledge, in this paper, we first present a brief introduction to cloud computing, describe its fundamentals, and present a detailed comparative study of some popular cloud computing simulation tools.

Keywords Cloud computing · Simulation tools · CloudSim · iCanCloud

1 Introduction

In the last decades, many computing paradigms have been developed to provide computing services as a utility. Cloud computing is a developing computing model that promises to simultaneously deliver on-demand computing resources over the Internet to many consumers. It has progressed at a remarkable pace in the past ten years. Cloud computing grew in popularity as a result of its ability to shift the IT industry's physiognomies through the use of virtualization. In contrast to the cloud

T. Umar (✉) · M. Nadeem · M. Sajid
Department of Computer Science, Aligarh Muslim University, Aligarh, India
e-mail: talhaumar.amu@gmail.com

S. Das et al. (eds.), *Advances in Data-driven Computing and Intelligent Systems*, Lecture Notes in Networks and Systems 653, https://doi.org/10.1007/978-981-99-0981-0_19

desires framework listed in Table 1, it has been adopted successfully by businesses since it satisfies their needs [1]. Services delivered in cloud computing managed by providers, and it requires users to have only a computer system and Internet to access those services. Data centers, service providers, applications, data catalogs, metadata, and information models are all components of a cloud computing system. Cloud computing services, unlike traditional computing models, are pay-as-you-go and can be scaled up or down based on your needs. A wide range of deployment and cloud service models have been made available to meet the needs of customers. Cloud environment is a very large and complex system as it is composed of several components such as data centers, service providers, applications, data catalog, metadata, and information model. All cloud-based systems require unique configurations; hence, cloud computing researchers must undertake actual tests. Cloud-based experimentation is often expensive, both in terms of setup and execution. To model cloud settings and complete the required testing, cloud simulation tools are an enticing alternative.

As there is a vast number of simulation tools present in the market which are different in terms of their features and variants, users need to choose a suitable simulator for specific research to their needs and utility. To fully comprehend the situation, it is important to examine available cloud computing simulation tools in one place. As a result, we concentrated our efforts to provide a deeper understanding of cloud computing, and its simulation tools. We present a detailed study on some popular cloud simulators such as CloudSim, GreenCloud, iCanCloud, and others. Here, we focus on their execution time, platform portability, simulation type, and many more and present a comparative analysis among them.

The remainder of this paper is structured as follows. Section 2 gives an overview on cloud computing in which we describe cloud computing definition and its key features which contribute to its successful adoption in industries. Section 3 presents a detailed

Table 1 Key desires of cloud computing paradigm

Cloud desires	Description
Security	The desire to receive a technological service which is secure as that which may be obtained from a locally run server
Availability	The desire to use services as reliable as it is available from a locally installed server
Latency	The desire to receive a technical service with a latency that is at least comparable to that of a locally running server
Variety	The desire to obtain services with a level of complexity (variety) that is appropriate for the operating environment
Abstraction	The urge to abstract and hide non-essential complexity, particularly the complexity of managing the underlying IT infrastructure and applications
Scalable	A desire for a scalable service that can satisfy demand
Efficiency	A desire for a service that assists users in becoming more cost-effective
Simplicity	The desire to make available services simple to understand and use

overview on CloudSim, GreenCloud, iCanCloud, secCloudSim, CloudSched, and SCORE. Section 4 concludes the chapter.

2 Cloud Computing

Cloud computing has been evolved as a utility entity that derives every business task in today's world. In 1961, John McCarthy suggested the idea of computing utility. He said that computing may be used as a utility, as now we are doing with water, electricity, and gas [2]. That vision of computing as a utility predicts a transformation in the computing paradigm where services can be delivered on demand anywhere. With the origin of the Internet, the vision of computing utility became clear, many computing paradigms like distributed computing, grid computing, and utility computing have emerged to achieve the vision of computing as a utility. Cloud computing is a lately emerging technology in this direction [3].

Cloud computing is an outcome of improvements made in the area of distributed computing, Internet, and hardware virtualization concepts to make IT services elastically available through the Internet anywhere[4]. The cloud computing paradigm creates an infrastructure that allows computing resources such as computing facilities, networks, applications, and storage to be accessed over the Internet from anywhere in a pay-as-you-go model. These services can be provisioned and released without any human interaction.

Cloud computing can be defined as a computing paradigm which offers an easily accessible resource pool over the Internet where the user just needs a computer system and Internet to access those services at a very convenient cost. The services are provisioned on the basis of an established service-level agreement (SLA) between a cloud service provider and consumers [5]. Cloud computing provides four types of deployment models as per user's requirements. These are Private Cloud, Public Cloud, Community Cloud, and Hybrid Cloud presented in Fig. 1.

2.1 Key Features of Cloud Computing

Shared Resource Pooling. Cloud computing infrastructure offers a resource pool from which computing resources can be provisioned to multiple users dynamically at a time, which maximizes the resource utilization and also minimizes the cost.

Geo-Distribution and Ubiquitous Network Access. Cloud services are available and accessible through internetworking. Any device with Internet access from anywhere in the world can use cloud services at any time. Nowadays, a large number of companies (cloud providers) are establishing many data centers across the globe to achieve high performance and localization [6].

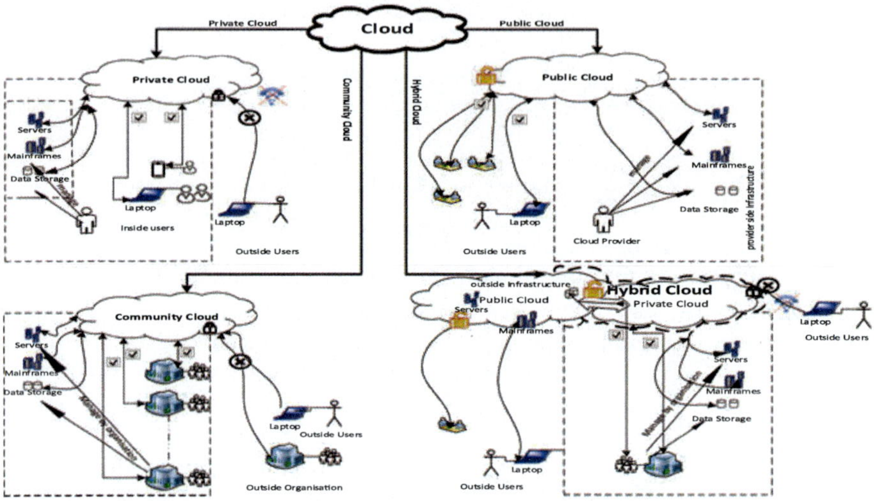

Fig. 1 Schematic representation of cloud infrastructures based on deployment models

Service-Oriented. Cloud computing architecture adopts a service-driven model which places strong importance on service delivery. In a cloud system, the service provider has to provide the services and resources according to the established service-legal agreement (SLA) between the consumer and the provider.

Dynamic/Elastic Resource Provisioned. Cloud computing services are so elastic; cloud infrastructure offers unlimited resources where users can use any number of resources. In a cloud environment, users can also add or release resources anytime as per their requirements.

Self-Organization. In the cloud computing paradigm, services are available on-demand where users can use any service provided by a cloud provider whenever they require it. Cloud users want on-demand and instant access to resources. This paradigm allows users to use any delivered service based on their demand without the intervention of the service provider.

Utility-Based Pricing. Cloud infrastructure allows a pay-per-use manner of billing. There is no need for up-front commitment by users, and it offers users to use a required number of services on the basis of the user's need. Service charges will be based on the usage of resources [7].

3 Cloud Simulation Tools

3.1 CloudSim

CloudSim, one of the most widely used cloud simulators, was created by Cloud Laboratory at the University of Melbourne in Australia. It was created in Java and based on the GridSim platform. It is an open-source simulation tool that allows for continuous cloud computing and application provisioning modeling, simulation, and testing [8]. CloudSim is an event-driven simulation tool because its components retain a message queue and produce messages, which they then transmit to other entities [9].

CloudSim simulator is not a ready-to-use solution. It necessitates the creation of a Java program that uses its components to create the desired scenario and gather the data for cloud application performance and security analysis [10].

The simulation tools in CloudSim use a layered approach. Figure 2. The CloudSim core simulation engine, which is responsible for queueing and communication among components, is located at the simulator's lowest layer. The middle layer contains all of the sublayers that contain main components such as sensors, data centers, and so on. User code is at the top of the stack, and it provides configuration parameters for hosts, cloudlets, virtual machines (VMs), user count, and broker scheduling methods [11].

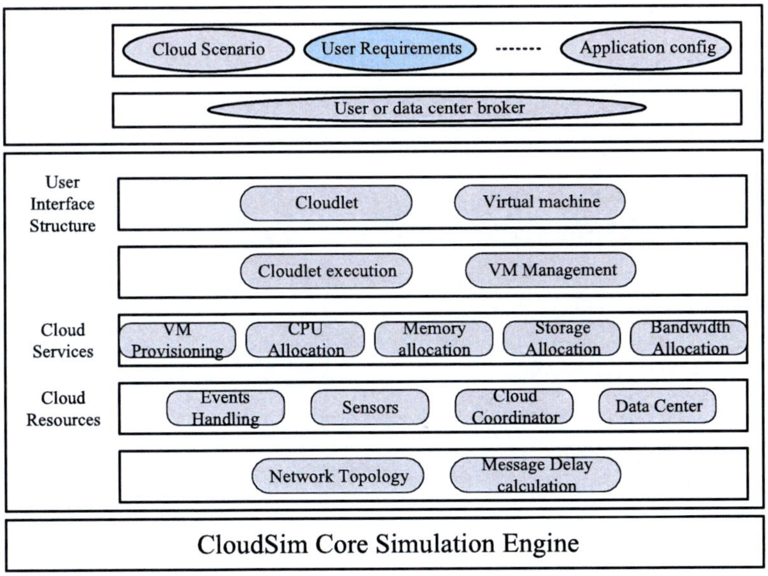

Fig. 2 Architecture of CloudSim

3.2 GreenCloud

GreenCloud simulator is a packet-level simulator which was developed as an extended version of the networking simulation tool NS-2. Similar to the NS2 simulator, it is also written in C++ with OTcl libraries. This simulator was created to conduct advanced energy-conscious investigations of cloud computing data centers in a realistic environment [12].

In actual circumstances, GreenCloud measures the amount of energy utilized by data center components (switches, servers, and networks), as well as packet-level communication patterns. It can also look at how the network's workload is spread [13].

The architecture of the GreenCloud simulator shown in Fig. 3 consists of three layers. First layer is the network layer which represents the core network of the data center. Second layer is the aggregation layer which represents the aggregation network. Third layer is the access layer which connects to the servers.

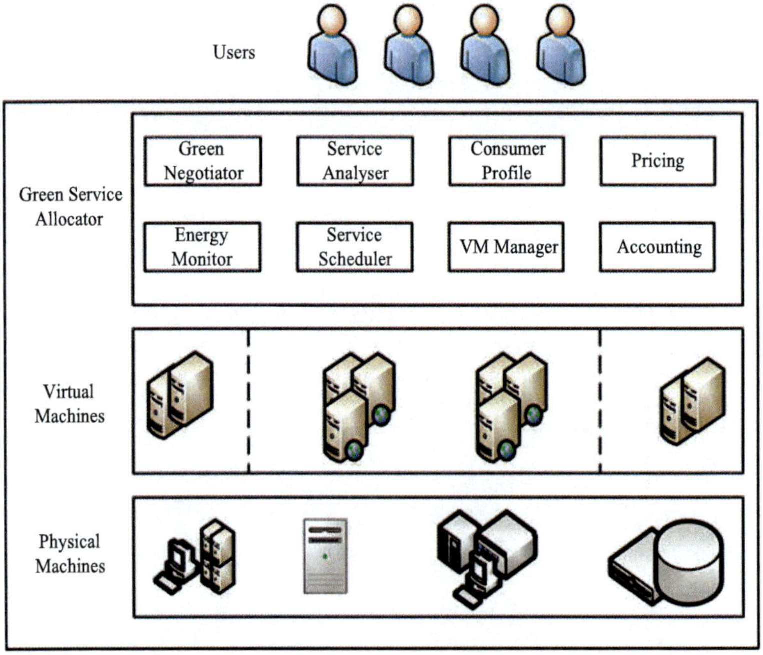

Fig. 3 Architecture of GreenCloud

3.3 ICanCloud

iCanCloud was created in response to the shortcomings of previous simulators such as CloudSim and GreenCloud. It is based on OMNET platform. iCanCloud was designed to conduct large experiments so it was written in C++ language. It is designed to simulate Amazon Elastic Compute Cloud (EC2) cloud resources, while its developers believe that it may be modified to simulate other environments using the given API [14].

The key benefit of iCanCloud is that it provides a full hypervisor for various brokering schemes. It facilitates the generation and deployment of a flexible hypervisor module that makes it simple to integrate existing and new cloud brokering policies [15]. It can forecast the cost-performance trade-offs of applications running on simulated hardware.

Bottom of the architecture is the hardware model layer. This layer contains the models responsible for simulating the hardware components of a system. Upper layer consists of VMs repository. A user-defined set of virtual machines (VMs) is stored in those VMs repositories. Similarly, application repository contains a collection of predefined applications in the layer above VMs repository layer. Upper layer is the hypervisor layer that controls the execution of all tasks and virtual machines (VMs). The cloud system module is the topmost layer. This module contains a definition of the whole cloud system, which comprises mostly the hypervisor and virtual machine that makes up the system [11, 14]. The architecture of iCanCloud is depicted in Fig. 4.

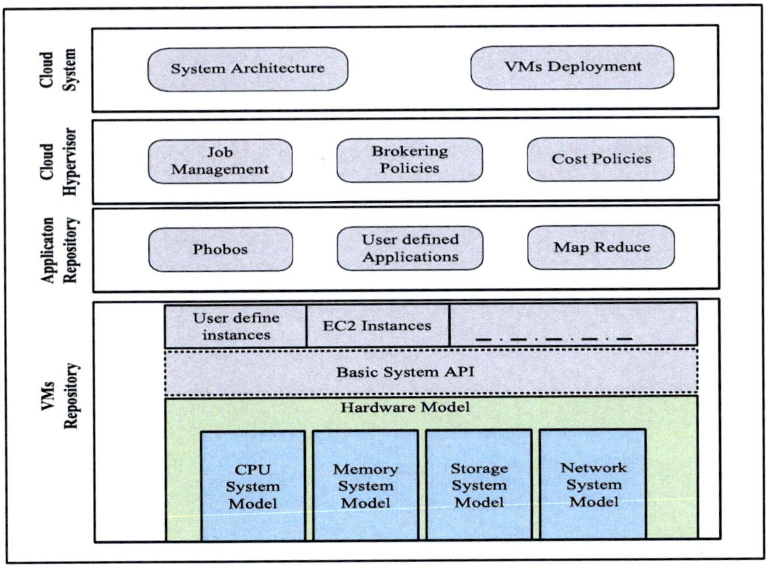

Fig. 4 Architecture of iCanCloud

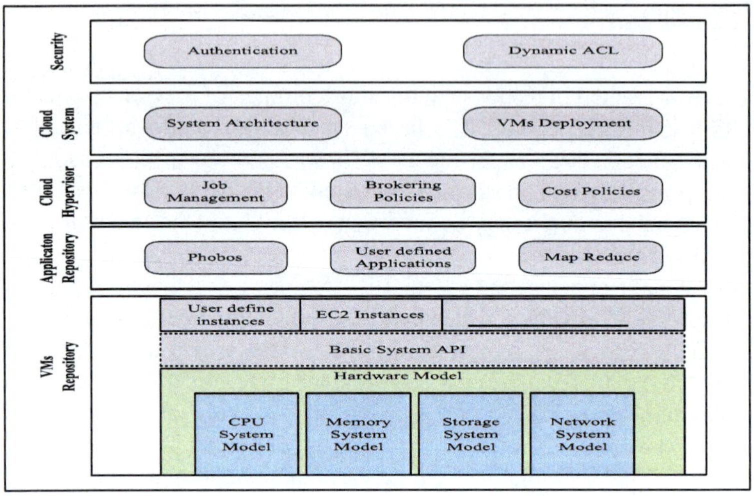

Fig. 5 Architecture of secCloudSim

3.4 *SecCloudSim*

secCloudSim simulator has been developed as an extension of iCanCloud simulator which provides basic security features such as authentication and authorization.

In SecCloudSim, security features are coded in C language [16]. Users can create security features including encryption, decryption, encapsulation, authentication, and privacy guarantee using secCloudSim.

secCloudSim architecture Fig. 5 is the same as iCanCloud simulator with one new security layer. New security layer contains two components: an authentication module and a dynamic access control list module. Authentication module is responsible to authenticate users with Challenge-Handshake Authentication Protocol (CHAP). The dynamic access control list module stores a list of users, as well as a list of cloud services such as virtual machines, network systems, storage, and processing power, and also their access privileges. The authorization sets of privileges for each user can be customized based on their needs.

3.5 *CloudSched*

The CloudSched simulation program was created to model and simulate the cloud environment, with an emphasis on comparing various IaaS scheduling techniques in terms of servers and workload. It is a lightweight simulator which is written in the Java programming language [17].

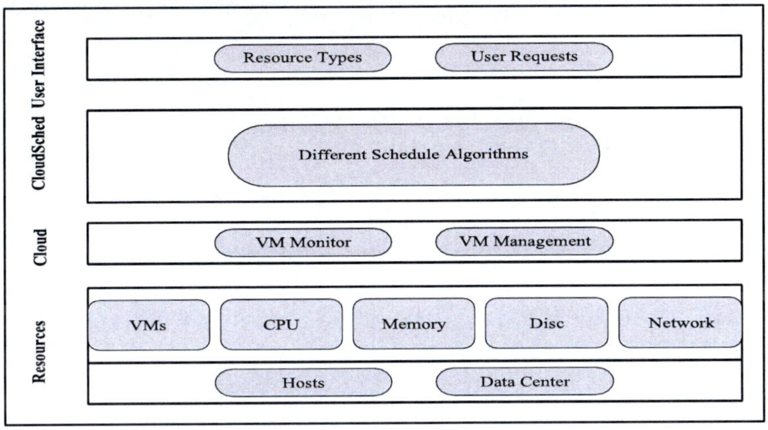

Fig. 6 Architecture of CloudSched

Users can use the CloudSched simulator to identify and explore the best solution for a variety of resource scheduling policies and algorithms [15]. It is in charge of providing a policy for modeling and estimating the implementation of various IaaS-based arrangement techniques.

CloudSched simulator architecture consists of three main layers as shown in Fig. 6. First layer is the user interface which allows users to choose resources and send requests as there are a few types of VMs that are preconfigured for users to choose. Second layer is the core layer of scheduling which is responsible for choosing the appropriate data center and physical machine as per user's request. The lowest layer is cloud resources which contain physical resources and VMs.

3.6 SCORE

SCORE is based on the Google Omega lightweight simulator, which has been improved with the addition of resource utilization, discrete-event, and multi-agent scheduling models. It is a high-performance cloud simulation tool which focuses on data center energy efficiency [18].

SCORE is an easily extensible tool that includes multiple techniques for various scheduling models, allocation, and shut-down and power-on procedures that are already established and ready to use. SCORE is a trustworthy tool to analyze and simulate security, energy efficiency, and scheduling techniques in cloud environments [11].

The architecture of SCORE simulator Fig. 7 is composed of two modules named core simulator and the efficiency module. Core simulator model is inherited from omega lightweight simulator, and its architecture has three layers. The first layer is

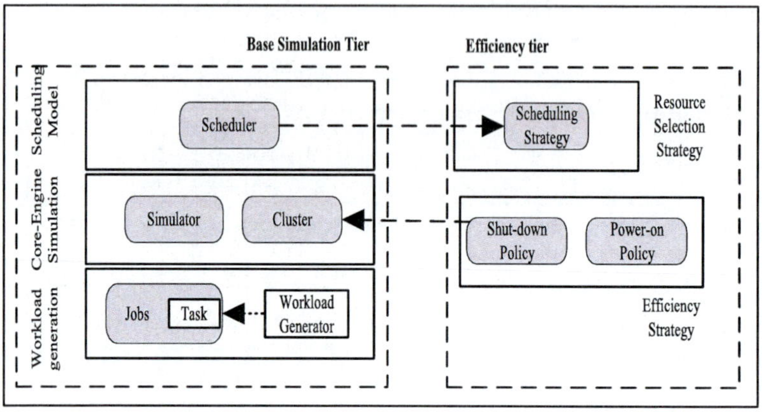

Fig. 7 Architecture of SCORE

the workload generation layer, which is in charge of creating cloud computing workloads. The second layer is made up of core engine simulations, which are responsible for workload characteristics, and the third layer is made up of scheduling models, which are in charge of implementing various scheduling methods. The efficient module's second component is made up of two efficiency layers that are responsible for executing different energy-saving and resource-selection strategies.

In contrast to other simulation tools, CloudSim offers a migration policy. Other simulation tools lack such a strategy. GreenCloud simulation required a longer simulation period and a considerable quantity of memory. secCloudSim's security layer is the only distinction between iCanCloud and secCloudSim, and secCloudSim does not release open-source code. SCORE simulation tools are the most recent simulators developed using the Scala programming language. This simulator is incompatible with communication modeling. Table 2 compares cloud simulators based on numerous factors, including platform, programming language, memory space, graphics, parallel experiment, simulation time, and migration approach, as all simulation programs differ in terms of their features and versions [9–11, 15, 19].

4 Conclusion

A variety of computing paradigms have emerged in response to the idea of computing as a utility; in this journey, cloud computer is the most recently created computing paradigm. Cloud computing emerged as a multi-service provider which shares information, software, and computing resources within an Internet-based environment. This new paradigm has a wide range of effects on diverse areas which are related to information technology (IT). Cloud computing systems are very complex and large, so researchers need to evaluate and analyze the different models somewhere.

Table 2 Comparison between CloudSim, GreenCloud, iCanCloud, secCloudSim, CloudSched, and SCORE cloud simulation tool

Parameters	CloudSim	GreenCloud	iCanCloud	secCloudSim	CloudSched	SCORE
Publication year	2009	2010	2011	2014	2015	2018
Platform	GridSim	NS2	OMNET	iCanCloud	Any	Google Omega
Programming language	Java	C++/OTCL	C++, MPI, OMNeT	C++	Java	Scala
Memory space	Small	Large	Medium	NA	Small	NA
Platform portability	Yes	No	Yes	Yes	Yes	Yes
Graphics	No	No	Yes	No	No	No
Parallel experiment	No	No	Yes	Yes	No	Yes
Energy consumption	Yes	Yes	No	NA	Yes	Yes
Congestion control	No	No	No	No	Yes	No
Simulation time	Seconds	Tens of minutes	Seconds	Seconds	Seconds	Seconds
Modeling of public cloud providers	No	No	Yes (Amazon EC2)	No	Yes (Amazon EC2)	NA
Availability	Open source	Open source	Open source	Not Available	Open source	Open source
Migration policy	Yes	No	No	No	No	No
Communication modeling support	Limited	Yes (full)	Yes (full)	Yes (full)	Limited	Not support
Energy modeling support	Yes	Yes (servers and network)	No	No	Yes (servers and network)	Yes
Simulator type	Event-based	Packet level	Event-based	Event-based	Event-based	NA

It is a very difficult task to analyze the performance of various cloud models in the real cloud environment. So, cloud computing simulation tools are very useful for evaluating the system behavior of various cloud systems under different scenarios. There is a wide range of cloud simulations available but every simulator is different in its features and utility. This paper explains the foundations of cloud computing and conducts a thorough examination of some popular cloud simulators. CloudSim, GreenCloud, iCanCloud, secCloudSim, CloudSched, and SCORE are some of the most popular cloud simulators these days. This study delves into all of these simulators, as well as their structures. This chapter also includes a comparative study that gives a thorough overview of various simulators. As we examined and adjusted all of the objectives, we discovered that no single simulator is ideal. Users must select a suitable simulator that meets their requirements.

References

1. Venters W, Whitley EA (2012) A critical review of cloud computing: researching desires and realities. J Inf Technol 27:179–197. https://doi.org/10.1057/jit.2012.17
2. Gourisaria MK, Samanta A, Saha A, Patra SS, Khilar PM (2020) An extensive review on cloud computing. Presented at the (2020). https://doi.org/10.1007/978-981-15-1097-7_6
3. Buyya R, Yeo CS, Venugopal S, Broberg J, Brandic I (2009) Cloud computing and emerging IT platforms: vision, hype, and reality for delivering computing as the 5th utility. Futur Gener Comput Syst 25:599–616. https://doi.org/10.1016/j.future.2008.12.001
4. Buyya R, Broberg J, Goscinski A (2011) Cloud computing: principles and paradigms. In: Cloud computing: principles and paradigms. https://doi.org/10.1002/9780470940105
5. Buyya R, Yeo CS, Venugopal S (2008) Market-oriented cloud computing: vision, hype, and reality for delivering IT services as computing utilities. In: 2008 10th IEEE international conference on high performance computing and communications. IEEE, pp 5–13. https://doi.org/10.1109/HPCC.2008.172
6. Mahmoud MS, Xia Y (2019) Networked control systems: cloud control and secure control. In: Networked control systems
7. Sajid M, Raza Z (2013) Cloud computing: issues AND challenges. In: International conference on cloud, big data and trust, pp 13–15
8. Goyal T, Singh A, Agrawal A (2012) Cloudsim: simulator for cloud computing infrastructure and modeling. Procedia Eng 38:3566–3572. https://doi.org/10.1016/j.proeng.2012.06.412
9. Fakhfakh F, Kacem HH, Kacem AH (2017) Simulation tools for cloud computing: a survey and comparative study. In: 2017 IEEE/ACIS 16th international conference on computer and information science (ICIS). IEEE, pp 221–226. https://doi.org/10.1109/ICIS.2017.7959997
10. Sinha U, Shekhar M (2015) Comparison of various cloud simulation tools available in cloud computing. IJARCCE 171–176. https://doi.org/10.17148/IJARCCE.2015.4342
11. Mansouri N, Ghafari R, Zade BMH (2020) Cloud computing simulators: a comprehensive review. Simul Model Pract Theory 104:102144. https://doi.org/10.1016/j.simpat.2020.102144
12. Kliazovich D, Bouvry P, Audzevich Y, Khan SU (2010) GreenCloud: a packet-level simulator of energy-aware cloud computing data centers. In: 2010 IEEE global telecommunications conference GLOBECOM 2010. IEEE, pp 1–5. https://doi.org/10.1109/GLOCOM.2010.5683561
13. Ahmad MO, Khan RZ (2018) Load balancing tools and techniques in cloud computing: a systematic review. Presented at the (2018). https://doi.org/10.1007/978-981-10-3773-3_18
14. Núñez A, Vázquez-Poletti JL, Caminero AC, Castañé GG, Carretero J, Llorente IM (2012) iCanCloud: a flexible and scalable cloud infrastructure simulator. J Grid Comput 10:185–209. https://doi.org/10.1007/s10723-012-9208-5
15. Khalil KM (2017) Cloud simulators–an evaluation study. Int J Inf Model Anal
16. Rehman UU, Ali A, Anwar Z (2014) secCloudSim: secure cloud simulator. In: 2014 12th international conference on frontiers of information technology. IEEE, pp 208–213. https://doi.org/10.1109/FIT.2014.47
17. Tian W, Zhao Y, Xu M, Zhong Y, Sun X (2015) A toolkit for modeling and simulation of real-time virtual machine allocation in a cloud data center. IEEE Trans Autom Sci Eng 12:153–161. https://doi.org/10.1109/TASE.2013.2266338

18. Fernández-Cerero D, Fernández-Montes A, Jakóbik A, Kołodziej J, Toro M (2018) SCORE: simulator for cloud optimization of resources and energy consumption. Simul Model Pract Theory 82:160–173. https://doi.org/10.1016/j.simpat.2018.01.004
19. Shishira SR, Kandasamy A, Chandrasekaran K (2018) Comparative study of simulation tools and challenging issues in cloud computing. Presented at the (2018). https://doi.org/10.1007/978-981-10-7635-0_1

Identification of Ischemic Stroke Origin Using Machine Learning Techniques

Rajan Prasad and Praveen Kumar Shukla

Abstract An ischemic stroke arises as a result of atherosclerosis. It promotes tissue and cholesterol accumulation in the vascular system. When a large amount of plaque collects in one region, it has the potential to reduce blood supply to crucial organs. Carotid arteries are blood vessels that carry blood to the brain and are found in the neck. A stroke may develop if plaque plugs a carotid artery (carotid artery disease). Atherosclerosis is mostly asymptomatic. As a result, many people are unaware they have atherosclerosis until they have a stroke or suffer the additional repercussions of clogged arteries near other organs, such as a heart attack. The primary goal of this project is to identify cases of ischemic stroke by utilizing blood clot imaging datasets with CNN and CatBoost classifiers. The experimental results show that the CNN achieved 97.5% accuracy, whereas the CatBoost achieved 73.5% accuracy. This study will aid healthcare providers in properly detecting the sources of blood clots in fatal strokes, allowing doctors to deliver the best post-stroke treatment and minimize the risk of a second stroke.

Keywords Ischemic stroke · Deep learning · Classification · CNN · CatBoost

1 Introduction

An ischemic stroke is still the world's second biggest cause of mortality. Over 15 million people in the world report an ischemic stroke every year as a result of a blood clot obstructing an artery to the brain [1]. A second stroke (23% of total occurrences are recurrent) reduces the patient's odds of survival. However, if physicians can pinpoint stroke aetiology, which determines therapeutic therapy following stroke episodes, additional strokes may be minimized. For most of the previous decade, fibrinolysis therapy was the standard treatment for acute stroke caused by a major arterial blockage. As a consequence, collected clots may be investigated [2].

R. Prasad (✉) · P. K. Shukla
Artificial Intelligence Research Center, Department of Computer Science and Engineering, School of Engineering, Babu Banarasi Das University, Lucknow, India
e-mail: rajan18781@gmail.com

S. Das et al. (eds.), *Advances in Data-driven Computing and Intelligent Systems*, Lecture Notes in Networks and Systems 653, https://doi.org/10.1007/978-981-99-0981-0_20

Healthcare professionals are currently attempting to apply machine learning techniques to predict ischemic stroke aetiology and clot origin. To decrease the chances of subsequent strokes, the Mayo Clinic Neurovascular Research Laboratory (MCNRL) [3] encourages scientists to develop an artificial intelligence-based aetiology classification so that physicians are better equipped to prescribe the correct treatment [4]. AI-based approaches for the classification of blood clots could help save the lives of stroke survivors and help us better understand the world's second-leading cause of death. Figure 1 shows the sample dataset collected from the Striped-AI blood clot dataset [4] accessed in June 2022. Convolutional neural networks (CNNs) are artificial neural networks [5] that are designed to describe comprehensive input image datasets. CNNs are commonly used in visual applications such as image segmentation and categorization. They have also achieved success in a variety of disciplines such as agriculture, health care, microbiology, all disciplines of engineering [6], and many more. It has the ability to identify picture patterns such as lines, gradients, and circles. CNN design principles are based on feed-forward neural networks [7] with 25–30 layers. The power of CNN is derived from a sort of layer known as a convolutional layer. The use of convolutional layers in convolutional neural networks mimics the human visual brain, in which a series of layers evaluate an input image and acknowledge increasingly complex aspects [8].

Gradient boosting [9] is another excellent machine learning technique that delivers ground-breaking outcomes in a variety of complex problems. It has long been the preferred approach for learning problems with noisy data, complicated relationships, and heterogeneous characteristics, like online search, recommender systems, and weather forecasting [10]. It is supported by strong hypothetical implications that illustrate how powerful predictors can be developed by repeatedly integrating lesser prototypes using a greedy technique related to gradient descent in a function space. Because gradient boosting only accepts numeric inputs, categorical characteristics must be modified. The one-hot encoding (OHE) [11] technique is widely used for converting categorical to numeric data, but it has a few drawbacks. CatBoost [12] employs a novel approach to converting categorical to numerical data. To begin with,

Fig. 1 Sample dataset

the data is randomly sorted. Then, for each category variable, new numerical features are generated by taking the mean of the goal but only using historical observations to eliminate bias. CatBoost has the potential to outperform XGBoost [13]. The main disadvantage is that it requires more time to run. As basis predictors, decision trees are commonly used in gradient boosting solutions. Although decision trees can be useful for numerical variables, many datasets contain categorical characteristics that are essential for prediction. Converting categorical characteristics to numbers before training is the most generally utilized approach for dealing with them in gradient boosting.

This chapter is structured as follows: Sect. 2 presents literature reviews; Sect. 3 covers the system descriptions and datasets; Sect. 4 includes the blood clot classification models; Sect. 5 includes the findings and discussion; and Sect. 6 presents the conclusion and future research directions.

2 Literature Review

Sultan et al. [14] suggested a brain tumour classification model based on a multi-layer CNN that uses two publicly accessible datasets to classify three kinds of brain tumours. First, the deep learning model defines tumours as glioma, meningioma, and pituitary tumours, whereas the second model classifies the three glioma grades as 96.14%, 98.6%, and 96.14%, respectively.

Ismael et al. [15] suggested a model to classify brain tumours. The proposed model used an MRI dataset to classify glioma, meningioma, and pituitary. The author used the Gabor filter to transform the domain and construct feature characteristics. A back-propagation (BP) algorithm was used to train the model and achieve 91.9% accuracy.

Baddran et al. [16] presented a brain tumour classification model using MRI. As a confirmation phase, a CNN is used to classify tumours as benign or malignant, and the tumour region is identified using ROI methods. To create an image binary, segmentation was performed at the level two steps of pre-processing using canny edge detection. The Harris approach is then used to extract distinguishing traits. Two neural networks are trained using these characteristics. The first NN is to classify brains as having no tumours or having tumours that are benign or malignant. The neural model is trained using the Levenberg–Marquardt approach. The model achieved 75%.

Chen et al. [17] investigated the ischemic stroke using various machine learning algorithms, including the CatBoost model. The CatBoost model has better accuracy as compared to random forest, AdaBoost, and XGBoost. A study proposed by Ayyanar [18] investigated a cardiovascular dataset by linking risk factors in order to provide various analytical insights that reflect the risk population prone to heart disorders. The study used multiple boosting algorithms to predict cardiac illness, including light gradient, AdaBoost, XGBoost, and CatBoost.

Fuzzy systems are also utilized for the identification of ischemic stroke [19–21]. Interpretability is an important feature of fuzzy systems [22–24], which should be taken into account while implementing the prediction systems. It would help in interpreting the results or predicting the outcomes of the designed system.

3 System Description

The proposed system is divided into different blocks. From the input block to the classification block, followed by the performance evaluation block, each block has a specific task. Both experiments follow the same system architecture. Figure 2 shows the block diagram of the proposed system.

The classification task begins with image extraction from the dataset. In this process, the images are extracted from the dataset. On the other hand, the corresponding label is fetched from the csv file of the dataset. After that, image labels are renamed from the respective class (CE or LAA), and the dataset is divided into training and testing parts. The architecture of sequential CNN model starts with the setting of hyper-parameters. Furthermore, the model is trained using a training dataset. In this process, a supervised learning approach has to be followed. After that, apply the testing dataset to test the model. Finally, the performance evaluation matrices are used to compute accuracy, recall, and precision.

3.1 Dataset Descriptions and Experimental Setup

The Mayo Hospital Blood Clots Dataset contains 756 high-resolution TIF images [3]. Figure 3a depicts a blood clot from cardioembolic atherosclerosis (CE), while Fig. 3b depicts a blood clot from large artery atherosclerosis (LAA). The following experiments were carried out with an Intel Core processor (TM) i5-1035G1 CPU @ 1.99 GHz 8, 16 GB RAM, and GeForce GT 730 GPU, as well as the required Ubuntu 16.04 LTS and Python 3.8 packages.

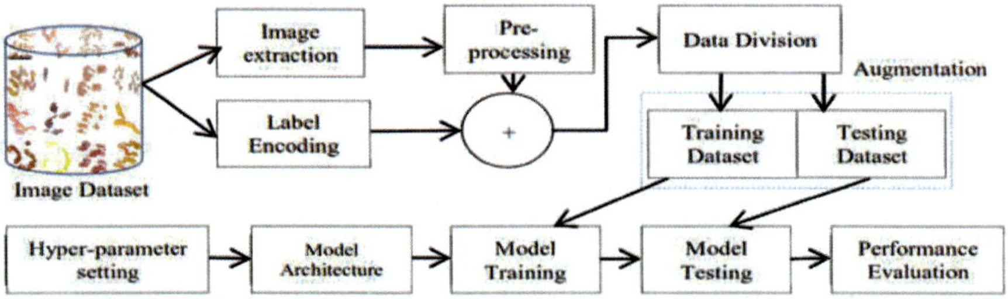

Fig. 2 Blood clot classification model

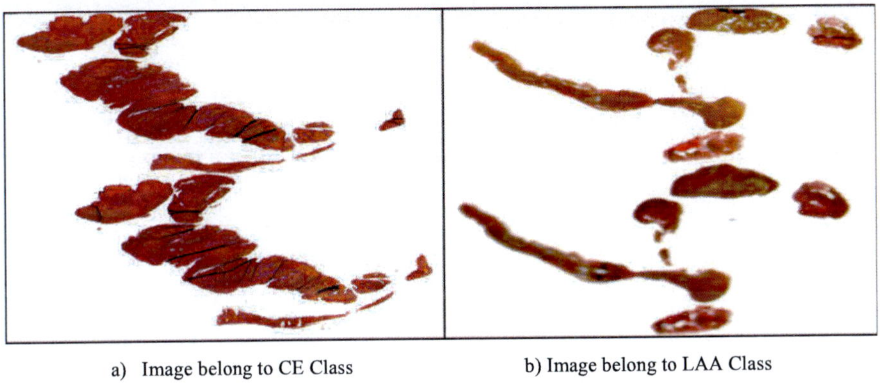

Fig. 3 Sample data of CE class

3.2 *Data Pre-Processing*

TIF images come in a variety of sizes and dimensions. However, they must first be down-sampled from TIF images to.jpg format with a set size of 200 × 200 pixels. Actually, this approach reduces the complexity of calculations while preserving picture information, which is subsequently tagged and processed.

Figure 4 shows the histogram of the sample data selected for the experiments. It shows that the 512 images belong to the CE class and 206 images belong to the LAA class.

Data augmentation techniques are used to avoid over-fitting problems. In this strategy, the training dataset is modified in various ways, such as horizontal and vertical flips and random cropping, without modifying the class label. The data augmentation aids in the improvement of model performance.

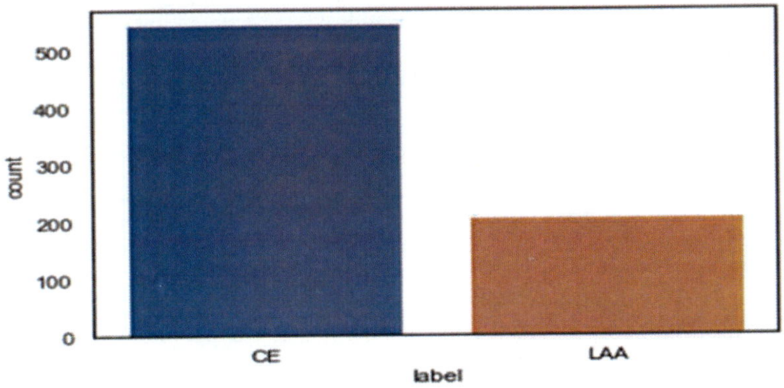

Fig. 4 Histogram of class label data

4 Identification of Ischemic Stroke Origin

In this section, the authors implemented the two prominent classification models along with the different values of hyper-parameters. The first experiment is based on the CNN model, and the second experiment is performed with the CatBoost classifier as given below.

4.1 Convolution Neural Network

CNN is a popular neural model used for object classification and identification in particular. CNN is particularly adept at extracting complex features for categorization. CNN is made up of neurons that can learn weight and bias. Each neuron gets input; the weight total is then applied to the activation function. A CNN extracts important features with specified dimensions using consecutive convolution layers and a nonlinear ReLU function. The maxpool layer is used to shrink the size of the feature map. Every neuron in the fully connected layer is associated with every other neuron in the dense layer that comes before it. Gradient descent and back-propagation algorithms are employed to train the network. To keep all output values between 0 and 1, the Softmax function applies a posterior probability. Based on hidden layer depth, CNN extracts information that lets neural networks learn minor image attributes. Figure 5 depicts the architecture of the proposed model.

CNN has three levels: pooling, convolution, and fully connected layers. CNN architecture is produced when these layers are laid out. Figure 6 displays a simplified CNN architecture for classifying blood clots.

The key functions of the proposed CNN model can be classified as:

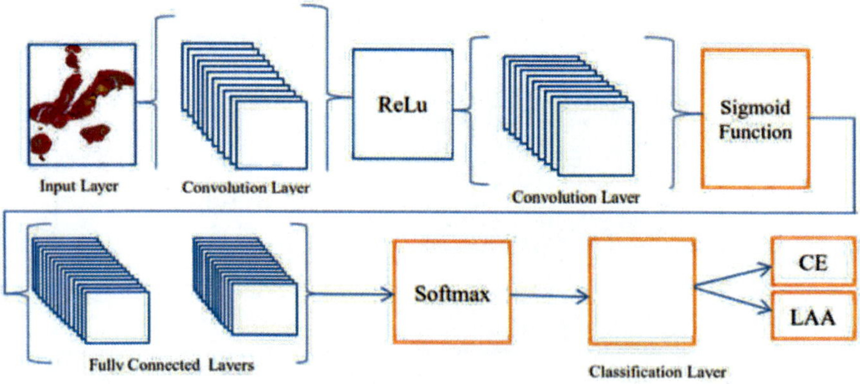

Fig. 5 Proposed CNN architecture

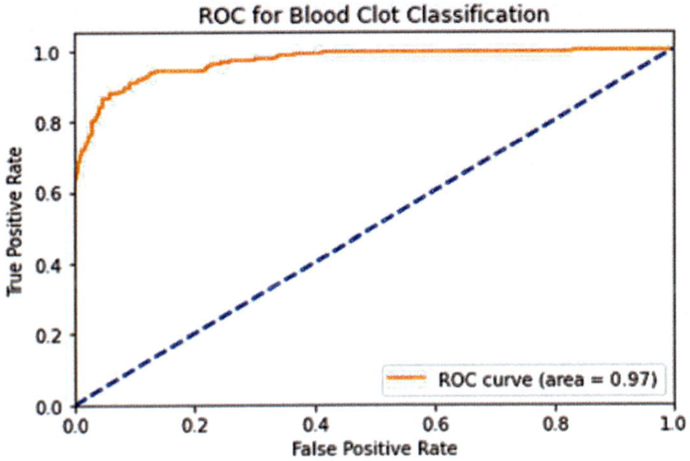

Fig. 6 ROC of blood clot classification using CNN model

1. The task of the input layer is similar to that in other types of ANN; it is used to hold the pixel value of the image input dataset.
2. The dot product of the individual weights and the region corresponding to the input volume will be computed by the convolution layer.
3. ReLU attempts to apply an element-wise activation function, and a sigmoid is used in the activation output from the previous layer.
4. The pooling layer is used to sample down the spatial dimensions of the supplied input to lower the number of parameters inside that activation.
5. The dense layers will function similarly to typical ANNs, trying to calculate the final scores for classification from the activation. The ReLU layer could be used between these levels to improve performance.

CNN may use convolution processes to modify the original input in each layer to find class scores for classification problems. Individual layer tasks are defined as:

CNNs rely heavily on convolutional layers. The layer configuration is centred on the use of trainable kernels and is typically low in spatial dimensions yet spans throughout the entire strength of the input. Each filter in a convolutional layer is convolved across the spatial dimensions of the input to form a 2D activation map as data enters it. As we traverse through the inputs, the dot product for every element in that kernel is calculated. The network learns kernels that "fire" when a specific feature is detected in a given input bounded area. It is frequently referred to as "activations".

Each kernel will be assigned a distinct activation map. It will be overlaid along the depth dimension to reflect the convolutional layer's whole output level. Training on such images yields models that are too complex to learn because of the fully connected nature of the traditional ANN neurons. To compensate, neurons in a convolutional layer are only coupled to a small portion of the input volume. The depth amplitude of

Table 1 Parameter setting of the CNN model

Name	Type	Activations	Learnable parameters
Input layer	Input image	200 × 200 × 3	–
Convolution layer-1	Convolution	200 × 200 × 32	Weight 3 × 3 × 3 × 32 Bias 1 × 1 × 32
ReLU	ReLU	200 × 200 × 32	–
Convolution layer-2	Convolution	200 × 200 × 28	Weight 3 × 3 × 28 Bias 1 × 1 × 28
Sigmoid	Sigmoid	200 × 200 × 28	–
Dense layer-1	Dense	1 × 1 × 200	Weight 200 × 1,120,000 Bias 200 × 1
Dense layer-2	Dense	1 × 1 × 2	Weight 2 × 200 Bias 2 × 1
Softmax	Softmax	1 × 1 x 2	–
Classification layer	Classification output	0	0

connections has always been equivalent to the depth of the input, and the dimensionality of this area is generally referred to as the neuron's receptive size of the field. The suggested CNN input image dimensions of 200 × 200 × 3 (for RGB image) and set the receptive field size as 32 × 32, resulting in a total of 896 weights for each neuron inside convolution layer-1. To put this into context, a conventional neuron observed in other kinds of ANN would include 8092 weights in the second convolution layer. There are three hyper-parameters (depth (9), stride (1-1), and zero-padding (1-1-1-1)) that are adjusted to appropriate values to reduce the model's complexity. Table 1 shows the setup summary of the hyper-parameters. The depth of the convolution layers' output volume can be manually adjusted by adjusting the number of neurons inside the layer in the same area as the input. Every neuron in the hidden layer is already connected to every other neuron. Reduced depth reduces the overall number of neurons in the network, but it also reduces model accuracy.

CatBoost

The CatBoost classifier is another machine learning technique that is effective at predicting categorical features. It is a gradient boosting method that uses binary decision trees (BDT) as base predictors. Yandex recommended (2017). Gradient boosting is an effective machine learning strategy for dealing with complex problems that involve heterogeneous features, inaccurate data, and complex dependencies. The classification of blood clots, parameter setting, and reducing the losses as discussed below.

Suppose the input dataset $D = \{(X_j, y_j)\}._{j=1,2,\ldots,m}$, where $X_j = x_j^1, x_j^2, \ldots, x_j^n$ represent a vector of n features and output feature vector $y_j \in R$ which can be modelled as binary class (CE/LAA) and prearranged with numerical features (1 and 0). Samples (X_j, y_j) are independently, and the learning task's goal is to train a function $A : R^n \rightarrow R$ using identically distributed based on a certain random distribution $p(.,.,.,)$, which minimizes the estimated loss shown in Eq. (2).

$$L(A) := \mathrm{EL}(y, A(X)), \tag{2}$$

where $L(., ., .)$ denote function loss and (X, y) denotes the test dataset.

The gradient boosting approach [24] iteratively generates a sequence of approximations $A^t : R^m \rightarrow R, t = 0, 1, \ldots$ based on greedy style. By using additive process, the value of A^t can be obtained from Eq. (2).

$$A^t = A^{t-1} + \alpha g^t \tag{3}$$

Here, α denotes the step size and $g^t : R^n \rightarrow R$ is the function of the base predictor.

The variables listed above are chosen from a collection of functions G in order to minimize the predicted loss, as described by Eq. (5).

$$g^t = \arg \arg \min_{g \in G} L\left(A^{t-1} + g\right) \tag{4}$$

$$g^t = \arg \arg \min_{g \in G} \mathrm{EL}\left(y, A^{t-1}(X) + g(X)\right) \tag{5}$$

For the experiment purpose, the hyper-parameters of the CatBoost model are indexed in Table 2. Depth of the decision tree is set to be 6, the value of step size is 0.01 describing the best learning rate and the model executed up to 100 iterations provides the better results. The outcome of the proposed model is explained in the next section.

Table 2 Hyper-parameters of CatBoost model

Parameter name	Value
Depth	6
Learning rate	0.01
No. of iteration	100
Loss function	Binary class

5 Results and Discussion

The proposed study builds two classifiers: the CNN classifier and the CatBoost classifier. The dataset contains 754 images, and both classifiers were trained utilizing 80% (567) of the dataset and tested with the remaining 20% (207). It is ensured that both experiments follow the same division of the dataset for the results analysis. For the performance analysis, the authors used the confusion matrix and ROC to measure the accuracy, precision, and recall values of the proposed models as given below:

5.1 *Confusion Matrix*

As per the experimental results of both classifiers, the confusion matrix is shown in Table 3, which summarizes the classifier's performance. The X-axis and Y-axis represent the expected and true labels, respectively. It shows the precision, accuracy, and recall values of the classifiers computed by given Eqs. (6), (7), and (8). As results obtained by the confusion matrices clearly show, the accuracy of the CNN model is better than the CatBoost model.

$$\text{Precision} = \frac{\text{TP}}{\text{TP} + \text{FP}} \tag{6}$$

$$\text{Recall} = \frac{\text{TP}}{\text{TP} + \text{FN}} \tag{7}$$

$$\text{Accuracy} = \frac{\text{TP} + \text{TN}}{\text{TP} + \text{TN} + \text{FP} + \text{FN}} \tag{8}$$

Here, TP denotes the total number of positive predicted cases that actually belong to Positive. TN indicates negative predictions that actually belong to the negative category. FP indicates positive predicted, whereas they actually belong to negative, and FN denotes negative predicted cases where they actually belong to positive.

Table 3 Confusion matrix

Method	Actual status	Predicted status			
		Training		Testing	
		CE	LAA	CE	LAA
CNN	CE	526	19	15	5
	LAA	100	43	4	30
CatBoost	CE	440	30	16	5
	LAA	120	67	3	22

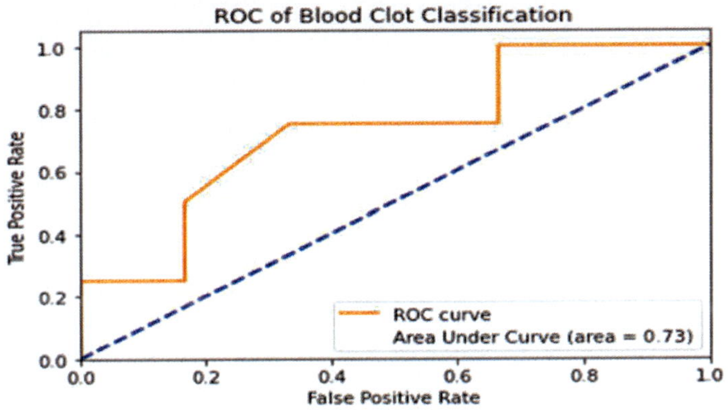

Fig. 7 ROC of blood clot classification using CatBoost classifier

Table 4 Performance of CNN and CatBoost classifier

Model	Accuracy	Recall value	precision
CNN	0.9741246	0.75	0.85
Cat boost	0.731576	0.77	0.87
ANN	0.94325	0.87	0.79
SVM	0.863452	0.74	0.81

5.2 Receiver Operating Characteristic (ROC)

The receiver operating characteristic (ROC) curve demonstrates the model's capacity to differentiate across classes. The ROC area under the curve is useful in determining how well a model performs in classification. Figure 6 shows the curve of the proposed CNN model. It covers 0.97% of the plot. It was interpreted as meaning the proposed CNN model achieved 97% accuracy. In contrast to Fig. 7, shown the ROC of the CatBoost classifier, it shows the cover area of the plot is 0.73% of the whole and accuracy can be determined as 73%. The area under the curve score (ROC-AUC Score) for each class in comparison with both classes is calculated using pairwise analysis. It is very clear that the ROC-AUC for CNN is found to be higher than that of CatBoost. CNN has a higher AUC value than CatBoost classifier.

Furthermore, artificial neural networks and support vector machines are used for testing. The results are shown in Table 4.

6 Conclusion

Accurate blood clot classification lowers the risk of a second ischemic stroke and allows the neurologist to better treat the patient. On the other hand, it will minimize the pathological test cost. The experimental results obtained from the two classifiers in Table 4 showed that the accuracy of the CNN model is better than CatBoost for the classification of blood clots using TIF pictures. CNN has the greatest recall, precision, and accuracy values. The accuracy of the classifiers across three parameters shows that CNN outperforms the CatBoost classifier by a wide margin. As a result, CNN is the best alternative for the most precise and reliable classification due to much-improved performance across several measures such as ROC, precision, and accuracy recall value. In future studies, appropriate feature choices and optimization techniques can be used to improve the accuracy of the CNN and CatBoost classifiers. Blood clot classification is also possible with enough datasets. Or other hybrid classifiers can be employed for more accurate analysis. This experimental investigation will assist neurologists in determining the exact origin of blood clots in order to prevent subsequent ischemic stroke and improve the treatment of this dangerous condition.

References

1. Feigin VL, Norrving B, Mensah GA (2017) Global burden of stroke. Cir Res 120(3):439–448
2. Van der Worp HB, van Gijn J (2007) Acute ischemic stroke. New Engl J Med 357(6):572–579
3. https://www.mayo.edu/research/labs/neurovascular-research/overview
4. https://www.mayoclinic.org/diseases-conditions/stroke/symptoms-causes/syc-20350113
5. Wang S-C (2003) Artificial neural network. In: Interdisciplinary computing in java programming. Springer, Boston, pp 81–100
6. Zhang Y, Sung WJ, Mavris DN (2018) Application of convolutional neural network to predict airfoil lift coefficient. In: 2018 AIAA/ASCE/AHS/ASC structures, structural dynamics, and materials conference, p 1903
7. Bebis G, Georgiopoulos M (1994) Feed-forward neural networks. IEEE Potentials 13(4):27–31
8. Wan J, Wang D, Hoi SC, Wu P, Zhu J, Zhang Y, Li J (2014) Deep learning for content-based image retrieval: a comprehensive study. In: Proceedings of the 22nd ACM international conference on multimedia, pp 157–166
9. Friedman JH (2002) Stochastic gradient boosting. Comput Stat Data Anal 38(4):367–378
10. Gneiting T, Raftery AE (2005) Weather forecasting with ensemble methods. Science 310(5746):248–249
11. Potdar K, Pardawala TS, Pai CD (2017) A comparative study of categorical variable encoding techniques for neural network classifiers. Int J Comput Appl 175(4):7–9
12. Dorogush AV, Ershov V, Gulin A (2018) CatBoost: gradient boosting with categorical features support. arXiv preprint arXiv:1810.11363
13. Chen T, He T, Benesty M, Khotilovich V, Tang Y, Cho H, Chen K (2015) Xgboost: extreme gradient boosting. R package version 0.4–2 1(4):1–4
14. Sultan HH, Salem NM, Al-Atabany W (2019) Multi-classification of brain tumor images using deep neural network. IEEE Access 7:69215–69225
15. Ismael MR, Abdel-Qader I (2018) Brain tumor classification via statistical features and back-propagation neural network. In: 2018 IEEE international conference on electro/information technology (EIT). IEEE, pp 0252–0257
16. Badran EF, Mahmoud EG, Hamdy N (2010) An algorithm for detecting brain tumors in MRI images. In: The 2010 international conference on computer engineering and systems. IEEE, pp 368–373

17. Chen SD, You J, Yang XM et al (2022) Machine learning is an effective method to predict the 90-day prognosis of patients with transient ischemic attack and minor stroke. BMC Med Res Methodol 22:195. https://doi.org/10.1186/s12874-022-01672-z
18. Ayyanar M, Jeganathan S, Parthasarathy S, Jayaraman V, Lakshminarayanan AR (2022) Predicting the cardiac diseases using select best method equipped light gradient boosting machine. In: 2022 6th International conference on trends in electronics and informatics (ICOEI)
19. Shukla PK, Tripathi SP (2012) A review on the interpretability-accuracy trade-off in evolutionary multi-objective fuzzy systems (EMOFS). Information 3(3):256–277
20. Shukla PK, Tripathi SP (2011) A survey on interpretability-accuracy (IA) trade-off in evolutionary fuzzy systems. In: 2011 5th international conference on genetic and evolutionary computing. IEEE, pp 97–101
21. Shukla PK, Tripathi SP (2014) A new approach for tuning interval type-2 fuzzy knowledge bases using genetic algorithms. J Uncertainty Anal Appl 2(1):1–15
22. Shukla PK, Tripathi SP (2012) Interpretability issues in evolutionary multi-objective fuzzy knowledge base systems. In: Proceedings of 7th international conference on bio-inspired computing: theories and applications (BIC-TA 2012). Springer, India, pp 473–484
23. Shukla PK, Tripathi SP (2012) On the design of interpretable evolutionary fuzzy systems (I-EFS) with improved accuracy. In: 2012 international conference on computing sciences. IEEE, pp 11–14
24. Prasad R, Shukla PK (2022) A review on the hybridization of fuzzy systems and machine learning techniques. In: Bansal JC, Engelbrecht A, Shukla PK (eds) Computer vision and robotics. Algorithms for intelligent systems. Springer, Singapore. https://doi.org/10.1007/978-981-16-8225-4_32

Comparative Study of Pre-trained Language Models for Text Classification in Smart Agriculture Domain

Sargam Yadav and Abhishek Kaushik

Abstract Disruptive technologies (DTs) have infiltrated and revolutionized several fields. The data collected from Internet of Things (IoT) devices can be leveraged for performance review, analysis, and predictive modelling. The agricultural industry has benefited greatly from the use of DTs and information and communication technologies (ICTs). However, the eventual success of an emerging technology is greatly dependent on general knowledge and perceived benefits of its use. This study aims to analyse user perception towards smart farming technologies by performing sentiment analysis on data collected from YouTube using 3 transformer models: generative pre-trained transformer 2 (GPT-2), bidirectional encoder representations from transformers (BERT) for agricultural domain, and DistilBERT. Analysis of the results and statistical testing implies that transformers are a viable option for classifying technical and agricultural text, with the domain-specific pre-trained agriculture BERT outperforming others based on macro-F1-score and accuracy.

Keywords Smart farming · Bidirectional encoder representations from transformers · Sentiment analysis · Transformers · Multiclass text classification · Natural language processing

1 Introduction

Smart farming (SF), or smart agriculture, refers to the practice of integrating ICTs as well as DTs into farming practices. Although there have been several successful implementations of SF projects all over the world, a critical challenge towards widespread adoption remains in the form of lack of technical knowledge about such

S. Yadav (✉)
Dublin Business School, Dublin, D02 WC04, Ireland
e-mail: sargam.yadav@edhec.com

A. Kaushik
Dundalk Institute of Technology, Dundalk A91 K584, Ireland
e-mail: Abhishek.Kaushik@dkit.ie

S. Das et al. (eds.), *Advances in Data-driven Computing and Intelligent Systems*, Lecture Notes in Networks and Systems 653, https://doi.org/10.1007/978-981-99-0981-0_21

topics [1]. Monitoring user sentiment towards a product, service, or a topic is a challenging task. It is important that there is a level of general public awareness towards major technological advancements, particularly if the project could be potentially life-changing in terms of ensuring food sustainability [2]. Agriculture is an industry that is essential for human survival, and innovation is constantly desired in the field. There have been several videos posted explaining or demonstrating innovative technologies in smart agriculture, and user interaction on these videos is a vital source for mining opinion towards these tools. YouTube is the one of the largest media sharing platforms, with over 2.6 billion users monthly worldwide, and it is host is to nearly every type of content [3]. Sentiment analysis is a natural language processing (NLP) technique which determines opinion or emotion in text [4, 5] and can provide a crucial functionality to government bodies and other concerned parties that want to monitor how the public is responding to new projects being undertaken. Manual analysis of user sentiment is not feasible for data collected from such large platforms. For a more automated approach, NLP models and toolkits have shown tremendous potential in correctly identifying sentiment from text. Since its inception, the BERT model [6] has consistently set new state-of-the-art results at various NLP tasks such as question answering [7], language inference, and text classification [4].

In this study, we implement 3 transformer models, DistilBERT, GPT-2, and agriculture BERT, on data collected through channels about smart farming. This work serves as an extension of the experiment done by Yadav et al. [8], where a preliminary sentiment analysis on the same dataset using the pattern Python library[1] was performed. The classification results are compared based on several metrics such as accuracy, F1-score, precision, and recall, and an appropriately trained model is selected. Friedman statistical testing is also performed to determine if the results obtained by the experiment are statistically significant. The rest of the article is structured as follows: Sect. 2 discusses the motivation behind conducting this study and the hypothesis to be explored. Section 3 discusses the relevant literature for SF technologies and state-of-the-art language models, Sect. 4 explains the methodology employed in the experiment, and Sect. 5 analyses the results of our experiments along with determining statistical significance. The discussion and conclusion that follow the study are covered in Sect. 6 and Sect. 7, respectively.

2 Motivation

Smart farming has proven to be an invaluable tool to help farmers and the agricultural industry prepare for the uncertainty arising from climate change, decrease in the water supply and availability of arable land, and market fluctuations in the farming sector [9]. However, the concerned authorities need to consider the opinion of citizens before committing to a project, as it requires large-scale infrastructural changes and could have an extreme long-term impact on the area. To address this challenge, we conduct

[1] https://github.com/clips/pattern.

sentiment analysis on a dataset that consists of comments collected from YouTube from 16 videos explaining different smart farming practices. Three transformer-based models have been selected for this study, namely DistilBERT, GPT-2, and agriculture BERT, which has been specifically trained on data from the agricultural domain. Transformers have provided excellent results on various benchmark NLP tasks [4, 5]. This study aims to analyse the performance of the 3 models and draw insights from it.

The hypothesis that is explored throughout the study is as follows:

Hypothesis: *Transformer models can be used to classify technical text relevant to the agricultural domain and smart farming technologies.*

The following research questions have been crafted for further exploration of the hypothesis:

1. What is the performance of transformers in classifying data containing user opinion on smart farming?
2. Does a BERT-based model pre-trained on agricultural text perform better classification on the smart farming dataset than other transformer models?
3. What are the appropriate evaluation metrics used for quantifying model performance on imbalanced datasets?
4. Are the results obtained from the experiment statistically significant?

3 Literature Review

In this section, we discuss the current state-of-the-art technologies that are being used to automate and improve farming practices, and an overview of the previous work that has been done with transformers.

3.1 Smart Farming

In recent years, there has been a conscious shift towards the utilization of ICTs and DTs in the agricultural sector [8, 9]. This shift has been mainly due to the availability of low-cost sensors of small sizes [10], popularity of predictive and analytical tools such as big data and machine learning [11], and the need for sustainable approaches to agricultural practices to ease the strain on the global food supply from the drastically increasing world population [12]. Sensors that can measure various external factors such as moisture and temperature can be installed on unmanned aerial vehicles (UAVs), unmanned ground vehicles (UGVs), or around the agricultural land. UAVs can also be utilized to provide medical assistance and drug delivery [13], and continuous research is being conducted to improve navigation and determine the shortest navigation path [14, 15]. Wireless sensors networks (WSNs) and Internet of Things (IoT) can allow for the collection, storage, and analysis of the data collected from sensors [16]. The European Union has undertaken several smart agriculture projects in the past decade that employ technologies like cloud computing (CC), image processing, big data, etc., to perform useful functions such as crop monitoring, harvesting, remote imaging, and more [1, 8, 17]. Other countries, such India and Israel, are also funding projects that implement and refine technology for improving farming practices [18].

3.2 Pre-trained Language Models (PLMs)

The BERT model is a PLM introduce by Google AI language, and it is trained on a large English corpus [6]. The transformer architecture was introduced by Vaswani et al. [19] and is capable of handling long-range dependencies in text due to the self-attention mechanism. BERT was pre-trained with masked language modelling (MLM) and next sentence prediction (NSP) objectives and can be fine-tuned for any downstream task by implementing an additional output layer without undergoing any modifications in its architecture. BERT can also be pre-trained on a domain-specific corpus such as FinBERT for finance domain [20], BioBERT for biomedical corpora [21], and patentBERT for patent classification [22]. DistilBERT is a distilled version of BERT that preserves 97% of its language understanding capabilities through knowledge distillation, while reducing its size by 40% and speeding up the training process by 60% [23]. As opposed to BERT which has 12 layers, DistilBERT only has 6 hidden layers in the transformer encoder and is computationally cheaper. GPT-2 [24] is a large model based on the transformer architecture and is trained on an English corpus with the objective of predicting the next word in a sequence. Unlike BERT, GPT-2 and its predecessor GPT [25] were trained with a casual language modelling (CLM) objective. BERT for agricultural domain[2] is a BERT-based uncased language model pre-trained from the SciBERT starting point on a corpus of 1.2 million paragraphs from National Agricultural Library (NAL) from the US Gov. and 5.3 million paragraphs from books and common literature from the agriculture domain. SciBERT is a pre-trained BERT-based language model trained on large multi-domain corpus of scientific publications [26]. Thus, the BERT model for agriculture domain should have good representations for text from both scientific and agricultural domain.

4 Methodology

This section describes the methodology used in our study. Details of the dataset are discussed, along with the steps involved in the experiment. The evaluation metrics used to evaluate model performance are also explained.

[2] https://huggingface.co/recobo/agriculture-bert-uncased.

4.1 Dataset

The dataset is composed of comments gathered from 16 videos on YouTube that featured content relevant to smart farming tools and technologies. A total of 7334 comments are divided into 6 categories by 2 human annotators, and agreement is reached between discrepancies. The comments are divided into 6 categories: praising (970 comments), opinion (3224 comments), suggestion (231 comments), undefined (1386 comments), queries (1007 comments), and hybrid (516 comments). Details about the videos are provided in the previous work [8]. The dataset is highly imbalanced, with the 'opinion' class having the highest number of comments.

4.2 Experiment

Pre-trained models from the HuggingFace library[3] are fine-tuned for the purpose of text classification using PyTorch. We have selected the DistilBERT,[4] GPT-2,[5], and agriculture BERT models for our experiment. Hyperparameter tuning is carried out by varying the following: batch size, learning rate, Adam's epsilon, and number of epochs. Batch size is the total number of training samples that is passed through the model in one iteration. Learning rate is the rate at which the model updates its weights with respect to the loss gradient. The number of epochs determines the total number of times the entire training set is passed forward and backward through the neural network. The AdamW optimizer implements the Adam algorithm for optimization, but decouples the weight decay from the optimization step [27]. The epsilon hyperparameter for the AdamW optimizer, which is to avoid divide by zero error in the Adam algorithm, is tuned in this study. Table 1 provides all the 15 configurations tested in our study.

Evaluation Metrics In order to evaluate model performance, it is necessary to employ suitable evaluation metrics taking into account the class imbalance in the dataset. Thus, accuracy score is an unreliable measure of model performance, as the model can simply score higher by predicting the class with the highest number of samples [28]. The model will thus not generalize well to new data. Precision (macro-P), recall (macro-R), and F1-score (macro-F1) are calculated as a macro-average, which means that each class will be considered equal and a greater penalization occurs if the model incorrectly predicts a minority class [29]. The weighted average of F1 (weighted-F1) and precision (weighted-P) are recorded but not considered for evaluation, as the weighted average is useful when we want to favour the majority class.

---

[3] https://huggingface.co/.

[4] https://huggingface.co/distilbert-base-uncased.

[5] https://huggingface.co/gpt2.

Table 1 Configuration table

Config no.	Epochs	Train batch size	Eval batch size	Adam's epsilon	Learning rate
1.	1	32	32	1e−5	4e−5
2.	1	16	16	1e−5	4e−5
3.	2	16	16	1e−5	4e−5
4.	1	8	8	1e−5	4e−5
5.	1	16	16	1e−5	3e−4
6.	1	32	32	1e−5	3e−4
7.	1	16	16	1e−5	3e−5
8.	3	8	8	1e−8	5e−5
9.	1	16	16	1e−8	3e−4
10.	1	16	16	1e−8	3e−5
11.	1	16	16	1e−8	5e−5
12.	2	16	16	1e−8	5e−5
13.	1	8	8	1e−8	5e−5
14.	2	8	8	1e−8	5e−5
15.	2	32	32	1e−5	4e−5

Matthews correlation coefficient (MCC) is computed between the observed and predicted classifications, ranges from +1 (perfect prediction) to −1 (inverse prediction) and is considered a viable metric for imbalanced multi-class text classification [30].

5 Results and Analysis

In this section, we discuss the results of the experiment and analyse model performance on the evaluation criteria.

5.1 Empirical Results

Tables 2, 3, and 4 list the results of our experiment for DistilBERT, GPT-2, and agriculture BERT, respectively. The highest score of each metric is marked in bold. Agriculture BERT achieves the highest accuracy of 77.04% out of the 3 models, with a training and evaluation batch size of 16, learning rate of 4e−5, Adam's epsilon value of 1e−5, and 2 training epochs. DistilBERT gives the highest MCC of 0.6819, with a training and evaluation batch size of 8, learning rate of 5e−5, Adam's epsilon value of 1e−6, and 2 training epochs. Agriculture BERT gives the best macro-F1-score of 0.6963 out of the 3 models, followed by DistilBERT's score of 0.6953, and GPT-2's

Table 2 Experiment results for DistilBERT

Config no.	Accuracy	Weighted-F1	Weighted-P	MCC	Macro-P	Macro-R	Macro-F1
1.	65.86	0.5919	0.6014	0.5186	0.4610	0.4437	0.4256
2.	72.41	0.6950	0.7001	0.6102	0.6015	0.5312	0.5469
3.	76.00	0.7497	0.7504	0.6648	0.7026	0.6299	0.6470
4.	74.75	0.7317	0.7417	0.6475	0.7242	0.6025	0.6080
5.	71.37	0.6986	0.6916	0.5984	0.5768	0.5535	0.5596
6.	72.68	0.7050	0.7123	0.6157	0.6078	0.5548	0.5707
7.	70.77	0.6650	0.6917	0.5876	0.6126	0.5078	0.5037
8.	**76.99**	**0.7678**	**0.7671**	**0.6819**	0.7082	**0.6863**	**0.6953**
9.	70.39	0.6866	0.7224	0.5964	0.7415	0.5661	0.5566
10.	72.79	0.7002	0.7045	0.6157	0.6089	0.5385	0.5557
11.	74.75	0.7286	0.7222	0.6435	0.6199	0.5755	0.5883
12.	76.88	0.7590	0.7604	0.6773	0.7144	0.6466	0.6668
13.	75.19	0.7388	0.7498	0.6535	0.7508	0.6194	0.6327
14.	76.55	0.7609	0.7616	0.6737	0.7065	0.6709	0.6860
15.	73.44	0.7086	0.7434	0.6276	**0.7799**	0.5653	0.5656

score of 0.6742. Thus, based on accuracy and macro-F1-score, BERT for agriculture domain is the best performing model out of the three. Training and evaluation batch sizes of 8 and 16 give the best model performance, which decreases with an increase in batch size. Based on the results, the optimal learning rate is 5e−5 for DistilBERT and GPT-2 and 4e−5 for agriculture BERT. GPT-2 achieves the highest accuracy of 75.29% for a training and evaluation batch size of 8, learning rate of 5e−5, Adam's epsilon value of 1e−8, and 2 training epochs. GPT-2 achieves the lowest best scores on all metrics out of the 3 models.

Figure 1 displays the confusion matrix for config no. 13 in Table 1 for agriculture BERT. It allows us to calculate the true positive, true negative, false positive, and false negative counts for each class. The training and validation loss for model training of DistilBERT, GPT-2, and agriculture BERT are graphed in Fig. 2, respectively, for 5 epochs, with the training and evaluation batch size of 8, Adam's epsilon value of 1e−8, and learning rate of 5e−5. In Fig. 2, the training and validation loss become much closer in value at 2 training epochs for all 3 models, after which validation loss gradually increases indicating that the model is overfitting on the data. Thus, it follows that convergence occurs at 2 training epochs. The values of the MCC for all 15 configurations are plotted in Fig. 3 for the 3 pre-trained models selected in the study. It suggests that with the exception of GPT-2, DistilBERT and BERT for agricultural domain have MCC close in value to each other.

Table 3 Experiment results for GPT-2

Config no.	Accuracy	Weighted-F1	Weighted-P	MCC	Macro-P	Macro-R	Macro-F1
1.	43.56	0.3683	0.4304	0.1363	0.3324	0.2271	0.2207
2.	59.97	0.5489	0.5509	0.4181	0.4134	0.3769	0.3693
3.	73.55	0.7173	0.7384	0.6292	**0.7623**	0.5866	0.5909
4.	69.13	0.6696	0.6886	0.5764	0.6331	0.5387	0.5231
5.	71.42	0.6806	0.7016	0.5952	0.6269	0.5185	0.5246
6.	65.43	0.6132	0.6519	0.5094	0.5500	0.4426	0.4334
7.	60.90	0.5803	0.5782	0.4382	0.4723	0.4221	0.4272
8.	73.77	0.7306	0.7295	0.6364	0.6627	0.6288	0.6396
9.	70.88	0.6810	0.6813	0.5865	0.5818	0.5195	0.5295
10.	61.72	0.5607	0.6155	0.4433	0.5345	0.3854	0.3818
11.	66.08	0.6204	0.6627	0.5132	0.5940	0.4438	0.4447
12.	72.79	0.7097	0.7349	0.6174	0.7599	0.5774	0.5854
13.	71.86	0.6953	0.7259	0.6067	0.7486	0.5466	0.5436
14.	**75.29**	**0.7457**	**0.7475**	**0.6564**	0.6935	**0.6406**	**0.6566**
15.	61.61	0.5712	0.6093	0.4466	0.5192	0.4002	0.4238

Table 4 Experiment results for agriculture BERT

Config no.	Accuracy	Weighted-F1	Weighted-P	MCC	Macro-P	Macro-R	Macro-F1
1.	70.01	0.6700	0.6780	0.5811	0.5234	**0.7001**	0.5246
2.	73.71	0.7174	0.7199	0.6302	0.6169	0.5700	0.5860
3.	**77.04**	0.7623	0.7644	**0.6800**	0.7311	0.6583	0.6791
4.	75.19	0.7392	0.7449	0.6552	0.7087	0.6184	0.6259
5.	69.84	0.6895	0.7329	0.5773	0.7663	0.5471	0.5719
6.	73.00	0.7172	0.7278	0.6197	0.7004	0.5917	0.6260
7.	74.26	0.7246	0.7166	0.6376	0.6046	0.5802	0.5889
8.	76.06	0.7598	0.7603	0.6705	0.6915	0.6795	0.6839
9.	67.01	0.6516	0.6966	0.5414	0.7343	0.5142	0.5150
10.	73.93	0.7213	0.7489	0.6336	**0.7737**	0.5794	0.5948
11.	73.11	0.7194	0.7379	0.6198	0.7407	0.5785	0.6183
12.	76.66	0.7607	0.7617	0.6757	0.7140	0.6698	0.6860
13.	75.68	0.7474	0.7494	0.6797	0.6871	0.6191	0.6328
14.	76.77	**0.7643**	**0.7652**	0.6777	0.7119	0.6848	**0.6963**
15.	74.91	0.7317	0.7490	0.6492	0.7434	0.6011	0.6125

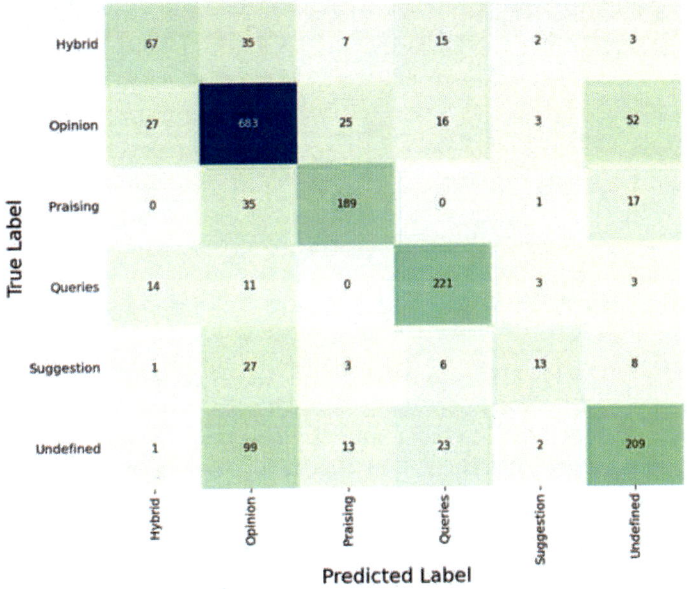

Fig. 1 Confusion matrix for agriculture BERT

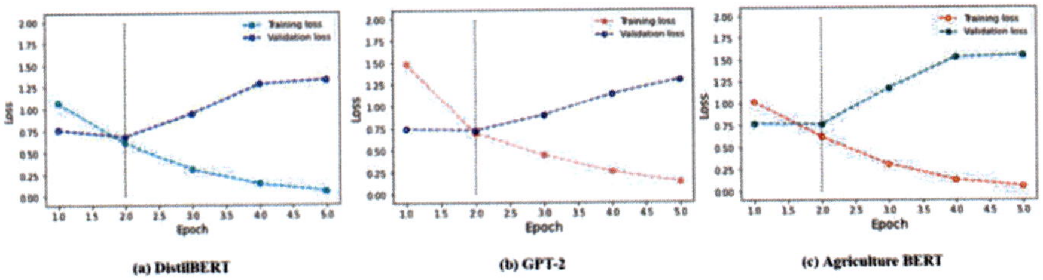

Fig. 2 Training and validation loss

Fig. 3 Matthews correlation coefficient for 3 pre-trained language models

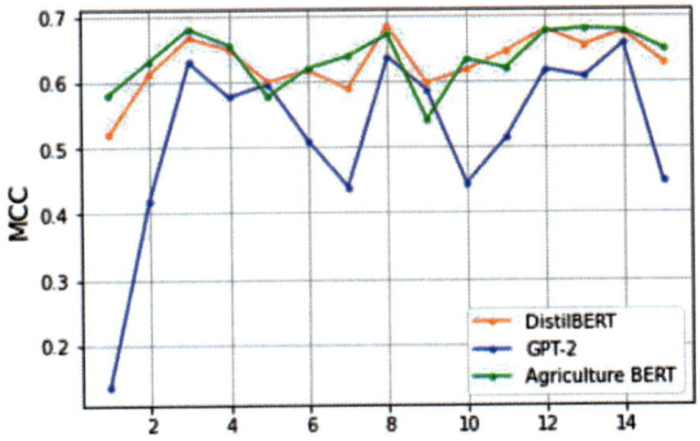

Table 5 Friedman statistical test

Metric	X_r^2 statistic	p-value
Accuracy	12.9333	0.00155
MCC	17.2	0.00018
Macro-precision	13.52	0.0011
Macro-recall	20.8	0.00003
Macro-F1	22.9333	0.00001

5.2 *Statistical Testing*

In order to ensure that the results produced by our study are significant and not random, Friedman statistical testing was conducted using the Scipy python library.[6] The results generated by all the 15 configurations were compared for the 3 models. The null hypothesis assumes that there is no statistically significant difference between the means of the results given by the three classifiers. The alternative hypothesis assumes that at least one population mean is different.

Table 5 lists the results of the statistical test for accuracy, MCC, macro-precision, macro-recall, and macro-F1-score, along with the p-value and the X_r^2 test statistic. The results are calculated for significance level of 0.01, 0.05, and 0.10. For all metrics, results are significantly different for all three models for the significance level of 0.01, 0.05, and 0.10. Thus, the null hypothesis is rejected and we can conclude that there is a significant difference between the results provided by the 3 models, and agriculture BERT outperforms the others.

6 Discussion

This study was conducted to analyse the performance of pre-trained language models on a dataset consisting of comments about smart farming technologies. In this section, the research questions which were crafted to explore the hypothesis have been discussed as follows:

1. What is the performance of transformers in classifying data containing user opinion on smart farming?
   The highest value of MCC for DistilBERT is 0.6819, for GPT-2 it is 0.6564, and for agriculture BERT it is 0.6800. Highest macro-F1 for DistilBERT is 0.6953, for GPT-2 is 0.6566, and for agriculture BERT is 0.6963. DistilBERT gives the best performance on the basis of MCC, and agriculture BERT gives the best performance on the basis of macro-F1-score and accuracy. The study by Yadav et al. [4] produced the highest MCC score of 0.8706 for mix-code text classification on a perfectly balanced dataset.

---

[6] https://docs.scipy.org/doc/scipy/reference/generated/scipy.stats.friedmanchisquare.html.

2. Does a BERT-based model pre-trained on agricultural text perform better classification on the smart farming dataset than other transformer models? The discussion in Sect. 5 suggests that the agriculture BERT provides the highest macro-F1 of 0.6963 and accuracy of 77.04%. This suggests that the agriculture BERT model performs better than the other two models trained on an English text corpus on the basis of accuracy and macro-F1-score. However, a higher accuracy could not be achieved because the dataset is highly imbalanced and the comments in the dataset do not contain too much technical terminology pertaining to the agricultural domain. This supports the initial assumption in the study [8] and the findings in study [1] that digital users are not well-informed about the technical details of smart agricultural operations and remain unclear about the perceived risks and benefits. Further exploration of the vocabulary used to pre-train the agriculture BERT model and the dataset used for training could provide better insight into the results.
3. What are the appropriate evaluation metrics used for quantifying model performance on imbalanced datasets? To evaluate model performance, the macro-average of the f1-score is considered, as it takes minority classes into consideration and treats all classes equally. Further study into the reliability of metrics such as MCC [30] and macro-F1 for imbalanced datasets could provide better insight on model performance.
4. Are the results obtained from the experiment statistically significant? The results obtained from Sect. 5.2 suggest that the null hypothesis is rejected and the results obtained by the experiment are statistically significant. Friedman statistical testing suggests that the model performances differ significantly from each other, and the results were not produced by chance, confirming that the agriculture BERT gives the best performance.

7 Conclusion and Future Work

Understanding user opinion about an area of interest can be a cumbersome task for any concerned authority. Smart farming, with its plethora of unexplained potential, remains vastly unexplored due to the various socio-economic, political, and other concerns regarding its implementation. In this study, we leverage state-of-the-art transformer models to analyse the perception of general public towards the new and emerging trends in the field of smart agriculture. Three transformer models are trained and optimized, and their performance is measured on the dataset on the basis of several metrics such as accuracy, macro-F1, and MCC. On the basis on MCC, DistilBERT gives the highest score of 0.6824. Agriculture BERT is the best classifier on the basis of macro-F1-score of 0.6963. Through Friedman statistical testing, we determine that there is significant difference between model results. Though the BERT model for agricultural domain is pre-trained specifically on agricultural literature, it only marginally performs better than other pre-trained models when evaluated on the macro-F1-score. This indicates that our dataset did not contain an abundance of technical terminology from the agricultural domain. For future exploration into the hypothesis, we can explore and analyse the vocabulary of agriculture BERT and compare the word representations and semantic similarities with other BERT models. Techniques for tackling class imbalance problems such as downsampling the majority class, upsampling the minority class, and synthetic minority oversampling technique (SMOTE) can also be implemented, since the dataset is highly skewed towards one class. Model performance on agricultural data can be further explored by gathering data from other social media websites.

References

1. Das VJ, Sharma S, Kaushik A (2019) Views of irish farmers on smart farming technologies: an observational study. AgriEng 1(2):164–187
2. Salim JN, Trisnawarman D, Imam MC (2020) Twitter users opinion classification of smart farming in indonesia. In: IOP conference series: materials science and engineering, vol 852, issue 1. IOP Publishing, p 012165
3. YouTube Statistics. https://www.omnicoreagency.com/youtube-statistics/. Accessed 2 July 2022
4. Yadav S, Kaushik A, Sharma S (2021) Cooking well, with love, is an art: Transformers on youtube hinglish data. In: 2021 international conference on computational performance evaluation (ComPE). IEEE, pp 836–841
5. Kaur G, Kaushik A, Sharma S (2019) Cooking is creating emotion: a study on hinglish sentiments of YouTube cookery channels using semi-supervised approach. Big Data Cogn Comput 3(3):37
6. Devlin J, Chang MW, Lee K, Toutanova K (2018) Bert: pre-training of deep bidirectional transformers for language understanding. arXiv preprint arXiv:1810.04805
7. Qu C, Yang L, Qiu M, Croft WB, Zhang Y, Iyyer M (2019) Bert with history answer embedding for conversational question answering. In: Proceedings of the 42nd international ACM SIGIR conference on research and development in information retrieval, pp 1133–1136
8. Yadav S, Kaushik A, Sharma M, Sharma S (2022) Disruptive technologies in smart farming: an expanded view with sentiment analysis. AgriEng 4(2):424–460
9. Bacco M, Barsocchi P, Ferro E, Gotta A, Ruggeri M (2019) The digitisation of agriculture: a survey of research activities on smart farming. Array 3:100009
10. Rajasekaran T, Anandamurugan S (2019) Challenges and applications of wireless sensor networks in smart farming-a survey. In: Advances in big data and cloud computing, pp 353–361
11. Wolfert S, Ge L, Verdouw C, Bogaardt MJ (2017) Big data in smart farming-a review. Agric Syst 153:69–80
12. Hunter MC, Smith RG, Schipanski ME, Atwood LW, Mortensen DA (2017) Agriculture in 2050: recalibrating targets for sustainable intensification. Bioscience 67(4):386–391
13. Jacob B, Kaushik A, Velavan P, Sharma M (2022) Autonomous drones for medical assistance using reinforcement learning. In: Advances in augmented reality and virtual reality. Springer, pp 133–156
14. Jacob B, Kaushik A, Velavan P (2022) Autonomous navigation of drones using reinforcement learning. In: Advances in augmented reality and virtual reality. Springer, pp 159–176
15. Velavan P, Kaushik A, Jacob B, Sharma M (2022) A reinforcement learning approach for shortest path navigation in automated guided vehicles for medical assistance. In: Advances in augmented reality and virtual reality. Springer, pp 193–212
16. Boursianis AD, Papadopoulou MS, Diamantoulakis P, Liopa-Tsakalidi A, Barouchas P, Salahas G, Karagiannidis G, Wan S, Goudos SK (2022) Internet of things (IoT) and agricultural

unmanned aerial vehicles (UAVs) in smart farming: a comprehensive review. Internet of Things 18:100187
17. Smart farming European Union. https://ec.europa.eu/eip/agriculture/en/find-connect/online-resources/smart-farming-platform. Accessed 2 July 2022
18. Village of Excellence. https://www.business-standard.com/article/economy-policy/india-israel-sign-3-year-work-programme-for-cooperation-in-agri-tomar-121052401072.html. Accessed 2 July 2022
19. Vaswani A, Shazeer N, Parmar N, Uszkoreit J, Jones L, Gomez AN, Kaiser Ł, Polosukhin I (2017) Attention is all you need. Adv Neural Inf Process Syst 30
20. Yang Y, Uy MCS, Huang A (2020) Finbert: a pretrained language model for financial communications. arXiv preprint arXiv:2006.08097
21. Lee J, Yoon W, Kim S, Kim D, Kim S, So CH, Kang J (2020) Biobert: a pre-trained biomedical language representation model for biomedical text mining. Bioinformatics 36(4):1234–1240
22. Lee, J.S., Hsiang, J.: Patentbert: Patent classification with fine-tuning a pre-trained bert model. arXiv preprint arXiv:1906.02124 (2019)
23. Sanh V, Debut L, Chaumond J, Wolf T (2019) Distilbert, a distilled version of bert: smaller, faster, cheaper and lighter. arXiv preprint arXiv:1910.01108
24. Radford A, Wu J, Child R, Luan D, Amodei D, Sutskever I et al (2019) Language models are unsupervised multitask learners. OpenAI blog 1(8):9
25. Radford A, Sutskever I (2018) Improving language understanding by generative pre-training. In: arxiv
26. Beltagy I, Lo K, Cohan A (2019) Scibert: a pretrained language model for scientific text. arXiv preprint arXiv:1903.10676
27. Loshchilov I, Hutter F (2017) Decoupled weight decay regularization. arXiv preprint arXiv:1711.05101
28. Kotsiantis S, Kanellopoulos D, Pintelas P et al (2006) Handling imbalanced datasets: a review. GESTS Int Trans Comput Sci Eng 30(1):25–36
29. Opitz J, Burst S (2019) Macro f1 and macro f1. arXiv preprint arXiv:1911.03347
30. Boughorbel S, Jarray F, El-Anbari M (2017) Optimal classifier for imbalanced data using Matthews correlation coefficient metric. PloS one 12(6):e0177678

Reinforcement Learning of Self-enhancing Camera Image and Signal Processing

Chandrajit Bajaj, Yunhao Yang, and Yi Wang

Abstract Current camera image and signal processing pipelines (ISPs), including deep-trained versions, tend to apply a single filter that is uniformly applied to the entire image. This is despite the fact that most acquired camera images have spatially heterogeneous artifacts. This spatial heterogeneity manifests itself across the image space as varied Moire ringing, motion blur, color bleaching, or lens-based projection distortions. Moreover, combinations of these image artifacts can be present in small or large-pixel neighborhoods within an acquired image. Here, we present a deep reinforcement learning model that works in learned latent subspaces and recursively improves camera image quality through patch-based spatially adaptive artifact filtering and image enhancement. Our *Recursive Self-Enhancement Reinforcement Learning*(RSE-RL) model views the identification and correction of artifacts as a recursive self-learning and self-improvement exercise and consists of two major sub-modules: (i) The latent feature subspace clustering/grouping obtained through variational auto-encoders enabling rapid identification of the correspondence and discrepancy between noisy and clean image patches. (ii) The adaptive learned transformation is controlled by a soft actor-critic agent that progressively filters and enhances the noisy patches using its closest feature distance neighbors of clean patches. Artificial artifacts that may be introduced in a patch-based ISP are also removed through a reward-based deblocking recovery and image enhancement. We demonstrate the self-improvement feature of our model by recursively training and testing on images, wherein the enhanced images resulting from each epoch provide a natural data augmentation and robustness to the RSE-RL training-filtering pipeline. Our code is publicly available at https://github.com/CVC-Lab/rse_rl.git.

Chandrajit Bajaj, Yunhao Yang, and Yi Wang contributed equally to this work.

C. Bajaj (✉) · Y. Yang
Department of Computer Science and Oden Institute of Engineering and Sciences, University of Texas, Austin, TX, USA
e-mail: bajaj@cs.utexas.edu

Y. Wang
Oden Institute of Engineering and Sciences, University of Texas, Austin, TX, USA

S. Das et al. (eds.), *Advances in Data-driven Computing and Intelligent Systems*, Lecture Notes in Networks and Systems 653, https://doi.org/10.1007/978-981-99-0981-0_22

1 Introduction

Image Processing Pipelines for cameras, known as camera ISPs, have been a hot topic in past decades. For instance, image denoising has been investigated due to the vast demand for noise pattern removal. In the early era, traditional methods would explore the case where they apply Gaussian blurring, TV regularization [40], or a coefficient transform in Fourier domain [42] to remove high-frequency signals, commonly identified as noise signals. However, the idea of non-local mean denoising from [8] truly starts heterogeneous noise removal. The non-local mean method is built upon self-similarity and redundant information over realistic images without a uniform noise pattern assumption, which is proved to lead to a performance boost. Later on, BM3D [12], as a non-local method, explore the sample idea and exploits the sparsity further. There are discussions with respect to the subdivision of image domains [17] as well. Nowadays, with the advent of deep learning architectures, more and more works are using Convolutional Neural Networks (CNNs) or even Generative Adversarial Networks (GANs) that beat most of the classical, sophisticated methods [18, 21, 36, 51, 56]. Nevertheless, as noise in large-scale realistic photos is obliged to an unknown form, most prior works depend on the assumption that noise comes from easily synthesized additional signals. The most common noise model assumption is the AWGN noise model. AWGN noise model cannot effectively assist either a machine learning model or a deep learning model to better improving real photo noise/artifact removal, as discussed in earlier benchmarks [37]. How to fill in the gap between the synthesized noise model and real-world noisy images remains an open question.

In recent years, more self-supervised learning or unsupervised image denoising models have arisen. These models aim to find agnostic noise modeling under real photographs rather than synthesizing its noisy version from instrument error or Gaussian/Poisson noise distortions. Among all works, N2V [30] and N2N [36] are typical examples of such learning schemes. Later works [24, 50] explores a more generative, latent space representation of noise model equipped with more flexibility of learning dependently from image signals. Even though self-redundancy and realistic noise generation are partly solved in these proposed models, these models only assume a single type artifact per RAW image needs to be diagnosed and filtered. Heterogeneous camera image artifacts, especially mixtures, are insensitive to pixel locations, and thus it is hard to disentangle them from one pass. For a camera ISP, however, this assumption fails to capture spatial heterogeneous acquired corruption caused by a mismatch of acquisition settings and environmental lighting in realistic scenes. Moreover, one can hardly tell if the artifacts come from sensor limitations, environmental changes, or post-processing, such as lossy compression. Such varied real-world image processing issues motivate us to adopt a recursive self-improving learning approach for the next generation of camera ISPs. Although reinforcement learning-based image improvement pipeline was invented in parallel [15, 22, 54, 57], none of them take care of spatial heterogeneity compared with the unsupervised or self-supervised scheme. It is worthwhile to combine the idea of recursive improvement together with a spatial subdivision, and, if possible, control and understand patch-based latent space representation of realistic photos.

In this paper, we present a deep reinforcement learning model that recursively improves camera image quality through patch-based spatially adaptive artifact filtering. Our Recursive Self-Enhancement Reinforcement Learning(RSE-RL) model is introduced in Sect. 3. The main contribution of the RSE-RL model is a two-fold patch-based denoising filtering. On the one hand, we learn latent space feature disentanglement by projecting the encoding of corrupted patches into subspaces. With the discrepancy being identified through a transformation from subspaces, one can assign the decoding filters, where each filter is a neural network to recover the clean image. On the other hand, we iteratively refine our denoising result by changing the transformation through actions and optimizing the performance per test input. We demonstrate, in Sect. 4, our RSE-RL model provides a natural image enhancement throughout training-filtering pipeline, resulting in the quality improved photo-realistic image from each epoch.

2 Related Work

2.1 Image Denoising

Generative Noise Patterns Current image denoising research focuses more on model-free noise generators, no matter globally or locally. Liu et al. [33] adapts a Wavelet Transform as a separator for high-frequency signals and then applies the design of the invertible blocks to recover and refine low-resolution image signals. Xia et al. [49] discusses the contrastive learning under flashing/no-flashing pairs to cancel out noise under low-light conditions. Ren et al. [39] propose the consistency prior, which measures the correlation and tries to linearize the image noise model. Khademi et al. [25] extend the blindspot model for self-supervised denoising to handle Poisson-Gaussian noise specifically. Noise2same [50], as a recent attempt at self-supervised learning, takes both the full noisy image and the masked image as inputs and produces two outputs. The reconstruction loss is computed between the full noisy image and its corresponding output. The invariance loss is computed between the two outputs to align with a standard denoising filter. Among all these works, the generative model of noise is assumed as a global modification.

Latent Subspace Denoising One of our major proposed contributions is to learn a latent encoding for maximizing the usage of self-similarity in natural images via probing subspaces. Earlier work, such as [43], proposes a Gaussian mixture model for global noise. Inspired from latent space clustering using VAE [13, 53] introduces a latent mixture subspace and optimizes under multiple distinguished denoising filters. The variational inference under the latent space disentanglement yields better separability and approximation power toward a non-uniform, signal-dependent noise removal algorithm. Compared with [32], where they study the influence of Gaussian noise on parameters of the Gaussian Mixture Model. The result can adapt to any noise prior. However, for unseen data, its generalizability remains a question, and the real noisy image might fall into a simple filter's scope. Another attempt at latent space representation of noise modeling is NBNet [10], which learns latent space noise basis under projection with principal component analysis. In addition, for patch-based subspace noise learning, [24] concatenates the whole image with its noise level map taken as the encoder input. Given a query patch, similar context patches will be found by cross-patch sampling. It comprises two branches: the correlation of the inner patch grasp for the local feature while a cross-patch GCN extracts patch-wise consistency. Then the model aggregates the local and non-local features as the input of the decoder and applies residual blocks for fine-detailed denoise image output. PS-VAE [53] proposed to impose Gaussian Mixture Model in the latent space encoding for patch-based denoising. Soft labeling was introduced to assign different filters for patches. The model has the potential of denoising heterogeneously. Nevertheless, the prior of noise is intractable in a real photo, leading to an overly strong assumption that is difficult to generalize.

2.2 Camera ISP Meets Deep Learning Model

Deep learning approaches have progressively replaced the image and signal processing applied in conventional computational photography tasks. For instance, with low-level details and hierarchical structures of neural network implemented, one could achieve superior performance for image denoising (e.g., [3, 45, 46, 52, 55]) and deblurring (e.g., [31, 44]) tasks. There are numerous works related to camera ISP as well, where different types of articulated modelings are applied to different vision tasks. Among all of them, some related works such as color demosaicing(e.g., [26]), auto white balance correction (e.g., [4]), and removing lossy compression artifacts (e.g., [16]), etc., have been separately discussed under deep learning settings.

Within the scope of image denoising, CycleISP [55] develops a generative model to generate synthesized realistic image data that is both forward and reversed. Moreover, solving multi-task image enhancement is possible: [41] is an early work that focuses on jointly solving demosaicing and denoising problems by applying deep learning schemes from the source sampling of sensors. A recent work [23] presented a single deep learning model containing 5 parallel learning levels to replace the entire camera ISP pipeline, referred to as PyNET. The input RAW image from a cellphone is aligned with a DSLR camera output as the supervised training data and PyNET outputs a visually high-quality sRGB image. To solve noise pattern generation issues, [9] focus on synthesizing noise pattern which is dependent on RAW image signals and iteratively update the denoised image from an unpaired set of noisy and clean images to adapt and finetune the pre-trained denoiser.

2.3 Reinforcement Learning-Based Image Enhancement

Reinforcement learning methods have been applied in different low-level vision tasks, especially in image processing. Furuta et al. [15] has proposed a pixel-wise A3C (Asynchronous Advantage Actor Critic [35]) scheme to view the entire image domain as a multi-agent system but provide their filtering action and reward map convolution to efficiently compute mean reward. This early system has enlightened the later research to formulate the image processing task as a multi-agent, or multi-task learning problem, which could be controlled via reinforcement learning. A global image enhancement approach was introduced in [22], where different filters are applied over the RAW image with a decision-making procedure, i.e., a policy gradient descent. The network's input is a low-res, fixed 64 by 64 patch where the patch is used to determine which filter and the weight of filters applying over the high-res image. Yu et al. [54] extends the idea of filter decision, but the decision-making is based on Long Short-Term Memory(LSTM) rather than a conventional CNN as stated in [22].

3 Recursive Self-enhancing Camera ISP

In this section, we describe the architecture of our reinforcement learning-based camera ISP model that recursively improves and enhances acquired images by filtering heterogeneous image artifacts in a spatially adaptive manner. Figure 1 displays the overall framework of RSE-RL.

We target the problem of resolving image artifacts, specifically on image denoising, but our approach can be extended to solve other comprehensive image processing tasks, such as generating RAW to sRGB images [23, 59]. We assume the observed image I_{obs} is obtained via the following mixture model:

$$f(I_{\text{obs}}) = I_{\text{gt}} + \sum_{s=1}^{S} \Sigma_s \odot M_s. \tag{1}$$

The function f is an identity function performing sRGB to sRGB image denoising tasks. The noise Σ_s and the mask matrix M_s are independent and blind to the model ($\odot$ refers to the element-wise product). Moreover, we do not rely on the underlying distribution Σ_s. Except for the synthesized dataset, we cannot obtain the accurate number of artifact types S, and S is a hyper-parameter in most scenarios. We would like to present and discuss how to disentangle and filter from (1) using our RSE-RL. We will first train distinct filters in separated latent subspaces with supervision. Then we impose self-enhancement for testing data, by learning per-image latent transformation under a policy learning scheme.

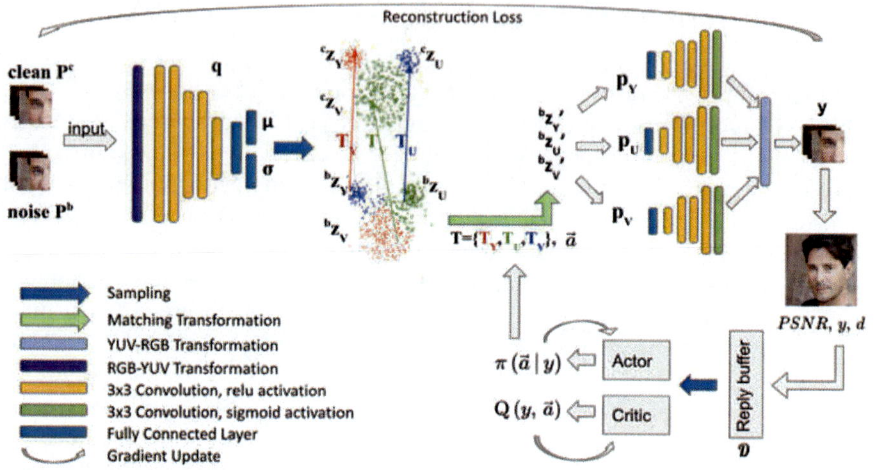

Fig. 1 The overall pipeline of our RSE-RL: For each captured (and training) image, we split the image into local patches and feed every patch as a stack into the encoding network. We divide the latent space into three subspaces Z_y, Z_u, and Z_v which preserve the YUV features of the patches. We apply the encoder to the clean and noisy patches and project them as well onto the same three latent subspaces. Then, we learn a set of transformations T that transforms the latent representation of the noisy patches to a corresponding representation of the clean patches, in all three subspaces. We send individually sampled transformed noisy representations to the decoders to reconstruct the image. Note that we have three decoders to reconstruct YUV features from the three subspaces. Then we merge and reconvert the YUV features to reconstruct the final RGB image. Once we reconstruct the images, we train a soft actor-critic reinforcement learning algorithm to further maximize the image PSNR. The RL algorithm uses 1-norm distance between a target PSNR and the actual PSNR as the reward to finetune the trainable weights in the set of transformations T. Application of the RL self-enhancing image denoising network. More details are presented below and in the Appendix

Overall Pipeline of RSE-RL Our recursive self-improving camera ISP in Fig. 1 consists of a multiple latent subspace variational autoencoder developed from [53]. For every input image $I_{\text{obs}} \in \mathbb{R}^{H\times W\times C}$, we first divide the input image into D by D patches $P \in \mathbb{R}^{D\times D\times C}$ with overlaps allowed. For every D by D patch P in the image, we denote its location mask H_n in the original image domain. P relates to the observed image in the form $P \stackrel{\text{def}}{=} I_{\text{obs}} \odot H_n$. Even though we do not estimate the magnitude of the artifacts, we approximate it within a patch with the single dominant model, namely:

$$f(P) = f(I_{\text{obs}} \odot H_n) = I_{\text{gt}} \odot H_n + \sum_{s=1}^{S} \Sigma_s \odot M_s \odot H_n, \approx (I_{gt} + \Sigma_i \odot M_i) \odot H_n. \quad (2)$$

Our RSE-RL network initially learns from clean image patches $P^c = I_{gt} \odot H_n$ and observed noisy patches P^b. We feed both clean patches P^c and noisy patches P^b into the variational encoder $q(P|\vec{\theta})$ parametererized by $\vec{\theta}$. The encoder projects the patches into three latent subspaces Z_y, Z_u, and Z_v and we draw samples in three latent spaces via:

$$\vec{z}_s \sim \mathcal{N}(q_s^{\mu}(P|\vec{\theta}), \mathrm{diag}[(q_s^{\sigma}(P|\vec{\theta}))^2]), \quad s \in \{y, u, v\}. \tag{3}$$

We denote $\{\vec{z}_y^b, \vec{z}_u^b, \vec{z}_v^b\}$ and $\{\vec{z}_y^c, \vec{z}_u^c, \vec{z}_v^c\}$ as the latent encodings (samples) for noisy patches P^b and clean patches P^c respectively. To obtain latent subspaces which learn to identify imminence and chrominance components, we apply an RGB to YUV transformation per pixel for each input patches P^b and P^c as a pre-processing layer in encoder q. We state the RGB-YUV transformation matrix in the Appendix.

Within the three subspaces Z_y, Z_u, and Z_v, we construct a set of three transformations $T = \{T_y, T_u, T_v\}$. We train the transformations so that the transformed latent encoding approaches the clean patch latent encoding: $T(\vec{z}^b) \approx \vec{z}^c$. Then, we set up three decoders $P_y := p_y(T_y(\vec{z}_y^b)|\vec{\psi}_y)$, $P_u := p_u(T_u(\vec{z}_u^b)|\vec{\psi}_u)$, $P_v := p_v(T_v(\vec{z}_v^b)|\vec{\psi}_v)$ with parameters $\vec{\psi}_s$, $s \in \{y, u, v\}$ respectively for reconstructing YUV channels from three disentangled latent subspaces. After the YUV features are constructed, we transform the patches' YUV features into RGB patches under a patch transformation PT, which is a pixel-wise YUV to RGB mapping:

$$PT(P_y, P_u, P_v) = \mathrm{vec}^{-1}_{D\times D\times C}\left(A\left[\mathrm{vec}(P_y), \mathrm{vec}(P_u), \mathrm{vec}(P_v)\right]^T\right). \tag{4}$$

Here the $\mathrm{vec}^{-1}_{D\times D\times C}$ is the reshaping operator that reshapes the patch back to D by D by C. The YUV-RGB patch transformation matrix A is the inverse of RGB-YUV linear transformation matrix, whose form is given in the Appendix. We reconstruct each patch independently and merge them after the decoding procedure. The output is the combined reconstructed patches and image that approximate P^c.

Loss Function In practical implementation, our training process optimizes different loss functions. For training the network, our training loss $\mathcal{L}_{\mathrm{vae}}$ consists of three parts: the evidence lower-bound(ELBO) [29], including the data fitting term and the KL loss, the transformation constraint loss, and the regularization term. The gradient computed from $\mathcal{L}_{\mathrm{vae}}$ is used to update all the parameters in the network, including the encoder, decoders, and transformations.

$$\mathcal{L}_{\mathrm{vae}} = \mathcal{L}_{\mathrm{MSE}} + \mathcal{L}_{\mathrm{KL}} + \mathcal{L}_{\mathrm{reg}} + \mathcal{L}_{\mathrm{tran}}, \tag{5}$$

where

$$\mathcal{L}_{\mathrm{MSE}} = \|PT(P_y, P_u, P_v) - P^c\|_F^2 + \|PT(\tilde{P}_y, \tilde{P}_u, \tilde{P}_v) - P^c\|_F^2, \tag{6}$$

$$\tilde{P}_s := p_s(\vec{z}_s^c|\vec{\psi}_s), \quad s \in \{y, u, v\}, \tag{7}$$

$$\mathcal{L}_{\mathrm{KL}} = -\frac{1}{2}\sum_{s\in\{y,u,v\}}\sum_j (1 + 2\log q_{s,j}^{\sigma}(P|\vec{\theta}) - (q_{s,j}^{\sigma}(P|\vec{\theta}))^2 - (q_{s,j}^{\mu}(P|\vec{\theta}))^2), \tag{8}$$

$$\mathcal{L}_{\mathrm{reg}} = \lambda_{\mathrm{reg}}(\|\vec{\theta}\|_2^2 + \|\vec{\psi}\|_2^2), \tag{9}$$

$$\mathcal{L}_{\mathrm{tran}} = \|\vec{z}_y^c - T_y(\vec{z}_y^b)\|_2^2 + \|\vec{z}_u^c - T_u(\vec{z}_u^b)\|_2^2 + \|\vec{z}_v^c - T_v(\vec{z}_v^b)\|_2^2. \tag{10}$$

On each latent subspace, we compute a loss $\mathcal{L}_{tran}$ from the noisy patch projection and clean patch projection and update the corresponding transformation accordingly.

Patch Assembly and Block Artifacts Upon getting the reconstructed blocks from the decoders, we merge these blocks back into the original image. If one naively concatenates two adjacent patches without any tolerance of the overlaps generated, one could observe block artifacts if these two patches pass different decoders. We apply post-processing to remove block artifacts generated from concatenation. We average the pixel output for overlapping regions based on the distance between the pixel and the overlapped patch centers via linear interpolation.

Reinforcement Learning New images acquired in the test or field operational stage, are reutilized to retrain the model. Since here we do not have access to the corresponding clean target patches, the trained transformations are fine tuned based on a reward function comprising of a target PSNR larger than the acquired image PSNR.

We use a Soft Actor-Critic (SAC) reinforcement learning algorithm [19, 20] as the self-enhancing mechanism. In our setup, the SAC algorithm learns to iteratively enhance from the current state of image patches P by updating the latent space transformation T using an weight vector $\vec{a}$. $\vec{a}$ can be viewed as a vectorial action, comprising of three component actions $\vec{a}_y$, $\vec{a}_u$ and $\vec{a}_v$, in the YUV latent subspaces. If we parameterize our transformation network $T_s, s \in \{u, v, w\}$ with parameter ξ_s, the weight vector update action can be expressed as an element-wise multiplication $\vec{\xi}_s' = a_s \odot \vec{\xi}_s$. Here $\vec{\xi}_s'$ is the updated weight, and $\vec{\xi}_s$ is the previous weight. The algorithm continuously updates these trainable weights to enhance the transformation from noisy to clean patch representations, consequently improving the final enhanced and filtered result.

We briefly explain SAC here and how it applies to our problem. SAC is an off-policy algorithm with entropy regularization. SAC trains an RL model to maximize the trade-off between expected return and entropy, a measure of randomness in the policy. SAC can prevent the policy from prematurely converging to a bad local optimum. In SAC, an entropy bonus is reflected in Q^π: $Q^\pi(P, a) \approx r(P, a) + \gamma(Q^\pi(P', \tilde{a}') - \alpha \log \pi(\tilde{a}'|P'))$, $\tilde{a}' \sim \pi(\cdot|P')$, where P is an input patch and P' is the resulting patch when the transformation in the network has been updated using $\tilde{a}'$, $\alpha > 0$ is the trade-off coefficient, $r(P, a, P')$ is the reward function and $\log \pi(\tilde{a}'|P')$ is the defined entropy.

SAC simultaneously learns a policy π and two Q functions Q_{ϕ_1}, Q_{ϕ_2}, with ϕ_1, ϕ_2 the network parameters. The actor-network learns the Q functions that minimizes a Mean-Squared Bellman Error(MSBE) $L(\phi_i, D)$, and SAC sets up the MSBE loss for each of the two Q-functions:

$$\mathcal{L}(\phi_j, D) = E\left[((Q_{\phi_j}(P, a) - f(r, P', d))^2\right], \quad (P, a, r, P', d) \sim D \tag{11}$$

where d is the done signal to set a terminating state, and f is given by

$$f(r, P', d) = r + \gamma(1 - d)(\min_{j=1,2}(Q_{\phi_j}(P', \tilde{a}') - \alpha \log \pi_\theta(\tilde{a}'|P'))), \quad \tilde{a}' \sim \pi(\cdot|P') \tag{12}$$

In the SAC algorithm, we consider patches from an image as state observations. The reward r is given by computing the L1 distance between the actual patch-PSNR and a target patch-PSNR for each of a minibatch of samples. Namely,

$$r(P, a, P') = b\left(\text{PSNR}(PT(P_y, P_u, P_v), P') - \text{PSNR}_t\right) + c, \tag{13}$$

where b and c are constants for adjusting the reward scale. PSNR is the quality metric, N is the total number of patches, and PSNR_t is the target PSNR of the entire patch-assembled image. The SAC trains an RL model to enhance the transformation and reduce the blocking artifacts. Thus, the optimized RL model can enhance images PSNR. We present more details and hyper-parameter settings in the Appendix.

The critic network learns the policy π_ω by maximizing a value function $V^\pi(I)$ to evaluate the current policy and computes gradients for the actor to find the optimal policy updating a accordingly via sampling from π_ω.

$$\begin{aligned}\mathcal{V}^\pi(P) &= E_{a\sim\pi_\omega}[Q^\pi(P, a)] + \alpha H(\pi_\omega(\cdot|P)) \\ &= E_{a\sim\pi_\omega}[Q^\pi(P, a) - \alpha \log \pi_\omega(\cdot|P)].\end{aligned} \tag{14}$$

Here Q^π takes the minimum value between Q_{ϕ_1} and Q_{ϕ_2}. We refer to readers the overall algorithm of our RSE-RL pipeline stated in Algorithm 1.

Algorithm 1 RSE-RL algorithm

Input: Training patches $\{P_n^b, P_n^c\}_{n=1}^N$, training steps MAX_TRAIN_ITER, minibatch size M, target noisy image I_{test}, gradient descent step size η, target PSNR $PSNR_t$.
Output: Trained VAE with three distinct decoders, denoised image $\tilde{I}_{test}$.
% Training Pipeline
for $i = 1$ to MAX_TRAIN_ITER **do**
  Sample a minibatch $\{P_m^b, P_m^c\}_{m=1}^M$ from $\{P_n^b, P_n^c\}_{n=1}^N$.
  Update network parameters via gradient descent of (5) using $\{P_m^b, P_m^c\}_{m=1}^M$.
end for
% Self-Enhancement Pipeline, which can be a decoupled procedure from training.
Collect patches $\{P_n\}$ from I_{test}.
while $PSNR < PSNR_t$ **do**
  Sample $P \sim \{P_n\}$, the observed state and $a \sim \pi_\omega(\cdot|P)$, the action.
  $P' \leftarrow PT(P_y, P_u, P_v)$.
  Compute $PSNR$ and $r(P, a, P')$
  Set done signal $d \leftarrow (PSNR < PSNR_t)$ AND the trajectory is less then 40 steps.
  Update replay buffer $D \leftarrow D \cup \{P, a, P', r, d\}$.
  for $i = 1$ to MAX_EXPLORE_ITER **do**
    Sample a minibatch $\{P_m, a_m, P_m', r_m, d_m\}_{m=1}^M$ from D.
    Compute $f(r_m, P_m', d_m)$ defined in (12).
    $\phi_j \leftarrow \phi_j - \frac{1}{M}\sum_{m=1}^M \eta\nabla_{\phi_j}\mathcal{L}(P_m, a_m, P_m')$ with $\mathcal{L}$ defined in (11).
    $\omega \leftarrow \omega - \frac{1}{M}\sum_{m=1}^M \eta\nabla_\omega\mathcal{V}^\pi(P_m)$ with V^π defined in (14).
  end for
  Assemble $\tilde{I}_{test}$ from patches by applying learnt policy and the training networks
  Return: $\tilde{I}_{test}$.
end while

4 Experiments

In the experiment section, we validate our RSE-RL algorithm on two datasets: Synthesized Noisy CelebFaces Attributes (CelebA) Dataset [34] and Smartphone Image Denoising Dataset (SIDD) [2]. CelebA dataset consists of a set of faces of celebrities, and we apply Gaussian noise to each image in the dataset to form a synthesized noisy dataset. SIDD dataset consists of images with real artifacts generated by smartphone cameras under various light conditions. We apply our RSE-RL architecture to remove the artifacts in the image and expect that our architecture outperforms other denoising models. The baseline model is an ordinary variational autoencoder trained under identical settings with our RSE-RL; we denote it as *Single Decoder VAE.*

4.1 CelebA Synthesized Artifacts Denoising

Dataset Construction We collect images from the CelebA_HQ/256 dataset, where the size of each image is 256×256 pixels. We apply uniform Gaussian noise to each image in the CelebA dataset (using OpenCV [6]) to form a synthesized noisy image corresponding to the original clean image. We consider the clean image and its noisy image as an image pair. We divide the image pairs into a training set and a validation set. The training dataset consists of 2250 image pairs, and the validation dataset has 11,250 image pairs. We divide each image into 16×16 pixels patches, with 4 pixels overlapping with the surrounding patches, and then feed these patches into our network.

Experimental Setup Our encoder projects the patch-based images into the latent space using 5 convolutional layers and 2 fully-connected layers in a subsequent order. The decoder has an inverse structure to the encoder. It has 2 upsampling layers and 5 transposed convolutional layers. We implement the networks in Keras [11] and Tensorflow [1] and use a single 12GB NVIDIA Tesla K80 GPU for training and testing the synthesized noisy CelebA dataset. We set the training batch size to 128, training epochs to 50, regularization coeficient $\lambda_{reg} = 0.01$, optimizer to Adam [28] with $\beta_1 = 0.9$ and $\beta_2 = 0.999$. The learning rate is set by an exponential decay learning rate scheduler with an initial rate of 0.001, decay factor of 0.95, and decay step 1000 (Fig. 2).

For self-enhancement weight finetuning, we implement the SAC algorithm using Stable-Baseline3 [38] from OpenAI [7]. We use OpenAI Gym for setting up the environment. The reward in the environment is computed as stated in Eq. (13) with $b = 1.25$, $c = 5$, and $\text{PSNR}_t = 30.0$. We define the terminating condition d as the following: the SAC will terminate when the actual PSNR reaches the target psnr PSNR_t or the number of epochs reaches the maximum steps allowed, which is 40 in our experiments. The model tries to maximize the reward by optimizing the actions that adjust the trainable weights in the three transformation functions. The action space is a set of weight vectors within the range (0.999, 1.001). In the synthesized

Fig. 2 CelebA Denoising Results: images on the top row are images that contain the synthesized artifacts (Gaussian Noise). The images on the bottom row are the denoising result from our RSE-RL

Table 1 CelebA results comparison

	PSNR	SSIM	UQI
Image with artifacts	16.64	0.5835	0.7327
N2V [30]	21.66	0.7242	0.9249
N2N [36]	26.60	/	/
VAE	26.81	0.7621	0.9604
RSE-RL(before)	28.83	0.8322	0.9721
RSE-RL	**29.03**	**0.8339**	**0.9731**
RSE-RL(S)	28.79	0.8317	0.9717

CelebA dataset, the learning rate for the model is set to 0.001. The results before and after the recursive self-enhancing procedure are indicated in Table 1, denoted *RSE-RL(before)* and *RSE-RL*, respectively. We also present a set of recursively enhanced images in Fig. 3.

Denoising results We evaluate our result using PSNR, SSIM [48], and UQI [47] scores. Some baseline denoising methods, Noise2Void(N2V) and Noise2Noise(N2N) are applied to the synthesized artifacts dataset, and its denoising results are compared with our networks as well. For N2V, the default model for 2D RGB images is trained with 400 noisy images for 50 epochs and tested with 1575 noisy images. In the default model, each image is divided into 128 $16 \times 16 \times 3$ patches, so a total of 51,200 patches are fed into the network. For N2N, the pre-trained N2N model for Gaussian noise is tested on 1575 noisy images. Table 1 provides the results obtained using 2250 training images and 11250 testing images. The training images are divided into 1 million $16 \times 16 \times 3$ patches (441 patches per-image) and fed into the network. The result of the baseline method is compared to the results of our RSE-RL. We can observe a significant improvement in all the quality metrics using RSE-RL. A visualization of our denoised image results after self-enhancement can be found in Fig. 2. After applying the RL model for self-improvement, we can observe enhancements on all three quality metrics. This demonstrates that our self-enhancing model can be optimized during the testing iterations. Another observation is that the training data size does not significantly affect RSE-RL's denoising performance. We perform

Fig. 3 Recursive Self-enhancing RL Visualization: the figure shows how the test images are recursively enhanced in a 50-epoch RL training. We observe a performance boost when iterating the weights of decomposed transformations under three latent subspaces. With the reinforcement learning agent, the network converges to a better result compared to the case where only a solo VAE framework can achieve. The last line also specifies the difference between the starting image we fed into the RL agent and the final result after recursive learning

an experiment using 450 training images (0.2 million patches) and the same set of testing images to demonstrate this feature (Fig. 4). We present the result in Table 1, denoted as *RSE-RL(S)*. This result can be compared with *RSE-RL(before)* to observe how the training data size affects the performance. This observation provides an effective way of utilizing this network to largely reduce training time (Fig. 5).

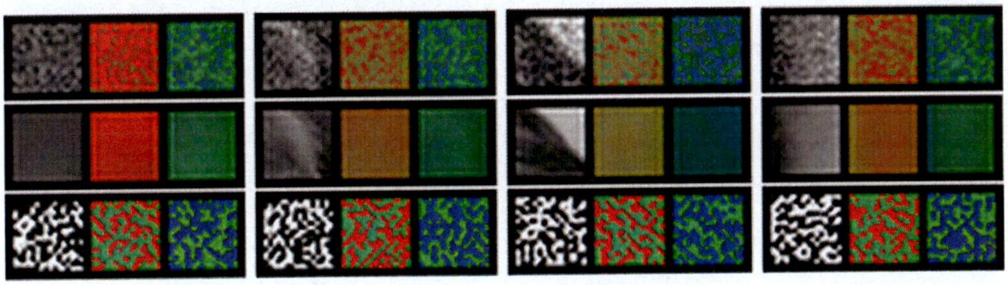

Fig. 4 Patch-based matching results on YUV spaces: first row shows the noisy patches, the second row is the clean patches that match the noisy patches, and the third is the contrast, representing the noise we are removing. Columns from left to right–three columns as a group–are the images on YUV spaces. The patches are scaled to [0, 255] for all the channels. The result justifies the correctness of our patch transferring scheme within each patch locally. The details are also preserved. Hence, our patch-based method is amenable to scale up to any size of the image

YUV Latent Space Visualization The results indicate that learning the transformations of the latent spaces is effective. By comparing the three latent subspaces Z_Y, Z_U, and Z_V, we can observe that the noise has the largest impact on the Y space Z_Y, which represents the luminance (brightness) of the image. There is no significant difference between the noisy and clean patch representations on the other two subspaces. This indicates that Gaussian noise significantly impacts brightness compared to chrominance (represented by U and V). The visualized Y latent subspaces is shown in Fig. 6. We also present the reconstructed patches from the YUV subspaces in Fig. 4

4.2 SIDD Denoising Result

Dataset Construction Smartphone Image Denoising Dataset (SIDD) is a benchmark for denoising algorithms. It contains about 30,000 noisy images captured by five representative smartphone cameras under 10 lighting conditions and their ground truth. Each noisy image features a mixture of artifacts generated under realistic scenarios. Artifacts caused by ISO levels, illumination, lighting conditions, and signal-dependent noise can all be seen within the dataset.

To test our network, we sample 320 sRGB images as the training data and used the SIDD Benchmark Data, which contains 40 noisy sRGB images and their ground truth, as the testing set. For each benchmark image, we sample 32 patches instead of taking an evaluation measure on the entire image. Besides, we preprocess the training dataset and divide each image, both noisy and growth truth, into $24 \times 24 \times 3$ patches, with 8 pixels overlapping. Consequently, there is a total of 11.19 million patches used for training.

Table 2 SIDD sRGB to sRGB results (small scale)

	PSNR	SSIM		PSNR	SSIM
Noisy image	31.18	0.831	BM3D [12]	25.65	0.685
NLM [8]	26.75	0.699	KSVD [5]	26.88	0.842
DANet [14]	39.25	0.955	RDB-Net [58]	38.11	0.945
VAE	31.89	0.874			
RSE-RL(before)	32.38	**0.891**	RSE-RL	**32.53**	0.887

Fig. 5 SIDD Denoising Result. To better visualize the denoising result, we zoom in at the area covered by the green rectangle in each figure on the left column. The figures from the left column to the right column are: original noisy images from SIDD, zoomed noisy patches, and zoomed denoising patches, respectively

Experiment Setup Our encoder (with approximately 2.2 million parameters) and decoder (with 1.6 million parameters) structures are the same as the one defined in the Experimental Setup of Sect. 4.1. Again, we use a single 12GB NVIDIA Tesla K80 GPU for training and testing on SIDD, and the training batch size is 128. The same optimizer is chosen as we did for the Synthesized Noisy CelebA dataset, defined in Sect. 4.1. Our model's parameters were optimized after 20 epochs of training. The SAC setting is identical to Sect. 4.1, except for we set $PSNR_t = 34.0$ in Eq. (13). The results before and after self-enhancement are shown in Table 2, denoted as *RSE-RL(before)* and *RSE-RL*, respectively.

Denoising Results The results show that our self-enhancing RL model contributes a small enhancement to PSNR, demonstrating that our RL model can improve the denoising results. Since we are only involving PSNR in the reward function, we can only observe some improvements in terms of PSNR.

Table 2 also lists benchmark denoising methods and deep learning methods used to compare against our network. In the table, noisy images are the images before denoising procedures; BM3D, NLM, and KSVD are the benchmark non-DL results; DANet and RDB-Net are two of the state-of-art deep learning methods used for comparison with our method. The performance of our model is significantly better than traditional methods. Figure 5 shows the visualized results of self-enhanced denoised images. As for efficiency comparison, our RSE-RL only contains 2.5 million parameters in total, whereas DANet contains $\sim$ 60 million parameters, leading our network train much faster than the state-of-the-art structure.

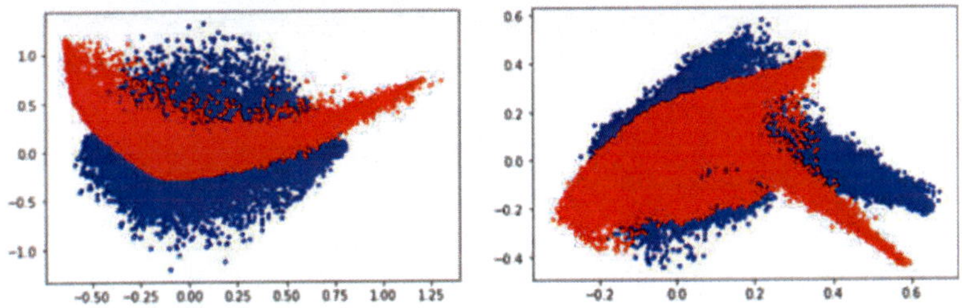

Fig. 6 Y latent subspaces display: the figure shows the difference between noisy and clean patches our network captured in Y latent subspaces. We show the Z_Y spaces in CelebA(left) and SIDD(right) datasets, respectively. Blue points are noise patch projections, and red points are clean patch projections. Principle Component Analysis is applied to reduce the dimension of the latent subspaces into 2 for visualization

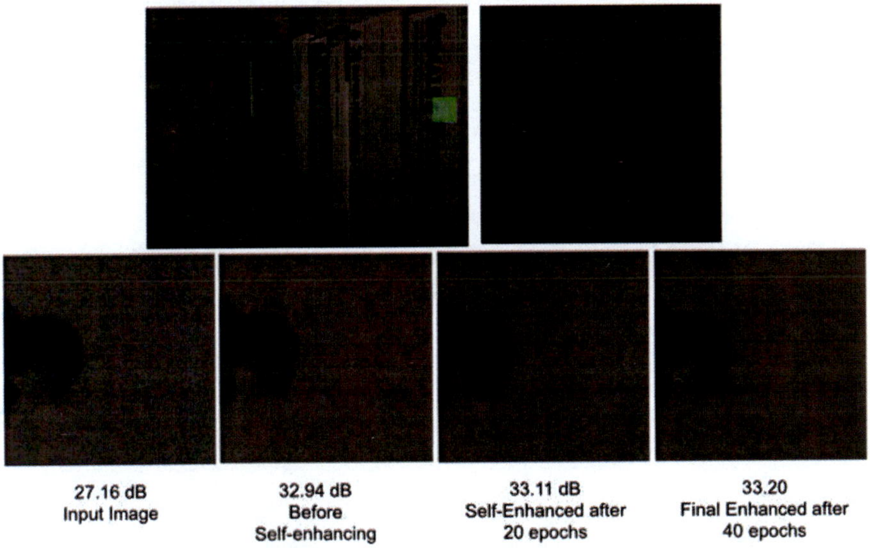

Fig. 7 Patch enhancement on SIDD: the top left figure shows where we select the patch; the top right figure shows the difference between this patch before and after the self-enhancement. The bottom row shows the patch enhancement over epochs

The images from SIDD consist of the same realistic artifacts generated by smartphone cameras. This leads to the same transformation in the latent space for every patch since each patch consists of the same types of noises. From Fig. 6 we can observe a transformation between the noisy patch projections and the clean patch projections on the latent space.

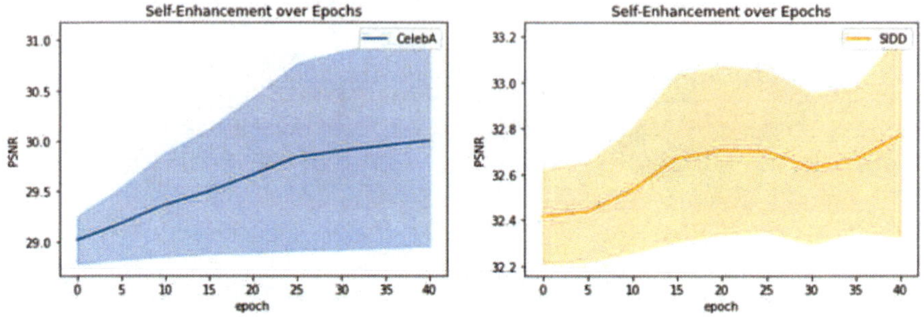

Fig. 8 Self-enhancement in PSNR over epochs: the left and right figures present the enhancement results on CelebA and SIDD, respectively. We also indicate the variance of our enhancement quality due to the RL algorithm. We perform the RL algorithm multiple times on the same set of images and compute the variances. As the figures presented, the PSNR tends to increase while the number of epochs increases

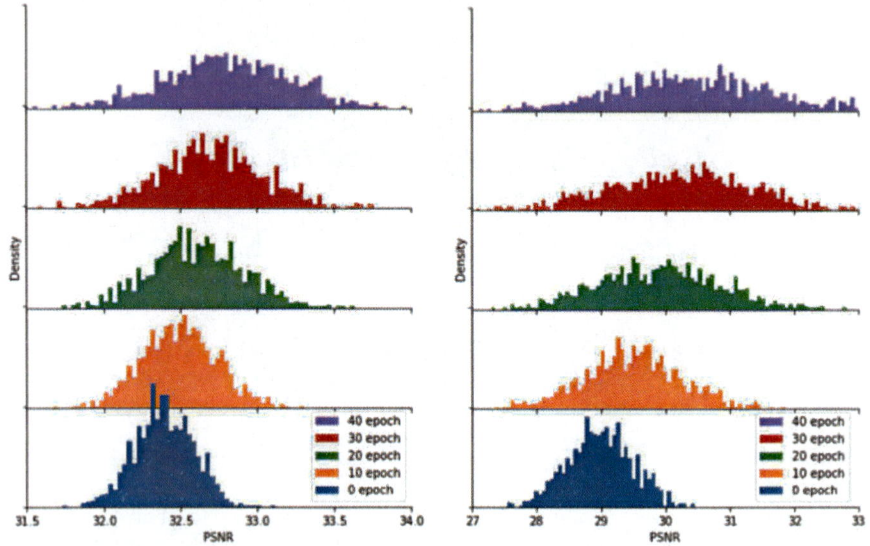

Fig. 9 Patch-PSNR distribution at each epoch: the left and right figures present the enhancement results on CelebA and SIDD test patches, respectively

Self-enhancing Visualization We then select a patch from an image in SIDD to visualize the self-enhancement on this zoomed region. We show the enhancement visualization of the zoomed region in Fig. 7. We can observe a small improvement in this patch before and after applying the RL algorithm. To quantify this improvement, we compute the PSNR scores for some sample images and show the relationship between the PSNR and the number of epochs of the RL algorithm in Fig. 8. We also compute the PSNR score for the patches and present the distribution of patch-PSNR at each epoch of self-enhancing in Fig. 9. We can observe a slight increase in the average PSNR of all the patches. Simultaneously, the variance of the patch-PSNR distribution also grows while the number of epochs increases.

5 Conclusions

We have presented a Recursive Self-Enhancing Reinforcement Learning (RSE-RL) model, for a self-improving camera ISP built upon adaptive and heterogeneous image filtering and patch specific policy learning. The patch-based transformations are progressively trained in multiple latent subspaces to identify and rectify, spatially heterogeneous and camera specific lens-color acquisitional image artifacts. We define our action spaces and reward function for a self-enhancement framework and further discuss its potential for real-word Camera ISPs. Nonetheless, our work is an early-stage exploration and exploitation solution. We are moving toward considering optimizing patch scan ordering specific protocols for more efficient processing, and more complicated environmental settings, to further strengthen our RSE_RL Camera ISP.

Acknowledgements The research was supported in part from the Peter O'Donnell Foundation. Further CB was funded in part from NIH DK129979, and from a grant from the Army Research Office accomplished under Cooperative Agreement Number W911NF-19-2-0333.

Appendix

Detailed Setup of Our RSE-RL Framework

Our network transforms the image patches from RGB to YUV channels before encoding. The RGB-YUV transformation is defined as

$$\begin{bmatrix} P_y \\ P_u \\ P_v \end{bmatrix} = \begin{bmatrix} 0.299 & 0.587 & 0.114 \\ -0.147 & -0.289 & 0.436 \\ 0.615 & -0.515 & -1.000 \end{bmatrix} \begin{bmatrix} P_r \\ P_g \\ P_b \end{bmatrix}$$

An encoder $\boldsymbol{q}$ encodes the YUV channels respectively and projects the patch information on three latent subspaces Z_y, Z_u, and Z_v. The dimension of one subspace is set to 72 for both sets of experiments, hence the latent space dimension is 216. In each of the latent subspaces, both clean and noisy patch representations are projected, and we want to learn a transformation that matches noisy patch representations to clean patch representations. The transformations T_y, T_u, T_v are defined and operated in their corresponding latent subspaces. Each of the transformation T_s ($s \in \{y, u, v\}$) is a three-layer MLP, with identical dimension layers and ReLU activation. Each transformation is trained to match from a noisy patch representation z_s^b to a clean patch representation z_s^c within its latent subspaces, using the loss function is defined in Eq. (8).

Additional Experimental Results of Our RSE-RL Framework

Decomposed Subspace Visualization The following sets of figures show the denoising results on each of the Y, U, and V spaces, which further demonstrate how the noises are removed from each space. Figures 10 and 11 present the examples on both our synthesized CelebA dataset and SIDD dataset. These figures give several specific patches as examples for demonstrating how the noisy patches map to the clean patches. It also shows the noises on the patches specifically, and we can observe the noises on the Y, U, and V spaces.

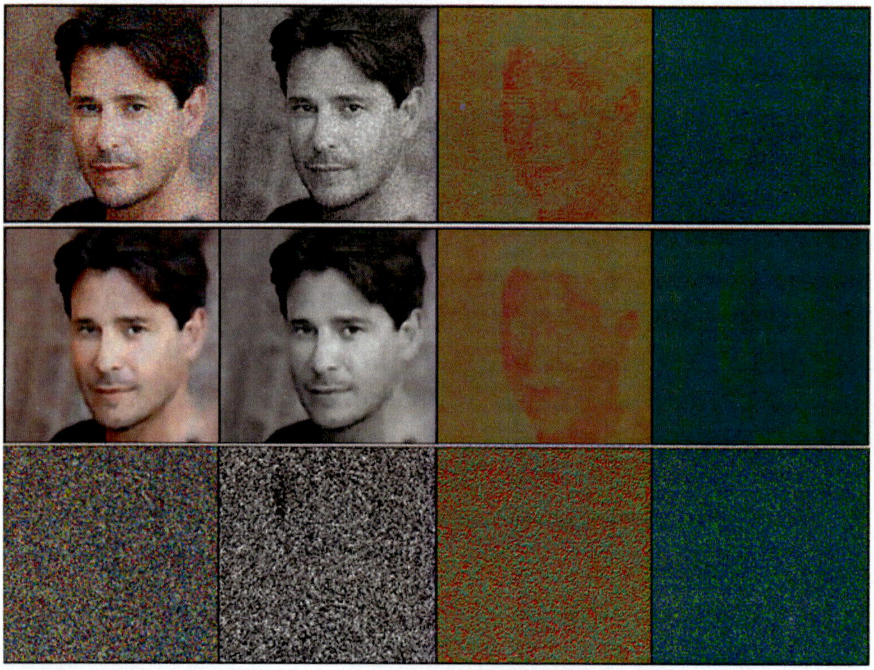

Fig. 10 CelebA Denoising Visualization in YUV Spaces: images on the top row are images that contain the synthesized artifacts (Gaussian Noise). The images on the second row are the denoising result from our RSE-RL. And the images on the bottom row show the difference between the first two rows, which are the expected noises we removed by the network. The images are scaled to [0, 255] for all channels. Columns from left to right show the images on RGB channels, Y space, U space, and V space, respectively. Our method reveals and remove the noise decomposed in three channels, respectively

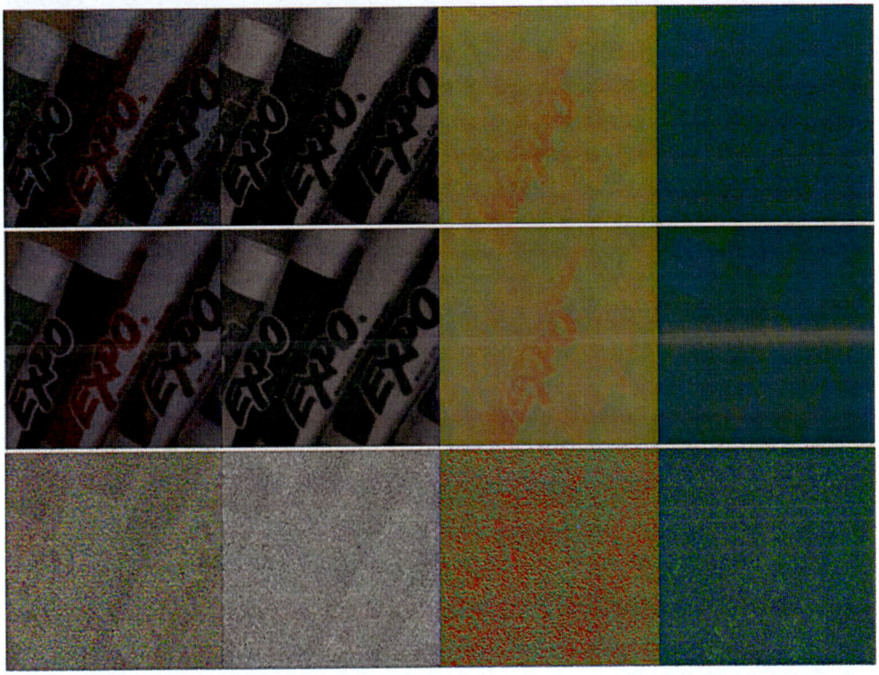

Fig. 11 SIDD Denoising Visualization in YUV Spaces: specifications are identical to Fig. 10. The figure demonstrates a noise removal over channels and show our patch-based method can apply to large-scale, realistic image as well

Fig. 12 Deblocking Results: this figure shows the results of a deblocking method [27], as well as our overlapping patch smoothing alternative. It shows that our overlapping patch smoothing method can remove the block artifacts that may be created by our patch-based scheme. The columns from left to right show the noisy image, image composed by patches without overlapping, non-overlapping patches with deblocking enhancement, image with overlapping patch smoothing, and image with overlapping patch smoothing + deblocking enhancement. The PSNR scores for these images from left to right are 19.89, 29.51, 29.50, 30.31, and 30.31

Justification of Removing Block Artifacts There might be the case where our filter generates patch-based enhancement result locally while ignoring the neighboring patches. The one-to-one correspondence from noisy patches to clean patches might cause additional block artifacts. We propose the post-processing using overlapping patch smoothing, or additional deblocking algorithm to correct the newly introduced artifacts. Below we show an ablation study under the influence of overlapping patch selections and the use of deblocking artifacts.

In general testing, we compare the qualities between the non-overlapping patches and overlapping patches, as well as the qualities before and after using the deblocking method [27]. The average PSNR for images composed of non-overlapping patches is 27.8214. And we can observe obvious blocking artifacts on the edge of the patches (in Fig. 12). When we apply the deblocking method, the average PSNR is slightly reduced to 27.8212 and the block artifacts can still be visualized.

By comparison, after we apply the overlapping patches, there is a smooth transition on each of the edges between the two blocks. The average PSNR for images composed of overlapping patches is 28.84, which is a significant improvement. We can also observe the enhancement in Fig. 12. However, we applied the deblocking method to the images with overlapping patches and there is no observable improvement, while the average PSNR stays the same.

References

1. Abadi M et al (2015) TensorFlow: large-scale machine learning on heterogeneous systems. Software available from https://www.tensorflow.org/
2. Abdelhamed A, Lin S, Brown MS (2018) A high-quality denoising dataset for smartphone cameras. In: 2018 IEEE/CVF conference on computer vision and pattern recognition, pp 1692–1700
3. Abdelhamed A, Timofte R, Brown MS, Yu S, Park B, Jeong J, Jung SW, Kim DW, Chung JR, Liu J, Wang Y, Wu CH, Xu Q, Wang C, Cai S, Ding Y, Fan H, Wang J, Zhang K, Zuo W, Zhussip M, Park DW, Soltanayev S, Chun SY, Xiong Z, Chen C, Haris M, Akita K, Yoshida T, Shakhnarovich G, Ukita N, Zamir SW, Arora A, Khan S, Khan FS, Shao L, Ko SJ, Lim DP, Kim SW, Ji SW, Lee SW, Tang W, Fan Y, Zhou Y, Liu D, Huang TS, Meng D, Zhang L, Yong H, Zhao Y, Tang P, Lu Y, Schettini R, Bianco S, Zini S, Li C, Wang Y, Cao Z (2019) Ntire 2019 challenge on real image denoising: methods and results. In: 2019 IEEE/CVF conference on computer vision and pattern recognition workshops (CVPRW), pp 2197–2210
4. Afifi M, Brown MS (2020) Deep white-balance editing. In: Proceedings of the IEEE/CVF conference on computer vision and pattern recognition, pp 1397–1406
5. Aharon M, Elad M, Bruckstein A (2006) K-svd: an algorithm for designing overcomplete dictionaries for sparse representation. IEEE Trans Sig Process 54(11):4311–4322
6. Bradski G (2000) The OpenCV Library. Dr. Dobb's, J Softw Tools
7. Brockman G, Cheung V, Pettersson L, Schneider J, Schulman J, Tang J, Zaremba W (2016) Openai gym. arXiv preprint arXiv:1606.01540
8. Buades A, Coll B, Morel JM (2005) A non-local algorithm for image denoising. In: 2005 IEEE computer society conference on computer vision and pattern recognition (CVPR'05), vol 2. IEEE, pp 60–65
9. Cao Y, Wu X, Qi S, Liu X, Wu Z, Zuo W (2021) Pseudo-isp: learning pseudo in-camera signal processing pipeline from a color image denoiser. arXiv preprint arXiv:2103.10234

10. Cheng S, Wang Y, Huang H, Liu D, Fan H, Liu S (2021) Nbnet: noise basis learning for image denoising with subspace projection. In: Proceedings of the IEEE/CVF conference on computer vision and pattern recognition (CVPR), pp 4896–4906
11. Chollet F et al (2015) Keras. https://keras.io
12. Dabov K, Foi A, Katkovnik V, Egiazarian K (2007) Image denoising by sparse 3-D transform-domain collaborative filtering. IEEE Trans Image Process 16(8):2080–2095
13. Dilokthanakul N, Mediano PA, Garnelo M, Lee MC, Salimbeni H, Arulkumaran K, Shanahan M (2016) Deep unsupervised clustering with gaussian mixture variational autoencoders. arXiv preprint arXiv:1611.02648
14. Fu J, Liu J, Tian H, Li Y, Bao Y, Fang Z, Lu H (2019) Dual attention network for scene segmentation. In: Proceedings of the IEEE/CVF conference on computer vision and pattern recognition, pp 3146–3154
15. Furuta R, Inoue N, Yamasaki T (2019) Pixelrl: fully convolutional network with reinforcement learning for image processing. IEEE Trans Multimedia 22(7):1704–1719
16. Galteri L, Seidenari L, Bertini M, Del Bimbo A (2017) Deep generative adversarial compression artifact removal. In: Proceedings of the IEEE international conference on computer vision, pp 4826–4835
17. Ghimpeţeanu G, Batard T, Seybold T (2016) Bertalmío M (2016) Local denoising applied to raw images may outperform non-local patch-based methods applied to the camera output. Electron Imaging 18:1–8
18. Guo S, Yan Z, Zhang K, Zuo W, Zhang L (2019) Toward convolutional blind denoising of real photographs. In: Proceedings of the IEEE conference on computer vision and pattern recognition, pp 1712–1722
19. Haarnoja T, Zhou A, Abbeel P, Levine S (2018) Soft actor-critic: off-policy maximum entropy deep reinforcement learning with a stochastic actor. In: International conference on machine learning. PMLR, pp 1861–1870
20. Haarnoja T, Zhou A, Hartikainen K, Tucker G, Ha S, Tan J, Kumar V, Zhu H, Gupta A, Abbeel P et al (2018) Soft actor-critic algorithms and applications. arXiv preprint arXiv:1812.05905
21. Hou Y, Xu J, Liu M, Liu G, Liu L, Zhu F, Shao L (2020) NLH: a blind pixel-level non-local method for real-world image denoising. IEEE Trans Image Process 29:5121–5135
22. Hu Y, He H, Xu C, Wang B, Lin S (2018) Exposure: a white-box photo post-processing framework. ACM Trans Graph (TOG) 37(2):1–17
23. Ignatov A, Van Gool L, Timofte R (2020) Replacing mobile camera isp with a single deep learning model. In: Proceedings of the IEEE/CVF conference on computer vision and pattern recognition workshops, pp 536–537
24. Jang G, Lee W, Son S, Lee KM (2021) C2n: practical generative noise modeling for real-world denoising. In: Proceedings of the IEEE/CVF international conference on computer vision (ICCV), pp 2350–2359
25. Khademi W, Rao S, Minnerath C, Hagen G, Ventura J (2021) Self-supervised poisson-gaussian denoising. In: Proceedings of the IEEE/CVF winter conference on applications of computer vision (WACV), pp 2131–2139
26. Khashabi D, Nowozin S, Jancsary J, Fitzgibbon AW (2014) Joint demosaicing and denoising via learned nonparametric random fields. IEEE Trans Image Process 23(12):4968–4981
27. Kim SD, Yi J, Kim HM, Ra JB (1999) A deblocking filter with two separate modes in block-based video coding. IEEE Trans Circ Syst Video Technol 9(1):156–160
28. Kingma DP, Ba J (2015) Adam: a method for stochastic optimization. In: 3rd International conference on learning representations, ICLR 2015, San Diego, CA, USA, May 7–9, 2015, conference track proceedings. http://arxiv.org/abs/1412.6980
29. Kingma DP, Welling M (2014) Auto-encoding variational bayes. In: 2nd International conference on learning representations, ICLR 2014, Banff, AB, Canada, April 14–16, 2014, Conference Track Proceedings
30. Krull A, Buchholz TO, Jug F (2019) Noise2void-learning denoising from single noisy images. In: Proceedings of the IEEE/CVF conference on computer vision and pattern recognition, pp 2129–2137

31. Kupyn O, Martyniuk T, Wu J, Wang Z (2019) Deblurgan-v2: deblurring (orders-of-magnitude) faster and better. In: Proceedings of the IEEE international conference on computer vision, pp 8878–8887
32. Liu H, Liu X, Lu J, Tan S (2021) Self-supervised image prior learning with gmm from a single noisy image. In: Proceedings of the IEEE/CVF international conference on computer vision (ICCV), pp 2845–2854
33. Liu Y, Qin Z, Anwar S, Ji P, Kim D, Caldwell S, Gedeon T (2021) Invertible denoising network: a light solution for real noise removal. In: Proceedings of the IEEE/CVF conference on computer vision and pattern recognition (CVPR), pp 13365–13374
34. Liu Z, Luo P, Wang X, Tang X (2015) Deep learning face attributes in the wild. In: Proceedings of international conference on computer vision (ICCV), pp 3730–3738
35. Mnih V, Badia AP, Mirza M, Graves A, Lillicrap T, Harley T, Silver D, Kavukcuoglu K (2016) Asynchronous methods for deep reinforcement learning. In: International conference on machine learning. PMLR, pp 1928–1937
36. Moran N, Schmidt D, Zhong Y, Coady P (2020) Noisier2noise: learning to denoise from unpaired noisy data. In: Proceedings of the IEEE/CVF conference on computer vision and pattern recognition, pp 12064–12072
37. Plotz T, Roth S (2017) Benchmarking denoising algorithms with real photographs. In: Proceedings of the IEEE conference on computer vision and pattern recognition, pp 1586–1595
38. Raffin A, Hill A, Ernestus M, Gleave A, Kanervisto A, Dormann N (2019) Stable baselines3. https://github.com/DLR-RM/stable-baselines3
39. Ren C, He X, Wang C, Zhao, Z (2021) Adaptive consistency prior based deep network for image denoising. In: Proceedings of the IEEE/CVF conference on computer vision and pattern recognition (CVPR), pp 8596–8606
40. Rudin LI, Osher S, Fatemi E (1992) Nonlinear total variation based noise removal algorithms. Phys D Nonlinear Phenom 60(1–4):259–268
41. Schwartz E, Giryes R, Bronstein AM (2018) Deepisp: toward learning an end-to-end image processing pipeline. IEEE Trans Image Process 28(2):912–923
42. Simoncelli EP, Adelson EH (1996) Noise removal via bayesian wavelet coring. In: Proceedings of 3rd IEEE international conference on image processing, vol 1. IEEE, pp 379–382
43. Strela V, Portilla J, Simoncelli EP (2000) Image denoising using a local gaussian scale mixture model in the wavelet domain. In: Wavelet applications in signal and image processing 8, vol 4119. International Society for Optics and Photonics, pp 363–371
44. Suin M, Purohit KRajagopalan A (2020) Spatially-attentive patch-hierarchical network for adaptive motion deblurring. In: Proceedings of the IEEE/CVF conference on computer vision and pattern recognition, pp 3606–3615
45. Tian C, Fei L, Zheng W, Xu Y, Zuo W, Lin CW (2020) Deep learning on image denoising: an overview. Neural Netw 131
46. Valsesia D, Fracastoro G, Magli E (2020) Deep graph-convolutional image denoising. IEEE Trans Image Process 29:8226–8237
47. Wang Z, Bovik AC (2002) A universal image quality index. IEEE Sig Process Lett 9(3):81–84
48. Wang Z, Bovik AC, Sheikh HR, Simoncelli EP (2004) Image quality assessment: from error visibility to structural similarity. IEEE Trans Image Process 13(4):600–612
49. Xia Z, Gharbi M, Perazzi F, Sunkavalli K, Chakrabarti A (2021) Deep denoising of flash and no-flash pairs for photography in low-light environments. In: Proceedings of the IEEE/CVF conference on computer vision and pattern recognition (CVPR), pp 2063–2072
50. Xie Y, Wang Z, Ji S (2020) Noise2same: optimizing a self-supervised bound for image denoising. In: Larochelle H, Ranzato M, Hadsell R, Balcan MF, Lin H (eds) Advances in neural information processing systems, vol 33. Curran Associates, Inc., pp 20320–20330. https://proceedings.neurips.cc/paper/2020/file/ea6b2efbdd4255a9f1b3bbc6399b58f4-Paper.pdf
51. Xu J, Huang Y, Cheng MM, Liu L, Zhu F, Xu Z, Shao L (2020) Noisy-as-clean: learning self-supervised denoising from corrupted image. IEEE Trans Image Process 29:9316–9329
52. Xu X, Li M, Sun W (2019) Learning deformable kernels for image and video denoising. arXiv preprint arXiv:1904.06903

53. Yang Y, Zheng Y, Wang Y, Bajaj C (2021) Learning deep latent subspaces for image denoising. arXiv preprint arXiv:2104.00253
54. Yu K, Dong C, Lin L, Loy CC (2018) Crafting a toolchain for image restoration by deep reinforcement learning. In: Proceedings of the IEEE conference on computer vision and pattern recognition, pp 2443–2452
55. Zamir SW, Arora A, Khan S, Hayat M, Khan FS, Yang MH, Shao L (2020) CycleISP: real image restoration via improved data synthesis. In: Proceedings of the IEEE/CVF conference on computer vision and pattern recognition, pp 2696–2705
56. Zhang K, Zuo W, Chen Y, Meng D, Zhang L (2017) Beyond a gaussian denoiser: residual learning of deep CNN for image denoising. IEEE Trans Image Process 26(7):3142–3155
57. Zhang R, Zhu J, Zha Z, Dauwels J, Wen B (2021) R3l: connecting deep reinforcement learning to recurrent neural networks for image denoising via residual recovery. In: 2021 IEEE international conference on image processing (ICIP). IEEE, pp 1624–1628
58. Zhang Y, Tian Y, Kong Y, Zhong B, Fu Y (2018) Residual dense network for image super-resolution. In: Proceedings of the IEEE/CVF conference on computer vision and pattern recognition, pp 2472–2481
59. Zhang Z, Wang H, Liu M, Wang R, Zhang J, Zuo W (2021) Learning raw-to-srgb mappings with inaccurately aligned supervision. In: Proceedings of the IEEE/CVF international conference on computer vision (CVPR), pp 4348–4358

PBDPA: A Task Scheduling Algorithm in Containerized Cloud Computing Environment

Himanshukamal Verma and Vivek Shrivastava

Abstract The container is a lightweight and agile virtualization solution that effectively addresses VM issues in cloud infrastructures such as overhead and compatibility. Users' tasks are assigned to containers in the form of cloudlets to ensure efficient management of cloud resources. Task scheduling with large job sizes and strict deadlines is a challenging job in cloud environments. Several jobs may need to be effectively assigned on several containers while minimizing makespan and optimizing resource efficiency. This paper presents a PROMETHEE-II-Based Dynamic Priority Algorithm (PBDPA) that incorporates multi-criteria-based optimization for dealing with multi-objective task scheduling issues. The suggested technique is evaluated using the CloudSim simulator. Experimental results on 3000 tasks successfully demonstrate that the proposed PBDPA algorithm maximizes performance in terms of time, revenue, resource utilization, and quality of service (QoS).

Keywords Cloud computing · Container · Scheduling · PROMETHEE · MCDM · PBDPA

1 Introduction

Cloud computing (CC) is a novel approach to internet service hosting. The cloud computing concept has emerged with the fast advancement of processing and storage technology over the Internet. Users may lease, manage, and release software applications and infrastructure over the Internet from anywhere at any time using CC services.

H. Verma (✉)
School of Computer Science & IT, Devi Ahilya University, Indore, Madhya Pradesh 452001, India
e-mail: himanshuk.verma1990@gmail.com

V. Shrivastava
International Institute of Professional Studies, Devi Ahilya University, Indore, Madhya Pradesh 452001, India
e-mail: vivek.shrivastava@iips.edu.in

S. Das et al. (eds.), *Advances in Data-driven Computing and Intelligent Systems*, Lecture Notes in Networks and Systems 653, https://doi.org/10.1007/978-981-99-0981-0_23

CC provides users with personalized, scalable, and QoS-assured computing environments that are accessible from anywhere. Users just have to invest in subscribing to computing services instead of purchasing and maintaining in-house computing resources. As a result, cloud consumers pay only for the services that they utilize on a pay per usage. These benefits make cloud computing a favorable model for fulfilling users' computing demands [1].

Container-based virtualization exploits the host's kernel to prevent hardware simulation as an alternate virtualization strategy for deploying applications and compute operations [2, 3]. Despite hypervisor-based virtualization, which runs a virtual machine instance with the entire operating system, containers are built on operating system images and share a safe set of system libraries [4]. The fundamental advantage of containers is that they are lightweight and use physical resources considerably more efficiently than VM-based hardware virtualization [5]. The more VMs that are deployed on a server, the worse the server's performance. Containers, on the other hand, may be deployed in the thousands on a server with hardly any effect [6].

Task scheduling has a major area of research in the cloud environment, gaining a lot of attention in the academic literature [7]. Task scheduling activities centered on the priority concept are still a crucial and challenging subject. It may not only improve cloud consumers' and providers' satisfaction, but also accomplish efficient resource utilization, maximize profit, and high-performance computing. To address the scheduling issues, several studies have been published. However, almost all of those papers treated all the tasks as similar, ignoring the fact that the tasks varied significantly in nature and requirements. They might potentially have various levels of importance.

2 Related Study

Shi et al. [8] introduced a BMin algorithm to improve the efficiency of the min-min method. The proposed technique is evaluated using CloudSim, with minimized completion time, maximized throughput, and improved resource load balancing.

Zong et al. [9] proposed a combination of a dynamic fusion mission planning technique, a genetic approach, and an ant colony system. Their work reduced the amount of time and energy consumption used by cloud computing data and storage facilities. Yao et al. [10] suggested a "three-stage selection procedure," and the "total-division-total" genetic approach for improving genetic strategy. The CloudSim tool's results showed that their algorithm beats a simple genetic algorithm (SGA) in terms of job completion time.

In the paper by Gupta et al. [11], the authors suggested a CC system simulation of the earliest deadline first scheduling based on priorities. The jobs are assigned according to their importance, with the highest priority tasks being scheduled first. The suggested approach improves memory consumption while simultaneously increasing performance. Similarly, Alla et al. [12] the authors presented task scheduling with energy awareness and deadline consideration in cloud computing environment. Prioritized tasks with deadline constraints are scheduled first. The suggested approach optimizes memory consumption while increasing performance.

The authors Bala et al. [13] suggested a strategy with task scheduling based on priorities in their study, which prioritizes tasks depending on the length of the instructions. Using six sigma control charts, their chapter prioritized numerous separate jobs in a workflow. The tasks have been divided into multiple levels, as well as the resources were distributed in accordance with the needs of the various work levels. The results of the experiments showed that the suggested method may significantly minimize makespan and execution time.

In business cloud architecture, Khan et al. [14] and Alla et al. [15] suggested priority-based service scheduling mechanisms for different sets and sizes of tasks. The suggested methods are based on the AHP method (Analytical Hierarchy Process). Before the services are handled in a specified sequence, they are given a predetermined based on a list of criteria for prioritization. The criterion weights of the submitted tasks were calculated using the AHP approach. An eigenvector of the matrix and a pairwise comparisons matrix are both used to calculate the criteria weights. The suggested method also helps to reduce service wait times in scheduler and maximize service quality before it is delivered to the client.

Although the majority of the works mentioned above are concerned with energy usage and makespan, there are still certain restrictions. Researchers enhanced task schedules by using energy consumption and revenue as targets, ignoring essential performance measures such as the computational time and load balancing. Similarly, higher energy efficiency should result from a resource allocation approach that considers resource consumption. However, scheduling a job with a single purpose in mind, such as energy consumption, might degrade overall performance without a flexible and optimum resource allocation methodology. Despite how crucial the deadline is, few just several works have taken this limitation into consideration. Some works are based on MCDM like ELLECTREE [16] and AHP [17–19], which consider the multiple criteria for efficient scheduling.

This work implements the scheduling scheme concatenated prioritization of tasks to solve the aforementioned challenges. Different users' requests might have varying criteria according to their requirements. In order to deal with the classification of user task, this research also presents a solution based on an MCDM that involves grouping with an evolutionary algorithm, such as differential evolution. As a first stage, this method ranks these tasks based on their importance and categorizes them into several types. It assigns every task a dynamic priority before scheduling them to the relevant resources. In order to create an intelligent, flexible, and unique system that can choose the optimum resources to manage and deploy, these resources are also categorized and given a priority. Several tests reflecting various scenarios were carried out to evaluate the efficacy of our proposed algorithm. The performance study is carried out under a variety of complicated circumstances, including both fixed and dynamic priority assignments.

The rest body of the chapter is structured as follows: Sect. 2 summarizes essential research. The scheduling problem is described in Sect. 3, followed by a discussion of the suggested solution in detail. Section 4 discusses the experimental setup and simulation findings. Section 5 concludes the paper.

3 Proposed Work

In CC, several factors should be taken into consideration in task scheduling. Prioritization of tasks is a critical problem that must be addressed throughout the scheduling process since certain jobs must be completed earlier than others that may be left unattended for an extended time. As a result, a good task scheduling algorithm should consider the priority of tasks according to multiple measures. As a result, the cloud broker should implement an optimum work scheduling algorithm based on these measures. The following are the primary objectives of the proposed research, based on the issues mentioned above:

- Using MCDM PROMETHEE-II [20], determine the priority of tasks.
- Distribute jobs across the dynamic Priority-Queue based on the distribution of decisions and the priority level.
- To obtain high performance, schedule, and execute the jobs held in the dynamic queue using a meta-heuristic method.

We develop PBDPA model representing above mention objectives as shown in Fig.1.

Another type of MCDM tool that can rank order choice alternatives is the preference function-based outranking method. Brans and Vincke created the PROMETHEE (preferred ranking organization method for enrichment evaluation) approach in 1985. The PROMETHEE-I approach may offer a partial organization of the choice options, but the PROMETHEE-II method can obtain the entire ranking.

Step I: The following equation is used to normalize the decision matrix.

For beneficial criteria

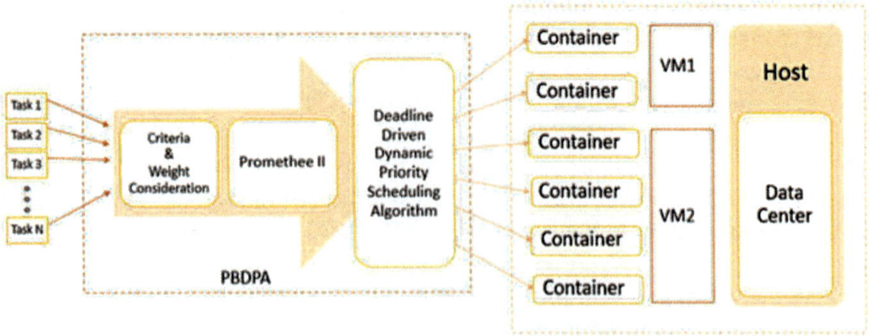

Fig. 1 PBDPA model

$$A_{ij} = \frac{[X_{ij} - \text{Min}(X_{ij})]}{[\text{Max}(X_{ij}) - \text{Min}(X_{ij})]} \quad (1)$$

For non-beneficial criteria

$$A_{ij} = \frac{[\text{Max}(X_{ij}) - X_{ij}]}{[\text{Max}(X_{ij}) - \text{Min}(X_{ij})]} \quad (2)$$

where X_{ij} is the performance measure of ith alternative concerning jth criterion.

Step II: Calculate the qualitative difference between the ith choice and others. This step includes calculating the variations in criterion values between options pairings.

Step III: Calculation of the preference function $P_J(i, i')$.

$$P_J(i, i') = 0 \text{ if } A_{ij} \leq A_{i'j} \quad (3)$$

$$P_J(i, i') = (A_{ij} - A_{i'j}) \text{ if } A_{ij} > A_{i'j} \quad (4)$$

Step IV: Calculation of aggregated preference function (APF) according to criteria weights.

$$\text{APF}(i, i) = \frac{\left[\sum_{j=1}^{m} W_j X P_j(i, i')\right]}{\sum_{j=1}^{m} W_j} \quad (5)$$

Step V: Calculate the leaving outranking (LR) and entering outranking (ER) flows.

$$\text{LR}(i) = \frac{1}{n-1} \sum_{i'=1}^{n} \text{APF}(i, i')i \neq i' \quad (6)$$

$$\text{ER}(i) = \frac{1}{n-1} \sum_{i'=1}^{n} \text{APF}(i', i)i \neq i' \quad (7)$$

Step VI: Net outranking (NR) calculation for each alternative.

$$\text{NR} = \text{LR} - \text{ER} \quad (8)$$

Step VII: Determine the order in which all of the options are ranked based on the values of F. The better the alternative, the higher the value of NR. As a result, the best option is the one with the highest NR value.

The PROMETHEE-II technique is a multi-criteria interactive decision-making strategy that can handle both quantitative and qualitative criteria with discrete options. Other MCDM methodologies, such as multi-attribute utility theory (MAUT) and AHP, have major disadvantages when compared to the PROMETHEE-II method.

The PROMETHEE-II approach can categorize non-comparable options as alternatives that are difficult to compare due to a trade-off relationship between assessment standards [20].

Further, in the next step of the algorithm based on decision distribution and priority level, distribute jobs across dynamic Priority-Queue. Then tasks are allocated to required resources on the basis of their deadline constraints.

The proposed algorithm PBDPA, which incorporates PROMETHEE-II, may be exploited for task scheduling with optimal resource utilization.

4 Experimental Setup

The suggested technique is simulated and evaluated using the CloudSim simulator. Anyone may model and simulate single and linked clouds with this simulator. CloudSim makes it possible to model, test, and simulate cloud computing systems and its applications in a seamless environment. The simulation is run on 2 Datacenters, 2 hosts, and 30 Container VMs, with CPU MIPS 1000, RAM 512 MB, and bandwidth is 1000. We acquire tasks from workload data as the source of the task.

PBDPA Algorithm
Input: Number of Tasks
Output: Scheduled Tasks
Let T_i is the number of tasks such that $i = 0$ to $i = n - 1$
And C_j is the number of criteria such that $j = 0$ to $j = m - 1$
Assume that the weight for each criterion C_j is W_j
Initialize TaskCriteria[m][n] such that Criteria[i][j] := [A_i][C_j]
for $i = 0$ to $n - 1$ & $j = 0$ to $m - 1$
Call ImposeCriteria (TaskCriteria, C_j,, flag) for $j = 0$ to $j = m - 1$
where flag := 0 for beneficial criteria, flag := 1 for non-beneficial criteria
Assign *NormalizeSpace[n][n]* by Performing following steps:
Initialize l=0
for $i = 0$ to $i = n - 1$
Initialize k=i
for $j = 0$ to $j = n - 1$
Assign (TaskCriteria[j][k]-TaskCriteria[i][l])W_l to NormalizeSpace[i][j] for Non-negative value;*
Assign 0 otherwise;
End for
l=l+1
End for
Call TaskSchedule (T_i)T_i with all attributes to Schedule tasks according to higher weight factor first.
Allocate Resources to T_i for $TaskBurst_i$ < $Deadline_i$
Continue with T_{i+1} otherwise;

5 Experimental Result

Several tests were conducted with different parameter values to evaluate the effectiveness of the suggested strategy. The comparison of the four makespans of four methods is shown in Fig. 2. The suggested algorithm PBDPA outperformed FCFS, EDF, and random in terms of makespan in simulations compared to the other three algorithms. Makespan of PBDPA is superior due to the fact that the tasks are sorted into various levels and submitted in accordance with the priorities. PBDPA has a competitive edge in terms of performance and resource utilization.

The comparison of PBDPA, FCFS, EDF, and random in terms of throughput is shown in Fig. 3. Figure showing the number of tasks executed successfully in PBDPA is greater than the other three algorithms. It represents that QoS and performance of PBDPA are better as compared to the other three algorithms.

To evaluate the throughput of the presented method PBDPA, a number of experiments were performed with strict deadlines and limited time. The comparison of the throughput of various algorithms is shown in Fig. 4. When compared to the other three algorithms, PBDPA executes a greater number of tasks at a particular time.

As the paper discussed earlier, the proposed algorithm PBDPA is multi-objective and focused on several criteria. One of the important criteria is to generate better revenue for a cloud service provider with other objectives. The experimental results show that the proposed algorithm PBDPA is much better performed than FCFS, EDF, and random in terms of revenue generation in simulations. The comparison of the revenues is shown in Fig. 4. When compared to the other three algorithms, the revenue of PBDPA is higher than the other three algorithms.

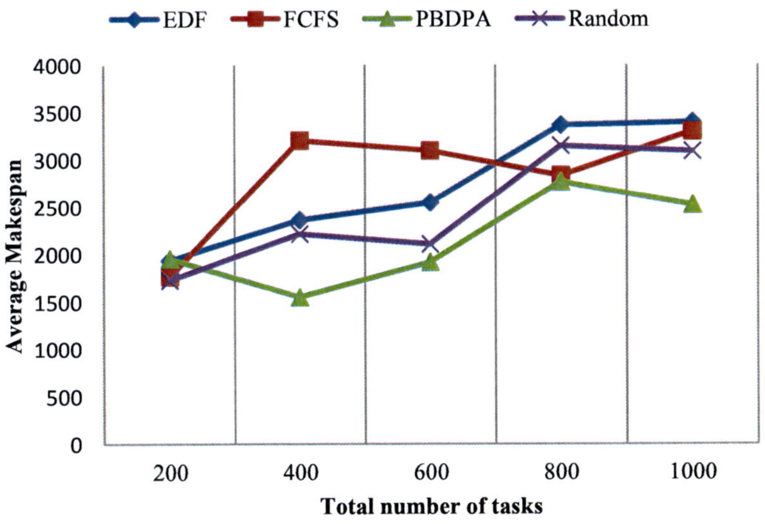

Fig. 2 Average makespan comparison

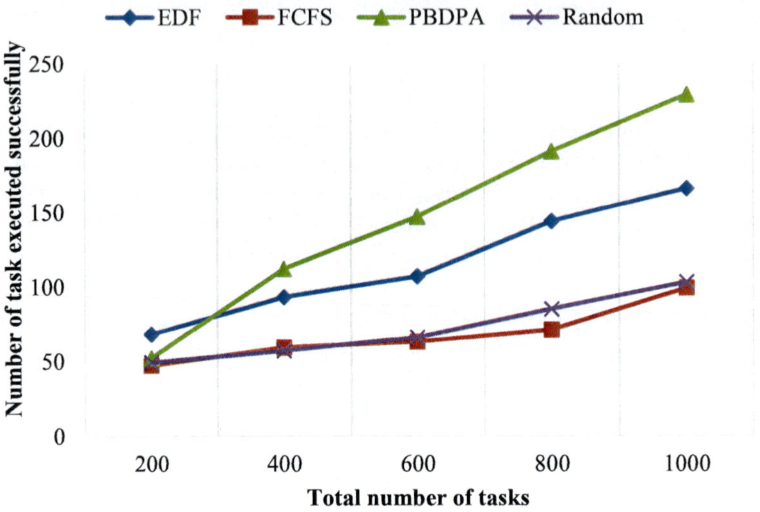

Fig. 3 Throughput comparison

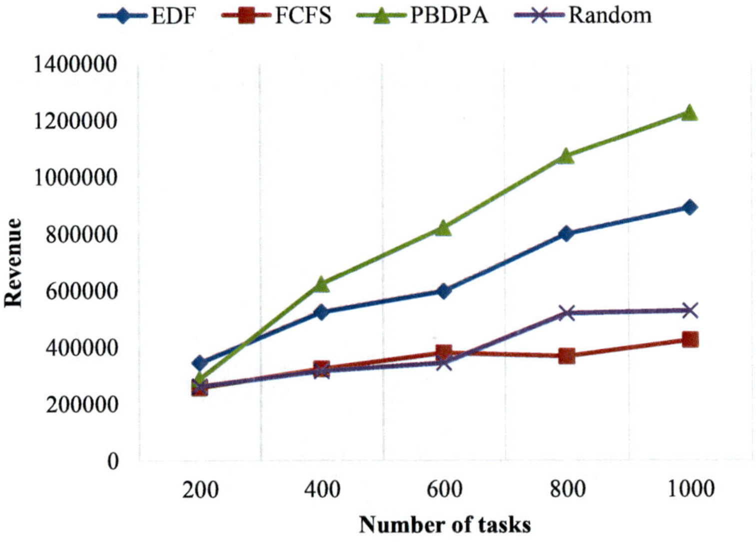

Fig. 4 Revenue comparison

6 Conclusion

In today's era of cloud computing, scheduling is an interesting problem. Task prioritization is an essential consideration in task scheduling. This work offers an effective prioritization for scheduling tasks in cloud computing based on a multiple criteria decision-making approach. This work proposes an effective approach for dealing with task scheduling with a deadline constraint. While fulfilling the deadline restriction, the recommended method aims to shorten the makespan while increasing throughput and revenue. Following that, each job is scheduled to the appropriate container VM to maximize resource utilization. When PBDPA is equated to the earliest deadline first (EDF), First Come First Served (FCFS), and random algorithms, this unique PBDPA algorithm produces a better scheduling with less makespan and better revenue. In the future, we may concentrate on adding more parameters to improve the algorithm. Energy awareness and efficiency may be another direction of future work.

References

1. Buyya R, Yeo CS, Venugopal S, Broberg J, Brandic I (2009) Cloud computing and emerging IT platforms: vision, hype, and reality for delivering computing as the 5th utility. Futur Gener Comput Syst 25(6):599–616
2. Han R, Guo L, Ghanem MM, Guo Y (2012) Lightweight resource scaling for cloud applications. In: 2012 12th IEEE/ACM international symposium on cluster, cloud and grid computing (ccgrid 2012), pp 644–651
3. He S, Guo L, Guo Y, Wu C, Ghanem M, Han R (2012) Elastic application container: a lightweight approach for cloud resource provisioning. In: 2012 IEEE 26th international conference on advanced information networking and applications, pp 15–22
4. Soltesz, S., Pötzl, H., Fiuczynski, M. E., Bavier, A., & Peterson, L. (2007). Container-based operating system virtualization: a scalable, high-performance alternative to hypervisors. In Proceedings of the 2Nd ACM SIGOPS/EuroSys European conference on computer systems 2007, pp 275–287
5. Guo Y, Lama P, Rao J, Zhou X (2013) V-cache: towards flexible resource provisioning for multi-tier applications in iaas clouds. In: 2013 IEEE 27th international symposium on parallel and distributed processing. IEEE, pp 88–99
6. Yin L, Luo J, Luo H (2018) Tasks scheduling and resource allocation in fog computing based on containers for smart manufacturing. IEEE Trans Industr Inf 14(10):4712–4721
7. Ibrahim IM (2021) Task scheduling algorithms in cloud computing: a review. Turk J Comput Math Educ (TURCOMAT) 12(4):1041–1053
8. Shi Y, Suo K, Kemp S, Hodge J (2020) A task scheduling approach for cloud resource management. In: 2020 fourth world conference on smart trends in systems, security and sustainability (WorldS4)
9. Zong Z (2020) An improvement of task scheduling algorithms for green cloud computing. In: Paper presented at the the 15th international conference on computer science & education (ICCSE 2020)
10. Hu Yao XF, Li H, Dong G, Li J (2020) Cloud task scheduling algorithm based on improved genetic algorithm. Int J Performability Eng 13:1070–1076
11. Gupta G, Kumawat V, Laxmi P, Singh D, Jain V, Singh R (2014) A simulation of priority based earliest deadline first scheduling for cloud computing system. In: 2014 first international conference on networks & soft computing (ICNSC2014)
12. Ben Alla S, Ben Alla H, Touhafi A, Ezzati A (2019) An efficient energy-aware tasks scheduling with deadline-constrained in cloud computing. Computers 8(2):46
13. Bala A, Chana I (2016) Multilevel priority-based task scheduling algorithm for workflows in cloud computing environment. In: Satapathy SC, Joshi A, Modi N, Pathak N (eds) Proceedings of international conference on ICT for sustainable development. AISC, vol 408. Springer, Singapore, pp 685–693

14. Khan G, Sengupta S, Sarkar A (2016) Priority based service scheduling in enterprise cloud bus architecture. In: IEEE international conference on foundations and frontiers in computer, communication and electrical engineering (C2E2 2016), SKFGI, pp 363–368
15. Ben Alla H, Ben Alla S, Ezzati A (2016) A novel architecture for task scheduling based on Dynamic Queues and Particle Swarm Optimization in cloud computing. In: 2016 2nd international conference on cloud computing technologies and applications (CloudTech)
16. Figueira J, Mousseau V, Roy B (2016) Electre methods. In: International series in operations research & management science, pp 133–153
17. Alla HB, Alla SB, Touhaf A, Ezzati A (2018) A novel task scheduling approach based on dynamic queues and hybrid meta-heuristic algorithms for cloud computing environment. Cluster Comput 21(4):1797–1820
18. Figueira J, Mousseau V, Roy B (2022) ELECTRE methods. Retrieved 26 March (2022)
19. Gawali MB, Shinde SK (2018) Task scheduling and resource allocation in cloud computing using a heuristic approach. Journal of Cloud Computing 7(1), 1–16
20. Brans JP, Mareschal B, Vincke PH (1984) PROMETHEE: a new family of outranking methods in multicriteria analysis. Operational Research IFORS 84, Amsetrdam

Application of Fuzzy-Based Multi-criteria Decision-Making Technique for Highway Concrete Bridge Health Assessment

Sudha Das Khan, Aloke Kumar Datta, and Pijush Topdar

Abstract Bridges are vital components of highway networks. Decision-makers often face challenges in selecting maintenance and repair plans for a massive backlog of deteriorated bridge structures due to limited funds. In this context, a prioritization scheme is required for making necessary decisions. So, evaluating the health of bridges is a crucial activity. However, it is cumbersome to evaluate all the influencing parameters quantitatively. Several tests are available, but these are time-consuming, costly, and uncertain. In this regard, visual inspection is quick, economical, and easy to evaluate the overall bridges' condition qualitatively. In the inspection, bridge inspectors visually inspect and evaluate the present condition of bridges. So, these data are subjective and inherently involve imprecision and uncertainty. Hence, this work suggests a combined fuzzy set theory-based multi-criteria decision-making model for health evaluation and prioritization of concrete bridges. Here, fuzzy set theory is employed for the approximate categorization of inspected information and converting those into quantifiable information. Next, this quantifiable information is used as input in a multi-criteria decision-making technique, i.e., VIseKriterijumska Optimizacija I Kompromisno Resenje, for prioritization of alternatives. A numerical example on a concrete bridge is presented to demonstrate the application of the suggested method. As a result, a prioritization list is obtained according to the potential of alternatives to damage the bridge's structural integrity. The author believes that this ranking will help asset managers to decide on mitigation measures and fund allocation.

Keywords Multi-criteria decision-making · Fuzzy set theory · VIseKriterijumska Optimizacija I Kompromisno Resenje

S. Das Khan (✉) · A. K. Datta · P. Topdar
National Institute of Technology Durgapur, Durgapur, West Bengal 713209, India
e-mail: sdk.18ce1510@phd.nitdgp.ac.in

S. Das Khan
BIT Sindri, Dhanbad, Jharkhand 828123, India

S. Das et al. (eds.), *Advances in Data-driven Computing and Intelligent Systems*, Lecture Notes in Networks and Systems 653, https://doi.org/10.1007/978-981-99-0981-0_24

1 Introduction

Bridge infrastructure is a vital part of a highway network. The social and economic growth of a country depends on bridge infrastructure. However, the health of a bridge deteriorates over time. To keep this vital infrastructure in operational and safe mode, a huge amount of funds is required for health inspection and maintenance purposes. Further, there is huge confusion regarding the distribution of limited funds among the massive backlog of deteriorated bridge structures, especially in developing countries like India. The number of deteriorated bridges is huge, but there is a lack of reliable procedures for prioritizing bridges based on their deteriorated condition, which can help allocate funds accordingly.

The most common in-practice health assessment technique is visual inspection. A bridge inspector evaluates the present condition of the bridge as per the country's guidelines. In India, we have 'Guidelines for inspection and maintenance of bridges' and 'Bridge Inspector's Reference Manual' for present condition assessment of highway bridges (IRC-SP:35 [1], IRC-SP:52 [2]). The information gathered using visual inspection technique is subjective, incomplete, imprecise, and uncertain. Further, this technique cannot be neglected because it can quickly and economically collect global information about bridge health. Therefore, a computational technique is required to tackle the imprecision and uncertainty of bridge inspection data and provide reliable management information.

In this context, fuzzy set theory (FST) is observed to be effectively applied in various fields. Elevli [3] applied fuzzy technique in decision-making for determining center location for logistic freight. Chen [4] used fuzzy techniques in the evaluation of weapon systems. Chen and Hwan [5] applied multi-attribute decision-making methods in a fuzzy environment. Yong [6] employed fuzziness in selecting the location of the plant. Jain and Bhattacharjee [7] deployed fuzzy techniques in deteriorating reinforced concrete structures. Furuta [8] applied fuzziness in the analysis of reliability. Yadav and Barai [9] deployed a fuzzy concept in the bridge management system.

FST, developed by Zadeh [10], has the capability to handle imprecise, uncertain, and insufficient data. A review work by Das Khan et al. [11] showed that several research works were carried out using fuzzy techniques for the health evaluation of bridges. Tee et al. [12] presented a fuzzy-based approach for the complete condition rating of bridge structures. Tarighat and Miyamoto [13] employed a fuzzy-based condition rating method using a fuzzy inference system (FIS) for bridge decks for practical application. Sasmal and Ramanjaneyulu [14] used analytic hierarchy process (AHP), fuzzy synthetic evaluation (FSE), and resolution identity techniques combined to calculate the priority ranking of bridges. Omar et al. [15] developed an index for complete condition of a bridge deck using weighted fuzzy union (WFU) method by combining the fuzzy membership function of the condition classes. It is observed that the outcome of most of the existing models is as fuzzy sets. Defuzzification of the fuzzy sets and mapping with a rating scale is an additional complex computational process. Besides, multi-criteria decision-making techniques (MCDM) are capable of delivering output as a ranking list. Therefore, it is worthwhile to apply a hybrid FST and MCDM model for condition evaluation and prioritization of concrete bridges. Further, this fuzzy-based MCDM technique can be helpful for ranking a bridge health condition considering different evaluated parameters, which can be used for mitigation measures and fund allocation.

In this context, a recent study by Das Khan et al. [16] used AHP and fuzzy technique for order of preference by similarity to ideal solution (TOPSIS) techniques for the ranking of defects of concrete bridges. However, the use of various MCDM approaches hasn't been extensively researched in this field. This study attempts to explore another popular MCDM technique, i.e., VIseKriterijumska Optimizacija I Kompromisno Resenje (VIKOR) in this field. In this regard, VIKOR is one of the popular techniques which is successfully used in many decision-making problems. VIKOR technique was established as an MCDM technique for resolving distinct decision issues with non-commensurable and conflicting attributes (Opricovic and Tzeng [17], Ogrizovic [18]). This technique is established on the closest to the ideal solution principle for ordering alternatives and determining the compromised solution (Opricovic and Tzeng [17], Ogrizovic [18]). It provides a compromise solution based on maximizing group benefit and minimizing personal regret.

Wide application of this technique is observed in the literature. Liu et al. [19] used fuzzy VIKOR for failure mode and effect analysis of industries. Lam et al. [20] used a unified entropy–fuzzy VIKOR method for performance assessment of construction companies. Mete et al. [21] used Pythagorean fuzzy VIKOR for calculating the occupational risk during constructing a natural pipeline. Liu et al. [22] proposed a VIKOR-based fuzzy MCDM technique for assessing healthcare waste disposal. Awasthi et al. [23] employed fuzzy AHP and VIKOR methods for selecting sustainable global suppliers. However, VIKOR was used to order various bridge repair projects by Gao et al. [24]. But the evaluation indices have not been considered in a fuzzy environment. Therefore, no study has used fuzzy VIKOR in condition assessment of bridges based on bridge inspection data for health assessment as per our literature survey. Hence, this study employs fuzzy VIKOR for the health evaluation of concrete bridges.

2 Methodology

Bridge is a complex structure. It is cumbersome to evaluate all the influencing parameters quantitatively. Several tests are available, but these are time-consuming, costly, and uncertain. In this regard, visual inspection is quick, economical, and easy to assess the global condition of bridges qualitatively. To overcome the imprecision and uncertainty in visual inspection data, it is worth applying FST to approximately categorize the inspection information and convert it into quantifiable information. Next, this quantifiable information is further used as input in fuzzy VIKOR technique to prioritize the alternatives.

In this study, fuzzy multi-criteria optimization (FUMCO) concepts presented by Opricovic and Tzeng [25] are considered. FUMCO is conceptualized in two phases: CFU phase to convert fuzzy data into crisp data and MCO phase to determine compromise ranking using the results of CFU phase. VIKOR technique [18] is applied in this study for ranking purposes in a fuzzy environment. We consider this approach to reduce unnecessary complexity due to over-fuzzification.

2.1 *Fuzzy Set Theory*

The FST concept was developed to deal with 'classes' that do not possess clearly defined membership criteria (Zadeh [10]). As a result, there are no clear boundaries between any two 'classes.' Fuzzy linguistic variables are useful in approximate characterization of any complex situation using linguistic terms which are challenging to describe in conventional quantitative ways. A fuzzy set is represented as follows (Chen [4], Zadeh [10], Kaufmann and Gupta [26]):

Suppose, U is the universe of discourse with $\{u_1, u_1, \ldots, u_n\}$ elements. A fuzzy set $\tilde{A}$ of U is defined by a set of pairs of $\left\{(u_1, \mu_{\tilde{A}}(u_1)), (u_2, \mu_{\tilde{A}}(u_2)), \ldots, (u_n, \mu_{\tilde{A}}(u_n))\right\}$, where $\mu_{\tilde{A}}$ represents the membership function of $\tilde{A}$ and $\mu_{\tilde{A}}(u_i)$ associates with the membership value of u_i in $\tilde{A}$. The membership function value varies from 0 to 1 that indicates the distribution of an element to the fuzzy set.

In this study, the alternatives are assessed considering criteria using fuzzy linguistic scales. Linguistic scales articulated in natural or artificial language encompasses fuzzy numbers as shown in Table 1. In this study, triangular membership functions are considered. A triangular fuzzy number (TFN) is articulated as $\tilde{a}(l\alpha, m\beta, u\gamma)$. Now, defuzzification is the procedure which converts a fuzzy linguistic value into crisp value. Center of area (COA) technique is mostly employed for the defuzzification of input data. Suppose, a fuzzy rating of an alternative i with respect to criteria j is articulated in terms of a TFN, $\tilde{a}(l\alpha, m\beta, u\gamma)$. Then the defuzzified value of $\tilde{a}$ can be calculated by:

$$a = l\alpha + \left[\frac{(u\gamma - l\alpha) + (m\beta - l\alpha)}{3}\right] \tag{1}$$

The defuzzified values of TFN/ BNP values are employed in the evaluation process using classical VIKOR technique for ranking of alternatives.

2.2 *VIKOR*

The step-by-step process of VIKOR technique (Opricovic [18]) is outlined below:

Step 1. Formulation of the performance rating matrix:

Table 1 Fuzzy linguistic scale

Linguistic term and TFN
Very low (0.0, 0.0, 0.2)
Low (0.0, 0.2, 0.4)
Medium (0.2, 0.4, 0.6)
High (0.4, 0.6, 0.8)
Very high (0.6, 0.8, 1.0)
Extreme (0.8, 1.0, 1.0)

At first, an expert evaluates the performance of alternatives against each criterion. Suppose, m number of alternatives are evaluated against n number of criteria by a decision-maker. A decision matrix is formulated based on the expert's rating (a_{ij}).

$$DM = \left[a_{ij}\right]_{m \times n}; i = 1, 2, \ldots, m; j = 1, 2, \ldots.., n \tag{2}$$

Step 2. Identification of the best and worst values of all criteria:

The best and worst value of each criterion $(j = 1, 2, \ldots., n)$ for all alternatives $(i = 1, 2, \ldots, m)$ is evaluated as follows:

$$\text{Best value } (f_j^*) = \max_i\left\{a_{ij}\right\} \text{ for benifit criteria or } \min_i\left\{a_{ij}\right\} \text{ for cost criteria} \tag{3}$$

$$\text{Worst value}\left(f_j^*\right) = \min_i\left\{a_{ij}\right\} \text{ for benefit criteria or } \max_i\left\{a_{ij}\right\} \text{ for cost criteria} \tag{4}$$

Step 3. Computation of S_i and R_i values:

The value of utility measure (S_i) and regret measure (R_i) are calculated using the following equations, respectively:

$$S_i = \sum_{j=1}^{n} w_j \frac{f_j^* - a_{ij}}{f_j^* - f_j^-} \tag{5}$$

$$R_i = \max_i w_j \frac{f_j^* - a_{ij}}{f_j^* - f_j^-} \tag{6}$$

where w_j = criteria weight. It shows the expert's significance of one criterion over other criteria.

Step 4. Computation of VIKOR index:

The VIKOR index (Q_i) values are determined by the following:

$$Q_i = v\frac{S_i - S^*}{S^- — S^*} + (1 - v)\frac{R_i - R^*}{R^- - R^*} \tag{7}$$

$$\text{where, } S_i = \min_i S_i,\ S^- = \max_i S_i$$

$$R^* = \min_i R_i;\ R^- = \max_i R_i$$

$$\nu = \text{the weight for the strategy of max group utility} (= 0.5 \text{ considered})$$

$$(1 - \nu) = \text{the weight for individual regret}$$

Step 5. Ranking of alternatives:

The ranking of alternatives is done by sorting the values of S, R, and Q in increasing order. There are three ranking lists for attributes (RoA) considering individual values of S, R, and Q, i.e., $(RoA)_S$, $(RoA)_R\, and (RoA)_Q$, respectively. The VIKOR index signifies the distance measured from the best solution for an alternative.

Step 6. Proposal of compromise solution:

Finally, a compromise solution $(A^{(1)})$ is obtained based on the ranking $(\text{RoA})_Q$as Q minimum, if the subsequent two requirements are fulfilled.

Condition I. The 'acceptable advantage':

$$Q\left(A^{(2)}\right) - Q\left(A^{(1)}\right) \geq \text{DQ} \tag{8}$$

$$\text{DQ} = \frac{1}{J - 1} \tag{9}$$

where $(A^{(1)}) = \text{first ranked alternative in} (\text{RoA})_Q$

$$(A^{(2)}) = \text{second ranked alternative in} (\text{RoA})_Q$$

$$J = \text{total number of alternatives}$$

Condition II. 'Acceptable stability in decision-making':

Alternative $(A^{(1)})$ must also be the rank one in $(\text{RoA})_S$, and $(RoA)_R$.

If one of the above prerequisites fails to fulfill, then the compromise solutions are as follows:

- Alternatives $(A^{(1)})$ and $(A^{(2)})$ if only condition II fails to satisfy
- Alternatives $(A^{(1)}), (A^{(2)}), \ldots, (A^{(N)})$ if Condition I fail to satisfy; and $(A^{(N)})$ is measured by the relation $(QA^{(N)}) - (QA^{(1)}) < DQ$, the positions of these alternatives are 'in closeness.'

3 Numerical Example

A numerical example is performed to explain the ranking process. Defects present in a concrete bridge are evaluated and ranked based on evaluation indices as in

Table 2 List of alternatives, criteria, and criteria weights

Alternatives	Symbol	Criteria	Symbol	Weight
Cracks	A-1	Temperature	Crt-1	0.0371
Potholes	A-2	Shrinkage-induced distress	Crt-2	0.0371
Scaling	A-3	Settlement-induced distress	Crt-3	0.3335
Delamination	A-4	Impact	Crt-4	0.0349
Spalling	A-5	Abrasion	Crt-5	0.0160
Corrosion	A-6	Erosion	Crt-6	0.0349
		Overload	Crt-7	0.2122
		Fatigue	Crt-8	0.2122
		Carbon dioxide and chloride	Crt-9	0.0348
		Sulfates	Crt-10	0.0043
		Carbonation	Crt-11	0.0384
		Alkali-aggregate reaction	Crt-12	0.0048

the study done by Das Khan et al. [16]. Six defects and twelve evaluation criteria are considered and shown in Table 2 [16, 27, 28]. Now, the defects present in the concrete bridge are evaluated using the linguistic scale (Table 1) in a field inspection. During the inspection, a bridge inspector assesses the condition of defects based on the evaluation indices and scores them using the linguistic scale. Based on the assessment, a fuzzy evaluation matrix is formulated (Table 3). Next, each fuzzy data is converted into crisp data using Eq. (1). Criteria weights are considered from the study, Das Khan et al. [16] and presented in Table 2.

Next, the values of Si, Ri, and Qi are obtained by Eqs. (5), (6), and (7), respectively, and presented in Table 4. The three ranks of alternatives are based on ascending order of Si, Ri, and Qi, as shown in Table 4. Though 'Potholes (A-2)' consists minimum Q value, it could not meet the 'acceptable advantage' condition means $Q(2) - Q(1) = 0.159 - 0.00 = 0.159$ is less than $0.2\left[\frac{1}{j-1} = \frac{1}{6-1} = 0.2\right]$. Besides, 'Potholes (A-2)' satisfies 'acceptable stability in decision-making' condition, means best ranked as per Si and Ri value also. Because only condition 1 is not fulfilled, therefore, the set of compromised alternative lists will be 'Potholes (A-2)' and 'Spalling (A-5)' for the following inequities:

$$Q(2) - Q(1) = 0.159 - 0.00 = 0.159 < 0.2$$

$$Q(3) - Q(1) = 0.326 - 0.000 = 0.326 > 0.2$$

Table 3 A fuzzy evaluation matrix

	A-1	A-2	A-3	A-4	A-5	A-6
Crt-1	(0.0, 0.2, 0.4)	(0.4, 0.6, 0.8)	(0.4, 0.6, 0.8)	(0.4, 0.6, 0.8)	(0.0, 0.2, 0.4)	(0.2, 0.4, 0.6)
Crt-2	(0.2, 0.4, 0.6)	(0.4, 0.6, 0.8)	(0.4, 0.6, 0.8)	(0.4, 0.6, 0.8)	(0.0, 0.2, 0.4)	(0.2, 0.4, 0.6)
Crt-3	(0.2, 0.4, 0.6)	(0.4, 0.6, 0.8)	(0.2, 0.4, 0.6)	(0.4, 0.6, 0.8)	(0.2, 0.4, 0.6)	(0.0, 0.2, 0.4)
Crt-4	(0.2, 0.4, 0.6)	(0.4, 0.6, 0.8)	(0.2, 0.4, 0.6)	(0.4, 0.6, 0.8)	(0.0, 0.2, 0.4)	(0.2, 0.4, 0.6)
Crt-5	(0.0, 0.2, 0.4)	(0.4, 0.6, 0.8)	(0.0, 0.2, 0.4)	(0.0, 0.2, 0.4)	(0.4, 0.6, 0.8)	(0.0, 0.2, 0.4)
Crt-6	(0.0, 0.2, 0.4)	(0.2, 0.4, 0.6)	(0.0, 0.2, 0.4)	(0.0, 0.2, 0.4)	(0.4, 0.6, 0.8)	(0.0, 0.2, 0.4)
Crt-7	(0.2, 0.4, 0.6)	(0.2, 0.4, 0.6)	(0.0, 0.2, 0.4)	(0.0, 0.2, 0.4)	(0.4, 0.6, 0.8)	(0.0, 0.2, 0.4)
Crt-8	(0.4, 0.6, 0.8)	(0.2, 0.4, 0.6)	(0.0, 0.2, 0.4)	(0.2, 0.4, 0.6)	(0.4, 0.6, 0.8)	(0.4, 0.6, 0.8)
Crt-9	(0.2, 0.4, 0.6)	(0.2, 0.4, 0.6)	(0.2, 0.4, 0.6)	(0.2, 0.4, 0.6)	(0.2, 0.4, 0.6)	(0.4, 0.6, 0.8)
Crt-10	(0.2, 0.4, 0.6)	(0.2, 0.4, 0.6)	(0.4, 0.6, 0.8)	(0.0, 0.2, 0.4)	(0.4, 0.6, 0.8)	(0.0, 0.2, 0.4)
Crt-11	(0.2, 0.4, 0.6)	(0.0, 0.2, 0.4)	(0.4, 0.6, 0.8)	(0.0, 0.2, 0.4)	(0.2, 0.4, 0.6)	(0.0, 0.2, 0.4)
Crt-12	(0.0, 0.2, 0.4)	(0.0, 0.2, 0.4)	(0.4, 0.6, 0.8)	(0.2, 0.4, 0.6)	(0.4, 0.6, 0.8)	(0.4, 0.6, 0.8)

Table 4 Results of fuzzy VIKOR ($\nu = 0.50$)

Alternatives	Si	$(RoA)_S$	Ri	$(RoA)_R$	Qi	$(RoA)_Q$
A-1	0.458	4	0.167	2	0.326	3
A-2	0.310	1	0.106	1	0.000	1
A-3	0.694	6	0.212	4	0.733	5
A-4	0.449	3	0.212	5	0.414	4
A-5	0.330	2	0.167	3	0.159	2
A-6	0.694	5	0.334	6	0.999	6

4 Conclusion

Bridge is a complex structure. Decision-makers often face challenges in the selection of maintenance and repair plans for a massive backlog of deteriorated bridge structures due to limited funds. Also, it is cumbersome to evaluate all the influencing parameters quantitatively. Several tests are available, but these are time-consuming, costly, and uncertain. In this regard, visual inspection is quick, economical, and easy to qualitatively assess the global condition of bridge structures. However, imprecision and uncertainty are inherently involved in bridge inspection data. Therefore, a computational technique is required to tackle the imprecision and uncertainty of bridge inspection data and provide reliable management information. In this context, the use of various MCDM approaches hasn't been extensively researched in this field. This study attempts to explore a popular MCDM technique, i.e., VIKOR in this field.

Hence, this study has made an attempt to apply a hybrid FST-based VIKOR model for condition evaluation and prioritization of defects in concrete bridges. In the model, FST has been helped to approximately categorize the inspection information and convert those into quantifiable information. Next, this quantifiable information has been used as input in the VIKOR technique for prioritizing the alternatives. A numerical example on a concrete bridge is used to demonstrate the application of the proposed method. As a result, 'Potholes (A-2)' and 'Spalling (A-5)' have been identified as the top prioritized alternatives which need the most attention for that particular numerical example. As a result, a prioritization list has been obtained as per the potential of those defects to damage the bridge's structural integrity. The author believes that this defect ranking will help asset managers to decide on mitigation measures and allocation of limited funds.

References

1. IRC-SP:35 (1990) Guidelines for inspection and maintenance of bridges. The Indian Road Congress, New Delhi, India
2. IRC-SP:52 (1999) Bridge inspector reference manual. The Indian Road Congress, New Delhi, India
3. Elevli B (2014) Logistics freight center locations decision by using Fuzzy-PROMETHEE. Transport, 29:412–418. https://doi.org/10.3846/16484142.2014.983966
4. Chen SM (1996) Evaluating weapon systems using fuzzy arithmetic operations. Fuzzy Sets Syst 77:265–276. https://doi.org/10.1016/0165-0114(95)00096-8
5. Chen SJ, Hwang CL (1992) Fuzzy multiple attribute decision making methods. In: Lecture notes in economics and mathematical systems book series. Fuzzy multiple attribute decision making. Springer, Berlin, pp 289–486
6. Yong D (2006) Plant location selection based on fuzzy TOPSIS. Int J Adv Manuf Technol 28:839–844
7. Jain KK, Bhattacharjee B (2012) Application of fuzzy concepts to the visual assessment of deteriorating reinforced concrete structures. J Constr Eng Manag 138:399–408. https://doi.org/10.1061/(ASCE)CO.1943-7862.0000430
8. Furuta H (1995) Fuzzy logic and its contribution to reliability analysis. In: Rackwitz R, Augusti G, Borri A (eds) Reliability and optimization of structural systems. Springer, Boston, pp 61–76
9. Yadav D, Barai S (2005) Fuzzy inference driven internet based bridge management system. Transport 20:37–44. https://doi.org/10.1080/16484142.2005.9637993
10. Zadeh LA (1965) Fuzzy sets. Inf Control 8:338–353
11. Das Khan S, Topdar P, Datta AK (2020) Applicability of fuzzy-based visual inspection approach for condition assessment of bridges in developing countries: a state-of-the-art review. J Inst Eng India Ser A
12. Tee AB, Bowman MD, Sinha KC (1988) A fuzzy mathematical approach for bridge condition evaluation. Civil Eng Syst 5:27–24. https://doi.org/10.1080/02630258808970498

13. Tarighat A, Miyamoto A (2009) Fuzzy concrete bridge deck condition rating method for practical bridge management system. Expert Syst Appl 36:12077–12085. https://doi.org/10.1016/j.eswa.2009.04.043
14. Sasmal S, Ramanjaneyulu K (2008) Condition evaluation of existing reinforced concrete bridges using fuzzy based analytic hierarchy approach. Expert syst appl 35:1430–1443
15. Omar T, Nehdi ML, Zayed T (2017) Integrated condition rating model for reinforced concrete bridge decks. J Perform Constructed Facil 31:1–14
16. Das Khan S, Datta AK, Topdar P, Sagi SR (2022) A cause-based defect ranking approach for existing concrete bridges using Analytic Hierarchy Process and fuzzy-TOPSIS. Struct Infrastruct Eng. https://doi.org/10.1080/15732479.2022.2035407
17. Opricovic S, Tzeng G-H (2004) Compromise solution by MCDM methods: a comparative analysis of VIKOR and TOPSIS. Eur J Oper Res 156:445–455. https://doi.org/10.1016/S0377-2217(03)00020-1
18. Opricovic S (1998) Multicriteria optimization of civil engineering systems. PhD thesis
19. Liu H-C, You J-X, You X-Y, Shana M-M (2015) A novel approach for failure mode and effects analysis using combination weighting and fuzzy VIKOR method. Appl soft Comput 28:579–588. https://doi.org/10.1016/j.asoc.2014.11.036
20. Lam WS, Lam WH, Jaaman SH, Liew KF (2021) Performance evaluation of construction companies using integrated entropy–fuzzy VIKOR model. Entropy 23. https://doi.org/10.3390/E23030320
21. Mete S, Serin F, Oz NE, Gul M (2019) A decision-support system based on Pythagorean fuzzy VIKOR for occupational risk assessment of a natural gas pipeline construction. J Nat Gas Sci Eng 71:102979. https://doi.org/10.1016/J.JNGSE.2019.102979
22. Liu HC, Wu J, Li P (2013) Assessment of health-care waste disposal methods using a VIKOR-based fuzzy multi-criteria decision making method. Waste Manage 33:2744–2751. https://doi.org/10.1016/J.WASMAN.2013.08.006
23. Awasthi A, Govindan K, Gold S (2018) Multi-tier sustainable global supplier selection using a fuzzy AHP-VIKOR based approach. Int J Prod Econ 195:106–117. https://doi.org/10.1016/J.IJPE.2017.10.013
24. Gao Z, Liang RY, Xuan T (2019) VIKOR method for ranking concrete bridge repair projects with target-based criteria. Results Eng 3:100018. https://doi.org/10.1016/J.RINENG.2019.100018
25. Opricovic S, Tzeng G (2003) Fuzzy multicriteria model for postearthquake land-use planning. Nat Hazard Rev 4:59–64. https://doi.org/10.1061/(ASCE)1527-6988(2003)4:2(59)
26. Kaufmann A, Gupta MM (1986) Introduction to fuzzy arithmetic: theory and applications. Elsevier, New York
27. Joshi S, Sagi SR, Toraskar H et al (2020) Handbook: for implementing IBMS/UBMS, 1st edn. self, Mumbai
28. EN 1504 (2008) ES Products and systems for repair and protection of concrete structures

Modality Direct Image Contrast Enhancement for Liver Tumour Detection

S. Amutha, A. R. Deepa, and S. Joyal

Abstract Medical image analysis relies heavily on segmenting medical images. Numerous methods based on edge or area characteristics have been developed for segmenting medical images. These are reliant on the image's quality. When detecting a region of interest, such as a lesion, on a CT or MRI scan, contrast is crucial (s). Manual histogram adjustment techniques are used for improving the contrast of image where contrast enhancement is based on cross-modality in a 2D histogram specification technique. Our proposed multi-modality guided histogram specification approach improves the liver CT images by MRI images for input data. The optimization scheme is used in our technique which is based on guided picture enhancement as well as image quality management. In a two-step process, the suggested Improved Supervised Contrast Enhancement Technique (ISCET) scheme makes use of both structural and contextual information. Applying a two-dimensional histogram definition while utilizing contextual data from the matching guiding image, such as a magnetic resonance image, is the initial step (MRI). The second stage involves using an optimization strategy that keeps the original image's structural information while employing a structural similarity metric. The results show that the performance is improved by the increase in entropy and MIGLCM values.

Keywords Histogram · Enhancement · Optimization · Liver cancer · Contrast image · ISCET

S. Amutha (✉)
School of Computer Science and Engineering, SCOPE, Vellore Institute of Technology, Chennai, India
e-mail: amuthabenziker@gmail.com

A. R. Deepa
Department of Computer Science and Engineering, Koneru Lakshmaiah Education Foundation, Andhra Pradesh, India

S. Joyal
Electrical and Electronics Engineering, Saveetha Engineering College, Chennai, India

S. Das et al. (eds.), *Advances in Data-driven Computing and Intelligent Systems*, Lecture Notes in Networks and Systems 653, https://doi.org/10.1007/978-981-99-0981-0_25

1 Introduction

The upper right section of our belly is where the liver is located. There are many types of cancer in the liver, and the most common is hepatocellular carcinoma Lei et al. [1]. However, early cancer tumour discovery and efficient treatment approaches can increase the overall survival rate. Although rapid cancer diagnosis is facilitated by diagnostic imaging techniques like CT, their utility is constrained by low contrast and noise. Additionally, due to the low contrast of these images, tumour detection, and segmentation are tough, and it is solved by performing a contrast enhancement beforehand. It is important for pertinent structural data from the organs can be obtained by a medical imaging modality. Therefore, it would be intriguing to employ the extra data that the imaging modality (like MRI) captures for improving the other (like CT). The idea of enhancing an image from a single modality utilizing information from multiple modality images is not new; similar concepts have been successfully used to enhance photographs of the natural world.

A method for enhancing the image of tumours and arteries using MR images was proposed. Cross-modality guided enhancement methods outperformed classic single-image enhancement approaches. There are two main issues with image enhancement. First off, the majority of contemporary enhancement methods are designed solely for particular kinds of photographs. Second, it can be challenging to identify a reliable standard against which to compare the current enhancement techniques. Due to these factors, evaluations of the effectiveness of enhancement initiatives frequently focus on how they affect the underlying application. The general purpose of CE in medical imaging is for enhancing the visual appearance to improve diagnosis and treatment Banovac et al. [2]. In addition, research on picture quality enhancement to better segmentation of organ features Balakrishnan et al. [3], Clatz et al. [4], Celik [5]. As found in Rueckert et al. [6], by using CE as a pre-processing step, it is possible to achieve better segmentation for CT images. Combining a quality control system with the contrast enhancement strategy is one way to get around these restrictions. Additionally, optimization is used to avoid the saturation artefacts with histogram-based approaches. Similar suggestions for improving nature photos were made in Smistad et al. [7]. It requires employing an optimization strategy to preserve the structures of the input picture while mapping the input image's histogram to that of a reference image. In this study, we suggest a related strategy for medical pictures utilizing cross-modal data Chen et al. [8]. Later, the function of CE in aiding tumour segmentation is examined. To test the entire processing technique, a real dataset including CT and MRI images of the human liver, as well as segmentation, is employed. Two-dimensional histogram specification-based CE technique protects multi-modal medical imaging data. The proposed cross-modal guiding and quality control combination improve the picture by using contextual information. By applying segmentation results to actual multi-modal liver data and using objective quality criteria, a novel goal-oriented performance evaluation of suggested approach is carried out. Comparisons with single enhancement methods demonstrate the suggested method's higher performance.

2 Related Work

When rejecting outliers, the non-rigid registration method switches from approximation to interpolation formulation. This approach speeds up calculation while reducing error. During surgery, the resection is simple due to the clean margin between the diseased and healthy tissue Clatz et al. [4].

They have employed parallel and distributed computing because image guided neurosurgery on critical patients has to be done quickly and successfully. The feasibility of landmark tracking throughout the entire image volume for brain MRI near-real-time image fusion. This has improved speed, fault tolerance, simplicity of use, and execution time Chrisochoides et al. [9].

A 3D contrast-enhanced breast MRI method with non-rigid registration that can detect breast motion was developed. Affine transformation is used to simulate the global motion of the breast, while free-form deformation based on B-splines is used to model the local motion. Voxel-based contrast enhancement similarity measurement. However, low-resolution 3D breast images are insufficient. Rueckert et al. [6]. A comparable strategy has been offered Bao et al. [10].

Using free-form deformation to model the movements of the liver. The intensity-based rigid and non-rigid registration algorithm may be used to determine the deformation's intensity. This approach can provide 3D MRI performance that is superior to earlier 2D images. The accuracy rate for calculating organ motion and deformation is great Rohlfing et al. [11].

The thermal ablation therapy is carried out when the patient has liver metastases. The information from the ultrasound image has to be translated from preoperative to intraoperative during the procedure. The pixel or voxel map interacts with the pre-process in this method. Then it is transformed into a vessel probability value using the registration algorithm. The error rate has decreased because of this. In order to improve the image, the two-dimensional histogram equalization (2DHE) technique is developed. This algorithm is relatively simple to apply by just adding the spatial neighbourhood parameter, which determines and increases the difference in grey levels between two nearby pixels. It achieves good performance results, but one may also make minor adjustments to achieve alternative values Celik [5].

Cross-modality guided enhancement (CMGE), a new technique, converts data from one modality to another, for example, converting an MRI picture to a CT image. The image quality that produces the better results was determined using IEM and EME measures Naseem et al. [12].

Local structures are preserved via SSIM, while artefacts added during improvement are minimized Naseem et al. [12]. It has been determined that better categorization of important structures in CT images can be accomplished using CE as a processing step.

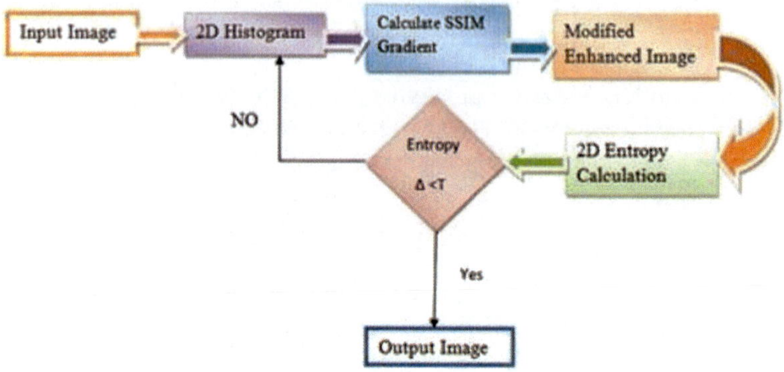

Fig. 1 Flowchart of proposed model

3 Proposed Method

Our proposed approach new CE strategy is based on

- Quality control and cross-modality guided medical image enhancement.
- Local structure preservation and artefact minimization are accomplished using the structural similarity index measure.
- Tumour is classified.

The overall processing strategy is evaluated by using an actual dataset with scans of the human body and its flowchart is shown in Fig. 1.

4 2D Histogram Specification

Two-dimensional (2D) histogram improves picture contrast. The output image's 2D histogram is different from the target histogram because of the 1D cumulative distribution function. A pairwise pixel-value mapping approach based on a 2D CDF. According to experimental findings, the proposed method's output image's 2D histogram closely resembles the 2D target histogram. The histogram threshold technique is a strong contender for segmenting grayscale images. It is based on the characteristics of the smoothed histogram's shape, including its peaks, valleys, and curves and improves the additional type of 2D grey-level histogram. A Cartesian sum of the original 1D grey-level histogram and the 1D local average grey-level histogram, for local window to each pixel in the image to averaging the grey levels within the window. The pixel value change in the horizontal or vertical directions seems sluggish when compared to a change in the diagonal direction, but the gradation change continuity appears powerful. The 1D histogram method fails to account for the fact that a colour cluster is not always present in each component of an RGB colour or multispectral image, and the combination of the various segmentations is unable to capture this spatial property of colours. Additionally, it disregards the relationship between the various elements. Multiple histogram-based thresholding is therefore necessary. The 3D-histogram approach, however, is limited in a fully multi-dimensional manner by data sparsity and complexity to search algorithms in a large memory space.

The usage of a 2D-histogram is an intriguing alternative technique that picks two colour bands simultaneously, bands in RGB space colour. This technique is acquired by projecting a 3D-histogram onto two colour planes. The number of pixels in an RGB colour image I is represented by the 2D histogram pn as $p(\times 1, \times 2)$. Here, we review the fundamental ideas and concepts covered in the 2D histogram specification methods. To increase contrast, guided contrast enhancement transforms an image (f) as input into an output image (fe). The reference or guide image in this process is [g], a higher perceptual quality image. The value of CON is calculated as per Eq. 1.

$$\mathrm{CON_{FRE}}(a, b) = \sum_{l=0}^{n-1} \mathrm{GLCM} \sum_{m=0}^{n-1} (f(i, j), f(o, p)) \tag{1}$$

Here, a and b denote the pixel values, (i,j) and (o,p) denote coordinates in the images, n denote the total number of grey levels, and 0 ≤ a; j ≤ n- 1;

$$\partial ij(q, r) = \{1, \text{ if } i = i \text{ and } j = b\, 0, \quad \text{otherwise}$$

The 2D normalized histogram's transition probability for grey levels is obtained from the HGM as follows (Eq. 2):

$$\mathrm{HGM}(a, b) = \frac{\mathrm{CON_{FRB(a,b)}}}{\sum_{l=0}^{n-1} x \sum_{m=0}^{n-1} y} \mathrm{CON_{FRE}}(a, b) \tag{2}$$

The grey-level mapping methodology uses the 2D-histogram. The mapping process was calculated using the two-dimensional Cumulative Distribution Function (CDF) which is represented in Eq. 3.

$$\mathrm{HIS}_D(a, b) = \sum_{l=0}^{n-1} XY \sum_{m=0}^{n-1} \mathrm{HGM}(a, b) \tag{3}$$

Similar calculations are utilized to construct and describe the guiding image's 2D-CDF expression as Hg. T maps the input and output signal, and it is obtained as

$$T(i, j) = \arg \arg |Hf(i, j) - Hg(k; l)| + \eta(|i - k| + |j - l|)a \tag{4}$$

The mapping is completed by locating the target pixel values $T(i,j)1$ and $T(i,j)2$, to match i and j in $[f]$. The difference between the two CDFs among the candidate pixel values is as minimal as possible, by Eq. 4; $|i-k| + |j-1|$ is a symbol that symbolizes a number, and it has the value 104. Equation 5 is used to calculate the intensity values for the enhanced picture [fe]:

$$\mathrm{fe}(m; n) = T(f(m; n); f(m; n+1)) \tag{5}$$

Each element's surrounding element influences the values in the original image and is changed into a new one with respect to Eq. 5. The method uses contextual information between the pixels to make use of the 1D histogram specification that takes into account distinct values when calculating the CDFs.

5 Gradient-Based Structural Similarity Measure

Histogram settings are widely used to boost visual contrast. Processing distortions can be controlled in two ways: by incorporating an objective function or halting criterion into the CE process, or by framing the entire problem inside a constrained optimization framework. Structural similarity changes between the original image and its enhanced form are controlled by applying global HS to a low-contrast image.

The SSIM is a well-known statistic for determining how similar two photographs are to one another. The index compares the quality of the picture under consideration to the quality of a reference image. The SSIM between matching local blocks in pictures [A] and [B] is computed to yield a single SSIM value for overall similarity index as calculated from Eqs. 6 and 7. Assume that x_a, y_b and x_a, y_b represent the corresponding blocks in each image; x_a, y_b and x_a, y_b represent the mean intensity values of a and b, respectively; and $\mathrm{x_a}$, $\mathrm{y_b}$ and x_a, y_b, respectively, represent the standard deviations. Then, we express the SSIM between the two blocks axe and bx as:

$$\mathrm{SSIM}\ (x_a, y_b) = \frac{(2\alpha x_{a\alpha_{yb+}} C1)2(2\alpha x_{a\alpha_{yb+}} C2)}{\left(\alpha^2 x_{a+\alpha^2_{yb+}} C1\right)\left(\alpha^2 x_{a+\alpha^2_{yb+}} C2\right)} \tag{6}$$

Overall images and its SSIM index are expressed, and its total number of the images is represented as N.

$$\mathrm{SSIM}(A, B) = \frac{1}{N} \sum_{\forall x} \mathrm{SSIM\ Index_{map}}(x_{a,b}, y; x) \tag{7}$$

The SSIM gradient-based optimization method is used in this work for the cross-modal medical picture improvement, as was before indicated. Here, the greater quality of MR images is utilized by 2D-HS to enhance CT images. The SSIM gradient gradually enhances the enhancing process when utilized inside an optimization framework. The addition of SSIM allows for the general morphology of the original image to be kept with the least amount of information loss.

$$\mathrm{SSIM}(x_e, y) = \frac{1}{N} \sum_{\forall x} \mathrm{INXSSIM}_{\mathrm{map}}\left(x_{xa}, x_{yb}; x\right) \tag{8}$$

5.1 *Contrast Enhancement with Quality Control*

Before discussing the OPTGCE methodology, this paragraph briefly discusses the 2D-HS and SSIM gradient approaches. Input CT image $[x]$ equal to $[x']$ and the guidance MRI as $[g]$. CDFs $[x]$ and $[g]$ are calculated in Eq. 4 for calculating the transformation matrix T. Using Eq. 5, the pixel values in $[x']$ are converted to new values to produce the enhanced image [xe]. In Eq. 7, change $[B]$ to $[f]$ and write: The suggested method next computes the SSIM gradient with respect to the image $[xe]$, denoted by fe, and uses Eq. 6 to determine the structural similarity between $[x\ e]$ and $[f]$. After that, $[x']$ is changed in accordance with step 5's description in method 1.

5.2 *Performance of Suitable Step Size*

In this part, we describe how to determine the best step size empirically so that the algorithm can achieve a higher SSIM with fewer repetitions. Mathematically speaking, the anticipated rise in SSIM at iteration t is expressed as Eqs. 9 and 10:

$$\nabla \mathrm{SSIM}(t) = \alpha Z \sum_{\forall X} (\partial_{\mathrm{fe}} \mathrm{SSIM}(f, f_e(t)))^2 \tag{9}$$

In SSIM(t) at several iterations, ΔSSIM(t) can be modelled by $\alpha r s^t$ and the final value of SSIM is expressed as

$$\mathrm{SSIM}_f = \mathrm{SSIM}' + \frac{r \propto Z}{1 - s} \tag{10}$$

Algorithm

Guided Image Contrast Enhancement Through an Optimized (GICEP)

IPI = Input CT image and

GI = Guidance image

Calculate 2D-CDF of guidance image

Set f′ = f, threshold = 0.05 and $t = 1$

While ∇ E > -threshold

do

Apply 2D histogram

Calculate structural similarity between f_e and f, SSIM(f, f_e)

Calculate E^t and ∇E

Increment thres as t = thres + 1

Decrement thres as t = thres -1

Modified f′ contents using SSIM gradient

Factors as: f′ = (f, f_e)

end while

Image enhanced f_e

Output Image

where $r = \forall x \partial f e\text{SSIM}(f, \text{fet})2s = \Delta\text{SSIM}11\text{SSim}1$, ΔSSIM2 and ΔSSIM(1) denote the increase in SSIM values at t = 2 and t = 1, respectively.

SSIM's starting point is determined by the first iteration. By this experiment, SSIM value changes faster in first iterations, and hence, the algorithm is performed three times to determine the values. The SSIM values between the original image and the corresponding enhanced image slices and the range are from 20 to 60.

In order to make tumour classification easier, we try to boost contrast while maintaining structural similarities with the input image. Along with ensuring structural similarity, an additional criterion is to evaluate the contrast that is enhanced at each iteration by employing 2D-HS. 2D entropy controls the degree of enhancement. The enhancement process needs a gain in two-dimensional entropy in order to be completed. Our methodology is intended to benefit from inter-pixel correlation, so we used 2D entropy to develop this criterion.

$$E_t = -\sum_{i=0}^{k-1}\sum_{j=0}^{k-1} h_{f_{e(t)}}(i, j)\ln\ln\left(h_{f_{e(t)}}(i, j)\right) \tag{11}$$

The value of the 2D entropy for an image of type f_e at iteration t is represented by E in Eq. 11, where t is a number between 1 and 10. The value h (f (e(t)) represents the chance of a transition occurring between two grey-level pairs (i, j). The rise in entropy of the augmented image is calculated for each cycle as per Eq. 12.

$$\Delta E = Et - Et - 1 \tag{12}$$

Additionally, when the method is used on our dataset, Fig. 2 shows the variation in entropy values over all rounds (normalized to fall within the range [0,1]). When applied to images that have been updated using the suggested technique, the segmentation accuracy is increased; further applications cause a modest entropy increase.

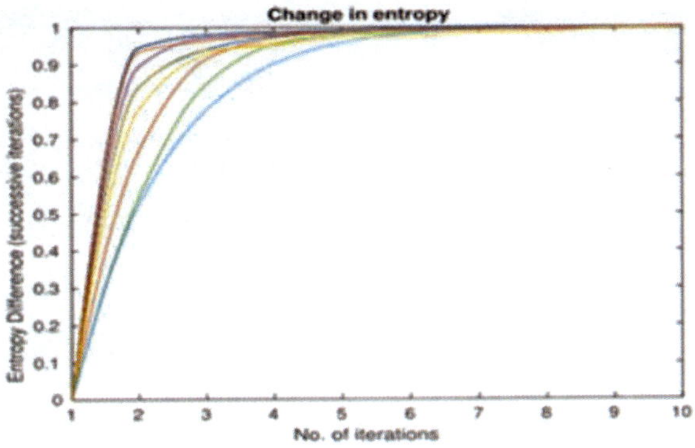

Fig. 2 Variation in entropy values with iteration

The application of the ISCET method's results is shown in the next section, along with a comparison to alternative methods.

6 Result

In this section Penney et al.[13], Satpute et al. [14], and Naseem et al. [12], the dataset used in the experiment and the findings from different approaches are described. More information on both the quantitative and qualitative ratings is provided in Figs. 3 and 4.

The subject of image quality assessment (IQA), particularly in the context of nature photographs, has drawn a lot of research attention. However, applying the present IQA criteria in a medical setting has some serious disadvantages. The objectives of CE are substantially different in the setting of medicine. In contrast to natural pictures, where the objective is to quantify the influence of various aberrations on the perceived quality of the image, the focus in the medical environment is instead on the diagnosis, despite the fact that some degradation may anger the radiologists. Because of this, extra care must be used while applying the current IQA metrics. Another challenging problem is how to evaluate the perceived quality of an algorithm for improving image quality. In the suggested system, place some weight on a few contrast enhancement evaluation (CEE) parameters. We've chosen three distinct CEE measures to evaluate the effectiveness of better photographs. A mutual information-based no reference metric is referred to as MIGLCM for the first metric. With the use of the GLCM, this measure offers numerical standards for contrasting the statistical characteristics, joint entropy, and mutual information of the original and modified images. In addition to MIGLCM, we also employed the more modern multi-criteria contrast enhancement metric.

Entropy, which is frequently employed in QA of medical picture improvement, was chosen as the final metric. Table 1 displays the MIGLCM and entropy median values. An improved CE algorithm's performance is shown by a higher MIGLCM

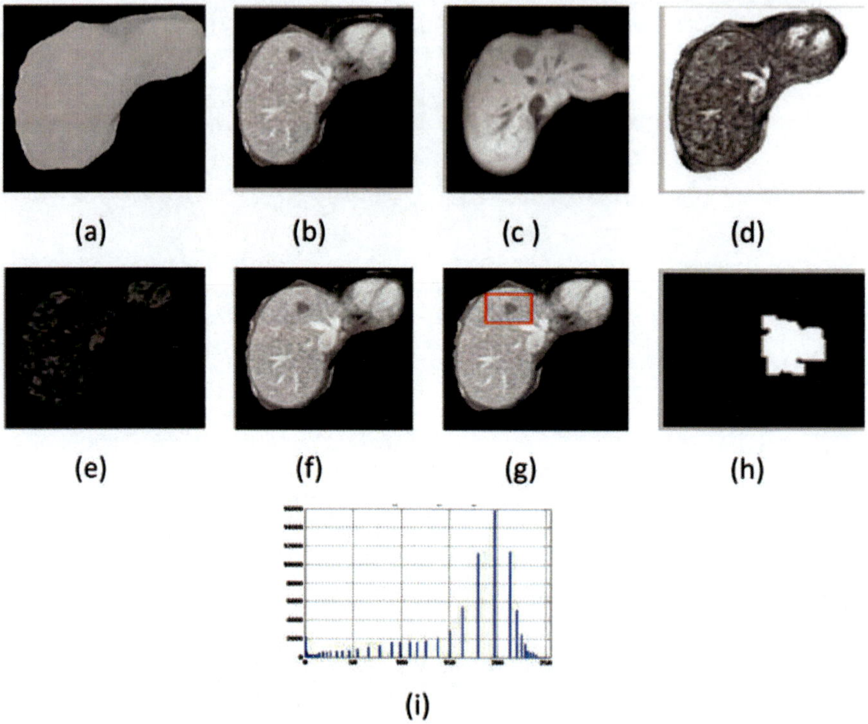

Fig. 3 **a** Input image 1. **b** Guidance. **c** 2D-Histogram. **d** SSIM. **e** SSIM gradient. **f** Enhanced. **g** ROI image. **h** ISCE image. **i** Histogram

value. Furthermore, higher entropy values have been linked to improved CE performance, but no specific range for this parameter exists. As can be seen from the tabular results, OPTGCE performs the best. Cross-modality guidance-based enhancement (CMGE) and histogram equalization with maximum intensity coverage (HEMIC) have poor overall results for MCCEE and entropy, according to the two QA measures shown in Table 2.

7 Conclusion and Future Work

This research suggests a guided contrast enhancement strategy for poor contrast images. The proposed technique extracts data from the enhanced perceptual quality guide picture using a context-aware 2D histogram-based strategy for global difference improvement, though local image structures are improved using SSIM-based measures. The morphological data of the image is preserved through enhancement to this combination's effective contrast improvement and minimization of the artefacts brought on by conventional histogram-based enhancement techniques. A qualitative

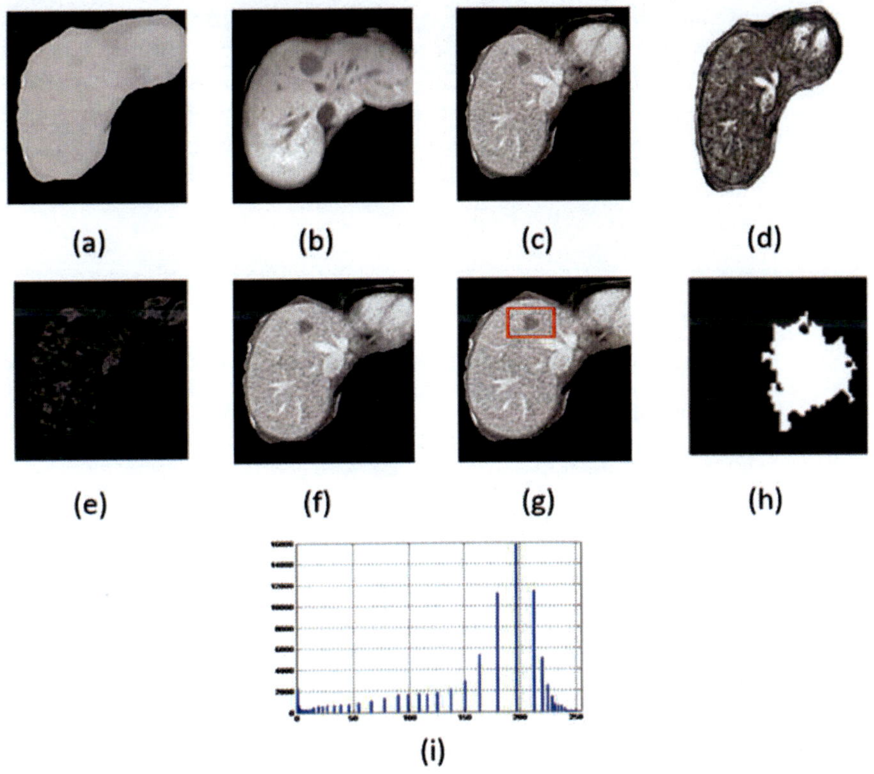

Fig. 4 **a** Input image 2. **b** Guidance image. **c** 2D-Histogram. **d** SSIM image specification. **e** SSIM gradient. **f** Enhanced image. **g** ROI image. **h** ISCET image. **i** Histogram

Table 1 Quantitative assessment of different enhanced methods

Image	Entropy				MIGLCM			
	SSIM	EM	ROI	Proposed	SSIM	EM	ROI	Proposed
1	2.33	2.31	2.71	3.16	1.3	1.09	0.97	1.10
2	1.59	1.6	1.9	2.7	0.97	0.8	0.88	1.14

Table 2 Median MCCEE values for different methods

Image	SSIM	EM	ROI	ISCET
1	0.23	**0.31**	0.33	0.36
2	0.25	**0.27**	0.23	0.29

and quantitative analysis shows that the suggested method beats surviving methods which do not include a supervision mechanism. An improved image is then subjected to a tumour segmentation algorithm to evaluate how well the suggested strategy facilitates tumour segmentation. Future research should focus on dynamic range stretching and adaptive tissue attenuation to improve the visual contrast. Users of the parametric model can quickly increase image contrast by changing the attenuation scale.

References

1. Lei Y, Fu Y, Wang T, Liu Y, Patel P, Curran WJ, Liu T, Yang X (2020) 4D-CT deformable image registration using multiscale unsupervised deep learning. Phys Med Biol 65(8):085003. https://doi.org/10.1088/1361-6560/ab79c4
2. Banovac F, Wilson E, Zhang H, Cleary K (2006) Needle biopsy of anatomically unfavorable liver lesions with an electromagnetic navigation assist device in a computed tomography environment. J Vasc Interv Radiol 17(10):1671–1675
3. Balakrishnan G, Zhao A, Sabuncu MR, Guttag J, Dalca AV (2019) VoxelMorph: a learning framework for deformable medical image registration. IEEE Trans Med Imag 38(8):1788–1800
4. Clatz O, Delingette H, Talos IF, Golby AJ, Kikinis R et al (2005) Robust nonrigid registration to capture brain shift from intraoperative MRI. IEEE Trans Med Imaging 24(11):1417–1427
5. Celik T (2012) Two-dimensional histogram equalization and contrast enhancement. Pattern Recogn 45(10):3810–3824
6. Rueckert D, Sonoda LI, Hayes C, Hill DL, Leach M O et al (1999) Nonrigid registration using free-form deformations: application to breast MR images. IEEE Trans Med Imag 18(8):712–721
7. Smistad E, Elster AC, Lindseth F (2014) GPU accelerated segmentation and centerline extraction of tubular structures from medical images. Int J Comput Assist Radiol Surg 9(4):561–575
8. Chen H, Zhang Y, Kalra MK, Lin F, Chen Y et al (2017) Low-dose CT with a residual encoder-decoder convolutional neural network. IEEE Trans Med Imaging 36(12):2524–2535
9. Chrisochoides N, Fedorov A, Kot A, Archip N, Black P et al (2006) Toward real-time image guided neurosurgery using distributed and grid computing. In: Proceedings of the 2006 ACM/IEEE conference on Supercomputing, pp 76-es
10. Bao P, Warmath J, Galloway R, Herline A (2005) Ultrasound-to-computer-tomography registration for image-guided laparoscopic liver surgery. Surg Endosc Other Interv Tech 19(3):424–429
11. Rohlfing T, Maurer CR Jr, O'Dell WG, Zhong J (2004) Modeling liver motion and deformation during the respiratory cycle using intensity-based nonrigid registration of gated MR images. Med Phys 31(3):427–432
12. Naseem R, Cheikh FA, Beghdadi A, Elle OJ, Lindseth F (2019) Cross modality guided liver image enhancement of CT using MRI. In: 2019 8th European workshop on visual information processing (EUVIP), pp 46–51
13. Penney GP, Blackall JM, Hamady MS, Sabharwal T, Adam A, Hawkes DJ (2004) Registration of freehand 3D ultrasound and magnetic resonance liver images. Med Image Anal 8(1):81–91
14. Satpute N, Naseem R, Pelanis E, Gómez-Luna J, Cheikh FA et al (2020) GPU acceleration of liver enhancement for tumor segmentation. Comput Methods Programs Biomed 184:105285

An Enhanced Deep Learning Technique for Crack Identification in Composite Materials

Saveeth Ramanathan, Uma Maheswari Sankareswaran, and Prabhavathy Mohanraj

Abstract Composite material consists of more than one substance, which inherit the superior properties of the constituent's elements for light weight and tensile strength. The identification of cracks in composite material at an earlier stage reduces the cost and effort. Deep learning is a complex model in machine learning to extract significant features from a large dataset. The suitable construction of convolutional neural network topology for the application will give better results. Our proposed model uses Bayesian optimization for hyper-parameter tuning which updates the hyper-parameter value for each iteration to improvise the accuracy. In this work real time, composite material images were taken as dataset captured by Silicon Electron Microscope and Evolution VF camera. Applied proposed model on Silicon Electron Microscope dataset, obtained accuracy over 88% and on Evolution VF dataset obtained accuracy over 91%. The integration of optimization techniques with CNN reduces the computational time.

Keywords Composite material · Convolutional neural network · Bayesian algorithm

1 Introduction

One of the primary justifications for using composite materials over traditional materials for components is weight savings. In addition to being stronger than other materials, composites can also be lighter. For instance, reinforced carbon fiber can be up

S. Ramanathan (✉)
Department of Computer Science Engineering, Coimbatore Institute of Technology, Coimbatore, India
e-mail: saveeth.r@cit.edu.in

U. M. Sankareswaran
Department of Electronics and Communication Engineering, Coimbatore Institute of Technology, Coimbatore, India

P. Mohanraj
Department of Artificial Intelligence and Data Science, Coimbatore Institute of Technology, Coimbatore, India

S. Das et al. (eds.), *Advances in Data-driven Computing and Intelligent Systems*, Lecture Notes in Networks and Systems 653, https://doi.org/10.1007/978-981-99-0981-0_26

to five times stronger than 1020 grade steel while weighing only one-fifth as much, making it ideal for structural use.

1.1 Deep Learning

Deep learning [9] is a class of machine learning approaches that employs several processing layers, to learn representations of data at successive levels of abstraction, the output from one layer is used as an input for the next. The excellency of the feature extraction procedure, which is dependent on the data and model types. When utilizing a trial-and-error method for feature extraction, traditional learning takes a long time, and the efficacy of the feature extraction process is determined by the user experience. The deep learning approach, on the other hand, will benefit from an automated feature extraction procedure based on the learning of a large number of nonlinear filters before making any decisions. As a result, DL combines feature extraction and decision-making into a single model, obviating the need for often-suboptimal human handcrafting.

1.2 Convolutional Neural Network

Convolutional neural network (CNN) is one of the most common deep learning frameworks, and it is mostly used for image classification. According to, it is useful for detecting patterns in photos that aid in the automated recognition of authentic physical objects. The selected patterns are recovered immediately from picture datasets by CNNs, eliminating the need to manually extract features, which is the most essential component that makes CNN so popular. CNN also has the ability to be retrained to conduct new recognition and based on previously produced models, and it gives very accurate recognition results. It provides an ideal model architecture that allows for advancements in object detection and recognition. As a result, it has become a crucial tool in automated facial recognition. Filters are used to extract information and can range from simple parameters such as brightness to more complicated ones that uniquely identify an item. CNN may contain hundreds of layers to assist find patterns in pictures. This filtering is done to each training picture, and the convolution result of each image may be utilized as an input to the next layer. An input layer, an output layer, and several hidden layers make up a CNN as shown in Fig. 1.

The layers that are not visible execute feature-learning tasks with the three most typical feature-learning layers:

- Convolution: It activates some features in pictures by applying convolutional filters, the nonlinear filters are represented by a matrix of weights that slide along the pixel brightness input matrix to create a feature map matrix using a specific dot product.

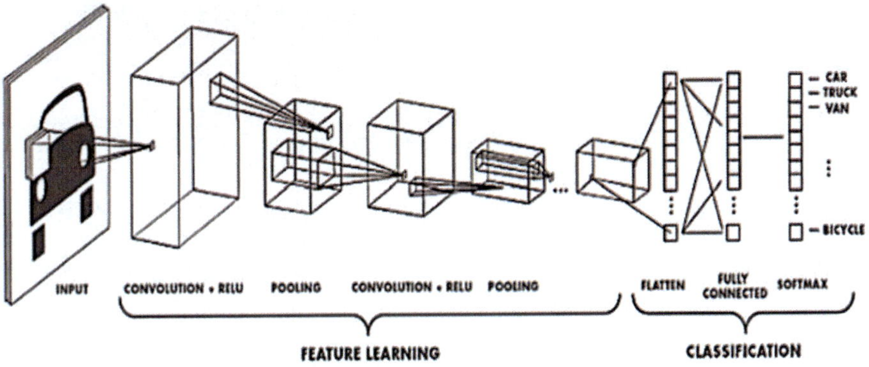

Fig. 1 General structure of CNN

- Batch normalization: It is employed after each convolutional layer as a supplement layer to lessen the risk of overfitting by normalizing the succeeding layer's input values (Yamashita et al. 2018).
- Pooling: It minimizes the dimensionality of the output volume (McDermott 2021) without compromising critical characteristics that contribute to decreasing the computational cost. It reduces the number of parameters by performing nonlinear down-sampling which simplifies the output. There are two forms of pooling: max pooling and average pooling. The max pooling takes the most activated feature, and average pooling, which takes the future's average presence. As a result, dark backgrounds benefit from maximum pooling, whereas white backgrounds benefit from average pooling.

CNN has two tiers of categorization:

- The first part consists of a convolution and pooling layer which merely do feature extraction.
- The second part consists of a fully connected layer and classifier. The extracted features will be fed into the fully connected layer, and processed weights will be given into the classifier for classification.

2 Literature

Li et al. [5] presented the concept of artificial intelligence (AI) based on the idea of having robots act like humans, the development of intelligent systems can boost productivity and efficiency. One of the most widely used AI approaches is a type of machine learning based on artificial neural networks (ANNs). Filippis et al. [2] explored an idea like, a large amount of data can be modeled using the computers, which allows for the discovery of intricate correlations. Wuest et al. [12] presented that ANNs are capable of handling high-dimensional data in real time, allowing implicit patterns to be extracted to predict the future state of complex systems. As a result of ANN's ability to cope with nonlinearity, it can also solve complex dynamic problems. ANNs can be trained quite easily using historical data by changing control parameters such as learning rate and momentum. Wang et al. [11] discussed the advancement of deep learning algorithms over machine learning in smart manufacturing and quicker reliable decision-making.

Geissbauer et al. [4] discussed the merits of big data analytics in the Industry 4.0 paradigm, also known as the Fourth Industrial Revolution, which aims to develop smart systems using the Internet of Things (IoTs), cyber-physical systems (CPSs), and cloud computing. (CC). Ciortea [1] proposed a stochastic system that examines the relationship between process and system performance and offers a quantitative analysis of the system's performance as the first step in modeling an IoT system. However, there are still a number of challenges when employing ANN, the most important of which is data collection because the accessibility of pertinent data is not guaranteed. Additionally, after obtaining a dataset, putting appropriate data mining into practice might be challenging, especially when a significant amount of irrelevant data has been gathered, which could harm the effectiveness of the ensuing models.

Sinha et al. [9] studied that deep learning (DL) approaches are now well established as an enhancement to ANN capabilities, generating greater learning potential. Prior to the decision-making step, they have the benefit of adopting automated feature extraction by learning a large number of nonlinear filters. One of the most popular deep learning networks is the convolutional neural network (CNN). The CNN is stacked with Bayes theorem (NB-CNN) to monitor the physical structure of the reactor in a nuclear power plant. Periodically, video will be taken over the nuclear plant and this video will be converted into frames. Each frame will be taken as input image and processed to remove the noisy pattern.

Sinha et al. [10] presented the work on the optimization of convolutional neural network parameters for image classification. He focused on optimal CNN architecture to achieve high accuracy value.

Fu-Chen et al. [3] presented the advancement of Bayes theorem to maximize the accuracy value. The Bayes algorithm will make decisions based upon the previous output. Ibrahim [8] improvised the stochastic gradient descent to obtain the high true positive value in MRI scan segmentation and also to locate the correct boundaries of inside and outside of contour. The manual inspection of concrete structure is tedious, therefore using image processing the monitoring of cracks in the concrete structures using multi-scale line filters with the Hessian matrix is proposed by Fujita and Hamamoto [14].

The composite materials are being used for insulating the steel pipelines in oil and gas industries. Over the period, inherent defects will occur in the pipeline, therefore to monitor the internal defects microwave signal will be passed which will penetrate the steel pipelines and highlight the defects. This microwave signal nondestructive testing (NDT) is united with k-means clustering and unsupervised learning for defect classification.

Saveeth and Maheswari [7] proposed the Haar cascade classifier to classify the crack images in real-time composite material images, and the computational complexity is minimized by the deployment of the cascade classifier.

Saveeth and Maheswari [6] deployed deep neural network (DNN), and it is being optimized by modified crow search algorithm (MCSA) for crack classification in composite material images. The deep learning algorithm combined with the optimization algorithm gives leading results in pattern recognition, image processing, and natural language processing as discussed by Xu et al. [13].

3 Proposed Novel Hybrid Algorithm

In this proposed method, the learning rate parameters are optimized using Bayesian optimization. The first step will be to use the design of experiments approach to look at the major elements that influence validation accuracy (VA). Then, by optimizing the significant factors, BO will be utilized to identify the ideal hyper-parameters for the network in order to minimize the classification error on the validation set, which is the objective function. Figure 2 states the overall pseudocode for BO_CNN.

The proposed work is organized as follows: (3.1) Image Acquisition, (3.2) Bayesian Based CNN Model (BO-CNN), (3.3) Bayesian Optimization, (3.4) Composite Crack Classification Based on CNN model, (3.4.1) CNN Topology, (3.4.2) Feature Extraction Layer, and (3.4.3) Training Algorithms.

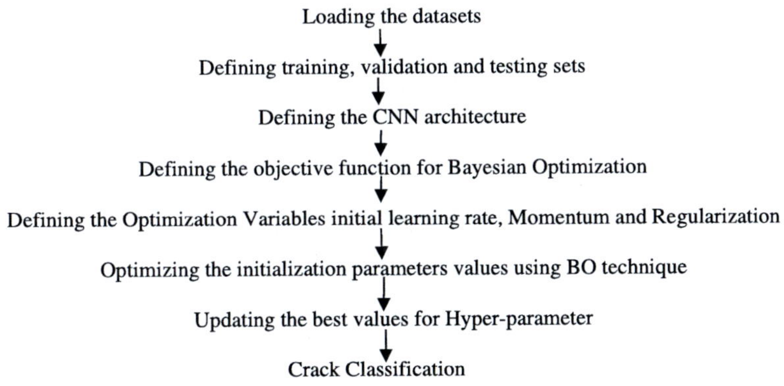

Fig. 2 BO-CNN pseudocode

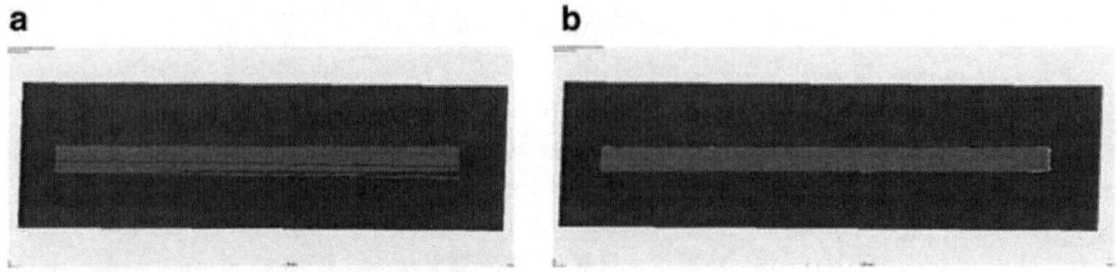

Fig. 3 EVF dataset: composite material image **a** cracks and **b** non-crack image

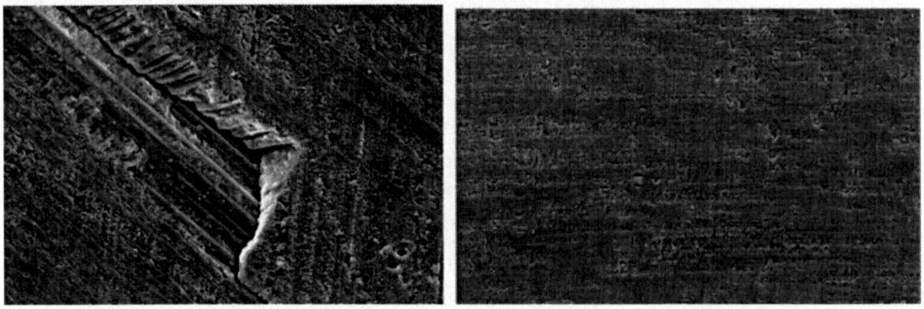

Fig. 4 SEM dataset: composite material image **a** cracks and **b** non-crack image

3.1 Image Acquisition

The dataset image been procured from composite lab, Indian Institute of Technology, Madras. In this work, two types of real-time datasets have been used, one datasets are being captured by Evolution VF (EVF) camera namely EVF datasets, and another being from Scanning Electron Microscope (SEM) namely SEM datasets shown in Figs. 3 and 4, respectively.

3.2 Bayesian Based CNN Model (BO-CNN)

Initially, the grid search method was in use to optimize the hyper-parameter value. In grid search, all random values have been arranged in a matrix form, each value will be taken sequentially and checked for optimization. The time complexity will be high in the grid search, then random search will be in use for optimizing the parameter value. The random search technique is not a promising one, so Bayesian optimization techniques are in use for turning the hyper-parameter value. The CNN architecture depends upon the hyper-parameter value for good accuracy results. The learning parameter value will get updated for each iteration in order to minimize the loss function. In CNN during the training process, many iterations will happen and hyper-parameters will get updated during back propagation to minimize the gradient descent.

3.3 Bayesian Optimization

In general, the model parameters such as weights and bias are set randomly by the machine learning algorithms, whereas the hyper-parameter can be optimized by the user which will drive the model parameter to get good validation results. In Bayesian optimization (BO), initially the objective function should be defined where hyper-parameter value will be evaluated. The probabilistic model which is of less expensive used to optimize the objective function is called surrogate function. The objective function formula is

$$Y^* = \arg\min_{y \in Y} f(y) \tag{1}$$

$f(y)$ is the objective function score to minimize the error rate on evaluation of the validation set. Y^* is the set of hyper-parameters, whereas y can take any values in the range of Y.

The algorithm for the Bayesian optimization is as follows:

a. Construct a probabilistic model called surrogate model
b. Choose the hyper-parameters that best performs on the surrogate model
c. The selected hyper-parameters been evaluated on the objective function
d. Update the hyper-parameter value until to get better optimization value
e. Repeat the steps II-IV for maximum iteration or until achieve the good optimization value.

3.4 Composite Crack Classification Based on CNN Model

The CNN model is a supervised learning algorithm, which will do automatic feature extraction. The model parameter values like weight and bias have been randomly initialized by the model. On each backward propagation, the weight value will be updated to minimize the loss function.

3.4.1 CNN Topology

In this work, CNN topology has five convolution layers, three max pooling, and two fully connected layers as shown in Fig. 5. The result of the CNN architecture depends upon how we design the CNN architecture and how each layer is being connected with each other layers. The output of the convolution layer is the feature map which is the

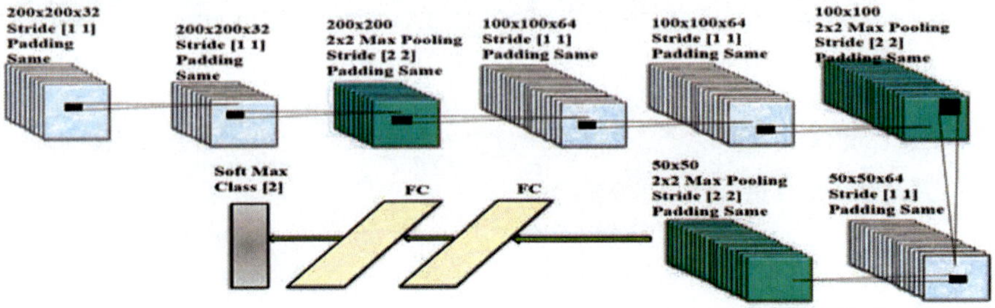

Fig. 5 CNN architecture

two-dimensional matrix. The main advantage of CNN is the adaptation of automatic feature extraction using convolution kernels. The convolution operation applied on set of parameterized filters called kernel with input image to get a new information feature map. The filters will slide from left top to right on the input image with a stride value of '1'. In practice, the value for stride will be set as 1 or 2, the chance of loss of prominent data if the value of stride is more than 2. The padding will be set to restore the border value with zero of the input image. After each convolution layer, the feature map will be fed into the activation layer. The ReLU activation function is the most commonly used function to improve the computational performance, and it will make the negative value as zero and keep the positive value as such in the feature map. The pooling layer will reduce the dimensional complexity. The last convolutional layer is followed by two fully connected layer and a softmax classifier. The CNN topology is depicted in Fig. 5.

3.4.2 Feature Extraction Layer

In order to have even distribution of the input image, all the dataset image has been resized to 200 × 200. The size of the input after each convolution layer is determined by the following formula:

$$\text{Convolution width} = \frac{\text{Composite Input Image Width} - \text{Convolution filter width} + (2 \times ZP)}{S_{\text{width}}} + 1, \tag{2}$$

$$\text{Convolution width} = \frac{\text{Composite Input Image Hight} - \text{Convolution filter hight} + (2 \times ZP)}{S_{\text{height}}} + 1 \tag{3}$$

Table 1 Convolution, pooling, and fully connected layer configurations of the proposed CNN

Layer	Kernel size	Kernel	Stride
Convolution1	3 × 3 × 1	32	1
Convolution2	3 × 3 × 1	32	1
Max pool 1	3 × 3	–	2
Convolution 3	3 × 3 × 1	64	1
Max pool 2	3 × 3	–	2
Convolution 3	3 × 3 × 1	64	1
Max pool 3	3 × 3	–	2
Fully connected 1	10	10	–
Fully connected 2	10	2	–

ZP is for 'zero padding' and *S* for stride value. After each convolution layer, feature map width and height size is calculated using Eqs. (2) and (3).

The max pooling formula is

$$MP = \frac{\text{Previous Image size}}{\text{Stride}} \tag{4}$$

The proposed CNN model layer details have been shown in Table 1. The last layer is softmax classifier which will decide crack or non-crack based on the final score.

3.4.3 Training Algorithm

The objective of the training is to minimize the loss function L_W by adjusting the value of the parameters with respect to the algorithm. The mostly used algorithm is gradient backpropagation. In this work, stochastic gradient descent method (SGDM) is used to minimize the loss function. The SGDM helps to reduce the difference between prediction and actual output, and it overcomes the problem arise in gradient descent. The SGDM is little faster in training the datasets. The crucial part in this method is setting the value for learning parameter. Here, the optimized value is obtained from the Bayesian optimization algorithm. The SGDM updates the value of *Y* for every iteration for the objective function *F*(*y*), and α is the learning parameter.

$$Y = y - \alpha \nabla y E[F(Y)] \tag{5}$$

4 Evaluation Function

Classification problem has discrete output value, and the three assessment parameters namely accuracy, precision, and recall are used to validate the classification method.

Accuracy value will gives the degree of reliability in the model

$$\text{Accuracy} = \frac{\text{Number of true positive predicted}}{\text{Total umber of predictions made}} \tag{6}$$

Precision value gives how far the true positive predicted from total positive prediction

$$\text{Precision} = \frac{\text{Number of true positive predicted}}{\text{Total number of positive predictions made}} \tag{7}$$

Recall value gives the value of true positive prediction from total number of actual positive samples

$$\text{Recall} = \frac{\text{Number of true positive predicted}}{\text{Total number of actual positive samples}} \tag{8}$$

5 Results and Discussions

In this work, a total of 620 EVF datasets and 630 SEM datasets have been taken for analysis and k-fold validation has been done with the value of 100. The momentum value set to be 0.9 and the optimized learning rate is 0.01. Each iteration is directly proportional to the value of assessment parameter value, therefore the graph gradually increases for each iteration. The values obtained for EVF and SEM datasets have been shown in Tables 2a and 2b.

According to Table 2a, the average values for accuracy, precision, and recall for EVF datasets are 91.35, 95.84 and 94.72, respectively. The average values obtained for accuracy, precision, and recall are 88.27, 92.8 and 93.53, respectively, for SEM datasets, as shown in Table 2b. Compared to SEM datasets, the EVF datasets exhibit slightly superior value. Since the work involves sensitive data, the recall parameter is the main focus.

Table 2a EVF datasets

Accuracy	Precision	Recall
94.59	97.14	97.14
86.49	93.94	91.18
89.19	94.12	94.12
94.59	97.06	97.06
91.89	96.97	94.12
91.35	95.846	94.724

Table 2b SEM datasets

Accuracy	Precision	Recall
93.1	96	96
93.1	96	96
86.21	92	92
82.76	88	91.67
86.21	92	92
88.276	92.8	93.534

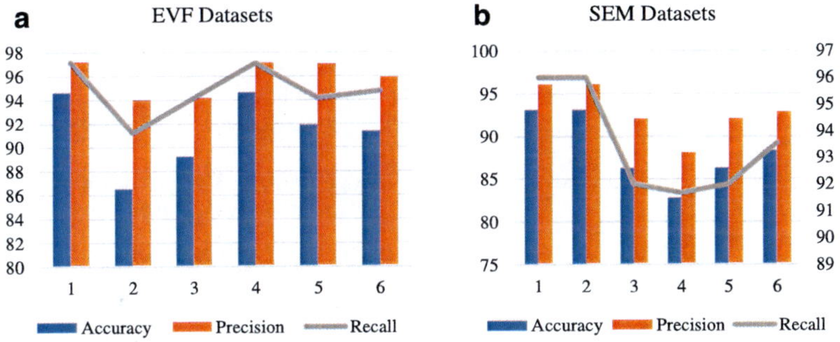

Fig. 6 **a** Performance analysis for EVF datasets. **b** Performance analysis for EVF datasets

The graphical representation of the performance analysis for EVF and SEM datasets has been shown in the Fig. 6a and b, respectively.

6 Conclusion

This work proposes a method called BO-CNN to polish up the classification task of composite material images. The work comprises CNN incorporated with Bayesian optimization algorithm to improvise the classification accuracy. The optimization algorithm is used to obtain the optimal learning rate. The weight of the kernel being adjusted for each iteration by the gradient descent algorithm. The updation of weight aids in reducing the mean square error. The proposed work has been validated for the assessment metrics of recall, accuracy, and precision. Applying the BO-CNN algorithm to EVF and SEM datasets, resulted in a validation accuracy of 91.53 and 88.27% and precision of 95.84 and 92.8% and the obtained recall value is of 94.72% and 93.53%, respectively. The new proposed BO-CNN method may be improved in the future by optimizing the weight regularization factor in the convolutional layers and fully connected layer utilizing BO to increase CNN performance.

References

1. Li BH, Hou BC, Yu WT, Lu XB, Yang CW (2017) Applications of artificial intelligence in intelligent manufacturing: a review. Front Inform Technol Electron Eng 18(1):86–96. https://doi.org/10.1631/FITEE.1601885
2. De Filippis LAC, Serio LM, Facchini F, Mummolo G (2017) ANN modelling to optimize manufacturing process. In: Advanced applications for artificial neural networks. pp 201–226
3. Wuest T, Weimer D, Irgens C, Thoben KD (2016) Machine learning in manufacturing: advantages, challenges, and applications. Prod Manuf Res 4(1):23–45. https://doi.org/10.1080/21693277.2016.1192517
4. Wang J, Ma Y, Zhang L, Gao RX, Wu D (2018) Deep learning for smart manufacturing: methods and applications. J Manuf Syst 48:144–56. https://doi.org/10.1016/j.jmsy.2018.01.003
5. Geissbauer R, Vedso J, Schrauf S (2016) Global industry 4.0 survey: building the digital enterprise, 1. Retrieved from PwC Website https://www.pwc.com/gx/en/industries/industries-4.0/landing-page/industry-4.0-building-your-digital-enterprise
6. Ciortea EM (2018) IoT analysis of manufacturing using petri nets. In: IOP conference series: materials science and engineering, vol 400(4)
7. Singh AK, Ganapathy Subramanian B, Sarkar S, Singh A (2018) Deep learning for plant stress phenotyping: trends and future perspectives. Trends in Plant Sci 23(10):883–98. https://doi.org/10.1016/j.tplants.2018.07.004
8. Sinha T, Verma B, Haidar A (2017) Optimization of convolutional neural network parameters for image classification. In: IEEE symposium series on computational intelligence (SSCI), Honolulu, HI, United States, IEEE
9. Chen F-C, Jahanshahi MR (2018) Deep learning-based crack detection using convolutional neural network and Naïve Bayes data fusion NB-CNN. IEEE Trans Indus Electron 65(5)
10. Ibrahim RW, Hasan AM, Jalab HA (2018) A new deformable model based on fractional wright energy function for tumor segmentation of volumetric brain MRI scans. Comput Methods Programs Biomed 163:21–28
11. Fujita Y, Hamamoto Y (2010) A robust automatic crack detection method from noisy concrete surfaces. Mach Vis Appl 22(2):245–254
12. Saveeth R, Uma Maheswari S (2019) HCCD: Haar based cascade classifier for crack detection on a propeller blade. In book: Sustainable technologies for computational intelligence, November 2019, Springer Publications. https://doi.org/10.1007/978-981-15-0029-9_33
13. Saveeth R, Uma Maheswari S (2022) Crack detection in composite materials using Mcrowdnn. Intell Autom Soft Comput 34(2):983–1000. https://doi.org/10.32604/iasc.2022.023455
14. Xu F, Pun CM, Li H, Zhang Y, Song Y, Gao H (2019) Training feed-forward artificial neural networks with a modified artificial bee colony algorithm. Neurocomputing

Vaccine-Block: A Blockchain-Based Prevention of COVID-19 Vaccine Misplacement

Swami Ranjan and Ayan Kumar Das

Abstract The World Health Organization has classified COVID-19 as a highly contagious disease. In order to provide a quick and prompt diagnosis, appropriate medical support is essential. Many researchers have already proposed intelligent schemes to identify the disease. However, vaccination is the only feasible way to prevent from COVID-19 virus. Many countries have already developed the vaccine and vaccination is going on till now. In this paper, a secure blockchain-based scheme is designed for vaccine management including the prevention of vaccines from theft and misuse. Blockchain is an emerging technology for the decentralized storage of data. The proposed scheme used blockchain to store the vaccination record and protect from fraud registration or misplacement of vaccines. The performance evaluation is done in terms of true generated response, altered data, and product loss with respect to the number of vaccine units. The proposed approach is evaluated in terms of true generated response, altered data, and product loss ratio with respect to number of vaccine units. The results are compared with other two existing approaches. The comparative analysis shows that the proposed scheme outperforms both the existing schemes.

Keywords Blockchain · COVID-19 · Vaccination · Security · Smart contract

1 Introduction

The epidemic spreading of the COVID-19 virus causes a huge loss in human society. Akshita et al. [1] stated that the quick contaminating nature of this virus not only caused a pandemic but also destroyed the economy and the daily routine of humans. The World Health Organization (WHO) has advised vaccination to prevent from COVID-19 virus. Rathee et al. [2] analyzed that coronavirus infection is caused due

S. Ranjan (✉) · A. K. Das
Birla Institute of Technology, Patna Campus, Mesra, Patna 800014, India
e-mail: phdcs10060.20@bitmesra.ac.in

A. K. Das
e-mail: das.ayan@bitmesra.ac.in

S. Das et al. (eds.), *Advances in Data-driven Computing and Intelligent Systems*, Lecture Notes in Networks and Systems 653, https://doi.org/10.1007/978-981-99-0981-0_27

to Severe Acute Respiratory Syndrome (SARS-CoV) and reduces the oxygen supply in the lungs. COVID-19 is found to be a highly contagious disease spreading rapidly from person to person. The whole world is engaged in giving their efforts to control this epidemic with all its strength and power. The researchers are encouraged to understand, establish, and test new devices and to provide different services to cure them in the current situation. In the meantime, COVID-19 has a profound effect on health facilities as its treatment of it is still unknown and an unlimited number of cases has increased rapidly. The WHO [3] has reported that the recent confirmed cases are about 440 million in the world; confirmed death is about 60 million, and 380 million recovered patients. The COVID-19 virus affects people in different ways, the most common symptoms like cough, fever, loss of taste, loss of smell, tiredness, etc., as a mild infection and will recover without hospitalization.

Vaccination is a very important tool to deal with this epidemic and thus it is required to ensure mass immunizations where possible. The idea of vaccinating people is used to fight infectious diseases through intensive vaccination. There are many types of COVID-19 vaccines in the markets of different developed countries and the vaccination of the people is already going on. MoHFW [4] has broadcasted that more than 179 crores people of India are vaccinated and among them, 80.4 Crores get fully vaccinated. This means 58.3% of Indians are vaccinated. Worldwide 1.09 trillion Crores doses are given to the people and 442 Crores are fully vaccinated.

Deka et al. [5] has implemented a successful COVID-19 vaccine supply chain system that requires a well-planned and organized system. Such a system will need to have effective storage of vaccines, management of stock, handling of the vaccine, and control of user records. Nature's article [6] has claimed that there are dangers of corruption in every policy of the shipping process. For example, vaccines may be stolen from a series of public service delivery during the transit process. The process may also transfer to a black market or be reserved for personal use. Farzanegan and Hofmann [7] finds that vaccine supplies are also at risk once they arrive at the hospital or public health facility or in the absence of reliable management measures. Employees of public health facilities may also steal vaccines, reselling in the black market, or their private activities. This risk is particularly significant if the supply is limited, and the need is high, as it was during the epidemic. A transparent and secure way for the distribution of vaccines efficiently to the vaccination centers needs to be explored.

Blockchain technology is a distributed network that keeps records as private or public ledgers of all the events. Antal et al. [8] stated that the blockchain can provide real-time visibility of policy distribution and retention chains from production to management, eliminating blind spots in the supply chain of vaccines. Rouhani and Deters [9] used the blockchain technology for solving some of the persistent challenges of supply chain management, such as participant accountability, accuracy in tracking items, problems in stock management, and stored the data in Ethereum blockchain network. However, the biggest advantage of the blockchain, in this case, would be the volatility of the end-to-end data. Beni and Zarko [10] analyzed that the blockchain can enhance the efficiency and transparency of the distribution of the COVID-19 policy by ensuring strict compliance and evaluation of storage and delivery conditions. In our view, blockchain-based solutions can provide complete automatic deployment of data accountability and tracking. Blockchain technology is used in the supply chain management system of COVID-19 vaccination, it improves the efficiency of the distribution to the vaccination center. Datta et al. [11] defined a transparent network for blockchain technology to assist in developing top-level administration to the bottom-level users and develop trust between them. Also, Datta et al. [12] analyzed that the secure transactions are required for each block that takes place for each and every transaction in that network and will be notified to all the chain members.

In this paper, we analyzed the proper distribution of the vaccine and the user who registered for the vaccination by proper means, the admin will ensure that only the registered user will get the vaccination at the registered vaccination center. After every vaccination, a block is created and added to the existing blockchain. This will ensure that no attacker or intruders will enter the existing block.

The work in this paper contributes to offer a blockchain-based application to overcome the risks and the challenges. The finest way to avoid variation in the data record is to keep using a distributed block.

2 Related Work

Hao et al. [13] proposed a secured and fine-grained self-controlled data removal system in the cloud-based Internet of Things. It permits data holders to exactly and forever delete the IoT-determined data that is warehoused in the cloud without trusting the third party. The proposed work demonstrates a foreign policy approach collectively with decryption techniques along with key schemes. The effectiveness of this method is demonstrated by complete comparisons with the parameters of deleting data from cloud servers. Another work has been projected by Hasanat et al. [14] in which the actual data monitoring system for the transmission and distribution of vaccines is defined. The proposed work provides an exclusive feature to manage and create a humidity system with carrier moisture and temperature. The approach has successfully improved surveillance of vaccine distribution in circulated contexts. After that Rathee et al. [15] proposed a blockchain inherited vaccine distribution system relying on IoT technology, where a level wise-blockchain architecture has been maintained throughout. Security is assured by enhancing vaccine distribution without any interruption and less complexity. As the proposed system uses the level wise-blockchain system therefore the requirements of storage maintenance are improved. Now from the above work and study, Frustaci et al. [16] proposed an IoT network using blockchain-based security techniques for social networks and allowing the people to interact and share the information between a secured networks. Another work has been proposed by Yong et al. [17] for a novel intelligent system for supervision of vaccines for the vaccine supply chain with the help of blockchain technology. They create a smart contract based on Ethereum for maintaining the records and vaccine circulation. The research focused on vaccine circulation was designed for consumers and the government to trace the vaccine operation records. The Good Manufacturing Practice (GMP) chain is used for vaccine production and distribution, the release chain is designed for the vaccine enterprises with GMP certification and the vaccination chain is used to store the immunization records of patients. Therefore, it is found that supply chain management is very important for the proper distribution of vaccines. Hence, Saberi et al. [18] discussed the blockchain technology adoption in the network supply chain of vaccines, they identified and categorized blockchain barriers to the adoption of the supply chain management, and the authors identified and categorized blockchain barriers and adopted the new technology for the supply chain. Now security is an important aspect for creating a block. Therefore, Pan et al. [19] proposed a blockchain technology based on a trust system, from these they implemented a value exchange and enterprise vaccine records as the improvement of enterprise, this paper revealed the existing model of blockchain, and designed a new operational capability for the improvement of enterprises operations. Nash and Olmsted [20] has proposed a web application programming-based distributed model for uploading, creating, and communicating information with their test models. This article focuses on the entropy comparisons using actual data for vaccine-preventable diseases as an example of a demonstrating program. Kamble et al. [21] encourage the policymakers to adopt blockchain technology for the supply chain management system. The framework in this model is a combination of interpretive structural modeling (ISM) and decision-making trial and evaluation laboratory (DEMATEL) methods, to visualize the relationship between blockchain technology enablers. Farrugia et al. [22] reviewed the characteristics of blockchain technology in supply chain management systems. The system simplifies a huge amount of data, i.e., collected about the products in the manufacturing industry which proves to be beneficial to a range of different organizations and researchers. Chen [23] has stated that the blockchain is considered a supply chain of storage in which every operation is verified, accountable, and immutable. These essential features make it a possible solution for health data systems that are concerned with both sharing and patient privacy. Therefore, a blockchain-based storage scheme and service framework for storing, sharing, and using medical data is proposed in this study. Table 1 describes the comparative analysis of state-of-the-art studies on vaccine management and distribution.

3 Proposed Work

In this section, we proposed our architecture for the vaccine distribution records using a blockchain-based IoT framework. The architecture of this process is shown in Fig. 1. We created an Ethereum Smart Contract; our proposed work is to maintain the records of vaccines and how much vaccination is done. The main mechanisms of the system are given as:

Administrator: The administrator is at the top level of the database; the administrator can sight all the information warehoused and can access the vaccination

Table 1 Comparative analysis

Author	Methodology	Description	Limitations
Hasanat et al. [14]	Real-time application monitoring system for vaccine chain	An app-based monitoring system for the supply of vaccine cold chains to ensure the transaction of vaccines in the health centers	Not user friendly
Rathee et al. [15]	Level wise-blockchain network for secure vaccine distribution	A secured COVID-19 vaccine distribution over IoT-based systems and artificial neural network (ANN)-based mechanism to decrease the complexity and computation	Requirement to maintain the ledger's time
Yong et al. [17]	Ethereum blockchain technology for vaccine supply chain	Development of a vaccine blockchain system integrating the blockchain with machine learning in order to trace vaccines and smart contract functions for addressing the problems of vaccine expiration and vaccine record fraud	Not much suitable for real-time environments for the vaccine supply chain
Saberi et al. [18]	Blockchain technology in supply chain management	A review on examination of blockchain and smart contracts for supply chain management	The study on the environmental and social/ humanity dimension of sustainability including the U.N. sustainable development areas
Pan et al. [19]	Trust relationship-based operation on vaccine distribution	Implementation of blockchain technology in vaccine management to build a trust relationship and enhance collaboration among vaccine supply chain members	Appropriate time to implement blockchain technology for expansion of enterprise asset scale is required

records. Blockchain technology has two options, i.e., private and public blockchain. The difference between them is the result of accepting/rejecting the entering of users in the chain without verification.

Users: The users will be registered to one center record by their UID and get a slot for the vaccination.

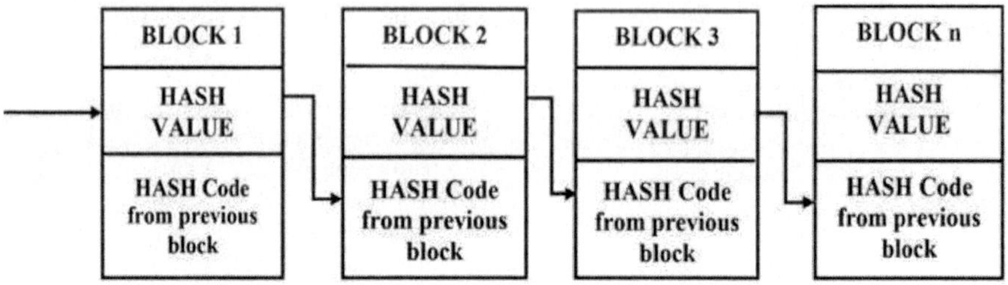

Fig. 1 Blockchain generation

Cloud Service: It is used to create a local network and store the data in cloud storage.

Blockchain Networks: The blockchain networks are used to keep data in blocks and it will be verified with the blockchain node.

Our method of tracking the vaccination is given in detail in this segment. We proposed a cloud-based system in which all the registration records are kept and accessed by the authorized/admin members of the vaccination centers. Blockchain network itself is a secured network, in this, the hash is used for the input of data and to keep the registration information secured. For every registration of the vaccination, a unique block is created for every user, the block contains user records such as name, address, previous vaccination details.

Hash functions are used during the entire Ethereum blockchain configuration process, transactions are kept as a hash. The hash code is generated based on the inputs of the blockchain administrator and other blockchain functions. So when a new block is added to the blockchain network, a new hash is created using the hash function and the code will be compared to the hashes of the remaining blocks in that network, if the code is correct, the block will added and the information will be verified, but if the code is incorrect, the block will be deleted as an entry error or fraudulent entry. This work was founded as smart contracts used in Ethereum networks, we created two smart contracts, the first for vaccination centers and the other for users who need access to vaccination records.

3.1 Formal Process

In our work, every time a user is vaccinated by the network at a vaccination center, the network creates a new block on the existing blockchain as a transaction record, and each transaction belongs to each member of the network, i.e., each member in the chain propagated to the node that is unique and cannot be changed.

The idea was conceived to be used as a protection for the COVID-19 vaccine at the vaccination center. In this module users can register by their Aadhar (UID) and book a slot at any vaccination center; the details of the registration are stored in cloud storage.

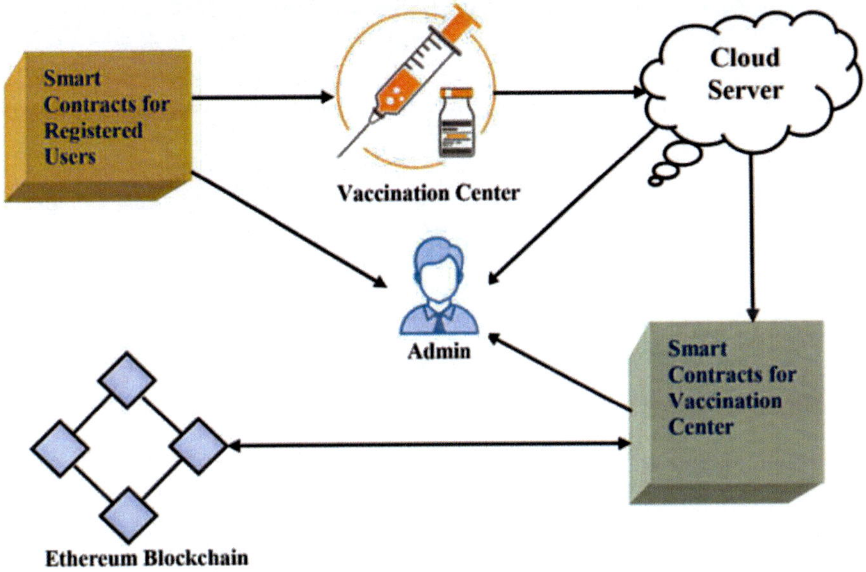

Fig. 2 Smart contract communication system

In the admin module, i.e., the vaccination center can have access to that storage, now when the registered users get vaccinated at the dedicated vaccination center, the admin will have to update the user as vaccination is done in their database, i.e., the cloud storage. The admin module stores the hash value of each transaction, this will make it easy to track all the transactions in the blockchain as shown in Fig. 2.

3.2 Smart Contracts

Khatoon et al. [24] stated that the smart contracts are just a collection of code or we can say a set of agreements followed by the users in the network while carrying out the processes in the blockchain network. Also, Datta et al. [25] discussed that for secure transactions, smart contracts are very important for authentication of the users in that network. In Fig. 2, the communication system of smart contract is shown and also with the help of smart contract the blockchain is created. In our work, the smart contract is used in vaccination centers that will create the blockchain network. The vaccination centers will keep records of vaccines and at the same time can store vaccine-related files to the cloud storage. Also, the vaccination centers can encrypt the data in such a way that only the individuals have proper access to it because the files can be read by government authorized persons or the person whose vaccination record is kept. On the other hand, individuals using the vaccine can interact with the blockchain network and access their earlier records when it is required. Confidentiality is a very important aspect in terms of medical things; thus, it is maintained because each individual has a different block address. The smart contract has activities to display specific vaccine records such as vaccination date and its variants. To keep the vaccine records safe, the authors Abeyratne and Monfared [26] found that the supply chain is very important for taking care of the records and it can also reduce the technical fault in that network.

Figure 3 depicts the flowchart for the storage of vaccination records. The user is registered by their UID and the data is stored in the cloud storage. The details of authenticated users are available for the admin. However, the algorithm will not allow to save any unauthenticated or duplicated data in the cloud storage. In Fig. 4, the flow diagram is depicted to describe the process of creation of the Ethereum blockchain network. If any users get vaccinated then a block will be created and joined to the blockchain network.

Fig. 3 Flowchart for storing vaccination records

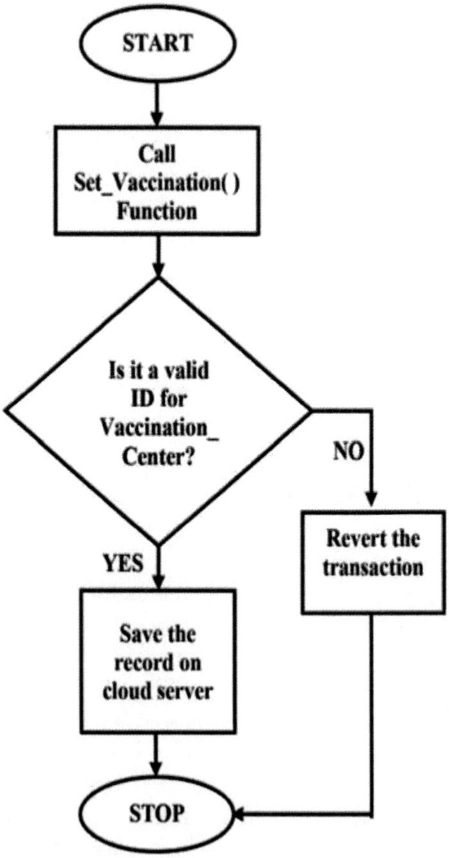

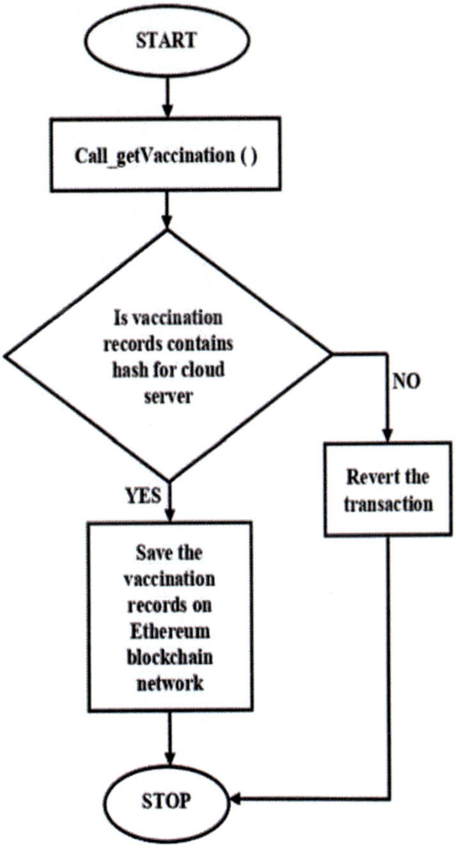

Fig. 4 Flowchart for creation of ethereum network

4 Results and Discussion

The performance of our proposed approach has been analyzed in terms of the true generated response, altered data and product loss ratio. The defined performance parameters, i.e., true generated response and altered data analysis results in preventing misplacement of vaccines and improving product loss. Figure 5 represents an assessment of two security measures, actual generated reports, and modified data in proposed and existing cases. The true generated response is defined as the total number of vaccinations done to the patients versus the number of vaccine units present in the vaccination centers. The graph determines the number of true generated responses produced by the proposed method. Our findings show a better response compared to other approaches in terms of vaccinations. The reports produced in the instance of the projected entity are always accurate and related to the methods available due to the maintenance of blockchain structures among a few organizations. Blockchain maintains transparency by ensuring accurate and precise reporting of IoT

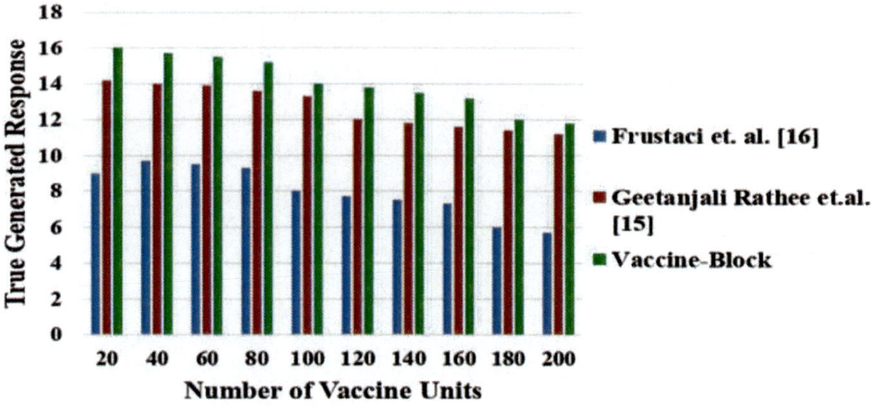

Fig. 5 True generated response versus number of vaccine units

devices during communication. The proposed work generated significant improvement as compared to the other two approaches due to the use of the blockchain network.

In Fig. 6, the change in the number of units produced by any malicious middle organization can be seen quickly and easily projected status. The altered data is defined as the number of true generated responses and altered by transitional units while doing the vaccination. Modified data production is very important in maintaining a blockchain database during the communication process. The percentage of altered data with respect to the number of vaccine units in Frustaci et al. [16], does not give the better results, it finds only about 15% of the data alteration record. But the findings of Rathee et al. [15] show that the percentage of data alteration is about 25% that is better compared to Frustaci et al. [16], whereas our proposed scheme finds the better percentage of data alteration, i.e., about 30% which is higher than Rathee et al. [15] and Frustaci et al. [16]. This proposed situation makes significant improvements compared to conventional methods due to the blockchain network.

Figure 7, shows the product loss ratio in our work is less compared with the other approaches. In our work, the percentage of the product loss ratio of the vaccine is decreased with the increase in the number of vaccinations as compared with other two approaches. Our findings show only 4% of product loss ratio, whereas Rathee et al. [15] finding is about 6% and the results of Frustaci et al. [16] is approximately 9%.

5 Conclusion

In this paper, we discussed the supply chain of the COVID-19 vaccine, and keep an eye on the vaccination center because of the misuse and theft of the vaccine. Blockchain technology is adopted for records and tracking of the supply chain. We

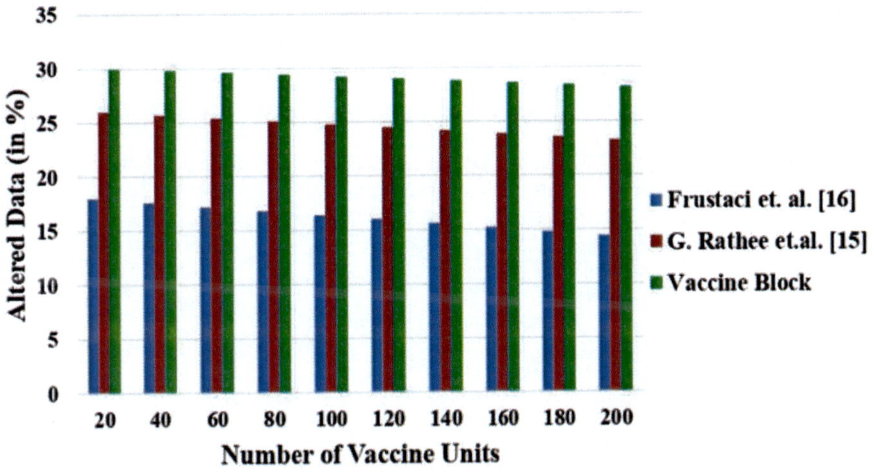

Fig. 6 Altered data versus number of vaccine units

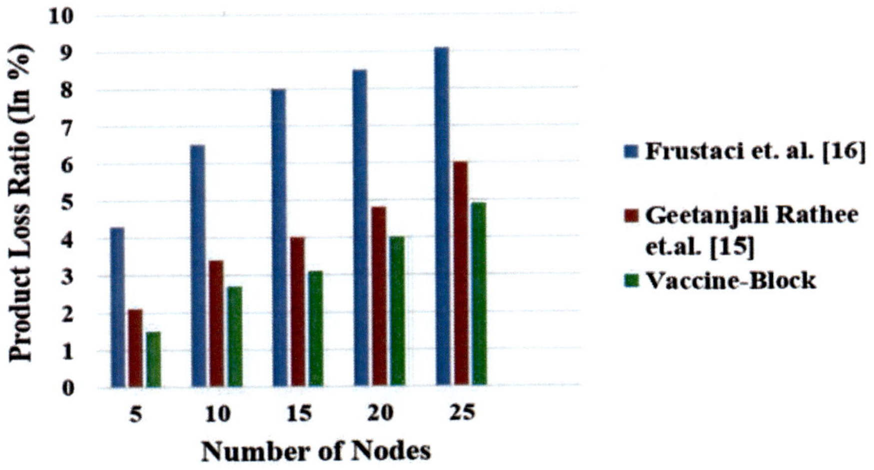

Fig. 7 Product loss ratio versus number of nodes

introduce a detailed study of the vaccine record-keeping model on the blockchain network and discuss various factors such as the design, maintenance, and feasibility of the proposed solution. The result analysis of the proposed method is presented in terms of various parameters such as true generated response, altered data, and product loss ratio. The proposed work in this paper aims to offer a standard effective solution for keeping vaccination records on the blockchain network safely and economically.

References

1. Akshita V, Dhanush JS, Dikahitha V (2021) Blockchain based Covid vaccine booking and vaccine management system. (ICOSEC)/IEEE. https://doi.org/10.1109/ICOSEC51865.2021.9591965
2. Rathee G, Balasaraswathi M, Chandra KP, Gupta SD, Boopathi CS (2017) A secure IoT sensors communication in industry 4.0 using blockchain technology. https://doi.org/10.1007/s12652-020-02017-8
3. https://www.who.int/data
4. https://www.mygov.in/covid-19
5. Deka SK, Goswami S, Anand A (2020) A Blockchain based technique for storing vaccination records. In: IEEE bombay section signature conference (IBSSC).https://doi.org/10.1109/IBSSC51096.2020.9332171
6. https://www.nature.com/articles/s41598-021-02802-1.pdf
7. Farzanegan MR, Hofmann HP (2021) Effect of public corruption on the COVID-19 immunization progress. https://doi.org/10.1038/s41598-021-02802-1
8. Antal CD, Cioara T, Antal M, Anghel I (2021) Blockchain platform for COVID-19 vaccine supply management. IEEE Open J Comput Soc 2:164–178. ISSN: 2644-1268. https://doi.org/10.1109/OJCS.2021.3067450
9. Rouhani S, Deters R (2017) Performance analysis of Ethereum transactions in private blockchain. In: 2017 8th IEEE international conference on software engineering and service science (ICSESS), pp 70–74
10. Beni FM, Zarko IP (2018) Distributed ledger technology: blockchain compared to directed acyclic graph. In 2018 IEEE 38th international conference on distributed computing systems (ICDCS), pp 1569–1570
11. Datta S, Kumar S, Sinha D (2022) BSSFFS: blockchain-based sybil-secured smart forest fire surveillance. J Ambient Intell Human Comput 13:2479–2510. https://doi.org/10.1007/s12652-021-03591-1
12. Datta S, Sinha D (2021) BESDDFFS: Blockchain and EdgeDrone based secured data delivery for forest fire surveillance. Peer-to-Peer Netw Appl 14:3688–3717. https://doi.org/10.1007/s12083-021-01187-2
13. Hao J, Liu J, Wu W, Tang F, Xian M (2015) Secure and finegrained self-controlled outsourced data deletion in cloud-based IoT. IEEE Internet of Things J 7(2):1140–1153
14. Hasanat R, Rahman T, Md. A, Mansoor N (2020) An IoT based real-time datacentric monitoring system for vaccine cold chain. In: IEEE East-West design and test symposium, 2020, pp 1–5
15. Rathee G, Sahil G, Georges K (2021) An IoT-based secure vaccine distribution system through a blockchain network. IEEE Internet of Things Magazine
16. Frustaci M, Pace P, Aloi G, Forting G (2017) Evaluating critical security issues of the IoT world: present and future challenges. IEEE Internet of Things J. https://doi.org/10.1109/JIOT.2017.2767291
17. Yonga B, Shenc J, Liua X, Lic F, Chenc H, Zhou Q (2019) An intelligent blockchain-based system for safe vaccine supply and supervision. https://doi.org/10.1016/j.ijinfomgt.2019.10.009
18. Saberi S, Kouhizadeh M, Sarkis J, Shen L (2019) Blockchain technology and its relationships to sustainable supply chain management. Int J Prod Res 57(7):2117–2135
19. Pan X, Pan X, Song M, Ai B, Ming Y (2019) Blockchain technology and enterprise operational capabilities: an empirical test. Int J Inform Managem. https://doi.org/10.1016/j.ijinfomgt.2019.05.002
20. Nash T, Olmsted A (2017) shinysdm: point and click species distribution modeling. In: 12th international conference internet technology and secured transactions. pp 450
21. Kamble SS, Gunasekaran A, Sharma R (2020) Modeling the blockchain enabled traceability in agriculture supply chain. Int J Inform Managem 52:101967. ISSN 0268-4012. https://doi.org/10.1016/j.ijinfomgt.2019.05.023

22. Farrugiaa S, Ellul J, Azzopardi G (2020) Detection of illicit accounts over the Ethereum blockchain. https://doi.org/10.1016/j.eswa.2020.113318
23. Chen Y, Ding S, Xu Z, Zheng H, Yang S (2019) Blockchain-based medical records secure storage and medical service framework. J Med Syst 43(1):5
24. Khatoon A (2020) A blockchain-based smart contract system for healthcare management. Electronics 9(1):94. https://doi.org/10.3390/electronics9010094
25. Datta S, Das AK, Kumar A, Khushboo, Sinha D (2020) Authentication and privacy preservation in IoT based forest fire detection by using blockchain—a review. In: Nain N, Vipparthi S (eds) 4th international conference on internet of things and connected technologies (ICIoTCT), 2019. ICIoTCT 2019. Advances in intelligent systems and computing, vol 1122. Springer, Cham. https://doi.org/10.1007/978-3-030-39875-0_14
26. Abeyratne SA, Monfared RP (2016) Blockchain ready manufacturing supply chain using distributed ledger. Int J Res Eng Technol 5(9):1–10

MRI-Based Early Diagnosis and Quantification of Trans-Ischemic Stroke Using Machine Learning—An Overview

R. Bhuvana and R. J. Hemalatha

Abstract A "Transient Ischemic Attack" (TIA) is a brief period of neurological impairment brought on by cerebral ischemia that is not accompanied by persistent cerebral infarction. This type of neurological impairment can be caused by a lack of blood flow to the brain. According to the World Health Organization (WHO), stroke is the leading cause of death and disability on a global scale. A transient ischemic attack (TIA) should be evaluated as soon as possible using imaging and laboratory tests in order to cut down on the possibility of further strokes. The primary objective of treatment for transient ischemic attack (TIA) is to lower the patient's chance of experiencing another TIA or stroke. Getting treatment after a transient ischemic attack can significantly lower the risk of having an early stroke. The severity of a stroke can be reduced if the numerous warning symptoms for a stroke are identified and treated promptly. When it comes to making a clinical diagnosis of TIA, the evaluation of the patient's medical history is still an essential component. In light of this, our primary objective was to provide an overview of the pathophysiology underlying transient ischemic attacks. In addition to this, we take a more in-depth look at the evaluation of the diagnostic procedures that can be used to identify transient ischemic attacks. The numerous works that have been done on the machine learning algorithms that are utilized to differentiate and identify the TIA and provide assistance with early diagnosis have been reviewed and discussed.

Keywords Trans Ischemic Attack (TIA) · CT · MRI · DWI · Deep learning · SVM

1 Introduction

A stroke is an urgent medical disease defined by acute vascular injury or cerebral perfusion. Stroke is the sixth most common cause of death and the leading cause of disability worldwide, and it can result in permanent brain damage and impairment. A

R. Bhuvana · R. J. Hemalatha (✉)
Department of Biomedical Engineering, VISTAS, Pallavaram, Chennai, Tamil Nadu 600117, India
e-mail: bhuvana2089@gmail.com

S. Das et al. (eds.), *Advances in Data-driven Computing and Intelligent Systems*, Lecture Notes in Networks and Systems 653, https://doi.org/10.1007/978-981-99-0981-0_28

stroke happens when a blood vessel in the brain bursts or ruptures or when the blood supply to a portion of the brain is cut off (ischemic stroke) (hemorrhagic stroke).

Transient ischemic attack (TIA) is a transient stage of symptoms that resembles the symptoms of a stroke [1]. This is a brief interruption of blood flow to a part of the brain, spinal cord, or retina that can cause symptoms like a temporary stroke, but it can damage brain cells or be permanent [2]. It does not cause any trouble. About one in three people who have had a TIA will have a subsequent stroke. Within 48 h of TIA, the risk of stroke is very high. Up to 80% of strokes after TIA are preventable and early diagnosis is the key to treatment. On average, the annual risk of future ischemic stroke after TIA or first ischemic stroke is 3–4% [3], up to 11% over the next 7 days, and up to 30% over the next 5 years [4].

2 Prevalence

The incidence of TIAs in a population is difficult to estimate because of other similar disorders. Internationally, the probability of a TIA is about 0.42 per 1000 in developed countries [5]. The incidence of TIA in the United States is probably about half a million people per year, and the probability is about 1.1 per 1000 people in the United States. The estimated overall prevalence of TIA in adults in the United States is about 2% [3]. TIAs increased with age, from 13 cases per 100,000 people under the age of 35–1500 cases per 100,000 people over the age of 85 [6]. Less than 3% of all major strokes occur in children. Strokes in children can often have a very different etiology than strokes in adults and are less frequent. The prevalence of TIA is significantly higher in men than in women [7]. One-third of TIAs worldwide occur in people under the age of 65, and stroke in women is becoming an ongoing epidemic. The transient ischemic attack is no longer considered a benign event but rather a critical alarm for stroke. If you don't recognize and respond to this warning sign quickly, you could miss your chance to prevent permanent injury or death [8]. The most common signs of TIA are glucose disturbances, migraine headaches, seizures, retrorectal states, and tumors (especially with acute bleeding).

Several studies have shown that the majority of people who have had a TIA have a rapid onset and peak intensity is usually reached within minutes. One- or two-second episodes of nonspecific symptoms such as fatigue, dizziness, and bilateral rhythmic tremor are likely indicative of acute penetrating ischemia. Most patients who report symptoms of a transient ischemic attack should be referred to the emergency room and undergo a thorough history and physical examination, as well as selected tests to determine whether the patient is on thrombolytic therapy or not [9].

3 Risk Factors for TIA

According to an evidence-based classification, the following are the major risk factors for TIAs: Age, sex, and significant family history are unchangeable. Other factors that can be changed include smoking, obesity, inactivity, cardiovascular and lipid disorders, coronary artery disease, myocardial infarction, valvular heart disease, atrial fibrillation, diabetes mellitus, arterial hypertension, and peripheral artery disease. Other factors that could be changed include age, sex, and significant family history (history of migraine, obstructive sleep apnea, high-risk sleeping, and alcohol habits).

Obesity, smoking, alcohol usage, diet, drug misuse, and inactivity were found to be modifiable lifestyle variables that increase the risk of stroke [10]. An independent risk factor is obesity [11]. Particularly in young persons with short sleep duration, sleep patterns are thought to be a potential risk factor [4]. Another factor is daytime nappy migraine, which is more common in older people and is associated with airflow obstruction and obstructive sleep apnea [4]. A poor way of life is not only associated with a higher risk of stroke but also increased all-cause mortality after stroke [3, 4] rate of TIA.

4 Review of Documentation

4.1 Materials and Methods

Imaging Modality

Immediate awareness of these disorders is required and may require neurosurgical intervention or specific therapeutic approaches. CT scans can identify conditions that mimic TIA, such as epilepsy-related tumors and other masses. Brain CT scans can detect early signs of brain injury or signs of a bold stroke [12, 13]. Magnetic Resonance Imaging (MRI) is generally the best method for structural analysis of the brain because it provides high-contrast images and high spatial resolution of soft tissues and poses no health risks [14]. Modality such as computer tomography (CT) and positron emission tomography (PET) are also used for brain examination, but MRI is the most common and this task focuses on MRI. The benefits of brain MRI overhead CT include better tissue imaging (i.e., higher sensitivity for early diagnosis), better imaging of the posterior fossa (including brain stem and cerebellum), and additional imaging [15]. Includes planes (sagittal, coronal, and diagonal). No radiation exposure [16]. CT is not sensitive to the assessment of brainstem or cerebellar disease due to the increased bone artifacts in the skull area. In these cases, magnetic resonance imaging (MRI) is the preferred test. For this reason, MRI is useful and can replace an immediate CT scan of the head during the initial evaluation of TIA patients. MRI or MRA is recommended if cerebral vascular malformations, aneurysms, cerebral venous thrombosis, or arthritis are suspected [17]. Table 1 describes landmark ischemic stroke research work over the past decade. It also represents the work being done in the detection of brain tumors and ischemic strokes related to the work we are interested in [18].

MRI-assisted detection can be performed using two-dimensional (2D) or 3D images, and it also provides the option to choose image registration methods, such as T1, crystallization economy, and weight (DW). Diffusion equilibrium imaging detects vascular mass as early as 10 to 15 min after symptom onset [7]. In addition, MRI-assisted diagnosis will help gather detailed information such as the severity, location, and volume of the affected part of the brain, which plays a key role in planning and performing treatment. In addition, the fusion modality also allows for improved diagnosis, when MRI is considered.

Table 1 Literature Overview of trans ischemic stroke in the past decade

Year/Journal	Article title	First author	Result
2022/Scientific report	Validation of CSR model to predict stroke risk after transient ischemic attack	Lu Zhao	The CSR model and ABCD3 I score have good predictive value for predicting 90-day stroke risk in definite TIA patients. In addition, the CSR model outperformed the ABCD3 I score in predicting 90-day stroke risk in DWI-positive TIA patients. Using the CSR model, we were able to further stratify high-risk TIA patients
2021/Pubmed	The role of urgent imaging in the diagnosis and management of patients with TIA and minor stroke	Negar Asdaghi	A significant proportion of patients with mild neurological symptoms or who have completely resolved show evidence of vascular or tissue abnormalities in acute neuroimaging studies. These factors have proven invaluable in risk stratification, treatment planning, and outcome prediction in these patients

(continued)

Table 1 (continued)

Year/Journal	Article title	First author	Result
2020/Springer	Posterior circulation stroke: machine learning-based detection of early ischemic changes in acute non-contrast CT scans [10]	Helge C. Kniep	The proposed artificial intelligence-based algorithm confirms the feasibility of automated, reproducible, reader-independent detection of ischemic lesions in the posterior circulation. Stroke patient. The prediction performance was excellent or Equivalent to a visual assessment by at least two neuro radiologists. The proposed approach extends traditional measurements (a) Do not integrate texture and filter-based image functionality Evaluable by the human eye, (b) use artificial intelligence algorithms for automated and standardized data interpretation. This system can facilitate reproducible analysis in future studies and provide supportive tools for clinical decision making, treatment planning, and outcomes
2019/Pubmed	Diagnosis and Management of Transient Ischemic Attack	Shelagh B. Coutts	This article reviews the diagnosis, investigation, and recommended the management of following a transient ischemic attack (TIA) and explains how to make an accurate diagnosis, including diagnosing a mimic TIA

(continued)

Table 1 (continued)

Year/Journal	Article title	First author	Result
2019/Biomedical journal	Artificial Intelligence in Diagnosis and Management of Trans Ischemic Stroke	Swati Gupta	This review highlights recently developed AI-based tools for stroke diagnosis and treatment. The application of AI in predicting and predicting outcomes is also summarized
2017/Pubmed	Deep Learning for Brain MRI Segmentation: State of the Art and Future Directions	Zeynettin Akkus1	This review evaluates current deep learning architectures used to segment anatomical brain structures and brain lesions. Then, the performance, speed, and characteristics of deep learning methods are summarized and discussed
2015/Indian journal of science and technology	Computer Reinforced Analysis for Trans Ischemic Stroke Recognition: A Review	R. Kanchana	This stroke survey was to study such a diagnostic system to do a better job of highlighting stroke lesions. Such automated methods that use brain MRI and CT images to classify stroke-active tumors from non-tumor images greatly assist researchers and physicians
2014/Pubmed	Magnetic resonance imaging in patients with transient ischemic attack	Mohamed Al-Khaled	In summary, brain imaging in patients with transient neurological symptoms may help answer the question of whether TIA patients with acute infarction should be classified as acute ischemic stroke.. Furthermore, he found that TIA patients with acute infarction due to DWIMRI or CCT had a higher risk of disabling stroke

(continued)

Table 1 (continued)

Year/Journal	Article title	First author	Result
2014/AIMS	A fast multiparameter MRI approach for acute stroke assessment on a 3 T clinical scanner: preliminary results in a non-human primate model with transient ischemic occlusion	Xiaodong Zhang	The application of parallel imaging technique significantly reduces the TA of the most time-consuming MRI measurements and enables rapid and/or repeatable examination of acute stroke lesions by multiparameter MRI. Established procedures are validated with a monkey stroke model in a clinical setting and can be used to assess the time course of stroke lesions in a PSN stroke model or possibly in a patients stroke patient. Applying this approach to stroke patients would benefit from an even larger human brain volume

As 3D and 4D imaging became commonplace and physiological and functional imaging increased, medical imaging data increased in size and complexity. Therefore, it is essential to develop tools that can help extract insights from these large data sets. Machine learning is a set of algorithmic techniques that allow a computer system to make data-driven predictions from big data. These techniques have many different applications that can be adapted to the medical field. Some other well-known and publicly available datasets for brain MRI are Brain Tumor Segmentation (BRATS), Ischemic Brain Injury Segmentation (ISLES), Mild Traumatic Brain Injury Outcomes Prediction (mTOP), Multiple Sclerosis Segmentation (MSSEG), Neonatal Brain Segmentation (NeoBrainS12), and MR Brain Imaging Segmentation [19].

5 Early Diagnosis and Machine Learning

In the literature, the researchers used biomedical data to perform several early detection procedures for stroke. The work of Johnson et al. (2019) confirmed that cerebral vascular accidents (stroke) are the second chief reason for death and the third principal reason of disability in humans. The earlier research work also confirmed that MRI-supported stroke detection is widely recommended to detect the location and harshness accurately. The work of Maier et al. (2018) presented a detailed assessment of the mining of Ischemic Stroke Lesions (ISL) in multispectral MRI using SVM. Similar work by the researchers presents a comparison of semi-automated /automated segmentation procedures for ISL using the images of the database. Table 2 presents the details of literature works done on the early diagnosis of ischemic stroke and the various diagnostic procedures and management techniques used in this process to overcome the disease.

Table 2 Literature of early diagnosis of ischemic stroke using CT and MRI

Title	Year/Journal	Modality	Processing technique
Characterization of Brain Stroke Using Image and Signal Processing Techniques *Abdullah Alamoudi et al.*	2021	CT and MRI	Sharpen features using various image processing algorithms, such as ROI and watershed-based segmentation methods
Brain stroke computed tomography images analysis using image processing: A review *Nur Hasanah Ali*	2021/International journal of AI	CT	Compare segmentation techniques such as Fuzzy C-mean, Threshold, Region Growth, k-means, and watershed segmentation techniques
MRI image analysis methods and applications: an algorithmic perspective using brain tumors as an exemplar *Vadmal et al.* [20]	2020	MRI brain	Filtering, surface fitting, segmentation, and histogram
Survey of Image Processing Techniques for Brain Pathology Diagnosis: Challenges and Opportunities *Martin Cenek et al.*	2018/Frontier in robotics and AI	MRI (T1, T2, FLAIR, and DTI)	The study aimed to precisely identify the active regions of the tumor and its extension in the brain and label the different regions of the tumor accordingly
Image Processing for Enhancement of Ischemic Stroke	2019/Research gateway	Non-enhanced computed tomography (CT) and nuclear magnetic resonance imaging (MRI)	This research proposes an approach based on a computational algorithm, highlighting regions of ischemic stroke

The related works of Maier et al. [1, 12, 13] presented segmentation of stroke lesion in MRI of chosen modality using different techniques. Subbanna et al. [21] demonstrated the evaluation of ISL in Flair MRI using the modifier Markov random field. Zhang et al. [21] presented multiplane information fusion-based segmentation from various MRI modalities. Singh et al. [22] discussed deep learning (DL) supported ISL detection. The work by Rajinikanth and Satapathy [23] presented joint thresholding and segmentation-based ISL assessment, and a similar attempt was presented in the research by Lin et al. [24]. The recent work by Hemanth et al. [25] implemented a multimodality fusion-based ISL examination. The review by Zhang et al. [26] confirmed the following limitations in earlier works: (1) modality-specific detection, (2) in most of the modalities, automated extraction and evaluation is quite difficult, and (3) less detection accuracy for T1 modality case [24].

The work carried out by Shivangi Gupta et al. stated that MRI images are influenced by various noises which are needed to be removed before any processing. The Gaussian filter turns out to be the one with the maximum SNR and at the same time, no valuable information is lost [27]. The fact that the brain is quasi-symmetric has been used to detect strokes [28]. Various parameters have been calculated and extracted that are used to train the neural network to classify normal and stroke victims [12]. In recent years, a collection of image review methods semi-automatic and full bioimaging has been discussed and performed by researchers [29]. Semi-automated computer tools are widely practiced for evaluating many types of medical images due to their accuracy [30]. Automated methods have also been implemented to improve outcomes during medical imaging evaluation [12].

Considerable efforts have been made to develop classical machine learning algorithms to segment normal (e.g., white and gray matter) and abnormal (e.g., brain tumors) brain tissue in MRI [18]. However, generating image features that enable such segmentation requires careful engineering and expertise. Despite the considerable effort of the medical imaging community, automated segmentation of brain structures and detection of anomalies remains an unresolved problem due to normal anatomical variations in imaging. Brain morphology, variations in acquisition parameters and MRI scans, image acquisition defects, and variations in the appearance of pathology. Table 3 describes various machine learning techniques for the early diagnosis of TIAs by MRI [19].

6 Discussion

TIA evaluation is essential for making an accurate diagnosis, and a good medical history is essential. Once a diagnosis of a TIA has been made, computed tomography can help inform potential etiologies and guide the initiation of evidence-based secondary stroke prevention strategies. Despite the significant impact of deep learning techniques in quantitative brain MRI, it is still difficult to have a single universal method robust enough for all variations of brain MRI images from different MRI scanners and institutions [31]. The performance of deep learning methods is

Table 3 Literature on MRI-based early detection of TIA using machine learning

S. No.	Title	Year/Journal	Methodology	Result
1	Early Identification of High-Risk TIA or Minor Stroke Using Artificial Neural Network *Ka Lung Chan*	2019/Frontier journal	They obtained the sensitivity, specificity, accuracy, and c-statistics of each ANN model from 5 rounds of cross-validation and compared it between Support Vector Machine (SVM) and Naïve Bayes classifier models in the analysis of patient risk strategies	The mean sensitivity, specificity, accuracy, and c-statistics of ANN models for predicting recurrent stroke at 1 year were 75%, 75%, 75%, and 0.77, respectively. The ANN model outperformed the SVM and Naïve Bayes classifiers in the data set in predicting recurrence after TIA or mild stroke
2	Transient ischemic attack analysis through non-contact approaches *Qing Zhang*	2020/Springer	The advantages of SVM include high precision, good theoretical guarantees on the overfitting, etc	The idea of RF is to build many decision trees to form a "forest" and make decisions by voting. Both theoretical and experimental studies show that this method can effectively improve the accuracy of classification…
3	Automatic detection of ischemic stroke using higher-order spectra features in brain MRI images U Rajendra Acharya	2019/Research gate way	The basic idea behind the system design is to find discriminative features which can be used to extract diagnostically relevant information from DWI images. Feature extraction is necessary because the machine learning algorithms, in our case the SVM, cannot readily address high-dimensional data, such as is typically found in biomedical imagery. The feature extraction projects the image into a lower-dimensional space, which is then input to a machine classifier	The first method was based on a first-order polynomial kernel, also known as a linear kernel. Two more methods were constructed by increasing the polynomial order to two and three, respectively. The fourth method employs the RBF kernel function which is based on the squared Euclidean distance between the feature vectors

(continued)

Table 3 (continued)

S. No.	Title	Year/Journal	Methodology	Result
4	Development and clinical application of a deep learning model to identify acute infarct on magnetic resonance imaging Christopher P. Bridge	2022/Scientific reports	This paper sought to develop a machine learning algorithm that would both detect and segment acute infarcts on MRI imaging. Then demonstrated the effectiveness of this algorithm in three clinical scenarios including two-stroke code test sets (at training and non-training hospitals) and an international test set	Classification is reported less frequently amongst published models; a recent study reported sensitivity of 91% and specificity of 75% while our model had sensitivity between 89.3 and 100.0%, and specificity between 86.6% and 98.0% for the test sets

highly dependent on several key steps such as pre-processing, initialization, and post-processing. In addition, the training dataset is relatively small compared to the large-scale Image Network dataset (e.g., millions of images) to achieve generalization across the dataset [32, 33]. Furthermore, current deep learning architectures are based on supervised learning and require a manual generation of basic truthy labels, which is tedious work for large-scale data [27, 30]. Therefore, deep learning models that are robust to brain MRI variants or capable of unsupervised learning with fewer requirements for underlying truth labels are needed [19].

Accurate detection of ischemic stroke is still very demanding because the appearance, size, shape, and structure of ischemic strokes can vary. Although ischemic stroke segmentation methods have shown great potential in the analysis and detection of ischemic stroke on MRI images, much improvement is still needed for segmentation and segmentation precise classification of the area of ischemic stroke. The present work presents limitations and challenges in identifying the background structures of ischemic stroke regions and classifying healthy and unhealthy images [10].

In summary, this survey covers all the important aspects and the latest work done to date with their limitations and challenges. It will be helpful for researchers to develop an understanding of doing new research in the short term and in the right direction. Deep learning has made a significant contribution but still requires a common technique. These methods yielded better results when training and testing were performed on similar acquisition characteristics (intensity range and resolution); however, a small variation in training and test images directly affects the robustness of the methods. In future work, research can be conducted to detect AIT and deep features can be merged to improve classification results. Likewise, lightweight methods like quantum machine learning play an important role in improving accuracy and efficiency, saving radiologists time, and increasing patient survival rates.

7 Conclusion

Trans Ischemic Stroke is the partial blockage of blood vessels in the brain that subsequently leads to brain tissue damage and can cause further brain damage, necrosis, and even death. The timely diagnosis of Trans Ischemic Stroke is vital for functional recovery and to minimize mortality. The advancement in imaging technology for stroke diagnosis led to the availability of a large volume of scattered neuroimaging information. AI has been employed in several ways to extract the most coherent information which can be used as an identifier or marker for stroke diagnosis and for analyzing its severity. The ability of AI to provide clinically relevant output information solely depends on the correctness of the input data and the machine learning method used to train the AI model. Therefore, researchers are focusing on the development of a new learning algorithm that can handle a large volume of information and provide more precise output information in a reasonable time frame.

References

1. Dey N, Rajinikanth V (2022) Automated detection of ischemic stroke with brain MRI using machine learning and deep learning features. Elsevier BV
2. Bridge CP, Bizzo BC, Hillis jm, Chin JK et al (2022) Development and clinical application of a deep learning model to identify acute infarct on magnetic resonance imaging. Sci Rep
3. Przelaskowski A et al (2007) Improved early stroke detection: wavelet-based perception enhancement of computerized tomography exams. Comput Biol Med 37:524–533 (Science Direct)
4. Kniep HC, Sporns PB, Broocks G, Kemmling A, Nawabi J, Rusche T, Fiehler J, Hanning U (2020) Posterior circulation stroke: machine learning-based detection of early ischemic changes in acute non-contrast CT scans. J Neurol
5. Zhu G, Chen H, Jiang B, Chen F, Xie Y, Wintermark M (2022) Application of deep learning to ischemic and hemorrhagic stroke computed tomography and magnetic resonance imaging. Sem Ultrasound CT MRI
6. Balasooriya U, Perera MUS (2012) Intelligent brain hemorrhage diagnosis using artificial neural networks. In: 2012 IEEE business, engineering & industrial applications colloquium (BEIAC)
7. Hemalatha RJ, Vijaybaskar V, Thamizhvani TR (2018) Performance evaluation of contour based segmentation methods for ultrasound images. Adv Multimedia
8. Reed DM, Resch JA, Hayaski T, MacLean C, Yano K (1988) A prospective study of cerebral atherosclerosis. Stroke 19:820–825
9. Crinion J, Holland AL, Thompson CK, Hillis AE (2013) Neuroimaging in aphasia treatment research: quantifying brain lesions after stroke
10. Bacchi S, Oakden-Rayner L, Zerner T, Kleinig T, Patel S, Jannes J. Deep learning natural languageprocessing successfully predicts the cerebrovascular cause of transient ischemic attack-like presentations. Stroke
11. Lee Y, Takahashi N, Tsai DY, Fujita H (2006) Detectability improvement of an early sign of acute stroke on brain CT images using an adaptive partial smoothing filter. In: Proceedings of the society of photo optical instrumentation engineering and medical imaging, vol 6144, pp 2138–2145
12. Zhang Q, Li Y, Al-Turjman F, Zhou X, Yang X (2020) Transient ischemic attack analysis through non-contact approaches. Human-Centric Comput Inf Sci
13. Prabhu Das I, Baker M, Altice C, Castro KM, Brandys B, Mitchell SA (2020) Neural Comput Appl 32:15897–15908. Outcomes of multidisciplinary treatment planning in US cancer care settings. Cancer 2018, 124:3656–3667 [PubMed]
14. Ren Z (2022) Chapter 3 complications of aneurysm embolization and prevention. Springer
15. Hankey GJ, Warlow CP, Sellar RJ (1990) Cerebral angiographic risk in mild cerebrovascular disease. Stroke 21:209–222
16. Zhang L, Song R, Wang Y, Zhu C, Liu J, Yang J, Liu L (2020) Ischemic stroke lesion segmentation using multi-plane information fusion. IEEE Access 8:45715–45725
17. Willinsky RA, Taylor SM, Ter Brugge K, Farb RI, Tomlinson G, Montanera W (2003) Neurologic complications of cerebral angiography: prospective analysis of 2,899 procedures and review of the literature. Radiology 227:522–528
18. Hemalatha RJ, Vijayabaskarin V (2018) Histogram based synovitis scoring system in ultrasound images of rheumatoid arthritis. J Clin Diagnostic Res
19. Akkus Z, Galimzianova A, Hoogi A, Rubin DL, Erickson BJ (2017) Deep learning for brain MRI segmentation: state of the art and future directions. J Digit Imaging
20. Johnston CS, Gress DR, Browner WS, Sidney S (2000) Short-term prognosis after emergency department diagnosis of TIA. JAMA 284:2901–2906
21. Mousa AE, Elrakhawy MM, Zaher AA (2013) Multimodal CT assessment of acute ischemic stroke. Egypt J Radiol Nuclear Med 71–81

22. Vadmal V, Junno G, Badve C, Huang W, Waite KA, Barnholtz-Sloan JS (2020) MRI image analysis methods and applications: an algorithmic perspective using brain tumors as an exemplar. Neuro Oncol Adv
23. Zhang S, Xu S, Tan L, Wang H, Meng J (2021) Stroke lesion detection and analysis in MRI images based on deep learning. J. Healthc Eng
24. Zhao L, Cao S, Pei L, Fang H, Liu H, Wu J, Sun S, Gao Y, Song B, Xu Y (2022) Validation of CSR model to predict stroke risk after transient ischemic attack. Sci Rep
25. Maier O, Menze BH, von der Gablentz J, Häni L, Heinrich MP, Liebrand M, Reyes M ISLES 2015—A public evaluation benchmark for ischemic stroke lesion segmentation from multispectral MRI. Med. Image Anal. 35:250–269 [PubMed]
26. Subbanna NK, Rajashekar D, Cheng B, Thomalla G, Fiehler J, Arbel T, Forkert ND (2019) Stroke lesion segmentation in Flair MRI datasets using customized Markov random fields. Front Neuroanat 10:541
27. Kanchana R, Menaka R (2015) Computer reinforced analysis for ischemic stroke recognition: a review. Ind J Sci Technol
28. Chan T (2007) Computer-aided detection of small acute intracranial hemorrhage on computer tomography of the brain. Comput Med Imaging Graph
29. Gupta S, Mishra A, Menaka R (2014) Ischemic stroke detection using image processing and ANN. In: 2014 IEEE international conference on advanced communications, control and computing technologies
30. Hemalatha RJ, Thamizhvani T, Dhivya AJA, Joseph JE, Babu B (2018) Active contour based segmentation techniques for medical image analysis. Med Biol Image Anal
31. Alazawee WS, Naji ZH, Ali WT (2022) Analyzing and detecting hemorrhagic and ischemic strokebased on bit plane slicing and edge detection algorithms. Indonesian J Electr Eng Comput Sci
32. Ambrosini RD, Wang P, O'Dell WG (2010) Computer-aided detection of metastatic brain tumors using automated three-dimensional template matching. J Magn Reson Imaging 31:85–93 [PubMed]
33. Coutts SB, Simon JE, Eliasziw M, Sohn C-H et al. Triaging transient ischemic attack and minor stroke patients using acute magnetic resonance imaging

A Federated Learning Approach to Converting Photos to Sketch

Gowri Namratha Meedinti, Anannya Popat, and Lakshya Gupta

Abstract Worldwide, communication frequently involves the use of images. Numerous applications use images, such as medical imaging, remote sensing, educational imaging, and electronic commerce. In the age of information and the introduction of big data, we require a greater level of security protection for our photographs. Since public networks are used to transfer and view our photographs, in order to train models using input from several clients without forcing each client to provide all of their data to a central server, federated learning (FL), a decentralized, privacy-preserving machine learning technique, is used. In this paper, we provide a novel method that employs machine learning (ML) to train the camera filter and provide generic filter capabilities while taking privacy concerns into account and using the concepts of federated learning and auto-encoding with the help of the CUHK Face Sketch Database (CUFS) dataset to transform any given image into a sketched representation of itself.

Keywords Federated learning · CUHK Face Sketch Database (CUFS) · Auto-encoding · Decentralized · Photo · Sketch

1 Introduction

One of the common types of data that we encounter is vision data. Almost every industry, including fashion, streaming platforms, medicine, law, and finance, uses them for different use cases. One of the most prominent examples is social media.

G. N. Meedinti (✉) · A. Popat · L. Gupta
School of Computer Science and Engineering, Vellore Institute of Technology, Vellore, Tamil Nadu, India
e-mail: gowri.namratha2019@vitstudent.ac.in

A. Popat
e-mail: anannyarajesh.popat2019@vitstudent.ac.in

L. Gupta
e-mail: lakshya.gupta2019@vitstudent.ac.in

S. Das et al. (eds.), *Advances in Data-driven Computing and Intelligent Systems*, Lecture Notes in Networks and Systems 653, https://doi.org/10.1007/978-981-99-0981-0_29

Image data has greatly benefited from AI, which has now taken over everything on the globe.

The disadvantages of earlier image processing methods include their inability to adequately reveal high-level dimensionality. These issues have been resolved, and deep learning algorithms have been shown to be far more dependable. The list of activities they are employed for nowadays is nearly endless and includes things like object identification, object tracking, picture classification, image segmentation and localization, 3D pose estimation, video matting, and many more. But they still have a primary concern–Privacy.

Images published on well-known social media and content-sharing platforms pose more of a privacy and security hazard than ever before. Finding a solution to these issues has become increasingly important as our digital footprints expand exponentially. Numerous studies on the protection of image privacy through privacy policy guidelines and configurations have been conducted in order to address these issues [1]. The public has been forced to reconsider the need for individual privacy while sharing images on the Internet as a result of recent image breaches from well-known Internet services and the exploitation of private photos utilizing cutting-edge algorithms (such as DeepFake) [2]. However, the act of sharing images is comparatively complex, and the systems in place to maintain privacy in daily life are labor-intensive but fall short of offering individualized, precise, and adaptable privacy protection. As a result, there is a need for a more intelligent setting for image sharing that respects privacy [2]. As a result of advancements in deep learning technology, Ho Bae et al. highlighted in their survey that AI-based applications have become widespread in a number of industries, but that current deep learning models are vulnerable to a number of security and privacy problems [3].

The most significant cyberattacks, such as poisoning, inference, and generative adversarial network attacks, were highlighted in an overview of initiated attacks employing multiple computer players to establish a unique threat classification. There is still a gap between the current state of federated AI and a future in which widespread adoption is feasible, as they show in their research that present protocols do not always provide enough security when it comes to managing multiple attacks from both clients and servers [4]. FL has drawn a lot of attention since it protects users' privacy by separating the local data stored on each end-user device and aggregating machine learning model parameters, like neural network weights and biases. The sole aim of FL is to jointly train a global model without compromising data privacy. FL offers considerable privacy advantages over data center training on a data set. Even when "anonymized" data is kept on a server because it is linked to other data sets, client privacy may still be in danger. On the other hand, the information given for FL consists of minor tweaks to increase a machine learning model's accuracy. Other problems like network unavailability in edge devices, another area where Federated Learning excels, may prevent businesses from combining datasets from many sources. Even when data sources can only interact occasionally, federated learning makes it simpler to obtain a variety of data. Since, models in FL are continuously improved by utilizing client input, there is no need to aggregate data for continuous learning. The federated learning technique requires less complicated hardware since federated learning models don't need a single complicated central server to analyze data.

A component of computer vision is image processing. Computer vision systems enable the computer to comprehend and derive meaning from images, while image processing systems concentrate on changing images from one form to another. Image Processing algorithms are widely used in Computer Vision systems. A face improvement software, for instance, might utilize computer vision algorithms to identify faces in a picture before applying Image processing methods like smoothing or grayscale filters. In order to transform images for a variety of tasks, such as applying artistic filters, adjusting an image for the best quality, or enhancing particular image details to maximize quality for computer vision tasks, many advanced image processing techniques use machine learning models like deep neural networks. Convolutional Neural Networks (CNN) learn to do tasks like object detection, image segmentation, and classification by taking in an input image and applying filters to it. Recent advances in machine learning enable engineers to enhance image data in addition to doing image alteration. A machine learning model is only as good as the dataset when there is not enough training data.

It is possible to generate entirely new datasets rather than search for and label additional datasets. Simple image transformation techniques like horizontal flipping, color space augmentations, zooming, and random cropping can be used to achieve this, or deep learning algorithms like Feature Space Augmentation and Auto-encoders, Generative Adversarial Networks (GANs), and Meta-Learning.mize quality can be used to achieve this.

Our methodology uses an auto-encoder and an auto-decoder that have been pre-trained for the source domain and the target domain, respectively, and cross-connects them by mapping their feature spaces between the two domains, in contrast to commonly used feature space learning approaches [5–8].

Also, there have been a lot of advancements in the FL framework as well over the past few years [9, 10]. In addition, much has been accomplished in achieving a fairer contribution from the training of clients in architecture [11, 12]. Federated learning has made it possible to achieve robust and generalizable analytical techniques using healthcare data which is often quite fragmented and private [13]. FL is also a solid technique for increasing communication effectiveness [14].

2 Background

2.1 Federated Learning

By separating the capacity to do machine learning from the requirement to put the training data in the cloud, federated learning enables mobile devices to cooperatively develop a shared prediction model while maintaining all of the training data on the device. Google launched federated learning, commonly referred to as decentralized learning, as a relatively new machine learning technology in 2016. Not requiring them to submit the data to a central site like the Google Cloud gives end-users control over the information they have gathered. Only the global model, which is sent to several client devices throughout each communication loop, is kept on the central server. Contrarily, traditional machine learning methods (also referred to as centralized learning) cannot start the training process until all the data has been gathered in one spot, which not only compromises the privacy of the users but also drives up the cost of holding this data.

An essential part of FL is the server's inclusion of these parameters from the client models into the global model. Numerous techniques have been created, such as CO-OP and Federated Stochastic Variance Reduced Gradient (FSRVG) [15]. It has been demonstrated that the most often used algorithm, FedAvg, performs better than the others [16].

In each communication round, the Federated Averaging (FedAvg) method computes the weighted total of all the weight updates from the training clients in order to maintain the shared global model. The shared global model weights are hosted by a central server, which also controls the training. On the other hand, the actual optimization is done locally by the client using a variety of optimization algorithms such as SGD, Adam, or Adagrad.

The FedAvg algorithm uses hyper-parameters to control the percentage of training clients (C), the number of epochs (E), the batch size for client data (B), and the learning rate. B is heavily used when training with SGD. E, a commonly utilized parameter in most optimization algorithms including SGD, Adagrad, and Adam [16], is the total number of iterations that are performed over the same data while training is done on the client's federated data.

In each communication round, the Federated Averaging (FedAvg) method computes the weighted total of all the weight updates from the training clients in order to maintain the shared global model. The shared global model weights are hosted by a central server, which also controls the training. On the other hand, the actual optimization is done locally by the client using a variety of optimization algorithms such as SGD, Adam, or Adagrad.

The global model's weights are originally assigned at random by the FedAvg algorithm. The same technique is used in each cycle of communication between the clients and server. Firstly, a random subset of training clients St, $|St| = C \cdot K \geq 1$, are selected by the server for global model update purposes. These clients are provided with the weights wt of the global model, who thereafter update their local weights w_t^k to the global weights, $w_t^k \leftarrow w_t$. Thereafter, each of these clients splits their federated data into batches of size B and trains their local models E a number of times on the total number of batches. Lastly, the clients communicate the updated weights, w_{t+1}^k, from the training process back to the server, where aggregation of the weight updates is carried out by calculating their weighted sum, which subsequently leads to an updated global model, w_{t+1} [16].

2.2 Auto-Encoder

An artificial neural network called an auto-encoder is used for unsupervised data encoding. The goal of an auto-encoder is to train the network to capture the most crucial elements of the input image in order to learn a lower-dimensional representation (encoding) for higher-dimensional data, often for dimensionality reduction.

Three components make up auto-encoders:

- Encoder: A component that shrinks the input data from the train-validate-test set into an encoded representation that is often several orders of magnitude less.
- Bottleneck: A module that is the most crucial component of the network because it includes the compressed knowledge representations.
- Decoder: A component that aids the network in "decompressing" knowledge representations and recovering the data from its encoded state. Next, the output is contrasted with a source of truth.

2.3 Dataset

The dataset [17]: CUHK Face Sketch Database (CUFS) is used for research on face sketch synthesis and face sketch identification. It contains 295 faces from the XM2VTS database, 123 faces from the AR database, and 188 faces from the student database of the Chinese University of Hong Kong (CUHK). There are a total of 606 faces. There is a sketch created by an artist for each face based on a frontal position, natural lighting, and neutral expression photograph (Fig. 1).

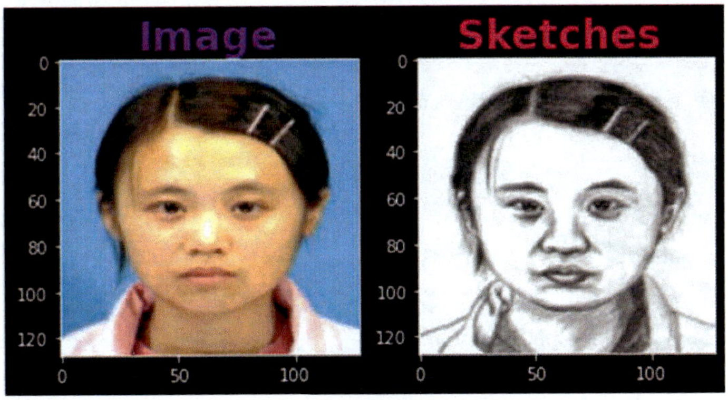

Fig. 1 Example dataset (Photo versus corresponding sketch)

3 Methodology

3.1 Federated Learning Architecture

We utilized the centralized architecture as shown in Fig. 2, which was superior to the hierarchal, regional, decentralized, and centralized architectures [18]. With the adoption of a centralized architecture and the communication of all client updates to a client-server, it maintained its accuracy consistency across numerous datasets [18]. The client-server performs the duty of aggregating model changes and keeps the overall model up to date using federated averaging. Because it has a single central location that is in charge of managing every participant edge device as well as computing model aggregation, the centralized architecture in FL is simpler to set up and administer than other solutions [18].

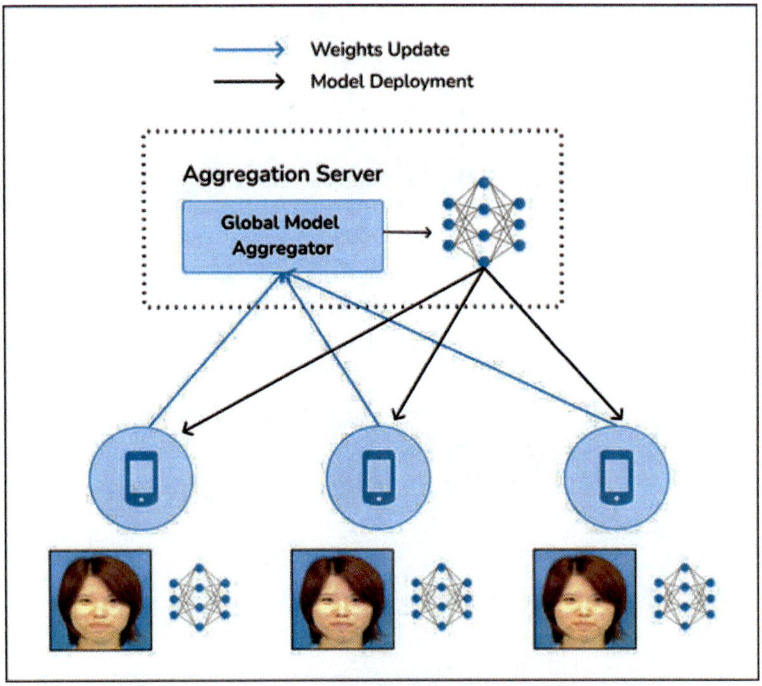

Fig. 2 Federated learning architecture

3.2 Data Pre-processing and Feature Extraction

Different folders for images and sketches are included in the dataset. A blank list for pictures and sketches is made first. All photographs are initially in BGR format. The next step is to convert all of the photos from BGR format to RGB format. One of the most important reasons for converting a BGR image to RGB is that various image processing packages have varying pixel orderings.

Lastly, we normalized all photos into pixel values between the range of 0 and 1, resizing the images from the shape (256,256,3) to (128,128,3) for faster processing. By lowering values between 0 and 1, the normalization of images primarily aims to increase computation efficiency.

To expand the number of images from 188 to 1504, the images are then inverted horizontally and vertically and saved in the photos and drawing arrays, respectively.

Additionally, a pandas data frame with a variety of photographs and doodles as its columns is constructed. The conventional dataset must first be transformed into a federated dataset for the Federated Learning operations before continuing. Local data that is kept up to date by each client is referred to as federated data. As a result, in our instance, we use the original dataset and randomly assign 6 clients to it for additional preparation.

Federated data frequently contains non-i.i.d data, which presents a special set of difficulties [16, 19]. To make testing simpler, we gave each image a client ID.

The combined data was then divided into training and testing groups. The first four clients were utilized for training, and the next five and a half client IDs were used for testing (Fig. 3). For this simulation, Tensorflow Federated (TFF) was the selected framework. Now we create our deep CNN model using an auto-encoder and a decoder. The encoder network passes our image through a succession of convolution and max-pooling layers, downsampling it from 128 by 128 to a 16 by 16 latent vector. This 16×16 latent vector that has been downscaled is upscaled by running it through several convolution and upsampling layers. The final decoder output and our encoder input are identical. The reconstruction loss is measured after comparing the upsampled decoder output with our sketches. By updating the weight and bias of the network through back-propagation, this loss is reduced to a minimum.

The training and validation loss metrics are then obtained by running this model for a number of epochs in each of the training clients over the 16-sized batches. We then map the real photographs against the anticipated designs. Lastly, we plot the real photographs against the anticipated designs.

3.3 Proposed Model

Figure 4 shows the architecture being implemented in our approach.

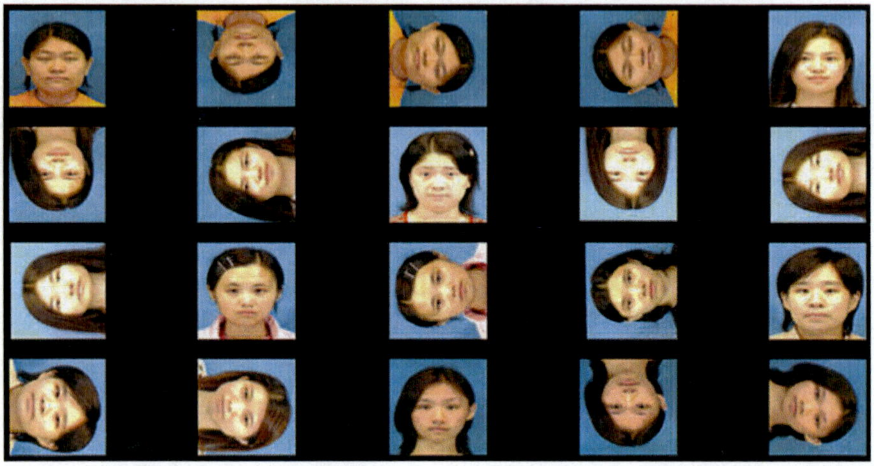

Fig. 3 Heterogeneity of data

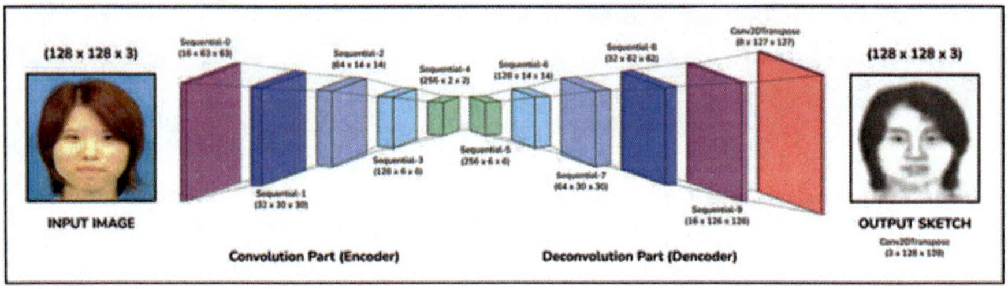

Fig. 4 Proposed model

a. Encoder (Down-Sampling)

The encoder compresses the input image by reducing its dimensions and we start working on the distorted version of the original image.

Firstly, features are extracted from the input image using the following operation

$$y^{j(r)} = \max\left(0, b^{j(r)} \sum_{i} k^{i,j(r)} * x^{i(r)}\right) \tag{1}$$

Filters of sizes 16, 32, 64, 128, and 256 were applied to each layer along with a MaxPool layer of size 2×2 ($s \times s$) and a stride of 2. The max-pooling layer takes the maximum value of the mentioned kernel size, as seen in Eq. (2)

$$y_{i,j(k)} = \max\left(x_{i,j*s+m,k*s+n}\right) \tag{2}$$

Then we perform batch normalization to standardize the current data:

$$x_{\text{normalized}} = \frac{x - \text{mean}}{x_{\max} - x_{\min}} \tag{3}$$

The features obtained from the fully connected layer are passed through an activation function to get the output values. The activation function used here is the 'tanh' function. This is portrayed in Eq. (4).

$$x = \frac{e^{x} - e^{-x}}{e^{x} + e^{-x}} \tag{4}$$

b. De-Encoder (Up-Sampling)

The decoder reconstructs the encoded image back to the original dimension. Here, we use Conv2DTranspose. The Transpose Convolutional layer, also called the Deconvolution layer, is an inverse convolutional layer that will both upsample input and learn how to fill in details during the model training process. Filters of sizes 256, 128, 64, 3,2, and 16 were applied to each layer with a kernel size of 3 × 3. The array is obtained after each layer undergoes a dropout of 0.3. The features obtained from the transposed convolution layer are passed through an activation function to get the output values. The activation function used here is the 'tanh' function. The final output obtained from this is in the shape of the original image. Auto-encoding helps regenerate a sketch of the input photo of the same dimensions.

4 Results

Figure 5 gives a visual representation of the frequency distribution of images in each of the training clients. It can be seen that every client has a varied distribution, which is a property of Federated Learning wherein we often deal with non-i.i.d data.

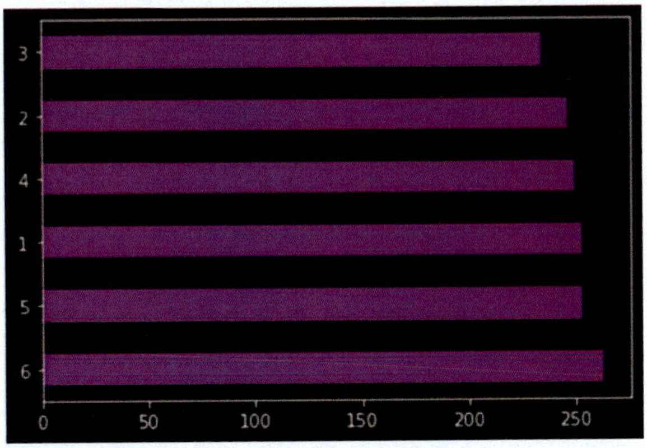

Fig. 5 Distribution of client IDs

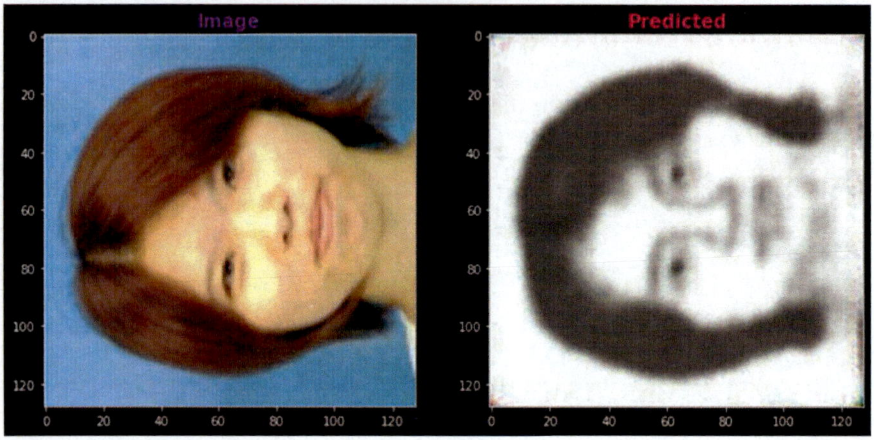

Fig. 6 Sketch output

Figure 6 shows our model's sketch outputs for a given input image. In comparison with the centralized approach, a marginal quality deterioration can be seen. However, it has the added functionality of privacy preservation which is the main focus of the approach and this is something that should outweigh the slight performance reduction.

Lastly, Fig. 7 shows the training and testing loss for the model. The loss maintains a constant decline throughout the training as expected, though the testing did have a few abnormal patterns in the beginning which eventually started decreasing consistently.

5 Conclusion

In this paper, we use the concept of federated learning and auto-encoding to convert any given image into a sketched version of itself. The proposed technique allows a central server to develop a global model that learns from the individual training of different clients on their own independent data without having to communicate this data to a central location. It also significantly helps in cutting storage costs as an added benefit apart from ensuring privacy. The motive of carrying out this research is to aid the developers of applications that use filters on the user's face to generate interesting outputs. Instead of extracting user data from their public or private profiles, federated learning provides a more secure approach for developers to train their models on user images without actually breaching privacy policies on sensitive data.

Our research shows a modest deviation in results from the centralized approach carried out for this problem statement, but it holds a significant advantage over existing solutions due to the previously stated benefits of privacy and storage cost reduction. The marginal performance decrease is present due to the use of non-i.i.d type data, which can be improved upon in future research.

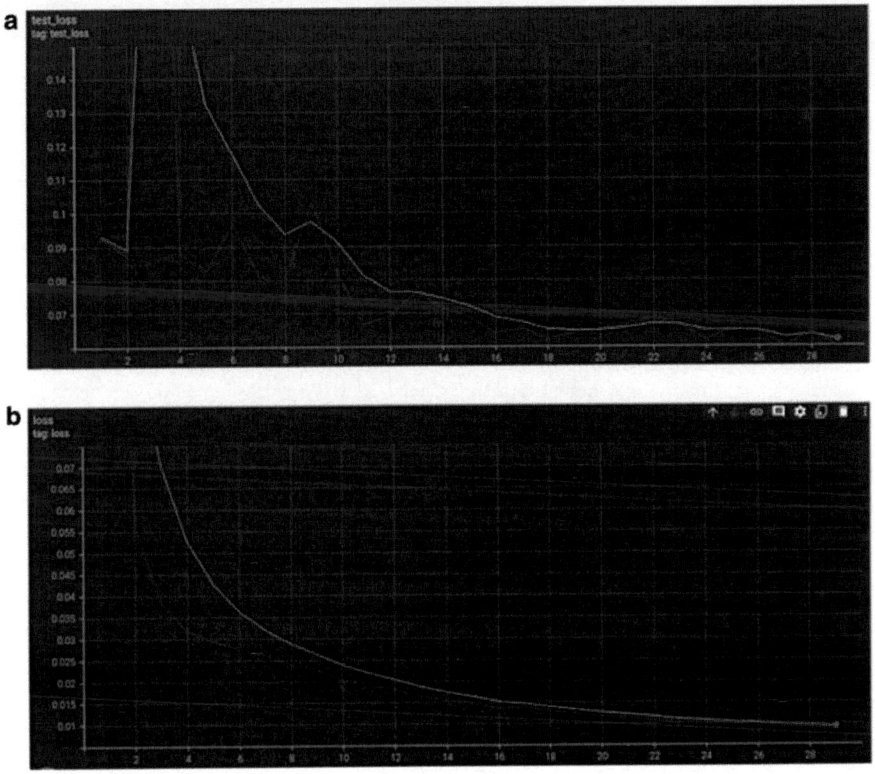

Fig. 7 FL model **a** testing loss, **b** training loss

References

1. Rakhmawati L, Wirawan, Suwadi (2018) Image privacy protection techniques: a survey. In: TENCON 2018—2018 IEEE
2. Liu C, Zhu T, Zhang J, Zhou W (2020) Privacy intelligence: a survey on image sharing on online social networks
3. Bae H, Jang J, Jung D, Jang H, Ha H, Lee H, Yoon S (2018) Security and privacy issues in deep learning. arXiv.
4. Mothukuri V, Parizi RM, Pouriyeh S, Huang Y, Dehghantanha A, Srivastava G (2021) A survey on security and privacy of federated learning. Future Gener Comput Syst 115
5. Vincent P, Larochelle H, Lajoie I, Bengio Y, Manzagol PA (2010) Stacked denoising autoencoders: learning useful representations in a deep network with a local denoising criterion. J Mach Learn Res 11:3371–3408
6. Masci J, Meier U, Cire San D, Schmidhuber J (2011) Stacked convolutional auto-encoders for hierarchical feature extraction. In: Proceedings of the 21st international conference on artificial neural networks, Espoo, Finland, 14–17 June 2011, pp 52–59
7. Rifai S, Vincent P, Muller X, Glorot X, Bengio Y (2011) Contractive auto-encoders: explicit invariance during feature extraction. In: Proceedings of the 28th International Conference on Machine Learning, Washington, DC, USA, 28 June–2 July 2011; pp. 833–840.

8. Nguyen A, Do T, Tran M, Nguyen BX, Duong C, Phan T, Tjiputra E, Tran QD (2021) Deep federated learning for autonomous driving. arXiv
9. Wei K et al (2020) Federated learning with differential privacy: algorithms and performance analysis. IEEE Trans Inf Forensics Secur 15:3454–3469
10. Qi T, Wu F, W, C, Lyu L, X, T, Yang Z, Huang Y, Xie X (2022) FairVFL: a fair vertical federated learning framework with contrastive adversarial learning. arXiv
11. Wei S, Tong Y, Zhou Z, Song T (2020) Efficient and fair data valuation for horizontal federated learning. In: Yang Q, Fan L, Yu H (eds) Federated learning. Lecture notes in computer science, vol 12500. Springer, Cham
12. Sheller MJ, Edwards B, Reina GA et al (2020) Federated learning in medicine: facilitating multi-institutional collaborations without sharing patient data. Sci Rep 10:12598
13. Xu J, Glicksberg BS, Su C et al (2021) Federated learning for healthcare informatics. J Healthc Inform Res 5:1–19
14. Konecný J, McMahan HB, Yu FX, Richtárik P, Suresh AT, Ba-con D (2016) Federated learning: strategies for improving communication efficiency. In: NIPS workshop on private multi-party machine learning
15. Nilsson A, Smith S (2018) Evaluating the performance of federated learning. Master's thesis, University of Gothenburg
16. Zhang W, Wang X, Tang X (2011) Coupled information-theoretic encoding for face photo-sketch recognition. In: 2011 IEEE conference on computer vision and pattern recognition (CVPR)
17. Zhang H, Bosch J, Olsson H (2020) Federated learning systems: architecture alternatives. 385–394. 10.1109
18. Darba A, Alferaidi Y, Kusum A, Yasser V, Wattan K, Sandee D, Gaurav (2022) Federated learning algorithms to optimize the client and cost selection, 2022 AB
19. Zhu H, Xu J, Liu S, Jin Y (2021) Federated learning on non-IID data: a survey. arXiv

Corrosion Behaviour of Spray Pyrolysis Deposited Titanium Oxide Coating Over Aluminium Alloy

Dalip Singh, Ajay Saini, and Veena Dhayal

Abstract This study is to investigate how a spray pyrolysis process can be used to produce a corrosion resistant titania sol–gel coating on an aluminium alloy sample. In order to deposit a stable titania coating on an aluminium alloy, oxime-modified titania(IV) isopropoxide was used as the sol–gel precursor. Scanning electron microscopy (SEM) photographs of the coated sample show that a uniform and crack-free coating was applied. The corrosion resistances of uncoated and coated aluminium alloy samples in a 3.5 wt% aqueous NaCl solution were investigated using electrochemical impedance spectroscopy and potentiodynamic polarization. The values of the coated samples' corrosion current density (i_{corr}) and equilibrium corrosion potential (E_{corr}) revealed a noticeable change. These findings clearly show that titania coatings are corrosion resistant.

Keywords Sol–gel · Coating · Spray pyrolysis · Corrosion

1 Introduction

Applications for titanium oxide thin films with nanostructures include many different ones. These applications consist of UV filters for packing and optics material Ivanova et al. [1], Bonini et al. [2] antireflection coatings for solar industries Perera et al. [3], photo catalysts for water and air filtration and treatment Mao et al. [4], keshmiri et al. [5], batteries anode terminal Huanga et al. [6] electro chromic displays Aliev and shin [7] self-cleaning window and tile coatings, transparent conductors Fretwell and Douglas [8], sensors used to measure humidity Tai and Oh [9], Koarunagaran

D. Singh (✉)
Department of Automobile Engineering, Manipal University Jaipur, Jaipur 303007, India
e-mail: dalip08784@gmail.com

A. Saini
Central Analytical Facilities, Manipal University Jaipur, Jaipur 303007, India

V. Dhayal
Department of Chemistry, Manipal University Jaipur, Jaipur 303007, India

S. Das et al. (eds.), *Advances in Data-driven Computing and Intelligent Systems*, Lecture Notes in Networks and Systems 653, https://doi.org/10.1007/978-981-99-0981-0_30

et al. [10] and anti-corrosion coatings Shen et al. [11]. It has been demonstrated that a nanostructured TiO_2 phase benefits several applications significantly. Indeed, numerous approaches have recently been used to produce nanostructured TiO_2 thin films Chow et al. [12], Carotta et al. [13], Bally et al. [14]. Several processes have been used to produce titania coatings on metallic surfaces, including plasma spray and electrophoretic deposition Cho et al. [15], sputtering Boyadzhiev et al. [16], sol–gel dip coating Tang et al. [17], Garzall et al. [18] and spray pyrolysis deposition Nakaruk et al. [19]. Spray pyrolysis deposition is one of these processes that has various advantages: cheap operating expenses, basic facilities, ambient operation Okayu et al. [20], Kavitha et al. [21], chakraborty et al. [22], mass production potential Aukkaravittayapum et al. [23], film reproducibility, rapid coating growth (up to 100 nm/s) Sun et al. [24], and coverage of a vast area Murakami et al. [25], Shinde et al. [26], Reina et al. [27].

A more general method to increase corrosion resistance is to apply protective films or coatings that are not cracked. With increasing current density, the presence of cracks and flaws in the coating will result in localized corrosion Gluszek et al. [28]. To improve corrosion and abrasion resistance, a sol–gel-based titania coating has recently been created and described on mild steel and aluminium alloy Ruhi et al. [29], Hawthorne et al. [30]. Due to the numerous applications of aluminium alloy in industry, this study demonstrates an attempt to improve the corrosion resistant qualities of an aluminium alloy substrate by spray pyrolysis deposition of TiO_2 nanoparticle coating.

We have presented a method for getting anti-corrosion behaviour of titanium dioxide coatings on aluminium alloys using the spray pyrolysis deposition process in this paper. Using oxime-modified titanium isopropoxide, a protective titania coating was formed over an aluminium substrate. The coating that was produced as a result is dense, which increases the anti-corrosion capabilities of the substrate that was coated.

2 Materials and Methods

In this experiment, an aluminium alloy substrate was used. It has a high magnesium concentration, and its composition is measured in weight per cent being aluminium 92.45%, Silicon 0–0.4%, Copper 0.1%, magnesium 4–4.9%, Zinc 0.25%, Iron 0.4%, manganese 0.4–1%, chromium 0.05–0.25%, Tin 0.15%. The test pieces were fabricated to a size of $20 \times 30 \times 3\ mm^3$.

For the purpose of this investigation, an aluminium alloy substrate was utilized. It has a high magnesium concentration, and its make-up, as measured in weight per cent, is as follows: tin 0.15%, chromium 0.05–0.25%, manganese 0.4–1%, iron 0.4%, zinc 0.25%, copper 0.1%, silicon 0–0.4%, magnesium 4–4.9% and aluminium 92.45%. The test pieces were manufactured to the dimensions of 20 mm by 30 mm by 3 mm.

2.1 *Sample Preparation*

Before being dip coated to create aluminium oxide coatings, the specimens were mechanically ground down to 220–2000 grit with silicon carbide (SiC) paper. Chemical pre-treatment included acetone rinsing, ultrasonic rinsing in acetone for 15 min, acid treatment of the cleaning substrate by immersion in an alkaline solution containing 50 g L^{-1} sodium hydroxide for 10 s, and etching in a 0.8 mol L^{-1} nitric acid solution for 10 min to prevent surface defects and achieve uniform surface hydroxides/oxides. The specimens were cleaned and activated in an acidic solution before being rinsed with deionized water.

2.2 *Titania Sol Synthesis Process*

The disclosed approach was used to produce and analyse the oxime-modified titanium(IV) isopropoxide Saini et al. [31].

$$\left[Ti\left(O^{i}Pr\right)_4\right] + 2\left[HONC(CH_3)_2\right] \quad [Ti(OiPr)_2\{ONC(CH_3)_2\}_2$$

Titanium(IV) Isopropoxide — Acetoxime — Acetoxime − modifiedtitanium(IV) isopropoxide

3 Results and Discussion

3.1 *Coating Deposition and Their Characterization*

The thermogravimetric analysis (TGA) of $[Ti(OiPr)_2(ONC(CH_3)_2)_2$ is shown in Fig. 1. The TG curve manifests in weight loss below 250 °C, which is when TiO_2 is produced stoichiometrically. The oximato-complex can be used as a spray pyrolysis precursor for TiO_2 coating deposition on aluminium alloy at 400 °C because of its low temperature conversion to titania (250 °C). Titanium coating was applied to the specified aluminium sample using an automated spray pyrolysis machine with optimized deposition parameters (distance between sample and nozzle: 150 mm, air pressure: 1.5 bar, nozzle diameter: 0.7 mm, spray produce: 900, solution flow rate: 1.0 mL min^{-1} at 400 °C). This titanium(IV) isopropoxide was modified with an oxime, and the temperature was 400 °C.

SEM–EDX and AFM techniques were used to further characterize the deposited titania coatings. SEM analysis of the coated surface's morphology reveals the presence of compact and crack-free surfaces, as illustrated in Fig. 2. Utilizing EDX, the composition of the coated sample was analysed, revealing the deposition of a

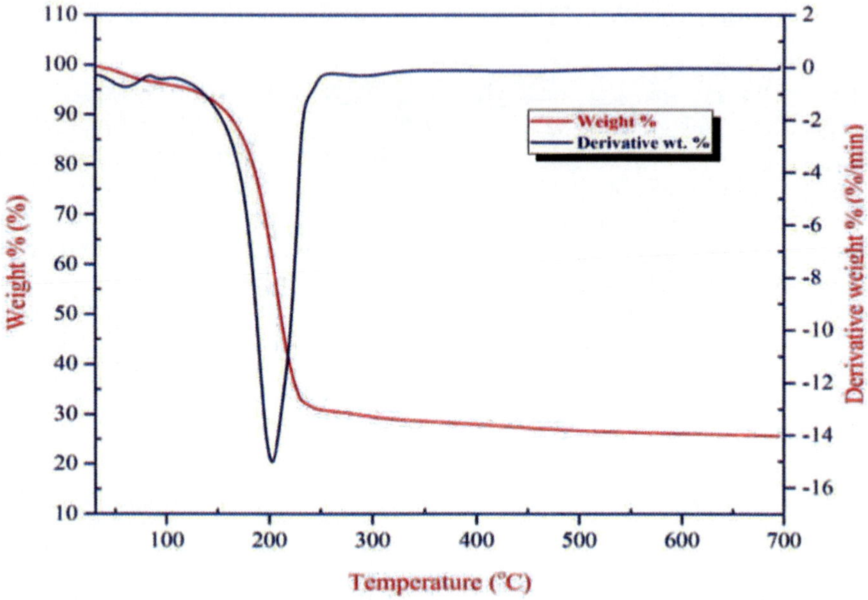

Fig. 1 TG curve of $[Ti(OiPr)_2(ONC(CH_3)_2)_2]$

titania coating on an aluminium alloy substrate. In the case of a coated substrate, the titanium weight percentage affects the presence of film, as illustrated in Fig. 3 Ienei et al. [32], Novakoric et al. [33].

The titania spray coated aluminium substrate is micrographed using atomic force microscopy (AFM) in Fig. 4. The obtained values, 3.06 and 4.21 nm, are the Ra (arithmetic average deviation) and Rq (root mean square deviation), respectively, Saini et al. [34], Rahimi et al. [35]. These values were found to be lower than those published in the literature and indicate smooth coated surfaces Hamid et al. [36], Jungsuwattananon et al. [37].

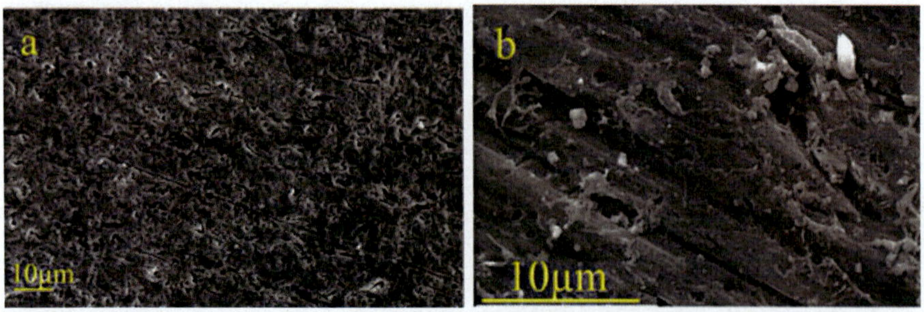

Fig. 2 SEM images of **a** bare and **b** titania coated aluminium substrates

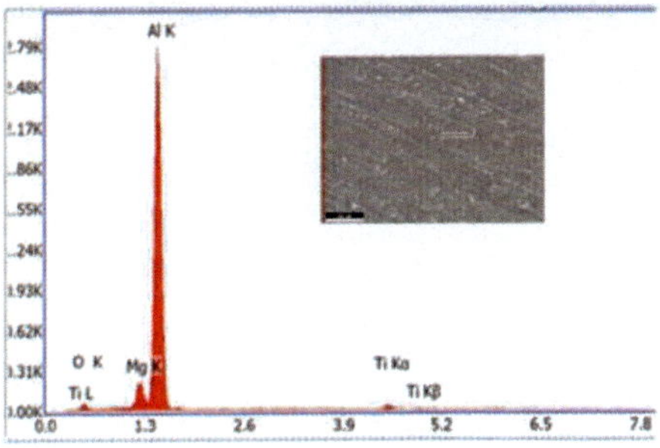

Element	Weight %	Atomic %	Net Int.	Error %
O K	5.87	9.56	11.56	19.1
MgK	5.56	5.95	68.37	5.86
AlK	84.96	81.98	943.89	3.06
TiK	2.18	1.18	10.6	18.73

Fig. 3 SEM–EDX of titania coated aluminium alloy

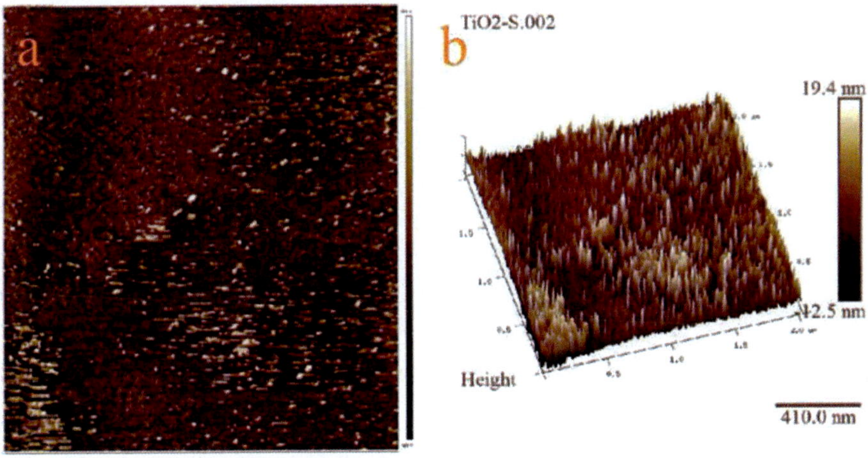

Fig. 4 Titanium-coated aluminium alloy AFM pictures **a** in 2D and **b** in 3D

3.2 *Electrochemical Measurements*

In 3.5 wt% aqueous NaCl solution, electrochemical measures such as potentiodynamic polarization curves were performed to assess the corrosion behaviour of bare and coated aluminium surfaces Onofre et al. [38]. Figure 5 displays the potentiodynamic polarization (PD) curves for substrates with 0.05 M precursor concentration and 30 mL volume.

Corrosion potential (E_{corr}) and corrosion current density (i_{corr}) were obtained by extrapolating cathodic and anodic branches to the point of junction in PD curves (i_{corr}) Saini et al. [39], Singh et al. [40]. All corrosion parameters are summarized in Table 1. In coated titania substrates, the corrosion potential, E_{corr}, changed to the positive side (− 0.732 V) as compared to bare substrate (− 1.33 V), and the corrosion current density (i_{corr}) reduced dramatically when compared to bare substrate Saini

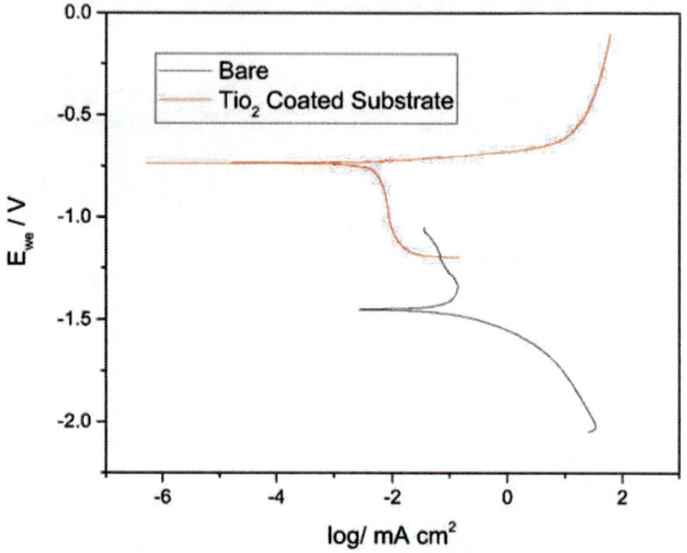

Fig. 5 Tafel plot for bare and titania coated aluminium substrate

Table 1 Comparison values for E_{corr}, i_{corr} and corrosion resistance for Bare, titania coated substrate

Sample	E_{corr} (mV)	I_{corr} (uA/cm^2)	Corrosion rate (mmpy)
Bare	− 1362.293	66.640	$7.47939 \cdot 10^{-3}$
Titania coated	− 732.897	5.822	$7.294 \cdot 10^{-6}$

et al. [41], Singh et al. [42]. This suggests that titania coatings have anti-corrosion properties [43].

4 Conclusion

Titanium coatings were produced via spray pyrolysis on aluminium substrates with a 0.05 M concentration and 30 mL volume. Their morphological, structural and corrosion resistance characteristics have all been studied. According to SEM findings, a crack-free coating has been applied to an aluminium substrate, and EDX results indicate that titania is present as a coating. AFM readings show smoother substrate-overcoatings. Titania coatings' ability to prevent corrosion was evaluated using electrochemical tests PD. These findings indicate that titania coatings have corrosion-protective qualities since they demonstrate a shift in equilibrium corrosion potential (E_{corr}) towards positive potential and a reduction in corrosion current density (i_{corr}).

References

1. Ivanova T, Harizanova A, Surtchev M, Nenova Z (2003) Investigation of sol–gel derived thin films of titanium dioxide doped with vanadium oxide. Sol Energy Mater Sol Cells 76:591–598
2. Bonini N, Carotta MC, Chiorino A, Guidi V, Malagù C, Martinelli G, Paglialonga L, Sacerdoti M (2000) Doping of a nanostructured titania thick film: structural and electrical investigations. Sens Actuators B Chem 68:274–280
3. Perera VPS, Jayaweera PVV, Pitigala PKDD, Bandaranayake PKM, Hastings G, Perera AGU, Tennakone K (2004) Construction of a photovoltaic device by deposition of thin films of the conducting polymer polythiocyanogen. Synth Met 143:283–287
4. Mao D, Lu G, Chen Q (2004) Influence of calcination temperature and preparation method of TiO_2–ZrO_2 on conversion of cyclohexanone oxime to ε-caprolactam over B_2O_3/TiO_2–ZrO_2 catalyst. Appl Catal A 263:83–89
5. Keshmiri M, Mohseni M, Troczynski T (2004) Development of novel TiO_2 sol-gel-derived composite and its photocatalytic activities for trichloroethylene oxidation. Appl Catal B 53:209–219
6. Huanga SY, Kavana L, Exnarb I, Grätzela M (1995) Rocking chair lithium battery based on nanocrystalline TiO_2 (anatase). J Electrochem Soc 142:142–144
7. Aliev AE, Shin HW (2002) Image diffusion and cross-talk in passive matrix electrochromic displays. Displays 23:239–247
8. Fretwell R, Douglas P (2001) An active, robust and transparent nanocrystalline anatase TiO_2 thin film—preparation, characterisation and the kinetics of photodegradation of model pollutants. J Photochem Photobiol A Chem 143:229–240
9. Tai WP, Oh J-H (2002) Fabrication and humidity sensing properties of nanostructured TiO_2–SnO_2 thin films. Sens Actuators B Chem 85:154–157
10. Karunagaran B, Uthirakumar P, Chung SJ, Velumani S, Suh E-K (2007) TiO_2 thin film gas sensor for monitoring ammonia. Mater Charact 58:680–684
11. Shen GX, Chen YC, Lin CJ (2005) Corrosion protection of 316 L stainless steel by a TiO_2 nanoparticle coating prepared by sol–gel method. Thin Solid Films 489:130–136
12. Chow LLW, Yuen MMF, Chan PCH (1996) A novel method for the preparation of nanosized TiO_2 thin films. Adv Mater 8:334–337
13. Carotta MC, Ferroni M, Guidi V (1999) Martinelli preparation and characterization of nanostructured titania thick films. Adv Mater 11:943–946
14. Bally AR, Korobeinikova EN, Schmid PE, Lévy F (1998) Bussy, structural and electrical properties of Fe-doped thin films. J Phys D Appl Phys 31:1149–1154
15. Cho J, Schaab S, Roether JA, Boccaccin AR (2008) Nanostructured carbon nanotube/TiO_2 composite coatings using electrophoretic deposition (EPD). J Nanopart Res 10:99–105
16. Boyadzhiev S, Georgieva V, Rassovska M (2010) Characterization of reactive sputtered TiO_2 thin films for gas sensor applications. J Phys Conf Ser 253:1–6
17. Tang H, Prasad K, Sanjinés R, Lévy F (1995) TiO_2 anatase thin films as gas sensors. Sens Actuators B Chem 26:71–75
18. Garzella C, Comini E, Tempesti E, Frigeri C, Sberveglieri G (2000) TiO_2 thin films by a novel sol–gel processing for gas sensor applications. Sens Actuators B Chem 68:189–196
19. Nakaruk A, Sorrell CC (2010) Conceptual model for spray pyrolysis mechanism: fabrication and annealing of titania thin films. J Coat Technol Res 7:665–676
20. Okuya M, Prokudina NA, Mushika K, Kaneko S (2000) TiO_2 thin films synthesized by the spray pyrolysis deposition (SPD) technique. J Eur Ceram Soc 19:903–906
21. Kavitha R, Meghani S, Jayaram V (2007) Synthesis of titania films by combustion flame spray pyrolysis technique and its characterization for photocatalysis. Mater Sci Eng B 139:134–140
22. Chakraborty A, Mondal T, Bera SK, Sen SK, Ghosh R, Paul GK (2008) Effects of aluminum and indium incorporation on the structural and optical properties of ZnO thin films synthesized by spray pyrolysis technique. Mater Chem Phys 112:162–166

23. Aukkaravittayapun S, Wongtida N, Kasecwatin T, Charojrochkul S, Unnanon K, Chindaudom P (2006) Large scale F-doped SnO_2 coating on glass by spray pyrolysis. Thin Solid Films 496:117–120
24. Sun H, Wang C, Pang S, Li X, Tao Y, Tang H, Liu M (2008) Photocatalytic TiO_2 films prepared by chemical vapor deposition at atmosphere pressure. J Non-Cryst Solids 354:1440–1443
25. Murakami K, Nakajima K, Kaneko S (2007) Initial growth of SnO_2 thin film on the glass substrate deposited by the spray pyrolysis technique. Thin Solid Films 515:8632–8636
26. Shinde PS, Sadale SB, Patil PS, Bhosale PN, Brüger A, Neumann- Spallart M, Bhosale CH (2008) Properties of spray deposited titanium dioxide thin films and their application in photoelectrocatalysis. Sol Energy Mater Sol Cells 92:283–290
27. Reina A, Jia X, Ho J, Nezich D, Son H, Bulovic V, Dresselhaus MS, Kong J (2009) Layer area, few-layer graphene films on arbitrary substrates by chemical vapor deposition. Nano Lett 9:3087
28. Gluszek J, Jędrkowiak J, Markowski J, Masalski J (1990) Galvanic couples of 316L steel with Ti and ion plated Ti and TiN coatings in Ringer's solutions. Biomaterials 11:330–335
29. Ruhi G, Modi OP, Singh IB, Jha AK, Yegneswaran AH (2006) Wear and electrochemical characterization of sol-gel alumina coating on chemically pre- treated mild steel substrate. Surf Coat Technol 201:1866–1872
30. Hawthorne HM, Neville A, Troczynski T, Hu X, Thammachart M, Xie Y, Fu J, Yang Q (2004) Characterization of chemically bonded composite sol–gel based alumina coatings on steel substrates. Surf Coat Technol 176:243–252
31. Saini A, Jat SK, Shekhawat DS, Kumar A, Dhayal V, Agarwal DC (2017) Oxime-modified aluminium(III) alkoxides: potential precursors for γ-alumina nano-powders and optically transparent alumina film. Mater Res Bull 93:373–380
32. Ienei E, Milea AC, Duta A (2014) Influence of spray pyrolysis deposition parameters on the optical properties of porous alumina films. Energy Procedia 48:97–104
33. Novaković T, Radić N, Grbić B, Dondur V, Mitrić M, Randjelović D, Stoychev D, Stefanov P (2008) The thermal stability of porous alumina/stainless steel catalyst support obtained by spray pyrolysis. Appl Surf Sci 255:3049–3055
34. Saini A, Dhayal V, Agarwal DC (2018) Evaluation of corrosion protective behaviour of alumina film deposited by oxime-modified aluminium(III) alkoxide precursor. Surf Coat Technol 335:241–247
35. Rahimi H, Mozaffarinia R, Hojjati A (2013) Corrosion and wear resistance characterization of environmentally friendly sol–gel hybrid nanocomposite coating on AA5083. J Mater Sci Technol 29:603–608
36. Hamid M, Rahman I (2003) Preparation of titanium dioxide (TiO_2) thin films by sol gel dip coating method. Malays J Chem 5:86–98
37. Jungsuwattananon K, Saesoo S, Pimpha N, Negishi N (2008) Characterization and bactericidal activity of thin-film TiO_2 photocatalyst. J Nat Sci Spec Iss Nanotechnol 7:25–31
38. Onofre-Bustamante E, Domínguez-Crespo M, Torres-Huerta A (2009) Characterization of cerium-based conversion coatings for corrosion protection of AISI-1010 commercial carbon steel. J Solid State Electrochem 13:1785–1799
39. Singh D, Dhayal V, Agarwal DC (2019) Corrosion performance of alumina coatings over anodized aluminum alloys by dip coating method. Surf Eng Appl Electrochem 55436–55442
40. Saini A, Singh D (2021) The effect of coating morphology on anti-corrosion behavior of modified alumina coating over aluminum alloy. Prot Met Phys Chem Surf 57(5):995–1001
41. Saini A, Singh D (2022) Structural and optical properties of titania nanostructures obtained from oxime-modified titanium(IV) precursor. Mater Res Innovat 26(5):275–284
42. Singh D, Saini A, Dhayal V, Agarwal DC (2019) Oxime-modified aluminum(III) isopropoxide: a promising sol-gel precursor for corrosion resistive nano-alumina coating on an aluminum alloy. Prot Met Phys Chem Surf 55:682–688
43. Dhayal V, Singh D, Saini A, Sonewane S, Agarawal DC (2020) Vibration and corrosion analysis of modified alumina coating over aluminium alloy. J Fail Prevent 21:130–137

Topologies of Shear and Strain Promote Chaotic Mixing in Helical Flow

Priyam Chakraborty

Abstract Physics aids explainable artificial intelligence. The inherent topology of a chaotic system is often a boon to learning algorithms. Helical or screw flows are chaotic. Their velocity and rotational fields are parallel to each other, typically hosting coherent structures that contain (either strain or shear) barriers which resist fluid flow across them. Here, we apply perturbation to coherent fluid particles to construct a criterion governing the topological changes in their mixing across barriers, which we define using the macroscopic statistical measure of finite-time Lyapunov exponent. Our findings demonstrate that the rigid coherent structures essentially support mixing in purely helical flows. These findings have far-reaching implications in diverse fields of applications, ranging from dynamos in growing magnetic field, classical turbulence in superfluid helium to supercell atmospheric tornadoes.

Keywords Chaos · Helicity · Lyapunov exponent · Streamline

1 Introduction

The topology of chaotic signals around equilibria is an active area of interest in data-centric computing and neural networks. A chaotic signal is deterministic and bounded while being sensitive to initial conditions. Chaotic signals have been recently represented as artificial neurons in the novel ChaosNet architecture to improvise machine-enabled classification [1]. To add, asymptotic perturbations have defined constriction factors which ensure the diversity of solutions in multi-objective optimizations in learning algorithms [2]. Recent studies have affirmed that the Langevin [3] and Lagrangian [4] approaches unravel the underlying physics of neural networks in data science. Since the application of machine learning is gaining traction to comprehend the physics of fluid flow [5], the topological examination of a certain flow field is the subject matter of this article.

P. Chakraborty (✉)
Happymonk AI Labs, Bengaluru, Karnataka 560078, India
e-mail: priyam@happymonk.co

S. Das et al. (eds.), *Advances in Data-driven Computing and Intelligent Systems*, Lecture Notes in Networks and Systems 653, https://doi.org/10.1007/978-981-99-0981-0_31

Helicity is a flow property which measures the extent of alignment between a vector and its curl. Kinetic helicity quantifies that alignment between vorticity (rotationality) and velocity in the flow. Rotating column of fluid mass in tornado may be an ideal use case of purely helical flow. By virtue of conservation [6, 7], helicity is a fundamental quantity in explaining the dynamo which is a spontaneously growing magnetic field in an electrically conducting Earth's liquid core [8, 9]. This has interdisciplinary implications, ranging from examining the astrophysical magnetism to assessing the stability of magnetic field in thermonuclear fusion [10]. It is known that a geometrically constrained advection of helical flow hosting a repeated sequence of stretch, twist and fold can be described by a superposition of three orthogonal components of velocity also known as ABC flow, which is attributed to the contributions by Arnol'd, Beltrami and Childress [11]. ABC flow is an exact solution to the steady Euler equations of motion. The modeled flow comprises compartments or cells due to the alternate presence of saddles and foci which is evident from the application of critical point theory [12]. The cells may be associated with rigid coherent structures which modulate fluidic mixing. Rigidity is a measure of increase in timescale of coherence in the mixing phenomenon. An ideal rigid coherent structure neither grows nor decays with time.

Fluidic mixing in purely helical flows is an outstanding area of interdisciplinary interest. Chaotic advection facilitates mixing which in turn depends on repelling stationary points [8]. ABC flow is elusively simple but the pathlines stretch and fold owing to bounded helicity and space periodicity [13]. Recently, we have examined the topology and transport of mixing in ABC flows using the instantaneous and Lagrangian descriptors, and reported that the contours of finite-time Lyapunov exponent signify mixing due to transport barriers [14]. The physical connection between the time-resolved (Eulerian) and time-averaged (Lagrangian) description of mixing in helical flow is the subject matter of this article.

Here, we claim that a topological change in fluid mass that is advecting between locally minimized shear and strain in rigid coherent structures promotes mixing in purely helical unsteady flow. In order to test the hypothesis, we examine the topology of unsteady streamlines and find an asymptotic match between material curves that satisfies smooth mixing under constrained shear and strain. The assessment begins with velocity. To this end, Haller [15] illustrated the field of Lyapunov exponents on ABC-type flows. Further, Barenghi et al. [16] found concentrated vortices using a template of ABC flow for the normal component of turbulent ^{4}He. Taking cue from these earlier works on unsteady helical rotationalities, a kinematic template of velocity (u, w) in the incompressible helical (ABC) flow may be given as:

$$\begin{aligned} u &= A\,\sin(k(z + \sin(A\,t/\lambda_{ABC}))) + C_0\,\cos(k(y + \sin(A\,t/\lambda_{ABC}))) \\ w &= C_0\,\sin(k(y + \sin(A\,t/\lambda_{ABC}))) + B_0\,\cos(k(x + \sin(A\,t/\lambda_{ABC}))) \end{aligned} \tag{1}$$

where the unsteady parameter A is given as $A_0 + 0.5 \times t \times \sin(\pi\, t)$, t denotes time, the constant parameters A_0, B_0 and C_0 obey the constraint $A_0^2 + B_0^2 + C_0^2 = 3$, $k = 2\pi/\lambda_{ABC}$ and λ_{ABC} is the length scale of the domain. The flow field in Eq. 1 has dimensions in SI. For illustrations in this article, $A_0 = B_0 = \sqrt{3/2}$, $C_0 = 0$ and $\lambda_{ABC} = 1$.

The significance of the present work is that it unifies the underlying mixing due to helical rotationalities in apparently distinct physical systems, such as the Earth's magnetic dynamo, superfluid helium and tornado. A dynamo comprises the nonlinear interactions between the chaotic flow and the magnetic field within the Earth's liquid core [10]. Chaos modulates the separation of nearby fluid parcels which is often measured with the finite-time Lyapunov exponent (FTLE) that has interdisciplinary applications. For instance, the Lyapunov exponent is frequently used to diagnose nonlinear signals from the human brain [17]. Alexakis [18], in his search for sustained growing dynamos, computed FTLE in laminar ABC flow and reported certain intervals of the model parameters A, B and C that define chaos. However, this finding alone does not adequately explain the underlying mixing, as attributable to a recent study by the present authors that deliberates the topology of mixing in generalized helical flows by using FTLE in the coordinate space rather than the parametric space of the model [14]. Chaotic motions are essential elements of fast dynamos, i.e., dynamos that operate over timescales much shorter than the turnover timescales generating the fluid flow. Within the Earth's liquid core, a relaxing magnetic field is topologically equivalent to a state of magnetostatic equilibrium which is analogous to Euler flow. Core flow, being turbulent, is asymmetric (that is, chiral with nonzero helicity) [10]. This allows for solving the magnetic field with a tractable idealized ABC flow. In the study of turbulence in superfluid helium, ^{4}He II, Angstrom (Å)-thick superfluid vortex lines have been modeled [19], their existence and growth within normal fluid vortex cores have been quantified [20] and the mechanisms of superfluid–normal fluid vortex matching have been identified [16]. It is well-known that discrete turbulent normal fluid vortices are ideally analogous to those appearing in ABC flow, subject to the condition that the turbulent vortices in ^{4}He II are not closely spaced to initiate any topological change or reconnection [21]. From a geophysical view, the study of tornadoes is crucial to the climatic modeling of the Earth's heat imbalance [22]. Supercell storms originate near swirling intersection of warm inflow and cool outflow. Strong wind shear around bulk updraft orients the vortical flow to be optimally helical in order for the storm to be a sustainable tornado [23], which supercell storm transforms into tornado is an open problem [24]. This article complements the progress in these challenges. Rest of the article is organized as follows. We examine the instantaneous flow topology and identify characteristic timescales in the modeled flow. We probe the topology using Lagrangian descriptors such as pathlines and FTLE to emphasize the fundamental differences between steady and unsteady ABC flow. We then perturb the deformation of fluid parcels to identify the conditions of smooth mixing between minimized strain and shear in the flow where FTLE vanishes. These regions are associated with rigid coherent structures and occupy bulk of the flow. Finally, we discuss the implications of our results on interdisciplinary physical systems and conclude this article.

2 Topological Regimes in Helical Flow

We find that the temporal probes of the modeled flow field (Eq. 1) at different locations in the domain (Fig. 1) suggest that the model is bounded which is essential for the representation of physical flows. We now consider the topological evolution of the model to examine the categorically distinct mixing phenomenon of unsteady helical flow. Figure 2 illustrates the topological bifurcations as a function of time in the $x - z$ plane at $y = 0$ where the Cartesian coordinates have their origin at the center of the plane. To aid the visualization, we superpose the streamlines on the vorticity field (ω_y) of ABC flow in Fig. 2. We find three topological regimes where saddles and vortices undergo unique deformations. First, locally maximum vorticity traces the domain of the unsteady flow with time ($t < 1.15$ s, 1.5 s $< t < 3.25$ s, 3.27 s $< t < 3.62$ s and 4.2 s $< t < 5$ s). Please see Fig. 2a–c for illustration. This is reported in the context of superfluid turbulence as well [16]. Second, there are two occurrences of abrupt topological bifurcation, 3.25 s $< t < 3.27$ s (Fig. 2d–f) and 3.77 s $< t < 3.79$ s. The bifurcations are evident from the evolving pattern of streamlines. Third, there are regions of vorticity which evolve and grow while being stationary on certain time intervals (1.15 s $< t < 1.5$ s, 3.62 s $< t < 3.77$ s and 3.79 s $< t < 4.2$ s). Please see Fig. 2g–i for illustration. Thus, the repeated sequence of the three regimes is a candidate that facilitates mixing.

Probing further, we examine the continuum deformation that is associated with flow topology. Two neutrally buoyant neighboring fluid particles separate expo-

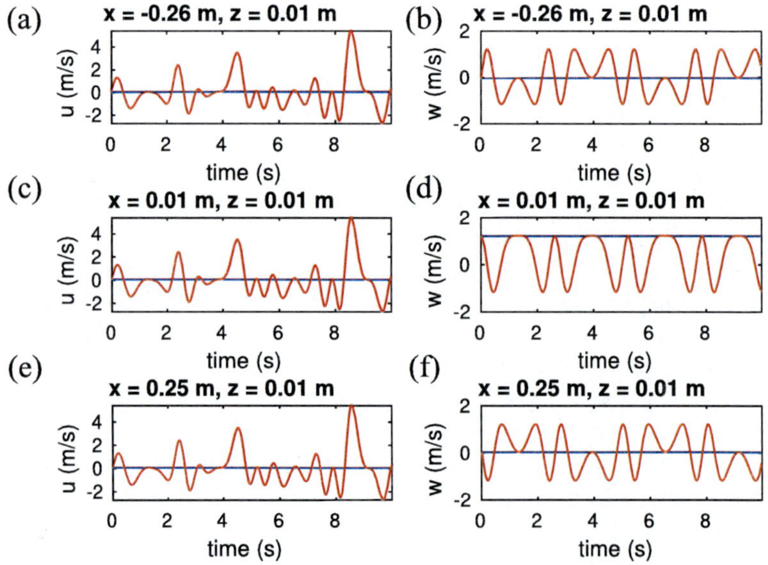

Fig. 1 **a–f** Variation of flow components (u, w) with time for steady (horizontal blue) and unsteady (fluctuating red) ABC flow at three specified locations

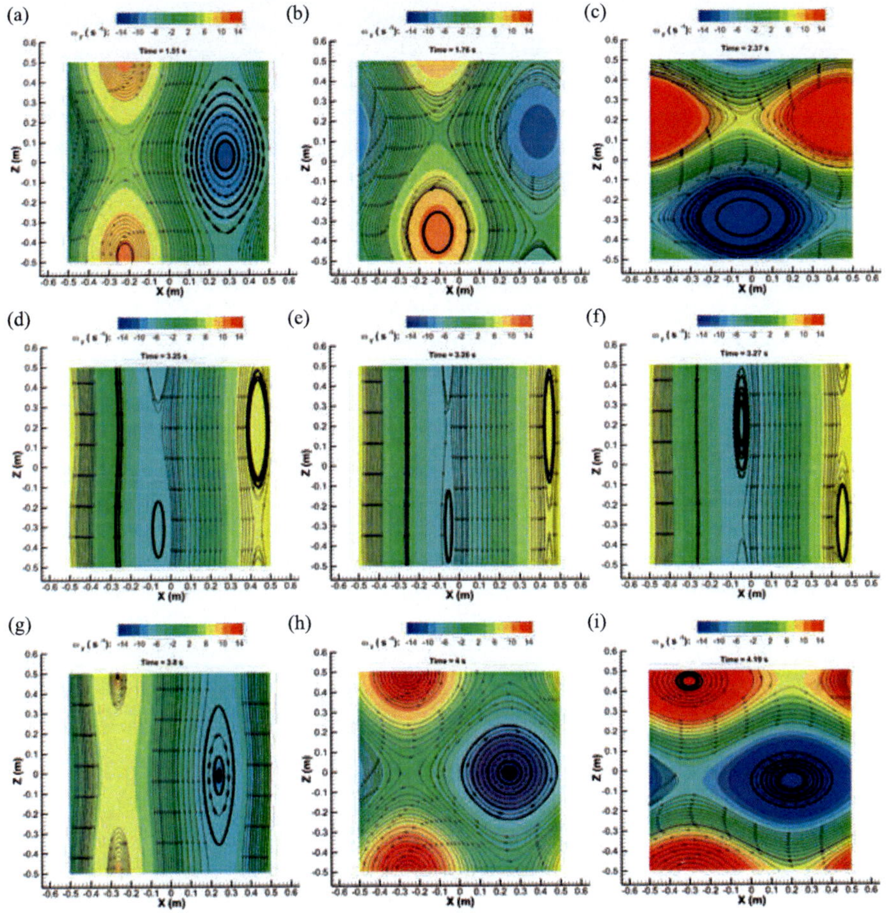

Fig. 2 Topological regimes in unsteady ABC flow. The vorticity field (ω_y) and the interplay of superimposed streamlines suggest that there are three distinct topological regimes: **a**–**c**, **d**–**f** and **g**–**i**. The illustrative streamlines are equal in number from **a** to **i**

nentially over time as elucidated by a first-order Taylor series expansion of the velocity field about a spatial coordinate. This leads to computing the growth rate of a multi-dimensional neighborhood. Consider a two-dimensional flow map $F : (\bar{x}_0, t_0) \rightarrow (\bar{x}, t)$. Initial neighborhood $\delta(\bar{x}_0)$ acted upon by the deformation operator $\bar{\nabla} F$ results in $\delta(\bar{x}) = (\bar{\nabla} F)\,\delta(\bar{x}_0)$. Hence $|\delta(\bar{x})|^2 = \langle(\bar{\nabla} F)\,\delta(\bar{x}_0), (\bar{\nabla} F)\,\delta(\bar{x}_0)\rangle = \langle\delta(\bar{x}_0), (\bar{\nabla} F)^T \bar{\nabla} F\,\delta(\bar{x}_0)\rangle = \langle\delta(\bar{x}_0), C\,\delta(\bar{x}_0)\rangle$. Here $\langle\cdot,\cdot\rangle$ denotes inner product and C is the Cauchy-Green strain tensor, which is positive-definite. Accordingly, all its eigenvalues ($\lambda_i, i = 1, 2, 3$) are positive. Mass conservation between instants t_0 and t for an incompressible flow requires the determinant of C to be unity (§14 of [25]). This implies that, in two dimensions (2D), $\lambda_1 \times \lambda_2 = 1$, that is, λ_1 and λ_2 must satisfy the inequality: $0 < \lambda_1 \le 1 \le \lambda_2$ [26]. We note that, since the stretch, twist and fold of vortices in helical systems do not constitute the subject matter of this article, and there is no loss of generality in examining a 2D modeled ABC flow. The 2D velocity field (u, w) exhibits chaos the way it exists in a three-dimensional system [27]. Moreover, the imposed length scale in the third dimension is far greater than the mixing scale of the helicity-driven dynamics. For instance, the size of the Earth and troposphere is orders of magnitude larger than the cross sections of the Earth's liquid core and tornado, respectively. Thus, a 2D model enables a conservative understanding of mixing.

The extent of deformation of an n-dimensional fluid mass may be defined in terms of n Lyapunov exponents along principal directions which are given by the eigenvectors of the tensor C. Since the separation of neighbors in the flow is exponential, the largest Lyapunov exponent overwhelms the remaining $(n-1)$ exponents. Mathematically, the time-averaged logarithmic largest Lyapunov exponent is known as FTLE, that is, FTLE $= \frac{1}{\Delta t}\log(\lambda_n)$. Thus, the fluid particles attract and repel along λ_1 and λ_2, respectively, in 2D flow. When the eigenvalues of C are each unity, FTLE is zero which implies coherence due to no net exponential growth or decay. Starting with the velocity of the modeled ABC flow (Eq. 1), we computed FTLEs that denote the repulsion of fluid parcels from these structures forward in time. We sampled the model at every 0.01 second for 1000 time steps and customized an open-source software for the computations. We employed a fourth-order Runge–Kutta scheme to discretize time.

3 Perturbation of Coherent Fluid Parcels

To elucidate the topological changes in coherent structures as shown in Fig. 2, we may quantify the deformation of a neutrally buoyant material curve in the flow (Fig. 3). Mathematically, a length-averaged property $Q(\gamma) = \frac{1}{\sigma}\int_0^{\sigma}\mathcal{L}(\bar{r}(s), \bar{r}'(s))ds$ of curve γ in ϵ-neighborhood is invariant with $O(\epsilon^2)$-accuracy if the differential of Q vanishes (that is, when Q is indifferent to change in γ). Here, s denotes the length along γ. $\bar{r}$ is a position on γ and $\bar{r}'$ is the tangent to γ at that point (Fig. 3). Minimizing Q gives a set of Euler equations subject to which the total derivative of a function (known as the first integral) vanishes when $\mathcal{L}$ is independent of s. The first integral is invariant for the curve γ which is traveling with the flow given that the Q-minimizing condition is true (known as Noether's theorem) [28].

Physically, Q may be constructed to imply that either strain or shear governs the flow. Firstly, when Q is defined as a strain ratio (l_t/l_0) by comparing the parametric lengths of curve γ at a later and initial time instants, the dimensional scaling of s with respect to $\bar{r}'$ transforms $\mathcal{L}$ to $\mathcal{L}'(\bar{r}, \bar{r}') = \langle \bar{r}', \frac{1}{2}(C - \lambda^2 I)\bar{r}' \rangle$, which is a function of strain energy (Supplementary material of [29]) and defines an energy integral. The tensor $\frac{1}{2}(C - \lambda^2 I)$ in 2D flow has oppositely signed eigenvalues $\lambda_1 - \lambda^2$ and $\lambda_2 - \lambda^2$, and is known as a symmetric Lorentzian metric in the geometric sense. The net deformation due to the tensor has two components: isotropic due to λ, and anisotropic due to λ_1 and λ_2 (Fig. 3). The energy integral ensures a minimum-strain

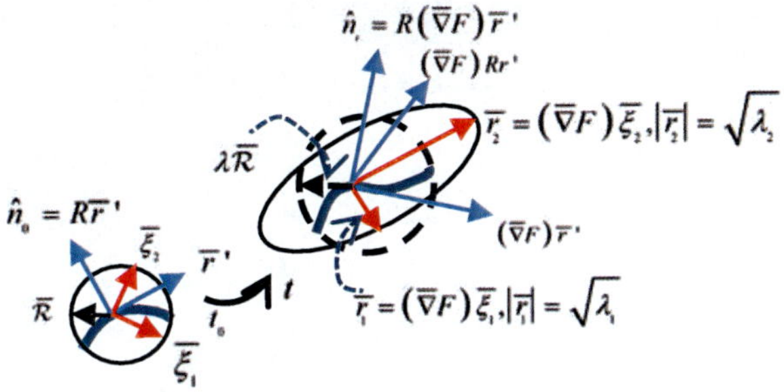

Fig. 3 Deformation of a material curve (solid blue) and the vectors attached to it. $\bar{\xi}_1, \bar{\xi}_2$ comprise principal directions in an illustrative 2D deformation

deformation of curve γ in the Lorentzian metric space which confirms that FTLE vanishes at least once within a coherent vortical region enclosed by curve γ in the flow [29]. The vanishing FTLE is a fundamental observation in the analysis of ABC flow presented in this article (Fig. 4).

Secondly, when Q is defined as a shear due to the projection of the advected unit normal vector on the tangent vector, $\mathcal{L}$ transforms into a new $\mathcal{L}'$ which contains $\frac{1}{2}(CR - RC)$ that is again a symmetric Lorentzian metric tensor because its eigenvalues in 2D, $\pm\sqrt{C_{12}^2 + \frac{(C_{22}-C_{11})^2}{4}}$, have opposite sign. Accordingly, the energy integral ensures a minimum-shear deformation of curve γ. Here, $C = \begin{bmatrix} C_{11} & C_{12} \\ C_{12} & C_{22} \end{bmatrix}$, and $R = \begin{bmatrix} 0 & -1 \\ 1 & 0 \end{bmatrix}$ rotates a vector counter-clockwise by $\frac{\pi}{2}$.

Figure 4 shows that the coherent FTLEs divide the domain into 'cells'. Moreover, we attribute mixing to the sensitive pathlines of fluid particles which are spread over greater parts of the domain in the unsteady modeled flow than in the steady flow. To this end, we have shown earlier that the transport in generalized helical flows is a function of topology and FTLE [14]. Here, we examine the conditions which underlie the topological differences during the transport of a material curve through shear- and strain-driven cells of the domain under the influence of vanishing FTLE. Since the definition of the length-averaged property Q minimizes either strain or shear of a deforming curve γ, it is incumbent that there are separate corresponding geometric descriptions of the phenomena. Accordingly, a tangent $\bar{r}'$ to a point on the closed material curve γ that is advecting in the flow with a minimum-strain deformation [26] may be given as Eq. 2, which is a linear combination of orthonormal eigenvectors, $\bar{\xi}_1$ and $\bar{\xi}_2$, of the tensor C. Similarly, $\bar{r}'$ for a curve γ under minimum shear [30] may be given as Eq. 3.

$$\bar{r}'(s) = \frac{\sqrt{\lambda_2 - \lambda^2}}{\sqrt{\lambda_2 - \lambda_1}}\bar{\xi}_1(\bar{r}) \pm \frac{\sqrt{\lambda^2 - \lambda_1}}{\sqrt{\lambda_2 - \lambda_1}}\bar{\xi}_2(\bar{r}) \tag{2}$$

$$\bar{r}'(s) = \frac{\sqrt{\sqrt{\lambda_2}}}{\sqrt{\sqrt{\lambda_1} + \sqrt{\lambda_2}}}\bar{\xi}_1(\bar{r}) \pm \frac{\sqrt{\sqrt{\lambda_1}}}{\sqrt{\sqrt{\lambda_1} + \sqrt{\lambda_2}}}\bar{\xi}_2(\bar{r}) \tag{3}$$

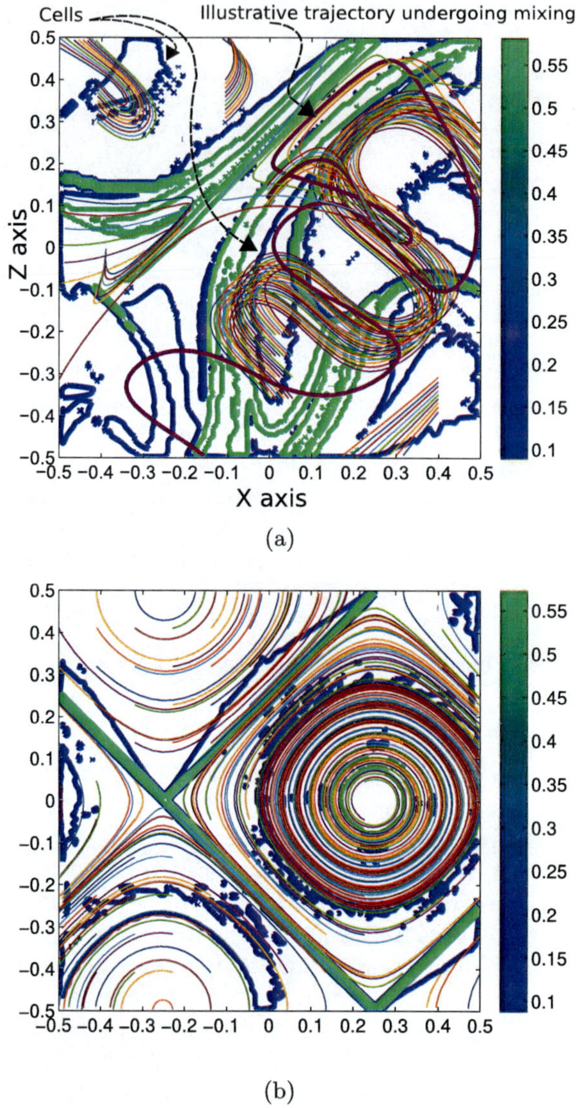

Fig. 4 Forward FTLE contour superimposed with trajectories

Since the transition of the curve γ from a minimizing strain to shear deformation requires that Eqs. 2 and 3 are linearly dependent, the corresponding coefficients of $\bar{\xi}_1$ and $\bar{\xi}_2$ in these two equations must be proportional to each other. Since the eigenvalues λ_1 and λ_2 of the tensor C approach unity when FTLE is vanishing, Eqs. 2 and 3 have characteristic differences. While the coefficient of either $\bar{\xi}_1$ or $\bar{\xi}_2$ is $\frac{1}{\sqrt{2}}$ in Eq. 3, we consider an asymptotic perturbation to avoid singularity in Eq. 2. Noting that $\lambda_2 = 1/\lambda_1$, we let $\epsilon = \sqrt{1 - \lambda_1^2}$ to be the perturbation when λ_1 approaches unity.

As $\epsilon \to 0$, a transition between shear- and strain-minimizing curves in the flow will occur when there is asymptotic matching between Eqs. 2 and 3 with respect to the coefficients of $\bar{\xi}_1$ and $\bar{\xi}_2$.

After some algebra, the matching between the coefficients of $\bar{\xi}_1$ occurs according to:

$$\frac{1}{\epsilon}\sqrt{1 - \lambda^2\sqrt{1-\epsilon^2}} \sim \frac{1}{\sqrt{2}} \tag{4}$$

which implies that $1 - \lambda^2(1 - \frac{\epsilon^2}{2} - \frac{\epsilon^4}{8}) \sim \frac{\epsilon^2}{2}$, and hence $\lambda \to 1^+$. Similarly, the coefficients of $\bar{\xi}_2$ match when:

$$\frac{1}{\epsilon}\sqrt{\lambda^2\sqrt{1-\epsilon^2} - (1-\epsilon^2)} \sim \frac{1}{\sqrt{2}} \tag{5}$$

which leads to $\lambda^2 \sim \frac{1-\frac{\epsilon^2}{2}}{\sqrt{1-\epsilon^2}}$, and hence $\lambda \to 1^+$. That is, λ is never less than one. Thus, we find that the presence of rigid coherent structures (λ on the order of but not less than one) is a necessary condition for the characteristic topological modulations in unsteady helical ABC flow.

4 Discussion

Our findings have implications on the study of terrestrial magnetic dynamos, superfluid turbulence and tornadoes. First, the equation of magnetic induction may be expressed as $\partial \bar{B}/\partial t = \bar{\nabla} \times (\bar{u} \times \bar{B}) + (\nabla^2 \bar{B})/R_m$, where $\bar{u}$ and $\bar{B}$ are velocity and magnetic fields, respectively, and $R_m = U_c L_c \mu_0 \sigma$ is non-dimensional magnetic Reynolds number with U_c and L_c being characteristic velocity and length, respectively, μ_0 being free space permeability and σ being electrical conductivity. Solution to the induction equation reveals the largest Lyapunov exponent (indicator of chaos) and unstable critical points, both of which can be used as metrics to find a relation that tweaks the parameters A, B and C of the unsteady ABC flow with varying R_m (and hence U_c and L_c) so that FTLE contours can bear equivalent metrics. Actual solutions of induction equation with ABC velocity field at lower values of R_m are required to establish this relation. This can provide apriori information about advection of magnetic field when R_m is high ($O(\sim 10^3)$ for Earth) and direct solution of the induction equation is delayed due to constraint on computing power. Moreover, unsteady ABC flow shows promise in being used as input to induction equation to find growth rates of both large scale and small scale dynamos over a wide range of R_m [31]. We do realize that anti-dynamo theorem [32] negates formation of sustained growing dynamo in 2D flow. However, our analysis in this article can be easily extended and is applicable to three dimensions. Second, particle image velocimetry and direct numerical solution of the equation of moving particles in fluid are potent tools for flow visualization of He II. However, they depend on properties of particles being injected as tracers [33]. FTLE contour of ABC flow displays presence of idealized discrete vortices of superfluid He II without requiring tracers. Instantaneous maximum vorticity regions in our unsteady model are found to linearly trace the domain as has been reported with other unsteady models [16]. Onset of vortex wave instability as mechanism for superfluid–normal fluid vortex matching [16] can be linked with FTLE contours. Moreover, there is scope to explore analogy between cell mixing in unsteady ABC flow and turbulent vortex interactions in superfluid He II at temperatures varying below critical 2.17 K. Third, observations from Doppler Radar [34] can be used to identify unsteady ABC flow field with matched critical stagnation points. This is to be followed by tagging the supercell storms, both objectively (largest FTLE) and subjectively (extent of cell mixing due to zero-level FTLE contours). Thus storm tagging will help build a database documenting the behavior of storms leading to tornadoes.

5 Conclusion

To summarize, we analyze the significance of macroscopic Lagrangian behavior and mixing characteristics of purely helical flow in understanding the first principles of interdisciplinary problems. In this regard, we consider an unsteady model of helical flow, and identify coherent structures in both steady and unsteady helical flow. Fluid mixing across cells is absent in the steady but evident in the unsteady flow. We identify distinct topological changes in the unsteady flow as well. These observations highlight the role of perturbation in unsteady field and affirms our new analytical procedure to understand how coherent fluid parcels transition between shear and strain barriers and hence induce cell mixing. Largest FTLE in the domain and the degree of cell mixing appear to emerge as two indicators that extend the scope of purely helical flow as a model for dynamo theory, mixing in superfluid He II and tagging of supercell tornadoes.

Acknowledgements The author is grateful to Professor Snehanshu Saha (CSIS and APPCAIR, BITS Pilani, K. K. Birla Goa Campus, Goa, India) for his insightful comments and review of the manuscript.

References

1. Balakrishnan HN, Kathpalia A, Saha S, Nagaraj N (2019) ChaosNet: a chaos based artificial neural network architecture for classification. Chaos: Interdiscip J Nonlinear Sci 29(11):113125
2. Bhattacharya A, Saha S, Nagaraj N (2021) SMPSO revisited: a theoretical analysis of exponentially-averaged momentum in multi-objective problems. https://doi.org/10.48550/ARXIV.2104.10040
3. Marceau-Caron G, Ollivier Y (2017) Natural Langevin dynamics for neural networks. In: International conference on geometric science of information. Springer, pp 451–459

4. Cranmer M, Greydanus S, Hoyer S, Battaglia P, Spergel D, Ho S (2020) Lagrangian neural networks. arXiv preprint arXiv:2003.04630
5. Vinuesa R, Brunton SL (2022) Enhancing computational fluid dynamics with machine learning. Nat Comput Sci 2(6):358–366
6. Enciso A, Peralta-Salas D, Torres de Lizaur F (2016) Helicity is the only integral invariant of volume-preserving transformations. Proc Natl Acad Sci USA 113(8):2035–2040. https://doi.org/10.1073/pnas.1516213113
7. Scheeler MW, Kleckner D, Proment D, Kindlmann GL, Irvine WTM (2014) Helicity conservation by flow across scales in reconnecting vortex links and knots. Proc Natl Acad Sci USA 111(43):15350–15355. https://doi.org/10.1073/pnas.1407232111
8. Moffatt HK (1989) Stretch, twist and fold. Nature 341:285–286. https://doi.org/10.1038/340301a0
9. Gilbert AD (1991) Fast dynamo action in a steady chaotic flow. Nature 350:483–485. https://doi.org/10.1038/353737a0
10. Moffatt HK (2014) Helicity and singular structures in fluid dynamics. Proc Natl Acad Sci USA 111(10):3663–3670. https://doi.org/10.1073/pnas.1400277111
11. Dombre T, Frisch U, Greene JM, Henon M, Mehr A, Soward AM (1986) Chaotic streamlines in the ABC flows. J Fluid Mech 167:353–391
12. Strogatz SH (1994) Nonlinear dynamics and chaos with applications to physics, biology, chemistry, and engineering. Perseus Books
13. Ottino J (1990) Mixing, chaotic advection, and turbulence. Annu Rev Fluid Mech 22(1):207–253. https://doi.org/10.1146/annurev.fluid.22.1.207
14. Chakraborty P, Roy A, Chakraborty S (2021) Topology and transport in generalized helical flows. Phys Fluids 33(11):117106
15. Haller G (2001) Distinguished material surfaces and coherent structures in three-dimensional fluid flows. Physica D 149(4):248–277. https://doi.org/10.1016/S0167-2789(00)00199-8
16. Barenghi CF, Samuels DC, Bauer GH, Donnelly RJ (1997) Superfluid vortex lines in a model of turbulent flow. Phys Fluids 9(9):2631–2643. https://doi.org/10.1063/1.869379
17. Mohanchandra K, Saha S, Murthy KS (2016) Evidence of chaos in EEG signals: an application to BCI. In: Advances in chaos theory and intelligent control. Springer, pp 609–625
18. Alexakis A (2011) Searching for the fastest dynamo: laminar ABC flows. Phys Rev E 84(2):026321(10). https://doi.org/10.1103/PhysRevE.84.026321
19. Schwarz KW (1982) Generation of superfluid turbulence deduced from simple dynamical rules. Phys Rev Lett 49(4):283–285. https://doi.org/10.1103/PhysRevLett.59.2117
20. Samuels DC (1993) Response of superfluid vortex filaments to concentrated normal-fluid vorticity. Phys Rev B 47(2):1107–1110
21. Paoletti MS, Fisher ME, Sreenivasan KR, Lathrop DP (2008) Velocity statistics distinguish quantum turbulence from classical turbulence. Phys Rev Lett 101(15):154501(4). https://doi.org/10.1103/PhysRevLett.101.154501
22. Diffenbaugh NS, Scherer M, Trapp RJ (2013) Robust increases in severe thunderstorm environments in response to greenhouse forcing. Proc Natl Acad Sci USA 110(41):16361–16366. https://doi.org/10.1073/pnas.1307758110
23. Lilly DK (1986) The structure, energetics and propagation of rotating convective storms. Part II: helicity and storm stabilization. https://doi.org/10.1175/1520-0469(1986)043<0126:TSEAPO>2.0.CO;2
24. Rotunno R (2013) The fluid dynamics of tornadoes. Annu Rev Fluid Mech 45(1):59–84. https://doi.org/10.1146/annurev-fluid-011212-140639
25. Lamb H (1975) Hydrodynamics. Cambridge University Press
26. Haller G (2015) Lagrangian coherent structures. Annu Rev Fluid Mech 47:137–162. https://doi.org/10.1146/annurev-fluid-010313-141322
27. Galloway DJ, Proctor MRE (1992) Numerical calculations of fast dynamos in smooth velocity fields with realistic diffusion. Nature 356:691–693
28. Gelfand I, Fomin SV (1963) Calculus of variations. Prentice-Hall, Inc.

29. Haller G, Beron-Vera FJ (2013) Coherent Lagrangian vortices: the black holes of turbulence. J Fluid Mech 731:R4-1–R4-10. https://doi.org/10.1017/jfm.2013.391
30. Haller G, Beron-Vera FJ (2012) Geodesic theory of transport barriers in two-dimensional flows. Physica D 241:1680–1702
31. Cameron A, Alexakis A (2016) Fate of alpha dynamos at large Rm. Phys Rev Lett 117(20):205105(5). https://doi.org/10.1103/PhysRevLett.117.205101
32. Zil'dovich IB (1957) The magnetic field in the two-dimensional motion of a conducting turbulent liquid. J Exp Theor Phys 31:460–462
33. Poole DR, Barenghi CF, Sergeev YA, Vinen WF (2005) Motion of tracer particles in He II. Phys Rev B 71(6):064514(16). https://doi.org/10.1103/PhysRevB.71.064514
34. Wurman J, Straka JM, Rasmussen EN (1996) Fine-scale doppler radar observations of tornadoes. Science 272:1774–1777. https://doi.org/10.1126/science.272.5269.1774

Comparative Study of Pruning Techniques in Recurrent Neural Networks

Sagar Choudhury, Asis Kumar Rout, Pragnesh Thaker, and Biju R. Mohan

Abstract In recent years, there has been a drastic development in the field of neural networks. They have evolved from simple feed-forward neural networks to more complex neural networks such as convolutional neural networks (CNNs) and recurrent neural networks (RNNs). CNNs are used for tasks such as image recognition where the sequence is not essential, while RNNs are useful when order is important such as machine translation. By increasing the number of layers in the network, we can improve the performance of the neural network (Alford et al. in Pruned and structurally sparse neural networks, 2018 [1]). However, this will also increase the complexity of the network, and also training will require more power and time. By introducing sparsity in the architecture of the neural network, we can tackle this problem. Pruning is one of the processes through which a neural network can be made sparse (Zhu and Gupta in To prune, or not to prune: exploring the efficacy of pruning for model compression, 2017 [2]). Sparse RNNs can be easily implemented on mobile devices and resource-constraint servers (Wen et al. in Learning intrinsic sparse structures within long short-term memory, 2017 [3]). We investigate the following methods to induce sparsity in RNNs: RNN pruning and automated gradual pruning. We also investigate how the pruning techniques impact the model's performance and provide a detailed comparison between the two techniques. We also experiment by pruning input-to-hidden and hidden-to-hidden weights. Based on the results of pruning experiments, we conclude that it is possible to reduce the complexity of RNNs by more than 80%.

Keywords Sparse neural networks · Deep learning · RNN compression · Automated gradual pruning · LSTM · GRU

S. Choudhury (✉) · A. K. Rout · P. Thaker · B. R. Mohan
National Institute of Technology Karnataka, Surathkal 575025, India
e-mail: sagarchoudhury.nitk@gmail.com
URL: http://www.nitk.ac.in/

S. Das et al. (eds.), *Advances in Data-driven Computing and Intelligent Systems*, Lecture Notes in Networks and Systems 653, https://doi.org/10.1007/978-981-99-0981-0_32

1 Introduction

Sparse neural networks are a viable option to make neural networks less complex and make them memory efficient. The benefit of sparse neural networks is that it leads to a reduction in inference and training time. There is a reduction in storage requirement as the connections between layers and weights are reduced [4, 5]. Such sparsity in neural networks can be induced by pruning the weights in the network.

In this work, we implement two different pruning techniques, RNN pruning and automated gradual pruning, to induce sparsity in recurrent neural networks. Recurrent neural networks result in better in text analysis tasks but often require a lot of memory to store their weights [6]. These large sizes add computational burden which in turn makes it difficult for their deployment on small devices [7]. We investigate the impact of these pruning techniques on the final accuracy of deep learning models and provide a comparison between the two techniques. We also experiment by implementing the pruning techniques on different datasets.

2 Literature Review

A few researchers have worked on pruning to induce sparsity in recurrent neural networks in the past couple of decades. Some of the notable works in the pruning of recurrent neural networks have been discussed briefly below:

- Zhang and Stadie [8] proposed a pruning method that was not dependent on any specific recurrent neural network architecture of choice. Their algorithm outperformed many random pruning techniques even at high sparsity levels. The performance of this algorithm increased with the increase in the network size. This can be attributed to the effective distribution of sparse connections across the entire weight matrix. But, this method could not perform well when the size of the network is small. Other models such as SNIP and Foresight performed better compared to this when the network size was small. This was not an iterative method, and one of the major trade-offs was accuracy.
- In 2019, Semionov [9] implemented a pruning technique on three different datasets and provided a comparison for the same. Her proposed methodology was simple to implement and can easily be implemented by modifying the basic hyperparameter settings. While the proposed method was relatively simple, the author used shallow networks that do not allow us to make solid conclusions about the method.
- Zhong [10] proposed an end-to-end pruning method that was capable of compressing a variety of network structures with comparable accuracy. The proposed method worked well on convolutional as well as fully connected networks. The experimental results revealed that the method was able to learn the sensitivity of each network layer. One of the major drawbacks of this method was that pruning through this method led to performance loss on deeper networks. Also, there was no experimentation on recurrent neural networks.

- Wen et al. [11] have proposed a new method to prune individual neurons of recurrent neural networks. In this method, they have introduced binary gates on recurrent and input units such that sparse masks for the weight matrix can be generated. Under sparsity constraints, this allowed for effective neuron selection under sparsity constraints. This method can be readily implemented for vanilla RNNs and GRU. But, the optimization of the introduced two random variables was computationally intractable. Combining the neuron selection with quantization algorithms is still to be experimented with for further reduction in model sizes.

After going through the research papers, we conclude that most of the pruning techniques were implemented for convolutional neural networks, and very few pruning techniques were implemented for LSTMs and GRUs. Only a few comparison studies have been done between the existing algorithms to find out which one outperforms the other. Most pruning techniques were not tested for a wide variety of datasets and test conditions.

3 Methodology

The methods proposed below consider the work done till now and build upon those works with a significant focus on pruning in recurrent neural networks.

3.1 RNN Pruning

There are various approaches to pruning neural networks. However, we chose RNN pruning for our research, as mentioned in [12]. The advantages of the proposed method:

- Computational simplicity
- No need for additional training of the model.

This method creates a binary mask corresponding to every weight in the network. Initially, they are all set to one. The weight is multiplied by the mask after every optimizer update step. The masks that are below a certain threshold are updated to zero. The threshold value is calculated by using hyperparameters.

The duration, rate, and frequency of pruning the parameters for each layer are controlled by the hyperparameters. For each layer, we have used a different hyperparameters set, resulting in a different threshold for each layer. The start_itr and stop_itr hyperparameters are used to specify the start and end of the pruning of the neural network, respectively. In the backpropagation step, the weights are updated using gradient descent. The weights that are larger than the threshold of that particular layer will be involved in the forward step, and those which are less are eliminated.

The heuristics to help determine start iteration (start_itr), ramp iteration (ramp_itr), and end iteration (end_itr) are generated by repeated testing. The ramp slope is adjusted to change the pruning rate, while freq specifies the number of iterations after which the threshold is updated. After picking these hyperparameters, we calculate start_slope (θ) using Eq. 1.

$$\theta = \frac{2 * q * \text{freq}}{2 * (\text{ramp\_itr} - \text{start\_itr}) + 3 * (\text{end\_itr} - \text{ramp\_itr})} \tag{1}$$

Parameters Frequency (freq), Ramp Iteration (ramp_itr), Start Iteration (start_itr), End Iteration (end_itr), and threshold are selected separately for each layer. The parameter q is the nineteeth percentile of the absolute values of the weights trained without model pruning as given in [12].

The calculated threshold decides the rate of pruning. The threshold value is increased more aggressively from Ramp Iteration (ramp_itr) to End Iteration (end_itr). We calculate threshold from Start Iteration (start_itr) to Ramp Iteration (ramp_itr) using Eq. 2.

$$\epsilon = \frac{\theta * (\text{current\_itr} - \text{start\_itr} + 1)}{\text{freq}} \tag{2}$$

The pruning rate from Ramp Iteration (ramp_itr) to End Iteration (end_itr) is decided by a new threshold. We calculate the new threshold from Ramp Iteration (ramp_itr) to End Iteration (end_itr) using 3.

$$\epsilon = \frac{\theta * (\text{ramp\_itr} - \text{start\_itr} + 1)}{\text{freq}} + \frac{\phi * (\text{current\_itr} - \text{ramp\_itr} + 1)}{\text{freq}} \tag{3}$$

3.2 *Automated Gradual Pruning*

In this approach, as proposed by Zhu and Gupta [2], pruning of the network's connections is carried out during the training phase itself. In the starting phase of the training, the redundant connections are more in number in the network. Hence, we prune these redundant connections rapidly in the starting phase. Gradually, we reduce pruning the network as the number of remaining weights in the network decreases [13].

Initially, in this approach, all the weights in the layer are sorted by their absolute values. A binary mask is maintained for each weight. This decides which of the weights in the layer will participate during the forward execution phase of the training. The smallest magnitude weights are masked to zero until the desired sparsity level(s) for the layer is achieved. During the backpropagation phase, the back-propagated gradients flow through the binary masks, and the weights that were masked in the forward execution phase do not get updated. In this gradual pruning algorithm, the sparsity is increased over a span of n pruning steps from an initial sparsity (s_i) value

0 to a final sparsity (s_f) value with pruning frequency Δt. The final sparsity is calculated using Eq. 4.

$$s_t = s_f + (s_i - s_f)\left(1 - \frac{t - t_0}{n\Delta t}\right)^3 \quad \text{for } t \in \{t_0, t_0 + \Delta t, \ldots, t_0 + n\Delta t\} \tag{4}$$

As the network is trained, after every Δt step, the binary weight masks are updated to increase the sparsity of the network gradually for effective pruning [14]. Once the model achieves the target sparsity, the weight masks are no longer updated. With a small learning rate, we observed that it is difficult for subsequent training steps to recover from the loss in accuracy, which is caused by forcing the weight to zero. In comparison, going with a high learning rate results in pruning the weights even before they converged to a good solution. So it is crucial to choose the pruning schedule closely with the learning rate schedule [15].

All the layers are pruned using the same sparsity function, and to allow the network to heal from the pruning-induced loss of accuracy, pruning occurs in the regime where the learning rate is high.

3.3 Datasets

IMDB Dataset: The IMDB Movie Reviews dataset is a binary sentiment analysis dataset consisting of 50,000 reviews from the Internet Movie Database (IMDb). The dataset contains positive and negative reviews. There are 25,000 movie reviews for training and 25,000 for testing. The movie reviews are labeled by sentiments (positive/negative), where 0 indicates a negative review and 1 indicates a positive review. There are a maximum of 30 reviews included per movie.

PTB Dataset: The English Penn Treebank (PTB) corpus, and in particular the section of the corpus corresponding to the articles of Wall Street Journal (WSJ), is one of the most known and used corpus for the evaluation of models for sequence labeling. The task consists of annotating each word with its Part-of-Speech tag. In the most common split of this corpus, sections from 0 to 18 are used for training (38,219 sentences, 912,344 tokens), sections from 19 to 21 are used for validation (5527 sentences, 131,768 tokens), and sections from 22 to 24 are used for testing (5462 sentences, 129,654 tokens). The corpus is also commonly used for character-level and word-level language modeling.

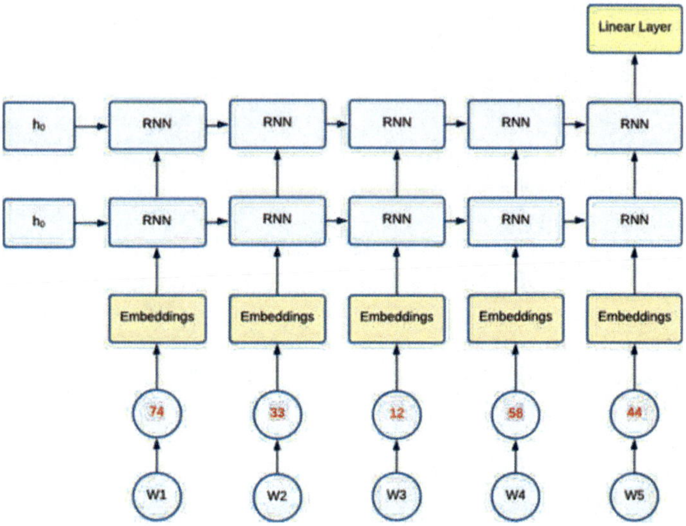

Fig. 1 Visualizing training of a recurrent network with 2 recurrent layers

3.4 Model Architecture

The base model consists of a regular RNN model followed by a linear layer and stacked upon an embedding layer. The input layer feeds the data into the model. The same is seen in Fig. 1.

Our base model starts with an input layer that feeds data to the model. Since our data is in text format, we must convert each word to a corresponding integer value. Here, we assign integers starting from 0 to each word in the dataset and store them in a dictionary. The model takes sequences of word indexes as an input.

In Fig. 2, after the input, the second layer is an embedding layer used to develop word embeddings. It transforms each input word into a fixed-length vector. The value of an embedding dimension must be equal to the number of neurons in the next layer.

After the embedding layer, the subsequent two layers are recurrent, each layer containing (250/500/1000) neurons based on whether the model is small, medium, or large. These layers start with an initial hidden state of shape (batch size, hidden size), comprising all zeros. In LSTM, these layers begin with one initial hidden state and one initial cell state, both of similar shape and comprising of all zeros. Each neuron in recurrent layers is an individual LSTM unit in the case of the LSTM recurrent network and GRU unit in the case of GRU recurrent network. Dropout is applied after the embedding layer and the Recurrent layer to prevent the neural network from overfitting [16].

These recurrent layers are followed by a linear layer that takes input from the second recurrent layer and applies the following linear transformation to the input:

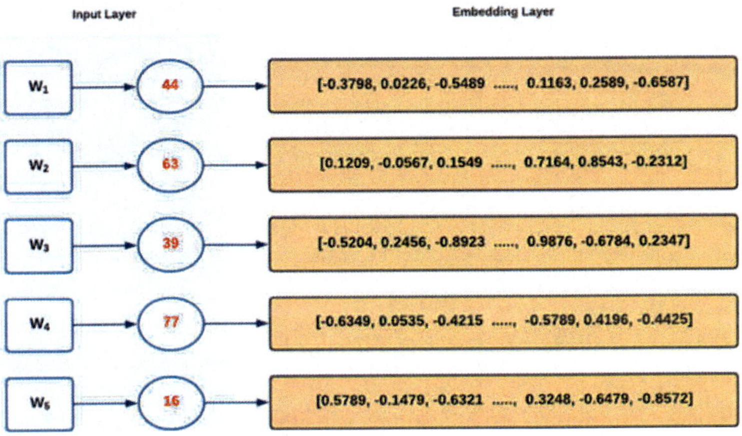

Fig. 2 Generating embeddings after converting each word of this sequence into its corresponding integer value

$$\hat{y} = xW^T + b \tag{5}$$

where $\hat{y}$ is the output, W represents weight matrices of shape (output size = number of word tokens, i.e., 9066 in case of word language model and 1 in sentiment analysis, input size = number of hidden units in a neural network), x is the input, and b is bias.

Since the second recurrent layer's hidden size is 250, the linear layer's input is of size 250, and output is of size 9066 because our target has 9066 different words, and for each word, the corresponding probability distribution is found in the case of word language model. In sentiment analysis, the output is of size 1, which denotes a fractional value from 0 to 1.

After getting the output of the probability distribution, the model calculates the cross-entropy loss for word language model and BCEwithLogitsLoss for the sentiment analysis model and optimizes it via backpropagation through time to a minimum value to achieve better performance.

We train this model on our training dataset of 19,044 sequences for 20 epochs, and to check its performance, we evaluate it on the test dataset of 8243 sequences in the case of the PTB dataset. Similarly, for the IMDB dataset, we train it for 25,000 reviews for ten epochs and evaluate it on a test dataset of 25,000 reviews. This training and evaluation employ the hyperparameters shown in the following table.

For training and evaluation in the IMDB dataset, we have used BCEwithLogits Loss with Adam optimizer and a batch size of 65, while in the PTB dataset, we used cross-entropy loss with stochastic gradient descent optimizer and a batch size of 20.

Since we work with LSTM and GRU, we also perform this training multiple times by varying the model size for both LSTM and GRU Models.

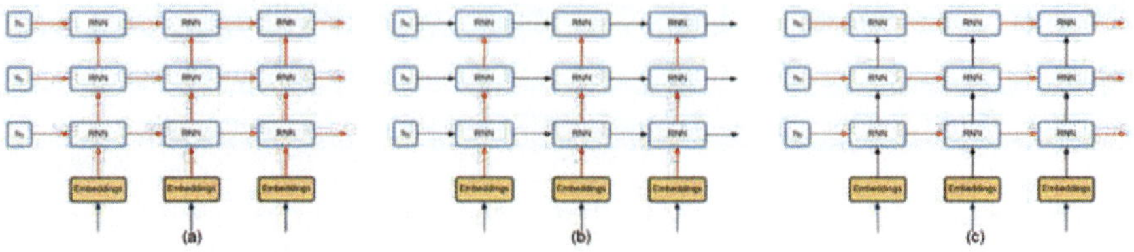

Fig. 3 **a** Simultaneously pruning input-to-hidden and hidden-to-hidden weights, **b** pruning input-to-hidden weights, **c** pruning hidden-to-hidden weights

3.5 Pruned Recurrent Neural Networks

One of the ways to introduce sparsity in recurrent networks is to prune weights below a certain threshold. We have three different types of weights; input-to-hidden weights, hidden-to-hidden weights, and hidden-to-output weights. In our experiment, we individually and simultaneously prune input-to-hidden and hidden-to-hidden weights as envisioned in Fig. 3.

In both the pruning algorithms, we calculate the threshold. After calculating the threshold, we create a binary tensor called a mask, where one corresponds to the absolute weight value above the threshold, and 0 corresponds to the absolute weight value below the threshold in the corresponding weight matrix. Element-wise multiplication of this binary mask with the corresponding weight matrix zero-outs the absolute weight values below the threshold value. After pruning, we evaluate pruned model's performance on the test dataset to see how such pruning affects an already trained model's overall accuracy. These pruning and retraining steps are repeated three times per RNN variant (i.e., LSTM and GRU), the first time for pruning both input-to-hidden and hidden-to-hidden weights, the second time for pruning only input-to-hidden weights, and the third time for pruning only hidden-to-hidden weights.

4 Results and Analysis

Here, we present our results from the experiments conducted. We present base model performance, which we then compare with pruning results. We further show the results by pruning only input-to-hidden and hidden-to-hidden weights. The experiments are conducted on IMDB and PTB datasets for LSTM and GRU recurrent units.

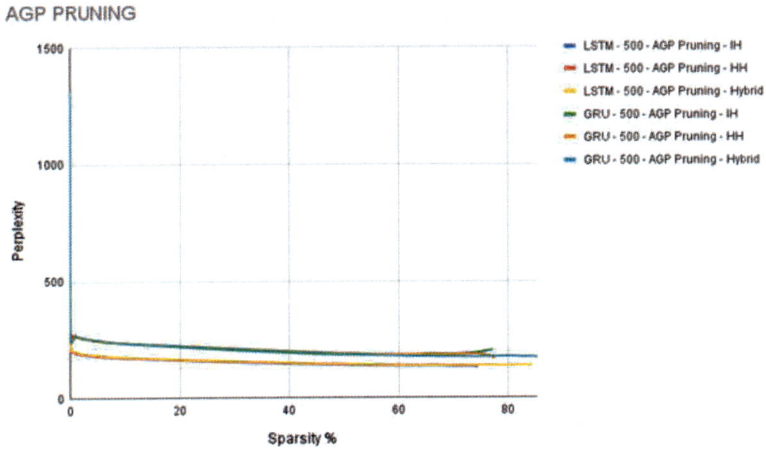

Fig. 4 Performance of LSTM model and GRU model by AGP pruning

4.1 PTB Dataset—Word Language Model

4.1.1 Base Model Performance

When the base LSTM model with 250, 500, and 1000 units in the hidden layer is trained for 30 epochs, it achieves test perplexity 140.28, 96.18, and 57.35 on the training dataset and 115.05, 108.69, 77.52 on the test dataset, respectively. The base GRU model with 500 units in a hidden layer after training for 30 epochs achieved a perplexity of 110.03 in the training data and 104.86 perplexities in test data.

4.1.2 Pruning of Models by Automated Gradual Pruning

Pruning Input-to-Hidden Weights and Hidden-to-Hidden weights We pruned the input-to-hidden weights and hidden-to-hidden weights of the medium LSTM model and medium GRU model. The medium LSTM model with 500 units in hidden layer has perplexity of 144.06 in 75.15% sparsity by pruning input-to-hidden weights and perplexity of 144.49 in 75.15% sparsity by pruning hidden-to-hidden weights. There was not much difference in pruning the weights separately in automated gradual pruning.

Hybrid Pruning In Fig. 4, the medium LSTM model has perplexity of 141.455 in 79.74% sparsity by AGP pruning. The large LSTM model has perplexity of 103.45 perplexity in 84.58% sparsity by AGP pruning.

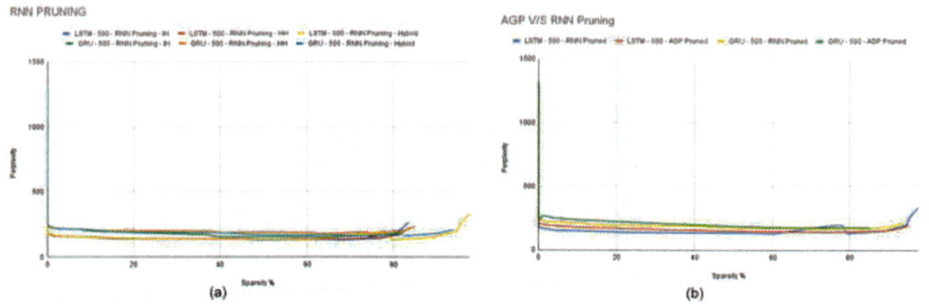

Fig. 5 **a** Performance of LSTM model and GRU model by RNN pruning, **b** performance of LSTM model and GRU model by AGP and RNN pruning

4.1.3 Pruning of Models by RNN Pruning

Pruning Input-to-Hidden Weights and Hidden-to-Hidden weights The medium LSTM model with 500 units in the hidden layer has perplexity of 140.85 perplexity in 75.97% sparsity by pruning input-to-hidden weights and perplexity of 157.49 in 74.72% sparsity by pruning hidden-to-hidden weights, Fig. 5a. For hybrid pruning, we got a perplexity of 137.18 at 76.16% sparsity as evident in Fig. 5b.

In RNN pruning, input-to-hidden pruning is more effective as compared to hidden-to-hidden pruning.

Hybrid Pruning The medium LSTM model has perplexity of 132.72 in 79.49% sparsity by RNN pruning. The large LSTM model has perplexity of 93.80 in 82.88% sparsity by RNN pruning. RNN pruner performed better as compared to AGP pruning. The model at 89% sparsity in the case of RNN pruning and 81% sparsity in the case of AGP pruning has a similar perplexity to that of the base model in the test dataset.

4.2 IMDB Dataset

4.2.1 Base Model Performance

The base LSTM model with 250 units in the hidden layer after training for ten epochs achieved an accuracy of 0.92 in the training set, performing steadily in the last few epochs. The test accuracy obtained is 0.85. The base LSTM model with 500 units in a hidden layer after training for ten epochs achieved an accuracy of 0.95 in the training set. The test accuracy obtained is 0.85. The model performed the same as compared to the small model. The base LSTM model with 1000 units in a hidden layer after training for ten epochs achieved an accuracy of 0.92 in the training set. The test accuracy obtained is 0.82. The model did not perform better as compared to both the small and medium models. The base GRU model with 500 units in a hidden layer after training for ten epochs achieved an accuracy of 0.95 in the training data and an accuracy of 0.85 in test data.

4.2.2 Pruning of Models by Automated Gradual Pruning

Pruning Input-to-Hidden Weights and Hidden-to-Hidden weights We pruned the input-to-hidden weights and hidden-to-hidden weights of the medium LSTM model and medium GRU model. The medium LSTM model with 500 units in the hidden layer has an accuracy of 0.93 in 63% sparsity and an accuracy of 0.85 in 63% for the test dataset by pruning input-to-hidden weights. The GRU model with 500 units in the hidden layer has an accuracy of 0.95 in 65% sparsity. The model has a test accuracy of 0.86 after 65% pruning in the case of pruning only input-to-hidden weights.

The medium LSTM model with 500 units in the hidden layer has an accuracy of 0.92 in 52% sparsity and 0.94 accuracy in 63% sparsity by pruning hidden-to-hidden weights. The GRU model with 500 units in the hidden layer has an accuracy of 0.95 in 65% sparsity. The model has a test accuracy of 0.85 after 65% pruning in the case of pruning only hidden-to-hidden weights.

Hybrid Pruning In Fig. 6, we take a look at the performance of LSTM model and GRU model by AGP pruning. The medium LSTM model with 500 units in the hidden layer has an accuracy of 0.94 in 66% sparsity. This model observes a better performance as compared to the small model. The model has an accuracy of 0.83 in the test dataset after 66% pruning. The large LSTM model with 1000 units in the hidden layer has an accuracy of 0.90 in 36% sparsity and 0.95 accuracy in 67% sparsity. The model has an accuracy of 0.84 in test dataset after 63% pruning. The GRU model with 500 units in the hidden layer has an accuracy of 0.93 in 49% sparsity and 0.95 accuracy in 67% sparsity. The model has an accuracy of 0.85 in the test dataset after 67% pruning.

4.2.3 Pruning of Models by RNN Pruning

Pruning Input-to-Hidden Weights and Hidden-to-Hidden weights The medium LSTM model with 500 units in the hidden layer has an accuracy of 0.88 in 44% sparsity and 0.93 accuracy in 90% sparsity. The model has an accuracy of 0.86 in the test dataset in 90% sparsity by pruning input-to-hidden weights. The GRU model with 500 units in the hidden layer has an accuracy of 0.94 in 92% sparsity and an test accuracy of 0.83 with the same sparsity in the case of pruning only input-to-hidden weights.

The medium LSTM model with 500 units in the hidden layer has an accuracy of 0.85 in 45% sparsity and 0.93 in 90% sparsity. The model has a test accuracy of 0.86 in 90% sparsity by pruning hidden-to-hidden weights. The GRU model with 500 units in the hidden layer has an accuracy of 0.95 in 92% sparsity. The model has a test accuracy of 0.86 in 92% sparsity in the case of pruning only hidden-to-hidden weights.

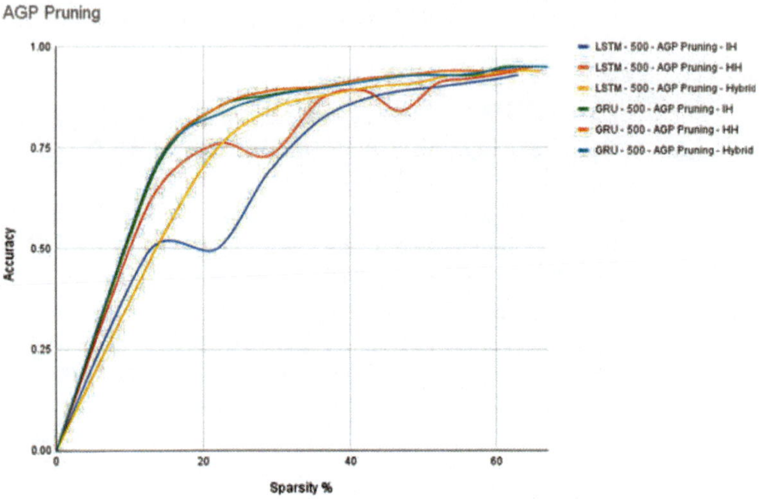

Fig. 6 Performance of LSTM model and GRU model by AGP pruning

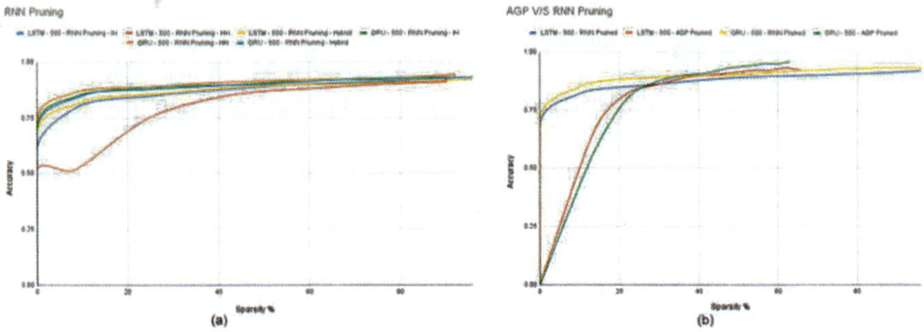

Fig. 7 **a** Performance of LSTM model and GRU model by RNN pruning, **b** performance of LSTM model and GRU model by AGP and RNN pruning

In Fig. 7a, we look at the performance of LSTM model and GRU model by RNN Pruning, while in Fig. 7b, we compare the performance of LSTM model and GRU model by AGP and RNN Pruning.

Hybrid Pruning The small LSTM model with 250 units in the hidden layer has an accuracy of 0.88 in 48% sparsity and 0.91 accuracy in 96% sparsity. The medium LSTM model with 500 units in the hidden layer has an accuracy of 0.89 in 49% sparsity. Further pruning of the model did not have a negative effect. The accuracy increased to 0.92 at 96% sparsity. The model has an accuracy of 0.86 in the test dataset after 96% pruning. The large LSTM model with 1000 units in the hidden layer has an accuracy of only 0.50 in 54% sparsity. At the same time, further pruning and training led to an increase in the accuracy to 0.89 in 98% sparsity. The model has a test accuracy of 0.79 at 98% sparsity, which is less compared to the small and medium models. The GRU model with 500 units in the hidden layer has an accuracy of 0.90 in 48% sparsity and 0.94 accuracy in 96% sparsity. The model has a test accuracy of 0.85 in 96% sparsity.

4.3 Analysis

4.3.1 Base Model Performance

For our experiment purpose, we developed custom recurrent models using PyTorch such that we can easily modify the weights based on our requirements. In the case of the PTB dataset, the large model outperformed both the small and medium models, while in the case of the IMDB dataset, the medium and small LSTM models outperformed the larger model.

4.3.2 Pruning Input-to-Hidden and Hidden-to-Hidden Weights

Our pruning experiment was divided into three separate sub-experiments: pruning input-to-hidden weights, pruning hidden-to-hidden weights, and pruning input-to-hidden and hidden-to-hidden weights simultaneously.

In automated gradual pruning, pruning input-to-hidden and hidden-to-hidden weights separately in the medium LSTM model produced nearly similar results. In GRU, pruning input-to-hidden weights had a better perplexity at 77.31% pruning as compared to pruning hidden-to-hidden weights. In RNN pruner, pruning input-to-hidden weights is much more effective than hidden-to-hidden weights pruning for both LSTM and GRU models. For the IMDB dataset, pruning through RNN pruner resulted in similar results as compared to pruning through automated gradual pruner for all the models at a sparsity of more than 90%. Pruning input-to-hidden weights was much more effective than pruning hidden-to-hidden weights.

4.3.3 Hybrid Pruning

Later, we experimented with our results by pruning input-to-hidden weights and hidden-to-hidden weights simultaneously. In the case of automated gradual pruning, the large LSTM model after pruning 81% has a similar perplexity to that of the base model in the test dataset, while the medium LSTM model after pruning 85% has a similar perplexity to that of the base medium model. The smaller model is not able to perform as well as the large medium models after pruning. For the IMDB dataset, we see the test accuracy for the small model is 0.85 while for the medium and large model, it is 0.83 and 0.84, respectively, with around 65% sparsity. We see the small model was able to perform well compared to the medium and large models with around 65% sparsity.

In the case of RNN pruner, the large model can be pruned to 89%, while the medium model can be pruned to 81%. In the smaller model, pruning is effective till 86%. RNN pruner is more effective than automated gradual pruning. The algorithm gave a better perplexity in both training and test dataset at 80% sparsity. For the IMDB dataset, all the models had a similar performance as the models, which were pruned through automated gradual pruning but were able to be pruned for more than 90%. The small model has a test accuracy of 0.84 at 96% sparsity, while it is 0.86 and 0.79 for the medium and large model with around 96% sparsity. Pruning input-to-hidden and hidden-to-hidden weights simultaneously gave better results than pruning the input-to-hidden or hidden-to-hidden weights individually.

5 Conclusion and Future Work

5.1 Conclusion

The main objective of our work was to investigate the impact of sparsity in recurrent neural networks. Based on the results of pruning experiments, we conclude that it is possible to reduce the complexity of RNNs by more than 80%. This can be observed both for IMDB and PTB datasets. RNN Pruner algorithm outperforms automated gradual pruning in most of the models. Pruning input-to-hidden and hidden-to-hidden weights simultaneously gave the best results. Pruning input-to-hidden weights are much more effective than pruning hidden-to-hidden weights.

5.2 Future Work

AGP and RNN pruning techniques can be further implemented for Vanilla RNN. The pruning techniques can be combined with regularization techniques to penalize network weights [17] further to check whether the sparsity could be decreased without compromising the accuracy or not. A comparative study of L1-Regularization, L2-Regularization, and Group lasso Regularization with the pruning techniques can be carried out.

Notes and Comments. The first author and the second author have done the same amount of work.

References

1. Alford S et al (2018) Pruned and structurally sparse neural networks. CoRR abs/1810.00299. arXiv: 1810.00299

2. Zhu M, Gupta S (2017) To prune, or not to prune: exploring the efficacy of pruning for model compression. arXiv preprint arXiv:1710.01878
3. Wen W, He Y, Rajbhandari S, Zhang M, Wang W, Liu F, Hu B, Chen Y, Li H (2017) Learning intrinsic sparse structures within long short-term memory. arXiv preprint arXiv:1709.05027
4. Furuya T, Suetake K, Taniguchi K, Kusumoto H, Saiin R, Daimon T (2021) Spectral pruning for recurrent neural networks. arXiv:2105.10832v1
5. Mao H et al (2017) Exploring the regularity of sparse structure in convolutional neural networks. arXiv preprint arXiv:1705.08922
6. Ekaterina Lobacheva E, Nadezhda Chirkova N, Dmitry Vetrov D (2017) Bayesian sparsification of recurrent neural networks. arXiv:1708.00077v1
7. Wen L, Zhang X, Bai H, Xu Z (2019) Structured pruning of recurrent neural networks through neuron selection. arXiv:1906.06847v2
8. Zhang MS, Stadie BC (2019) One-shot pruning of recurrent neural networks by Jacobian spectrum evaluation. arXiv:1912.00120v1
9. Semionov N (2019) Pruning of long short-term memory neural networks
10. Narang S, Undersander E, Diamos G (2017) Block-sparse recurrent neural networks. arXiv:1711.02782v1
11. Zhong J, Ding G, Guo Y, Han J, Wang B (2018) Where to prune: using LSTM to guide end-to-end pruning (IJCAI-18)
12. Narang S, Elsen E, Diamos G, Sengupta S (2017) Exploring sparsity in recurrent neural networks. arXiv:1704.05119v2. 2017 potential of RFID in anti-counterfeiting. In: Proceedings of ACM symposium on applied computing, pp 1607–1612
13. He K et al (2016) Deep residual learning for image recognition. In: 2016 IEEE conference on computer vision and pattern recognition (CVPR), pp 770–778. https://doi.org/10.1109/CVPR.2016.90
14. Sun Y, Wang X, Tang X (2016) Sparsifying neural network connections for face recognition. In: Proceedings of the IEEE conference on computer vision and pattern recognition, pp 4856–4864
15. Shiwei L, Iftitahu N, Vlado M, Mocanu DC, Pechenizkiy M (2021) Efficient and effective training of sparse recurrent neural networks
16. Zaremba W, Sutskever I, Vinyals O (2014) Recurrent neural network regularization. arXiv:1409.2329
17. Christos C, Dimitrios K, Dimitropoulos K, Daras P (2021) Recurrent neural network pruning using dynamical systems and iterative fine-tuning

Issues, Challenges, and Opportunities in Advancement of Factory Automation System (FAS)

Janhavi Namjoshi and Manish Rawat

Abstract In the current scenario of the factory automation system (FAS), it is obligatory to automate the factory for the upgradation of manufacturing technologies and processes. FAS incorporates cyber-physical technology with conventional manufacturing systems, making the convoluted technologies more precise and complex than they used to be. This paper discusses the evolution and recent developments of the FAS in the last decade. The key technologies are part of the network layer, the data application layer, and/or the physical resource layer. Additionally, key emerging technologies and issues associated with them are explored. This includes real-time communication, coordination, undiagnosed bottleneck, breakdowns, and safety issues, which are implanted in the manufacturing process. With the FAS evolution, efforts have mainly been focused on network, wireless communication and connectivity, efficiency and productivity, transportation, safety, security, and surveillance. This paper provides a review of the evolution of the FAS and its scope in the future. The detailed state-of-the-art conclusion gives an idea to the researchers of potential issues and challenges.

Keywords Factory automation system · Industry 4.0 · Smart factory · IIoT

1 Background

Factory automation system is the deployment of computers, robots, information technologies, and control systems to operate several different machinery and processes in a factory to replace humans. It is the next step after mechanization in the scope of industrialization. This results in lowering costs, improving quality, increasing flexibility, improving reliability and efficiency, and reduction in environmental impact. The strengths of FAS include increased productivity, improved quality, consistency, and minimizing labor costs. In the later period of the twenty-first century, FAS was

J. Namjoshi · M. Rawat (✉)
Department of Mechatronics Engineering, School of Automobile, Mechatronics and Mechanical Engineering (SAMM), Manipal University Jaipur (MUJ), Jaipur, Rajasthan 303007, India
e-mail: manish.rawat@jaipur.manipal.edu

S. Das et al. (eds.), *Advances in Data-driven Computing and Intelligent Systems*, Lecture Notes in Networks and Systems 653, https://doi.org/10.1007/978-981-99-0981-0_33

developed very rapidly to satisfy the requirements of mass production and reduce product costs. The global FAS is segregated based on control and safety systems, components, industry-based technology, and connectivity. Based on the control system, is categorized into the programmable logic controllers, distributed control system, manufacturing execution system, supervisory control and data acquisition system, safety instrumented system, and human–machine interface. The components comprise sensors, actuators, transmitters and receivers, and other mechatronic components. The technology and connectivity used differ from factory to factory as per their requirement. From a common perspective, the following goals of the FAS can be identified:

- For high productivity.
- Increase in safety.
- Lower labor and employee cost.
- Increased repeatability and accuracy.
- Reduction in operation and lead time.
- Improve workers' safety and help accomplish tasks not possible manually.
- High production quality.

We will need to establish certain parameters to give a useful survey. Product and process design, supply chain, and steady-state optimizations are not discussed in this work. Even though they fall under the umbrella of FAS, these are enormous topics in and of themselves. Technologies important to FAS are briefly discussed in the context of the integration of various concepts; however, due to the large body of research on this topic, a comprehensive discussion of such technologies is not included in the study. The literature to be examined will span the years 2009 through 2019. (With very few exceptions). In the past decade, the FAS has changed dramatically. As a result, the purpose of this paper is to assess and review the progress of the FAS as well as to offer recommendations for future research. Our argument is two fold:

1. The challenges faced in the FAS and the new technologies and methodologies were highlighted in the last decade, i.e., 2009 onwards
2. The evolution of the FAS in the last decade, i.e., 2009 onwards and perspective of future scope in the FAS.

The rest of the paper is organized as follows Sect. 2 describes the challenges in the factory automation related to real-time application, breakdowns and undiagnosed bottlenecks, coordination and scheduling, and safety in FAS. Where Sect. 3 provides the details of technologies and methodologies used in the FAS. Section 4.

2 Challenges in Factory Automation

2.1 Real-Time Application

In the era of FAS, it is necessary for the system to be quick and more precisely real time. Real-time communication and network refer to that there is very tight timing required between a computer or controlling system receiving information and dealing with it. New technologies for FAS raise important problems that need to be solved, one of these is real-time applications. Unfortunately, because of several reasons that have been debated in the literature, wireless systems cannot be thought of as fully replacing wired networks in factories, particularly when a real-time application is a key issue. Field buses have been the sole option to reduce engineering efforts, improve dependability, and simplify the installation procedure at the field level for almost a decade. Setting up wireless communication for a dependable real-time communication system that can work in harsh settings where FASs are used is required. Implementation of real-time capable FAS software that increases the dependability of FAS using soft sensors. For wireless real-time communication, a concept of Wireless Interface for Sensors and Actuators (WISA), with two sub-concepts of WISA-com and WISA-power, is introduced, by Kjellsson et al. [1]. They implemented a real-time communication system that is reliable and able to deal with the extreme environment where the FAS is implemented. The efficient integration of the WISA concept and the wired field network is discussed for FAS. Process field network and process fieldbus decentral peripherals and concepts for the integration of WISA into the field network of process field network/process fieldbus decentral peripherals were introduced for the improvement of coexisted other wireless technologies. Seno et al. [2] considered wireless extensions of Ethernet Powerlink, which is a real-time ethernet network, executed using the IEEE 802.11 WLAN. They worked on hybrid networks, i.e., wired + wireless. They focused on a widespread network configuration, addressed two types of add-ons for gateway, and bridge devices and provided a perusal of hybrid networks focused on analyzing the most admissible performance indexes for real-time communication. Bibinagar and Kim [3] developed an architecture using multi-client–multi-server on LAN by applying user-datagram protocol as the communication protocol. They implemented real-time feedback control of various components connected to one or more control systems over the network. Suto, et al. [4] introduced a system that can take feedback in real-time and has the potential to collect, manage, and process the data in real time. They named it wireless computing system, which higher capability of collecting data and processing it. For real-time feedback, wireless computing system has a sustainable delay in data collection. Patti and Bello [5] proposed priority-aware multichannel adaptive framework which is based on the low latency deterministic network protocol. This model can work with low latency and can be applied to large networks also. It enhances network reliability and has the potential to handle real-time data traffic. FAS produces a huge amount of data that needs to be transmitted through large-scale wireless networks in real time. For that, Jin et al. [6] proposed a fusion of centralized and distributed data transmission schemes, i.e., hierarchical data transmission framework with all the advantages of both centralized and distributed data transmission schemes for the reliability and real-time performance of the FAS.

2.2 Breakdowns and Undiagnosed Bottlenecks

FAS aims to minimize human errors in a process. These errors prevent systems from achieving their maximum output and hamper their productivity. FAS helps overcome these delays by identifying situations like machine breakdowns before they occur. Automation provides solutions to avoid delays and breakdowns while keeping up the system's quality. Vrba and Marik [7] worked on a control system that can alter its behavior to cooperate with abnormal situations like brisk orders or breakdown in machinery hardware. They introduced the manufacturing agent simulation tool, which can cooperate with dynamic conditions in the material flow control system. Hoshino et al. [8] focused on problems related to batch production systems in which the handling and processing of the material by robots are to be done at a fixed time cycle. So, to resolve this bottleneck issue, they proposed opposite operational techniques for the proper coordination of the multi-robot system to operate relatively among themselves. Zhao et al. [9] introduced an aggregation-based iterative course of action to examine the downtime, setup-time reduction, and overall performance of the system. They recommended equipping indicators to correct current bottlenecks for machine and product types, with shorter setup times resulting in higher line production rates.

2.3 Coordination and Scheduling

In the past, scheduling has been a major issue in FAS. As factories become more reliant on their processes, scheduling and coordination of processes become increasingly important to adapt rapidly to changing market conditions. The rise of FAS technology has provided manufacturers with an opportunity to coordinate and schedule their process and control systems to increase their responsiveness. Effective coordination and scheduling approaches have been created to seize the opportunity. The essential concerns for scheduling and coordination are focused on architecture, solution concept, and scalability, and approaches for scheduling and coordination are presented. Hoshino et al. [8] focused on problems related to batch production systems in which the handling and processing of the material by robots are to be done at a fixed time cycle. So, to resolve this coordination and scheduling issue, they proposed opposite operational techniques for the proper coordination of the multi-robot system to operate relatively among themselves. Sanchez and Bucio [9] had their final assignment of the course of discrete event systems, i.e., to devise and impose a flexible hierarchical discrete-event controller for the coordination of FAS assembled by LEGO blocks. They introduced an approach to bind the nomenclature and the criteria of design. They also justified it by implementing their model to programmable logic controller.

2.4 *Safety in FAS*

FAS safety is critical since it protects human life, particularly in high industries like nuclear, aerospace, chemical, oil and gas, and mining, as mistakes might be fatal. Hence, safety in FAS plays a very crucial role. Some standards are made for safety in designing FASs. Sanchez and Bucio [10] worked on the IEC 61499 standard to enable the construction of complicated FAS based on models, where a prototype of the controlled physical processes known as a plant is co-developed with the controller. The safety guidelines provided by the IEC for managing and estimating the menace using quantitative and qualitative analysis techniques for both hardware and software, respectively. Jiang et al. [11] worked on FAS's stability and safety measures. They promoted the fault diagnosis and the monitoring of the system by a collection of data from previous operations and past observations online in real time. The collected data is used to analyze the fault and error detection, and to perform all the control tasks and operations safely and with stability in the current or future instance for making the FAS more reliable.

3 Technologies and Methodologies

3.1 *Adaptive Automation*

Depending on the operator's behavior and the system's workload, automation can adapt to changes in its level of control. Recent systems have started using psychophysiological measures and neuro-ergonomics approaches to decide on changes in automation levels. This way, FAS systems are more integrated into the environment, so they are seen as 'coworkers' rather than machinery, going over the normal human–computer interaction levels. Sheridan TB and Leitão et al. [12, 13] revisited various concepts to differentiate between supervisory control, allocation authority, adaptive control, and adaptive automation. They explained human supervisory adaptation from the control engineering outlook and suggested relevant classification for adaptive automation for both supervisory and direct control in a FAS. Sheridan TB [12] discussed the problems of industries embracing adaptive and more flexible systems to compete with opponents in technology. To solve this problem, they introduced a multiagent system technology that provides another way to fabricate these intricated systems by decentralizing the control system over distributed units. They focused on the integration of process control and quality of product and contributed to the hike in the factory profitability by employing self-adaption methods at global and local levels also.

3.2 *Human Collaboration and Interaction*

In FAS, human interaction in manufacturing systems is an important consideration in achieving higher flexibility and adaptability. Designing manufacturing systems including human operators is considered a valuable resource. Busogi and Kim [14] proposed an approach of human in the loop to scrutinize the complexity of the operator's choice in the mixed-model assembly line. They identified the important factors that influence the complexity of a human operator's decision. A simulation model was created with various degrees of choice complexity to train and evaluate the human reaction time. This model, along with a detailed case study, gave quantitative insight into evaluating how operator errors can be mitigated without affecting overall supply chain manufacturing. Steinmetz et al. [15] discussed the problem of applying suitable workflow among experts and shop floor workers in a FAS. They proposed a robot-level task programming framework named RAZER in which both parameter interface and proficient programming were blended. They coined the term of human–robot interface which is a graphical user interface running in a browser and providing access to several other man–machine interfaces. The proposed framework serves the workflow among both the experts and the shop floor workers for creating the skills and using the skills respectively. Small and medium-sized enterprises need to improve their competitive capacity to withstand the developed industries. While implementing FAS in those small- and medium-sized enterprises, human labor can be utilized for performing tasks that require capabilities like sensing, cognitive skills, and flexibility. Reimann and Sziebig [16] introduced the concept of intelligent factory space for interacting humans with the FAS. The IFS is comprised of different layers and various compatible components, according to the user's demand. Two-way communication provides feedback to human users from machines to industrial standards.

3.3 *Augmented Reality (AR)*

Augmented reality is being used to track changes in a system, detect machine breakdowns, and visualize a finished product among many other things. Manufacturers are using it to display text, statistics, and information relevant to the worker's current job in addition to digital letters, photos, or material. This technology is taking over traditional prototypes as they are very time-consuming. AR technologies must speed up the decision-making process and saved a lot of time taken to bring modifications to prototypes. By augmenting the task and simplifying coordination and communication between the parties, AR can help speed up this process. A corporate executive can use augmented reality to see the actual product being designed and built-in real time. AR can be used to provide remote help. Factory planning can benefit from AR. Data from digital planning can help make better decisions. Since augmented reality, smart glasses are gaining popularity and are considered a vital technology for shop floor workers in the smart FAS. Syberfeldt et al. [17] tried to facilitate and accelerate the espousal of this augmented reality, smart glasses technology by the FASs. A variety of products based on this augmented reality, smart glasses technology was available in the market, so it was difficult and time-consuming to select the best alternative. To overcome this problem, they introduced an efficient way of the step-by-step process for the evaluation of ARSG and listed the minimum values and guidelines to select the best product. With their contribution, it is easier for factories to make an optimal decision quickly and its implementation on the shop floors. AR, which when employed in an industrial environment is called industrial augmented reality, is such technology that can be applied in various circumstances to serve useful and attractive interfaces. Blanco-Novoa et al. [18] presented the state-of-the-art on industrial augmented reality implementation on shipbuilding and FAS. They listed the most suitable industrial augmented reality software and hardware tools which are based on fog computing architecture. Their input aided shipyard operators in obtaining information about their tasks and interacting with the environment.

3.4 Robotics

Robots are the reason behind the changing face of manufacturing and FAS. They were designed and deployed for changing the orientation and moving the materials and can also perform various programmed tasks in FAS. They were also deployed in some unfavorable environments which are unsuitable for humans and repetitional tasks which can cause boredom and lead to an accident due to the inattentiveness of humans (workers). Robots were used in FAS for maximizing the productivity and efficiency of the process or factory. Repetitive processes and applications with precision could only be accomplished using a robot, as this level of reliability is very difficult to achieve using any other methodology. Robots can be upgraded accordingly, and it also increases the workplace's safety. Fukukawa et al. [19] worked on peg-in-hole task (precise insertion processing) for efficient and flexible assembly in FAS. Many difficulties were faced by the robotic manipulator to insert the peg inside the hole. To overcome this difficulty, they proposed a technique based on the passive alignment principle to solve the problems of deformation and accuracy. This passive alignment principle technique corrects the ring position placement in the nano order and eliminates the ring deformation. Liu et al. and Von Drigalski et al. [20, 21] discussed the multi-robot pairwise transportation technique, in which two levels of hybrid planning methodology are proposed. One is a single robot-level planner based on the incidental delivery process, and the second is a group robot-level planner based on simulated tampering. The single-level planner configures each robot's transportation path and plans to reduce transportation costs, while the group-level planner uses predetermined actions to explore the task's solution space. In dynamic conditions, the multi-robot pairwise transportation methodology can be updated to tackle online task allocation or reallocation difficulties. Li and Savkin [22] proposed a self-navigating algorithm based on wireless sensor networks for micro-flying robots in FAS. Since the micro-flying robots cannot be deployed with some heavy obstacle sensors for navigating the paths. They introduced a three-dimensional range finder based on wireless sensor networks, used to trace the dynamic and static hindrances in the factory environment, and pilot micro flying robots to evade any collision.

3.5 Artificial Intelligence and Industrial Internet of Things (IIOT)

The IIoT is the most popular topic in both academia and the factories of Industry 4.0. In the IIoT system, the transmission of data with different delay flows among various smart components. A centralized edge computing and software-defined networks in IIoT of FAS are proposed in [23]. Authors subjugated the traditional method's limitations and for solving the problem of delays and latency in data transmission in IIoT system. IIoT is an unfolding technology that is distinct from the consumer IoT in properties, types of components employed, networking, and QoS requirements. A profound study on IIoT from the FAS perspective is presented by Xu et al. [24]. Authors explained every possible difference between IIoT and consumer Internet of Things in depth. In the modern FAS, the number of nodes increases, and the size of the network also became larger; thus, the conventional IIoT architecture no longer can bear such an extensive system. J. Wan, J. Li, M. Imran, D. Li, and Fazal-e-Amin, 2019 introduced a three-dimensional architecture for designing distributed networks to make the conventional IIoT system more effective. The authors analyzed the problems faced in the conventional IIoT system and proposed a model based on blockchain architecture for better privacy and security. They dismantled the traditional IIoT architecture to create a partially decentralized design for a new multi-center. Then, to strengthen and optimize the new architecture introduced certain essential security technologies. The IIoT is created to help FAS execute more agile and efficient automation and control while also improving energy efficiency. The negative consequences of exploiting unprotected IIoT have been frequently shown and publicly published. There have been numerous stories of invasive and hackable devices, with the well-known Stuxnet worm specifically targeting IIoT systems [25]. The aging process, defined as a modest degradation of the IIoT physical system, is proposed in [26]. In the framework of aging in IIoT, they presented a concept for diagnosing abnormal behavior in aging in IIoT.

3.6 Issues and Challenges in FAS

Evolution is the change of characteristics and properties over successive times. We have seen this evolution in the FAS also over a long period. Since we are surveying only the last decade that is from 2009 onwards, we will discuss only this timeline. Factory automation has progressed from basic pneumatic and hydraulic systems to today's advanced robotics. Most of the factory processes are automated to increase production and efficiency while lowering labor expenses. Since its introduction, FAS has made significant progress in automating activities that were previously performed manually. A factory that fully automates its processes with the latest technology produces high-quality products and has a high production rate, greater efficiency, and lower labor and production costs. The world benefits from high-quality goods and greater energy, resource, and raw material utilization because of the invention and evolution of FAS in manufacturing industries. The purpose of this state-of-the-art is to research and understand the evolution of automation in factories. So, for this understanding of evolution in FAS, we are taking it part by part, by categorizing the technologies and methodologies in different fields. For this categorization, we are using broad fields. We are dividing these fields of technologies because then we will get to know what happened in this evolution and what will be its scope by seeing the trends of these fields. We are categorizing the fields in:

- Software and programming up-gradation: In this field of categorization, we count all those papers and articles which are based on only software development and its up-gradation in the FAS. In this FAS, software and programming are the basic units. This software is the interface between the operator and the machines. As this software is becoming more advanced, the operators are also facing less difficulty, or we can say the operation process is being eased. By the programming language, the operators give the command to the machines; hence, we can say that the programming language is the language of command between the operator and the machines. To work accordingly, both the software and the programming are essential for the operator to operate the machines.
- Hardware and technical upgradation: In this field of categorization, we count all those papers and articles which are based on hardware and machinery enhancement in the FAS. In factories, machines and hardware play the most crucial role as they are the only tools that make the process done. By this FAS, we must automate this hardware part to work accordingly. So, the papers and articles reflecting the upgradation in hardware and machinery are in this categorization field.
- Optimization and energy (power) saving: In this field of categorization, we took those papers and articles which are based on optimizing the process and the technologies to save energy in the FAS. Since the factories consume a huge amount of energy to run, it is always tried to consume less power, and many technologies have been developed to do so. So, the papers and articles that belong to optimization and power saving come under this field of categorization.
- Wireless networks and communications (Connectivity): In this field of categorization, we look up the papers and articles which are on wireless networks and

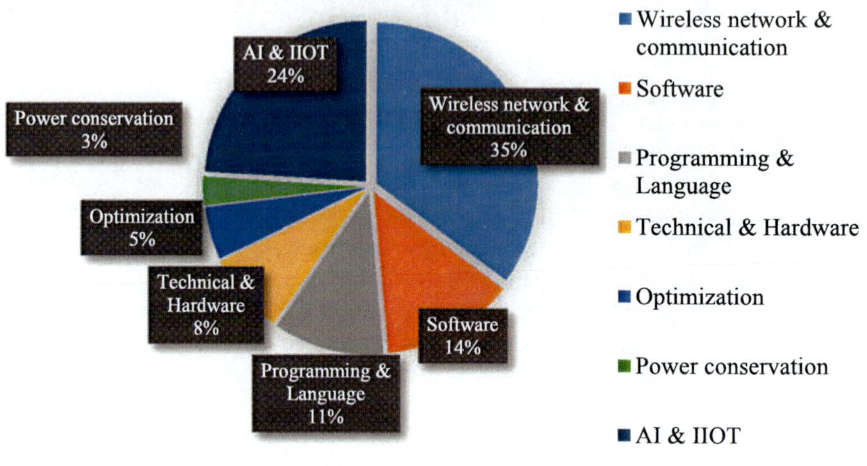

Fig. 1 Evolution of different fields in factory automation system

communications between all the components present in the factory. IIOT and other technologies also come under connectivity technology. For automating a factory, connectivity and communication among different components are necessary to work rhythmically

- According to our study, we prepared an evaluated chart that shows the evolution in different fields. By Fig. 1, the wireless network and communication are 46%, software and programming are 32%, hardware and machinery have 11%, and the optimization and power saving holds 11%. Therefore, we came to know that most of the evolution part is done in the wireless networks and the communication area is the connectivity part.

4 Conclusions

Application of FAS is the highest priority in the manufacturing industries as they increase productivity and reduces the manufacturing cost per product. However, this survey illustrated that the link between the evolution of FAS and the connectivity of components is deeper:

- Wireless network communication in factories has a high rate of evolution from the beginning; approximately half of the inventions, research, and articles are based on these fields of technology.
- As the basic field level devices, i.e., sensors, actuators, etc., are getting advanced and upgraded, accordingly the above controlling and governing levels are also getting advanced and automated.

- In the latest era, human–robot collaboration, AI-based systems, merging of OT and IT, and such integrated approaches are observed.
- FAS has no boundaries for the technology to dig in, and in the future, technologies like 5G connectivity will also rise.

All the above suggests strongly that the future of FAS will be more in the field of connectivity, integrated and adaptive automation, human–machine collaboration, and artificially intelligent systems. Technologies like 5G network and connectivity, holographic, and virtually designed architectures will also crop up shortly of FAS. Even though FAS has been attacked for potentially causing massive unemployment, the future of FAS is quite bright. Engineers are working to improve current industrial robots to produce future robots that can perform several tasks in a short period. As a result, a single machine can perform multiple industrial duties. Furthermore, industrial robots must have more human-like characteristics, such as the ability to make decisions and work independently. Self-diagnostic capabilities and predictive maintenance are also available. As FAS advances, future factories will be more efficient in their use of raw materials, human resources, and, most importantly, energy.

References

1. Kjellsson J, Vallestad AE, Steigmann R, Dzung D (2009) Integration of a wireless I/O interface for PROFIBUS and PROFINET for factory automation. IEEE Trans Ind Electron 56(10):4279–4287
2. Seno L, Vitturi S, Zunino C (2009) Analysis of Ethernet powerlink wireless extensions based on the IEEE 802.11 WLAN. IEEE Trans Ind Inf 5(2):86–98
3. Bibinagar N, Kim WJ (2011) Switched Ethernet-based real-time networked control system with multiple-client–server architecture. IEEE/ASME Trans Mechatron 18(1):104–112
4. Suto K, Nishiyama H, Kato N, Huang CW (2015) An energy-efficient and delay-aware wireless computing system for industrial wireless sensor networks. IEEE Access 3:1026–1035
5. Patti G, Bello LL (2016) A priority-aware multichannel adaptive framework for the IEEE 802.15. 4e-LLDN. IEEE Trans Ind Electron 63(10):6360–6370
6. Jin X, Kong F, Kong L, Wang H, Xia C, Zeng P, Deng Q (2017) A hierarchical data transmission framework for industrial wireless sensor and actuator networks. IEEE Trans Ind Inf 13(4):2019–2029
7. Vrba P, Marik V (2009) Capabilities of dynamic reconfiguration of multiagent-based industrial control systems. IEEE Trans Syst Man Cybern Part A: Syst Hum 40(2):213–223
8. Hoshino S, Seki H, Naka Y, Ota J (2010) Multirobot coordination for flexible batch manufacturing systems experiencing bottlenecks. IEEE Trans Autom Sci Eng 7(4):887–901
9. Zhao C, Li J, Huang N, Horst JA (2016) Flexible serial lines with setups: Analysis, improvement, and application. IEEE Robot Autom Lett 2(1):120–127
10. Sanchez A, Bucio J (2011) Improving the teaching of discrete-event control systems using a LEGO manufacturing prototype. IEEE Trans Educ 55(3):326–331
11. Jiang Y, Yin S, Kaynak O (2018) Data-driven monitoring and safety control of industrial cyber-physical systems: basics and beyond. IEEE Access 6:47374–47384
12. Sheridan TB (2011) Adaptive automation, level of automation, allocation authority, supervisory control, and adaptive control: Distinctions and modes of adaptation. IEEE Trans Syst Man Cybern Part A: Syst Hum 41(4):662–666

13. Leitão P, Rodrigues N, Turrin C, Pagani A (2015) Multiagent system integrating process and quality control in a factory producing laundry washing machines. IEEE Trans Ind Inf 11(4):879–886
14. Busogi M, Kim N (2017) Analytical modeling of human choice complexity in a mixed model assembly line using machine learning-based human in the loop simulation. IEEE Access 5:10434–10444
15. Steinmetz F, Wollschläger A, Weitschat R (2018) Razer—A HRI for visual task-level programming and intuitive skill parameterization. IEEE Robot Autom Lett 3(3):1362–1369
16. Reimann J, Sziebig G (2019) The intelligent factory space–a concept for observing, learning and communicating in the digitalized factory. IEEE Access 7:70891–70900
17. Syberfeldt A, Danielsson O, Gustavsson P (2017) Augmented reality smart glasses in the smart factory: product evaluation guidelines and review of available products. IEEE Access 5:9118–9130
18. Blanco-Novoa O, Fernandez-Carames TM, Fraga-Lamas P, Vilar-Montesinos MA (2018) A practical evaluation of commercial industrial augmented reality systems in an industry 4.0 shipyard. IEEE Access 6:8201–8218
19. Fukukawa T, Takahashi J, Fukuda T (2016) Precise assembly of ring part with optimized hollowed finger. Robomech Journal 3(1):1–8
20. Liu Z, Wang H, Chen W, Yu J, Chen J (2016) An incidental delivery-based method for resolving multirobot pairwised transportation problems. IEEE Trans Intell Transp Syst 17(7):1852–1866
21. Von Drigalski F, El Hafi L, Eljuri PMU, Ricardez GAG, Takamatsu J, Ogasawara T (2017) Vibration-reducing end effector for automation of drilling tasks in aircraft manufacturing. IEEE Robot Autom Lett 2(4):2316–2321
22. Li H, Savkin AV (2018) Wireless sensor network-based navigation of micro flying robots in the industrial internet of things. IEEE Trans Ind Inf 14(8):3524–3533
23. Li X, Li D, Wan J, Liu C, Imran M (2018) Adaptive transmission optimization in SDN-based industrial Internet of Things with edge computing. IEEE Internet Things J 5(3):1351–1360
24. Xu H, Yu W, Griffith D, Golmie N (2018) A survey on industrial Internet of Things: a cyber-physical systems perspective. IEEE Access 6:78238–78259
25. Wan J, Li J, Imran M, Li D (2019) A blockchain-based solution for enhancing security and privacy in smart factory. IEEE Trans Ind Inf 15(6):3652–3660
26. Genge B, Haller P, Enăchescu C (2019) Anomaly detection in aging industrial internet of things. IEEE Access 7:74217–74230

Investigation of Low-Cost IoT Device for Health Monitoring

Fariya Oyshi, Mushrafa Jahan Suha, Jawaad Rashid, and Farruk Ahmed

Abstract While the whole world has gone through the phase of fighting against the fatal coronavirus, monitoring so many COVID-19 patients admitted in the hospital or staying in home isolation has become a challenge. Moreover, many patients who are unable to commute or prefer to stay back home and take treatment from there finds it difficult to do so, as there is no system which could be used by the doctors to monitor patients remotely. In order to meet the challenge, we investigated a portable system which could monitor the health condition of a COVID-19 patient by measuring temperature, blood oxygen saturation, and heart rate of the patient, and provided live data through cloud feed as well as data logging. We analyzed the feasibility of such a device that can also help fighting upcoming hazards from outbreak of similar pandemic. Besides, we provide design implications for future research and conclude with discussing our limitations.

Keywords IoT · Remote health monitoring · COVID-19 patient monitoring

1 Introduction

Wireless technology has advanced in recent years to meet the needs of numerous industries [13, 32]. In recent years, IoT has dominated the industrial sector, particularly sensing, automation, and control [7, 20, 29]. One current trend in providing better health care and biomedical sensing [1, 4, 7]. IoT technology has opened up not only hospitals but also personal healthcare facilities. As a result of having a smart system, numerous metrics that use power, cost, and boost efficiency are detected. While the doctors can barely afford monitoring every patient admitted in a hospital, they can hardly support patients under home quarantine. Since IoT enables mass sensing with high reliability and low latency, one possible solution is the utilization of IoT-based solution to assist doctors in monitoring high influx of patients in such

F. Oyshi (✉) · M. Jahan Suha · J. Rashid · F. Ahmed
Independent University Bangladesh, Dhaka 1229, Bangladesh
e-mail: oyshifariya.fo@gmail.com

S. Das et al. (eds.), *Advances in Data-driven Computing and Intelligent Systems*, Lecture Notes in Networks and Systems 653, https://doi.org/10.1007/978-981-99-0981-0_34

a situation [12, 19]. However, IoT-based solution requires a complex set of prerequisite and system implementation. As a result, while top-notch institutes have the opportunity to implement such a system, we do not observe mass implementation of such systems in other cases.

With a mitigation plan in mind, we aim to incorporate a low-cost IoT-based health monitoring system. The goal of this work is to investigate the implementation and feasibility of a low-cost IoT-based health monitoring device to assist doctors in assessing pandemic affected patients, and with COVID-19 pandemic as a case study we implement heart rate, oxygen saturation, and temperature sensors [8, 16, 18] to assist the observation of the patient's condition as well as remote monitoring.

We developed a complete end-to-end system with commercially available, low cost, and easy to access sensors, controller, and software solution. Such a device can not only be handy at current pressing situation, but also can be utilized with a similar topology for future solution of similar pandemic situation.

2 Background Study

Our work cuts the field of IoT sensing and remote monitoring, and health assessment and monitoring sensors. We discuss the scope of our work in conjunction and contrast with prior works in the following subsections.

2.1 IoT Sensing and Remote Monitoring

Several prior works fused Android and IoT-based systems to perform sensing and monitoring of different parameters over the cloud, for instance gardening and automating the watering process [2, 10]. Thamaraimanalan et al. [27] used NodeMcu[1] to connect different sensors which collects the parameters of soil condition such as temperature, moisture and humidity, and transfer the information to Firebase[2] through inbuilt Wi-Fi. Sharma and Aarthy [23] used cloud feed to add RFID in attendance system. They combined the RFID with IoT when there is no need to take attendance by the lecturer, and cloud service is used here as storage for better performance.

Parmar et al. [21], Kiruthika and Umamakeswari [14], and Sunny et al. [25] proposed low cost and portable environment monitoring system. Chowdhury et al. [5] proposed a system by exploiting IoT architecture to monitor the current weather and air quality of the surrounding environment with a mobile and web application. The device can store data for an ample amount of time for future analysis of the changes over time. Moparthi et al. [17] used an IoT system to monitor water quality.

[1] https://www.nodemcu.com/.

[2] https://firebase.google.com/.

The device can measure the pH level and other irregularity. They used it in the various types of water resources and reservoir using the Arduino board and GSM module for message passing in addition to using LED display to have continuous observation on water parameters. The sensor data has been sent to the cloud for global monitoring of water quality. Our work is inspired by such prior works and similarly provisions a low cost and easy to implement IoT system. The concept of data classification and aggregations is utilized to avoid clogging the network with unnecessary data traffic [24]. A delay-aware routing metric is proposed to be used by the local cloud in its multi-hop communication to maintain the nature of the plants by continuously monitoring the parameters leading to the increased life of both plants and human being [3].

2.2 *Health Assessment and Monitoring System*

Wireless Body Area Networks (WBANs) [11, 30, 31] are one of the options for the remote health monitoring. However, such systems suffer from limitations due to the requirement of processing large amount data. Hence, WBANs are used with cloud computing which provides effective solutions and promote the performance [3] by dividing the cloud into local one, that includes the monitored users and local medical staff and global one that includes the outer world.

Gupta et al. [6] used a deep learning-based approach for ensuring social distancing in COVID-19 pandemic situation. Vaishya et al. [28] amalgamated IoT and AI system to provide seven critical implications for fighting such a situation. Kumar et al. [15] and Hassanalieragh et al. [7] proposed a smart device-based IoT health monitoring system for checking the status of SpO_2, ECG, heart rate, etc. However, such devices are not abundant and hard to integrate due to the cost and complexity of the system. Reference [9] is an existing work on wearable monitoring devices and respiratory support systems which are used to assist the coronavirus affected people discusses that the portable future trends is also down to select the best technology for COVID-19 affected people, and this work also analyzes the services they provide and the working procedure of the present situation of the merits and demerits with costs.

In a similar vein and with an attempt to resolve the system complexity and plausible solution for health monitoring, we propose a low cost, portable, and smart health monitoring system exploiting IoT infrastructure.

3 Methodology

We describe our methodology in the following subsections.

3.1 Sensor Reasoning

Fever is a major factor present in COVID-19 patients and is correlated to the severity of the inflammation of lungs [18]. Hence, temperature data is an essential for monitoring the health condition of a COVID-19 patient as well as lung infected patient. Oxygen saturation percentage is another crucial factor for continuous monitoring of COVID-19 patients [16] as well as severe lung infected patient and also for determining the type of care a patient requires [22]. Heart rate indicates the health condition of a patient suffering from severe lung infection and COVID-19 infected patient [8]. As a result, our system uses a temperature sensor and a pulse oximeter sensor to monitor the parameters—temperature, SpO_2, and heart rate, the three most critical factors for monitoring such a patient.

3.2 System Architecture

Our system is created to assist doctors in checking the patient's live report from any location utilizing an IoT feed. We utilized an oximeter, a temperature sensor, and a NodeMCU to create this device. To take the reading, the patient must first place their finger in the sensor, following which the sensor will detect the patient's heart rate, oxygen saturation, and temperature, which will be displayed on the onboard display and stored in the meantime on a web server. The live graph in the IoT feed can then be viewed by medical personnel. Figure 1 shows the overall system architecture.

3.3 Anomaly Detection

There is a fixed range of temperature, SpO_2 saturation, and BPM which is regarded as being the normal range, anything out of that range indicates deterioration of health condition. Monitoring of SpO_2 saturation is necessary for lung infected patients and using this system if the level goes below the fixed threshold, then a warning is generated. Similarly, there is a fixed threshold for temp and if temperature increases

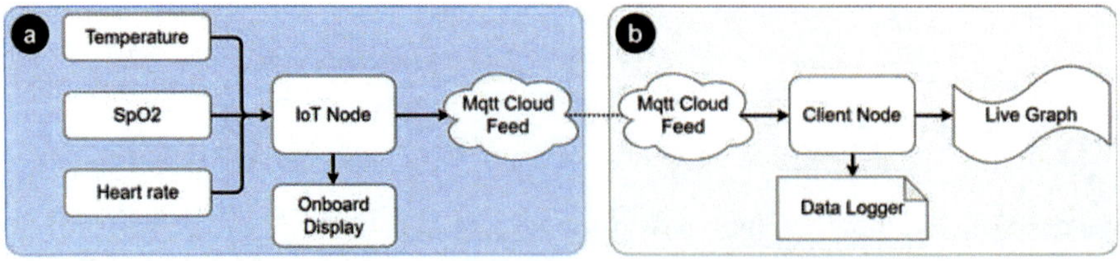

Fig. 1 System architecture **a** IoT node, and **b** client node

above that threshold, it refers to a hazardous scenario. Given that the data, the onboard controller and the data transferred over the cloud condition the threshold and hence an anomaly is detected using the thresholding method.

4 Implementation

In order to give functionality and utility to the user, IoT technology requires communication with the server. Various connection systems connect the server and IoT module to the Internet/cloud. AdafruitIo has been chosen as our cloud server because it eliminates the need for the developer to run or manage code manually. It is accessible over the internet to the user/developer. Its fundamental function is to store and retrieve data.

4.1 Hardware

The COVID-19 patient monitoring system necessitates the collection of several data points at the same time. We use a temperature sensor and a pulse oximeter, which combines pulse oximetry and heart rate monitor sensors. These are standardized off-the-shelf sensors that can be utilized with a variety of systems. The following is a list of sensors and their capabilities.

4.1.1 IoT Node (ESP32)

The open source low value NodeMCU has been accustomed sending the sensing element information to the cloud feed. The ESP8266, designed and created by Espressif Systems three, contains the crucial components of a computer: mainframe, RAM, networking (Wi-Fi), and even a recent code and SDK. It is a good choice for our projected system as, the firmware may be tailored according to the need.

4.1.2 MAX30100 Pulse Oximeter

The sensing element combines a pulse measuring instrument and heart rate monitor sensing element in one package. It detects pulse and pulse rate signals using two LEDs, a photo detector, improved optics, and low-noise analog signal process. It runs on one .8 to 3.3 V power sources and may be turned off by software package with little or no standby current, permitting the ability provide to be connected in the slightest degree times.

4.1.3 Temperature Sensor (DSB 18B20)

The DS18B20 temperature sensing element is associated with one-wire digital temperature sensing element. With the assistance of DS18B20 temperature sensor, we will live the temperature from -55 to 125 °C with accuracy of ± 5 °C.

4.1.4 Battery Management System (TP4056)

It is a complete constant-current and constant-voltage linear charger for single cell lithium-ion batteries. Its low external component count makes the TP4056 ideally suited to portable applications. Furthermore, the TP4056 can work using a USB and wall adapter. So, this has been chosen ideally for managing and charging the facility source of the system proposed by us.

4.1.5 Power Source (18650 Li-ion)

We utilize the popular and commercially available 18650 Li-ion battery. This battery supports the device during any interruption from the power outlet. Besides, the battery has the capacity to run the device up to 6 h without being re-charged. With improved sleep mode, this duration can also be increased in the future.

4.2 Software

4.2.1 Mqtt Brocker (AdafruitIo)

The data collected through the sensors is displayed on a live graph which is obtained from and stored in AdafruitIo.[3]

4.2.2 API (Python Bokeh)

Python Bokeh[4] API has been used to implement the live graph capturing the sensor data sent via NodeMCU.

4.2.3 Sensor Data Acquisition

The sensor data acquisition pseudocode can be found as depicted in Algorithm 1.

---

[3] https://io.adafruit.com/.

[4] https://bokeh.org/.

Algorithm 1: Senor data acquisition

```
initialize sensor and mqtt;
NULL-STATE = True;
while True do
    mqtt.run();
    if NULL-STATE is True then
        SampleCount = 0;
        wait until the sensor begins;
        update sensor data;
        if data is valid then
            set NULL-STATE to False;
        end
    else
        if CurrentTime - LastReportTime > REPORT-PERIOD and SampleCount >=
         MINIMAL-SAMPLE then
            SPO2 = sensor.getSpO2();
            HBEAT = sensor.getHeartRate();
            TEMP = sensor.getTemperature();
            print Sensor data in local LCD;
            mqtt.publish(Sensor data);
            LastReportTime = CurrentTime;
            SampleCount = 0;
        else
            update sensor data;
            if data is invalid then
                set NULL-STATE to True;
                goto NULL-STATE;
            end
            SampleCount++;
        end
    end
end
```

5 Result and Analysis

5.1 Feasibility

To take the reading, the patients must first place their finger in the sensor. Following which the sensor detects the patient's heart rate, oxygen saturation, and the temperature. This device has been tested through constant operation. The device takes three readings to confirm the exact measurement. The output shows after taking three successful inputs. First, it shows data in the LCD screen (see Fig. 4) and stored in the meantime on a web server. Then, it shows in the Adafruit screen (see Fig. 4). When the user interface is loaded, the live graph is displayed on the screen (see Fig. 2). The PCB design (see Fig. 3) is user friendly and also it is not heavy to carry anywhere.

Fig. 2 **a** PCB design and simulation for the test device. **b** Fabrication of the circuit. **c** Fabricated and assembled device

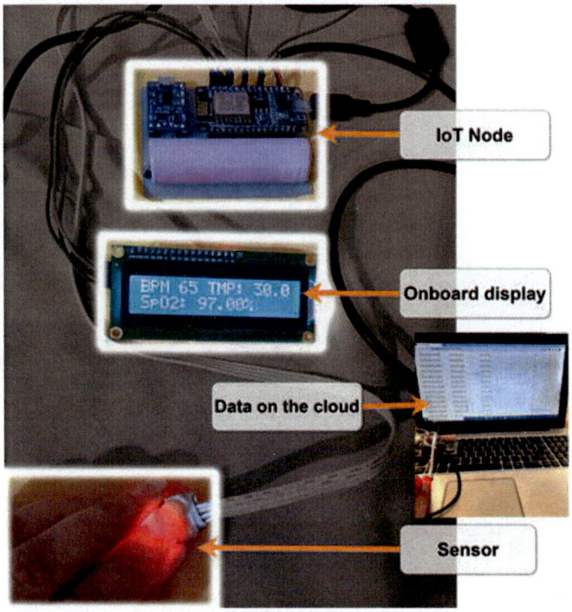

Fig. 3 Capturing data from the device and transmitting over the cloud server

Patients can move with this. The security of the web server is strong. Doctors can check their patients from any location utilizing an IoT feed.

5.2 *Cost Estimation*

Our system has been developed using commercially available but low-cost components, which is under $20. With fabrication and other passive element, we project the cost to $30. The cost breakdown has been given in (see Table 1) which displays that the costs of the components are as low as 1 dollar and a maximum of 6 dollar.

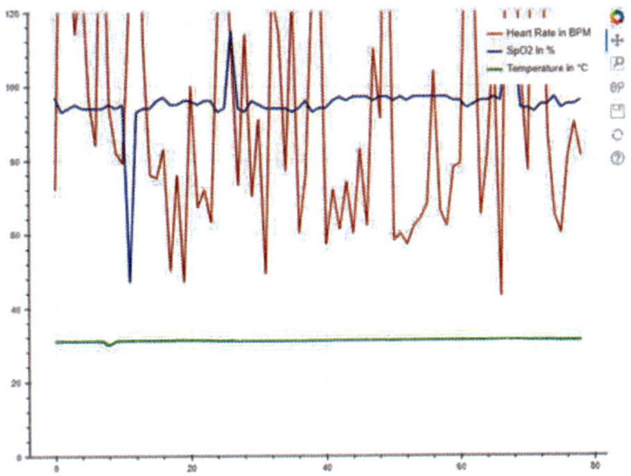

Fig. 4 Live graph

Table 1 Cost estimation of the fabricated device

	Component name	Cost (in USD)
i	DSB 18B20	3
ii	MAX30100 pulse oximeter	4.5
iii	NodeMCU ESP32	6
iv	BMS TP4056	1
v	Li-ion 18650	3
	Total cost	17.5

5.3 *Use-Case Scenario*

For any viral lung infected disease, a patient can be monitored with such a low-cost solution. This is a handy system, so patients can also monitor oneself at their home. This system suggests which patient needs more attention and a better monitoring approach for the doctors from a remote location.

5.4 *Design Implication*

Through our work, we investigated several design implications that can potentially help future researchers to find similar IoT-based solution.

- Reliability: Reliability is a crucial factor in sensitive issues like health monitoring. Before deploying a healthcare technology, the sensitivity and reliability of the device must be assessed during its clinical trial period [26]. While high-tech systems are being used for health monitoring, the use of low-cost solution could bring about so much disturbances in the reliability of the output data to be used.

- Latency: In this device on of the main sources of latency is the pulse oximeter which takes three inputs before displaying an output. Other sources of latency could be while the data is being received by the NodeMCU and send to the cloud feed.
- User Interface: A user friendly interface is an important factor in a health monitoring system, as it would not only save time but also will also help the doctors retrieve any necessary past data to reach pivotal decisions in the earliest manner.
- Storage: Larger storage is required to store a huge set of data acquired from a hospital which may contain a huge crowd of patient during the outbreak of such pandemic.

5.5 Limitations

The primary limitation of this work is that, it has not yet been implemented to take data from multiple patients. Also, it has not been tested on a large group of people to insure it's accuracy.

6 Conclusion and Future Work

A solution to one of the most challenging tasks of monitoring patients has been proposed through this research work. The portable device is able to measure temperature, blood oxygen saturation, and heart beat rate of a patient successfully and display it on the LCD screen of the device. The data is being sent to the cloud feed where it is stored and can be downloaded later on. The doctor can also view the data in the form of a graph in a web API. The doctor can be safe from any kind of virus attack or diseases for the benefit of the patient. The device is portable enough, hence, a patient staying in home isolation can also use this device to check all the three necessary parameters very easily. For future work, this system could be updated by adding more functionality for monitoring patients suffering from other health issues.

Acknowledgements We are thankful to Independent University Bangladesh for all the technical and financial support.

References

1. Aktas F, Ceken C, Erdemli YE (2018) IoT-based healthcare framework for biomedical applications. J Med Biol Eng 38(6):966–979
2. Ali M et al (2020) IoT based smart garden monitoring system using NodeMCU microcontroller. Int J Adv Appl Sci 7(8):117–124

3. Almashaqbeh G et al (2014) QoS-aware health monitoring system using cloud-based WBANs. J Med Syst 38(10):1–20
4. Banerjee A et al (2020) Emerging trends in IoT and big data analytics for biomedical and health care technologies. In: Handbook of data science approaches for biomedical engineering, pp 121–152
5. Chowdhury AH, Al Arabi A, Amin MA (2019) In search of a low-cost IoT system for real-time environment monitoring. In: 2019 IEEE international conference on robotics, automation, artificial intelligence and internet-of-things (RAAICON). IEEE, pp 87–92
6. Gupta M, Abdelsalam M, Mittal S (2020) Enabling and enforcing social distancing measures using smart city and its infrastructures: a COVID-19 use case. arXiv preprint arXiv:2004.09246
7. Hassanalieragh M (2015) Health monitoring and management using Internet-of-Things (IoT) sensing with cloud-based processing: opportunities and challenges. In: IEEE international conference on services computing. IEEE, pp 285–292
8. Hasty F et al (2021) Heart rate variability as a possible predictive marker for acute inflammatory response in COVID-19 patients. Milit Med 186(1–2):e34–e38
9. Islam M et al (2020) Wearable technology to assist the patients infected with novel coronavirus (COVID-19). SN Comput Sci 1(6):1–9
10. Jain RK et al (2020) IOT enabled smart drip irrigation system using web/Android applications. In: 2020 11th international conference on computing, communication and networking technologies (ICCCNT). IEEE, pp 1–6
11. Jegadeesan S et al (2020) EPAW: efficient privacy preserving anonymous mutual authentication scheme for wireless body area networks (WBANs). IEEE Access 8:48576–48586
12. Kamal M, Aljohani A, Alanazi E (2020) IoT meets COVID-19: status, challenges, and opportunities. arXiv preprint arXiv:2007.12268
13. Katiyar V, Chand N, Chauhan N (2010) Recent advances and future trends in wireless sensor networks. Int J Appl Eng Res 1(3):330
14. Kiruthika R, Umamakeswari A (2017) Low cost pollution control and air quality monitoring system using Raspberry Pi for Internet of Things. In: International conference on energy, communication, data analytics and soft computing (ICECDS). IEEE, pp 2319–2326
15. Kumar S et al (2020) A wristwatch-based wireless sensor platform for IoT health monitoring applications. Sensors 20(6):1675
16. Michard F, Shelley K, L'Her E (2021) COVID-19: pulse oximeters in the spotlight
17. Moparthi NR, Mukesh Ch, Vidya Sagar P (2018) Water quality monitoring system using IoT. In: 2018 fourth international conference on advances in electrical, electronics, information, communication and bio-informatics (AEEICB). IEEE, pp 1–5
18. Motta LP et al (2020) An emergency system for monitoring pulse oximetry, peak expiratory flow and body temperature of patients with COVID-19 at home: development and preliminary application. medRxiv
19. Ndiaye M et al (2020) IoT in the wake of COVID-19: a survey on contributions, challenges and evolution. IEEE Access 8:186821–186839
20. Neagu G (2017) A cloud-IoT based sensing service for health monitoring. In: E-Health and bioengineering conference (EHB). IEEE, pp 53–56
21. Parmar G, Lakhani S, Chattopadhyay MK (2017) An IoT based low cost air pollution monitoring system. In: 2017 international conference on recent innovations in signal processing and embedded systems (RISE). IEEE, pp 524–528
22. Shah S et al (2020) Novel use of home pulse oximetry monitoring in COVID-19 patients discharged from the emergency department identifies need for hospitalization. Acad Emerg Med 27(8):681–692
23. Sharma T, Aarthy SL (2016) An automatic attendance monitoring system using RFID and IOT using Cloud. In: 2016 online international conference on green engineering and technologies (IC-GET). IEEE, pp 1–4
24. Smys S, Kumar AD (2016) Secured WBANs for pervasive m-healthcare social networks. In: 2016 10th international conference on intelligent systems and control (ISCO). IEEE, pp 1–4

25. Sunny AI et al (2020) Low-cost IoT-based sensor system: a case study on harsh environmental monitoring. Sensors 21(1):214
26. Thabane L et al (2013) A tutorial on sensitivity analyses in clinical trials: the what, why, when and how. BMC Med Res Methodol 13(1):1–12
27. Thamaraimanalan T et al (2018) Smart garden monitoring system using IoT. Asian J Appl Sci Technol (AJAST) 2(2):186–192
28. Vaishya R et al (2020) Artificial intelligence (AI) applications for COVID-19 pandemic. Diabetes Metab Syndr: Clin Res Rev 14(4):337–339
29. Verma A et al (2019) Sensing, controlling, and IoT infrastructure in smart building: a review. IEEE Sens J 19(20):9036–9046
30. Yan S, Soh PJ, Vandenbosch GA (2018) Wearable ultrawideband technology—a review of ultrawideband antennas, propagation channels, and applications in wireless body area networks. IEEE Access 6:42177–42185
31. Yeh K-H (2016) A secure IoT-based healthcare system with body sensor networks. IEEE Access 4:10288–10299
32. Yen DC, Chou DC (2001) Wireless communication: the next wave of Internet technology. Technol Soc 23(2):217–226

Blockchain of Medical Things: Security Challenges and Applications

Namrata Singh and Ayan Kumar Das

Abstract Blockchain of Medical Things is the fusion of blockchain technology and Internet of Medical Things that introduces smart contracts in the network enabling medical devices to function anonymously in a secured way by leveraging cryptographic techniques. Internet of Medical Things integrated with the cloud environment has a centralized architecture that suffers with single point of failure. The privacy and authentication of data are also the major threats to the Internet of Medical Things. It is prone to various security attacks categorized as physical attacks, network attacks and software attacks. Blockchain design, its architecture, consensus methods and cryptographic techniques are supplemented with the Internet of Medical Things to deal with security issues in healthcare system. This integration is a favourable and distributed approach to provide transparency, privacy, immutability and failure resilience in remote healthcare monitoring. This paper presents a systematic review of Internet of Medical Things, security requirements and security attacks. An integrated architecture of blockchain and Internet of Medical Things, named as blockchain of Medical Things, is presented after a detailed analysis of blockchain technology for the purpose. The consensus algorithms, its compatibility with the Internet of Medical Things and its applicability for different security attacks has been illustrated. The paper also contains the applications of Blockchain of Medical Things in electronic health records, pharmaceutical supply chain and vaccine management for better administration of healthcare data.

Keywords Blockchain · Internet of things · Security attacks · Healthcare

N. Singh (✉) · A. K. Das
Birla Institute of Technology, Mesra, Patna Campus, Patna 800014, India
e-mail: phdcs10061.20@bitmesra.ac.in

A. K. Das
e-mail: das.ayan@bitmesra.ac.in

S. Das et al. (eds.), *Advances in Data-driven Computing and Intelligent Systems*, Lecture Notes in Networks and Systems 653, https://doi.org/10.1007/978-981-99-0981-0_35

1 Introduction

The healthcare is very crucial part of everyone's life. High quality healthcare services help in preventing the diseases and improving the quality of life. Healthcare industry is directly connected with the social welfare and lives of people. The population of developing countries, particularly in India, mostly lives in rural areas where the healthcare services are not accessible in real-time basis. Although, healthcare systems have experienced a significant growth, early diagnosis and cure of diseases is not possible in remote areas. Budida and Mangrulkar [1] specify the requirement of continuous and regular health monitoring for early prediction of diseases as it is considered a best way for early diagnosis and cure. Baker et al. [2] analysed that healthcare systems has become overcrowded with the increase of population and chronic diseases. It requires a large number of healthcare providers such as doctors, nurses and medical staffs as well as healthcare resources such as hospital beds, equipment for assessing patient health condition, drugs and vaccines. People's healthcare requirements are met through the creation and design of cost-effective smart healthcare monitoring systems and industries with advancement of technology. Baker et al. [2] states that Internet of Things (IoT) has been shown as an effective solution to reduce the burden on a healthcare system and lowering the costs. According to Yaacoub et al. [3], IoT is the Internet-based interconnection of multiple devices that allows for data gathering and sharing inside the network. Anything is regarded as a smart object in IoT that is connected to other objects through Internet.

Healthcare industry integrated with IoT evolves a new terminology namely Internet of Medical Things (IoMT). Yaacoub et al. [3] and Gatouillat et al. [4] describe IoMT as a network that connects a range of personal medical devices, servers and healthcare providers such as hospitals, doctors and private firms. IoMT comprises of wearable and non-wearable sensor devices for medical data collection. It consists of gateways to provide interconnection among different medical devices and servers present in the network. The servers are to provide storage of data, computation required and analysis of unusual behaviour. The interconnected and distributed network of IoMT demonstrates an excellent solution for continuous and regular health monitoring. It reduces doctor visits, which is especially beneficial for elderly people who are unable to see a doctor on a daily basis. Budida and Mangrulkar [1] demonstrates that IoMT is primarily focussed on remote health monitoring of different health parameters such as body temperature, blood glucose level, heart rate and other such health conditions. IoMT has a number of potentials in healthcare such as reduced medical cost, improved diagnosis and treatment, elderly care and vaccine management. However, it consists of a number of flaws as well in terms of security and privacy. This paper first describes the security issues of IoMT systems illustrating various security requirements as well as possible security attacks. The security techniques to deal with various types IoMT security attacks is then described. This is followed by illustrating an integrated environment of blockchain with in order to overcome the challenges of IoMT. The main objective of this review work is not only to provide different security challenges of IoMT and blockchain-based solutions, but also to analyse the compatibility of different consensus mechanism with healthcare systems. This papers also demonstrates different blockchain-based IoMT applications such as electronic health record management, organ donation and transplantation, pharmaceutical supply chain and vaccine management.

2 Security Issues of IoMT

The record and data shared among smart devices in this Internet era and the devices itself are very prone to malicious attacks. Papaioannou et al. [5] illustrated the security issues of the IoMT architecture as it is centralized and is vulnerable to many security threats. Wazid et al. [6] described another security risk with the IoMT as unauthorized access by an attacker without being discovered. Unauthorized access to this type of database can result in the theft of sensitive and confidential medical data and also their identity. Attackers can also tamper with the data resulting in the incorrect diagnosis and medication to the patients which may also lead to their death. Yaacoub et al. [3] considers security as one of the most important requirements that cannot be compromised. Gatouillat et al. [4] states that explicit verification and validation techniques are required to be incorporated with IoMT that has unique requirements associated with security and privacy due to additional legal requirements to protect medical information of a patient. Moreover, Papaioannou et al. [5], Wazid et al. [6] and Ghubaish et al. [7] describe the basic requirements related to security in IoMT as: (a) *Authentication*—It means verifying the user's identity, i.e. identity of patients and healthcare providers before giving access to the healthcare system. (b) *Authorization*—It refers to the user's access for any network services or information such as access for IoMT devices or collected medical data of any patient. (c) *Confidentiality*—The patient medical data shared to the physician, therapist and other medical practitioners are very sensitive data and must not be disclosed to any third unauthorized party. (d) *Non-repudiation*—It means that each authorized user is accountable for his or her activities. It prevents an authorized entity from denying previous commitments or actions in a communication. (e) *Integrity*—In IoMT perspective, integrity preserves the accuracy of patient personal medical data, health record, clinical prescriptions and test reports. (f) *Anonymity*—It means keeping the identities of the doctors and patients anonymous or private from the unauthorized users. In IoMT, there are three types of possible security attacks classified based on the attacking behaviour provided in Table 1 and described as below.

2.1 Physical Attacks

The hardware components of the IoMT environment are the target of these type of attacks. The attacker should be close or inside to the IoMT system for successful

Table 1 Security attacks of IoMT and security techniques

Authors	Security attacks	IoMT layer	Attack behaviour	Security techniques
Yanambaka et al. [8]	Node tampering and injection of malicious code	Perception layer	Obtain access to the confidential communication of users	PMsec: physical unclonable function (PUF) for authentication of medical devices
Porambage et al. [10]	Injection of fake node	Perception layer	Control the flow of information	PAuthKey technique
Sicari et al. [11]	Permanent denial of service (PDoS)	Perception layer	Destruction of the network or the system	Middleware named as networked smart object (NOS)
Hei et al. [12]	Sleep denial attack	Perception layer	Shutdown of the system or the node	Support vector machine (SVM)
Airehrour et al. [13]	Sybil attack	Network layer	Network performance reduction	SecTrust RPL
Liu et al. [14]	Traffic analysis	Network layer	Breach of the confidential and sensitive data	Privacy preserving traffic obfuscation framework
Cervantes et al. [15]	Sinkhole attack	Network layer	Breach of the confidential and sensitive data	INTI (intrusion detection system) using concept of hash chain
Shukla [16]	Wormhole attack	Network layer	Diverting data packets towards themselves by showing low latency	Intrusion detection system using machine learning algorithms
Yin et al. [17]	DoS /DDoS	Network layer	Crash or shutdown of the server or system	The SDx paradigm with cosine similarity of vectors technique
Liu et al. [18]	Spyware, Trojan horse, virus, worm, adware	Application layer	Destroy the network or system by increasing traffic	mutual auditing scheme using lightweight and distributed framework

execution of physical attack. Different types of physical attacks have been described as follows:

Node tampering: Yanamka et al. [8] defines tampering as the direct altering of an equipment or a communication link is known as tampering. An attacker destroys the sensor node by manually replacing all or part of the node hardware and acquires the access of sensitive information of users such as cryptographic keys used for encryption and decryption of confidential data.

Injection of fake node: Ratta et al. [9] illustrates that this type of is performed by inserting or placing one or more nodes between the two nodes by an attacker in the IoMT system. The data flows between the two nodes under the control of attacker as data flows via the malicious nodes. The attacker either tampers the data or just obtain access to the confidential information without tampering. The nodes of IoMT system are completely unaware about this malicious activity and trust that the data is coming from the source node. Porambage et al. [10] named this attack as man-in-the-middle attack.

Permanent Denial of Service (PDoS): Papaioannou et al. [5] and Sicari et al. [11] describes PDoS as a sort of Denial of Service (DoS) attack in which an IoMT device is entirely destroyed due to hardware tampering. The assault is launched by utilizing malware to delete firmware or upload a corrupt Basic Input or Output System (BIOS) that is used to bootstrap the system. The sufferer has no choice in this case, but to repair or replace the equipment in order to resume normal functioning.

Sleep denial attack: Hei et al. [12] states that the sensors used in the IoMT system are largely powered by batteries. The batteries of sensor nodes are set to rest when they are not in use to extend the battery life. The attackers in this type of attack keep the nodes busy by providing incorrect inputs which causes the nodes to use more power and shut down.

2.2 *Network Attacks*

Network attacks are focussed on the Internet of Medical things (IoMT) system network and the attacker can be away from the system in order to execute the attack on the system or network. Network attacks are illustrated as follows:

Sybil attack: Airehrour et al. [13] demonstrate that one malicious node acquires numerous identities in this type of attack and traverses throughout the network. This node is known as Sybil node. The attacker is able to gain excessive control over the network as sybil nodes are like a mask that appears to be honest users. The presence of destructive behaviour in a network can have an impact on data integrity, resource consumption and also on the entire system performance. Yanamka et al. [8] has presented its results in a critical security defence in place to ensure that the IoMT system continues to operate effectively and efficiently.

Traffic analysis: Liu et al. [14] defines traffic analysis attack. In this type of attack, the attacker continuously monitors the traffic flowing through the network with the aim to discover communication patterns and the users involved while monitoring in order to learn about the network thoroughly. They launch an attack on the system once they have obtained all of the necessary information. Sicari et al. [11] describes that monitoring is done using applications such as port scanning. Port scanning applications are used to see which ports are active.

Sinkhole attack: Cervantes et al. [15] defines sinkhole attacks that they are particularly dangerous because they make the hacked node appear desirable in terms of routing algorithms, implying that it is the shortest path. It stops the base station from receiving precise and correct sensitive data by diverting traffic away from the base station towards itself. Sicari et al. [11] analysed that this sort of attack compromises data confidentiality and disables network service by discarding all the data packets instead of sending those to the final destination.

Wormhole attack: Shukla [16] illustrated that an attacker creates a low latency connection with the intent of routing data packets from one point to another to perform wormhole attack.

Denial of Service (DoS)/Distributed Denial of Service (DDoS): Ghubaish et al. [7] and Yin et al. [17] presented that the fundamental goal of a DoS attack is to make the server of a given target unavailable. This is accomplished by sending massive amounts of traffic from multiple systems. This massive traffic keeps the server constantly busy serving fraudulent or fake requests. This massive traffic requires a lot of processing power and computational power which leads the server to be slow down or even crash in some of the cases.

2.3 Software Attacks

In software attack, attacker uses the software or security vulnerabilities exposed by an IoT system to perform the attack. Software attacks are carried out using malwares. Malware is a malicious software that is intended to harm or destroy the network or computer. Liu et al. [18] classified the Malwares as:

Spyware: It is a kind of malware that monitors user actions without their permission. It takes advantage of software flaws and attaches itself to a regular programme.

Trojan Horse: It is a malware that mimics a genuine programme in attempt to fool people into downloading and installing it on their systems. It gains access to the system and confidential data once it has been downloaded. It also uploads other viruses into the system in order to carry out destructive tasks.

Virus: It has the ability to replicate itself and propagate to other machines. It is used by hackers to infect multiple nodes in the network. They use these devices to perform a DDoS attack on a target after they have gained access to them.

Worm: It is a harmful software that multiplies itself and spreads copies throughout its network. Attackers frequently exploit this to steal personal information, erase files, or set up a botnet.

Adware: It resides in the system and displays advertisements. Some adware tracks the activity in order to serve the personalized advertisements to the users.

3 Blockchain of Medical Things (BoMT)

Blockchain is a novel and evolving technology that can be used in conjunction with IoMT to increase IoMT security. Blockchain is a relatively new decentralization tool that provides solutions to healthcare system's security and efficiency challenges. McGhin et al. [19] and Hasselgren et al. [20] has presented an idea of implementing blockchain in healthcare inspired by the requirements of security, interoperability, verification and validation. BoMT is defined as the integration of blockchain technology and IoMT which enables medical devices to function anonymously by introducing smart contracts in the network and using cryptographic techniques. BoMT provides a data structure of virtually immutable interconnecting blocks where patient data are securely stored using some cryptographic hash function. Figure 1 describes a general architecture of BoMT. The patients are remotely monitored continuously using some wearable and non-wearable medical sensory devices and these real-time health data are continuously updated to blockchain network. The healthcare providers registered in the network are also present physically at some remote location and authorized to access the patient health data from the network. The health report and treatment of every patient registered to network is shared on the distributed ledger to the authorized healthcare providers. The doctor consults the patients according their condition and other caretakers are also responsible to provide their services.

The information from every medical sensor device is continuously get monitored in BoMT environment using the unique identifiers of each device. It provides

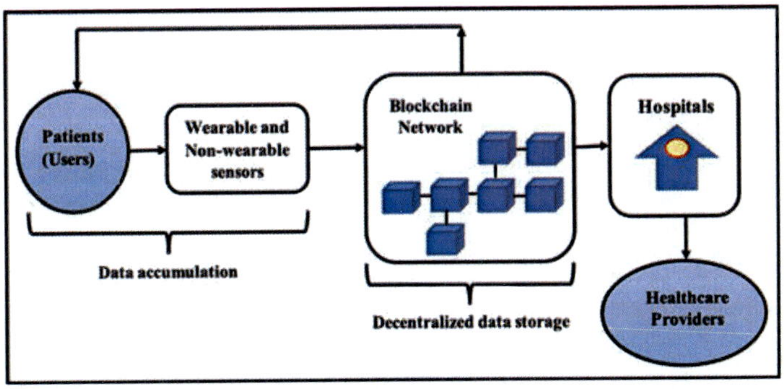

Fig. 1 A general architecture of BoMT

reliable aggregated certification with identity verification. BoMT architecture also ensures the security of healthcare ecosystem as each information is cryptographically hashed. It provides an immutable ledger containing the history patient information which builds a confidence in patients as they are able to monitor every transaction without fear of manipulation. The decentralized network facilitates higher security, transparency, interoperability, integrity and traceability in efficient data sharing and storage. It helps in increasing the trust among healthcare providers, accurate and timely diagnosis, sophistic clinical trials and other beneficial services in healthcare environment. Hospitals registered with the BOMT framework shares the patient data using web application in the front-end securely. Once the shared data in blockchain network goes in charge of patients, they are free to share them with other hospitals or healthcare providers.

4 Analysis of Consensus Algorithms for BoMT

Consensus algorithm is a method for arriving at or agreeing on a common data value or common state of the blockchain. There are various types of consensus algorithms designed is described below. Table 2 illustrates the compatibility of different consensus algorithms with IoMT.

Proof of Work (PoW): Arul and Renuka [21] explains PoW in which every node of the blockchain network reaches an agreement where miners compete each other to solve a mathematical problem and discover a nonce for a given block. The person who solves the challenge first is the one who broadcasts a block to the entire network. The block is then added to the blockchain network, once it is confirmed and validated by other blocks. The person is then rewarded with some amount. Zheng et al. [22] defines the procedure of selecting miners in PoW where the node with higher processing power should be selected as miner as it has to update the nonce value regularly. Ray et al. [23] and Sharma et al. [24] analysed that PoW is not considered ideal for IoMT due to its high requirement of bandwidth and computational power.

Proof of Stake (PoS): Arul and Renuka [21] shows PoS as an alternative technique of PoW and the second most popular consensus algorithm. It requires less energy, less time to process, less computational power and is cheaper than PoW. The node that gets a chance to create a new block in the blockchain are called validators. All the nodes stake their wealth for a set length of time in order to get a chance to add a new block. The node with the greatest stake has the highest opportunity of becoming a validator. Becoming a validator also largely depends on the period of time for which the money has been staked by the user. The validators are given the transaction fee associated with the block instead of reward. Zheng et al. [22] analysed that this algorithm not only saves the computational power and wealth of the validator but also the nodes that compete for becoming a validator. This strategy leads the quicker blockchains with lower electricity consumption and a lesser risk of a 51% attack. PoS-based algorithms are widely employed in public blockchains

Table 2 Compatibility of consensus algorithms with IoMT

Consensus algorithm	Year	Technique used	Benefits	Issues	Compatibility
Proof of work	1993	Mining based	Security	High computational power	Poor
Proof of stake	2012	Validation based	Fast and energy-efficient	Security vulnerabilities	Average
Proof of elapsed time	2016	Lottery based	Energy-efficient, high throughput, reduced latency and cost effective	Platform incompatibility	Good
Proof of burn	2012	Burn coins based	Low energy consumption	Security vulnerabilities, resource wastage	Poor
Stellar consensus	2014	Voting based	Distributed control, low latency, security, flexible trust	Inefficient in terms of message sent	Good
Proof of authority	2015	Identity or reputation based	Risk resilient, high throughput, less resource wastage	Complex to reach the reputation	Average

where validators are anonymous and untrusted. The most widely used blockchain platform Ethereum is going to switch from PoW to PoS algorithm. Ray et al. [23] and Sharma et al. [24] demonstrate that PoS could be a viable option for e-healthcare applications [21, 23, 24].

Proof of Elapsed time (PoET): It is lottery-based system and is developed mainly for sawtooth software allowing it to achieve better throughput. Every node that interacts within the blockchain network is assumed to be a loyal member in PoET. A timer is generated by the individual nodes randomly. The node whose timer expires first becomes the leader and the new block is added by them. Zheng et al. [22] analysed that it increases the efficiency and reduces the network delay as it does not require any computational power. Sharma et al. [24] show that PoET overcome the issue of high computational power faced by PoW algorithm and make the mining process cost effective. This protocol is mainly used in the private blockchain networks. Every node in the network has an equal chance of winning as it is an impartial lottery system. Ray et al. [23] and Sharma et al. [24] illustrate that this algorithm can be used in e-healthcare system due to the high throughput and reduced latency.

Proof of Burn (PoB): Ray et al. [23] defines PoB algorithm as PoW without wasting energy. Proof of burn is based on the idea of burning coins to reduce energy wastage in proof of work. Miners are supposed to present proof that they burnt some money.

This algorithm is combined with proof of work and proof of stake to enable block creation and network security. Burning coins implies transferring the coins to a particular address where cryptographic processes prevent the coins from being used. A transaction takes place to the burn address when coins are burned. A burn hash is determined after checking and confirming the burn transaction. The calculated burn hash is then checked or compared with the pre-defined target. If the burn hash is less than the pre-defined target then the proof of burn block is created. Zheng et al. [22] describes PoB as a mechanism for verifiably destroying cryptocurrency. The burn proof technique is still vulnerable to a 51% attack. A node with 51% hash power can attack the system, but it's unclear what that power is and how to measure it. It's suitable for bitcoin design. Sharma et al. [24] doesn't consider it ideal for e-health applications due to its unpredictable burning method.

Stellar Consensus Protocol (SCP): Ray et al. [23] explained that the nodes in the network don't need to trust the entire network in this decentralized system. They can choose the nodes to trust them. The collection of nodes that trusts one another in the network are known as quorum slice. A quorum is a group of nodes that work together to achieve consensus and a quorum slice is a subset of that group that aids a node in its agreement process. The two steps of SCP are as: nomination protocol and ballot protocol. The nomination protocol runs first. New values namely, candidate values are submitted for agreement at this time. Each node that receives these values will vote for one of them. It results in that one value will win with a majority of votes. The nodes deploy the ballot protocol when the nomination procedure has been successfully executed. This will involve voting to make decision to whether commit or abort the values obtained through the nomination procedure. The votes that were not counted are considered irrelevant. However, nodes can get trapped in states where they can't decide whether to abort or commit a value. This can be prevented by shifting it to a higher valued ballot and incorporating it into a next ballot protocol. SCP consensus protocol is considered as a suitable option when creating healthcare system.

Proof of Authority (PoAuthority): Sharma et al. [24] explains PoAuthority as a reputation-based consensus algorithm where the miner's reputation is at stake rather than coins. A validator takes on the job of a miner. Authorities are the people who validate the data. An authority should have a very good reputation to be a validator that prevents them from committing fraud. Each validator will generate a block using round robin. If a validator is malicious and offers an incorrect block, it is given a bad reputation. The validation nodes have complete control over new block decisions in a PoAuthority blockchain. The main benefits of using PoAuthority consensus are high risk tolerance except in cases where 51 per cent of validators are acting fraudulently, predictable block generation time in comparison to PoW or PoS, high transaction rate and no wastage of resources on computation as in PoW. However, this consensus mechanism is considerably not suitable for healthcare systems due to some dishonest action by third party as well as evaluation of same size stakes differently by different validators.

5 Applications of BoMT

BoMT can aid in the better administration of healthcare data. Various applications of BoMT are described as follows:

Electronic Health and Medical Records (EHR): EHR are the collection of patient's health data and are used by medical professionals to diagnose and treat patients. This data is considered extremely sensitive and confidential as it consists private information of patients and healthcare providers as well as medical information. BoMT can be employed in this case since it allows for efficient, secure and immutable data transmission and provide quick access to medical records. Usman et al. [25] implemented EHR utilizing a permissioned blockchain network such as Hyperledger that first registers users and health care providers. The certificate authority supplies the patient with a certificate and a private key. The health information of authenticated patients is maintained in the blockchain. Authorized doctors have access to this information.

Organ donation and transplantation: Organ transplantation is a very crucial part of the healthcare industry. The departed donors are the primary supply of organs. Ranjan et al. [26] and Niyigena et al. [27] has analysed that the kidney transplant waiting list has the greatest percentage of individuals waiting for a donor. The number of people donating an organ has been discovered to be significantly smaller than the number of patients in need of the organ. The communication gap between donors and receivers is the major cause of this. Organ donation does not have a transparent system. It results in medical professionals to be involved in the unlawful sale and acquisition of organs. There is an urgent need for a system that is fair and transparent. BoMT is one of the most effective solutions for organ donation system due to its decentralized, secure, distributed, transparent and immutable features. The blockchain technologies could be used to build a distributed and fully decentralized organ donation and transplantation system that acts as a link between donor networks in different nations. BoMT enables methods to make contracts between enterprises or governments open and trustworthy. Ranjan et al. [26] has presented a technique for development of decentralized application for organ donation and tissue transplantation using blockchain technology. It is developed using Ethereum blockchain platform that makes the system distributed, secure and decentralized. Smart Contracts are used to implement the application's logic which encompasses everything from registering the donor to performing the transplant.

Pharmaceutical Supply Chain (PSC): The pharmaceutical sector provides vital healthcare services by giving individuals with life-saving medicines and drugs. Zheng et al. [22] illustrates that medical drugs require a more precise management system due to their sensitive process of production and lifespan as well as their sensitivity to conservation conditions. Bocek et al. [28] presents another concern about the globalization of the PSC including the illegal trade in pharmaceuticals, which has increased public distrust. The PSC's complexity raises the necessity for product lifecycle transparency and drug traceability. BoMT for supply chain management provides the system's transparency and traceability keeping track of the medications. The structure of BoMT supply system can be divided into two tiers. (1) Backend includes the smart contract consisting of the conditions and rules and the blockchain network (2) The front-end comprises of sensors for determining the condition of the pharmaceutical product. Storing the data on the blockchain also ensure the immutability, illegal actions like manufacturer changing the expiry date of the medicine can be controlled with this feature of blockchain.

Vaccine management: Vaccine safety and management is a very critical issue. It involves people's life and health safety. Cui et al. [29] describes the risks involved with the supply chain of vaccines and quality measures of vaccines which are the major concern for researchers. The manufacturers may add some additives without any clinical trials to improve production records. The intermediary suppliers don't have any business qualification and may collude with the vaccination centres to replace the quality vaccines with the inferior quality and use transportation which does not meet the requirements of cold chain logistics. The front-end suppliers sell these inferior quality vaccines to gain profit. BoMT-based vaccine management system provide transparency in vaccine production and quality assurance. Only the registered manufactures and suppliers are authorized for production and supply of vaccines which develops a safe and trustworthy environment for vaccine management system.

6 Conclusion

Smart healthcare services are growing rapidly with the advancement of IoT-based technologies. Although, IoMT system has improved the quality and efficiency of treatment in tremendous way, it is exposed to various privacy and security threats. Blockchain technology has shown a great potential in mitigating these security challenges of IoMT devices. This paper addresses the different security issues associated with IoMT devices and the security requirements that should be met to overcome these issues. A blockchain enabled IoMT system namely, BoMT has been illustrated to overcome these security challenges that provides a transparent, secure and distributed healthcare environment. The compatibility of different consensus algorithms such as PoW, PoS, PoET, PoB, stellar and PoAuthority with IoMT system has also been analysed. PoET and Stellar are supposed to be good compatible with IoMT systems and can be considered best consensus mechanism for BoMT. The application of BoMT systems has been also presented in the area of EHR management, organ donation and transplantation, pharmaceutical supply chain management and vaccine management.

References

1. Budida DAM, Mangrulkar RS (2017) Design and implementation of smart healthcare system using IoT. In: 2017 international conference on innovations in information, embedded and communication systems (ICIIECS). https://doi.org/10.1109/iciiecs.2017.82759
2. Baker SB, Xiang W, Atkinson I (2017) Internet of things for smart healthcare: technologies, challenges, and opportunities. IEEE Access 5:26521–26544. https://doi.org/10.1109/access.2017.2775180
3. Yaacoub JPA, Noura M, Noura HN, Salman O, Yaacoub E, Couturier R, Chehab A (2019) Securing internet of medical things systems: limitations, issues and recommendations. Futur Gener Comput Syst. https://doi.org/10.1016/j.future.2019.12.028
4. Gatouillat A, Badr Y, Massot B, Sejdic E (2018) Internet of medical things: a review of recent contributions dealing with cyber-physical systems in medicine. IEEE Int Things J 1. https://doi.org/10.1109/jiot.2018.2849014
5. Papaioannou M, Karageorgou M, Mantas G, Sucasas V, Essop I, Rodriguez J, Lymberopoulos D (2022) A survey on security threats and countermeasures in internet of medical things (IoMT). Trans Emerg Telecommun Technol 33. https://doi.org/10.1002/ett.4049
6. Wazid M, Das AK, Rodrigues JJPC, Shetty S, Park Y (2019) IoMT malware detection approaches: analysis and research challenges. IEEE Access 7:182459–182476. https://doi.org/10.1109/ACCESS.2019.2960412
7. Ghubaish A, Salman T, Zolanvari M, Unal D, Ali A, Jain R (2021) Recent advances in the internet-of-medical-things (IoMT) systems security. IEEE Int Things J 8(11):8707–8718. https://doi.org/10.1109/JIOT.2020.3045653
8. Yanambaka V, Mohanty S, Kougianos E, Puthal D, Rachakonda L (2019) PMsec: PUF-based energy-efficient authentication of devices in the internet of medical things (IoMT). In: 2019 IEEE international symposium on smart electronic systems (iSES) (formerly iNiS), pp 320–321. https://doi.org/10.1109/iSES47678.2019.00079
9. Ratta P, Kaur A, Sharma S, Shabaz M, Dhiman G (2021) Application of blockchain and internet of things in healthcare and medical sector: applications, challenges, and future perspectives. J Food Qual 20. Article ID 7608296. https://doi.org/10.1155/2021/7608296
10. Porambage P, Schmitt C, Kumar P, Gurtov A, Ylianttila M (2014) PAuthKey: a pervasive authentication protocol and key establishment scheme for wireless sensor networks in distributed IoT applications. Int J Distrib Sens Netw 10(7):357430. https://doi.org/10.1155/2014/357430
11. Sicari S, Rizzardi A, Miorandi D, Coen-Porisini A (2018) REATO: reacting to denial-of-service attacks in the internet of things. Comput Netw 137:37–48. https://doi.org/10.1016/j.comnet.2018.03.020
12. Hei X, Du X, Wu J, Hu F (2010) Defending resource depletion attacks on implantable medical devices. In: 2010 IEEE global telecommunications conference GLOBECOM 2010. https://doi.org/10.1109/glocom.2010.5685228
13. Airehrour D, Gutierrez JA, Ray SK (2018) SecTrust-RPL: a secure trust-aware RPL routing protocol for internet of things. Futur Gener Comput Syst. https://doi.org/10.1016/j.future.2018.03.021
14. Liu J, Zhang C, Fang Y (2018) EPIC: a differential privacy framework to defend smart homes against internet traffic analysis. IEEE Internet Things J 5(2):1206–1217. https://doi.org/10.1109/jiot.2018.2799820
15. Cervantes C, Poplade D, Nogueira M, Santos A (2015) Detection of sinkhole attacks for supporting secure routing on 6LoWPAN for internet of things. In: 2015 IFIP/IEEE international symposium on integrated network management (IM). https://doi.org/10.1109/inm.2015.7140344
16. Shukla P (2017) ML-IDS: a machine learning approach to detect wormhole attacks in internet of things. In: 2017 intelligent systems conference (IntelliSys). https://doi.org/10.1109/intellisys.2017.83242

17. Yin D, Zhang L, Yang K (2018) A DDoS attack detection and mitigation with software-defined internet of things framework. IEEE Access 6:24694–24705. https://doi.org/10.1109/access.2018.2831284
18. Liu C, Cronin P, Yang C (2016) A mutual auditing framework to protect IoT against hardware Trojans. In: 2016 21st Asia and South Pacific design automation conference (ASP-DAC). https://doi.org/10.1109/aspdac.2016.7427991
19. McGhin T, Raymond Choo K-K, Liu CZ, He D (2019) Blockchain in healthcare applications: research challenges and opportunities. J Netw Comput Appl. https://doi.org/10.1016/j.jnca.2019.02.027
20. Hasselgren A, Kralevska K, Gligoroski D, Pedersen SA, Faxvaag A (2020) Blockchain in healthcare and health sciences—a scoping review. Int J Med Informatics 134:104040. https://doi.org/10.1016/j.ijmedinf.2019.10404
21. Arul P, Renuka S (2021) Blockchain technology using consensus mechanism for IoT-based e-healthcare system. IOP Conf Ser Mater Sci Eng 1055:012106. https://doi.org/10.1088/1757-899X/1055/1/012106
22. Zheng Z, Xie S, Dai H, Chen X, Wang H (2017) An overview of blockchain technology: architecture, consensus, and future trends. In: 2017 IEEE international congress on big data (BigData congress). https://doi.org/10.1109/bigdatacongress.2017
23. Ray P, Dash D, Salah K, Kumar N (2020) Blockchain for IoT-based healthcare: background, consensus, platforms, and use cases. IEEE Syst J 1–10. https://doi.org/10.1109/JSYST.2020.2963840
24. Sharma A, Kaur S, Singh M (2021) A comprehensive review on blockchain and Internet of Things in healthcare. Trans Emerg Telecommun Technol. https://doi.org/10.1002/ett.4333
25. Usman M, Qamar U (2020) Secure electronic medical records storage and sharing using blockchain technology. Proc Comput Sci 174:321–327. https://doi.org/10.1016/j.procs.2020.06.093
26. Ranjan P, Srivastava S, Gupta V, Tapaswi S, Kumar N (2019) Decentralised and distributed system for organ/tissue donation and transplantation. In: 2019 IEEE conference on information and communication technology. https://doi.org/10.1109/cict48419.2019.906622
27. Niyigena C, Seol S, Lenskiy A (2020) Survey on organ allocation algorithms and blockchain-based systems for organ donation and transplantation. In: 2020 international conference on information and communication technology convergence (ICTC). https://doi.org/10.1109/ictc49870.2020.928942
28. Bocek T, Rodrigues BB, Strasser T, Stiller B (2017) Blockchains everywhere—a use-case of blockchains in the pharma supply-chain. In: 2017 IFIP/IEEE symposium on integrated network and service management (IM). https://doi.org/10.23919/inm.2017.7987376
29. Cui L, Xiao Z, Wang J, Chen F, Pan Y, Dai H, Qin J (2021) Improving vaccine safety using blockchain. ACM Trans Int Technol 21(2). Article 38. https://doi.org/10.1145/3388446

Simulation and Synthesis of SHA-256 Using Verilog HDL for Blockchain Applications

Jitendra Goyal, Deeksha Ratnawat, Mushtaq Ahmed, and Dinesh Gopalani

Abstract Hash algorithms play a major role in ensuring data authenticity and integrity in many applications such as digital signature verification, password hashing, SSL handshake, integrity checking, blockchain applications. Although many popular hash algorithms exist such as MD-5, SHA-0, SHA-1, SHA-2. Among these SHA-2 is the most important hash family. Nowadays, blockchain technology is becoming more and more popular in the Internet world which uses SHA-256 (an algorithm of SHA-2 family). The key factor behind the success of the blockchain is the *Proof-of-Work* (PoW) consensus algorithm in which a hash value of a specific complexity is computed using SHA-256. It requires extensive computing resources and consumes a lot of power. Therefore, in this paper, FPGA-based hardware architecture for SHA-256 computation is synthesized and simulated on Verilog HDL for better resource management. Along with this the power and time consumed in different FPGA families have been analyzed.

Keywords Hashing · SHA-256 · FPGA · Verilog HDL · Blockchain

1 Introduction

Secure hash algorithms (SHAs) are cryptographic hash algorithms consisting of several hash functions that are used to ensure the authenticity and integrity of the data. SHAs take random input data lengths and convert them into fixed-length data called as message digests (hash values). Initially, in 1993, SHA was designed by NIST NSA based on the design of MD4 with significant changes. In 1995 a modified version was released as FIPS 180-1 and introduced as SHA-1 which generated hash values of 160-bit size. SHA-1 had a weak *collision avoidance* and less strength against *brute-force* attacks. Therefore SHA-1 was not widely used. After that in 2002, the National Security Agency (NSA) and the National Institute of Standards

J. Goyal (✉) · D. Ratnawat · M. Ahmed · D. Gopalani
Department of Computer Science and Engineering, Malaviya National Institute of Technology Jaipur, Jaipur 302017, India
e-mail: 2018rcp9080@mnit.ac.in

S. Das et al. (eds.), *Advances in Data-driven Computing and Intelligent Systems*, Lecture Notes in Networks and Systems 653, https://doi.org/10.1007/978-981-99-0981-0_36

and Technology (NIST) released the SHA-2 hash function; the successor to the SHA-1 family which was more secure. SHA-224 and SHA-256 are members of the SHA-2 family whose message digest lengths are 224 and 256 bits, respectively. All SHA-2 family algorithms are iterative in nature [1].

SHA-256 is used in various applications such as digital signatures, SSL handshakes, integrity checks, blockchain applications. Due to the strong security, high collision resistance, and computation time of SHA-256 it is used in many other important applications [2]. So far SHA-256 has been adopted by various blockchain projects including bitcoin and many other cryptocurrencies which have forked from the original bitcoin source code. For faster computation blockchain technology requires tools that are faster and re-configurable. FPGAs can be used to implement cryptographic functions that require high computational power. Examples of cryptographic functions are hash, encryption, and digital signature [3].

A wide range of applications require high-performance implementation of hash functions. For this Field Programmable Gate Array (FPGA) is a good choice because FPGA is flexible, physically secured and re-configurable. FPGA is an integrated circuit designed in such a way that a particular can configure it in a specific way after its manufacturing because it is *field-programmable*. FPGA includes multiple logic blocks that are programmable and can be used for complex combinational functions. It can work as a logic gate. Most of the FPGAs have some memory in form of flip-flops. A large number of FPGAs can implement various logic functions which make it reusable configuration similar to computer programs. Typically FPGAs are specified by a hardware description language (HDL) and to be used for an application-specific integrated circuit (ASIC). Nowadays, electronic design automation tools are used to define these HDL configurations. Verilog and VHDL are the most well-known HDLs. Verilog is a programming language for defining digital systems such as network switches, microprocessors, memory and flip-flops. It means that we can describe any digital hardware at any level using an HDL. HDL-described designs are technological-independent, highly easy to design and debug and typically more helpful than schematics; especially for the complex circuits.

1.1 Blockchain Technology

Blockchain technology is a *peer-to-peer* decentralized Distributed Ledger Technology (DLT). It allows multiple users to work together to maintain a secure ledger that is distributed over the blockchain network. A node (user) initiates a transaction over a blockchain network by signing it with his own private key. After that the transaction is broadcast over the blockchain network for performing the verification. Various methods can be used by the blockchain platform to verify whether the transaction is valid or not. These methods are called consensus algorithms. Once the transaction is verified by the nodes that the transaction is authentic, the transaction will find its place in the ledger. To protect the transaction from any modification the transaction has a timestamp and a unique ID. Then the transactions created are combined to

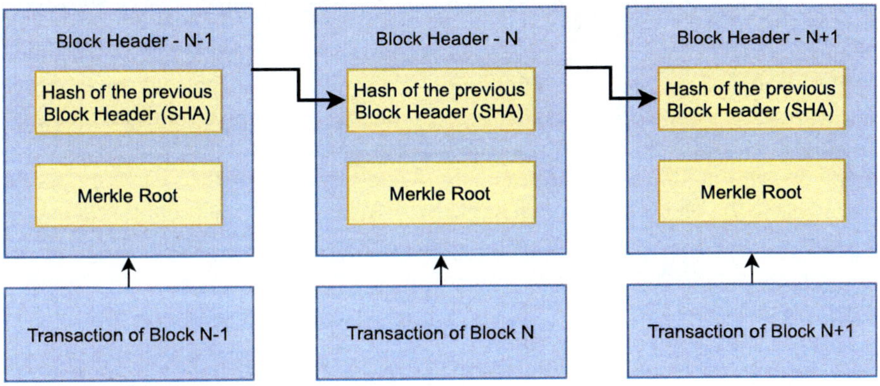

Fig. 1 SHA used in blockchain

form a block. As shown in Fig. 1. To generate a new block, a SHA-256 of this block data is computed [4]. After that connection of blocks happen in such a way that a block will get connected to its previous block via previous block hash and then a new block will get connected to this block and so on. In this the digital signature is generated from its private key using SHA-256 which is unique and no one is able to modify the digital signature which must be guaranteed. If someone tries to change the transaction details then the digital signature will get changed completely and after that verification of the digital signature will not be possible hence it will be rejected. So SHA-256 is the main principle used in blockchain architecture to convey security and privacy in the system [5].

The structure of this paper is: Sect. 2 has literature review; Sect. 3 describes SHA-256 algorithm and its properties; Sect. 4 describes the experimental setup used for the synthesis and simulation of SHA-256; and then Sect. 5 shows the simulation result and elaborates design results in RTL analysis.

2 Literature Review

In various authentication mechanism which are present in different security systems; the hashing is the important mechanism. SHA-256 is a cryptographic hashing algorithm that is utilized from IoT micro-devices to high-performance computers [6] and in various other security fields like blockchain [7], MACs [1] etc. SHAs are becoming resistant and robust against potential attacks [8]. Increasing the number of computation rounds increases the efficiency (safety) but this leads to computation complexity. That's why so many researchers are researching how to optimize secure hash function complexity.

While moving toward hardware implementation of cryptographic functions (using FPGA), hardware acceleration is ideal for implementing cryptographic algorithms, accomplishing tasks more quickly with greater flexibility and allowing for design optimization with lower costs. FPGA-based encryption is approximately 20 times faster than dual-core processor encryption while utilizing 85 percent lesser CPU. Hardware accelerators offer reasonable compromises regarding throughput and field cost [2]. Specifically hardware implementations of SHA-256 result in lower costs with better throughput and more secure computing platforms than its counterparts in software [9]. Pham et al. in their paper [10] proposed multimode processing architecture and three-stage arithmetic logic unit pipeline architecture to reduce the critical paths for block mining and observed the improved performance and hardware efficiency. Area and performance are two of the most essential design parameters to consider when implementing hash functions on FPGAs. Hash functions are used in many applications so their field-performance requirements can vary depending on the application. It is worth doing cryptographic calculations and working out plans that allow more productive.

In the IoT community, billions of devices communicate with each other worldwide and store data separately. This massive amount of data generated by these IoTs should be completely secured. Due to various challenges in security and scalability issues, centralized solutions are not suitable for these concerns. Here blockchain technology is an efficient solution. Its *distributed* method is used to solve these security challenges and scalability problems by optimizing data sharing methods and data privacy simultaneously [11]. The combination of blockchain technology and FPGA architecture provides significant solutions for decentralized applications such as IoT applications, supply chain management. [12].

In bitcoin the *PoW* consensus algorithm is used. Consensus requires a 32-bit number. For this many rounds of SHA-256 computation are needed for desired output. Merkle trees which help the blockchain network to maintain transactional integrity are also built using SHA-256 [10]. So doing SHA-256 computation capacity in bitcoin is the primary research focus [3]. The performance of SHA-256 can be improved by parallel computation on the hardware and software. Xiaoyong Zhang et al. in their paper [13] proposed a parallel hardware architecture for SHA-256 computation. The authors rescheduled SHA-256 calculations based on hardware characteristics. They found three independent paths in the round function and parallelized them by replacing with three parallel pipelines and shortened the long critical path. For synthesis and simulation they used *Intel 14nm* technology and achieved a 3-fold improvement in throughput. Unrolled architecture, parallel counters, pipeline and delay balancing are the various techniques used in hardware implementation of various cryptographic algorithms such as SHA-256 [1]. Jahan et al. in their paper [14] have proposed a software-based parallel blockchain technology that is used to store land-related documents. A satellite chain formation algorithm is used to digitally store land-related documents. In that algorithm there are satellite chains which are sub-chains created in parallel.

Table 1 Features of SHA-256 hash function

Feature of SHA-256	Value
Output length of hash value	256 bits
Number of round constants (K_t)	64
Message block size	512 bits
Word size	32 bits
Number of words	8 words
Number of iterations	64 rounds

3 Secure Hash Algorithm (SHA-256)

Secure hash functions are essential for computing Signatures and Message Authentication Codes (MACs) which enable users to authenticate the authenticity and integrity of data [15]. NSA and NIST jointly introduced SHA-256. The SHA-2 family includes the algorithms SHA-224, SHA-256, SHA-384, and SHA-512 which give them 224, 256, 384, or 512-bit output hash values, respectively. They are named according to the size of the digest. The size of the output message digest is always fixed regardless of the size of the input message. Due to SHA-2's advanced certificates it is more accurate and used in most of the applications. In SHA-256; 256 means the size of the final message digest is 256 bits. Other algorithms of SHA-2 family are similar to SHA-256 with some differences. Table 1 shows the characteristics of SHA-256.

3.1 SHA-256 Properties

For a cryptographic hashing function to be completely secure, it must have some specific properties: Compression, Avalanche effect, Determinism, Pre-image resistance (one-way function), Collision resistance, and Efficiency (Faster calculation).

- **Compression** Regardless of the input message size, the output hash will have a fixed number of bits which means that if the input size is 2 bits then the size of the output message digest is 256 bits. Even if the input size is 1000 bits, the output message digest size will also be 256 bits.
- **Avalanche Effect** As shown in Fig. 2, only a minimal change in the input completely changes the output hash value. This property prevents hackers from predicting the output hash value through trial and error.
- **Determinism** As shown in Fig. 3: if the input is same, the output will always be the same on different machines. Any machine in the world that understands a hashing algorithm should be able to generate the same output hash value for the same input message.

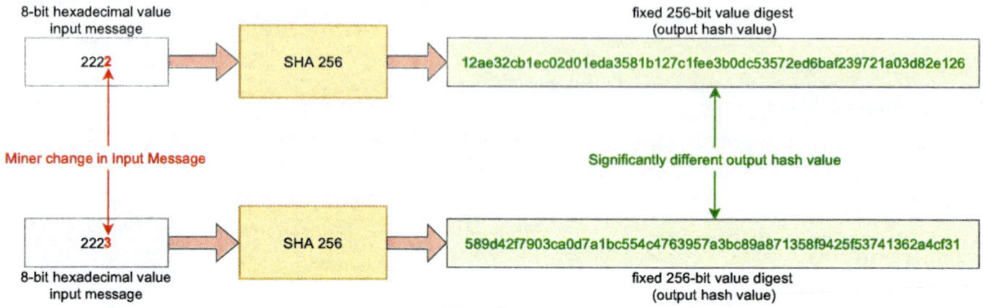

Fig. 2 Avalanche effect of SHA

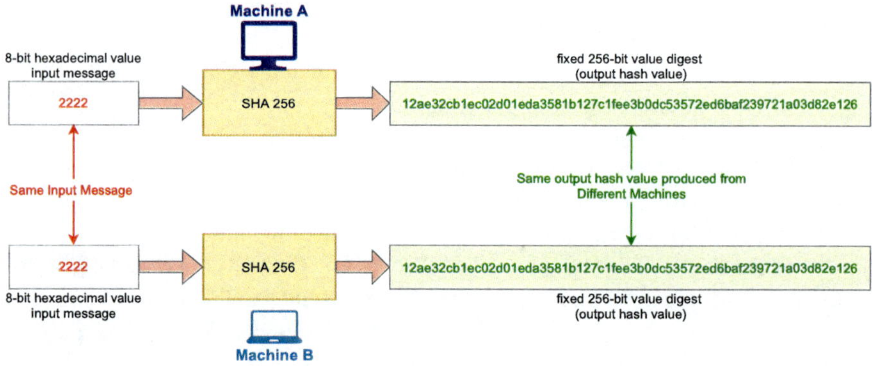

Fig. 3 Determinism property of SHA

- **Pre-Image Resistant (One-Way)** As shown in Fig. 4, the secure hashing algorithm should be a one-way function. This means that it should not be possible to get back the original message from the output hash value. The whole concept would fail if the input message retrieved from the output hash value.
- **Efficiency** The output hash value should be generated rapidly and should not require high computational power. It also shouldn't require supercomputers or high-end machines to generate the hash.
- **Collision Resistance** As shown in Fig. 5, it is practically impossible to find two different inputs that produce the same output. It must withstand collision. It is not possible to find two inputs that have hash value for the same output a and b such that $H(a) = H(b)$; given that $a \neq b$.

3.2 SHA-256 Algorithm

General Assumptions: The input message must be of $\leq 2^{64}$ bits and the processing of the input message is done sequentially in blocks of size 512 bits. This will contain

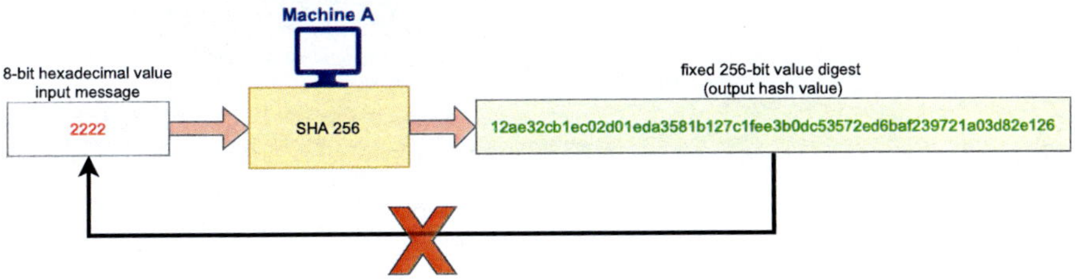

Fig. 4 SHA-Pre-image resistant (one-way function)

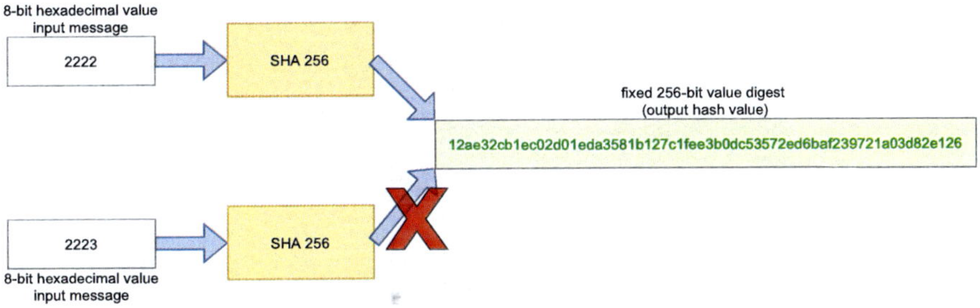

Fig. 5 Collision resistance property of SHA

the message digest or output hash value which must be of the size 256 bits. Figure 6 shows the entire SHA-256 algorithm process and its phases are shown below in the steps.

Step 1: **Convert the message block into 512 bits each by appending some bits**

- Convert *Message* into binary M.
- Append padding bit "1" at the end of the binary message.
- Divide the message M into the block of 512 bits $M_0, M_1, \ldots, M_{j-1}$, and in the remaining message M_j append padding bits 0's, i.e., the message length of last block is congruent to 448.
- Append the length of message in 64 bits (big endian) at the end of the message.

Step 2: **Initializing Hash Value** Initialize 8 words (32-bits each) message digest (MD) buffers, i.e., pre-defined constant values that represent the initial 32 bits of the fractional parts of the square roots of the first 8 primes: 2, 3, 5, 7, 11, 13, 17, 19. Table 2 shows these initial hash values represented in hexadecimal format.

Step 3: **Message Processing**

- Process each 512-bit block from $M_0, M_1, \ldots, M_j$ sequentially.
- Inputs are:

W_t is a 32-bit word from the message
K_t is a constant array (refer Table 3)

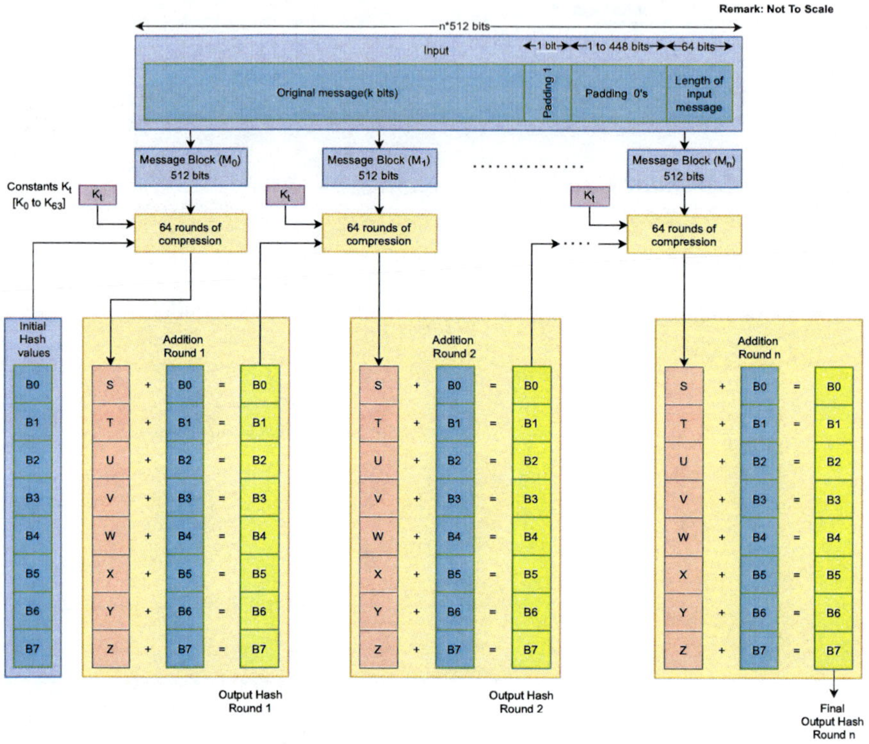

Fig. 6 SHA-256 algorithm used for our simulation

Table 2 Initial hash values

Buffer	Hash value (in hexadecimal)
B_0	0x6a09e667
B_1	0xbb67ae85
B_2	0x3c6ef372
B_3	0xa54ff53a
B_4	0x510e527f
B_5	0x9b05688c
B_6	0x1f83d9ab
B_7	0x5be0cd19

Input hash values are B_0, B_1, B_2, B_3, B_4, B_5, B_6, B_7 (refer Table 2)

– Initialize for each $M_i(S, T, U, V, W, X, Y, Z) = (B_0, B_1, B_2, B_3, B_4, B_5, B_6, B_7)$

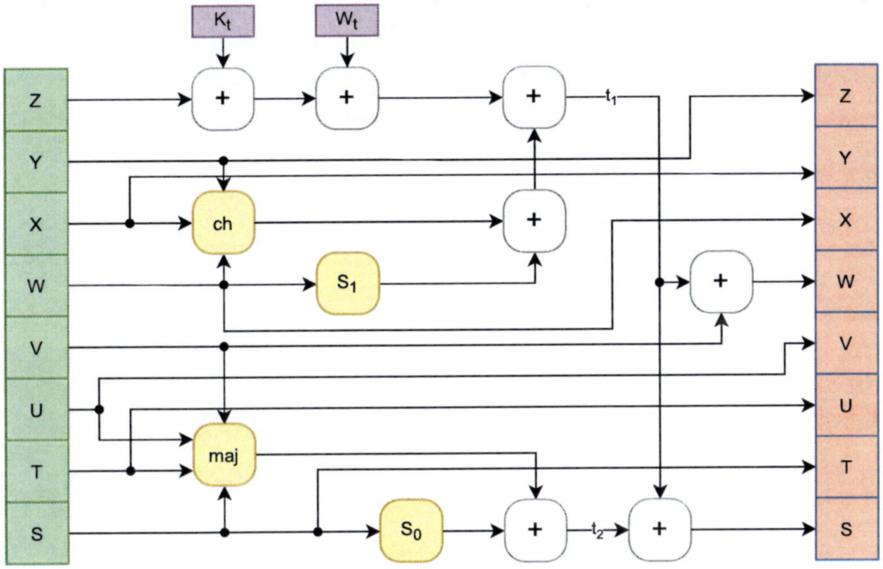

Fig. 7 SHA-256 compression function

Step 4: **Word Expansion** for $t = 0$ to 63

$$W_t = \begin{cases} t\text{th 32-bit word of block } M_j & \text{if } t \leqslant 16 \\ W_t - 16 + s_0 + W_t - 7 + s_1 & \text{if } 16 \leqslant t \leqslant 63 \end{cases}$$

where

- $s_0 = (W_t\text{-}15 \text{ rotateright } 7) \oplus (W_t\text{-}15 \text{ rotateright } 18) \oplus (W_t\text{-}15 \gg 3)$
- $s_1 = (W_t\text{-}2 \text{ rotateright } 17) \oplus (W_t\text{-}2 \text{ rotateright } 19) \oplus (W_t\text{-}2 \gg 10)$

Step 5: **Message Compression** Figure 7 illustrates the message compression for each step t $(0 \leqslant t \leqslant 63)$:

- $S_0 = (S \text{ rotateright } 2) \oplus (S \text{ rotateright } 13) \oplus (S \text{ rotateright } 22)$
- $\text{maj} = (S \cdot V) \oplus (S \cdot U) \oplus (T \cdot U)$
- $t_2 = S_0 + \text{maj}$
- $S_1 = (W \text{ rotateright } 6) \oplus (W \text{ rotateright } 11) \oplus (W \text{ rotateright } 25)$
- $\text{ch} = (W \cdot X) \oplus ((\neg W) \cdot Y)$
- $t_1 = Z + S_1 + \text{ch} + K_t + W_t$
- $(S, T, U, V, W, X, Y, Z) = (t_1 + t_2, S, T, U, V + t_1, W, X, Y)$

Step 6: **Output**: The 256-bit hash of M is available in B_0, B_1, B_2, B_3, B_4, B_5, B_6, B_7 once all M_j have been processed.

Initial value of round constants K_t are the first 32 bits of the fractional parts of the cube roots of the first 64 primes (2–311) make up these 64 constants. As shown in

Table 3 Initial values of round constants

Round constant	Value	Round constant	Value	Round constant	Value
K_0	428a2f98	K_{22}	5cb0a9dc	K_{43}	c76c51a3
K_1	71374491	K_{23}	76f988da	K_{44}	d192e819
K_2	b5c0fbcf	K_{24}	983e5152	K_{45}	d6990624
K_3	e9b5dba5	K_{25}	a831c66d	K_{46}	f40e3585
K_4	3956c25b	K_{26}	b00327c8	K_{47}	106aa070
K_5	59f111f1	K_{27}	bf597fc7	K_{48}	19a4c116
K_6	923f82a4	K_{28}	c6e00bf3	K_{49}	1e376c08
K_7	ab1c5ed5	K_{29}	d5a79147	K_{50}	2748774c
K_8	d807aa98	K_{30}	06ca6351	K_{51}	34b0bcb5
K_9	12835b01	K_{31}	14292967	K_{52}	391c0cb3
K_{10}	243185be	K_{32}	27b70a85	K_{53}	4ed8aa4a
K_{11}	550c7dc3	K_{33}	2e1b2138	K_{54}	5b9cca4f
K_{12}	72be5d74	K_{34}	4d2c6dfc	K_{55}	682e6ff3
K_{13}	80deb1fe	K_{35}	53380d13	K_{56}	748f82ee
K_{14}	9bdc06a7	K_{36}	650a7354	K_{57}	78a5636f
K_{15}	c19bf174	K_{37}	766a0abb	K_{58}	84c87814
K_{16}	e49b69c1	K_{38}	81c2c92e	K_{59}	8cc70208
K_{17}	efbe4786	K_{39}	92722c85	K_{60}	90befffa
K_{18}	0fc19dc6	K_{40}	a2bfe8a1	K_{61}	a4506ceb
K_{19}	240ca1cc	K_{41}	a81a664b	K_{62}	bef9a3f7
K_{20}	2de92c6f	K_{42}	c24b8b70	K_{63}	c67178f2
K_{21}	4a7484aa				

Table 3: these values of 64 round constants are represented in hexadecimal format. Since SHA-256 supports until 2^{64} input message size, 64 bits are required to append message length.

Suppose we have a original message L with the length of 640 bits. Since the message block size is 512 bits, it would require two message blocks ($n = 2$) to fit a 640-bit original message into 512-bit chunks. The first block M_0 will contain the first 512 bits without padding and the second block M_1 will contain the remaining 128 bits of the original message with some padding bits. Thus 512 bits of block M_1 are: 128 bits of original message + one bit for appending “1” + 319 bits of 0’s + 64 bit message length (decimal value of 640).

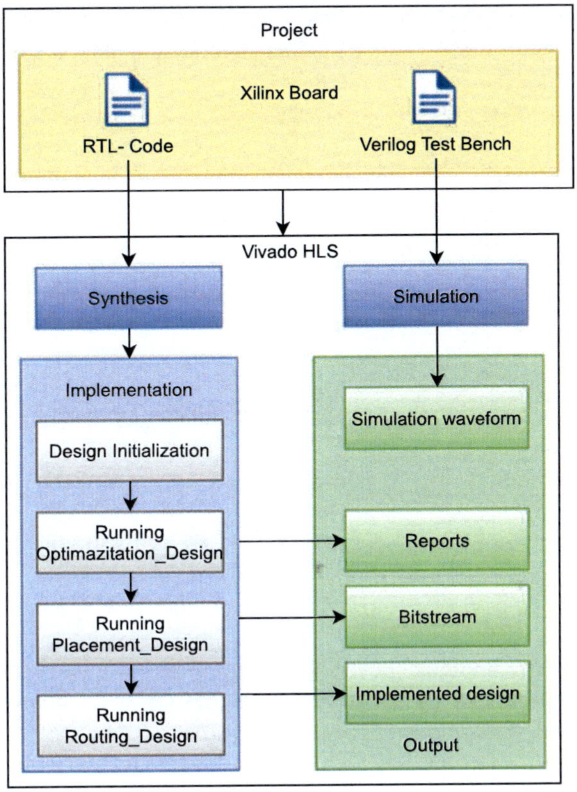

Fig. 8 Vivado architecture for synthesis and simulation

4 Experimental Setup

We are using Xilinx Vivado® Design Suite for simulation and synthesis of SHA-256. It is typically used to create a new design from a given programming source code. At each stage the design is analyzed and verified. It has many characteristics for high-end synthesis and chip development systems. In the Verilog code of SHA-256 we have used 8 registers; each of size 32-bits to store the initial hash values and a 512-bit register to store the output hash value. 64 registers are used to store the initial rounding constant where the size of each register is 32 bits. We have used Xilinx Artix-7 board, Kintex-7 board, and Spartan-7 board for synthesis of SHA-256.

Figure 8 shows the Vivado architecture showing the various stages of the simulation, synthesis, and implementation process. The process of converting a Register Transfer Level specified design (RTL-code) into a gate-level macro (Netlist) is known as synthesis and the process of creating models that mimic the device's behavior is known as simulation. We are also preparing a model (test-bench) along with a simulation model to test the device.

Table 4 Simulation results of SHA-256 from synthesis model

Name	Value (in hex)	Name	Value (in hex)
INPUT [0:31]	88888888	sum0[0:31]	3cf19125
H0 [0:31]	4d58bbbf	sum1[0:31]	469ec056
H1 [0:31]	fc4a5b54	Ma [0:31]	40e2b62f
H2 [0:31]	0d52aa5f	Ch [0:31]	32ae32f4
H3 [0:31]	ede84c0c	t1 [0:31]	657a8d64
H4 [0:31]	f2c3e43c	t2 [0:31]	7dd447f4
H5 [0:31]	4049675b	S [0:31]	e34ed558
H6 [0:31]	4fba0c9f	T [0:31]	40e2accf
H7 [0:31]	8f8e500a	U [0:31]	d0e3b6ed
k [0:63] [0:31]	Refer Table 3	d [0:31]	489856d2
w [0:63] [0:31]	88888888, 80000000, 00	V [0:31]	a1b591bd
i [31:0]	00000040	W [0:31]	a543fecf
s0 [0:31]	9469ea99	X [0:31]	303632f4
s1[0:31]	b90cbcff	Y [0:31]	33ad82f1

The implementation process consists of four steps: Initialization design, Running Optimize_design, Running Place_design, and Running Root_design on the available device resources of the target device.

- **Opt_design**: Optimizes the logical design to fit on the target Xilinx device more easily.
- **Place_design**: To optimize timing, it places the design on the target Xilinx device and does fanout replication.
- **Route_design**: The design is routed to the target Xilinx device.
- **Write Bitstream**: Creates a bitstream that can be used to configure Xilinx devices. Bitstream creation is usually done after implementation.

5 Result Analysis

5.1 SHA-256 Simulation

As shown in Table 4, We have given 32-bit input (in hexadecimal format), our input is 88888888. We have received the 256-bits output in the form of hash digest. The output is $4d58bbbf$ $fc4a5b54$ $0d52aa5f$ $ede84c0c$ $f2c3e43c$ $4049675b$ $4fba0c9f$ $8f8e500a$. Following steps are taken to view the simulation result:
⇒**Flow Navigator** → **Simulation** → **Run simulation**.

5.2 *Synthesized Design*

The Vivado Design Suite takes the RTL-Specific (Verilog HDL) Code and applies physical and time constraints to the target part. For this some related elements are loaded into memory where they can be analyzed and modified as per the need to accomplish the design. Modifications are made to these constraint files, netlists, debug cores. After that the configuration can be saved. The following steps are taken to synthesize a design:
⇒**Flow Navigator** → **Synthesis section** → **Run synthesis** → **Open Synthesized Design**.

5.3 *Implemented Design*

During implementation Vivado IDE takes the synthesized design (netlist) and applies physical and time constraints to the target part then runs the optimization, placement and routing process, and generates the implemented netlist. In the case when an implemented design is reopened, Vivado opens the implemented netlist and then applies the same physical and time constraints as applied during implementation. After that the placed logic of the applied design and the routed connections are loaded to complete the design.

All the generated constraint files, netlists, implementation results and design configuration can be saved for further analysis. Several analysis such as time analysis, power consumption analysis, visualization of utilization statistics can be performed to implement the design to determine whether the design converges on required performance goals. Following are the steps to open implemented design:
⇒**Flow Navigator** → **Implementation** → **Run implementation** → **Open Implemented Design**.

5.4 *Elaborated Design Results in RTL Analysis*

During this phase Vivado re-compiles the RTL netlists and plots physical and time constraints against the target part. During this many components of the elaborate design are loaded into memory. During RTL expansion there is no FPGA technology mapping. The following steps are taken to open an elaborated design:
⇒**Flow Navigator** → **RTL Analysis**→ **Open Elaborated Design**.

As shown in Table 5, IO utilization of all three product family is same which is 64%. Kintex-7 product family is better than Artix-7 and Spartan-7 for synthesis of SHA-256 because Kintex-7 has low total on-chip power.

Table 5 Elaborated design results in RTL analysis

Product family	Total on-chip power (W)
Artix-7	0.122
Kintex-7	0.084
Spartan-7	0.09

6 Conclusion

As compared to the software implementation, hardware implementation of cryptographic hash function has higher performance and more physical security because there is a physical separation of the cryptographic hash function from the main processor.

In this paper we have simulated the SHA-256 algorithm and synthesized it over three different FPGA families like Kintex-7, Artix-7, and Spartan-7. We have also calculated on-chip power consumption and junction temperature for these families. Based on the comparison we have found that Kintex-7 FPGA board consumes lesser on-chip power than Artix-7 and Spartan-7. So for the block mining in blockchain, which requires a faster hash rate, our synthesized SHA-256 on the Kintex-7 FPGA board would be a better choice. For future work most of the designs are focused on maximizing throughput but there is also some suggestions which are more concerned with power or area savings.

References

1. Chen Y, Li S (2020) A high-throughput hardware implementation of SHA-256 algorithm. In: 2020 IEEE international symposium on circuits and systems (ISCAS), pp 1–4
2. Gad AH, Abdalazeem SEE, Abdelmegid OA, Mostafa H (2020) Low power and area SHA-256 hardware accelerator on Virtex-7 FPGA. In: 2020 2nd novel intelligent and leading emerging sciences conference (NILES). IEEE, pp 181–185
3. Thomas A, Bhakthavatchalu R (2021) Implementation of SHA 256 using MATLAB and on FPGA by the application of block chain concepts. In: 2021 international conference on communication, control and information sciences (ICCISc), vol 1. IEEE, pp 1–5
4. binti Suhaili S, Watanabe T (2017) Design of high-throughput SHA-256 hash function based on FPGA. In: 2017 6th international conference on electrical engineering and informatics (ICEEI). IEEE, pp 1–6
5. Devika K, Bhakthavatchalu R (2019) Parameterizable FPGA implementation of SHA-256 using blockchain concept. In: 2019 international conference on communication and signal processing (ICCSP). IEEE, pp 0370–0374
6. Bensalem H, Blaquière Y, Savaria Y (2021) Acceleration of the secure hash algorithm-256 (SHA-256) on an FPGA-CPU cluster using OpenCL. In: 2021 IEEE international symposium on circuits and systems (ISCAS), pp 1–5
7. Kuznetsov A, Shekhanin K, Kolhatin A, Kovalchuk D, Babenko V, Perevozova I (2019) Performance of hash algorithms on GPUs for use in blockchain. In: 2019 IEEE international conference on advanced trends in information theory (ATIT), pp 166–170

8. Kammoun M, Elleuchi M, Abid M, BenSaleh MS (2020) FPGA-based implementation of the SHA-256 hash algorithm. In: 2020 IEEE international conference on design & test of integrated micro & nano-systems (DTS). IEEE, pp 1–6
9. Opritoiu F, Jurj SL, Vladutiu M (2017) Technological solutions for throughput improvement of a secure hash algorithm-256 engine. In: 2017 IEEE 23rd international symposium for design and technology in electronic packaging (SIITME). IEEE, pp 159–164
10. Pham HL, Tran TH, Le Duong VT, Nakashima Y (2022) A high-efficiency FPGA-based multimode SHA-2 accelerator. IEEE Access 10:11830–11845
11. Akarca D, Xiu PY, Ebbitt D, Mustafa B, Al-Ramadhani H, Albeyatti A (2019) Blockchain secured electronic health records: patient rights, privacy and cybersecurity. In: 2019 10th international conference on dependable systems, services and technologies (DESSERT), pp 108–111
12. Florin R, Ionut R (2019) FPGA based architecture for securing IoT with blockchain. In: 2019 international conference on speech technology and human-computer dialogue (SpeD). IEEE, pp 1–8
13. Zhang X, Ruizhen W, Wang M, Wang L (2019) A high-performance parallel computation hardware architecture in ASIC of SHA-256 hash. In: 2019 21st international conference on advanced communication technology (ICACT). IEEE, pp 52–55
14. Jahan F, Mostafa M, Chowdhury S (2020) SHA-256 in parallel blockchain technology: storing land related documents. Int J Comput Appl 975:8887
15. Kammoun M, Elleuchi M, Abid M, Obeid AM (2021) HW/SW architecture exploration for an efficient implementation of the secure hash algorithm SHA-256. J Commun Softw Syst 17(2):87–96

A Real-Time Graphical Representation of Various Path Finding Algorithms for Route Optimisation

Ravali Attivilli, Afraa Noureen, Vaibhavi Sachin Rao, and Siddhaling Urolagin

Abstract A country's economy relies heavily on transport—it connects us to the rest of the world and facilitates growth. There are many types of navigation systems for cars, including Google Maps, GPS, etc. These systems give the best route possible but are also susceptible to errors. Numerous algorithms have been developed and compared by researchers to determine the optimal route planning approach. A comparative analysis of various algorithms is included in this review, including Dijkstra, $A*$, uniform cost search, and Bellman-Ford (dynamic programming). These algorithms are compared to determine which one is most suitable for finding the optimal routes for the given dataset.

Keywords Route planning · $A*$ algorithm · Dijkstra algorithm · Uniform cost search algorithm · Dynamic programming

1 Introduction

According to Harvard University research done in 2019, people living in Los Angeles spend an average of 119 h a year stuck in traffic as the number of vehicles increases. We must incorporate technology to increase productivity and solve this problem. A transformation of the transportation industry is possible using the intersection of the digital and the physical realms. As a result, we are faced with the shortest path problem. This problem occurs quite often in graph theory and affects many aspects of our daily lives. Route planning and determining the shortest route are the

R. Attivilli (✉) · A. Noureen · V. S. Rao · S. Urolagin
BITS Pilani, Dubai, UAE
e-mail: f20190224@dubai.bits-pilani.ac.in

A. Noureen
e-mail: f20190239@dubai.bits-pilani.ac.in

V. S. Rao
e-mail: f20190151@dubai.bits-pilani.ac.in

S. Urolagin
e-mail: siddhaling@dubai.bits-pilani.ac.in

S. Das et al. (eds.), *Advances in Data-driven Computing and Intelligent Systems*, Lecture Notes in Networks and Systems 653, https://doi.org/10.1007/978-981-99-0981-0_37

primary aspects of this research. And the distance between the target destination and the starting point is minimised in the process. This problem is addressed by using algorithms that determine the shortest route. Time complexity, memory usage, search space, and accuracy are a few of the metrics used to evaluate algorithms that find the shortest route. These algorithms are used in GPS and Google Maps to determine the shortest route, which also considers external factors such as tolls, traffic, transportation modes, and adding extra stops.

In this paper, we aim to provide a practical understanding of shortest path algorithms so that relevant parties will be able to design applications that will assist their target audience in finding the shortest path. We utilise the concept of graphs to provide a visual representation of the area around Dubai International Academic City (DIAC) and then compare the paths that are generated using the Dijkstra algorithm, $A*$ algorithm, uniform cost search method, and Bellman-Ford algorithm.

2 Related Works

This section discusses the previous work proposed by various authors on finding shortest path and/or route optimisation. According to [1], the authors implemented shortest path algorithms like Bellman-Ford and Dijkstra's algorithm on roads of Bandung city to help users in navigation. They employed a five-step methodology: research, verification, comparison unit, adjustment analysis, and reconciliation. According to their results, the number of nodes was reduced and the negative weights could be handled using Bellman-Ford algorithm but not with Dijkstra's algorithm. Haria et al. [2] understood the working of Google Maps and how it provides the shortest route. They performed the analysis on real-time reports collected from smart devices and concluded that both Dijkstra's and $A*$ were highly effective than Bellman-Ford algorithm. In [3], Alzubaidi and Al-Thani modified Dijkstra's algorithm to identify a route in terms of road safety and quality. The proposed approach resulted in low accuracy as the dataset consisted of most popular cities in Yemen and data retrieval difficulties due to current situation in Yemen. The study involved the use of Dijkstra's, Bellman-Ford and Floyd Warshall algorithms. AbuSalim et al. [4] proposes a paper on research studies where Dijkstra's and Bellman-Ford algorithms are compared based on complexity and performance in terms of shortest path optimisation. While Dijkstra's is beneficial for large numbers of nodes, it takes up a great deal of storage space and has a high time complexity. In another paper [5], a comparison between Floyd Warshall and greedy algorithm was drawn to find out the optimal path. It was found that Floyd Warshall gave accurate results but took more time than greedy algorithm.

3 Dataset

For the purpose of this experiment, we took the area around Dubai International Academic City and Dubai Silicon Oasis, UAE and identified 48 important locations. Each location is considered as a node in the graph, and the connection between them is represented by an edge between the nodes. The latitude and longitude associated with each location are recorded as *x*–*y* coordinates and stored in our dataset in the form of a data dictionary so that they can be represented in the form of a graph. Figure 1 represents the actual GPS coordinates of the 48 locations that we have identified, in a real-time Google Maps simulator that was created for the sole purpose of this study.

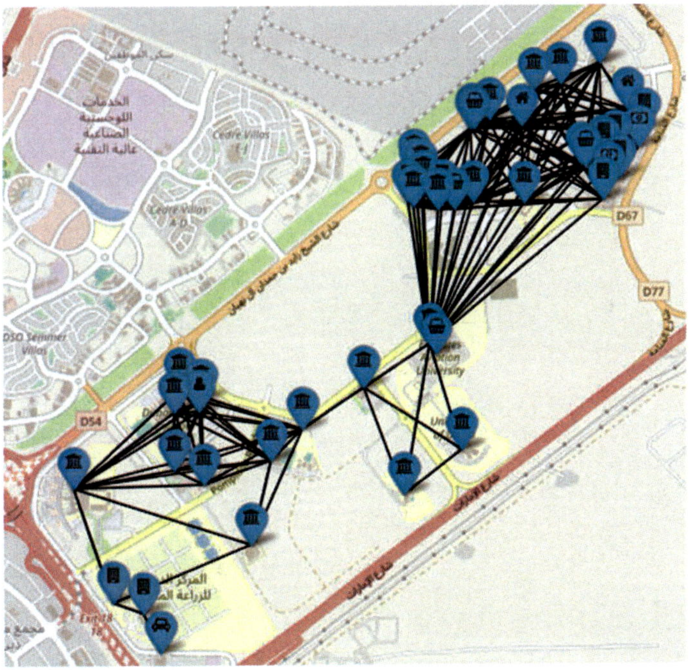

Fig. 1 Real-time map with all connections of the 48 locations represented using our Google Maps simulator

4 Methodology

In order to accomplish the objectives of our research, the study was divided into two different phases. In the first phase, we choose a starting point and a destination to compute the shortest path between them, the time taken to achieve the desired result was also noted. In the second phase, we simulate real-time traffic scenarios and make some roads in the path unavailable for use. We then run the simulation again to find the new paths taken and record the time. This trial is repeated for all the algorithms that were being studied.

Dijkstra's Algorithm—Greedy Method

Algorithm 1 Dijkstra's Algorithm

```
function dij(A,Q)
    for each vertex X in A
        dist[X] = infinite
        prev[X] = NULL
        if X != Q then add X to Priority Queue Y
    dist[Q] = 0
    while Y IS NOT EMPTY do
        P = Extract MIN from Y
        for each unvisited neighbour X of P
            tempDist = dist[P] + edge_wt(P,X)
            if tempDist < dist[X] then
                dist[X] = tempDist
                prev[X] = P
    end while
    return dist[ ], prev[ ]
```

Depending on how a graph is represented, Dijkstra's algorithm can be used to find shortest distances or the cheapest course of action. Essentially, find the shortest leg each time, starting at the end and working backwards.

*A** Algorithm

Algorithm 2 A* Algorithm

```
initialize
        opn_list = {initial}
        closd_list = {}
        g(initial) = 0
        h(initial) = heuristic_func(initial,final)
        f(initial) = g(initial) + h(initial)
while opn_list is not empty do
        x = Node on top of opn_list, with least f
        if x == final
                return
        remove x from opn_list
        add x to closd_list
        for each y in child(x)
                if y in closd_list then
                        continue
                cost = g(x) + dist(x,y)
                if y in closd_list and cost < g(y) then
                        remove y from opn_list as new path is better
                if y is closd_list and cost < g(y) then
                        remove y from closd_list
                if y is not opn_list and y is not closd_list then
                        add y to opn_list
                        g(y) = cost
                        h(y) = heuristic_func(y,final)
                        f(y) = g(y) + h(y)
return failure
```

An $A*$ begins by first calculating the cost of travelling to neighbouring nodes, then choosing the node with the lowest cost. $A*$'s efficiency depends on the heuristic value.

The calculation of the value can be done as follows:

$$f(y) = g(y) + h(y). \tag{1}$$

$g(y)$ Actual cost from initial node to y.
$h(y)$ Eestimation cost from y to goal node.

Uniform Cost Search Method

Algorithm 3 Uniform Cost Search Method

```
function UCS (Graph, start, target)
    Add the starting node to the open list. The node has zero distance value from itself
    while True do
        if open is empty then
            break
        selectedNode = remove from open list, the node with min distance value
        if selectedNode == target then
            calculate path
            return path
        add selectedNode to closed list
        newNodes = get the children of the selectedNode
        if the selected node has children then
            for each child in children
            calculate the distance value of child
            if child not in closed and open lists then
                child.parent = selectedNode
                add the child to open list
            else if child in open list then
                if the distance value of the child is lower than the corresponding node in open list then
                    child.parent = selectedNode
                    add the child to open list
```

This algorithm traverses a weighted graph or tree. When the costs for every edge are different, this algorithm comes into play. Its primary objective is to establish a route to the goal node that has the lowest cumulative cost. Nodes are expanded based on the cost of the path distance from the root node. A graph or tree whose optimal cost needs to be determined can be solved using this method. Under UCS, the lowest cumulative cost is given maximum priority using the priority queue. When all the edges have the same path cost, UCS is equivalent to BFS.

Bellman-Ford—Dynamic Programming

Algorithm 4 Bellman Ford Algorithm

```
function algo_bellman(A,S)
    for each vertex X in A
        dist[X] = infinite
            prev[X] = NULL
    dist[S] = 0
    for each vertex X in A
        for each edge (Y,X) in A
            tempDist = dist[Y] + edge_wt(Y,X)
            if tempDist < dist[X] then
                dist[X] = tempDist[X]
                prev[X] = Y
    for each edge (Y,X) in A
        if dist[Y] + edge_wt(Y,X) < dist[X]
            print 'Negative Cycle Exists'
    return dist[], prev[]
```

A dynamic programming solution, for example, breaks down optimisation problems into smaller pieces and reuses previous calculations to be more efficient than a recursive approach. The algorithm approximates function $j()$ which gives the shortest path from initial to final node.

The Bellman-Ford equation is written as

$$j(\text{node } A) = \min[C(\text{node } A, \text{node } B) + j(\text{node } B)] \text{ over all node } B \\ \times \text{ which are neighbours of node } A, \quad (2)$$

where $C() =$ transition cost and $j() =$ cost from initial to final node.

5 Results and Discussion

Our experiment was divided into two phases. In each phase, we took a total of three test cases (paths) and recorded the route it takes to reach the destination. We then tested each path with four algorithms, i.e. Dijkstra's, $A*$, uniform cost search, and Bellman-Ford, and the time taken to execute each algorithm was recorded. For the first phase, we assumed that there was no traffic and recorded the results. In the second phase, we simulated traffic and found out the time taken to find an alternative route and the new traversing path. All the obtained results were then plotted into the map.

5.1 Finding Shortest Path from Set El Sham (Node 0) to McDonald's (Node 46)

For Table 1, we chose the starting node as Set El Sham (0) and the destination node as McDonald's (0). The shortest route calculated in phase 1 is [0, 7, 46], and the shortest route in phase 2 (*paths 0–7 and 7–46 are unavailable*) is [0, 19, 7, 45, 46] (Figs. 2 and 3).

As seen from the results in phase 1 and phase 2, the execution time of each algorithm arranged in increasing order is

$$A^* \ll \text{Dijkstra's} \ll \text{Uniform cost search} \ll \text{Bellman-Ford}.$$

Table 1 Experimental results for the route from Set El Sham (0) to McDonald's (46)

	Without traffic management (phase 1)				With traffic management (phase 2)			
	Dijkstra's	*A**	Uniform cost search	Bellman-Ford	Dijkstra's	*A**	Uniform cost search	Bellman-Ford
Traversing path	[0, 7, 46]	[0, 7, 46]	i. [0, 7, 46] ii. [0, 7, 45, 46]	[0, 7, 46]	[0, 19, 7, 45, 46]	[0, 19, 7, 45, 46]	[0, 19, 7, 45, 46]	[0, 19, 7, 45, 46]
Time taken to execute (in sec)	1.0057	1.0056	1.0133	1.0439	1.00680	1.00434	1.01254	1.03711

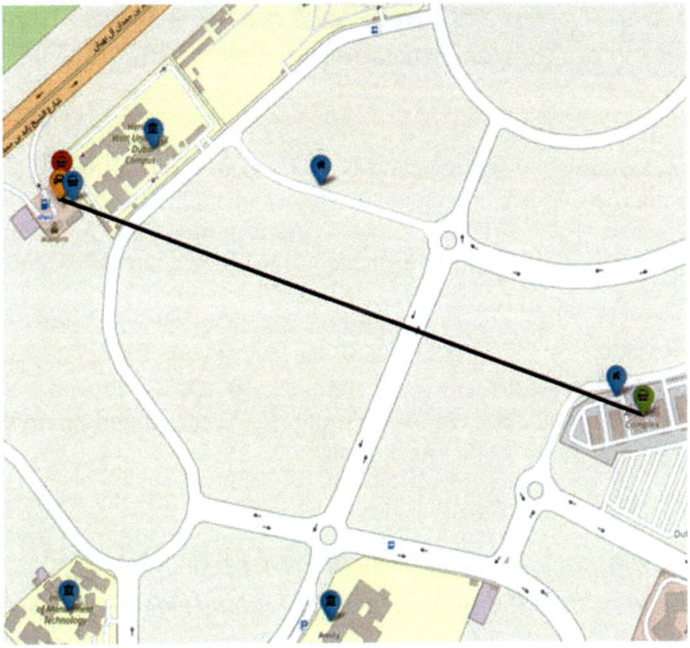

Fig. 2 Graphical representation of the route [0, 7, 46] (without traffic simulation)

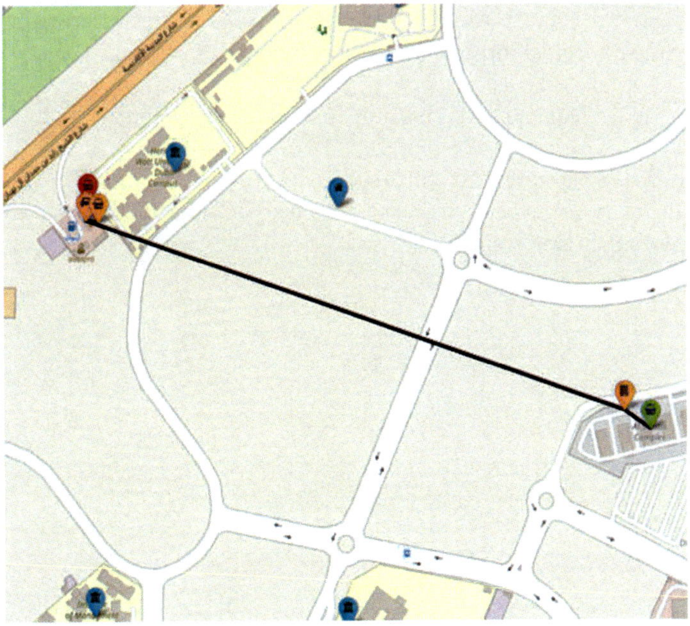

Fig. 3 Graphical representation of the route [0, 19, 7, 45, 46] (with traffic simulation)

5.2 *Finding Shortest Path from IMT Business School Dubai (Node 9) to German International School Dubai (Node 29)*

For Table 2, we chose the starting node as IMT Business School Dubai (9) and the destination node as German International School Dubai (29). In phase 1, the shortest route calculated is [9, 27, 29] (refer to Fig. 4). As seen from the experimental results in this phase, the execution time of each algorithm arranged in increasing order is

$$\text{Dijkstra's} \ll A^* \ll \text{Uniform cost search} \ll \text{Bellman-Ford}.$$

In phase 2, the shortest route calculated is [9, 10, 27, 28, 29] (refer to Fig. 5). Similarly, arranging the observed execution times for each algorithm in this phase in increasing order produces the same result.

5.3 *Finding Shortest Path from BITS-Dubai (Node 1) to Golden Pak Restaurant FZ LLC (Node 44)*

For Table 3, we chose the starting node as BITS-Dubai (1) and the destination node as Golden Pak Restaurant FZ LLC (44). In phase 1, the shortest route calculated is [1, 19, 44] (refer to Fig. 6). As seen from the experimental results in this phase, the execution time of each algorithm arranged in increasing order is

$$A^* \ll \text{Dijkstra's} \ll \text{Uniform cost search} \ll \text{Bellman-Ford}.$$

In phase 2, the shortest route calculated is [1, 0, 44] (refer to Fig. 7). Similarly, arranging the observed execution times for each algorithm in this phase in increasing order produces the same result.

Table 2 Experimental results for the route from IMT Business School Dubai (9) to German International School Dubai (29)

	Without traffic management (phase 1)				With traffic management (phase 2)			
	Dijkstra's	*A**	Uniform cost search	Bellman-Ford	Dijkstra's	*A**	Uniform cost search	Bellman-Ford
Traversing path	[9, 27, 29]	[9, 27, 29]	i. [9, 27, 29] ii. [9, 27, 28, 29]	[9, 27, 29]	[9, 10, 27, 28, 29]	[9, 10, 27, 28, 29]	[9, 10, 27, 28, 29]	[9, 10, 27, 28, 29]
Time taken to execute (in sec)	1.00547	1.00691	1.01035	1.10689	1.00823	1.00534	1.00899	1.11679

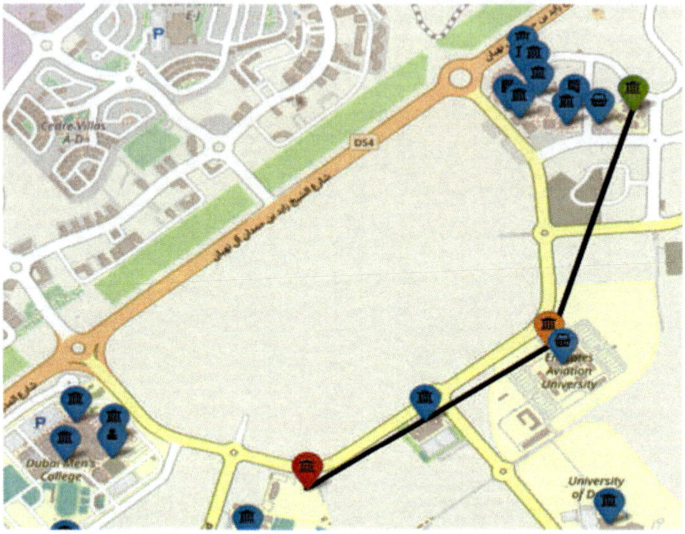

Fig. 4 Graphical representation of the route [9, 27, 29] (without traffic simulation)

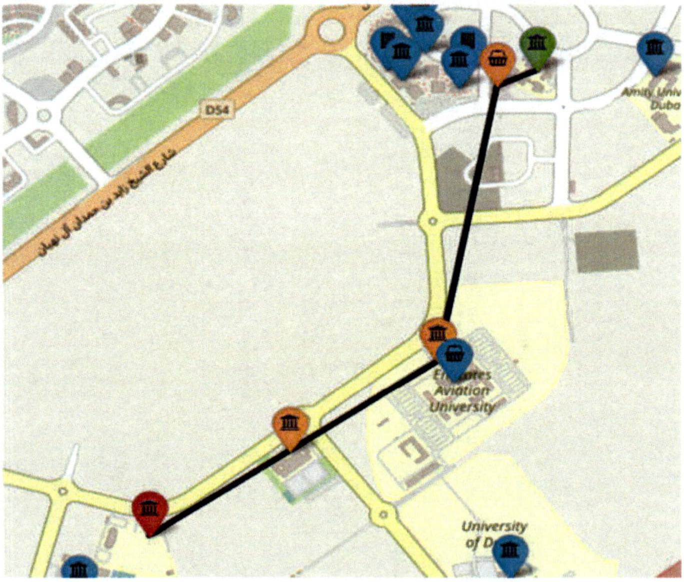

Fig. 5 Graphical representation of the route [9, 10, 27, 28, 29] (with traffic simulation)

Table 3 Experimental results for the route from BITS-Dubai (1) to Golden Pak Restaurant FZ LLC (44)

	Without traffic management (phase 1)				With traffic management (phase 2)			
	Dijkstra's	*A**	Uniform cost search	Bellman-Ford	Dijkstra's	*A**	Uniform cost search	Bellman-Ford
Traversing paths produced	[1, 19, 44]	[1, 19, 44]	i. [1, 19, 44] ii. [1, 0, 44] iii. [1, 19, 20, 44] iv. [1, 19, 20, 24, 44] v. [1, 19, 20, 25, 44]	[1, 19, 44]	[1, 0, 44]	[1, 0, 44]	i. [1, 0, 44] ii. [1, 0, 20, 44] iii. [1, 0, 20, 24, 44] iv. [1, 0, 20, 25, 44]	[1, 0, 44]
Time taken to execute (in sec)	1.00711	1.00275	1.01324	1.04941	1.00561	1.00095	1.01173	1.05492

Fig. 6 Graphical representation of the route [1, 19, 44] (without traffic simulation)

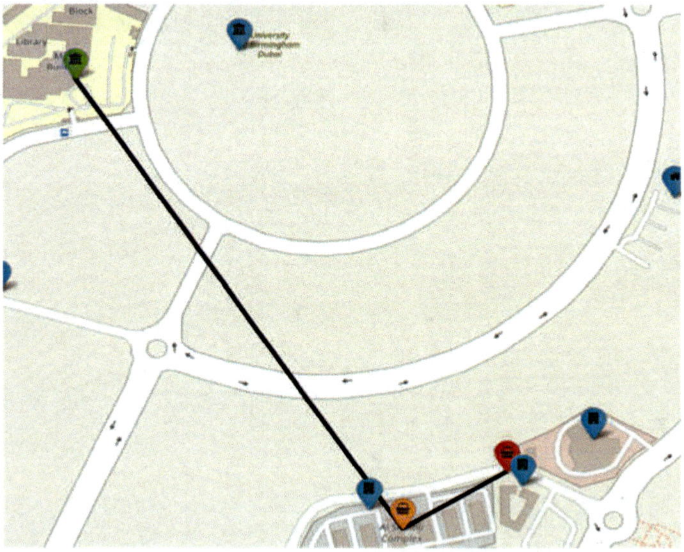

Fig. 7 Graphical representation of the route [1, 0, 44] (with traffic simulation)

6 Conclusion and Future Scope

This paper reviews multiple shortest path algorithms. Each of these algorithms was tested to find the optimal path with and without the simulation of traffic. While all the algorithms produced the same route for all the test cases, their execution time varied greatly. In Sect. 5.1, we observe that $A*$ performs the best (with execution times 1.0056 and 1.00434 s). Similarly in Sect. 5.2, $A*$ performs the best (with execution times 1.00691 and 1.00534 s). Finally in Sect. 5.3, we observe $A*$ outperforming the other algorithms (with execution times 1.00275 and 1.00095 s). Once the route was obtained, the results were plotted on the map—where the start node was marked green, the nodes traversed through were marked orange and the destination node was marked red. This was done to ease the comprehension of the results obtained and to view them in a real-life scenario. Thus, we were able to conclude that the $A*$ algorithm produces the most optimal results in the shortest time possible, followed by the greedy algorithm (Dijkstra's), then the uniform cost search method and lastly the Bellman-Ford algorithm.

The future work of this study will focus on obtaining the results in real time and on a larger dataset for deeper analysis.

References

1. Pramudita R, Heryanto H, Trias Handayanto R, Setiyadi D, Arifin R, Safitri N (2019) Shortest path calculation algorithms for geographic information systems. In: 2019 fourth international conference on informatics and computing (ICIC)
2. Haria V, Shah Y, Gangwar V, Chandwaney V, Jain T, Dedhia Y (2019) The working of google maps, and the commercial usage of navigation systems. IJIRT 6
3. Alzubaidi M, Al-Thani D (2021) Finding the safest path: the case of Yemen. In: 2021 international conference on information technology (ICIT)
4. Abu Salim S, Ibrahim R, Zainuri Saringat M, Jamel S, Abdul Wahab J (2020) Comparative analysis between Dijkstra and Bellman-Ford algorithms in shortest path optimization. IOP Conf Ser Mater Sci Eng 917:012077
5. Azis H, Mallongi R, Lantara D, Salim Y (2018) Comparison of Floyd-Warshall algorithm and greedy algorithm in determining the shortest route. In: 2018 2nd East Indonesia conference on computer and information technology (EIConCIT)

MS3A: Wrapper-Based Feature Selection with Multi-swarm Salp Search Optimization

Shathanaa Rajmohan, S. R. Sreeja, and E. Elakkiya

Abstract Feature selection is crucial in improving the effectiveness of classification or clustering algorithms as a large feature set can affect classification accuracy and learning time. The feature selection process includes choosing the most pertinent features from an initial feature set. This work introduces a new feature selection technique using salp swarm algorithm. In particular, an improved variation of the salp swarm algorithm is presented with modifications done to different stages of the algorithm. The proposed work is evaluated by first studying its performance on standard CEC optimization benchmarks. In addition to this, the applicability of the introduced algorithm for feature selection problems is verified by comparing its performance with existing feature selection algorithms. The experimental analysis depicts that the proposed methodology achieves performance improvement over existing algorithms for both numerical optimization and feature selection problems and reduces the feature subset size by 39.1% when compared to the traditional salp swarm algorithm.

Keywords Feature selection · Classification · Salp swarm algorithm · Multi-swarm · Wrapper approach

1 Introduction

Data mining is a computational task involving analysis of large datasets to find common patterns or trends. This process utilizes various techniques such as machine learning. The amount of data has been expanding quickly in recent years, and data mining has important applications such as knowledge retrieval from the data, prediction, and decision making. Finding knowledge from datasets that are articulated in

R. Shathanaa (✉) · S. R. Sreeja
Indian Institute of Information Technology, Sri City, India
e-mail: shathanaaraj@gmail.com

E. Elakkiya
SRM University, Amaravati, Andhra Pradesh, India

S. Das et al. (eds.), *Advances in Data-driven Computing and Intelligent Systems*, Lecture Notes in Networks and Systems 653, https://doi.org/10.1007/978-981-99-0981-0_38

a clear framework is the goal of data mining [1]. However, when the feature count is more than the number of patterns, an occurrence called as curse of dimensionality, it results in a huge number of classifier parameters, which is a significant issue with data mining applications like pattern recognition [2]. As a result, the classifier's performance can suffer and the data processing's computing complexity might increase. Feature selection, i.e., deleting unnecessary and repetitive features and choosing a subset of features from the input feature set, is a frequent method for solving this challenge. Feature selection is considered as an essential preprocessing step for classification and clustering problems [3]. Finding useful and non-redundant characteristics with the aid of feature selection techniques helps to minimize the system's complexity in terms of space and time. The fewer essential features also offer improved performance.

An assessment metric to determine the quality of subsets and a search approach to produce suitable feature subsets are needed in the search for the best feature subset [3]. The assessment algorithms can be categorized into filter and wrapper methods [4]. The filter approach estimates the quality of the selected features using metrics that are ingrained in the data and is not reliant on specific classification algorithms. Wrapper approaches measure the quality of the candidate subsets by evaluating it based on the performance of a classification algorithm. When comparing these two approaches, filter methods take less computation time when compared to wrapper methods. However, the accuracy of classification results is better with wrapper-based methods because of the built-in weight the classification adds to the selection approach.

Feature selection falls under the NP-hard set of problem category, and the optimal subset can be found by doing an exhaustive search that takes into account every conceivable subgroup. However, even with feature sets of a reasonable size, it is not feasible to do an exhaustive search. A suitable methodology for solving such problems involves using evolutionary algorithms, which are population-based optimization techniques that can be applied for single and many objective optimization problems. These algorithms are a popular choice for solving different problems from varying domains and have shown superior results.

This work focuses on building a feature selection process using the salp swarm algorithm (SSA), which is a contemporary evolutionary algorithm.

The subsequent sections are arranged as follows: Sect. 2 presents related work done on wrapper-based feature selection, Sect. 3 gives background on SSA, Sect. 4 describes the proposed work, Sect. 5 gives the experimental setup and findings, and Sect. 6 concludes the paper.

2 Related Work

Wrapper methods utilize a classifier to assess the efficiency of the selected feature subset. The classifier is used as a black box, and its prediction performance is integrated to the objective function. An exhaustive search entails evaluating 2^n subsets for a n-dimensional feature set. Therefore, a heuristic search algorithm could be used to find optimal/sub-optimal solutions in a feasible amount of time. The search approach usually proceeds in a sequential or random manner [5]. In a sequential search, features are either added to or deleted from the original set iteratively. However, this method has a high chance of getting stuck in the local optimum. Instead, randomness can be included in the search for avoiding getting stuck with sub-optimal solutions.

Under random methods, the application of evolutionary algorithms for feature selection problems has been investigated previously and has shown promising results. A genetic algorithm (GA) is a popular choice for feature selection and has been used in [6, 7]. An updated genetic algorithm, called CHCGA, was used in [6] with updates in the crossover and reproduction steps. A multi-objective genetic algorithm was proposed in [7] and was applied to feature selection for handwriting recognition problem. Particle swarm optimization (PSO) was integrated with GA for feature selection and was presented in [8] and applied to optimize investment portfolios. Apart from traditional evolutionary algorithms like GA and PSO, other modern evolutionary algorithms have also been widely used for feature selection. A detailed survey on different feature selection methods can be found in [5]. The research on optimizing the feature selection is still a growing field as constant research is being done to improve the accuracy of classification by finding the best feature subset.

3 Salp Swarm Algorithm

Salp swarm algorithm (SSA) is a recent evolutionary algorithm introduced in 2017 [9]. SSA simulates the behavior of salps during ocean scrounging. Salps typically create a structure called as salp chain in dense waters when searching for food. The salp in the first position is the leader in the SSA algorithm, while the remaining salps are referred to as followers. The salp positions are described as D-dimensional vectors where D denotes the problem dimension. Each salp position depicts a candidate solution in the search space, and the salps evolve and move in the search space through multiple generations. Eventually, the salp swarm will converge on the optimal solution and the algorithm can be terminated based on convergence criteria.

The steps in SSA are explained as follows: A population of salps is created initially with N salps each with D dimensions which can be represented as

$$S = \begin{bmatrix} s_1^1 & s_2^1 & \cdots & s_D^1 \\ s_1^2 & s_2^2 & \cdots & s_D^2 \\ \vdots & \vdots & \vdots & \vdots \\ s_1^N & s_2^N & \cdots & s_D^N \end{bmatrix}. \tag{1}$$

The leader salp can change position by using the next equation

$$s_j^1 = \begin{cases} G_j + k_1\left((u_j - l_j)k_2 + l_j\right) & k_3 \geq 0.5 \\ G_j - k_1\left((u_j - l_j)k_2 + l_j\right) & k_3 < 0.5 \end{cases}. \tag{2}$$

Here, s_j^1 is the jth dimension value of the leading salp. G_j is value of the global best solution (in current iteration) for dimension j. l_j and u_j are the lower and upper limit of jth dimension respectively. k_2 and k_3 are random numbers from [0, 1]. k_1 is calculated using Eq. (3).

$$k_1 = 2\mathrm{e}^{-\left(\frac{4\text{iter}}{\max\text{Iter}}\right)^2} \tag{3}$$

Here, iter denotes the current iteration of the population evolution and max Iter is total number of generations to evolve the population. To change position of the rest of the population, Eq. (4) is utilized.

$$s_j^i = \frac{s_j^i + s_j^{i-1}}{2}, \tag{4}$$

where s_j^i is the value of ith salp in jth dimension for $i > 1$ to exclude the leader.

The SSA algorithm is different from other evolutionary algorithms in its concept of a leader salp which guides the exploration, and it has shown good results for different applications. The existing works in the literature employ SSA for various applications such as COVID-19 and other related medical analysis [10, 11], engineering optimization [12, 13], and various other applications. In addition to these, SSA algorithm has been utilized for feature selection problem previously [14].

Although the simple steps of SSA reduce the exploration time, for complex problems the existing exploration is not sufficient to converge on the global optima quickly. To improve the exploration mechanism and add additional exploitation capability to aid quick convergence, this work designs an improved variation of the salp swarm algorithm namely multi-swarm salp search algorithm (MS3A).

4 Proposed Work

The proposed algorithm named as multi-swarm salp search algorithm (MS3A) addresses the issues in the traditional SSA by introducing the following concepts:

- A multi-swarm salp approach to explore maximum regions of search space thoroughly thus addressing the poor exploration problem.
- A novel swarm leader movement approach to quickly guide the subswarms toward convergence.
- A β-hill optimization applied to the leader of each swarm to balance exploration and exploitation.

4.1 Problem Formulation

To apply the salp swarm algorithm for feature selection, the population is encoded as binary vectors (s) representing the inclusion or exclusion of a feature in the selected feature subset. The size of the binary vector will be equal to number of features. To evaluate the candidate solutions and the selected feature subsets using the wrapper approach, the following fitness function to be minimized is used.

$$\text{fitness}(s) = w(1 - \text{acc}(s)) + (1 - w)\frac{d}{D}. \tag{5}$$

Here, d gives the count of ones in binary vector s indicating the size of feature subset. D is the feature count in original feature set. $\text{acc}(s)$ gives the classification accuracy achieved by the feature subset s, and $0 < w \leq 1$ is the weight assigned.

4.2 Initialization

The initial salp population is generated randomly, with X number of subswarms. Every subswarm contains N salp individuals representing candidate solutions. Each individual has D dimensions representing the total number of features

$$s_j^i(x) = \begin{cases} 1 & p_1 \geq 0.5 \\ 0 & p_1 < 0.5 \end{cases}. \tag{6}$$

Here, p_1 is a random value from [0, 1]. The proposed work follows this simple encoding approach to formulate the binary SSA. $s_j^i[x]$ denotes the jth dimension value of salp individual i from subswarm s. If it is set to 1, it implies that the jth feature is added to candidate feature subset, otherwise it will not be included.

4.3 Multi-swarm Exploration

After initialization, fitness of salp individuals is determined using Eq. (5). The leader in each subswarm is selected based on the fitness, and the salp with the smallest fitness score is taken as the global best.

The first step in exploration involves movement of the leader salps. For each pair of leader salps from subswarm x and y, the movement is calculated as follows:

$$s_j^1(x) = s_j^1(x) + \left(s^1(x) - s^1(y)\right)\alpha\delta, \tag{7}$$

where $s_j^1(x)$ and $s_j^1(y)$ are the leaders of the subswarms x and y, respectively, and $s_j^1(y)$ has better fitness out of the two. α is the step size, and δ is drawn from the normal distribution. $\left(s^1(x) - s^1(y)\right)$ is the Euclidean distance between the two leaders. This step creates the global movement of the subswarms and ensures quick convergence.

The next step is applicable for each subswarm. The leader of each subswarm is updated using Eq. (2) from the original salp swarm algorithm which was presented earlier. Once the subswarm leaders are updated, the rest of the individuals follow their respective leaders and are updated using Eq. (3).

At the end of exploration, two additional steps are performed. Each swarm will be merged with the global best swarm with a probability of 0.25. This is simply done by making the global leader as the leader of the swarm selected for merging. When the leader's position is updated, the rest of the subswarm will follow suit. This is done to ensure that the population converges eventually to the global optimum. After this local search using β-hill climbing is performed.

An important step in the proposed algorithm is the conversion of the candidate solutions to the binary form. After each generation, the population should be converted to binary form using Eq. (6) to aid fitness calculation.

4.4 Exploitation with Local Search

β-hill climbing is a local search technique introduced in [15]. When compared with other hill climbing algorithms, it has shown promising results. The proposed algorithm integrates β-hill climbing as an exploitation step to further improve the positions of subswarm leaders. For each subswarm leader $s^1(x)$ from every swarm x, the position of a randomly chosen dimension j is updated as follows:

$$s_j^1(x) = s_j^1(x) \pm b_1 \times bw, \tag{8}$$

where bw is the bandwidth parameter and b_1 is a random quantity from [0, 1].

After this, the value in each dimension j is updated as

$$s_j^1(x) = \begin{cases} s_r & b_2 \le \beta \\ s_j^1(x) & b_2 > \beta \end{cases}. \tag{9}$$

s_r is a random number from $\left[l_j, u_j\right]$, the limits of jth dimension. b_2 is uniform random value from [0, 1]. Figure 1 illustrates the flow of the presented approach.

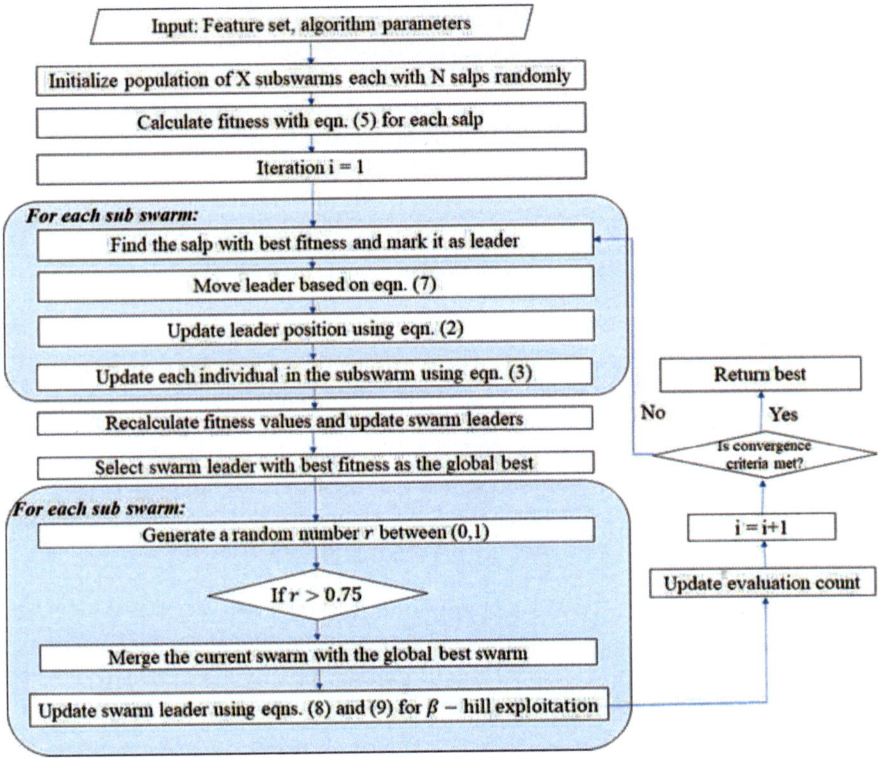

Fig. 1 Flow diagram showing steps of the proposed MS3A algorithm

5 Results and Discussion

The performance of the proposed approach is analyzed on two fronts: on 13 CEC benchmark functions selected from the literature [16] and 10 datasets for the feature selection task [17]. Every algorithm is executed for 20 times for evaluation. The experiments were carried out in an 11th Gen Intel i7 processor with 8 GB primary memory.

5.1 Results for Global Optimization Problems

MS3A is studied against original salp swarm optimization [9] and 5 other widely used evolutionary algorithms. Table 1 provides the list of algorithms compared along with their parameter settings based on the literature. The parameter settings for the proposed algorithm were finalized based on sensitivity analysis. Other common parameters were set as: the number of iterations was set to 500, population size as

Table 1 Parameter settings for different evolutionary algorithms

Algorithm	Parameter values
Particle swarm optimization (PSO)	$c1 = c2 = 1.8$; $w =$ from 1 to 0
Symbiotic organisms search (SOS)	No parameters to be tuned
Harris Hawks optimization (HHO)	$E_0 = (-1, 1)$; $\beta = 1.5$
Gray wolf algorithm (GWO)	$a = [2, 0]$
Whale optimization (WOA)	$a =$ linearly decreased from 2 to 0; $b = 1$
Multi swarm salp search algorithm (MS3A)	$w = 0.99$; $\alpha =$ linearly decreased from 1 to 0; $\beta = 0.05$; $bw = 0.5$

50 and the algorithms were tested for 20 runs. Table 2 shows the fitness values for different optimization benchmarks ($f1$–$f13$). The names of the functions are given in the second row, and the benchmark details are available in [15]. The overall best values for best and average fitness and standard deviation are highlighted in bold. The proposed MS3A algorithm converges on global optima for all benchmarks. Also, the average fitness of MS3A is superior to other algorithms for most of the benchmarks. Similarly, the standard deviation value for MS3A is minimal for most benchmarks, thus verifying the robustness of the algorithm.

Figure 2 presents the convergence plot of different algorithms for selected benchmarks. As shown by the plot, proposed algorithm shows early movement to better fit solutions and converges on best fitness values when compared to other algorithms.

5.2 Feature Selection Results

MS3A is studied against existing works for the feature selection task based on ten standard datasets considered from the UCI repository [17]. The properties of the benchmarks are available in Table 3. The selected features are evaluated on K-nearest neighbor (KNN) classifier with fivefold cross validation. MS3A algorithm is compared to other recent state-of-the-art works done on wrapper-based feature selection using evolutionary algorithms viz. BWOA [18], PSOC [19], AOA [20], and SSA [14]. Tables 4 and 5 show the fitness values obtained by the proposed and existing algorithms. The fitness values are calculated using Eq. (5), introduced earlier. The results with lower fitness values are considered better. The best fitness values obtained for each dataset are presented in bold. The observation shows that the proposed MS3A algorithm finds the best fitness solutions for the maximum number of datasets. As given by Eq. (5), the fitness is determined by both the classification accuracy and feature count.

Table 6 gives the average accuracy obtained from classification while using the feature subsets selected by different algorithms. The findings from Table 6 show that the proposed MS3A achieves better accuracy by selecting the optimal subset when contrasted against other existing algorithms. The overall average accuracy obtained

Table 2 Average, best, and standard deviation of fitness values obtained by proposed and existing methodologies

Alg	Value	$f1$	$f2$	$f3$	$f4$	$f5$	$f6$	$f7$	$f8$	$f9$	$f10$	$f11$	f12	f13
		Sphere	Bent cigar	Elliptic	Drop wave	Rosenbrock	Rastrigin	Alpine 1	Griewank	Penal 1	Penal 2	Branin RCOS	Hartmann	Zakarov
MS3A	Avg	**0**	**0**	**0**	**– 1**	**3.8e4**	**0**	9.3e–7	**0**	**1.0e–6**	**8.3e–6**	**0.39**	**– 3.8**	**0**
	SD	**0**	**0**	**0**	**0**	**4.9e4**	**0**	2.8e–6	**0**	**1.5e–6**	**8.1e–6**	1.2e–05	0.0003	**0**
	Best	**0**	**0**	**0**	**– 1**	**1.e–14**	**0**	**4.3e–263**	**0**	**9.4e–14**	**9.6e–8**	**0.39**	**– 3.8**	**0**
WOA	Avg	3e–94	8.3e–85	4.5e–96	– 0.98	0.6	**0**	4.9e–61	**0**	0.0016	0.01	**0.39**	– 3.8	**0**
	SD	5.4e–94	2.4e–84	9.8e–96	0.02	0.9	**0**	1.1e–60	**0**	0.0017	0.01	4.3e–05	0.05	405.9
	Best	2e–103	9.4e–98	4e–102	**– 1**	0.004	**0**	5.5e–64	**0**	2.17e–5	0.001	**0.39**	– 3.8	171.1
PSO	Avg	3329.7	3.0e09	26.3	– 0.99	845.6	8.9	18.9	33.6	6828.1	409.1	**0.39**	– 3.8	670.2
	SD	386.7	3.1e08	5.09	0.0002	665.7	1.03	2.5	3.8	9137.6	152.4	1.6e–05	3.6e–5	93.3
	Best	2544.5	2.6e09	19.8	**– 1**	7278	7.1	16.1	27.4	46.8	191.8	**0.39**	– 3.8	91.1
SOS	Avg	6e–148	6.4e–144	1.3e–151	– 1	25.8	**0**	**8.9e–77**	**0**	0.01	0.12	**0.39**	– 3.8	28.8
	SD	2e–147	1.7e–143	3.4e–151	0	0.521	**0**	**8.3e–77**	**0**	0.03	0.12	**0**	**8.8e–16**	132.2
	Best	1e–152	2.8e–146	8.6e–155	**– 1**	25.0	**0**	4.2e–78	**0**	6.6e–11	0.019	**0.39**	– 3.8	48.3
SSA	Avg	2.2e–8	586.6	2.6	– 1	130.7	5.2e–9	2.1	0.01	4.9	3.915	**0.39**	– 3.8	5.75E–61
	SD	6.5e–0	439.2	1.1	2.6e–13	123.8	1.2e–9	1.1	0.01	1.5	7.01	1.6e–14	7.8e–14	1.5E–60
	Best	8.3e–9	1.2	1.3	**–1**	22.7	2.6e–9	0.7	7.4e–6	1.7	3.5e–6	**0.39**	– 3.8	5.2E–60
HHO	Avg	2e–102	4.0e–98	4.7e–102	– 1	0.002	**0**	3.3e–07	**0**	2.3e–6	3.5e–5	**0.39**	– 3.8	1.3E–67
	SD	6e–102	8.3e–98	1.4e–101	0	0.002	**0**	9.9e–07	**0**	4.1e–6	2.7e–5	4.4e–08	1.5e–05	13.5
	Best	1e–115	1.3e–108	2e–119	**– 1**	0.001	**0**	3.1e–60	**0**	6.8e–8	7.0e–7	**0.39**	– 3.8	5.7
GWO	Avg	4.9e–36	2.6e–30	8e–39	– 1	26.7	**0**	0.0003	**0**	0.02	0.2	**0.39**	– 3.8	24.2
	SD	6.0e–36	3.0e–30	6.89e–39	0	0.7	**0**	0.0004	0.01	0.009	0.09	6.9e–07	0.002	5.6

(continued)

Table 2 (continued)

Alg	Value	*f*1	*f*2	*f*3	*f*4	*f*5	*f*6	*f*7	*f*8	*f*9	*f*10	*f*11	f12	f13
		Sphere	Bent cigar	Elliptic	Drop wave	Rosenbrock	Rastrigin	Alpine 1	Griewank	Penal 1	Penal 2	Branin RCOS	Hartmann	Zakarov
	Best	3.7e−37	1.1e−31	5.37e−40	**− 1**	25.7	**0**	4.6e−20	**0**	0.006	0.2	**0.39**	− 3.86	2.7E−65

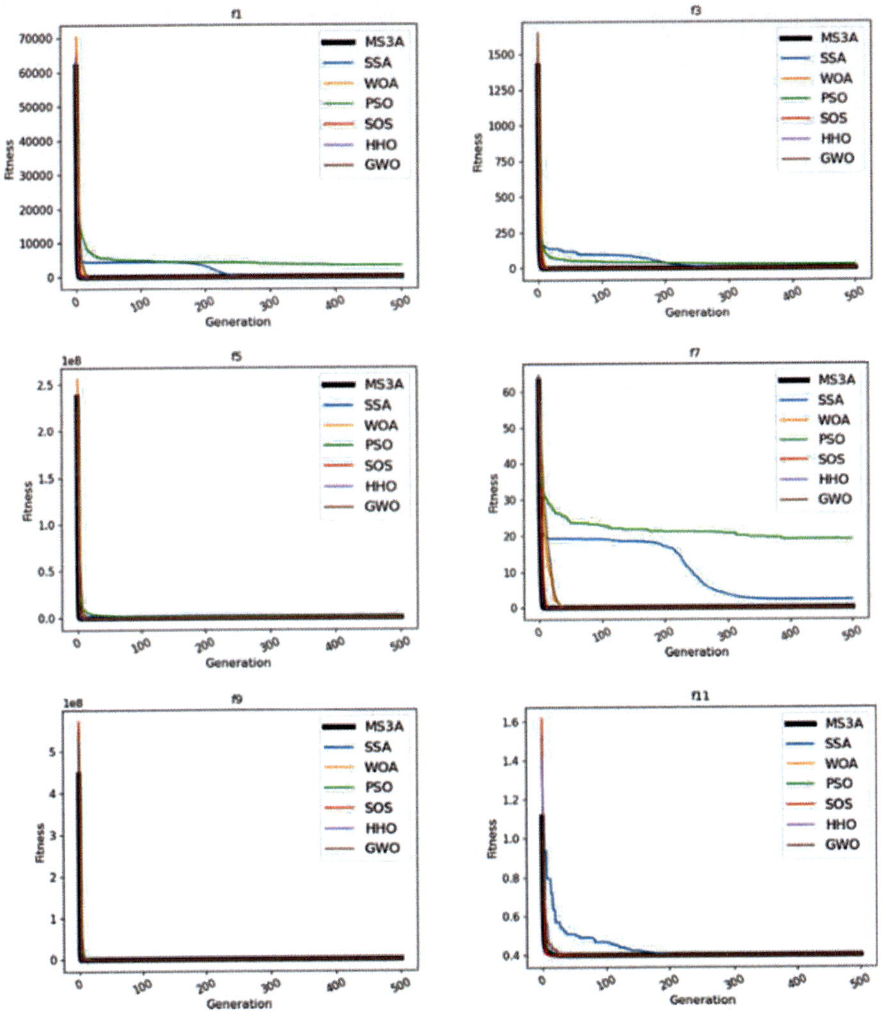

Fig. 2 Convergence plots for proposed and existing algorithm for selected benchmarks

by MS3A for all benchmarks is 0.904. Although BWOA and PSOC algorithms closely follow this with an overall average accuracy of 0.88, they use more features and take additional time for exploration.

The detailed comparison of feature count selected by different algorithms is given in Table 7. The average percentage of features included in the feature subset by each algorithm over 20 runs is presented in Table 7. The proposed algorithm picks minimal number of features when compared to existing algorithms that add maximum information to the classification process. The number of features selected can be modified by adjusting the weight factor w used in Eq. (5). The weight assigned to feature count is $1-w$ and a good value for w will in the range [0.6–1.0]. Increasing w

Table 3 Description of the feature selection datasets used

Dataset	Feature count	Instance count
Breastcancer1 (original)	10	699
Breastcancer2 (diagnostic)	32	569
Congress	16	435
Exactly	3	1000
Statlog (heart)	13	270
Ionosphere	34	351
Lymphography	18	148
Tic tac toe	9	958
WaveformEW	40	5000
Wine	13	178

Table 4 Best, mean, and standard deviation of fitness of the solutions found by proposed and existing algorithms for feature selection problem

	Proposed			BWOA			AOA		
	Best	Avg	SD	Best	Avg	SD	Best	Avg	SD
Breastcancer1	**0.024**	0.027	0.002	0.027	0.03	0.002	0.028	0.033	0.002
Breastcancer2	**0.045**	0.047	0.001	0.047	0.05	0.001	0.051	0.057	0.006
Congress	**0.03**	0.034	0.003	0.031	0.035	0.003	0.042	0.046	0.003
Exactly	0.005	0.011	0.01	0.005	0.005	0	0.019	0.256	0.086
Statlog (heart)	**0.135**	0.141	0.003	0.136	0.149	0.007	0.202	0.23	0.02
Ionosphere	**0.069**	0.082	0.01	0.069	0.093	0.014	0.126	0.132	0.004
Lymphography	0.107	0.131	0.013	**0.106**	0.146	0.021	0.158	0.178	0.011
Tic tac toe	**0.154**	0.157	0.002	0.156	0.183	0.02	0.158	0.219	0.032
WaveformEW	**0.166**	0.171	0.002	**0.166**	0.171	0.003	0.177	0.198	0.011
Wine	**0.031**	0.04	0.006	0.036	0.043	0.006	0.048	0.068	0.012

will decrease the weight given for number of features and will allow the exploration process to include more number of features in the subset without affecting the fitness.

Figure 3 gives the time taken by different algorithms which includes the classification time. The existing SSA algorithm takes the minimal time; however, it achieves sub-optimal results. The proposed algorithm takes a comparatively lesser amount of time than the other three existing algorithms, as it converges quickly on the global optima.

Table 5 Best, mean, and standard deviation of fitness of the solutions obtained by existing algorithms for feature selection problem

	PSOC			SSA		
	Best	Avg	SD	Best	Avg	SD
Breastcancer1	0.027	0.029	0.001	0.03	0.032	0.002
Breastcancer2	0.047	0.053	0.006	0.047	0.052	0.004
Congress	0.031	0.038	0.005	0.038	0.042	0.003
Exactly	**0.001**	0.076	0.101	0.189	0.259	0.026
Statlog (heart)	0.143	0.183	0.023	0.153	0.189	0.021
Ionosphere	0.099	0.119	0.014	0.08	0.104	0.009
Lymphography	0.121	0.159	0.019	0.144	0.172	0.014
Tic tac toe	0.157	0.153	0.001	0.155	0.16	0.003
WaveformEW	0.167	0.172	0.003	0.178	0.188	0.005
Wine	0.039	0.045	0.006	0.042	0.056	0.006

Table 6 Accuracy values obtained for the feature subsets selected by proposed and existing algorithms

Dataset	Algorithm				
	Proposed	BWOA	AOA	PSOC	SSA
Breastcancer1	0.97	0.969	0.966	0.968	0.968
Breastcancer2	0.945	0.945	0.941	0.942	0.94
Congress	0.96	0.96	0.955	0.957	0.957
Exactly	0.999	0.992	0.724	0.914	0.726
Statlog (heart)	0.836	0.782	0.742	0.783	0.791
Ionosphere	0.898	0.891	0.859	0.87	0.888
Lymphography	0.822	0.78	0.816	0.801	0.782
Tic tac toe	0.843	0.776	0.802	0.837	0.836
WaveformEW	0.827	0.827	0.8	0.823	0.812
Wine	0.936	0.942	0.916	0.934	0.928
Overall avg	0.904	0.886	0.852	0.883	0.863

6 Conclusion

This work studied the feature selection problem where a subset of features needs to be selected to shrink complexity of classification task. To solve this challenge, an evolutionary algorithm-based approach was presented. The proposed algorithm was an improved variation of the salp swarm algorithm which was modified on three fronts: multipopulational swarm, leader movement for exploration, and β-hill climbing for exploitation. The proposed algorithm was evaluated by first studying its performance on CEC numerical optimization benchmark. The experimental study analyzed the

Table 7 Average percentage of features selected by different algorithms in best fitness solutions

Dataset	Algorithm				
	Proposed	BWOA	AOA	PSOC	SSA
Breastcancer1	0.206	0.544	0.622	0.467	0.589
Breastcancer2	0.203	0.387	0.53	0.244	0.173
Congress	0.356	0.344	0.413	0.138	0.213
Exactly	0.485	0.462	0.615	0.143	0.938
Statlog (heart)	0.423	0.415	0.369	0.149	0.346
Ionosphere	0.121	0.262	0.479	0.175	0.15
Lymphography	0.439	0.467	0.494	0.179	0.45
Tic tac toe	0.257	0.833	0.689	1	1
WaveformEW	0.22	0.538	0.55	0.653	0.823
Wine	0.369	0.531	0.469	0.134	0.377
Overall avg	0.308	0.478	0.523	0.328	0.506

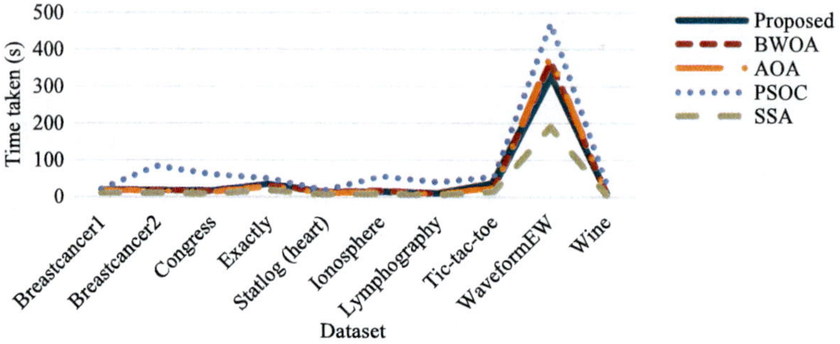

Fig. 3 Time taken by different algorithms for feature selection

proposed algorithm against other state-of-the-art evolutionary algorithms, and it was proved that the proposed algorithm achieves consistently good results against other algorithms under study. To analyze the proposed algorithm for feature selection task, the comparison was done with other recent evolutionary algorithm-based feature selection works. The results showed that the proposed methodology outperformed other existing algorithms based on both the accuracy of the results and size of the feature subset. For future work, we recommend the proposed algorithm can be studied further in the direction of enhancing the cooperation between the multiple swarms as this will improve the accuracy by helping to reach the global optimum and will avoid poor convergence.

References

1. Liu H, Yu L (2005) Toward integrating feature selection algorithms for classification and clustering. IEEE Trans Knowl Data Eng 17:491–502
2. Köppen M (2000) The curse of dimensionality. In: 5th online world conference on soft computing in industrial applications (WSC5), pp 4–8
3. Guyon I, Elisseeff A (2003) An introduction to variable and feature selection. J Mach Learn Res 3:1157–1182
4. Xue B, Zhang M, Browne WN, Yao X (2016) A survey on evolutionary computation approaches to feature selection. IEEE Trans Evol Comput 20:606–626
5. Chandrashekar G, Sahin F (2014) A survey on feature selection methods. Comput Electr Eng 40:16–28
6. Sun Y, Babbs CF, Delp EJ (2006) A comparison of feature selection methods for the detection of breast cancers in mammograms: adaptive sequential floating search vs. genetic algorithm. In: IEEE engineering in medicine and biology 27th annual conference, pp 6532–6535
7. Oliveira L, Sabourin R, Bortolozzi F, Suen CY (2003) A methodology for feature selection using multiobjective genetic algorithms for handwritten digit string recognition. Int J Pattern Recognit Artif Intell 17:903–929
8. Kuo RJ, Hong CW (2013) Integration of genetic algorithm and particle swarm optimization for investment portfolio optimization. Appl Math Inf Sci 7:2397
9. Mirjalili S, Gandomi AH, Mirjalili SZ, Saremi S, Faris H, Mirjalili SM (2017) Salp swarm algorithm: a bio-inspired optimizer for engineering design problems. Adv Eng Softw 114:163–191
10. Zhang Q, Wang Z, Heidari AA, Gui W, Shao Q, Chen H, Zaguia A, Turabieh H, Chen M (2021) Gaussian barebone salp swarm algorithm with stochastic fractal search for medical image segmentation: a COVID-19 case study. Comput Biol Med 139:104941
11. Al-qaness MAA, Ewees AA, Fan H, Abd El Aziz M (2020) Optimization method for forecasting confirmed cases of COVID-19 in China. J Clin Med 9(3)
12. Dou J, Ma H, Zhang Y, Wang S, Ye Y, Li S, Hu L (2022) Extreme learning machine model for state-of-charge estimation of lithium-ion battery using salp swarm algorithm. J Energy Storage 52:104996
13. Abbassi R, Abbassi A, Heidari AA, Mirjalili S (2019) An efficient salp swarm-inspired algorithm for parameters identification of photovoltaic cell models. Energy Convers Manage 179:362–372
14. Faris H, Mafarja M, Heidari A, Aljarah I, Al-Zoubi A, Mirjalili S et al (2018) An efficient binary salp swarm algorithm with crossover scheme for feature selection problems. Knowl Based Syst 154:43–67
15. Al-Betar MA (2016) β-hill climbing: an exploratory local search. Neural Comput Appl 28:153–168
16. Jamil M, Yang X-S (2013) A literature survey of benchmark functions for global optimization problems. arXiv preprint arXiv: 1308.4008
17. Dua D, Graff C (2017) UCI machine learning repository. http://archive.ics.uci.edu/ml
18. Tawhid MA, Ibrahim AM (2020) Feature selection based on rough set approach, wrapper approach, and binary whale optimization algorithm. Int J Mach Learn Cybernet 11:573–602
19. Adamu AT, Abdullahi M, Junaidu SB, Hassan IH (2018) A hybrid particle swarm optimization with crow search algorithm for feature selection. Mach Learn Appl 6:100108
20. Ewees AA, Al-qaness MAA, Abualigah LM, Oliva D, Algamal ZY, Anter AM, Ibrahim RA, Ghoniem RM, Elaziz MA (2021) Boosting arithmetic optimization algorithm with genetic algorithm operators for feature selection: case study on cox proportional hazards model. Mathematics 9(18):2321

BugFinder: Automatic Data Extraction Approach for Bug Reports from JIRA-Repositories

Rashmi Arora and Arvinder Kaur

Abstract Issue tracking systems (ITS) are used by organizations to organize and document the work of their projects. GitHub, GitLab, and Bugzilla are among the ITSs that have been explored thus far. Jira is one of the most extensively used ITSs in practice. Jira includes a lot of information on many projects. Each project confronts a variety of difficulties, including bug reports, improvements to already-existing features, features for a newer version, and completed jobs. Since each sort of problem has a unique set of characteristics, gathering such a large amount of data manually would be laborious, potentially fault-prone, and time-consuming. To eliminate errors brought on by human error and enhance accuracy, our main goal is to automate compiling bug reports. This paper proposed an automated bug report extraction approach called BugFinder. Our technique is implemented in Python and pulls data from the Jira repository. We used an automated technique to retrieve issues or bug reports from 15 projects and extracted 176,953 issues from the Apache JIRA repository and out of 103,018 are bug reports. The produced reports include issue characteristics such as key, priority, resolution, summary, and numerous more attributes. These reports can be used to identify various sorts of defects, such as security, memory, aging defects, and so on. The result also demonstrates that the number of bugs in Brooklyn is 72%, which is the highest among all the given projects, and the Lucene project is 41%.

Keywords Issues tracking system · Bug report · Data extraction · Jira repository

R. Arora (✉)
Guru Tegh Bahdur Institute of Technology, New Delhi, India
e-mail: arora.rashmi@gmail.com

A. Kaur
Guru Gobind Singh Indraprastha University, New Delhi, India

S. Das et al. (eds.), *Advances in Data-driven Computing and Intelligent Systems*, Lecture Notes in Networks and Systems 653, https://doi.org/10.1007/978-981-99-0981-0_39

1 Introduction

Many businesses utilize issue tracking systems that handle and manage a wide range of issues inside agile methodologies to accelerate software maintenance and evolution. Such methods are propitious since developers, testers, and other project stakeholders may readily report various difficulties. Bug reports, modification requests, enhancements, and new tasks, for example, are all examples of issues. A bug is a weakness or defect in software that causes the software system to behave in an unanticipated way, Gegick et al. [1]. When confronted with a new defect, it is possible to identify a viable remedy by looking back through the ages of open software history. There is a good probability that yet another open-source application has previously addressed a similar problem or crash. The issue is that each open-source project has its data store and uses separate bug tracking and version control methods. Furthermore, these systems have several data access interfaces. The information is not presented invariably, either. The idea that bug tracking tools and version control systems are not always linked complicates things even further. The former is in charge of the bug's life cycle, while the latter is in charge of the repairs. As a rule, developers construct a connection between two objects. Bug tracking systems (BTSs) are used in software projects to gather and track issues such as bug reports and feature requests. Bugs are information units that typically include a title, a description, and several characteristics such as status, priority, and fix version. The evolutionary refinement of issues (also known as an iterative improvement) is a primary emphasis of BTSs, which implies that information is gathered and improved over time while developers and stakeholders work together to solve problems.

For empirical research, BTS data is a gold mine [2–4]. BTS has been frequently utilized for testing ideas about maintenance and generating statistical prediction models. The problem discussions, in which programmers debate issues by offering technical specifics intertwined with views, give valuable information about the "why" of some design decisions or the state of a project. It is possible to investigate how developers communicate as well as their feelings about the projects and their peers by looking at these remarks. As a result, BTSs are a valuable source of data for researchers studying the productivity of development teams or the effectiveness of developers [5, 6]. Beyond issue tracking, Jira is an innovative management tool that includes scrum boards, Kanban boards, and roadmap management. Jira is similar to GitHub, GitLab, and Bugzilla in terms of ticket-centric architecture. Jira has a history of issue modifications, sophisticated issue connecting networks, and a broad collection of custom field configurations across organizations, among other advantages over competing ITSs. Jira is the most widely used issue tracking and agile project management application. Despite this, software engineering research is under-represented. The dearth of available and diverse Jira data to examine, we feel, is the explanation for such a paucity of widespread Jira research. One of the most often used BTS technologies is the Jira repository. Jira incorporates basic ITS information as well as additional project management-related information. Jira, e.g., offers a particular board that is a virtual replica of the popular agile kanban boards.

Jira also allows you to keep track of the team's work, manage your backlog, and schedule sprints. Jira is a favored decision for issue tracking because of its following highlights:

- Project Management: With the Jira tracking software, users can surely track the progress of your organization's current projects at any stage. It transmits JQL (a modified variation of SQL) to expedite the resolution of problems. Furthermore, Jira provides a minimal simplified approach for making testing run consistently. These are the reasons why Jira was chosen.
- It provides a lot of reporting and is used to track issues, tasks, projects, and other things.
- It offers a powerful inquiry and can be handily coordinated with different applications like Testlink, Atlassian, etc.
- It is incredibly user-friendly, both in terms of style and functionality, and it enables a large number of custom fields.
- Jira presents information gathered from transaction processing and application development in graphics known as reports. Those reports are available in a variety of formats and punctuation for your review. They serve as an estimating tool from the initial stages of a project through product delivery and management.
- Security: Jira contains bug tracking programming that restricts access to groups that have been granted permission to deal with bug management. Furthermore, Jira's default authorization scheme adds an extra degree of security when projects are deployed to groups.

In this paper, we design a bug tracking system called BugFinder, which uses the Jira ITS to collect all aspects of issues from Apache projects. It gathers data about the problem's state, resolution, priority, component, affected version, keys, issue type, watches, summary, description, and so on. All of these features are retrieved, and reports are generated, implying that the data acquired is more useful and examined. Jira is a Cloud-based application that is well-known for its issue management and application development capabilities. The primary clients of this tool are organizations submitting projects on the iterative model of programming advancement. A membership-based support program aids organizations in achieving a common goal. JIRA allows you to monitor and perform the overall progress of your enterprise, in addition to important tasks like workload development, scheduling, and delivery process. Jira also collaborates with a variety of other tools to improve issue tracking and accelerate the integration of new programming applications.

2 Related Work

Mining software archives is a new field that focuses on obtaining application metadata such as program code, issue reports, patch logs, as well as other data from repositories. Various scholars, e.g., Bettenbur et al. [2] analyze problem submissions and extract useful data from the source coding, like structural data or related defects. The vast majority of users manually retrieve defect reports from multiple sources, with little effort spent on autonomous bug report retrieval from issue tracking platforms. BUMPER(Bug Meta repository for developers and researchers), an open-source web-based application tool developed by Nayrolles and Hamou-Lhadj [7], is introduced. It offers a framework for finding bugs, patches, and system software for four independent projects: Netbeans, Eclipse, Apache, and Gnome, all of which use query language and NLP. The program extracts the program as well as the problems that go along with it. By processing and improving bug report pages, Yuk ad Jung [6] established a strategy for gathering applicable information from documents. Crawled web pages with bug information are turned into data trees. After that, the trees are analyzed, and the final findings are saved in a local database. Mozilla has been used to test the approach. Kaur and Jindal [8] developed bug report collecting system (BRCS), a solution that connects to the Jira repository's REST APIs and gets issues within a certain range of issue Ids.

Our work is based on the automatic download the all the issues of the project based on a Jira library. Our work is implemented through a Python script through which we can download all the data at once irrespective of any pre-condition [9].

3 Methodology

One of the most often used BTS technologies is the Jira repository. Jira provides the above-mentioned basic ITS information and other project management-related information. Jira, e.g., includes a unique board that is a virtual replica of the rapid whiteboards popular among engineers [10]. Jira also helps track team progress, manage your backlog, and schedule sprints. Our application uses the Jira bug reporting system to collect bug reports from numerous projects and provide data as a result of numerous reports. Jira library is used in Python to connect with Jira remotely. It offers a consistent interface for engaging with Jira and its associated apps. In Python script, the input is the project name, and the output is a full bug report description. Figure 1 depicts the basic data extraction process from the JIRA repository. There are two methods for obtaining the data using Python Script:

- Using Jira library in Python
- Using JIRA REST API in Python.

Using the Jira bug reporting system, we collect bug reports from many Apache projects and create the reports as a result. Jira library is a tool that allows you to interact with Jira from afar. It provides a uniform user experience for engaging with Jira and the rest of its different applications of it.

Our approach is based on Jira library in Python as it is easy to use as compared to the Jira Rest API method. The only pre-requisite for the methods is to have an account on the Jira webpage. Figure 2 explains the steps of using the JIRA Python library for extracting the issues from the JIRA repository. These procedures are as follows:

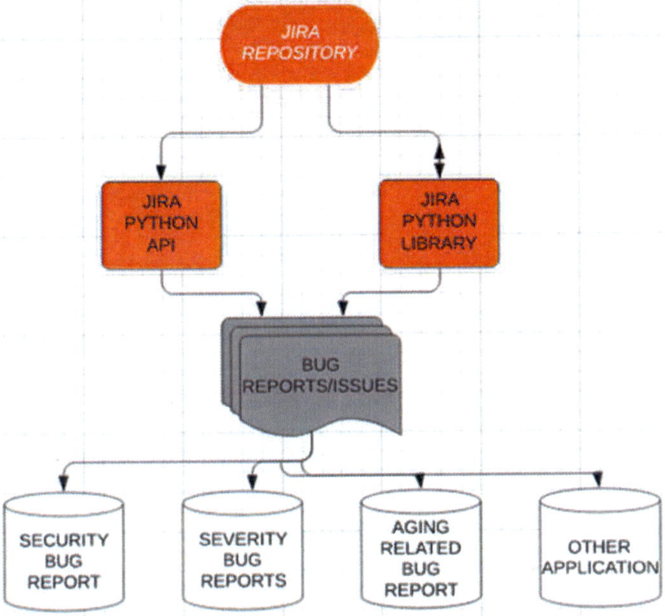

Fig. 1 JIRA

- Install the Jira module.
- Create a Jira client instance by creating the Atlassian account and the server key, which is also your domain name, is generated. Obtain in a JIRA client instance.
- Search all the issues against a project name that we want to extract from the repository.
- The process will fetch all the information stored in the project such as assignee, components, created, creator, description, resolution, resolution. date, status. description, issue type, key, priority, status, subtask, summary, updated versions, and watches as shown in Fig. 3.
- The process will continue to search until it will not get the details of every issue.

3.1 Elements of Bug Report

The quantity and quality of information supplied about the discovered problem in bug reports frequently differ. A thorough explanation of a failure is usual in a report. If the information provided there is not legitimate, developers have not been able to identify the exact position of the issue, which is insufficient to identify the source of the problem and the ability to start a forum about the report to address concerns.

Fig. 2 Data extraction approach

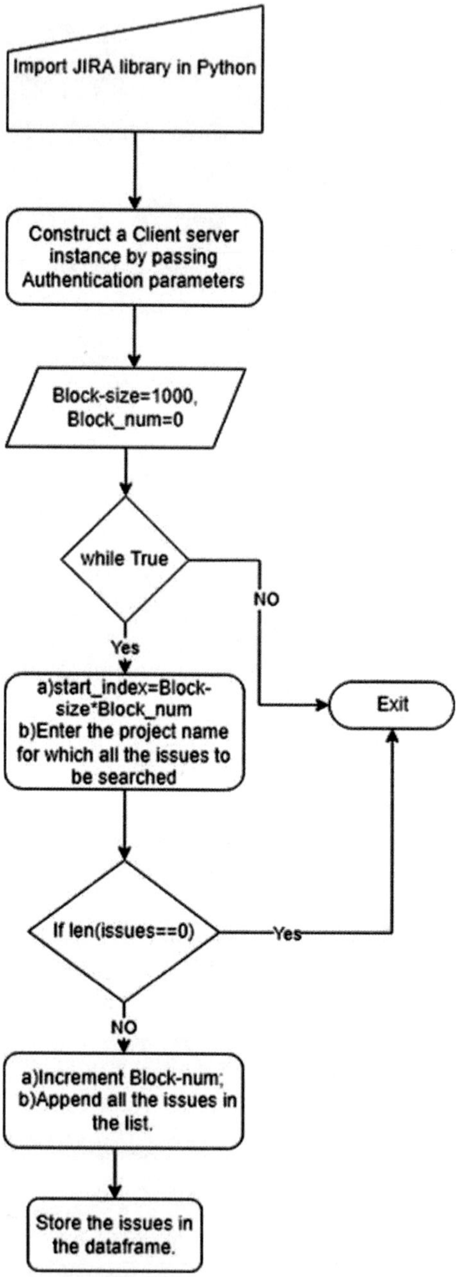

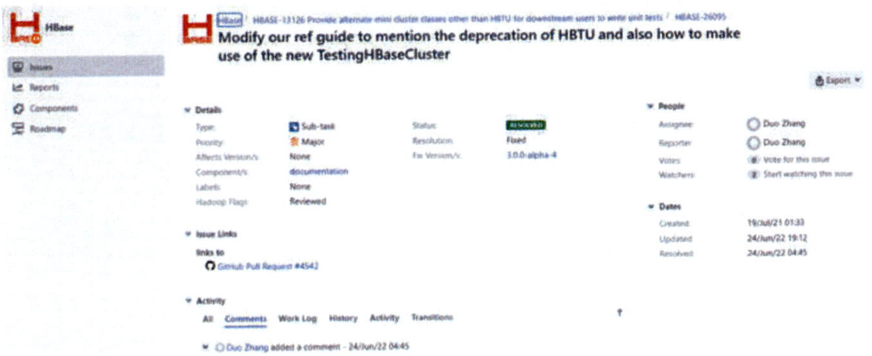

Fig. 3 Sample of bug report of HBase project in JIRA

- Assignee: a person who is assigned for resolving the issue.
- created. date: Data on which the issue is created.
- created: a person who created the issue.
- description: gives comprehensive issue details, including methods to replicate the problem and a potential solution.
- key: unique ID for every issue.
- resolution: status to tell whether the issue has been resolved or not.
- resolution. date: the date on which resolution is given or updated.
- summary: a concise yet informative explanation of the problem.
- issue type: represent whether it is a bug or improvement or task or document or proposal etc.
- status. name: represents whether the issue is open or resolved.

4 Experimental Setup

This section provides the experimental findings and attempts to evaluate our model's performance. Python was chosen because it is a multi-paradigm programming language with a large number of data science tools. The processor is an Intel® Core i5-9750H, while the computer system is an NVIDIA Geforce RTX 2060. We implement the Python script to extract all the project-related information stored in Java. First, we import the Jira library in the Python environment [9].

4.1 Dataset Description

We chose 15 distinct Apache projects from the Jira bug repository for implementing our approach, including Ambari, Qpid, Lucene, Cassandra, Hbase, Camel, Pig,

Table 1 Result of our approach for different projects in JIRA

Project name	Total # of issues	Total # of bugs	% of bugs in project
Zookeeper	4336	2282	52.6
Brooklyn	628	454	72.3
Wicket	6959	4183	60.1
Derby	7129	4165	58.4
HBASE	26,728	12,123	45.4
Groovy	9785	6495	66.4
Cassandra	15,927	8821	55.4
Lucene	9549	3965	41.5
Qpid	8358	5041	60.3
Ambari	25,242	17,866	70.8
Ignite	16,821	7093	42.2
Jclouds	1603	826	51.5
Sparks	38,554	26,589	69.0
Pig	5334	3115	58.4

Groovy, Zoo keeper, Brooklyn, Wicket, Derby, Spark, Ignite, and Jclouds. There is a large number of Apache projects from which to retrieve bugs. You may find the projects at https://issues.apache.org/jira/projects/projectname/issues [10].

5 Experimental Result

This section presents the experimental results of our approach. Python has been chosen because it is a multi-paradigm programming language with a comprehensive set of data science packages. We have a huge number of key values and issue types, which include enhancements to current features, new features added, new project tasks, and issues, however, the number of bugs is low. We gathered data from fifteen projects from Jira and computed the overall number of bug reports because we are just interested in bug reports. Table 1 summarizes our findings. This demonstrates that the number of bugs in Brooklyn is 72&, which is the highest among all the given projects, while the number of bugs in the Lucene project is 41%, which is the lowest of the fifteen projects. The findings indicate that bugs account for a portion of all difficulties in diverse projects.

6 Application of Bug Reports Gathered by BugFinder

This part discusses potential uses for bug data generated by our tool BugFinder. The different applications include the following.

6.1 Analysis of Bug Location

The automated process of locating potentially problematic files in a software project is known as bug localization. Bug localization allows developers to concentrate on crucial files. It has been suggested to use information retrieval (IR)-based technologies to help automatically uncover software flaws. For IR-based systems, however, some issue reports that are not semantically connected to the pertinent code are useless. Running a reporting system based on IR can produce erroneous positive findings. To resolve this issue, many researchers suggest a categorization approach for determining if a bug report is informative or not. By removing irrelevant data before running an IR-based issue location system, those methods reduce false positives and improve ranking performances. Those models are based on explicit characteristics set manually and implicit features discovered by neural network learning from bug reports, Fang et al. [11].

6.2 Bug Report Enrichment Using Machine Learning

When software maintenance tasks are carried out by developers, bug reporting is crucial. For instance, to comprehend how defects occur, developers must read bug reports. The information provided in bug reports can be used to help triage and assign the right developers to work on the associated problems. An issue report includes both text and non-textual data. Factors (such as products and services) are included in the non-textual data, whereas the summary (or title) and description are included in the primary free text. The descriptions, that make up the majority of a bug report, contain specific details about the provided problem, whereas the summaries offer an excellent overview of the bug. Bug reports are valuable tools that assist developers in understanding and fixing errors. To minimize developers' efforts, several recent research use machine learning approaches to assess the information in bug reports for handling various software maintenance activities such as automatic fixer recommendations. A machine learning method, for instance, was used by Anvik et al. [12] to identify a small group of engineers who would be most suited to address a newly discovered fault.

6.3 Bug Report Classification and Text Mining

A software bug is a fault. Because a fault can occur for a variety of causes, we have several categories of bugs such as security vulnerabilities, semantic defects, concurrency problems, and so on. Due to a lack of security domain expertise, security issues are frequently mislabeled as non-security defects in bug tracking systems. To address this issue, textual descriptions of defects must be retrieved by Gegick et al. [13], and a text mining technique must be created to identify bug reports as security or non-security vulnerabilities. To enhance security vulnerability assessments, similar work is done in the citation by Sadeghi et al. [14]. The Bayes theorem is utilized in text mining for categorization, and trials are done on Java and Android apps. Based on thorough descriptions of problem complaints from the bug tracking system, the concerns by Tan et al. [15] are further divided into storage, memory, and vulnerability bugs. The categorization is carried out using a two-level classification technique, various classification algorithms, and multiple assessment criteria. We can use BugFinder tool for the automated classification of Bug Reports.

7 Conclusion and Future Work

A bug global positioning framework is being developed that may be used to collect issues from the Apache project's issues global positioning framework and provide comprehensive data about all issue components. In this work, we extracted 176,953 issues from the Apache JIRA repository from fifteen projects and extracted out of the 103,018 bugs. The information obtained may be utilized for a variety of reasons, including the categorization of bugs based on one-line and extended descriptions, with bugs categorized as semantic bugs, aging-related bugs, vulnerability-related bugs; and predictions of severity level using machine learning methods. This work has been used for the Jira issue tracking system, however, in the future, not only JIRA ITS but also other open procurement issue tracking solutions such as Github, and Bugzilla, will be used. The structural information of the data can be helpful for various applications by applying various text mining techniques as well.

8 Data Availability

All accompanying codes, statistics, and output data are available on GitHub. https://github.com/rashmiarorakhera/BugFinder.

References

1. Gegick M, Rotella P, Xie T (2010) Identifying security bug reports via text mining: an industrial case study. In: 2010 7th IEEE working conference on mining software repositories (MSR 2010). IEEE, pp 11–20
2. Bettenburg N, Premraj R, Zimmermann T, Kim S (2008) Extracting structural information from bug reports. In: Proceedings of the 2008 international working conference on mining software repositories, pp 27–30
3. Young M (2002) Technical writer's handbook. University Science Books
4. Shokripour R, Kasirun ZM, Zamani S, Anvik J (2012) Automatic bug assignment using information extraction methods. In: 2012 international conference on advanced computer science applications and technologies (ACSAT). IEEE, pp 144–149
5. Lin T, Gao J, Fu X, Lin Y (2015) A novel bug report extraction approach. In: International conference on algorithms and architectures for parallel processing. Springer, Cham, pp 771–780
6. Yuk Y, Jung W (2013) Comparison of extraction methods for bug tracking system analysis. In: 2013 international conference on information science and applications (ICISA). IEEE, pp 1–2
7. Nayrolles M, Hamou-Lhadj A (2016) BUMPER: a tool for coping with natural language searches of millions of bugs and fixes. In: 2016 IEEE 23rd international conference on software analysis, evolution, and reengineering (SANER), vol 1. IEEE, pp 649–652
8. Kaur A, Jindal SG (2017) Bug report collection system (BRCS). In: 2017 7th international conference on cloud computing, data science & engineering-confluence. IEEE, pp 697–701
9. https://issues.apache.org/jira/projects/
10. https://pypi.org/project/jira/
11. Fang F, Wu J, Li Y, Ye X, Aljedaani W, Mkaouer MW (2021) On the classification of bug reports to improve bug localization. Soft Comput 25(11):7307–7323
12. Anvik J, Hiew L, Murphy GC (2006) Who should fix this bug? In: ICSE'06, pp 361–370
13. Gegick M, Rotella P, Xie T (2010) Identifying security bug reports via text mining: an industrial case study. In: 7th IEEE working conference on mining software repositories (MSR), pp 11–20
14. Sadeghi A, Esfahani N, Malek S (2014) Mining the categorized software repositories to improve the analysis of security vulnerabilities. In: International conference on fundamental approaches to software engineering, pp 155–169
15. Tan L, Liu C, Li Z (2014) Bug characteristics in open source software. Empir Softw Eng 19(6):1665–1705

Convolutional Neural Network for Parameter Identification of a Robot

Carlos Leopoldo Carreón Díaz de León, Sergio Vergara Limon, María Aurora D. Vargas Treviño, Jesús López Gómez, and Daniel Marcelo González Arriaga

Abstract Parametric identification is a crucial issue due to the possibility of emulating systems for motion control, collision detection, manufacturing, and other scientific areas. However, the conventional methodologies need an optimized trajectory to find all the parameters quickly. This paper shows an identification method based on a convolutional neural network to extract the dynamic parameters of a cartesian robot. First, the variables of position, velocity, acceleration, and torque with a set of parameters create an image with a conversion method. Second, the backpropagation algorithm trains the convolutional neural network to return the robot's parameters. Finally, a time-spectral evaluation distance gives the affinity of the results. The simulations show that the identification achieved 0.9916, and the experimental robot met 0.9196.

Keywords CNN · Robotics · Parameter identification · Similarity

1 Introduction

The dynamic model of robots plays an essential role in their design, where the parameters reflect the system's characteristics like inertia, gravity, and friction. Applications such as trajectory control [1], manufacturing [2], and model validation [3] require dynamic parameters to ensure acceptable results. In some dynamic param-

C. L. C. D. de León (✉) · D. M. G. Arriaga
Facultad de Ciencias de la Computación, Benemérita Universidad Autónoma de Puebla, Puebla, Mexico
e-mail: carlos.carreond@alumno.buap.mx

S. V. Limon · M. A. D. V. Treviño
Facultad de Ciencias de la Electrónica, Benemérita Universidad Autónoma de Puebla, Puebla, Mexico

J. L. Gómez
División Académica de Ingeniería y Arquitectura, Universidad Juárez Autónoma de Tabasco, Villahermosa, Mexico

S. Das et al. (eds.), *Advances in Data-driven Computing and Intelligent Systems*, Lecture Notes in Networks and Systems 653, https://doi.org/10.1007/978-981-99-0981-0_40

eters applications, the time required to make the parameter identification of the robot is an important limiting. When the mathematical model of the robot is linearly independent with respect to the dynamic parameters, the least-squares (LS) is a common solution [4]. The requirements of LS are the mathematical model of the robot corresponding to the real robot; i.e., if the dynamic parameters are known, the robot's behavior is wholly determined without any variations. LS can consider that the robot's signals are contaminated with a noise of gaussian distribution. Therefore, the mathematics founds of LS demonstrate that the dynamic parameters are easily determined, considering that the measures of position, velocity, and acceleration are known [5], i.e., these signals are continuous and are not quantized. However, only the position usually becomes from encoders or analogic-to-digital converters.

The conventional parameter identification works use methodologies that require all input and output signals to satisfy this methodologies theory. The experimental robots must contain position, velocity, acceleration, and torque sensors; this is expensive because industrial and scientific robots only use the position sensor. There are several methods to estimate the velocity with an encoder [6, 7], but they're still not the actual velocity of the robot. Usually, the parametric identification with LS needs an optimal trajectory to work appropriately, as described in [8].

The trajectory optimization step requires time because the experimental robot performs this trajectory. Consequently, the parametric identification can take more time to find the optimal motion than the LS algorithm execution. Some works split the parameter identification in conservative and friction models [9], gravity torque, and friction torque [10], and other works use a weighed version of LS (WLS) to accelerate the parametric identification [10]. This article proposes a Convolutional Neural Network (CNN) to extract the dynamic parameters of a robot using the position, velocity, acceleration, and torque without the optimization of the robot trajectory. The CNN does not require the exact measure signals to extract the parameters, as the LS-based methods need in its theory. The proposed methodology uses estimations of the velocity and acceleration to make an image and extract the dynamic parameters.

The neural networks (NNs) improve parameters extraction in methods based on LS, as [11] where a Long Short-Term Memory (LSTM) NN enhances the torque estimation. In this work, the LSTM learns the uncertainties of the measured torques to compensate for them and identify the parameters with LS. The NN extracts information that is not easily determined using equations in the robot's signals. The Convolutional Neural Networks (CNNs) is a NN architecture that analyses the input information by areas to extract features that are commonly processed by a Feed-Forward Neural Network (FFNN) to obtain the desired output [12]. The CNN is constructed with several convolutional layers with a non-linear activation function, pooling layers, and FFNN layers. The ConvNN has the advantage of not using combinational connections in the input layer where the size is large. Also, the convolutional filters named kernels extract the most relevant information from the input.

The CNN uses pooling layers that reduce the outputs of the convolutions to keep the execution time low. Because some relevant information can be extracted in reduced versions, CNN maintains its ability to learn features of the input data. In [13], a deep CNN is used to estimate the response of a mechanical system. This

Table 1 State of the art comparison

References	Method	Trajectory optimization	Neural network
[3, 5, 8, 9]	LS	Yes	–
[11]	LS and NN	Yes	LSTM
[13]	NN	No	CNN

work shows that the CNN applied to a 1-dimension input data overcomes the FFNN architecture. The CNN analyses the input in regions and learns the deep features that create the response estimation. In this case, the proposed CNN in [13] does not use an FFNN at the end. The comparison between the FFNN and the CNN demonstrates that the CNN works better for response estimation, even when the signals contain noise. Table 1 shows the conventional and neural networks articles for parameter identification.

The main contribution in this paper is a convolutional neural network applied to extract four dynamic parameters of an articulation of a cartesian robot. The signals of the robot assemble an image created with the dynamic parameters. The main idea is that the CNN extracts the vector residual from an initial set of parameters to the real dynamic parameters. An iterative algorithm creates an image and returns the dynamic parameters. The experimental torque is compared with the reconstructed torque with the dynamic model and the identified parameters. The paper is organized as follows: Sect. 1 shows the cartesian robot, Sect. 2 the proposed identification method, Sect. 3 displays the results, and Sect. 4 the conclusions.

2 The Cartesian Robot

The robot used in this paper is an articulation of a cartesian robot, as Fig. 1 displays. The joint consists of a motor coupled with a gearbox and an endless screw to move a base linearly. Equation 1 describes the dynamic model where x_{m}, x_{g}, x_{s} are the motor, gearbox, and screw angular positions, and x_{l} is the linear position.

The dynamic model considers the mechanical connections with springs and dampers. b_{m}, b_{g}, b_{s}, b_{l} model the viscosity, and J_{m}, J_{g}, and J_{s} model the inertia parameters. The parameter m_{l} models the base. The parameters b_{mg} and k_{mg} model the mechanical connection between x_{m} and x_{g}: b_{gs}, k_{gs}, b_{sl}, and k_{sl} follow the same notation. The parameters r and n denote the gearbox reduction and the nut-screw linear reduction.

$$\boldsymbol{u} = \boldsymbol{A}\ddot{\boldsymbol{x}} + \boldsymbol{B}\dot{\boldsymbol{x}} + \boldsymbol{s}(\boldsymbol{x}, \dot{\boldsymbol{x}}) + \boldsymbol{f}_{\boldsymbol{x}}(\dot{\boldsymbol{x}}) + \boldsymbol{e} \tag{1}$$

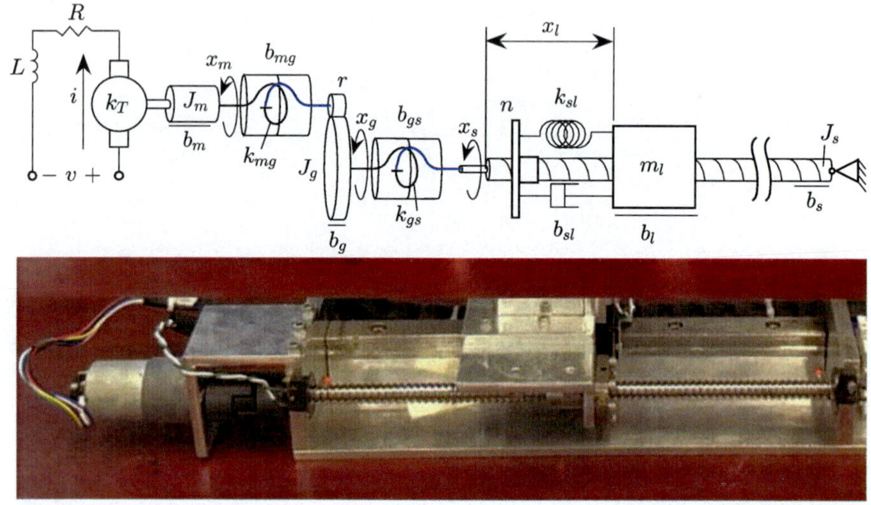

Fig. 1 Articulation of the cartesian robot

with $\boldsymbol{u} = [\tau, 0, 0, 0]^{\mathrm{T}}$, $\boldsymbol{x} = [x_{\mathrm{m}}, x_{\mathrm{g}}, x_{\mathrm{s}}, x_{\mathrm{l}}]^{\mathrm{T}}$, $A = \mathrm{diag}(J_{\mathrm{m}}, J_{\mathrm{g}}, J_{\mathrm{s}}, m_{\mathrm{l}})$,
$B = \mathrm{diag}(b_{\mathrm{m}}, b_{\mathrm{g}}, b_{\mathrm{s}}, b_{\mathrm{l}})$, $\boldsymbol{s} = [s_1, s_2, s_3, s_4]^{\mathrm{T}}$,
$w_1 = k_{\mathrm{mg}}(x_{\mathrm{m}} - r x_{\mathrm{g}}) + b_{\mathrm{mg}}(\dot{x}_{\mathrm{m}} - r\dot{x}_{\mathrm{g}})$,
$w_2 = k_{\mathrm{gs}}(x_{\mathrm{g}} - x_{\mathrm{s}}) + b_{\mathrm{gs}}(\dot{x}_{\mathrm{g}} - \dot{x}_{\mathrm{s}})$,
$w_3 = k_{\mathrm{sl}}(x_{\mathrm{s}} - n x_{\mathrm{l}}) + b_{\mathrm{sl}}(\dot{x}_{\mathrm{s}} - n\dot{x}_{\mathrm{l}})$,
$s_1 = w_1$, $s_2 = w_2 - r w_1$, $s_3 = w_3 - w_4$, $s_4 = -n w_3$,
$\boldsymbol{f}_{\mathrm{c}}(\dot{\boldsymbol{x}}) = [0, 0, 0, k_{\mathrm{s}}\mathrm{sign}(\dot{x}_{\mathrm{l}})]^{\mathrm{T}}$,
$\boldsymbol{e} = [0, 0, 0, o_{\mathrm{f}}]^{\mathrm{T}}$,
$v = Ri + L\frac{\mathrm{d}}{\mathrm{d}t}i$,

where $\boldsymbol{u}$ is the input, A contains the inertia and mass parameters, $\boldsymbol{s}$ is the mechanical connection between the components, $\boldsymbol{f}_{\mathrm{c}}$ is the Coulomb friction, and $\boldsymbol{e}$ is the input offset, $\tau = k_{\mathrm{T}} i$ is the input torque, k_{T} is the torque constant of the motor, R is the motor electrical resistance, L is the inductance of the motor, k_{c} is the parameter of Coulomb friction, and o_{f} is the offset value. The real robot has an encoder of 7.4800e−04 radians per count. Considering that the mechanical connection $\boldsymbol{s}$ has values under the resolution of the encoder, Eq. 2 simplifies Eq. 1 to four dynamic parameters:

$$\tau = J\ddot{x} + b\dot{x} + k_{\mathrm{c}}\mathrm{sign}(\dot{x}) + o_{\mathrm{f}} \tag{2}$$

J and b represent equivalent inertia and friction of Eq. 1 when $\boldsymbol{s}$ equals zero, and $\boldsymbol{\beta} = [J, b, k_{\mathrm{c}}, o_{\mathrm{f}}]^{\mathrm{T}}$.

3 Identification Method

The identification method requires an image to work in a short time. The robot signals of position, velocity, acceleration, and torque are not suitable for the algorithm due to their size. Also, these signals contain repetitive information in short areas. A conversion method transforms signals into an image.

First, $z = [x, \dot{x}, \ddot{x}, \tau]^{\mathrm{T}} \in \mathbb{R}^{N\times 4}$ is filtering with a gaussian low pass filter $z_{\mathrm{f}} = T^{-1}[T(z)\exp(-Py^2)]$ where $P = 3\pi^2 \ln(10)/20[h\omega_{\mathrm{c}}(N-1)]^2$, ω_{c} is the cut frequency, h is the sampling time set in 2.5 ms, $y = \{0, 1, \ldots, N-1\}$ is the spectrum frequency of the transform $T(.)$ of Eq. 3 where their inverse is $T(.)^{-1}$ [14], and the filtered signals are denoted by z_{f}. Second, z_{f} is cut with spectral frequency to 100 samples $z_{\mathrm{c}} = \{z_{\mathrm{f}} = T^{-1}[T(z_{\mathrm{f}}, y)], 0 \leq y \leq 99\}$.

$$T(x,k) = \sqrt{\frac{2}{N}} \sum_{i=1}^{N} a(k)x(i)\cos\left[\frac{\pi}{2N}(2i-1)(k-1)\right] \tag{3}$$
$$a = [1 + \delta(k-1)]^{-1/2},\ \delta(z) = 1,\ \text{if and only if}\, z = 0$$

With the filtered and cut signals, the next step is to construct the image with the vectors $\boldsymbol{p}_1 = 2\tau_{\mathrm{c}} - [\ddot{x}_{\mathrm{c}}, \dot{x}_{\mathrm{c}}, \mathrm{sign}(\dot{x}_{\mathrm{c}}), 1]^{\mathrm{T}}\boldsymbol{\beta}_{\mathrm{i}} \in \mathbb{R}^{100\times 1}$ and $\tau_{\mathrm{c}} \in \mathbb{R}^{100\times 1}$, where $\boldsymbol{\beta}_{\mathrm{i}} \in \mathbb{R}^{4\times 1}$ is the input set of parameters. Equation 4 determines the image $Z \in \mathbb{R}^{100\times 100}$ from the robot signals.

$$Z = [G - \min(G)]/[\max(G) - \min(G)],\ G = \boldsymbol{p}_1\boldsymbol{p}_2^{\mathrm{T}} \tag{4}$$

When the input parameters $\boldsymbol{\beta}_{\mathrm{i}}$ do not coincide with the torque and motion signals, the resulting image is different when $\boldsymbol{\beta}_{\mathrm{i}}$ is the actual parameters. Figure 2 shows two images constructed with Eq. 4: when $\boldsymbol{\beta}_{\mathrm{i}}$ is not the exact parameters, the image is denominated as Z_n, as Fig. 2b shows. If $\boldsymbol{\beta}_{\mathrm{i}}$ is the actual set of parameters, the image is called Z_{p}, as Fig. 2a shows.

The main idea behind the identification method is finding the actual parameters $\boldsymbol{\beta}$ from the input parameters $\boldsymbol{\beta}_{\mathrm{i}}$. If $\boldsymbol{\beta}_{\mathrm{r}}$ is known together with $\boldsymbol{\beta}_{\mathrm{i}}$, then the actual parameters $\boldsymbol{\beta}$ can be extracted: $\boldsymbol{\beta}_{\mathrm{r}} = \boldsymbol{\beta} - \boldsymbol{\beta}_{\mathrm{i}}$, where $\boldsymbol{\beta}_{\mathrm{i}}$ is randomly selected. Figure 3 shows a square set, where the real parameters $\boldsymbol{\beta}$, initial $\boldsymbol{\beta}_{\mathrm{i}}$, and the subtraction vector $\boldsymbol{\beta}_{\mathrm{r}}$ are located. The proposed CNN of Fig. 4 extracts from the image Z the vector $\boldsymbol{\beta}_{\mathrm{r}}$. Notice that $\boldsymbol{\beta}_{\mathrm{r}}$ is not directly measured, and the CNN returns from a condensed image Z constructed with the motion of the robot and $\boldsymbol{\beta}_{\mathrm{i}}$. It implies that the relevant information of the signals is in only 100 samples of position, velocity, acceleration, and torque.

The proposed convolutional neural network has three convolutional layers (Conv1, Conv2, Conv3) and three feed-forward layers (FC1, FC2, FC3). After Conv1 and Conv2, a pooling layer is placed. The convolution is performed by Eq. 5, where Y_{o} is the output, f_{a} is the activation function, b_{o} is the bias, X_{d} is the input, W_{do} is the convolutional kernel, the index d is for the input channels, and the index o is the output channels [12].

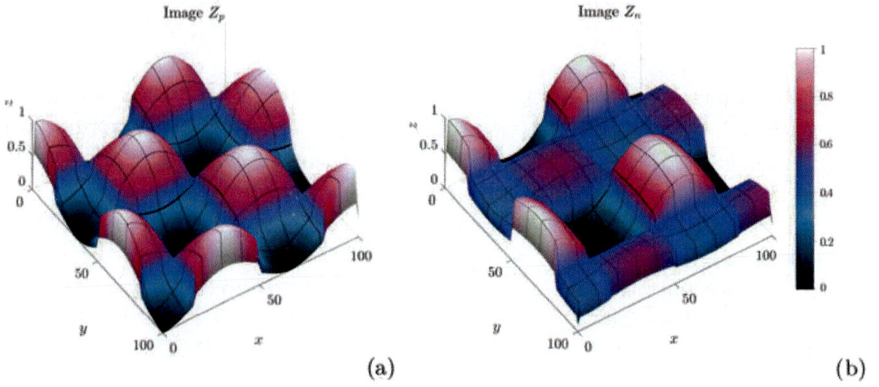

Fig. 2 Images created with the vectors $\boldsymbol{p}_1$ and $\boldsymbol{p}_2$: **a** $\boldsymbol{\beta}_\mathrm{i}$ is the correct set of parameters, **b** $\boldsymbol{\beta}_\mathrm{i}$ is not the actual parameters of the robot signals

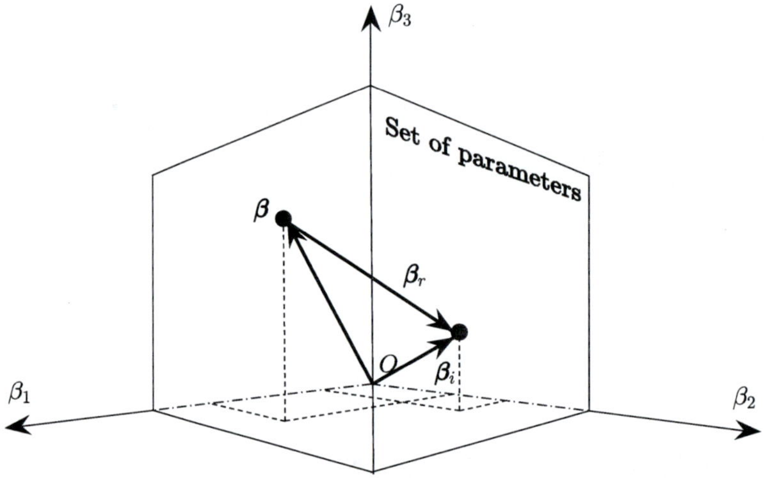

Fig. 3 Subtraction vector $\boldsymbol{\beta}_\mathrm{r}$ inside a limited set of three parameters

The kernels of the CNN are $9 \times 9 \times 1 \times 10$, $5 \times 5 \times 10 \times 10$, $3 \times 3 \times 10 \times 10$ for layers 1, 2, and 3. The feed-forward layers have 100, 100, and 4 neurons for layers 4, 5, 6. The activation functions in the CNN is $f_\mathrm{a}(X) = \max(0, X)$ for layers 1–5, and $f_\mathrm{a}(X) = [1 + \exp(-X)]^{-1}$ for the layer 6.

$$Y_\mathrm{o} = f_\mathrm{a}\left[b_\mathrm{o} + \sum_{d=0}^{D-1} \mathrm{conv}(X_\mathrm{d}, W_\mathrm{do})\right] \tag{5}$$

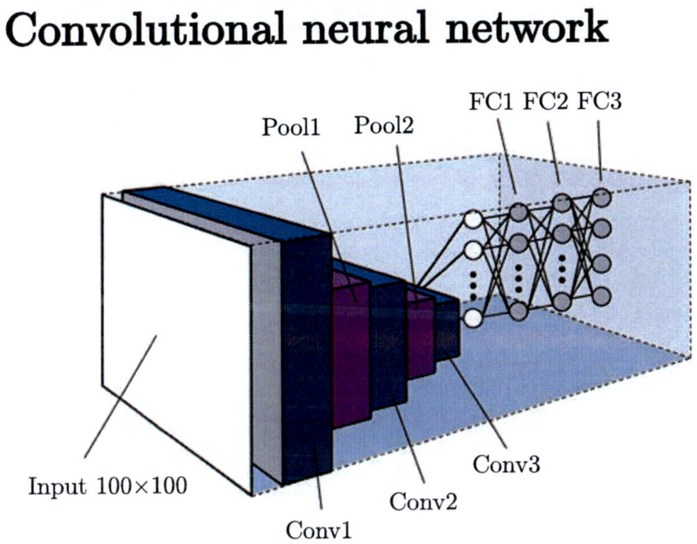

Fig. 4 Convolutional neural network used for the identification method

The CNN trains with images constructed with the signals of 10,000 simulations of Eq. 2, where the inertia, friction, and Coulomb parameters range randomly from 0.001 to 1, and the offset ranges from − 1 to 1. The backpropagation algorithm is used with a learning rate set in 0.14 and an error cost equal to $E = (1/8)\sum_{j=1}^{4}(Y_{d-i} - Y_i)^2$. Once the CNN training can extract $\boldsymbol{\beta}_{\mathrm{r}}$, it is put into the Flow chart of Fig. 5.

The identification method begins with the initialization of $\boldsymbol{\beta}_{\mathrm{i}}$, and the construction of the image Z using the motion of the cartesian robot. After that, the CNN extracts $\boldsymbol{\beta}_{\mathrm{r}}$, and by the addition of $\boldsymbol{\beta}_{\mathrm{i}}$, $\boldsymbol{\beta}$ is known. The model f of Eq. 2 reconstructs the input using the extracted $\boldsymbol{\beta}$ and the motion of the robot. The function distance d_{s} of Eq. 6 returns the affinity between the real τ and the reconstructed τ_{r}: If their value is over an umbral u, then the identification method returns $\boldsymbol{\beta}$.

$$d_{\mathrm{s}}(\tau, \tau_{\mathrm{r}}) = \left[\left(1 - \frac{|\sum\tau - \sum\tau_{\mathrm{r}}|}{\sum(|\tau| + |\tau_{\mathrm{r}}|)}\right)\left(1 - \frac{\sum|T(\tau) - T(\tau_{\mathrm{r}})|}{\sum(|T(\tau)| + |T(\tau_{\mathrm{r}})|)}\right)\right]^{1/2.5} \tag{6}$$

4 Results

4.1 Simulation Results

The trajectory used for the training data is a sinusoidal function that the Eq. 2 of the articulation of the cartesian robot follows, as Fig. 6 shows. The CNN has been implemented in Matlab with a GPU 1060 of NVIDIA© and an Intel© I7. The error cost

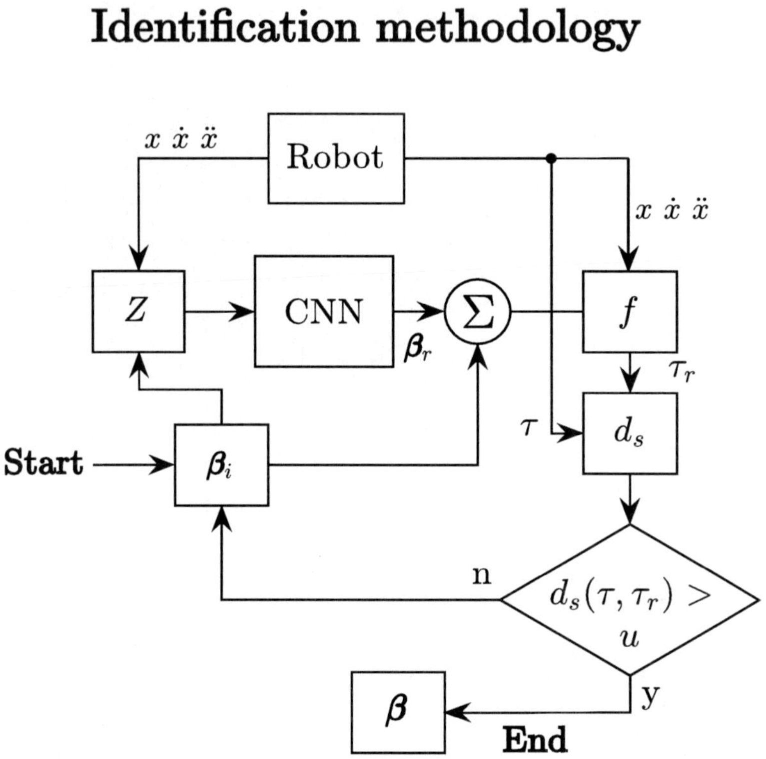

Fig. 5 Identification flow chart

has achieved a value of 2.8×10^{-3} with 4.3×10^5 training iterations. The identified method is tested with a simulation of Eq. 2.

Observer that the values of the simulation are outside of the range of training data, but with normalization and denormalization techniques in the input and output, the parametric identification method retrieval the actual set of parameters. The simulated and reconstructed torque of the simulation is in Fig. 7.

4.2 Experimental Results

For the real data of the robot of Fig. 1, the identified parameters are shown in Table 1. In this case, the articulation coincides with the horizontal, and the gravity effect does not appear in the offset parameter; However, there is a small value in o_f.

The experimental torque and their reconstruction with the parameters of Table 2, are displayed in Fig. 8. The error between the actual and reconstructed torque shows

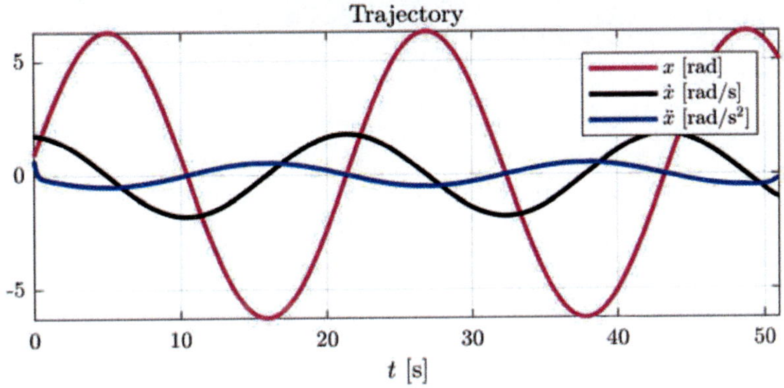

Fig. 6 Trajectory used for the parameter identification

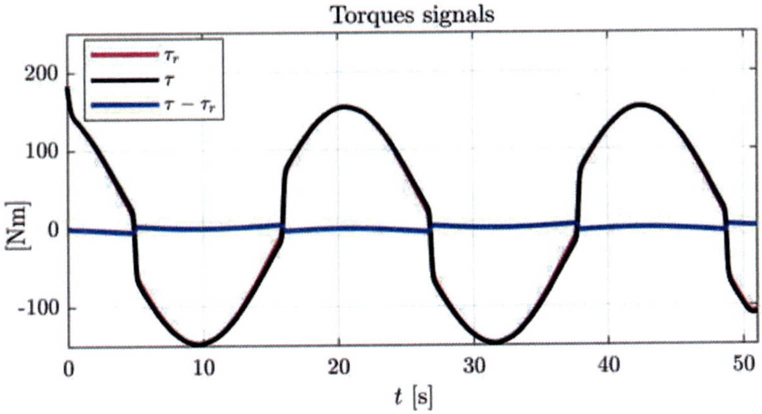

Fig. 7 Simulation torque and their reconstruction with the identified parameters

Table 2 Identification results for a simulation and experimental data, and training results of the CNN

Type test	J (kg m^2)	b (kg m^2 s^{-1})	k_c (Nm)	o_f (Nm)	CNN training error
Simulation	50	60	40	4	2.828×10^{-3}
Identified	51.4989	62.2023	35.85	4.1202	–
Experimental	0.0041	0.2302	0.0817	0.0001	–

that the model of Eq. 1 can be approximate with the simplification of Eq. 2. The affinity value for the experimental data is 0.9196 and 0.9916 for the simulation. Table 2 shows the simulation parameters and the identified parameters.

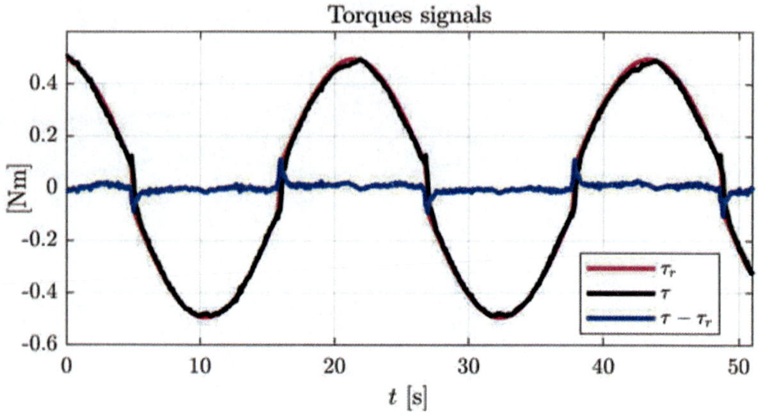

Fig. 8 Experimental torque and their reconstruction with the identified parameters

5 Conclusions

In this paper, a convolutional neural network for dynamic parameters extraction of a robot has been described. The common methods for parameter identification spend time optimizing a trajectory that performs the areal robot. Therefore, the parameter identification takes time. The identification method consists of extracting the dynamic parameters using an image generated by the robot signals and an initial set of parameters. The image construction uses a sinusoidal signal instead of an optimized trajectory and condenses in only 100×100 pixels the relevant information for the parametric identification.

The CNN trains with two classes of images, the Z_{p} and Z_{n}. When the parameters correspond to the torque and motion signals, the image is Z_{p}. Otherwise, it is a Z_{n} image. The tags of the training data are the subtraction vector of parameters $\boldsymbol{\beta}_{\text{r}}$, and the CNN can generalize their training to new images that are not included in the training. The main idea of the identification method described in this paper is identifying the parameters with the subtraction vector $\boldsymbol{\beta}_{\text{r}}$. With this vector, the parameters are determined by simple addition. The method starts with random parameter initialization ($\boldsymbol{\beta}_{\text{i}}$) and uses the CNN to extract the difference between the image's actual and $\boldsymbol{\beta}_{\text{i}}$. The results show that the identification method can return the parameters from a simulation and experimental data of the cartesian robot. Normalizing the simulation torque, the CNN can return the parameters outside their training range. The execution time for the simulation is 0.0476 s.

The experimental identification shows that the simplified model of Eq. 2 can accurately represent Eq. 1, which contains more parameters. The experimental parameters of inertia and friction correspond to the fact that the articulation tested does not contain the gravity effect; however, the offset parameter has a close zero value. The dynamic parameters can reconstruct the torque to achieve an affinity of 0.9196. The main conclusion is that the identification method can be suitable for parameter identification of experimental robots because, for the cartesian robot articulation, the method only takes 0.8126 s. The proposed methodology does not require an optimal trajectory of the robot to perform the parameter identification. Therefore, this article shows a methodology that has the potential to be used in many robots. The future work is applying the proposal to more robots.

Acknowledgements This work was support by the Consejo Nacional de Ciencia y Tecnología (CONACyT) of México by their national scholarship of postgraduate.

References

1. Zamora-Gómez GI, Zavala-Río A, López-Araujo DJ, Santibánez V (2019) Further results on the global continuous control for finite-time and exponential stabilisation of constrained-input mechanical systems: desired conservative-force compensation and experiments. IET Control Theory Appl 13(2):159–170. https://doi.org/10.1049/iet-cta.2018.5099
2. Petko M, Gac K, Góra G, Karpiel G, Ochoński J, Kobus K (2016) CNC system of the 5-axis hybrid robot for milling. Mechatronics 37:89–99. https://doi.org/10.1016/j.mechatronics.2016.03.001
3. Urrea C, Pascal J (2021) Design and validation of a dynamic parameter identification model for industrial manipulator robots. Arch Appl Mech 91:1981–2007. https://doi.org/10.1007/s00419-020-01865-2
4. Kelly R, Santibáñez V (2003) Control de Movimiento de Robots Manipuladores, 1st edn. Prentice Hall, Madrid
5. Jin J, Gans N (2016) Parameter identification for industrial robots with a fast and robust trajectory design approach. Robot Comput Integr Manuf 31:21–29. https://doi.org/10.1016/j.rcim.2014.06.004
6. Hace A, Čurkovič M (2018) Accurate FPGA-based velocity measurement with an incremental encoder by a fast generalized divisionless MT-type algorithm. Sensors 18(10):1–29. https://doi.org/10.3390/s18103250
7. Zhu WH, Lamarche T (2007) Velocity estimation by using position and acceleration sensors. IEEE Trans Ind Electron 54(5):2706–2715. https://doi.org/10.1109/TIE.2007.899936
8. Swevers J, Ganseman C, Bilgin-Tükel D, De Schutter J, Van Brussel H (1997) Optimal robot excitation and identification. IEEE Trans Robot Autom 13(5):730–740. https://doi.org/10.1109/70.631234
9. Benimeli F, Mata V, Valero F (2006) A comparison between direct and indirect dynamic parameter identification methods in industrial robots. Robotica 24:579–590. https://doi.org/10.1017/S0263574706002645
10. Cao P, Gan Y, Dai X (2019) Model-based sensorless robot collision detection under model uncertainties with a fast dynamics identification. Int J Adv Robot Syst 16(3):1–15. https://doi.org/10.1177/1729881419853713
11. Wang S, Shao X, Yang L, Liu N (2020) Deep learning aided dynamic parameter identification of 6-DOF robot manipulators. IEEE Access 2020:138102–138116. https://doi.org/10.1109/ACCESS.2020.3012196
12. Gua J, Wang Z, Kuen J, Ma L, Shahroudy A, Shuai B, Liu T, Wang X, Wang G, Cai J, Chen T (2018) Recent advances in convolutional neural networks. Pattern Recognit 77:354–377. https://doi.org/10.1016/j.patcog.2017.10.013

13. Wu RT, Jahanshahi MR (2019) Deep convolutional neural network for structural dynamic response estimation and system identification. J Eng Mech 145(1):1–25. https://doi.org/10.1061/(ASCE)EM.1943-7889.0001556
14. Matlab. Discrete cosine transform. mathworks.com. https://www.mathworks.com/help/signal/ref/dct.html. Accessed 10 Feb 2022

Traffic Jam Detection Using Regression Model Analysis on IoT-Based Smart City

D. H. Manjaiah, M. K. Praveena Kumari, K. S. Harishkumar, and Vivek Bongale

Abstract A well-mannered transportation system has an impact on the economy, well-being, and personal satisfaction of a country. The rate at which vehicle numbers increase is substantially faster than the rate at which the general population grows, resulting in increasingly congested and unsafe streets. This problem will no longer be solved by merely increasing the number of roadways. It is important to study and understand the traffic flow in thronged cities, and also it is necessary to mine traffic data and apply machine learning algorithms to it for the development of smart cities. We can cut transport delays, fuel consumption, traveler and freight movement costs, the frequency of crashes, tailpipe pollution, and improve city life by mining traffic data. Many cities in wealthy countries currently use a range of sensors to collect real-time traffic data, which are subsequently studied by using machine learning algorithms to improve traffic flow. The real-time traffic data of Aarhus city of Denmark is used in our work for exploring the traffic conditions of 449 junctions in the year 2014. Four regression models—linear, polynomial, lasso, and ridge are applied to the dataset to predict the flow of traffic in Aarhus city. The performance of these regression models is tested using statistical measures such as root mean square error and coefficient of determination. The experiment findings indicate that, for various routes in the Aarhus city of Denmark, lasso regression predictions tend to be the most accurate in predicting real traffic flow.

Keywords Traffic · Machine learning · Regression · IoT · ITS · Forecast

---

D. H. Manjaiah · M. K. Praveena Kumari (✉)
Department of Computer Science, Mangalore University, Mangalore, India
e-mail: narayan.praveena@gmail.com

D. H. Manjaiah
e-mail: drmdhmu@gmail.com

K. S. Harishkumar · V. Bongale
Department of Computer Science and Engineering, Presidency University, Bengaluru, India
e-mail: harishkumar@presidencyuniversity.in

V. Bongale
e-mail: vivek.bongale@presidencyuniversity.in

S. Das et al. (eds.), *Advances in Data-driven Computing and Intelligent Systems*, Lecture Notes in Networks and Systems 653, https://doi.org/10.1007/978-981-99-0981-0_41

1 Introduction

Significantly increased car traffic has been a key concern for municipal management agencies in recent years. Smart cities are a result of continual technological advancements aimed at improving the quality of life for their residents. Urban mobility is one of the most important features of smart cities. Urban overcrowding is becoming increasingly common as the number of vehicles in smart cities grows [1]. Currently, there have been one billion automobiles worldwide, but analysts believe that by 2050, there will be 2.5 billion [2]. People who relied on transportation systems have expanded in recent years, presenting transportation systems with many opportunities as well as difficulties [3]. First, as the frequency of vehicles grows, traffic congestion has become a more significant issue around the world. Second, the extension of transportation systems increases the danger of accidents, particularly in developing countries. Third, land resources are frequently depleted in many countries. Building new infrastructure, such as highways and freeways, is thus difficult. To govern traffic, each country has its own set of rules and guidelines. Conventional traffic control tactics were employed, such as the placement of traffic signals, traffic signs, traffic cops, and round-about [4]. In any case, these strategies are becoming antiquated and unsuitable in both developed and developing cities to enhance road traffic conditions. The transportation framework is not just a technology-driven independent system, but also a synchronized set of systems driven by information, as ITS generates huge volumes of data [5]. As a result, analyzing data obtained from a large number of auxiliary instruments in ITS, like inductive-loop detectors, cameras, GPS-oriented receivers, and microwave detectors, is the best technique for maximizing the utilization of the current transportation system. In addition to being transformed into valuable information, data can be used to develop new features and services in intelligent transportation systems (ITS). GPS data, for example, can be used to analyze and predict traffic trends. For future intelligent transportation systems (ITS) and smart city applications, knowledge mining from historical traffic big data is essential [6].

Intelligent computational researchers are currently using a variety of machine learning concepts and data mining methodologies and algorithms for mining traffic data to improve traffic management systems. However, the transportation issues are as yet rich in applying and utilizing machine learning strategies, and still more explores and approaches for examining and foreseeing road traffic are required because complex processes will be taking place almost every day on the road network of the cities and the circumstances and environment are distinctive in various parts of the world. Here, we have collected Aarhus city's real-time traffic data for exploring the traffic conditions of 449 junctions in the month of August and September of the year 2014. We have applied four regression models—linear regression, polynomial regression, lasso regression, and ridge regression to the real-time data used to forecast the city of Aarhus's daily traffic flow. To visualize, pre-process, and create regression models, we employed the Python high-level programming language. Finally, we compared the regression model's anticipated outcomes to the exact real-time traffic flow. The remainder of this paper is organized as follows: Sect. 2 presents related works on road traffic flow prediction. Section 3 delves into the study's scope and methodology, as well as the regression models employed to anticipate traffic flow. The performance criteria are presented in Sect. 4. Section 5 presents data visualization and presents the experimental results. Finally, the conclusion is discussed in Sect. 6.

2 Literature Review

Developing smart cities needs the ability to anticipate short- and long-term traffic patterns. Xiaobo et al. [5] used a dataset collected from 24 monitoring stations along Portland's highways to provide a short-term road traffic forecast model based on a least squared support vector regression (LSSVR) which is a hybrid genetic algorithm-based model. This approach can interpret the connection between the spatio-temporal parameters, which is one of its benefits.

Suguna and Neetha devised a method for forecasting traffic congestion by using machine learning algorithms to create models that may be utilized to make forecasts. They used the data collected after looking at the average speed of the road to evaluate their algorithm. The trials revealed that logistic regression was the most effective algorithm for their algorithms [6]. Dick et al. studied and developed efficient and easy-to-use linear and logistic regression models, for forecasting traffic volumes on low-volume roads in Wyoming, US. The response variable was log-transformed during the model construction process to address difficulties with error variance non-constancy. Both models were cost-efficient and highly accurate [7].

To forecast traffic flow, Nicholas et al. [8] suggested a unique architecture that is based on deep learning. The design was paired with a linear model that included a tenth layer sequence. Deep learning outperforms linear models in forecasting, according to their studies. Yalda et al. [9] used a technique that adapts to construct a real-time data flow forecast model. To represent variability, a two-step technique was adopted. PeMS, an open access database, was used to check the generality of their technique [10]. PeMS is a productivity measuring system that also includes many tools to analyze previous data. Their proposed method has been tested for missing data imputation, and the findings showed that the scheme is more effective than PPCA and k-NN. For short-term time series forecasting, Ling et al. [11] suggested a technique that is based on adaptive particle swarm optimization and multi-kernel support vector machine that is (APSO-MSVM). The technique had a substantially lower failure rate on both tollway and metropolitan roads, according to the study, which accumulated actual information from roadside units (RSU).

Wang et al. [12] used the deep learning method combined with nonparametric regression to predict nonlinearities spatio-temporal impacts. The deep learning algorithm's initial layer discovers spatio-temporal and other nonlinear correlations among predictors. When compared to other models, the deep learning approach using nonparametric regression performs far higher. The suggested methodology for forecasting the flow of traffic has better overall performance, according to the findings of the experiments.

Partial least squares, SVR, ARIMA, and kernel ridge regression were used by Zhan et al. to create an ensemble model [13]. Chen et al. [14] developed a radial basis function (RBF) neural network model for improved traffic forecasting in the big data environment on a modified artificial bee colony (ABC) algorithm. To be a reliable and interpretable traffic forecast, a Bayesian multivariate adaptive regression (MAR) technique was developed by Xu et al. [15]. These neural networks and Bayesian MAR techniques can estimate traffic with extreme accuracy, however, they come at a considerable cost in terms of computing, particularly when analyzing massive datasets.

3 Study Area and Methodology

3.1 Dataset Description

Aarhus, often known as the City of Smiles, is Denmark's second-largest city, with a population of over 300,000. It is situated on the Jutland Peninsula's east coast. Many visitors and internationals eager in settling in Aarhus were drawn to the city because of its rich culture and numerous opportunities. This paper's study is based on real-world traffic statistics from Aarhus, Denmark. The city administration has installed 449 sensor pairs along the city's key thoroughfares. Traffic data is gathered by counting the number of vehicles passing between two points during a period. Every five minutes, new observations are generated. A collection of datasets of automobile traffic was observed between two sites for a defined amount of time throughout nine months in 2014 (February, March, April, May, June, August, September, October, and November). Each pair of traffic sensors records the average vehicle speed, the number of vehicles on the route, and the expected travel time between the two stations. For the analysis, we have taken all 449 junctions traffic data of the August and September months and this sub-dataset contains more than 7,100,000 instances.

3.2 Models

For forecasting the traffic condition, we have utilized machine learning strategies, such as linear regression, polynomial regression, lasso regression, and ridge regression. Figure 1 shows the data source collection location. Figure 2 is the proposed architecture of the prediction models for forecasting the condition of traffic. The data contains nine columns. We have added a few new features such as year, month, and date in the supplied month, weekdays, and hour out from the timestamp column. The dataset contains August and September months traffic data. The filtered data is used to forecast traffic conditions once it has been pre-processed and enhanced. Machine learning algorithms such as linear regression, polynomial regression, lasso regression, and ridge regression are used to make predictions. Finally, cross-validation is utilized to choose the most appropriate model for predicting.

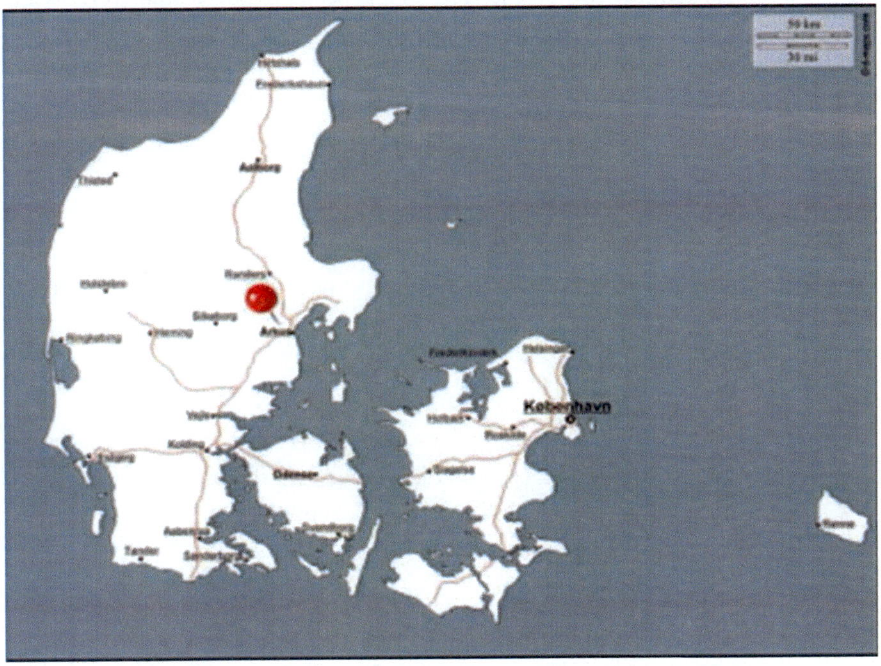

Fig. 1 Data source location

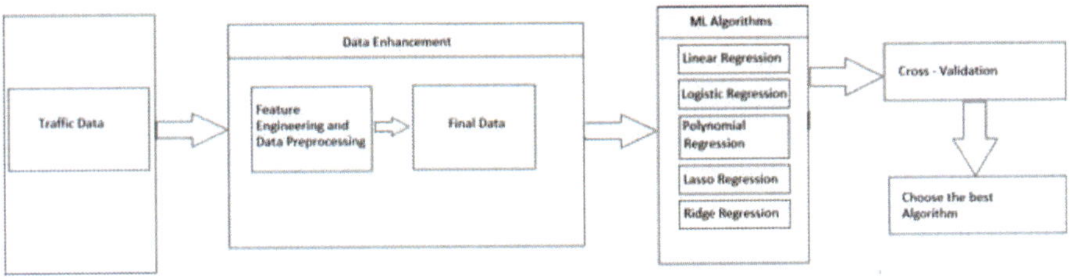

Fig. 2 Shows the proposed architecture of the prediction models for forecasting traffic conditions

3.2.1 Linear Regression Analysis

Linear regression is a well-known statistics technique. Linear regression is a standard statistical process in which a series of continuous predictor variables, also known as independent variables, covariates, or predictors, account for one continuous variable known as the dependent, outcome, or criterion variable [16]. It creates an estimating line based on the results of an explanatory variable by applying the equation of a line to forecast the outcome of a dependent variable. A simple linear regression's goal is to find the best fit line that reduces the sum of squared residuals. This is widely used to find a possible statistical correlation between two variables or to build a correlation between two variables. In linear regression, MSE or RMSE is used as the loss function. The linear regression model is represented by the equation below:

$$y = ax + b + e, \tag{1}$$

where a determines the slope of the line, b determines intercept, and e determines the error in the model.

3.2.2 Polynomial Regression

Because the change of the dependent variable is frequently influenced by several key elements in real-world problems, it is required to employ two or more influencing factors as independent variables to explain the change of the dependent variable, which is known as multiple regressions [17]. Multiple linear regressions are similar to polynomial regression, however, there are a few differences those are, in polynomial regression, the link between independent and dependent variables, X and Y, is denoted by the nth degree. The least mean squared approach is also used in polynomial regression. In polynomial regression, the best fit line is a curving line that runs through all of the data points using the power of X or the value of n. The polynomial regression is represented by the equation below:

$$y = \beta 0 + \beta 1x1 + \beta 2x2 + \beta 3x3 + \cdots + \beta nxn \tag{2}$$

3.2.3 Ridge Regression

Ridge regression is a statistical approach to creating a parsimonious model when the number of predictor variables in a set exceeds the number of observations, or when a dataset has multicollinearity (correlations between predictor variables) [18]. Ridge regression is typically employed when the independent variables have a strong correlation. This is because the least square estimations produce unbiased values in the case of multi-collinear data. However, if the collinearity is too high, there may be some bias. As a result, a bias matrix is included in the ridge regression equation. This is a strong regression strategy that reduces the likelihood of overfitting.

The ridge regression is denoted by the equation below, where the addition of (lambda) overcomes the problem of multicollinearity:

$$\sum_{i=0}^{x}(b_i - b_i')^2 = \sum_{i=1}^{x}\left(b_i - \sum_{j=0}^{y}\alpha_j * a_{ij}\right)^2 \tag{3}$$

3.2.4 Lasso Regression

The least absolute shrinkage and selection operator is a linear model for estimating sparse parameters that are particularly good at minimizing the number of parameters. As a result, in compressed sensing, the lasso regression model is commonly utilized [19]. Lasso regression is a type of machine learning regression that combines feature selection and regularization. It forbids the regression coefficient's absolute size. As a result, in ridge regression, the coefficient value approaches 0, which is not the case. As a result, the model is built using feature selection in lasso regression, which allows feature selection from the dataset. Only the required characteristics are used in lasso regression, while the others are set to zero. This prevents the model from becoming over-fit. When the independent variables are collinear, lasso regression chooses only one variable and shrinks the other to 0. The equation which is below that represents the lasso regression method:

$$\sum_{i=0}^{x}(b_i - b_i')^2 = \sum_{i=1}^{x}\left(b_i - \sum_{j=0}^{y}\alpha_j * a_{ij}\right)^2 + \lambda\sum_{j=0}^{y}\alpha. \tag{4}$$

4 Performance Criteria

Some of the statistical evaluations are used to evaluate the model performance such as Mean Absolute Error (MAE), Root Mean Square Error (RMSE), Mean Square Error (MSE), and coefficient of determination (R^2). The criteria formulas are shown in below:

Some of the statistical evaluations are used to evaluate the model performance such as Mean Absolute Error (MAE), Root Mean Square Error (RMSE), Mean Square Error (MSE), and coefficient of determination (R^2). The criteria formulas are shown in below:

Some of the statistical evaluations are used to evaluate the model performance such as Mean Absolute Error (MAE), Root Mean Square Error (RMSE), Mean Square Error (MSE), and coefficient of determination (R^2). The criteria formulas are shown in below:

Root mean square error (RMSE), mean square error (MSE), and coefficient of determination (R^2) are some of the statistical evaluations used to evaluate model performance. The formulas for the criterion are listed below.

$$\text{RMSE} = \sqrt{\frac{\sum_{t=1}^{n}(y_t - \hat{y}_t)^2}{n}}, \tag{5}$$

where m is the number of observations, $\hat{y}_t$ is the predicted value, and y_t is the actual value.

$$R^2 = \left[\frac{1}{M}\frac{\sum_{j=1}^{M}\left[(Y_j - \overline{Y})(X_j - \overline{X})\right]}{\sigma_y \sigma_x}\right]^2, \tag{6}$$

where M shows the number of observations, σ_x shows the standard deviation of the observation X, σ_y shows the standard deviation of Y, X_j shows the observed values, $\overline{X}$ is the mean of the observed values, Y_j shows the calculated values, and $\overline{Y}$ is the mean of the calculated values [20].

5 Data Visualization, Results, and Discussion

5.1 Data Visualization

We have plotted a time series graph to perceive information about the traffic condition of the city. Figure 3 shows that the data is consistent throughout all of the dates. Over the course of a day, as we expected, Fig. 4 shows that there are peaks in the morning and evening, as well as a fall during the night. Figure 5 shows that in terms of weekly trends, traffic is smoother on Sundays and Saturdays since there are fewer vehicles on the road and Monday through Friday, though, traffic is constant.

5.2 Results and Discussion

This paper looks into the usefulness and effectiveness of regression models for predicting the long-term traffic flow. The following regression models were taken into account: (a) linear regression, (b) polynomial regression, (c) lasso regression, and

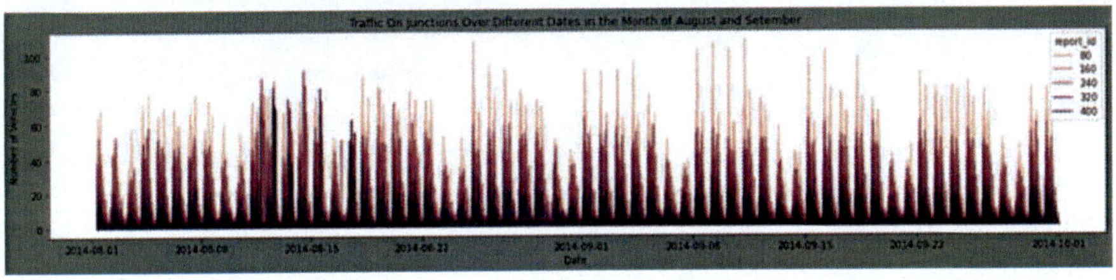

Fig. 3 Time series plot of date-wise vehicle count (August and September) at different junctions

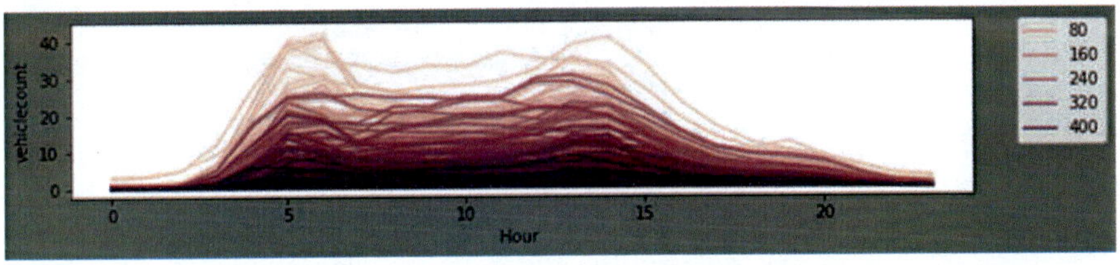

Fig. 4 Time series plot of vehicle count per hour (August and September) at different junctions

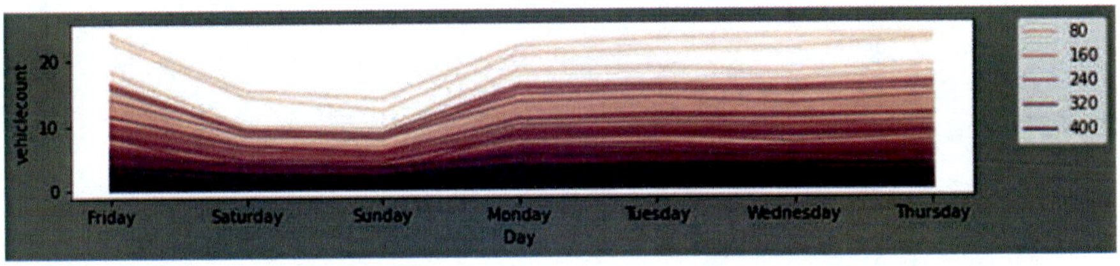

Fig. 5 Time series plot of vehicle count per day (August and September) at different junctions

(d) ridge regression. These models were employed in the trials, and cross-validation and performance criteria were used to assess the model's accuracy for each model.

The results of the prediction using the linear regression approach are shown in Fig. 6, where R^2 is 0.051. The actual numbers are represented by the blue color line, while the traffic prediction is represented by the red color line. Figure 7 shows the prediction results for the polynomial regression algorithm, R^2 is 0.141. Figure 8 shows the prediction results by using the ridge regression algorithm, R^2 is 0.141, and Fig. 9 determines the prediction results for the lasso regression algorithm, R^2 is 0.037. It is clear from Table 1 that the lasso regression model has ascertained its efficacy by having the lowest error rate than the linear, polynomial, and ridge regression models. Table 1 shows the results for all algorithms.

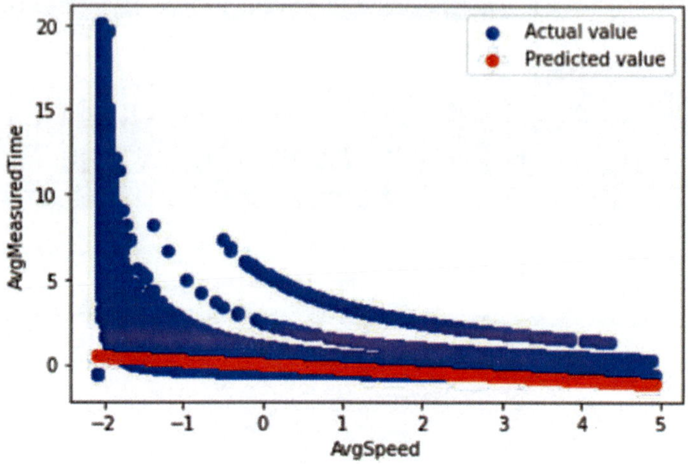

Fig. 6 Prediction model for linear regression

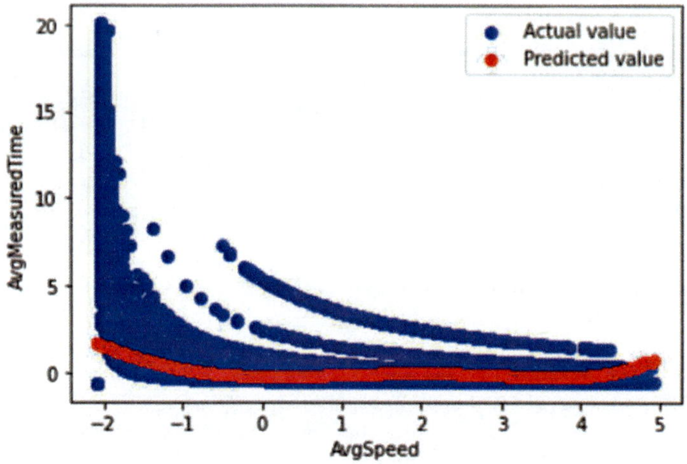

Fig. 7 Prediction model for polynomial regression

6 Conclusion

Intelligent transportation will make our everyday transit safer, greener, and more convenient in the not-too-distant future. ITS can help to minimize the number of traffic accidents, avoid injuries and fatalities, improve traffic flow, and save fuel costs. As a result, we must come up with innovative ways to strengthen and improve the future transportation system. To forecast future traffic flow for different roadways, we analyzed the effectiveness of four different regression models. Lasso regression model performed the best among them with the lowest error rate.

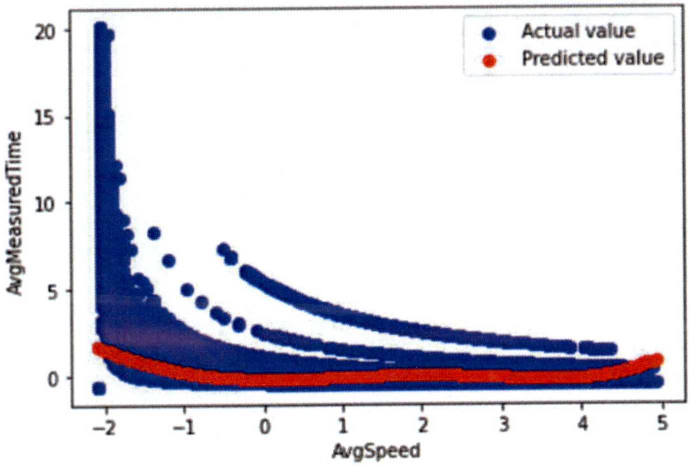

Fig. 8 Prediction model for ridge regression

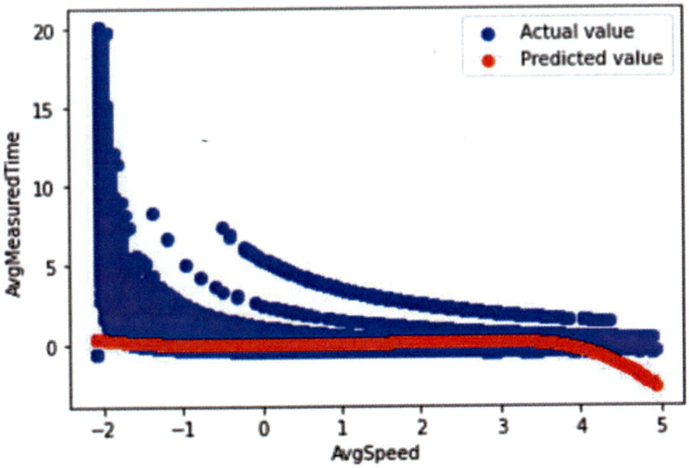

Fig. 9 Prediction model for lasso regression

Table 1 Best results for different machine learning models for traffic forecasting

Models	RMSE	R^2
Linear regression	0.96	0.051
Polynomial regression	0.87	0.141
Ridge regression	0.87	0.141
Lasso regression	0.97	0.037

References

1. Tekouabou SCK, Cherif W, Silkan H (2022) Improving parking availability prediction in smart cities with IoT and ensemble-based model. J King Saud Univ Comput Inf Sci 34(3):687–697
2. Sousanis J (2011) World vehicle population tops 1 billion units. Wards Auto 15
3. Zhang J, Wang FY, Wang K, Lin WH, Xu X, Chen C (2011) Data-driven intelligent transportation systems: A survey. IEEE Trans Intell Transp Syst 12(4):1624–1639
4. Alam I, Ahmed MF, Alam M, Ulisses J, Farid DM, Shatabda S, Rossetti RJ (2017) Pattern mining from historical traffic big data. In: 2017 IEEE region 10 symposium (TENSYMP). IEEE, pp 1–5
5. Chen X, Wei Z, Liu X, Cai Y, Li Z, Zhao F (2017) Spatiotemporal variable and parameter selection using sparse hybrid genetic algorithm for traffic flow forecasting. Int J Distrib Sens Netw 13(6):1550147717713376
6. Devi S, Neetha T (2017) Machine learning based traffic congestion prediction in a IoT based Smart City. Int. Res. J. Eng. Technol 4(5):3442–3445
7. Apronti D, Ksaibati K, Gerow K, Hepner JJ (2016) Estimating traffic volume on Wyoming low volume roads using linear and logistic regression methods. J Traffic Transp Eng (Engl Ed) 3(6):493–506
8. Polson NG, Sokolov VO (2017) Deep learning for short-term traffic flow prediction. Transp Res Part C Emerg Technol 79:1–17
9. Rajabzadeh Y, Rezaie AH, Amindavar H (2017) Short-term traffic flow prediction using time-varying Vasicek model. Transp Res Part C Emerg Technol 74:168–181
10. Harishkumar KS (2018) Multidimensional data model for air pollution data analysis. In: 2018 international conference on advances in computing, communications and informatics (ICACCI). IEEE. 2022 IEEE international conference on data science and information system, pp 1684–1689. 978-1-6654-9801-2/22/31.002022IEEE
11. Ling X, Feng X, Chen Z, Xu Y, Zheng H (2017) Short term traffic flow prediction with optimized multi-kernel support vector machine. In: 2017 IEEE congress on evolutionary computation (CEC). IEEE, pp 294–300
12. Arif M, Wang G, Chen S (2018) Deep learning with non-parametric regression model for traffic flow prediction. In: 2018 IEEE 16th international conference on dependable, autonomic and secure computing, 16th international conference on pervasive intelligence and computing, 4th international conference on big data intelligence and computing and cyber science and technology congress (DASC/PiCom/DataCom/CyberSciTech). IEEE, pp 681–688
13. Zhan H, Gomes G, Li XS, Madduri K, Sim A, Wu K (2018) Consensus ensemble system for traffic flow prediction. IEEE Trans Intell Transp Syst 19(12):3903–3914
14. Chen D (2017) Research on traffic flow prediction in the big data environment based on the improved RBF neural network. IEEE Trans Industr Inf 13(4):2000–2008
15. Xu Y, Kong QJ, Klette R, Liu Y (2014) Accurate and interpretable bayesian mars for traffic flow prediction. IEEE Trans Intell Transp Syst 15(6):2457–2469
16. Harish Kumar KS, Gad I (2020) Time series analysis for prediction of PM2. 5 using seasonal autoregressive integrated moving average (SARIMA) model on Taiwan air quality monitoring network data. J Comput Theor Nanosci 17(9–10):3964–3969
17. Dingen D, van't Veer M, Houthuizen P, Mestrom EH, Korsten EH, Bouwman AR, Van Wijk J (2018) Regression explorer: interactive exploration of logistic regression models with subgroup analysis. IEEE Trans Vis Comput Graph 25(1), 246–255
18. Liu K, Deng H (2021) The analysis of driver's recognition time of different traffic sign combinations on urban roads via driving simulation. J Adv Transp pp 1–11
19. Mei Y, Hu T, Yang LC (2020) Research on short-term urban traffic congestion based on fuzzy comprehensive evaluation and machine learning. In: International conference on data mining and big data. Springer, Singapore, pp 95–107
20. Harishkumar KS, Yogesh KM, Gad I (2020) Forecasting air pollution particulate matter ($PM_{2.5}$) using machine learning regression models. Proc Comput Sci 171:2057–2066

Conditional Generative Adversarial Networks for Image Transformation

C. N. Gireesh Babu, A. G. Guru Dutt, S. K. Pushpa, and T. N. Manjunath

Abstract Conditional adversarial networks are broadly used in picture-to-picture interpretation issues. These networks not only understand the mapping from input picture to the output picture, yet in addition gain proficiency with a misfortune capacity (also known as loss functions) to prepare this cross joining. This will in turn make it conceivable for the application of the very nonexclusive way to deal with issues which customarily would require totally different misfortune capacities (loss functions). The misfortune capacity (loss function) is pointed toward decreasing ancient rarities presented by GANs and guarantee better visual quality and accuracy of precision concerning the ground reality. The generator sub-network is built utilizing the U-Net engineering, though the discriminator is intended to use worldwide and neighborhood data to choose if a picture is genuine/counterfeit. Thus, the exhibition of this approach is compelling at orchestrating photographs from labeled maps and also recreating objects from edge maps.

Keywords Conditional adversarial networks · Convolutional neural networks · Image processing · KDE

1 Introduction

Picture interpretation involves creating a mapping between images from a source space and images from an objective time, and it has several uses, such as picture colorization, creating semantic markings from images, picture super goal, and space transformation. Many picture-to-picture interpretation techniques demand controlled working environments where sets of pertinent source and target images are available. The ability to "understand" an information picture into a comparison yield picture can be applied to a number of problems in computer vision, computer graphics, and image processing. A data picture can be "deciphered" into an image that is meant

C. N. Gireesh Babu (✉) · A. G. Guru Dutt · S. K. Pushpa · T. N. Manjunath
BMS Institute of Technology and Management, Bangalore Affiliated to Visvesvaraya Technological University, Belagavi, Karnataka, India
e-mail: gireeshbabu@bmsit.in

S. Das et al. (eds.), *Advances in Data-driven Computing and Intelligent Systems*, Lecture Notes in Networks and Systems 653, https://doi.org/10.1007/978-981-99-0981-0_42

to be looked at by yield picture because to various problems with picture handling, PC designs, and PC vision. Similar to how an English sentence can express an idea, an RGB image of a scene, a slant field an edge map, a semantic engraving map, and so forth. In relationship to tweaked language getting it, there is a depiction changed picture-to-picture interpretation as the errand of unraveling one likely portrayal of a scene into other, at whatever point outfitted with palatable readiness data.

The social class has recently taken gigantic steps in this course, with convolutional neural networks (CNNs) turning into the ordinary work horse behind a wide assortment of picture expectation issues. CNNs sort out some way to restrict a hardship (misfortune) limit an objective that scores the idea of results and but the learning strategy is programmed, a lot of manual effort really goes into arranging powerful adversities misfortunes).

When all is considered, we really need to tell CNN what we want it to be prohibited from doing. Whatever the case, Marcus Aurelius was right when he observed that our life is what we make it. If we choose a simplistic approach and ask the CNN to reduce the Euclidean distance between the predicted and actual pixels, the results will almost always be murky. This is due to clouding, which occurs when Euclidean distance is constrained by averaging a single extremely unlikely outcome. Writing accident works that compel CNN to accomplish what we actually need; such produce sharp, decent photographs, is a challenge and typically calls on expert data [1]. If we could just narrow it down, it would be incredibly alluring the result appears to be "indistinguishable from the real world" is a crucial level goal, and after that, develop expertise in a bad luck skill that will help you achieve this goal. Fortunately, the recently developed generative adversarial networks complete this exact task (GANs). GANs are successful at describing whether an output picture is authentic or fraudulent while concurrently constructing a generative model to limit this drawback. Hazy photos are impossible without major consequences since they look so obviously false. Since GANs can learn from errors that change as a result of the data, they can be used for a wide range of tasks that ordinarily call for extremely unusual types of error limit.

GANs have been predominantly focused on over the most recent two years and a significant number of the methodologies, we research in this paper have been as of late proposed. In any case, many papers focused in on unambiguous applications, and it has remained foggy how practical picture prohibitive GANs can be as a generally valuable reply for picture-to-picture understanding. In rundown, this paper makes the accompanying commitments:

(1) A contingent GAN-based system to address the difficult picture translation.
(2) A thickly associated U-Net generator sub-network that is explicitly planned by the expansion of skip associations between each layer of the engineering of the Generator.
(3) A patch GAN discriminator is proposed with fewer boundaries to decide whether the relating de-came down picture is genuine or counterfeit.

(4) Extensive analyses are led on freely accessible and combined datasets to exhibit the adequacy of the proposed technique as far as visual quality and quantitative execution.

This paper is coordinated as follows. A concise foundation on conditional generative antagonistic networks, related work of conditional generative adversarial networks is given in Sect. 2. The clarification of the proposed CGAN strategy is surrendered Area Sect. 3. Exploratory outcomes on the different datasets utilized are introduced in visual what's more, quantitative structures in Sect. 4.

2 Related Work

The previous work on the working of the GAN and their related work on how effectively they have been successfully implemented can be summarized as follows:

(a) **Choosing a solution for GAN-based culturization method**

GANs assessment is a difficult errand. GANs are prepared to arrive at a balance circumstance where the generator can cheat the discriminator. In any case, the previous, taken alone, is not related with an expense capability to be limited. Under these circumstances, it is difficult to anticipate when the generator will deliver tests that fit the objective likelihood conveyance at their best.

Thus, FID catches the closeness of produced pictures to genuine ones better than IS. On a fundamental level, we could look at the changed answers for picture culturization examined in the past area utilizing FID as finished in, which demonstrates the way that FID values can altogether fluctuate contingent upon explicit picture-to-picture interpretation task. Notwithstanding, we found it more fitting to analyze different GANs on a subjective premise to survey the degree to which every arrangement can meet the prerequisites for picture culturization (FID will assume a part in the following segment in tuning hyper parameters of the chose GAN arrangement). In particular, we saw that Pix2Pix, and comparative methodologies require matched datasets, which is an impediment since picture culturization ought to deal with exceptionally assembled datasets, e.g., made out of pictures downloaded from the web. InstaGAN did not uncover a possible answer for comparative reasons since it requires a cautiously planned dataset made out of pictures having a place with various social spaces and the relating veils. Following these initials contemplations, we carried out CycleGAN, attention GAN, and UNIT that do not present imperatives on the dataset, and afterward we subjectively assessed the outcomes. Tests affirmed that attention GAN has solid impediments when a picture-to-picture interpretation requires changing the picture shapes, frequently required in picture culturization. Both CycleGAN and UNIT may be great decisions: after subjective assessment, and since CycleGAN is the reason for the majority different methodologies (counting InstaGAN that may be rethought later on), we at last chosen it as the best possibility for picture culturization. Nonetheless, we do not see any deterrents in carrying out a similar technique depicted in the following sections with an alternate GAN [10].

(b) **Structured losses for image modelling**

Organized misfortunes for the picture showing image-to-picture understanding issues are often times arranged by per-pixel characterization or regression. These definitions treat the outcome space as "unstructured" as in each outcome pixel is seen as prohibitively independent from all others given the data picture. Restrictive GANs rather gain capability with a coordinated setback (misfortune) [6].

Organized disasters (misfortunes) rebuff the joint plan of the outcome. A fake colossal assortment of composing has contemplated hardships of this sort, with procedures including prohibitive sporadic fields, the SSIM metric, feature planning, nonparametric disasters, the convolutional pseudo-prior [8], and setbacks considering matching covariance estimations. The contingent GAN is different in that the adversity is learned, and can, on a fundamental level, rebuff any possible development that shifts among result and target. The principal approach utilizes a genuine informational collection from the Highway Safety Information System (HSIS) to assess the proposed CGAN-EB strategy.

Assessment considers (i) model fit to the accessible accident recurrence information; (ii) model execution for pre-ducting crash recurrence for some other time span; and (iii) network screening execution (e.g., distinguishing areas of interest). The fit execution of the models has been assessed utilizing three normal standards viz. coefficient of determination (R^2 score), mean outright mistake (MAE), and mean absolute percentage error (MAPE). A test set is used to assess the predictive performance of the models.

The evaluation of network screening performance uses four tests: site consistency test (SCT), rank difference test (RDT), method consistency test (MCT), and that are presented in Cheng and Washington [2], and the prediction difference test (PDT). All these tests assume that, in the absence of significant changes, detected hotspots remain hazardous over the time.

3 Proposed Method

The equation below expresses the conditional GAN objective:

$$\text{LcGAN}(M, N) = Ea, b\big[\log N(a, b)\big] + Ea, b[\log(1 - N(a, M(a, z))], \quad (1)$$

where M minimize this objective beside an adversarial D that tries to maximize it, i.e.,

$$M* = \arg\min M \max N \,\text{LcGAN}(M, N). \quad (2)$$

In testing the importance of conditioning the discriminator, there is a relation to an unconditional variant in which the discriminator does not observe x

$$LGAN(M, N) = Eb\big[\log N(b)\big] + Ea, z[\log(1 - N(M(a, z))]. \quad (3)$$

Past methodologies have found it valuable to blend the GAN objective in with a more conventional misfortune, for example, $L2$ distance [13].

The work of the discriminator remains constant. But the generator's tasks are far more comprehensive than that of a discriminator. It hinders the discriminator plus ensures it is close to the ground truth obtained from $L2$.

There is an investigation of additionally this choice, utilizing $L1$ distance instead of $L2$ as $L1$ supports less obscuring

$$LL1(M) = Ea, b, z[b - M(a, z)].$$

The final objective is

$$M* = \arg\min M \max N \,\text{LcGAN}(M, N) + \lambda LL1(M).$$

In starting tests, we did not find this methodology powerful—the generator just figured out how to disregard the commotion—which is predictable with [11]. All things considered, for the last models, there is a give commotion just as application of dropout that have been inculcated in our generator layers at both training and during the time of test. In spite of the dropout commotion, there is a perception just minor stochasticity in the result of our nets.

Past methodologies have found it useful to blend the GAN objective in with a more customary misfortune, for example, $L2$ many pictures managing, PC outlines, and PC vision issues could portray as "make an interpretation of" an information picture into a result picture. A scene can be tended to as a RGB picture, an edge map, a semantic name map, a point field and so on, comparably that a thought can be conveyed in the English or French. The preliminary of changing one likely portrayal of a scene in to another, given OK game plan information, is depicted as robotized picture-to-picture translate [2].

The preparation of a significant standard information grid to a significant standard yield cross section is an unmistakable property of picture-to-picture understanding hardships [5]. Moreover, the data and result for the issues we analyze have different surface appearances, yet they are the two depictions of a comparative fundamental development in like manner, the information structure is practically vague from the outcome structure [8].

These considerations support the generator designing. Ideally, we will rather give an irrefutable level objective, for instance, make the outcome dubious from this present reality", and the setback work for accomplishing that goal would be progressed naturally [13]. Fortunately, the actually proposed generative adversarial networks accomplish just that (GANs).

GANs are generative models that gains a planning [9] from irregular clamor vectors to a result picture b, G: $z \to b$. Conversely, restrictive GANs gain a preparation from saw picture x and inconsistent disturbance vector z, toy, G: $\{a, z\} \to b$ [11]. The generator G is prepared to produce results that cannot be discerned from "genuine" images by an adversarially arranged discriminator, N, which is prepared to perform similarly as expected in recognizing the generator's "fakes."

An adversarially prepared discriminator, N, is prepared to do as well as might be expected at differentiating the generator's "fakes", but the generator G is prepared to produce yields that cannot be distinguished from "real" photos.

Network Architecture of Generator

Generator with a U-Net-based design will be part of your network's architecture. The generator uses the input photos to provide predictions for the training region. The predictions are given into the discriminator, which evaluates prediction's resemblance to the ground truth to determine its accuracy.

If the forecast and the ground truth are too similar, the discriminator will be unable to tell them apart and identify the prediction as genuine. The GAN will be able to segment pictures with the same accuracy and precision as hand annotations with enough training.

As a result of preparing, the generator is continuously changing, and it has an inclination to fail to remember past tasks (with regards to the discriminator preparing, learning semantics, designs, and surfaces can be viewed as various undertakings). This discriminator is not compensated for holding an additional mind boggling of some.

information portrayal, which includes learning both worldwide and neighborhood visual differentiations [8]. The accompanying word is gotten from the state of the so-framed design, which informalize "U". By taking a gander at the construction and the many areas associated with the organization's creation, we can presume that it is a completely convolutional network [9]. They utilized no further layers, like thick or straighten, or anything like. In the realistic portrayal, a contracting course is trailed by an extending way. As demonstrated in the engineering, an information picture is communicated through the model, trailed by a couple convolutional layers utilizing the ReLU initiation work.

This is because of the use of unpadded convolutions (characterized as "genuine"), which brought about a decrease in general dimensionality. The encoder block use walks 2s maximum pooling layers to keep the picture size diminishing consistently [4]. We additionally have repeating convolutional layers with a rising number of channels in the encoder design. As we approach the decoder, the quantity of channels in the convolutional layers starts to diminish, joined by a continuous up inspecting in the progressive layers the whole way to the top. The utilization of skip associations with interface the past results to the decoder levels.

This skip interface is basic for keeping up with misfortune from past layers so it very well might be seen all the more plainly in complete qualities. They have likewise been logically displayed to create improved results and speed model assembly. In the last convolution block, we have a couple of convolutional layers followed by the last convolution layer [18].

This layer incorporates a second channel that has the suitable capacity for showing the outcomes. This last layer can be adjusted to meet the necessities of the current task. The U-Net design is one of the most huge and special advancements in profound

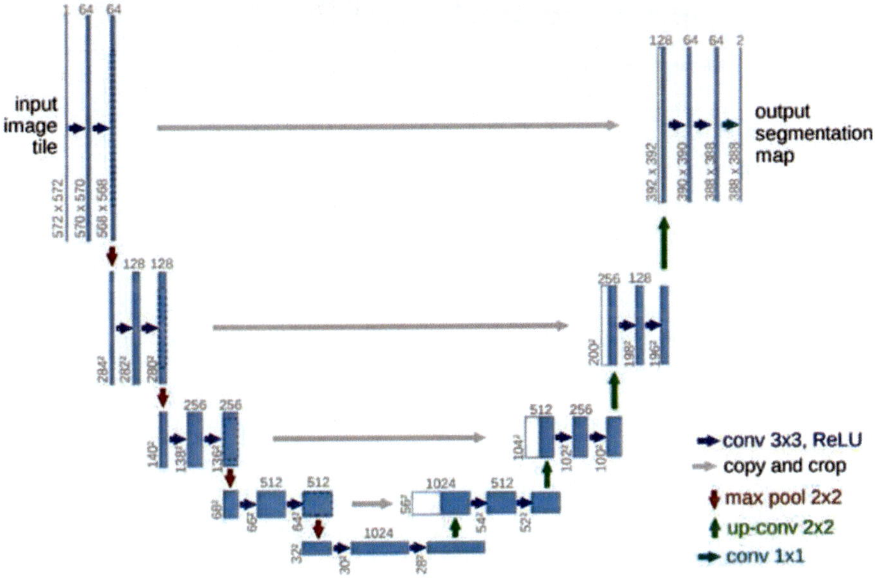

Fig. 1 Structure of U-Net framework

learning [14]. The U-Net design in Fig. 1 was at first proposed in an exploration paper to resolve the issue of biomedical image segmentation, though it was not restricted to that subject with connection to that application. The model has addressed and keeps on handling the most requesting profound learning issues.

Albeit a portion of the first design parts is as of now not accessible, there are different variations of this structure. Among them are U-Net with the attention, LadderNet, the intermittent and lingering convolutional U-Net, and more organizations.

Network Architecture of Discriminator

When an entire picture is sent into the deep convolution network to categorize, the discriminator network predicts whether a specific image is "fake" or "real" [4]. The cGAN-pix2pix model, on the other hand, employs a PatchGAN network, which classifies individual instead of the complete picture; sections of an input image are labeled as true or false.

The PatchGAN discriminator as shown in Fig. 2 classifies each $N \times N$ patch in an image as genuine or false, then runs convolutionally across the image to create a single feature map of true or false predictions, which may then beaveraged to get a singles core, which is the discriminator's final output D [9]. PatchGAN has the benefit of being able to employ any size picture with a fixed-size patch discriminator [15].

A source image and a target image are presented to the discriminator, and it must determine if the target image is a possible transformation of the source image [12].

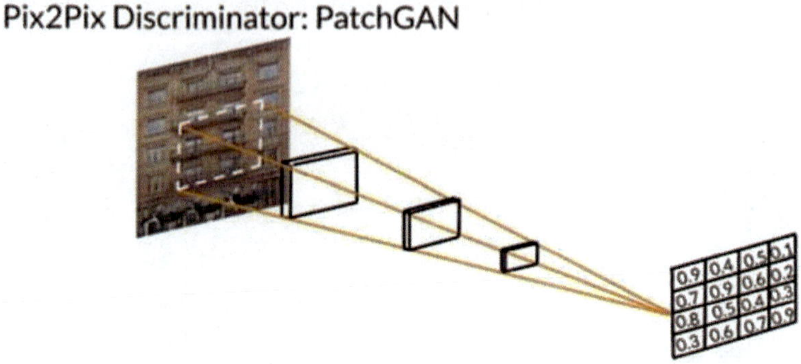

Fig. 2 Architecture of PatchGAN discriminator network

4 Evaluation Results

The evaluation results for the operation of cGANs used by us are summarized as follows:

(a) **Frechet Inception Distance and Kernel Inception distance**

It is described as the following: Bit Inception distance (KID) [9] and Frechet Inception distance (FID) [7] are the two most recognized estimations for evaluating picture-to-picture translation execution.

A lower distance score suggests the translated pictures are more comparative to those in the objective region. KID and FID for picture-to-picture understanding are difficult to recreate. In [9], most FID and KID scores of comparative task model settings shift. From this deduction, it may be perceived that to find the FID and KID scores of the different GAN_S utilized. The result of loss calculation FID and KID can be seen in Fig. 3.

For non-square CelebA pictures, first resize them to have a focal harvest of size 256 × 256, so a face has a similar viewpoint proportion [8]. All computations of KID and FID areas signed to the open-source light loyalty bundle.

All calculations of KID and FID scores from Table 1 are delegated to the open-source torch-fidelity package [7]. To faithfully reproduce existing models, use pre-trained models (if available). Otherwise, retrain them following the provided hyper parameter configurations.

(b) **Kernel Density Estimation**

KDE is broadly used to assess decoder-based models, and a variation was proposed in the setting of assessing Boltzmann machines [14].

It is described as the following:

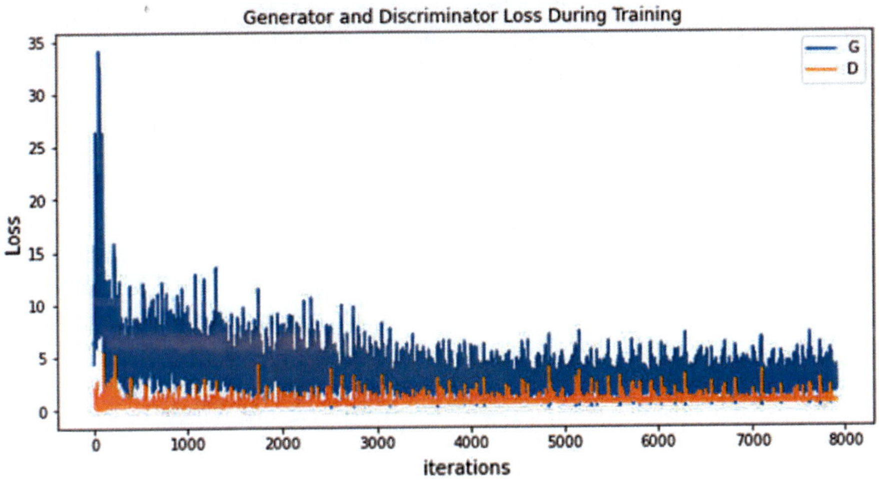

Fig. 3 Loss calculation using FID and KID

Table 1 FID and KID scores

GANS used	Selfie2Anime	Selfie2Anime	Anime2Selfie	Anime2Selfie
	FID	KID	FID	KID
DCGAN	99.8	3.22 ± 0.26	128.6	3.49 ± 0.33
CycleGAN	91.9	2.74 ± 0.26	126.0	2.57 ± 0.32
PatchGAN	82.8	7.34 ± 0.75	125.0	5.41 ± 0.41

$$f(x) = 1\frac{1}{nh}\sum_{k=1}^{n} K\left(\frac{x - xk}{h}\right).$$

Papers revealing KDE gauges frequently alert that the KDE is not anticipated to be applied in high-dimensional spaces and the outcomes could subsequently be mistaken. In any case, KDE stays the standard convention for assessing decoder-based models.

The precision of the KDE gauges was broken down by looking at against AIS. The two evaluations are stochastic lower limits on the genuine log-probability, so bigger qualities are ensured (with high likelihood) to be more exact. For every assessor, change was made to one boundary affecting the computational spending plan; for AIS, this was the quantity of middle of the road disseminations (looked over {100, 500, 1000, 2000, 10,000}), and for KDE, it was the quantity of tests (looked over {10,000, 100,000, 500,000, a million, 2,000,000}). Involving Self2Anime for delineation, plot both log-probability gauges 100 reproduced models as a capability of assessment time. Likewise plot the upper bound of probability given by running AIS in switch course. We see that hole ways to deal with nothing, approving the precision of AIS. We additionally see that the AIS assessor accomplishes significantly more

Table 2 Kernel density estimation of various AIS and IWAE bound values

Nats	AIS	AIS + encoder	IWAE bound	#dist AIS	#dist AIS + encoder
IWA	– 64.679	– 56.754	– 82.962	789	75
E	– 66.619	– 59.621	– 80.494	8559	888

precise appraisals during comparable assessment time. Moreover, the KDE gauges seem to even out off, recommending one cannot acquire exact outcomes in any event, utilizing significant degrees more examples. The KDE assessment mistake likewise influences the gauge of the perception commotion σ, since an enormous worth of σ is required for the examples to cover the full dissemination. On looking at the log-probabilities assessed by AIS and KDE with shifting decisions of σ on 100 preparation and approval instances of Selfie2Anime. On involving 1 million recreated tests for KDE assessment, which takes practically a similar time as running AIS assessment [16].

The log-probability of various AIS and IWAE is assessed by KDE and AIS as a capability of σ. Since the exactness of KDE declines pointedly for little σ values, it makes serious areas of strength for an inclination toward enormous σ as shown in Table 2.

(c) **FCN Per-Pixel Comparison**

Assessing the nature of created pictures is a difficult undertaking. Customary the measurements like the per-pixel mean-squared goof does not think about the joint bits of knowledge of the outcome and hence does not reflect development that coordinated mishaps try to depict it here.

Two systems are being utilized to evaluate all the more exhaustively work on the visual nature of our results. To start with, Amazon Mechanical Turk [5] offers "genuine versus misleading" discernment tests (AMT).

The extreme hold back nothing like colorization and in the photograph, creation is regularly trustworthiness to a human watcher. Subsequently, utilize this technique to assess our guide creation, airborne photograph age, and picture colorization. Second, we survey whether the engineered Cityscapes [3] are sensible enough for an off-the-rack object acknowledgment framework to perceive the things inside them. Perceptual examinations on AMT (AMT tests) Turkers were given a progression of preliminaries in which they needed to pick between a "veritable" picture and a "phony" picture made by the calculation. Each image showed for 1 s on every preliminary, and afterward disappeared, passing on Turkers with a boundless measure of time to conclude which was genuine [17].

Preparing for the guide flying picture was finished on 256 × 256 goal photographs, but testing was finished on 512 × 512 pictures utilizing completely the convolutional interpretation (made sense of above), which were accordingly down inspected and given to Turkers at 256 × 256 goal. The subjective outcomes of these changes on two marks photograph issues are displayed. When $L1$ is utilized alone, the outcomes are fair however foggy. The cGAN alone (setting = 0) delivers altogether better

Table 3 FCN-scores for various misfortunes, assessed on Cityscapes [3]

Loss	Per-pixel acc
Ground truth	0.80
L1 + cGAN	**0.66**
L1 + GAN	0.64
cGAN	0.57
GAN	0.22
L1	0.42

Bold represents results are better, compare to all other feasible combinations

outcomes, but for certain applications, it creates visual antiquities. These relics are decreased when the two terms are added as in Table 3.

The comparison results of the various losses of the GANs are given as follows in Fig. 4.

(d) **Patch Size Variations**

From a 1 × 1 "PixelGAN" to a complete 286 × 286 "ImageGAN", see what happens when we change the discriminator receptive fields' patch size *N*. Figure 5 depicts the analysis' qualitative findings, whereas Table 4 measures the effects utilizing the FCN-score.

The PixelGAN [12, 14] has little influence on spatial clarity, but it does make the output more colorful. At the point when the net is prepared with a $L1$ shortfall, the transport in Fig. 5 is dark, not red when the net is instructed with a PixelGAN deficit.

Fig. 4 Comparison of various losses induces different quality of results

Fig. 5 Analysis of PixelGAN to various pitch size variations

Table 4 FCN-scores for differed discriminator open field sizes in view of Cityscapes pictures

Discriminator receptive field	Per-pixel accuracy
1 × 1	0.39
16 × 16	0.65
70 × 70	**0.66**
286 × 286	0.42

Bold represents results are better, compare to all other feasible combinations

PixelGANs may be a possible lightweight answer for variety histogram coordinating, which is a predominant test in picture handling. Using a 16X16PatchGAN is enough to provide crisp outputs and strong FCN-scores, but it also causes tiling problems. Output vulnerability shows itself contrastingly for various misfortune capacities. Unsure districts become foggy and desiderated under $L1$ [13].

The 1 × 1 PixelGAN energizes variety yet lacking in spatial statistics. Locally, 16 × 16 PatchGAN delivers clear results, but it also introduces tiling artifacts that exceed the size it can detect. The spectrum (colorfulness) and spatial dimensions of the 70 × 70 PatchGAN demand clean, even if incorrect, outputs. According to our FCN-score criterion, the entire 286 × 286 ImageGAN produces visually equivalent but poorer results to the 70 × 70 PatchGAN. The results of the per-pixel accuracy with respect to discriminator receptive field are best suited for accuracy of 0.66 as shown in Table 4.

Scaling the results to the whole 286 × 286 Fig. 5 shows that ImageGAN does not seem to increment visual quality and on second thought results in an essentially lesser FCN-score. This perhaps could be on the grounds that the ImageGAN consists of a lot larger number of boundaries and profundity than the PatchGAN, making it way more difficult to train.

5 Conclusion

The outcomes in this paper propose that restrictive ill-disposed networks are a promising methodology for certain picture-to-picture interpretation errands, mainly those comprising profoundly organized graphical results. The articulate results put forth by the FID (lower scores) and KID show high promise of usage of GANs with good use for the image translation in medical diagnosis applications, traffic surveillance, and Cityscapes. Amazon Mechanical Turk [5] offers "genuine versus misleading" discernment tests (AMT).

The extreme hold back nothing like colorization and in the photograph, creation is regularly trustworthiness to a human watcher. The further analysis of FCN-scores technique to assess our guide creation, airborne photograph age, and picture colorization for the various types of GANs shows that the L1 + GAN has the highest per-pixel accuracy with respect to the candid original truth. According to our FCN-score criterion, the entire 286×286 ImageGAN produces visually equivalent but poorer results to the 70×70 PatchGAN. These organizations become familiar on a misfortune adjusted to the current errand and information, which makes them material in a wide assortment of settings across a plethora of industries ranging from image translations for medical applications, security measures of surveillance.

References

1. Chen L-C, Papandreou G, Kokkinos I, Murphy K, Yuille AL (2015) Semantic picture division with profound convolutional nets and completely associated NCRFs. In: ICLR
2. Chen T, Cheng M-M, Tan P, Shamir A, Hu SM (2009) Sketch2photo web picture montage. ACM Trans Graph (TOG) 28(5):124
3. Cordts M, Omran M, Ramos S, Rehfeld T, Enzweiler M, Benenson R, Franke U, Roth S, Schiele B (2016) The Cityscapes dataset for semantic metropolitans' cene understanding. In: CVPR
4. Denton E, Chintala S, Szlam A, Fergus R (2015) Profound generative picture models utilizing a Laplacian pyramid of ill-disposed networks. In: NIPS
5. Doersch C, Singh S, Gupta A, Sivic J, Efros A (2012) What compels Paris resemble Paris? ACM Trans Graph **31**(4)
6. Dosovitskiy A, Brox T (2016) Creating pictures with perceptual likeness measurements in light of profound organizations. In: NIPS
7. Efros AA, Freeman WT (2001) Picture sewing for surface union and move. In: SIGGRAPH
8. Efros AA, Leung TK (1999) Surface union by non-parametric examining. In: ICCV
9. Eitz M, Hays J, Alexa M (2012) How do people draw objects? In: SIGGRAPH, vol 4, p 12
10. Fergus R, Singh B, Hertzmann A, Roweis ST, Freeman WT (2006) Eliminating camera shake from a solitary photo. ACM Trans Graph (TOG) 25(3):787–794
11. Gatys LA, Ecker AS, Bethge M (2015) Surface combination utilizing convolutional brain organizations. In: NIPS
12. Gatys LA, Ecker AS, Bethge M (2016) Picture style move utilizing convolutional brain organizations. In: CVPR
13. Gauthier J (2014) Contingent generative antagonistic nets for convolutional face age. Class project for Stanford CS231N: convolutional neural networks for visual recognition. Winter Semester 5:2

14. Goodfellow I, Pouget-Abadie J, Mirza M, Xu B, Warde-Farley D, Ozair S, Courville A, Bengio Y (2014) Generative ill-disposed nets. In: NIPS
15. Hertzmann A, Jacobs CE, Oliver N, Curless B, Salesin DH (2001) Image analogies. In: SIGGRAPH
16. Hintonand GE, Salakhutdinov RR (2006) Lessening the dimensionality of information with brain organizations. Science 313(5786):504–507
17. Hwang S, Park J, Kim N, Choi Y, So Kweon I (2015) Multispectral passer by location: benchmark dataset and gauge. In: CVPR
18. Iizuka S, Simo-Serra E, Ishikawa H (2015) Let there be color!: joint end-to-end learning of global and local image priors. In: CVPR

Comparison of Data Collection Models in an Intelligent Tutoring System for the Inclusive Education of the Learning-Disabled

Sarthika Dutt and Neelu Jyothi Ahuja

Abstract Intelligent tutoring systems were developed to provide effective learning and educate learners without the human interventions. However, identifying learning disabilities is a challenge that requires an effective data collection process. As, the process of data collection affects the generated learner's details and their learning outcomes in an intelligent tutoring system. Therefore, in the present work, we have compared the data collection process to record the response of the learning-disabled learners for the developed ITS. Also, the machine learning (ML) techniques were explored to extract and select the features related to learning disabilities (dyslexia, dysgraphia, and dyscalculia). This study includes comparison of some popular and effective feature selection and classification algorithms. The feature selection methods utilized in the present study are relief, XGBoost (XGB), elastic net (EN), least absolute shrinkage and selection operator (LASSO), and gradient boosting decision tree (GDBT). The classification algorithms that have been compared based on these feature selection techniques for learning disability prediction are *K*-nearest neighbors (*K*NN), logistic regression (LR), linear discriminant analysis (LDA), classification and regression trees (CART), Naïve Bayes (NB), and support vector machine (SVM). The result proves both data collection methods were equally effective and accuracy of CART with the feature selection method LASSO yields highest performance for feature selection and classification process.

Keywords Classification · Feature selection · Learning-difficulties

---

S. Dutt (✉)
COER University, Roorkee, India
e-mail: Sarthikadutt@gmail.com

S. Dutt · N. J. Ahuja
University of Petroleum and Energy Studies (UPES), Dehradun, India
e-mail: Neelu@ddn.upes.ac.in

S. Das et al. (eds.), *Advances in Data-driven Computing and Intelligent Systems*, Lecture Notes in Networks and Systems 653, https://doi.org/10.1007/978-981-99-0981-0_43

1 Introduction

The identification of learning difficulties at an early stage of child development is important; learning disabilities are neurobiological disorder and difficult to be examined without psychologists and experts' interventions. The learning difficulties can be analyzed by the experts and proper therapies could be provided to the learners with learning difficulties. In education, with progress in adaptive e-learning, the learner is provided with a personalized learning environment according to the learner's requirement [1]. For learning disabled, the environment has to be more specific concerning the learners' characteristics. The advancements in artificial intelligence are focused to provide assistive technologies and support to the learners with learning difficulties.

Dyslexia disorder affects learners reading and learning skills. The learners with dyslexia find difficulties in literacy (LI), phonological awareness (PA), random naming (RAN), word fluency (WF), decoding (DE), and basic reading skills [2, 3]. On the other hand, dysgraphia difficulty affects the writing capabilities of the learners [4]. Also, the learners with dysgraphia show difficulties in motor skills (MS) [5], space knowledge (SK), and cognitive skills (CS) [6]. The learners with dyscalculia [7] difficulty are generally incapable to perform well with numbers and found difficulties in basic arithmetic operations [8]. The identification of these learning difficulties and providing assistive technologies at an early stage of child development can significantly improve these symptoms [9]. Several applications have already been developed utilizing artificial intelligence (AI) to identify disabilities among children and elderly people [10]. The machine learning algorithms were utilized on various data input in previous studies for learning difficulties identification [11].

An intelligent tutoring system (ITS) is an artificial intelligence (AI) application that primarily was developed for educating the students without teachers' interventions. The learner model of an ITS is responsible to collect data and identify learner characteristics from the collected data. The learner characteristics are mapped to provide learner specific content to the learners. The data collection is an integral process of an ITS learner model development. The data collection method affects the overall obtained accuracies of the utilized algorithms. As, the data plays an important role in overall accuracy improvement, the features related to dyslexia, dysgraphia, and dyscalculia prediction need to be carefully selected and analyzed before making the prediction. The feature selection techniques have been used for improving accuracies of machine learning algorithms [12, 13]. The data collection models for predicting dyslexia, dysgraphia, and dyscalculia were compared in this work. The objective of the study is to:

- Develop a learner database model into an ITS framework to collect dataset of learning-disabled learners.
- The two survey methods for data collection (in-person interviews and computer-based test) were compared for the measurement of learner's experience statistically.
- Additionally, the most important features were selected through the feature selection process and trained over machine learning models.

This paper is organized under the following sections. The section literature review discusses about the applications and methodologies used for LDs. The method section explains the data collection methods utilized for the development of learner database model of an ITS for LDs identification. The result and discussion section evaluates the learner's experience utilizing the two data collection methods into the ITS learner model, and finally, the paper conclude with the future possibilities of improvement in the developed ITS systems.

2 Literature Review

2.1 Advancements in Artificial Intelligence (AI)

The previous researches attempt to solve learning disabilities problems using various input features, features training through the machine learning algorithms. The machine learning approaches are widely used in communication, diagnosis, education, and other research domains. In a study, machine learning approach in the communication network is explored; unsupervised learning is applied to a real network to detect anomalies at multiple network layers and to identify the root of each anomaly [14]. In another study, long-term diabetes cause diabetic retinopathy (eye abnormality), micro-aneurysms (MAs) images have been analyzed to detect MA detectability, using machine learning classifiers [15]. The Wi-Fi signal strengths (RSSs) are sensitive to environmental factors and cause irregularity in signal strengths. An outlier detection technique is proposed using supervised, unsupervised, and ensemble machine learning models [16]. A study proposed an algorithm to determine the learning conflicts in the dataset. The algorithm is used in a real dataset of refrigeration, and results indicate it has improved the performance of machine learning algorithms [17]. Accurate respiratory distress syndrome is diagnosed following clinical and in each training label, this uncertainty of diagnosis is used as a graded weight of confidence associated with each label. To limit overfitting, a novel time series method is used to address the problem of intercorrelations among clinical data of each patient. The results indicate the improvement in the proposed algorithm when compared with the conventional SVM algorithm [18]. The learning automata algorithm is improved using adaptive fuzzy reinforcement learning to solve its stability problem. A numerical superposition algorithm is used to increase the learning rate and to improve stability [19]. The study proposed a geomagnetic data reconstruction approach using SVM, random forest, and gradient boosting, and RNN are explored further for improvement. The hyperplanes have been used for reconstructing data; result signifies the efficacy of the method proposed when compared with the traditional linear method [20].

2.2 Artificial Intelligence (AI) for Solving Learning Disabilities

In solving difficulties related to children and elderly people, machine learning has been used for the prediction of self-care problems of children with physical or motor disability. Several machine learning models are compared for accuracy; finally, *K*NN is proposed to predict the self-care problems [21]. Author has predicted learning disability in school-age children using ensemble techniques. The important parameters related to learning disabilities are focused while making predictions. The results of the ensemble and machine learning models are compared finally [22]. Autism detection using screening tools is time-consuming and expensive. The machine learning methodologies provide an effective way to diagnose autism spectrum disorder. The autism spectrum disorder is diagnosed using different machine learning models. The results have been compared based on the accuracy and efficiency of the models [23]. In order to predict Parkinson's disease based on voice data, different feature sets were compared and principal component analysis is used for selecting the most significant features. The classification algorithms on these feature sets have been trained and tested to compare their performance on several performance metrics [24]. Machine learning has been used for predicting learning difficulties, like dyslexia, dysgraphia, and dyscalculia. In studies, learning algorithms are used for game-based screening of learning disabilities and the output of this screening is used as input to the machine learning models like SVM, *K*NN, and random forest [25, 26]. A mobile application Pubudu based on deep learning has been developed for the screening of dyslexia, dysgraphia, and dyscalculia. The learners' handwritten letters and digits are included as input to CNN model in letter dysgraphia and numeric dysgraphia prediction. Learner's speech sample as audio clips has been included for classification of dyslexia from other disabilities [25, 27]. The disabilities have been predicted using the stacking ensemble technique. Logistic regression, *K*NN, and SVM results are calculated and are used in the stacking ensemble model along with the decision tree, and the final result is obtained. The result of machine learning algorithms is individually compared with the ensemble algorithm [22, 28].

3 Methodology

3.1 Design and Data Description

This study is focused to design and develop a model to collect data of learners with learning difficulties and provides the learners with personalized learning contents. As the different methods utilized in the previous studies for the interviews of learning-disabled learners involve bias in the responses [29]. Therefore, the two methods of surveys are compared in the present study-in-person interview and electronic response-based screening of LD learners through the developed ITS. In this study,

94 learning-disabled learners were interviewed in person. A pretest is conducted utilizing the learner module framework of an ITS on a sample of learners with and without learning disabilities. This recorded data has been quantitatively assessed to determine its impact on the learning of the learning-disabled learners. Pretest question then mapped to a particular skill which is either psychological or psychomotor ability related. For the development of questionnaires framework related to dyslexia, dysgraphia, and dyscalculia issues, schools and psychotherapeutics center have been collaborated. More than six schools and two psychotherapeutic centers were target for data collection and to develop questionnaire framework. The type of features used for dyslexia, dysgraphia, and dyscalculia discrimination and identification are related to each issue to be discriminated. These are the issues which impact reading skill, fluency, writing skills, motor skills, and reasoning and arithmetic skills of the learners. The speech features considered for classification process are phonological awareness, literacy, random naming, fluency, and loudness [30]. The motor skills, space knowledge, spellings, and legibility are features considered for writing analysis. These additional features were mapped to the cognitive strength (CI) and learning style (LS) of the learners [4, 6]. The system assisted screening is performed through training of learner's data to identify learning disabilities dyslexia, dysgraphia, and dyscalculia. The training data for learner model classification process included the data of learning disabled (dyslexia, dysgraphia, and dyscalculia) and data of non-learning-disabled learners. Intermediate output which is mapped with each pretest and sample of questionnaire is considered for profile development later. These intermediate outputs are problematic skills (PS), cognitive strength (CI), learning preference/style (LS). The problematic skills indicate the skills the person lacking in questionnaire. The cognitive strength represents the various mental activities performed by the learner mostly associated to the questions of learning and problem solving. It maps the strongest cognitive strength while performing questionnaire. The nine cognitive strengths are long-term memory, short-term memory, attention, concentration, orientation, abstraction/rational thinking, language ability, visual perception, and fluency [31]. The memory and attention, visual perception, and rational and logical thinking (processing) are the cognitive strength that are classified into groups for the proposed work. Also, learning style plays a crucial role in academic levels, and learning outcomes in many studies have included Felder-Silverman learning styles model in educational and tutoring models for adaptivity [32]. The learning style represents the style preferred by the learner during the learning process. This work frames out four learning styles reflective, visual, active and kinesthetic. The learning preference will give the most suitable preference of the learner according to the learner interest. In order to deliver the content according to the learning interest of the user, a learning style along with the cognitive abilities is required. The studies have shown the positive impact of using more than one learning style in learning process. The known learning models are of Klobe model (active/reflective/observant/experience), Felder Silverman model (active/reflective, sensory/intuitive, Visual/Verbal, Sequential/Global), and VARK model (visual, aural, read, and kinaesthetic) [33]. Final profile details include difficulties or disability type,

degree/severity, cognitive strength, learning style, and problematic skills involved with each participant (with or without learning disability).

3.2 Learner Data Preprocessing

The learner dataset utilized is collected through surveys, from psychotherapeutic centers and schools (rural and urban). Total of 94 learner data 41 were learning disabled with learning difficulties. The features for discriminating dyslexia difficulty include literacy, phonology, random naming, decoding, reading fluency, and loudness. The dysgraphia identification includes features such as visual-spatial response, spellings, sentence word expression, legibility, and visual-motor integration. For dyscalculia basic arithmetic skills, word problem, operator's difference, reasoning, and long/short-term memory features related to the learners were extracted from the questionnaires and samples input.

3.3 Feature Selection and Classification Methods

The five feature selection techniques were selected based on their effectiveness and complexity in previous researches [34, 35]. The features selection methods have been applied to text mining, image processing, bioinformatics, and industrial applications [36, 37]. Extra gradient boosting (XGB), relief, least absolute shrinkage and selection operator (LASSO), gradient boosting decision tree (GDBT), and elastic net (EN) are effective feature selection methods utilized in various applications [38, 39]. We have compared six machine learning algorithms—K-nearest neighbors (KNN), logistic regression (LR), linear discriminant analysis (LDA), classification and regression tree (CART), Naïve Bayes (NB), and support vector machine (SVM). These machine learning models have been selected based on their performance in learning difficulties identification [40]. In previous studies, SVM, KNN, and LR have been used for learning disability identification. LDA and CART are the machine learning models which are not been utilized before to identify learning difficulties. Therefore, KNN, SVM, and LR are selected in this study to compare their performance with the LDA and CART machine learning models. Additionally, these five machine algorithms are compared when they are combined with the best feature selection method as discussed above. Feature selection and classification algorithms are implemented utilizing a scikit-learn package (version 0.21) in Python (version 3.6.3).

4 Result and Discussion

Research question 1: To design and develop a learner database model in ITS to collect and store learning-disabled responses and to identify induced fatigue effect on the learning outcomes.

An intelligent tutoring system learner database model was created based on 19 pretest questions. The responses of the learners were recorded utilizing MYSQL database electronically, and responses to the same questions were recorded using in-person interview method. Further, these two data collection methods were compared to evaluate learner's experience. The learning experience through the manual and electronic mode of data collection has been compared. The performance of the learners in the developed ITS from the two survey modes has been depicted through Fig. 1. In pretest *A*, the survey method was in-person interviews with learning and non-learning-disabled learners. In pretest *B*, the electronic mode (developed ITS) has been selected for the conducted survey. It has been observed that the data collected from the two survey methods has induced fatigue in the learning process of the learners differently. The electronic mode has improved the learning outcomes as depicted through the result line of Pretest *B*, in Fig. 1. The average response rate of the learners in case of learning-disabled and non-learning-disabled learners has increased through computer-based evaluation in the developed ITS. However, the average response rate for the same learners was constant due to the induced fatigue or anxiety issues related with the in-person mode of survey (in-person interviews/questionnaires).

Research question 2: To compare the two survey methods to measure learner's experience.

To determine the impact of computer-based test and in-person method of interviews in the developed ITS of learning disabled, a statistical test is conducted. In order to, measure the impact of learner data collection method on the learner's performance,

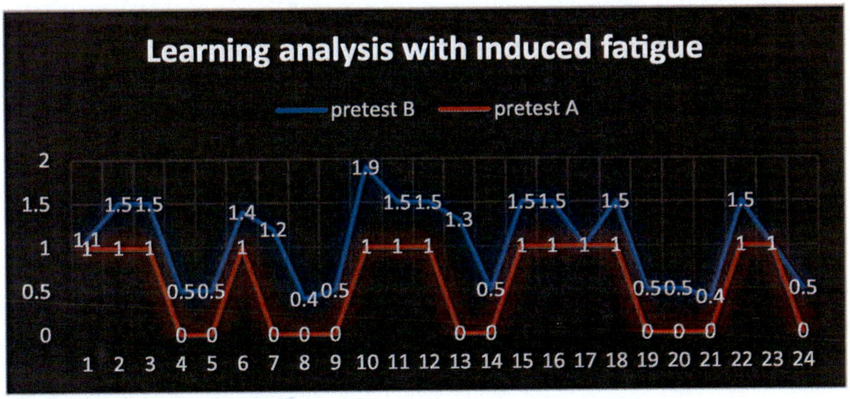

Fig. 1 Fatigue effect analysis on two modes of surveys

t-test is conducted. A *t*-paired test is conducted to find out the significance in pretest scores generated manually and electronically. The *t*-test paired sample for means is performed on pretest scores with unequal variance. The *t*-test (paired) is selected for evaluation as the same sample of population is used in the pretest.

Hypothesis: There is no significant difference present in the mean of each sample, with computer-based examination and in-person method of data collection into the ITS learner database model.

In Table 1, $P(T \leq t)$ two-tail (9.68406E−04) gives probability of t statistics (4.5282) which is greater in absolute value than the t critical (4.0128). The P value is greater than the alpha value (0.05) so accept the null hypothesis that there is no significant difference is present in the mean of each sample in pretest analysis. Hence, it is proved that the physical mode of survey conducted and electronic (computer-based test) method for interview in the ITS learner model for LDs identification are both significant and improve learner specific profiling. Therefore, the comparison of data collection method to create learner database model is equally significant and no difference in effectiveness found in either of the two methods utilized.

Research question 3: To compare feature selection and classification methods on the collected dataset to select the optimal feature selection and classification method for the development of learner classification module.

In the present study, the performance of feature selection technique and classification algorithm is evaluated based on accuracy, precision, recall, and $F1$-score. We observed six feature selection methods and selected the most important features for dyslexia, dysgraphia, and dyscalculia prediction. For comparison of machine learning models, the performance metrics considered are given in Table 2. In this comparison, feature selection method LASSO + classifier CART achieved the highest (accuracy:

Table 1 *t*-test for mean score

	In-person survey	Computer-based survey
Mean	0.5416	1.05
Variance	0.2590	0.2391
Observations	94	94
t-test		
Hypothesized mean difference	0	
Df	46	
t stat	4.5282	
$P(T \leq t)$ one-tail	0.0004	
t critical one-tail	1.5186	
$P(T \leq t)$ two-tail	0.0009	
t critical two-tail	4.4128	

Table 2 Obtained accuracies, precision, recall, and $F1$-score from feature selection and classifier modules combination

Classifier + feature selection method	Accuracy	Macro-average			Weighted average		
		Precision	Recall	$F1$-score	Precision	Recall	$F1$-score
KNN + XGB	98.254	0.73	0.73	0.73	0.97	0.98	0.97
LR + Relief	85.03	0.67	0.61	0.63	0.85	0.85	0.84
LDA + XGB	82.54	0.67	0.77	0.68	0.86	0.83	0.83
CART + LASSO	98.00	0.81	0.82	0.79	0.98	0.98	0.98
NB + LASSO	71.321	0.69	0.69	0.62	0.86	0.71	0.76
SVM + GDBT	85.03	0.62	0.58	0.59	0.86	0.85	0.84

0.98, precision: 0.81, recall: 0.82, $F1$: 0.79). Followed by XGB + KNN with (accuracy: 0.98, precision: 0.73, recall: 0.73, $F1$: 0.73), and relief + LR with (accuracy: 0.85, precision: 0.67, recall: 0.61, $F1$: 0.63), and GDBT + SVM with (accuracy: 0.85, precision: 0.62, recall: 0.58, $F1$: 0.59), XGB + LDA with (accuracy: 0.82, precision: 0.67, recall: 0.77, $F1$: 0.68), and LASSO + NB with (accuracy:0. 71, precision: 0.69, recall: 0.69, $F1$: 0.62).

Feature selection methods have slightly influenced the number of features selected and the overall performance of the machine learning model (please refer Table 2).

The accuracy of the LASSO feature selection method and classifier CART was considered as the best combination for prediction of learning disability. The performance of previous machine learning models used in learning disability prediction like KNN, SVM, and LR was compared with the performance of the LDA, and the CART machine learning model is found to be competitive. In previous studies for learning difficulties identification, KNN has been proven as an effective classifier module [28, 38]. In our study, the effectiveness of KNN has been noted, but CART with feature selection LASSO achieved the highest performance.

5 Conclusion

In this work, we compared two modes of surveys in-person interviews and electronic mode into an ITS for learning-disabled learners. Also, we have compared some popular feature selection and machine learning algorithms in learning difficulties prediction. In this comparison, we have found no difference in the effectiveness of data collection methods. Moreover, least absolute shrinkage and selection operator (LASSO) feature selection technique with classification and regression trees (CART) machine learning model yielded the highest performance. Meanwhile, other feature selection and classification algorithms like KNN, SVM, LASSO, and XGB demonstrated high performance for the identification of learning difficulties. This comparative investigation is important in predicting learning disability identification for attaining maximum performance metrics. For the future, these ML model performances can be compared with deep learning techniques, for learning difficulties identification. However, the limitation of the present work is that the less data is included for the experimental work. In future studies more, data of the participants could be involved and IoT devices could be incorporated in the developed ITS to examine user engagement, willingness, and system usefulness.

References

1. Awais Hassan M, Habiba U, Khalid H, Shoaib M, Arshad S (2019) An adaptive feedback system to improve student performance based on collaborative behavior. IEEE Access 7:107171–107178. https://doi.org/10.1109/access.2019.2931565
2. Palopoli L, Argyros A, Birchbauer J, Colombo A, Fontanelli D, Legay A et al (2015) Navigation assistance and guidance of older adults across complex public spaces: the DALi approach. Intel Serv Robot 8(2):77–92. https://doi.org/10.1007/s11370-015-0169-y
3. Rao C, Sumathi TA, Midha R, Oberoi G, Kar B, Khan M, Vaidya K, Midya V, Raman N, Gajre M, Singh NC (2021) Development and standardization of the DALI-DAB (dyslexia assessment for languages of India—dyslexia assessment battery). Ann Dyslexia. https://doi.org/10.1007/s11881-021-00227-z
4. Hebert M, Kearns DM, Hayes JB, Bazis P, Cooper S (2018) Why children with dyslexia struggle with writing and how to help them. Lang Speech Hear Serv Sch 49(4):843–863. https://doi.org/10.1044/2018_lshss-dyslc-18-0024
5. Biotteau M, Danna J, Baudou E, Puyjarinet F, Velay J-L, Albaret J-M, Chaix Y (2019) Developmental coordination disorder and dysgraphia: signs and symptoms, diagnosis, and rehabilitation. Neuropsychiatr Dis Treat 15:1873–1885. https://doi.org/10.2147/ndt.s120514
6. Dutt S, Ahuja NJ (2020) A novel approach of handwriting analysis for dysgraphia type diagnosis. Int J Adv Sci Technol 29(3):11812. Retrieved from http://sersc.org/journals/index.php/IJAST/article/view/29852
7. Skagerlund K, Träff U (2014) Number processing and heterogeneity of developmental dyscalculia. J Learn Disabil 49(1):36–50. https://doi.org/10.1177/0022219414522707
8. Rubinsten O (2015) Link between cognitive neuroscience and education: the case of clinical assessment of developmental dyscalculia. Front Hum Neurosci 9. https://doi.org/10.3389/fnhum.2015.00304
9. Kariyawasam R, Nadeeshani M, Hamid T, Subasinghe I, Samarasinghe P, Ratnayake P (2019) Pubudu: deep learning based screening and intervention of dyslexia, dysgraphia and dyscalculia. In: 2019 14th conference on industrial and information systems (ICIIS). https://doi.org/10.1109/iciis47346.2019.9063301
10. Dankovicova Z, Hurtuk J, Fecilak P (2019) Evaluation of digitalized handwriting for dysgraphia detection using random forest classification method. In: 2019 IEEE 17th international symposium on intelligent systems and informatics (SISY). https://doi.org/10.1109/sisy47553.2019.9111567
11. Rosenblum S, Dror G (2017) Identifying developmental dysgraphia characteristics utilizing handwriting classification methods. IEEE Trans Hum Mach Syst 47(2):293–298. https://doi.org/10.1109/thms.2016.2628799
12. Ashour AS, Nour MKA, Polat K, Guo Y, Alsaggaf W, El-Attar A (2020) A novel framework of two successive feature selection levels using weight-based procedure for voice-loss detection in Parkinson's disease. IEEE Access 8:76193–76203. https://doi.org/10.1109/access.2020.2989032
13. Mendes J, Freitas M, Siqueira H, Lazzaretti A, Stevan S, Pichorim S (2020) Comparative analysis among feature selection of sEMG signal for hand gesture classification by armband. IEEE Lat Am Trans 18(06):1135–1143. https://doi.org/10.1109/tla.2020.9099752

14. Côté D (2018) Using machine learning in communication networks [invited]. J Opt Commun Network 10(10):D100. https://doi.org/10.1364/jocn.10.00d100
15. Cao W, Czarnek N, Shan J, Li L (2018) Microaneurysm detection using principal component analysis and machine learning methods. IEEE Trans Nanobiosci 17(3):191–198. https://doi.org/10.1109/tnb.2018.2840084
16. Bhatti M, Riaz R, Rizvi S, Shokat S, Riaz F, Kwon S (2020) Outlier detection in indoor localization and Internet of things (IoT) using machine learning. J Commun Netw 22(3):236–243. https://doi.org/10.1109/jcn.2020.000018
17. Ledesma S, Ibarra-Manzano M, Cabal-Yepez E, Almanza-Ojeda D, Avina-Cervantes J (2018) Analysis of data sets with learning conflicts for machine learning. IEEE Access 6:45062–45070. https://doi.org/10.1109/access.2018.2865135
18. Reamaroon N, Sjoding M, Lin K, Iwashyna T, Najarian K (2019) Accounting for label uncertainty in machine learning for detection of acute respiratory distress syndrome. IEEE J Biomed Health Inform 23(1):407–415. https://doi.org/10.1109/jbhi.2018.2810820
19. Zhang B, He X, Ouyang F, Gu D, Dong Y, Zhang L, Mo X, Huang W, Tian J, Zhang S (2017) Radiomic machine-learning classifiers for prognostic biomarkers of advanced nasopharyngeal carcinoma. Cancer Lett 403:21–27. https://doi.org/10.1016/j.canlet.2017.06.004
20. Liu H, Liu Z, Liu S, Liu Y, Bin J, Shi F, Dong H (2019) A nonlinear regression application via machine learning techniques for geomagnetic data reconstruction processing. IEEE Trans Geosci Rem Sens 57(1):128–140. https://doi.org/10.1109/tgrs.2018.2852632
21. Islam B, Ashafuddula N, Mahmud F (2018) A machine learning approach to detect self-care problems of children with physical and motor disability. In: 2018 21st international conference of computer and information technology (ICCIT). https://doi.org/10.1109/iccitechn.2018.8631960
22. Mounica R, Soumya V, Krovvidi S, Chandrika K, Gayathri R (2019) A multi layer ensemble learning framework for learning disability detection in school-aged children. In: 2019 10th international conference on computing, communication and networking technologies (ICCCNT). https://doi.org/10.1109/icccnt45670.2019.8944774
23. Sharma A, Tanwar P (2020) Deep analysis of autism spectrum disorder detection techniques. In: 2020 international conference on intelligent engineering and management (ICIEM) (2020). https://doi.org/10.1109/iciem48762.2020.9160123
24. Aich S, Kim H, Younga K, Hui K, Al-Absi A, Sain M (2019) A supervised machine learning approach using different feature selection techniques on voice datasets for prediction of Parkinson's disease. In: 2019 21st international conference on advanced communication technology (ICACT). https://doi.org/10.23919/icact.2019.8701961
25. Kariyawasam R, Nadeeshani M, Hamid T, Subasinghe I, Ratnayake P (2019) A gamified approach for screening and intervention of dyslexia, dysgraphia and dyscalculia. In: 2019 international conference on advancements in computing (ICAC). https://doi.org/10.1109/icac49085.2019.9103336
26. Parmar C, Grossmann P, Bussink J, Lambin P, Aerts HJWL (2015) Machine learning methods for quantitative radiomic biomarkers. Sci Rep 5(1). https://doi.org/10.1038/srep13087
27. Dhamal P, Mehrotra S (2021) Deep learning approach for prediction of learning disability, pp 77–83. https://doi.org/10.1145/3461598.3461611
28. Sarker IH (2021) Machine learning: algorithms, real-world applications and research directions. SN Comput Sci 2:160. https://doi.org/10.1007/s42979-021-00592-x
29. Albreiki B, Zaki N, Alashwal H (2021) A systematic literature review of student' performance prediction using machine learning techniques. Educ Sci 11:552. https://doi.org/10.3390/educsci11090552
30. Foy JG, Mann VA (2011) Speech production deficits in early readers: predictors of risk. Read Writ 25(4):799–830. https://doi.org/10.1007/s11145-011-9300-4
31. Tsai R-C, Lin K-N, Wang H-J, Liu H-C (2007) Evaluating the uses of the total score and the domain scores in the cognitive abilities screening instrument, Chinese version (CASI C-2.0): results of confirmatory factor analysis. Int Psychogeriatr 19(06). https://doi.org/10.1017/s1041610207005327

32. Kumar A, Singh N, Jyothi-Ahuja N (2017) Learning styles based adaptive intelligent tutoring systems: document analysis of articles published between 2001 and 2016. Int J Cogn Res Sci Eng Educ 5(2):83–97. https://doi.org/10.5937/ijcrsee1702083k
33. Sharp JE (2004) A resource for teaching a learning-styles/teamwork module with the Soloman-Felder index of learning styles [the PACE report]. IEEE Antennas Propag Mag 46(6):138–143. https://doi.org/10.1109/map.2004.1396766
34. Yin P, Mao N, Zhao C, Wu J, Sun C, Chen L, Hong N (2018) Comparison of radiomics machine-learning classifiers and feature selection for differentiation of sacral chordoma and sacral giant cell tumour based on 3D computed tomography features. Eur Radiol 29(4):1841–1847. https://doi.org/10.1007/s00330-018-5730-6
35. Zhang X, Yan L-F, Hu Y-C, Li G, Yang Y, Han Y, Sun Y-Z, Liu Z-C, Tian Q, Han Z-Y, Liu L-D, Hu B-Q, Qiu Z-Y, Wang W, Cui G-B (2017) Optimizing a machine learning based glioma grading system using multi-parametric MRI histogram and texture features. Oncotarget 8(29). https://doi.org/10.18632/oncotarget.18001
36. Jovic A, Brkic K, Bogunovic N (2015) A review of feature selection methods with applications. In: 2015 38th international convention on information and communication technology, electronics and microelectronics (MIPRO) (2015). https://doi.org/10.1109/mipro.2015.7160458
37. Zhang G, Li H, Odbal (2019) Research on fuzzy enhanced learning model of multienhanced signal learning automata. IEEE Trans Ind Inf 15(11):5980–5987. https://doi.org/10.1109/tii.2019.2929086
38. Sun P, Wang D, Mok VC, Shi L (2014) Comparison of feature selection methods and machine learning classifiers for radiomics analysis in glioma grading. IEEE Access 7:102010–102020. https://doi.org/10.1109/access.2019.2928975
39. Uysal AK (2018) On two-stage feature selection methods for text classification. IEEE Access 6:43233–43251. https://doi.org/10.1109/access.2018.2863547
40. Huang Z, Yang C, Zhou X, Huang T (2019) A hybrid feature selection method based on binary state transition algorithm and reliefF. IEEE J Biomed Health Inform 23(5):1888–1898. https://doi.org/10.1109/jbhi.2018.2872811

Swarm Coverage in Continuous and Discrete Domain: A Survey of Robots' Behaviour

Banashree Mandal, Madhumita Sardar, and Deepanwita Das

Abstract This article analyses the behaviours of robot swarm that contribute to the solutions of various area coverage problems. Area coverage problems are broadly categorized into two domains such as continuous and discrete. To efficiently solve any given problem, different characteristics of the participating robots with respect to their structural and behavioural pattern, models of computation, synchrony, etc. play important roles. This paper discusses some of the significant related problems in both the domains and summarizes the behaviours of the constituent robots during their solutions.

Keywords Robot swarm · Area coverage · Distributed algorithm · Continuous domain · Discrete domain

1 Introduction

In nature, the small social insects like birds, fishes, bees, ants, etc. exhibit collective and collaborative behaviour. They work in a coordinated way. These behaviours have been imitated by the researchers to produce a group of simple, inexpensive, small robots to accomplish certain specific tasks. Tasks like: collaborative search and rescue, planetary exploration [27], forming patterns, exploring unknown terrains, gathering on a point, etc., are performed by robot swarm. In the literature, two types of environment have been studied while working with robot swarm, namely, *continuous* and *discrete* domain. In continuous domain, robots are placed over $2D$ environment [36] where they are able to move freely. However, discrete domain considers robots

B. Mandal (✉) · D. Das
National Institute of Technology, Durgapur, Durgapur, West Bengal 713209, India
e-mail: bm.20cs1101@phd.nitdgp.ac.in

D. Das
e-mail: deepanwita@cse.nitdgp.ac.in

M. Sardar
Haji Md. Serafat Mondal Government Polytechnic, Birbhum, West Bengal 731202, India
e-mail: madhumita.hmsm@wbscte.ac.in

S. Das et al. (eds.), *Advances in Data-driven Computing and Intelligent Systems*, Lecture Notes in Networks and Systems 653, https://doi.org/10.1007/978-981-99-0981-0_44

to be deployed on a graph. Here, robots are positioned on the vertices and are capable to traverse only through the edges of the graph. One of these problems that has been widely addressed in both the domains is *area coverage* and it has plenty of operations, like: painting, mowing a lawn, automated vacuum cleaning, surveillance as well as rescue operations, etc. Based on the domain, the definition of area coverage may differ. In continuous domain, robots work on $2D$ plane and all reachable portions of the considered area is to be explored by one robot (at least) of the swarm. However, in discrete domain, the area coverage problem is termed as *exploration* and it requires robots to travel over a graph in a manner that individual node of that particular graph is physically explored by one robot (at least) of the swarm. Moreover, this graph can be a tree, ring, grid, etc. This piece of work aims to study the behaviours of the robots that constitute the swarm and contribute to the solutions of variety of area coverage problems in both the domains.

To understand how coverage task is viewed in the two said domains, we consider two real life examples. Assume, a room with some windows and doors need to be painted by a robot swarm. The robots first divide the whole work into some tasks, such as painting all the windows, the walls and the doors, etc. and then divide themselves into small sub-groups which may carry out a particular task as a team. When all the sub-groups complete their given tasks, the whole room gets painted. During painting, an individual robot which may be responsible for painting a door, always tries to reach every accessible portion of that door. This is an example of area coverage in continuous domain. Consider another task in which a postman wants to deliver a letter to a particular house in an area. Assume that the area contains multiple number of streets and several similar houses like the target delivery location. The postman first tries to find out the specific street which leads to that particular house. Then it traverses through every house and checks for the targeted one. The postman, streets and the houses can be thought as a traversing robot, the edges and the nodes, respectively. This represents area coverage in discrete domain.

This paper studies the area coverage problem in both the domains. In particular, it highlights the variations in the models and characteristics/features of robots that have been considered to solve the area coverage problem. Section 2 discusses characteristics and models of robots based on which formulation of the problems are done. Section 3 includes different problems and their solutions. Finally, conclusion is discussed in Sect. 4 with future scope of research.

2 Characteristics/Features and Model

Robots' characteristics and the models they follow, play an important role to formulate a good solution for any specific task. This section discusses the characteristics of the robots and models that have been considered to achieve area coverage in both the domains.

Anonymous, Identical and Autonomous: Robots are anonymous in nature. Depending on shape, size, computational capability, visibility, etc. robots are of two types such as: homogeneous and heterogeneous. Moreover, robots are fully autonomous and there is no external control over them. Both the domains follow these features.

Memory: Robots are generally considered to be oblivious in both the domains. This implies that robots can not retain any information about their past observations. To ease this inadequacy, robots are sometimes equipped with some external memory support such as lights, pebbles.

Visibility: In continuous as well as discrete domains, robots are endowed with visual sensors by which they can observe their surroundings. There exist two variations of this capability, namely—*unlimited visibility* and *limited visibility*. In continuous domain, a robot having unlimited visibility, can view up to any distance. Conversely, with limited, it can view up to a limited distance. In discrete domain, with unlimited visibility, a robot can see the entire graph as well as the positions of all robots deployed on it. When the visibility is limited, a robot is able to see only those nodes which are located at one or two (limited) node distance from its current position.

Mobility: Robots are allowed to move freely over a region or in a graph. In continuous domain, there are two types of robot's movements: rigid and non-rigid. Rigid motion is undisturbed motion and robots reach the destinations without any halt. However, interruption occurs in non-rigid motion. In *discrete domain*, robots are restricted to traverse only through the edges of the graph. The moves of the robots are instantaneous and they can only traverse one edge at a time.

Communication: This plays a vital role among the robots, in continuous domain, robots can have direct communication where they can directly exchange information like their locations, states, etc. or have passive communication where robots can see other objects and take decision for movement. However, robots use lights as a mode of passive communication [15]. In discrete domain, robots can directly communicate among themselves or may passively communicate through visible lights or whiteboards [34].

Multiplicity Detection Capability: This capability equips a robot to discern whether a node contains one or more than one robot. A node that contains more than one robot is said to be host of a *tower*. Several variations of this capability exist. One variation equips a robot to distinguish a tower only at its current location (*local*) or perceive the same for all nodes of a graph (*global*). The other variation equips a robot to just sense the presence of a tower on any node (*weak*) or compute the exact number of nodes in that tower (*strong*).

The models considered during the solutions are as follows:

Computational Model: Robots follow various models such as Co-operative Distributed Asynchronous Model (CORDA) [31], SYm or ATOM [35], Broadcast-Listen-Process-Move [22] model. etc., however, CORDA is widely used standard model. In CORDA, throughout the process a robot executes computational cycles sequentially where a single cycle comprises of 3 basic phases, like: observe, compute and move. Individual robot repeatedly executes these 3 phases to perform a task. During $LOOK$, a robot observes the environment and gathers data. In $COMPUTE$ it calculates a goal point based on those data and move towards that point in $MOVE$ phase. Broadcast-Listen-Process-Move model is quite similar to CORDA model where broadcast and listen resemble $LOOK$ phase and process resembles $COMPUTE$ phase. However, robots are assigned with one unique ID and all the phases are having disjoint time span. In SYm or ATOM model all operations are instantaneous.

Models for Synchrony: In both domains three different timing models are considered, such as: Full-Synchronous, Semi-Synchronous and Asynchronous. Full-synchronous robots have same activation time; means all the robots follow a common global clock. Semi-synchronous model works same as the full-synchronous model. Here, all the active robots share a common clock. In asynchronous model robots do not follow same clock.

Models Based on Axes Agreements: In continuous domain, individual robot possess its own local coordinate system where current location of the robot is considered as the origin of the coordinate system. Robots do all calculations w.r.t. the local coordinate system. Based on axes agreements there are different models, such that: (i) *Full-compass*: All robots follow same directions and orientations of both the axes of the local coordinate systems means they agree on both direction and orientation. (ii) *Half-compass*: All robots follow same directions of axes but agree on the positive orientation of one axis only. (iii) *Direction-only*: All robots follow same directions of axes but the positive orientations of the axes may be dissimilar. (iv) *Axes-only*: Directions of both the axes are similar to all robots but the positive orientations of the axes may differ. Moreover, the robots disagree on positions of the two axes. (v) *No-compass*: There are no common axes. On the contrary, in discrete domain, robots generally do not have any coordinate system or a sense of direction.

3 Study Based on Area Coverage Problem

The study is broadly classified into the following sections: Sects. 3.1 and 3.2 present detailed analysis of area coverage problem in continuous domain and discrete domain respectively.

3.1 Area Coverage on Continuous Domain

Based on the count of robots, the coverage problem has been segmented into two sections: (A) single robot coverage and (B) multi-robot coverage.

(A) Single robot coverage. Single robot coverage does not require any kind of communication thus decreases the algorithmic complexity. Algorithm proposed by Zelinsky et al. [37] follows an approximate decomposition approach, in which the available place is divided into a number of cells of equal size and shape. Starting from the source cell, a robot tries to reach the goal cell by covering all the cells that comes along its path. The robot is synchronous, autonomous, non-oblivious with non-rigid motion and unlimited visibility and follows full-compass model. Later, Gabriely and Elon [21] have addressed one tool-based cellular decomposition where the tool is of square shape with size D and is used as a foot step of a robot. The entire working space of size $2D$ is subdivided with sub-cells of size D. During traversal, when the robot reaches a particular sub-cell, that is considered as covered. The robot bypasses cells partially covered by obstacles. It constructs a spanning tree by creating a graph joining the centre points of the cells. To cover the region, the robot follows the spanning tree and moves only along the counterclockwise orthogonal direction. Characteristics of robot are same as [37].

(B) Multi-robot area coverage. Multiple robots increase robustness as the work of a faulty robot can be taken up by the other robots [2]. There are two coverage techniques: (i) team-based technique and (ii) individual-based technique.

(i) Team-based area coverage: To work as a team, robots positioned alongside and move towards a particular direction to reach the goal. The members of a team stay together and explore the target region collectively. Choset [9] has discussed different area decomposition approaches, such as: *trapezoidal decomposition*, *exact cellular decomposition* and *boustrophedon cellular decomposition*, amongst all the last two are the most popular decomposition approaches. *Trapezoidal decomposition* requires the environment to be polygonal and decomposes the area with too many number of sub-cells. *Boustrophedon cellular decomposition* overcomes this disadvantage and the available space is broken down in fewer cells. During coverage inter-robot communication exists in the team, member robots are autonomous, synchronous and follow full-compass model. Rekletis et al. [33] has proposed an algorithm using *boustrophedon cellular decomposition* in which the team has 2 types of members: explorers and coverers. 2 robots from the team are selected as explorer to parallely explore the boundaries (both top and bottom) of the region and coverers work is to cover the free space between the boundaries. Robots communicate with each other using line-of-sight (LOS). During loss of LOS due to presence of obstacle, the team divides itself into two halves. Then newly two robots are selected as explorers from each of the sub-units. Moreover, one *reeb-graph* is used here to detect which cell is to cover next. Robots considered here are synchronous with unlimited visibility and follow full-compass model. The problem has been discussed under the same assumptions by Latimer et al. [28]. In this work, robots begin their execution at the same time and team members can communicate with each other. The team covers each cell by back and forth motion. Throughout their movements, once the team of robot detects a critical point, it segregates into 2 halves to cover the sub-cells. After covering sub-cells, the two sub-units reunited. Robot uses a common adjacency graph to differentiate between the information regarding both covered and undiscovered cells.

Whenever the team segregated or reunited the adjacency graph gets updated correspondingly for all team members. Assumed robots are autonomous, synchronous with non-rigid motion, having unlimited visibility and follow full-compass model.

(ii) Individual approach: Here, robots are placed arbitrarily all over the region as well as perform their task in an autonomous way. They break the entire task into several sub-tasks such that each individual task is performed by at least one robot. This approach comprises with two tasks namely, exploration and coverage. In exploration, robots collect data regarding the environment (such as obstacles, critical points etc.) and then depending on those data robots cover the area. Rekletis et al. [32] has introduced one such algorithm where depending on the number of robots, the bounded (rectangular) region is divided into a number of vertical stripes such that each stripe has a robot positioned at its upper left corner. Staring from there, a robot first scans the boundaries of the strip and then covers the whole strip by following a *cycle algorithm* which gives the robot a closed path to cover the entire strip and finally reaches back to its initial position. Both boundary exploration and coverage is done by following the same *cycle algorithm*. Besides that, one *reeb graph* used to gather the information regarding the critical points where the nodes denote the critical points and the edges denote the connections amid the stripes. During coverage once a robot encountered critical point, robot updates the reeb graph correspondingly. Here robots are synchronous with unlimited visibility, non-rigid motion and follow full-compass model. Algorithm proposed by Kong et al. [25] has been discussed under almost same assumptions as [32] however, robots are asynchronous in nature. This algorithm used *boustrophedon cellular decomposition* approach. Due to restricted communication, all robots have *adjacency graph* of the decomposed area to share the information regarding non-overlapped areas. Moreover, one *cycle algorithm* as in [32] is used by the robots for covering cells. During movements if a robot encounters critical point, it divides the entire cell into two dis-joined halves. Then the two disconnected parts will be added to the adjacency graph as new nodes. Karapetyan et al. [24] has proposed one *off-line* coverage algorithm where robots have a priory knowledge about the $2D$ environment which contains arbitrary shape of obstacles as well as the outline of the region to be covered. However, communication among the robots is strictly prohibited. Here two types of algorithms are used namely, *Efficient Complete Coverage* (EEC) and *Chinese postman problem* (CPP). In EEC, the *boustrophedon cellular decomposition* (BCD) algorithm is used to decompose the region and one reeb graph ($G_r = (V_r; E_r)$) is used to get the information about the obstacles as well as free uncovered cells. Here, the set of nodes (vertices) (V_r) are points on the corners of obstacles and the set of edges (E_r) denote single non-breakable obstacle-free cells. Then the CPP algorithm is introduced to figure out an optimal route over the complete graph. This CPP algorithm first generates an *Eulerian graph* (G_r) then calculates an Eulerian path, where robots can visit every edge of the graph only once. Finally, after applying *Eulerian tour*, it ends at the same vertex where it has started its exploration. Assumed robots are synchronous with non-rigid motion, unlimited visibility, restricted communication and follow full-compass model. Later, this same approach was improved under the same assumptions by Karapetyan to solve vehicle kinematic constraints

problem [23]. Das and Mukhopadhyaya [11, 12] have proposed one distributed painting algorithm of obstacle free area coverage using CORDA model. This model is a standard and suitable computational model for designing swarm based algorithms. In CORDA, *look-compute-move* phases are performed in a non-overlapping manner and observations are made on the basis of local coordinate systems. Based on relative ranking among the robots, the whole region is divided (virtually) into a number of non-overlapping cells where each robot is assigned one cell for painting according to their relative ranks. Once, each of the robots completes painting, the goal is achieved. This work considered, autonomous, oblivious, asynchronous robots with unlimited visibility that follow direction-only model. Additionally, there exists communication (passive) amid the robots. However, the robots maintain relative ranking among themselves to avoid crossover during their movements. To add on complexity, further, Das et al. [14] has addressed another problem where the region contains a random number of static horizontal line obstacles. To solve this particular problem, Das. et al. has introduced a prerequisite step, called *assembling* [13], where robots are assembled over the left boundary of the target area. Later, depending on the positions of the obstacles, the area decomposed into several rectangular blocks considering the positions of the obstacles such that the obstacles are located only along the boundaries of the blocks (not inside the blocks). At the beginning, all the robots are in explore state, and here they compute their blocks to paint. Homogeneous, oblivious, autonomous, synchronous robots with rigid motion, unlimited visibility are considered for this work which follows full-compass model. To make the algorithm more realistic and rational, Das and Mukhopadhyaya [10] addressed another coverage problem with a slight variation on the robots visibility capability. In this work, limited visibility of the robots are considered however, the target area is kept obstacle-free. In this work, robots are placed arbitrarily inside a rectangular region and they can view upto certain range. For solvability of the task, the robots are considered to be initially allied via a *visibility graph*. Each robot first identifies the strip to be painted by it while traversing from its current position to the upper boundary and from upper to lower boundary of the region. Once the strip boundaries are fixed, every robot paints its corresponding strips. However, it may possible that more than one robots to share a single strip. This is done via mutual agreement among them during the computation. The algorithm guarantees non-overlapped and complete coverage of the area. Robots are autonomous, oblivious, asynchronous with passive communication, rigid motion and follow full-compass, CORDA model. Fazli and Davoodi [18] addressed an algorithm that assumes the outline of the region where obstacles (static) are familiar to the robots and are line in nature. This algorithm places the static guards in the work area, applying the solving technique of the Art Gallery Problem. Constrained Delaunay Triangulation (CDT) is generated over the static guards as well as the beginning and ending points of obstacles so that individual obstacle has to arrive in the triangulation in the form of an edge. On CDT, a Floyd–Warshall algorithm is applied for calculating the shortest paths between every two points. Applying these information, one Reduced-CDT (R-CDT) is produced. To create R-CDT, the shortest path amongst all shortest paths (subjected to start and end points are static guards), including each and every vertex and edge is considered to be the fundamental element. The R-CDT

Table 1 List of papers in continuous domain based on different characteristics and model

Feature/model		Continuous domain												
		[37]	[25]	[33]	[24]	[28]	[12]	[14]	[18]	[10]	[32]	[23]	[21]	[9]
Visibility	Unlimited	✓	✓	✓	✓	✓	✓	✓			✓	✓	✓	✓
	Limited								✓	✓				
Communication	Active					✓					✓			✓
	Passive	✓	✓	✓	✓		✓	✓	✓	✓		✓	✓	
Mobility	Rigid							✓		✓				
	Non-rigid	✓	✓	✓	✓	✓	✓		✓		✓	✓	✓	✓
Autonomous		✓	✓	✓	✓	✓	✓	✓	✓	✓	✓	✓	✓	✓
Memory	Oblivious		✓	✓		✓	✓	✓	✓	✓	✓		✓	
	Non-oblivious	✓			✓							✓		✓
Synchrony	Full-synchronous	✓		✓	✓	✓		✓			✓	✓	✓	✓
	Semi-synchronous													
	Asynchronous		✓				✓		✓	✓				
Direction and orientation based model	Full-compass	✓	✓	✓	✓	✓		✓	✓	✓	✓	✓	✓	✓
	Half-compass													
	Direction-only						✓							
	Axis-only													
	No-compass													
Computational model	CORDA						✓	✓		✓				
	NON-CORDA	✓	✓	✓	✓	✓			✓		✓	✓	✓	✓

is progressively elongated by including the nearest static guard to the current element, counting the entire route through which it is attached to the element. Using Multi-Prims algorithm, the R-CDT graph is broken down into a forest of $|R|$ number of partial spanning trees (PSTs), where $|R|$ denotes the robots. To create a cycle on each PST, one Constrained Spanning Tour (CST) algorithm is applied on the set of PSTs. Then individual cycle is assigned to the respective robots to cover. As a result a complete area coverage is achieved. In the work, robots are asynchronous in nature and having limited visibility and follow full-compass model. Table 1 summarizes the coverage problems that are discussed above along with a detailed layout of the behaviours of the robots. In this table, literature [21, 37] follow single robot area coverage whereas [9, 10, 12, 14, 18, 23–25, 28, 32, 33] follow multi-robot area coverage algorithm. Further, [10, 12, 14, 18, 23–25, 32] follow individual approach. However, literature [9, 28, 33] follow team-based approach.

3.2 Area Coverage Problem in Discrete Domain

In discrete domain, the area coverage problem is termed as the *Exploration* problem. It is defined as a scenario where robots are randomly scattered on the vertices of a graph. Task of the robots is to move around the graph in a manner that individual vertex of the graph is physically explored by one robot (at least) of the swarm. Research has spread across both single as well as multi-robot exploration and are discussed in the following sub-portions separately.

(A) Single robot exploration. In case of single robot exploration, a robot is need to take a tour of all nodes and additionally all edges of the graph too. Here robots are assumed to have certain strong properties such as memory, sense of direction. None of the works cited below follow the CORDA model. Albers and Henzinger [1] have addressed an algorithm for exploring an unfamiliar directed strongly connected graph by a robot that has limited visibility, memory and a sense of direction. The robot tries to achieve exploration with minimum requirement of edge traversals. Starting from a vertex, the robot keeps on exploring new edges till it gets stuck at a vertex with no unvisited outgoing edge. Then it relocates to a vertex which has at least one unvisited outgoing edge and continues the exploration. During relocation, the explored edges get re-explored. Thus, to reduce the number of relocations, the robot keeps track of the nodes that have more visited outgoing edges than visited incoming edges, termed as *expensive* nodes. The robot tries to minimize its visit to these expensive nodes during relocation. The same problem was addressed by Panaite and Pelc [30] in an undirected connected graph. Robots are assumed to have limited visibility, memory and a sense of direction. Similar to [1], here also the robot explores new edges until it gets stuck at a vertex with no unvisited outgoing edge and then relocates to a vertex which has at least one unvisited outgoing edge. However, during relocation instead of just moving to any free node, it follows a dynamically constructed spanning tree of the explored graph. For exploration, the lower bound on the number of edge traversals is $|E(G)|$, and any excess number of edge traversals is termed as penalty. Panaite and Pelc [30] achieve exploration with a penalty of $O(V(G))$ where $V(G)$ and $E(G)$ denote the vertex set and edge set of the considered graph respectively. Bender et al. [5] addressed this problem in a strongly connected directed graph. However, certain additional assumptions were taken here such as, if the out-degree of a vertex is d then the outgoing edges of that vertex are numbered from 1 to d. Apart from having limited visibility, memory and sense of direction, a robot is assumed to have a pebble with which it can mark a node. It was shown here that if the number of vertices is known to the robot then, only a single pebble is enough for the robot is achieve exploration and build a map of the graph.

(B) Multi-robot exploration. With multiple robots, two variations of exploration have been addressed, namely, *perpetual* and *terminating*. Perpetual exploration requires that individual vertex of the graph is physically explored by a single robot infinite times. However, terminating exploration requires that each node has to be traversed at least once and all robots have to stop their movements once the task is complete. The discussion is broken down into two parts for presenting the two variations of exploration separately.

(i) Perpetual exploration: It has been studied across various graph topologies. All works cited below follow the CORDA model for computation and assume that robots do not communicate among themselves. Baldoni et al. [3] have addressed perpetual exploration in a graph while adding two constraints to the problem, that are—a node can host only one robot at a time and no two robots can visit the same edge at the same time. Incorporating the constraints, the problem is renamed as Constrained Perpetual Graph Exploration (CPGE). The objective here was to find out the maximal number of robots for that CPGE is feasible in that graph. The robots considered here have unlimited visibility, an unbounded amount of memory, sense of direction and are fully synchronous in nature. Moreover, robots have an outline about the graph. In the solution, the graph is converted into a labelled-tree that shows the paths that must be visited by a single robot at a time to achieve exclusiveness. The length (q) of the longest path is computed which determines the maximum number of robots (k). It was concluded that a CPGE is infeasible in a graph if $k > p - q$ where p denotes the number of vertices in the graph. However, same problem was reconsidered in case of a partial grid graph by Baldoni et al. [4]. Robots considered for the task have unlimited visibility, sense of direction and are fully synchronous in nature. Again, aim is to find out the maximal number of robots for which CPGE is feasible in a partial grid. Unlike [4], here robots do not have a map of the graph. Here the feasibility study is performed under various levels of visibility ρ ($\rho = 0, 1, \infty$) of the robots. The partial grid is first converted into a labelled-tree which shows the routes which have to be visited by a single robot at a time in order to achieve exclusiveness. The length (q) of the longest labelled path is computed. Similar to [4], here also q is the key parameter that places an upper bound on the number of robots. It was shown here that CPGE is infeasible when $\rho = 0$. Again, the maximum number of robots for which CPGE is feasible when $\rho = 1, \infty$ is $p - q$ where p is the number of nodes in the grid. Later Bonnet et al. [7] addressed the same problem for a grid with the aim of minimizing the count of robots. Here the considered robots are oblivious, asynchronous in nature and have no sense of direction. However, they do have unlimited visibility. A *configuration* denotes the locations of all robots in a graph at a instant of time. Starting from any random initial configuration, robots try to reach a special configuration. From this special configuration, robots execute a sequence of moves such that the special configuration is reached again. After a few iterations of executing the sequence, each robot has visited each node of the graph. The process is continued to achieve perpetual exploration. It was shown here that 3 robots are necessary and sufficient to perform a CPGE in a grid. In case of rings, Blin et al. [6] proposed an algorithm using oblivious, asynchronous robots that have unlimited visibility but no sense of direction. The objective was to find the minimum as well as maximum number of robots which are able to solve perpetual exploration while also achieving exclusiveness. The solution first introduces a few predefined configurations. Beginning with any initial configuration, robots first reach a predefined configuration. From there, they move in such a way that the predefined

configuration is obtained again. The process is repeated in a cyclic manner to achieve perpetual exploration.

(ii) Terminating exploration: Similar to perpetual, terminating exploration has been widely addressed in various types of graphs. In all works presented below, robots follow the CORDA model for computation. Until and unless specified, robots do not communicate among themselves. For trees, Flocchini et al. [20] have discussed a terminating exploration algorithm for a tree graph. The assumed robots are oblivious, asynchronous robots that have unlimited visibility and multiplicity detection capability. The main challenge of using oblivious robots is that at any stage robots do not remember how much of the tree is already explored and how much is left to explore. The solution in [20] first creates a specific configuration of robots called the *brain*. The brain is formed by a few robots and is a visible counter that indicates how much of the tree has been explored. The rest of the robots form a team of explorers that perform the actual exploration of the tree. Considering ring graphs, Flocchini et al. [19] have performed exploration using asynchronous, oblivious robots that do not communicate. Some strong results have been presented in [19] such as the problem is infeasible once the number of robots (say k) divides the number of nodes (n). From the initial configuration, robots put an effort to reach either a configuration where all robots are deployed on consecutive nodes of the ring, known as a *block* or a configuration where there are two such *blocks*. Once such configuration is reached, one or two towers are formed and one or two explorers are identified that explore the ring. Later the problem has been discussed with same robot characteristics by Lamani et al. [26] proving that a ring of even size cannot be explored by less than five robots. The proposed solution considers exactly five robots for exploration. Starting from a random initial configuration, five robots try to obtain a *block* configuration where all robots are located on consecutive nodes of the ring. The middlemost robot of the block forms a tower by moving to any of its neighbouring nodes. Then, an explorer robot is identified that explores the ring. Although unlimited visibility of the robots is an unrealistic assumption, all above works assume the same. This assumption has been removed by Datta et al. [16] who assumed myopic robots that can only see their neighbouring nodes. In addition to that, robots are oblivious, non-communicating, have multiplicity detection capability but have no sense of direction. Datta et al. [16] studied the problem in synchronous as well as asynchronous setting. They proposed algorithms for synchronous robots and showed that no deterministic algorithm can be designed for asynchronous setting. To explore a ring with myopic robots under the asynchronous setting, Ooshita and Tixeuil [29] have used luminous robots. Here, each robot is equipped with lights of two colours. Using lights, robots can have a passive communication among themselves. However, robots have no other constant memory as well as have no sense of direction. They show that in the fully synchronous model, 2 robots are necessary and sufficient to accomplish perpetual exploration whereas in terminating exploration the number is 3. Similarly, in the semi-synchronous and asynchronous models, the required number of robots are 3 and 4 for the two variations of exploration respectively. For finite grids, Devismes et al. [17] have addressed terminating exploration using asynchronous, oblivious robots

Table 2 List of papers in discrete domain based on different characteristics and model

Feature/model		Discrete domain													
		[1]	[30]	[5]	[3]	[4]	[7]	[6]	[20]	[19]	[26]	[16]	[29]	[17]	[8]
Visibility	Unlimited				✓	✓	✓	✓	✓	✓	✓			✓	✓
	Limited	✓	✓	✓								✓	✓		
Communication	Active														
	Passive/no communication	✓	✓	✓	✓	✓	✓	✓	✓	✓	✓	✓	✓	✓	✓
Autonomous		✓	✓	✓	✓	✓	✓	✓	✓	✓	✓	✓	✓	✓	✓
Memory	Oblivious						✓	✓	✓	✓	✓	✓		✓	✓
	Non-oblivious	✓	✓	✓	✓	✓							✓		
Synchrony	Full-synchronous				✓	✓						✓	✓		
	Semi-synchronous												✓		
	Asynchronous						✓	✓	✓	✓	✓		✓	✓	✓
Sense of direction	Yes	✓	✓	✓	✓	✓									
	No						✓	✓	✓	✓	✓	✓	✓	✓	✓
Computational model	CORDA				✓	✓	✓	✓	✓	✓	✓	✓	✓	✓	✓
	NON-CORDA	✓	✓	✓											
Multiplicity detection capability	Yes								✓	✓	✓	✓	✓	✓	✓
	No	✓	✓	✓	✓	✓	✓	✓							

equipped with unlimited visibility and multiplicity detection capability but with no sense of direction. It has been shown that three robots are necessary and sufficient to deal with exploration in a finite grid except for two cases. For the two exceptional cases, the number of robots required is four for a 2×2 grid and five for a 3×3 grid. In the general solution with three robots, robots execute collision-free moves to obtain a predefined configuration. On achieving that, two robots form a tower that remains immovable. Taking the tower as a reference, the third robot orients the grid and explores it. Chalopin et al. [8] present a discussion on the type of graphs that can be explored with asynchronous, oblivious robots that have unlimited visibility. Here they have assumed that any two edges of a node are uniquely labelled. However, the discussion does not consider the symmetric and/or periodic configurations that may exist in the graph. It was shown that it is the number of robots that decides whether a graph can be explored or not. It has also been proved that no graphs can be explored with less than three robots. Table 2 literature [1, 5, 30] follow single robot area coverage whereas [3, 4, 6–8, 16, 17, 19, 20, 26, 29] follow multi-robot area coverage algorithm. Further, [3, 4, 6, 7] follow perpetual exploration process whereas literature [8, 16, 17, 19, 20, 26, 29] follow terminating exploration process.

4 Conclusion and Future Scope

This paper summarizes some of the research works on area coverage problem and the corresponding solutions based on the behaviour of robot swarm in both the continuous and discrete domains. However, several other problems on area coverage exist, those are not included in this paper as flow of direction is little different from this survey. Studying the problems along with the solutions of those papers and forming new problems may be considered as the future scope of work. One can consider variations on the robots behaviours as well as environments, such as limited visibility of the robots, inclusion of external memory, restricted environment (convex/concave), inclusion of dynamic polygonal obstacles for new problem formation and consider those problems as future research challenges.

References

1. Albers S, Henzinger RM (2000) Exploring unknown environments. SIAM J Comput 29(4):1164–1188
2. Balch T, Arkin RC (1994) Communication in reactive multi-agent robotic systems. Auton Robots 1(1):27–52
3. Baldoni R, Bonnet F, Milani A, Raynal M (2008) Anonymous graph exploration without collision by mobile robots, vol 109, no 2. Tech report, IRISA, France, pp 98–103
4. Baldoni R, Bonnet F, Milani A, Raynal M (2008) On the solvability of anonymous partial grids exploration by mobile robots. In: Baker TP, Bui A, Tixeuil S (eds) OPODIS 2008. LNCS, vol 5401. Springer, Heidelberg, pp 428–445
5. Bender MA, Fernández A, Ron D, Sahai A, Vadhan S (2002) The power of a pebble: exploring and mapping directed graphs. Inf Comput 176(1):1–21
6. Blin L, Milani A, Potop-Butucaru M, Tixeuil S (2010) Exclusive perpetual ring exploration without chirality. In: Lynch NA et al (eds) Distributed computing, DISC 2010. LNCS, vol 6343. Springer, Berlin, Heidelberg, pp 312–327
7. Bonnet F, Milani A, Potop-Butucaru M, Tixeuil S (2011) Asynchronous exclusive perpetual grid exploration without sense of direction. In: Fernndez Anta A et al (eds) Principles of distributed systems. OPODIS 2011. LNCS, vol 7109. Springer, Berlin, Heidelberg, pp 251–265
8. Chalopin J, Flocchini P, Mans B, Santoro N (2010) Network exploration by silent and oblivious robots. In: Thilikos DM (ed) Graph theoretic concepts in computer science. WG 2010. Lecture notes in computer science, vol 6410. Springer, Berlin, Heidelberg, pp 208–219
9. Choset H (2000) Coverage of known spaces. The Boustrophedon cellular decomposition. Auton Robots 9(3):247–253
10. Das D, Mukhopadhyaya S (2018) Distributed algorithm for painting by a swarm of randomly deployed robots under limited visibility model. Int J Adv Robot Syst 15(5):17298814–18804508
11. Das D, Mukhopadhyaya S (2011) An algorithm for painting an area by swarm of mobile robots. In: International conference on control, automation and robotics (CAR) proceedings, pp 1–6
12. Das D, Mukhopadhyaya S (2013) Distributed painting by a swarm of robots with unlimited sensing capabilities and its simulation. arXiv preprint arXiv:1311.4952
13. Das D, Mukhopadhyaya S, Nandi D (2017) Multi-robot assembling along a boundary of a given region in presence of opaque line obstacles. In: Proceedings of 2nd international conference on intelligent computing and applications. Springer, Singapore, pp 21–29
14. Das D, Mukhopadhyaya S, Nandi D (2021) Swarm-based painting of an area cluttered with obstacles. Int J Parallel Emerg Distrib Syst 36(4):359–379

15. Das S, Yorozu Y, Hirano M, Oka K, Tagawa Y (2012) The power of lights: synchronizing asynchronous robots using visible bits. In: IEEE 32nd international conference on distributed computing systems, pp 506–515
16. Datta AK, Lamani AL, Larmore L, Petit F (2013) Ring exploration by oblivious agents with local vision. In: 33rd international conference on distributed computing (ICDCS), pp 347–356
17. Devismes S, Petit F, Tixeuil S (2010) Optimal probabilistic ring exploration by semi-synchronous oblivious robots. In: Kutten S et al (eds) Structural information and communication complexity. SIROCCO 2009. LNCS, vol 5869. Springer, Berlin, Heidelberg, pp 195–208
18. Fazli P, Davoodi A (2010) Multi-robot area coverage with limited visibility. In: Proceedings of the 9th international conference on autonomous agents and multi-agent systems, vol 1, pp 1501–1502
19. Flocchini P, Ilcinkas D, Pelc A, Santoro N (2013) Computing without communicating: ring exploration by asynchronous oblivious robots. Algorithmica 65:562–583
20. Flocchini P, Ilcinkas D, Pelc A, Santoro N (2008) Remembering without memory: tree exploration by asynchronous oblivious robots. In: Shvartsman A et al (eds) Structural information and communication complexity. SIROCCO 2008. LNCS, vol 5058. Springer, Berlin, Heidelberg, pp 33–47
21. Gabriely Y, Elon R (2001) Spanning-tree based coverage of continuous areas by a mobile robot. Ann Math Artif Intell 31(1):77–98
22. Ganguli A, Cortés J, Bullo F (2008) Distributed coverage of non-convex environments. In: Networked sensing information and control. Springer, Boston, MA, pp 289–305
23. Karapetyan N (2018) Multi-robot dubins coverage with autonomous surface vehicles. In: IEEE international conference on robotics and automation (ICRA). IEEE, pp 2373–2379
24. Karapetyan N, Benson K, McKinney C, Taslakian P, Rekleitis I (2017) Efficient multi-robot coverage of a known environment. In: IEEE/RSJ international conference on intelligent robots and systems (IROS). IEEE, pp 1846–1852
25. Kong CS, New AP, Rekleitis I (2006) Thin films and exchange anisotropy. Distributed coverage with multi-robot system on proceedings. In: IEEE international conference on robotics and automation, ICRA, pp 2423–2429
26. Lamani A, Potop-Butucaru MG, Tixeuil S (2010) Optimal deterministic ring exploration with oblivious asynchronous robots. In: Patt-Shamir B et al (eds) Structural information and communication complexity. SIROCCO 2010. LNCS, vol 6058. Springer, Berlin, Heidelberg, pp 183–196
27. Landis, Geoffrey A (2004) Robots and humans: synergy in planetary exploration. J Acta Astronaut 55(12):985–990
28. Latimer D, Srinivasa S, Lee-Shue V, Sonne S, Choset H (2002) Towards sensor-based coverage with robot teams. In: Proceedings IEEE international conference on robotics and automation, vol 1, pp 961–967
29. Ooshita F, Tixeuil S (2018) Ring exploration with myopic luminous robots. In: Izumi T et al (eds) Stabilization, safety, and security of distributed systems. SSS 2018. LNCS, vol 11201. Springer, Cham, pp 301–316
30. Panaite P, Pelc A (1999) Exploring unknown undirected graph. J Algorithms 33(2):281–295
31. Prencipe G et al (2001) Distributed coordination of a set of autonomous mobile robots. In: 4th European research seminar on advances in distributed systems, pp 185–190
32. Rekleitis I, New AP, Rankin Samuel E, Choset H (2008) Efficient Boustrophedon multi-robot coverage: an algorithmic approach. J Ann Math Artif Intell 52(2):109–142
33. Rekletis I, Lee Shue V, New AP, Choset H (2004) Limited communication, multi-robot team-based coverage. In: IEEE international conference on robotics and automation, pp 3462–3468
34. Sinha M, Mukhopadhyaya S (2018) Optimal tree search by a swarm of mobile robots. In: Information and communication technology. Springer, Singapore, pp 179–187
35. Suzuki I, Yamashita M (1994) Distributed anonymous mobile robots—formation and agreement problems. In: Proceedings of 3rd international colloquium on structural information and communication complexity (SIROCCO'96), pp 1347–1363

36. Xin B (2017) Distributed multi-robot motion planning for cooperative multi-area coverage. In: 13th IEEE international conference on control and automation (ICCA). IEEE, pp 361–366
37. Zelinsky A, Byrne RA, Yuta JCS (1993) Planning paths of complete coverage of an unstructured environment by a mobile robot. In: International conference on advanced robotics, pp 533–538

Defining Vibration Limits for Given Improvements in System Availability

L. G. Lasithan, P. V. Shouri, and V. G. Rajesh

Abstract Traditional industry is fast shifting to integrating automatic fault detection and correction. This includes online condition monitoring systems to make maintenance decisions based on the health or failure of the component machine. A major reason for the failure is unchecked vibrations of magnitude above their alert limit. When any of their component machines fail, the availability of the industrial plants is adversely affected. This research paper presents an experimental study to evaluate the impact of vibrations on availability of a system of Class-I machine of specific power. The methodology adopted includes accelerated life testing (ALT) procedures and analyzing the resultant system vibrations. The study results demonstrate when the state of the machine changes from the failure initiation (alert limit of vibration) to the potential failure limits (alarm limit of vibration), the values of system availability exhibit a unique pattern of variation. The results show that the definition of vibration limits for the given improvement in system availability and its significance on condition-based maintenance (CBM). Furthermore, an equation is established between allowable system load and system failure time.

Keywords Availability · Condition-based maintenance · Alert and alarm limits · Acceleration factor · Vibration limit

L. G. Lasithan (✉)
APJ Abdul Kalam Technological University, CET Campus, Thiruvananthapuram, Kerala 695016, India
e-mail: lasithanlg@geckkd.ac.in

P. V. Shouri
Department of Mechanical Engineering, Model Engineering College, Thrikkakara, Cochin, Kerala 682021, India
e-mail: pvshouri@mec.ac.in

V. G. Rajesh
Department of Mechanical Engineering, College of Engineering, Chengannur, Alappuzha, Kerala 689121, India
e-mail: rvg@mec.ac.in

S. Das et al. (eds.), *Advances in Data-driven Computing and Intelligent Systems*, Lecture Notes in Networks and Systems 653, https://doi.org/10.1007/978-981-99-0981-0_45

1 Introduction

The availability management plays a crucial role for the success of any company. Plant availability is a critical driver for the economic performance of an industrial process plant and maintaining higher value of availability is essential for its satisfactory performance [1–3]. The plant availability is determined by its component machines. If a component machine fails, its availability is badly affected which in turn influences the plant availability. Over the years, maintenance has advanced to adopt the idea of availability [4].

Machine availability can be defined as the probability that a system of machine will be available when required or as the proportion of total time the system will be available for use [1]. If there is a failure, the machine will not be available for use when required. If the failed machine is repaired, it will be available for its intended operation. Therefore, the availability involves the aspects of both reliability and maintainability. The machine reliability can be defined as the satisfactory performance of a system of machine over a specified operating period under specified operating conditions [1]. The machine maintainability can be defined as the probability of repairing a failed machine [1].

The system availability (A) can be calculated using the Eq. (1) [1]

$$A = \frac{\text{MTBF}}{\text{MTBF} + \text{MTTR}} \tag{1}$$

where MTBF = Mean Time Between Failures; MTTR = Mean Time to Repair

The objective of a CBM program is typically to determine a maintenance policy that optimizes system performance according to criteria like cost, availability, reliability, etc. [5]. Early failure detection is the responsibility of most CBM programs [6]. Li et al. [7] proposed a CBM model for assuring average system availability and plant safety.

For the CBM application, a potential failure-functional failure (P-F) curve can be plotted between the failure resistance or health condition of a machinery system and the time period of its failure [4]. It can be inferred from this curve that as time increases, failure resistance decreases to complete the system failure. This curve can be used to explain various stages of system failure, i.e., failure initiation (I), potential failure (P), and functional failure (F). Along this curve, a point "P" can be identified, where there is a potential to fail, and a point "F" beyond "P" can be detected, where the system will not perform as expected. After the point "P," the health condition decreases rapidly.

There is no hard data to define P-F intervals [4]. Therefore, the best strategy is to employ methods that could effectively ascertain the machine condition before the potential failure occurs, and this should permit scheduling the repair activity before the P-F interval.

Rotating machinery is extensively used in today's industries. Some of these are very critical for the successful operation of the plant [8]. The machine collapse may result in costly downtime of the plant [2]. The faults in rotating machinery such as machine out of balance and alignment, gear fault, resonance, bent shafts, bearing failures, mechanical looseness, etc., can be identified by measuring and analyzing the vibration generated by the machine [8, 9]. The rolling element bearings in the rotating machinery allow relative movement and bear shaft load [10].

An efficient diagnosis system is needed to predict the condition and consistent lead time of the machine. Vibration monitoring might be considered as the grandfather of condition/predictive maintenance, and it provides the foundation for most facilities' Condition Based Maintenance (CBM) programs [4]. An experimental study for CBM through vibration monitoring on submersible well pumps following vibration severity standards such as ISO 10816-7(2009) is conducted [11].

One of the main causes of failure of industrial bearing is due to rolling contact fatigue (RCF). The RCF wear mechanism involves false brinelling, characterized by plastically formed indentations due to overload generally caused by vibration. The cause of wear is that lubricant is squeezed out between the contact area of rolling elements and raceways, resulting in direct metal-to-metal contact. Vibration causes wear of the surfaces in contact and fine abrasive particles are produced rapidly that results in a characteristic groove with the oxide acting as an abrasive [12].

It is normally accepted that the vibration velocity recorded over the range of 10 Hz (600 RPM) to 1000 Hz (60,000 RPM) provides the finest indication of a vibration's severity on rotating machines [8, 9, 13]. As most of the rotating machines operate between 600 RPM and 60,000 RPM, the vibration velocity is the best candidate for vibration measurement and analysis [9, 13], whereas above 60,000 RPM vibration acceleration is only the fine indicator [9, 13]. The RMS value is directly related to the energy content of the vibration and thus its destructive capability [8, 9].

In industrial situations, it can often be seen that process plants are operated with certain machines which can be classified as class I, II, III, and IV based on vibration severity standards, ISO 10816 [8, 9, 14] whose failure is mainly due to severe vibrations in it. These machines often are subjected to more than the acceptable limits of vibration which invariably led to an increase in their failure rate, especially in the case of components with rotating parts. The study on the effect of high vibration (above acceptable limit) in gas turbine installed in Al Ghubra Power and Desalination plant [15] suggests using online condition monitoring methods to know the running condition of gas turbines in advance to plan maintenance activities to avoid sudden failure of the turbine.

The frequent failure of certain Class-I machines in an industrial process plant is mainly due to bearing wear [12], which causes vibrations in the machine and as the wear increases, the system vibration increases.

Electrical motors are integral parts of the majority of the machines installed in an industrial plant [4, 16]. The failure of any one of these motors degrades the machine performance, which, in turn, influences the overall plant availability. The paper reviewed machine learning (ML)-based fault detection techniques and predictive maintenance (PdM)/CBM strategies for electrical motors used in industrial plants [16]. Furthermore, the vibration data acquired from accelerometer sensors are extensively used for the data analysis. Kumar et al. [6] presented a comprehensive review of various faults in electric motors, failure causes, and advanced condition monitoring and diagnostic techniques. However, the availability aspects associated with the failure of electric motors and maintenance initiation points are not considered in these studies.

Usually, life testing under normal operating conditions of mechanical parts with high reliability is expensive in terms of both money and time. Hence, it is desirable to accelerate the testing procedure for gathering the failure data. The objective of accelerated testing procedures is to reduce the time required for life testing by strategies like intensified stress levels [1].

Physical acceleration or true acceleration means that operating a unit at high stress produces the same failures that would occur at typical use stresses, except that they happen much quicker. Then, by extrapolating the results suitably to the "normal use" conditions, we can obtain reasonably accurate estimates of the life of the component under the "normal use" conditions [1, 17]. However, the issue of prediction accuracy associated with extrapolating data outside the range of testing has not yet been fully addressed [17]. Acceleration factors show how time-to-fail at a particular operating stress level (for one failure mode or mechanism) can be used to predict the equivalent time to fail at a different operating stress level [18, 19].

Failure analysis of the bearing can be investigated by making artificial defects on various elements of the bearing and analysis is made with a vibration signature tool for monitoring its condition [20].

An availability-based feature in the vibration signal and its usability in CBM programs has not been reported widely by researchers [7, 21]. This work is an experimental study in that direction, and the results obtained are reported.

The remaining useful life (RUL) is the time measured from a specific time of operation of a process plant machine to its functional failure [4]. A plant engineer can schedule the maintenance activities based on the expected value of RUL which avoids unexpected machine breakdowns. The RUL prediction has high priority in CBM programs because of this reason [4]. Lot of studies can be found in literatures that predict machine RUL [22–24].

Usually, a maintenance engineer waits until the functional failure of the machine under operation for its maintenance activities. As per P-F curve, the RUL of an industrial plant machine is the time period up to its functional failure. It is not possible to predict the exact point of functional failure (F) [4, 24, 25]. The proposed study defines the exact point of potential failure (P) as alarm limit of vibration based on the vibration severity standards ISO 10816.

The proposed experimental study identifies that the machine availability declines much before potential failure point, and the maintenance initiation can be recommended before P. The study predicts a CBM initiation point for Class-I machines before P and at this point onwards the system reliability starts to decrease rapidly to zero [25]. The maintenance initiation at this point enhances over all life of the system. Furthermore, the study observes that the time period between P and F is very low and the maintenance initiation at a point before P will not affect much the economic performance of the plant.

In this experimental study, the accelerated life testing (ALT) procedures are used, and the failure of a system of Class-I machine is simulated by artificially wearing the bearing in the system. The wear in the bearing causes system vibrations, and the vibration velocity (RMS) values are measured as time series. The time recorded during the experiment is changed by a suitable transformation to the corresponding useful life in normal operating conditions of the machine [18, 19].

The analysis of the proposed work is performed between the alert and alarm limits of the vibration velocity. The quantification of the system availability and the vibration velocity is performed for CBM programs. The shrinkage pattern of availability with vibration velocity is assessed for a particular system load. The variation of percentage decreases in the vibration velocity with the given improvements in system availability for a particular system load is studied in this research article. The influence of system load on system failure time is also studied.

The rest of the paper is arranged as follows. Section 2 explains the experimental set up of a Class-I machine and its loading arrangement used in the study. The experimental procedure of accelerated life testing and vibration measurements are also explained in this section. Section 3 details the experimental data and failure analysis. A relation between system availability and vibration velocity is formulated in this section and vibration limits are defined for the given enhancements in system availability for CBM. Finally, Sect. 4 concludes the paper by enumerating the findings.

2 Experimental Set Up and Design of Accelerated Life Tests

As per vibration severity standards ISO 10816, Class-I machines are the integral part of the machines like blowers in an industrial process plant. The electrical motors of power up to 20 HP are usual examples of Class-I machines [8, 9]. The experimental set up consists of a delta-connected three-phase induction motor, a long shaft which is supported by two bearings mounted in bearing brackets and a mechanical loading mechanism with cooling accessories. The bearing brackets are numbered as 1 and 2 and the bearings inside these are named as bearing 1 and 2 respectively. The rated voltage, current, power, rpm, and frequency of the motor used are 415 V, 7.1A, 5HP, 1440 RPM, and 50 Hz respectively [25]. This is typically a Class-I machine.

The system of experimental set up is fabricated in such a way that, its failure is due to the excess wear in the bearing 2. The wear in the bearing is created deliberately using accelerated life testing (ALT) procedures [25]. The excess wear is assessed by measuring vibration levels as time series of vibration velocity (RMS). In the proposed study, the alert and alarm limits of vibration velocity for Class-I machines are defined respectively as 1.8 mm/s and 4.5 mm/s based on ISO 10816. The continuous measurement of time series vibration velocity levels is performed using a uni-axial accelerometer sensor which is mounted on the bearing bracket 2. As per P-F curve, the failure initiation and potential failures correspond to alert and alarm limits respectively [4].

The vibration data is gathered for a specific system load. The system loading arrangement used involves a brake drum, brake shoe, and a loading wheel. The loading is performed by turning the loading wheel which makes the brake shoe inside the brake shoe bracket to press against or release from the rotating hollow brake drum. The system load is noted from the load indicator dial from 14 to18 kg in steps of 1 kg. The trial for the experiment is started for a satisfactory level of vibration velocity, and the vibration velocity level increases as wear in the bearing 2 increases. The measurement continues until alarm limit is reached. Eight trials of experiments are conducted for a specific system load to assure repeatability, and a total of 40 trials are executed for the system loads 14–18 kg [25].

The analog signals of vibration coming from the accelerometer sensor are pre-processed, stored, and post-processed using CompactDAQ, PC, and the LabVIEW software (version 2017) respectively.

The methodology for the gathering experimental data from the system [25] is enumerated as:

(i) An experimental set up for the system of Class-I machine (delta-connected three-phase induction motor) is fabricated.
(ii) The experimental set up is fabricated in such a way that the failure of it is due to the excess wear in the bearing.
(iii) Accelerated life testing (ALT) is conducted by creating a severe environment of high stress resulted by the wear in the bearing that is close to an industrial situation.
(iv) The failure so created is assessed in terms of the induced excess vibration levels in the system by comparing with the vibration severity standards (ISO-10816).
(v) The induced vibration levels in the experimental set up are measured, preprocessed, and post-processed using an accelerometer sensor, CompactDAQ, and LabVIEW software, respectively.
(vi) The vibration data is stored in a PC as time series values of vibration velocity (RMS) for a particular system load.
(vii) Different trials are conducted to record the time series data for a particular radial system load applied on the experimental set up, and the trials are repeated for different system loads.

3 Failure Analysis, Discussion and Results

The failure analysis is performed between alert and alarm limits and the methodology used for the failure analysis is summarized [25] as:

- Error analysis of the recorded time series vibration data is conducted.
- The time corresponding to vibration velocity between predetermined limits (alert and alarm limits based on ISO 10816) are tabulated. Using the linear transformation of ALT, the time recorded was converted to the corresponding value in its normal use and the time obtained can be defined as *projected useful life period*.

- The MTBF values are calculated based on the projected useful life period. The system availability is calculated based on the MTBF and MTTR.
- A curve is plotted between the system availability and the vibration velocity for different system loads.
- The vibration limits are defined for the given improvement in system availability based on the relation between availability and vibration velocity and its significance on CBM is established.
- The influence of system load on system failure time is studied.

A suitable variable frequency drive (VFD) is provided between the motor and input three-phase supply to fix the motor speed at 1465 RPM [25]. Life L_2 of the bearing for a specific system load can be calculated using the Eq. (2) [12, 26]

$$\frac{L_1}{L_2} = \left(\frac{P_2}{P_1}\right)^3 \tag{2}$$

where L_1 is taken as the reference life 12,000 h since the expected life for bearings in industrial applications is in the range of 12,000 to 20,000 h [26]. It is assumed that the load P_1 acting corresponding to the reference life is 160 N (Maximum radial load on the bearing used in the experiment as per its data sheet). P_2 is the load acting on the bearing when the system load is applied in steps of 1 kg from 14 to 18 kg. P_2 is obtained by using equilibrium equations for the shaft configuration when a specific system load is applied [25].

The useful life period of the bearing 2 in the system during its normal operating conditions, L hr for a particular system load, m kg can be obtained using the Eq. (3) [18, 19]

$$(\mathrm{AF})^m \times t = L \times 3600 \tag{3}$$

where $(\mathrm{AF})^m$ is the acceleration factor for a specific possible system load m kg (varies from14 to 18 kg), and t is the observed failure time in seconds during experimentation.

Table 1 summarizes the measured failure data of vibration velocity (between alert to alarm limits), the time between failures, and the projected useful life period for the bearing under a system load 18 kg obtained for each of the five different trials.

In Table 1, the acceleration factor, 37,301.12 is used for the linear transformation of average time to the projected useful life period of the system. Similarly, the projected useful life period of the system for system loads 14–17 kg can be obtained using the acceleration factors 34,786.42, 32,723.24, 32,694.02 and 33,196.92 respectively [25].

Table 2 summarizes the measured failure data of vibration velocity (between alert to alarm limits), the projected useful life period, and the MTBF for the system loads 14–18 kg.

The system availability can be calculated using Eq. (1). The bearing failure can be resolved by occasional changing of shaft and replacement of the bearing. The average time calculated for all these activities (MTTR) is 47 min [25].

Table 1 The measured failure data and the projected useful life period for bearing 2 under the 18 kg system load

Vibration velocity (mm/h)	Time (s)						Projected time for bearing for its useful life period (h)
	Trial 1	Trial 2	Trial 3	Trial 4	Trial 5	Average value	
6480	800	808	804	806	807	805.00	8340.95
7200	870	874	872	874	874	872.80	9043.45
7920	944	942	944	943	943	943.20	9772.89
8640	998	990	994	992	990	992.80	10,286.82
9360	1066	1060	1062	1060	1059	1061.40	10,997.61
10,080	1116	1118	1118	1118	1117	1117.40	11,577.85
10,800	1156	1158	1159	1158	1158	1157.80	11,996.46
11,520	1192	1189	1190	1190	1190	1190.20	12,332.17
12,240	1220	1216	1217	1216	1215	1216.80	12,607.78
12,960	1246	1240	1242	1242	1240	1242.00	12,868.89
13,680	1272	1265	1268	1268	1265	1267.60	13,134.14
14,400	1298	1292	1294	1292	1289	1293.00	13,397.32
14,760	1310	1302	1304	1303	1301	1304.00	13,511.30
15,120	1322	1314	1316	1315	1312	1315.80	13,633.56
15,480	1330	1326	1329	1327	1326	1327.60	13,755.82
15,840	1342	1338	1339	1337	1340	1339.20	13,876.02
16,200	1351	1346	1352	1351	1354	1350.80	13,996.21

The system availability can be calculated for different values of vibration velocity between alert and alarm limits for system loads 14–18 kg. Based on the shrinkage pattern of system availability with vibration velocity, it is established that the maintenance initiation point of Class-I machines is 3.4 mm/s [25].

3.1 Defining the Vibration Limits for Given Improvements in Availability

For system load 14 kg, it can be found that, with the given enhancement of system availability from 0.001% to 0.004%, the percentage decreases in alert limit are − 10.23%, − 25.79%, − 31.16%, and − 36.10%, respectively (Table 3). Figure 1 shows the variation of percentage decreases in the vibration velocity (limits of a vibration velocity value) with the given improvements in system availability for a system load 14 kg.

It is found that, nearer to the alert limit, even for small improvements in availability, the decrease in vibration levels needed are considerably high (Fig. 1). The

Table 2 The vibration velocity, corresponding projected time and MTBF for different system loads

Vibration velocity (mm/h)	System load, 14 kg		System load, 15 kg		System load, 16 kg		System load, 17 kg		System load,18 kg	
	Projected time (h)	MTBF (h)	Projected time (h)	MTBF (h)	Projected time (h)	MTBF (h)	Projected time (h)	MTBF (h)	Projected time (hr)	MTBF (hr)
6480	16,773.18	12,998.20	11,348.60	12,863.57	10,347.66	9612.04	8629.663	7979.56	8340.945	5655.26
7200	18,278.98	11,492.40	12,899.93	11,312.24	11,333.93	8625.77	9341.245	7267.98	9043.45	4952.76
7920	19,515.83	10,255.55	14,102.81	10,109.36	12,311.12	7648.58	10,317.17	6292.05	9772.894	4223.32
8640	20,978.15	8793.23	15,242.06	8970.11	13,139.36	6820.34	11,030.29	5578.93	10,286.82	3709.39
9360	21,929.94	7841.44	16,393.44	7818.73	13,720.59	6239.11	11,703.45	4905.77	10,997.61	2998.60
10,080	22,938.1	6833.28	17,331.20	6880.97	14,229.16	5730.54	12,219.85	4389.37	11,577.85	2418.36
10,800	23,677.31	6094.07	18,296.23	5915.94	14,755.9	5203.80	12,656.33	3952.89	11,996.46	1999.75
11,520	24,360.16	5411.22	19,200.67	5011.50	15,329.86	4629.84	13,049.77	3559.45	12,332.17	1664.04
12,240	25,244.31	4527.07	20,026.32	4185.85	15,863.87	4095.83	13,618.42	2990.80	12,607.78	1388.43
12,960	26,415.13	3356.25	20,889.85	3322.32	16,303.42	3656.28	14,426.83	2182.39	12,868.89	1127.32
13,680	27,431.35	2340.03	21,788.23	2423.94	16,830.16	3129.54	15,490.36	1118.86	13,134.14	862.07
14,400	28,362.21	1409.17	22,532.07	1680.10	17,622.08	2337.62	16,092.82	516.40	13,397.32	598.89
14,760	28,716.51	1054.87	22,836.58	1375.59	18,337.71	1621.99	16,261.88	347.34	13,511.30	484.91
15,120	29,041.83	729.55	23,178.96	1033.21	19,325.8	633.90	16,361.78	247.44	13,633.56	362.65
15,480	29,318.83	452.55	23,556.19	655.98	19,687.25	272.45	16,427.87	181.35	13,755.82	240.39
15,840	29,571.68	199.70	23,927.36	284.81	19,839.82	119.88	16,504.71	104.51	13,876.02	120.19
16,200	29,771.38	0.00	24,212.17	0.00	19,959.7	0.00	16,609.22	0.00	13,996.21	0.00

Table 3 Defining vibration limits corresponding to the availability demand for the system load 14 kg

Vibration velocity (mm/h)	Availability	Percentage increase in availability	Percentage decrease in vibration velocity	Percentage increase in availability	Percentage decrease in vibration velocity	Percentage increase in availability	Percentage decrease in vibration velocity	Percentage increase in availability	Percentage decrease in vibration velocity
6480	0.999939765	0.001	– 10.227135	0.002	– 25.786922	0.003	– 31.161540	0.004	– 36.096586
7200	0.999931873	0.001	– 8.032799	0.002	– 22.276532	0.003	– 30.617801	0.004	– 34.990659
7920	0.999923657	0.001	– 6.793380	0.002	– 18.621259	0.003	– 27.697063	0.004	– 32.992783
8640	0.999910962	0.001	– 3.996752	0.002	– 11.902222	0.003	– 22.332215	0.004	– 29.120596
9360	0.999900156	0.001	– 2.963242	0.002	– 9.851964	0.003	– 17.964529	0.004	– 27.476523
10,080	0.999885427	0.001	– 2.135125	0.002	– 7.207529	0.003	– 13.135468	0.004	– 20.106223
10,800	0.999871531	0.001	– 2.035401	0.002	– 6.997669	0.003	– 11.462094	0.004	– 16.664084
11,520	0.999855322	0.001	– 1.897812	0.002	– 7.179121	0.003	– 10.501710	0.004	– 14.323177
12,240	0.999827070	0.001	– 1.731510	0.002	– 5.689282	0.003	– 7.881603	0.004	– 10.351849
12,960	0.999766758	0.001	– 1.315791	0.002	– 3.153222	0.003	– 3.283097	0.004	– 4.518192
13,680	0.999665501	0.001	– 1.002640	0.002	– 2.061292	0.003	– 2.575107	0.004	– 2.621646
14,400	0.999444663	0.001	– 0.901748	0.002	– 1.233610	0.003	– 1.059286	0.004	– 1.078428
14,760	0.999258276	0.001	– 0.861358	0.002	– 1.093979	0.003	– 0.999297	0.004	– 0.901984
15,120	0.998927884	0.001	– 0.713505	0.002	– 0.770380	0.003	– 0.726443	0.004	– 0.681668
15,480	0.998272775	0.001	– 0.671990	0.002	– 0.657505	0.003	– 0.639321	0.004	– 0.622699
15,840	0.996094422	0.001	– 0.578549	0.002	– 0.581649	0.003	– 0.584766	0.004	– 0.587899
16,200	0.000000000	0.001	0.000000	0.002	0.000000	0.003	0.000000	0.004	0.000000

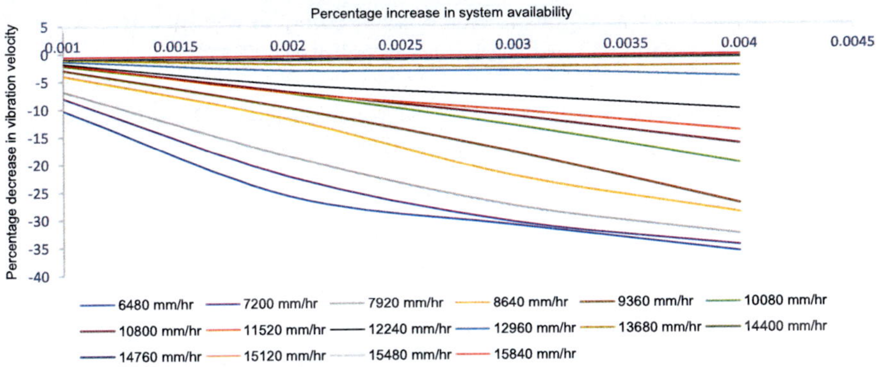

Fig. 1 Variation of % decrease in vibration velocity (between alert and alarm limits) with % enhancement in system availability for a system load 14 kg

relationship is according to the polynomial of order 2 for all the possible system loads. Therefore, the CBM initiation can be performed much before the alarm limit, and this will increase the overall life of the machine significantly. Moreover, it supports the established value of CBM initiation point (3.4 mm/s) for Class-I machines and after this point, there is not much time left to reach the functional failure point [25]. It can be detected that the percentage decrease in vibration velocity due to the percentage increase of system availability is not significant for the vibration velocity values nearer to alarm limit (Table 3).

In addition to the above findings, the following results are also identified. A curve can be plotted between projected potential failure time of the system, t_f hr, with the system load, m kg as shown in Fig. 2.

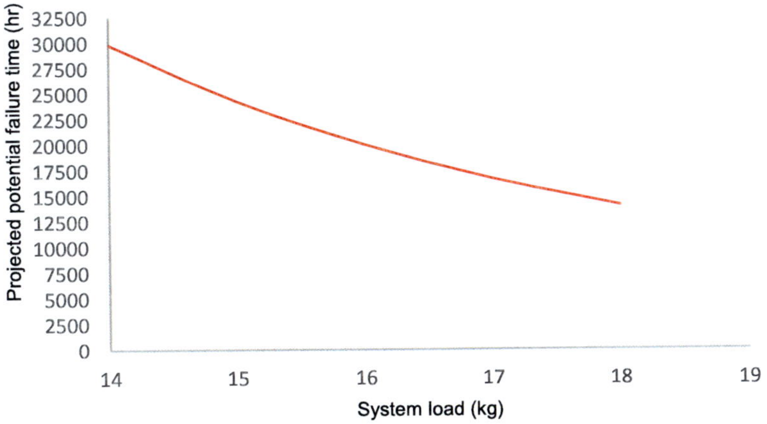

Fig. 2 Variation of potential failure time with the system load

It can be observed that the potential failure time of the system decreases exponentially as system load increases as given in Eq. (4),

$$t_f = 412751e^{-0.189m} \quad \left(R^2 = 0.9987\right) \tag{4}$$

3.2 *Error Report*

Eight trials of experiment are conducted for recording time series vibration velocity for each system load, and a total of forty bearings are failed during the experiment using ALT procedures. Using Grubbs' test at significance level $(\alpha) = 0.02$, the outliers in trials are detected, and the trials which include the outliers are removed from the rest for a specific system load. Out of eight trials, three trials are removed for the system loads 16 and 18 kg and two trials for the system loads 14, 15, and 17 kg.

The maximum value of percentage margin of error (at 95% confidence interval) in the average time is calculated as 3.5%, which corresponded to the average time of 1735.83 s and the vibration velocity 1.8 mm/s. The minimum value of the percentage margin of error in the average time is 0.013%, which corresponded to the average time of 2373.83 s and the vibration velocity 2.8 mm/s. Both errors happen for the system load 14 kg [25].

4 Conclusion

In this experimental study, the ALT procedures were implemented by building a high-stress environment in the bearing of a Class-I machine system. The wear in the bearing was enhanced quickly using a suitable external mechanism, and the resulting time series vibration velocity (RMS) data were gathered. The data acquisition was continued until the machine failure or the alarm limit of vibration was reached. A proper system loading mechanism was included in the experimental set up and different trials of data collection were conducted for a specific system load to ensure repeatability. The data collection was continued for different possible system loads. Using the linear transformation of ALT, the time recorded was transformed into the time corresponding to the normal operating conditions of the machine. The time obtained was named as *projected time,* and the MTBF was computed from it. The system availability was calculated from MTBF and MTTR, and the major contributions based on the analysis of system availability with vibration velocity can be enumerated as:

(i) The proposed study assures the suitability of vibration velocity (RMS) for the prediction of the maintenance initiation point in CBM.

(ii) The analysis of system availability enhancement and resulting vibration limits associated with each vibration velocity show the necessity of maintenance initiation for CBM much before alarm limit of vibration. Furthermore, the analysis of enhancement in system availability supports the established value of CBM initiation point, 3.4 mm/s for Class-I machines.
(iii) The study reveals that the potential failure time decreases exponentially as system load increases.

References

1. Rao SS (1992) Reliability-based design, 1st edn. McGraw-Hill Inc., New York, USA
2. Goel HD, Grievink J, Herder P, Weijnen MPC (2002) Integrating reliability optimization into chemical process synthesis. Reliab Eng Syst Saf 78(3):247–258
3. Devendra, C., Mayank, T., Ravi Shankar: Reliability, availability, and maintainability analysis of a cement plant: a case study. International Journal of Quality & Reliability Management 36(1) (2019).
4. Gulati R (2012) Maintenance best practices, 2nd edn. Industrial Press Inc., New York, USA
5. Alaswad S, Xiang Y (2017) A review on condition-based maintenance optimization models for stochastically deteriorating system. Reliab Eng Syst Saf 157:54–63
6. Kumar S, Mukherjee D, Guchhait PK, Banerjee R, Srivastava AK, Vishwakarma DN et al (2019) A comprehensive review of condition based prognostic maintenance (CBPM) for induction motor". IEEE Access 7:90690–90704
7. Li Y, Peng S, Li Y, Jiang W (2020) A review of condition-based maintenance: Its prognostic and operational aspects. Frontiers of Engineering Management 7(5):323–334
8. Mobley, R.K.: Root cause failure analysis. 1st ed. Elsevier (1999).
9. Mohanty AR (2014) Machinery condition monitoring: principles and practices. CRC Press, USA
10. Pattabhiraman S, Levesque G, Kim NH, Arakere NK (2010) Uncertainty analysis for rolling contact fatigue failure probability of silicon nitride ball bearings. Int J Solids Struct 47:2543–2553
11. Bianchini A, Rossi J, Antipodi L (2018) A procedure for condition-based maintenance and diagnostics of submersible well pumps through vibration monitoring. International Journal of System Assurance Engineering & Management 9(3):999–1013
12. Upadhyay, R.K., Kumaraswamy, L.A., Md. Sikandar, A.: Rolling element bearing failure analysis: A case study. Case Studies in Engineering Failure Analysis 1(1),15–17 (2013).
13. Adawi SKSA, Rameshkumar GR (2016) Vibration diagnosis approach for industrial gas turbine and failure analysis. British Journal of Applied Science & Technology 14(2):1–9
14. Sujatha, C.: Vibration and acoustics: measurement and signal analysis. Tata McGraw Hill, New Delhi, India (2009).
15. Sulaiman Khalifa SAA, Rameshkumar GR (2016) Vibration diagnosis approach for industrial gas turbine and failure analysis. Br J Appl Sci Technol 14(2):1–9
16. Manjare AA, Patil BG (2021) A review: condition-based techniques and predictive maintenance for motor. In: Proceedings of the international conference on artificial intelligence and smart systems. IEEE, JCT College of Engineering and Technology, Tamilnadu, India, pp 807–813
17. Zhang C, Chuckpaiwong I, Liang SY, Seth BB (2002) Mechanical component lifetime estimation based on accelerated life testing with singularity extrapolation. Mech Syst Signal Process 16(4):705–718
18. Bernstein JB (2014) Reliability prediction from burn-in data fit to reliability models, 1st edn. Elsevier, London, UK

19. NIST/SEMATECH: Engineering statistics handbook. https://www.itl.nist.gov/div898/handbook/ April 2012
20. Kulkarni S, Wadkar SB (2016) Experimental investigation for distributed defects in ball bearing using vibration signature analysis. Procedia Eng 144:781–789
21. Quatrini E, Costantino F, Di Gravio G, Patriarca R (2020) Condition-based maintenance—an extensive literature review. Machines 8(2):31
22. Coppe A, Pais MJ, Haftka RT, Kim NH (2012) Using a simple crack growth model in predicting remaining useful life. J Aircr 49(6):1965–1973
23. Kang Z, Catal C, Tekinerdogan B (2021) Remaining useful life (RUL) prediction of equipment in production lines using artificial neural networks. Sensors 21(3):932
24. Han X, Wang Z, Xie M, He Y, Li Y, Wang W (2021) Remaining useful life prediction and predictive maintenance strategies for multi-state manufacturing systems considering functional dependence. Reliab Eng Syst Saf 210(11):107560
25. Lasithan LG, Shouri PV, Rajesh VG (2022) Maintenance initiation prediction incorporating vibration and system availability. Adv Technol Innov 7(3):181–194
26. Bhandari VB (2017) Design of machine elements. McGraw Hill Education, New Delhi, India

A Study on Swarm-Based Approaches for Intrusion Detection System in Cloud Environment

Nishika, Kamna Solanki, and Sandeep Dalal

Abstract Cloud is an advanced Internet of Things enabled network. The layered and distributed nature of this network opens this network to the public environment and varied kind of users. Bulk of data is stored on centralized system, and large amount of information is communicated in the real environment. This public and large amount of shared data and communication opens the network for intruders. Various intrusion detection systems were designed and implemented on different layers of the cloud network to detect and prevent different kinds of internal and external attacks. In the recent years, various swarm-based methods were integrated within intrusion detection system to optimize the performance. In this paper, a detailed study and review of cloud- and Internet of Things-based intrusion detection systems is presented. The analytical and descriptive study of swarm-based intrusion detection system is provided in this paper. This study identifies that the swarm integration within intrusion detection and classification methods improved the performance and reliability of conventional systems.

Keywords Cloud computing · IoT · Swarm based · Intrusion detection · Optimization

1 Introduction

As the data communication and sensitivity increases, the chances of network penetration and intrusion also increase. The target of these intruders is to reveal sensitive information, degrading the network performance or destroying the communicating information. An intruder can attack different layers of a communication network. According to the type of users, type of information and application domain, the

Nishika (✉) · K. Solanki
University Institute of Engineering and Technology, MDU, Rohtak, India
e-mail: nishika510@gmail.com

S. Dalal
Department of Computer Science and Applications, MDU, Rohtak, India

S. Das et al. (eds.), *Advances in Data-driven Computing and Intelligent Systems*, Lecture Notes in Networks and Systems 653, https://doi.org/10.1007/978-981-99-0981-0_46

attack severity also changes. As the communication architecture becomes complex and the communication involves different kinds of hosts and controllers, the target for intruders also increases. An intrusion detection system (IDS) defines a security mechanism to detect the intruders and to improve the system reliability. An intrusion detection system is designed specific to the network architecture, communication type, security requirements, and network priorities. IDS can be applied on each node, centralized controller or randomly over the network to detect the intruder nodes. A firewall-based intruder detection systems are also popular in centralized and server-based networks and applications. The IDS can be applied to detect the intruder node or to take preventive decisions [1]. A standard functional architecture of IDS is shown in Fig. 1. Standard intrusion detection follows these steps to identify the intrusion at node or network level. The first task of IDS system is to track the network activity and communication. Based on this tracking, communication information and network activities-based data are collected. This collected information can be node or network or attack specific. Once the information is collected, the communication patterns are formed based on the available information, sequence, and communication schedule. These patterns are analyzed against different measures and methods to classify the attack patterns. Some rule-based decision methods can be applied to identify the suspicious or normal activity. As the suspicious activity is identified, the IDS system locates the source of the activity and treats the node as an intruder node. These suspicious nodes can be blocked or removed from the network that can ensure the reliability and security of the network.

Intrusion detection (IDS) is an intelligent mechanism that is capable to understand different attempts and methods that can be applied by the attackers to penetrate the security of the system. IDS should be capable of claiming security objectives including accountability, integrity, availability, integrity, and assurance. The IDS should detect all abnormal activities and security violations. The system should be capable enough to identify the insider's attack or behavior. The first priority of such a system is to predict the expected attack and stop it before the violation actually occurs. Even if the security violation happens, the IDS should be capable to deal with it with minimum loss. A quick detection of attacker or attacking behavior can prevent the system from bigger losses. Various statistical, pattern-based, machine learning, and artificial intelligence-based techniques were developed by researchers to enhance the functional and processing capabilities of IDS. The optimization algorithms can be integrated within the existing systems to improve the attack detection rate and early detection of attackers [2–5]. Various kinds of IDS systems that exist in different layers and use different methods are shown in Fig. 2. In the signature-based IDS system, the signatures of known attacks are maintained in the database. The pattern-based mapping is performed to identify malicious communication or intruders. Such systems require continuous update in the database to deal with advanced patterns and technologies. In anomaly-based IDS, the network is monitored using some rules or heuristic methods and classify the normal and malicious behavior of nodes or activities. These kinds of attacks are broadly classified as statistical, knowledge-based, and machine learning-based attacks. Figure 2 is showing various categories and subcategories of these kinds of attacks [6].

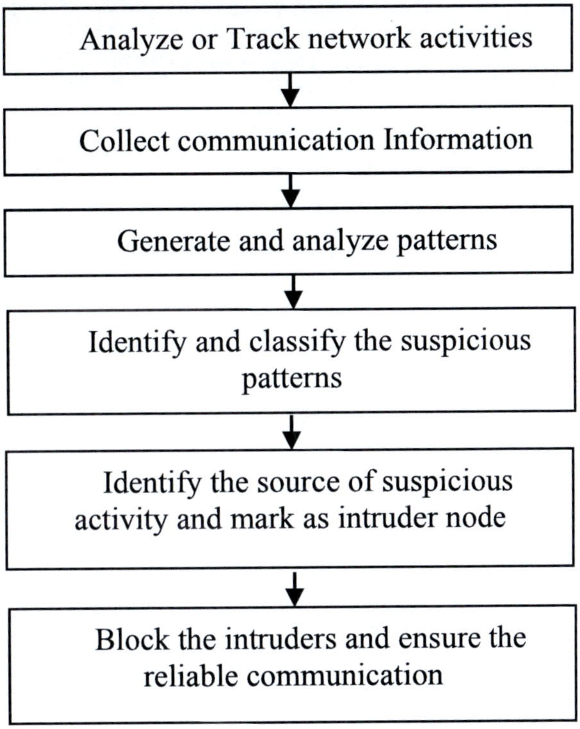

Fig. 1 Standard functioning of IDS

The statistical system performs a session-specific statistical check to observe the expected behavior of nodes and communication. The communication flow, delay, transmission rate, and other parameters can be analyzed over the session to identify the occurrence of attacks. The deviation based, time series based, and Markov methods are the most commonly used statistical methods. The knowledge-based intrusion detection systems capture the knowledge specific to attacks and vulnerabilities. The state transition, signature analysis, and expert system methods can be employed for detecting and categorizing the attacks. The effectiveness of such methods includes the availability of the knowledge. The regular updation of communication behavior and data transition is required to obtain effective results. Machine learning integrates artificial intelligence methods to utilize the knowledge, skills, and experience of nodes over the network. These systems are intelligent enough with resilience to noisy data and adaptive to historical facts [6].

2 Related Work

Mehanovic et al. [7] proposed a cloud-based parallel genetic algorithm to mitigate the intrusion in a cloud environment. Authors used the MapReduce implementation

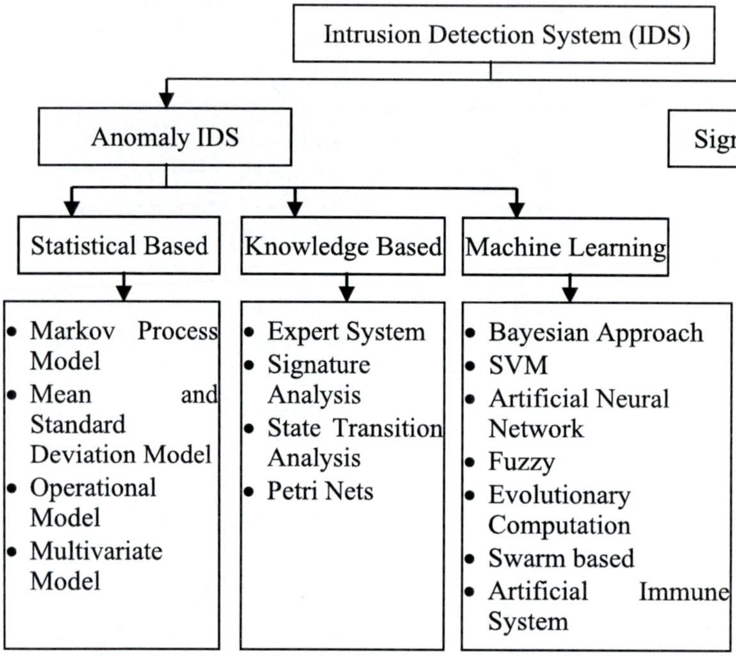

Fig. 2 Categorization of IDS [6]

with parallelism to improve the significance of the model. The genetic-based algorithm was integrated with machine learning methods including SVM, ANN, regression tree, naive Bayes, and logistic regression. These methods collectively identify the selective features and performed classification over it. The proposed model was applied on NSL-KDD dataset and achieved 90.45% accuracy. The data processing and outcomes were improved by the proposed model. Kalaivani et al. [8] proposed an artificial bee colony optimization-based method for optimizing the issues of intrusion in cloud environment. The model was defined against DOS, replay, and flood attack. The analysis results identify that the proposed model achieved 97% accuracy over existing Naive Bayes and decision tree classifiers. Sun et al. [9] proposed a KVM-based model to detect the intrusion in cloud-based network. The proposed model is defined as an improved back propagation network. The hybridization of the algorithm was defined by integrating the PSO algorithm within the classifier. The weights and threshold-based model was applied on adaptive learning method to detect intrusion effectively. The analysis results show that the proposed model achieved high accuracy rate for detection of intrusion in cloud environment. Zhou et al. [10] proposed a hybrid method using multi-objective PSO for detection of intrusion in a cloud environment. The attributes weights are computed in this model for identifying the optimal features. The proposed model resolved the issues of feature selection in intrusion detection system effectively and achieved higher accuracy and convergence in real environment.

Mayuranathan et al. [11] introduced a feature subset-based Random Harmony Search (RHS) optimization model to identify the best features. These selected and significant features were processed by the Restricted Boltzmann Machines (RBM)-based deep learning model for detection of DDOS attack. Author included seven extra layers to improve the performance and reliability of the system. The model was applied on KDD99 dataset, and accuracy up to 99.92%. Santos et al. [12] proposed a flow-based framework to detect the intrusion from an IoT network. The packet monitoring and flow features were analyzed at different layers of IoT architecture to recognize the type and severity of intrusion. The specification-based approach applied on captured traffic features. The IDS component and operations-based method was applied to optimize the usage of computational resources. The method achieved effective accuracy and lesser utilization of resources in comparison with conventional IDS. Kala et al. [13] combined the particle swarm optimization algorithm with a probabilistic neural network (PNN) for optimizing the performance of an intrusion detection system. The opposition-based PSO was defined to optimize the existing solution. The method was defined on NSL-KDD dataset and compared against neural network and opposition-based PSO models. The results show that the proposed model achieved 99.44% accuracy and outperformed existing methods and models.

3 IDS in Cloud and IoT-Based Environment

Cloud is a distributed environment that is deployed widely with all application domains. The reliable and broader scope of cloud computing make it popular to share the information, data, and services in a public and real environment. The public adaptation and access open this network for various security attacks. These attacks and vulnerability of this network are increasing day by day. Each layer and component of multi-layer architecture of cloud computing is under the threats of these attacks. The virtual machines, hypervisors, and attack nodes are the key concerns in this network. The adaptation of software-defined networks (SDN) with cloud computing is also growing these days. SDN is also prone to various kinds of security attacks including stealthy DDOS attack. The signature features of public nodes and usage of infrastructure as a service are the biggest concerns in security management of cloud and software-defined networks. The intrusion detection and prevention systems are integrated by the cloud service providers to ensure the reliable service delivery and data sharing. Various predictive models were designed by the service providers to enhance the security and reliability of the system. Figure 3 is showing a standard intrusion detection system for cloud network [14].

Figure 3 shows the IDS framework implied over the cloud system to monitor the network and to detect or prevent intrusion. The IDS system is integrated with the policy engine as the cover layer. The policy framework is defined and configured with some rules that can identify the normal and intrusion-specific communication. The interface library is also defined to accept the response respective to the queries.

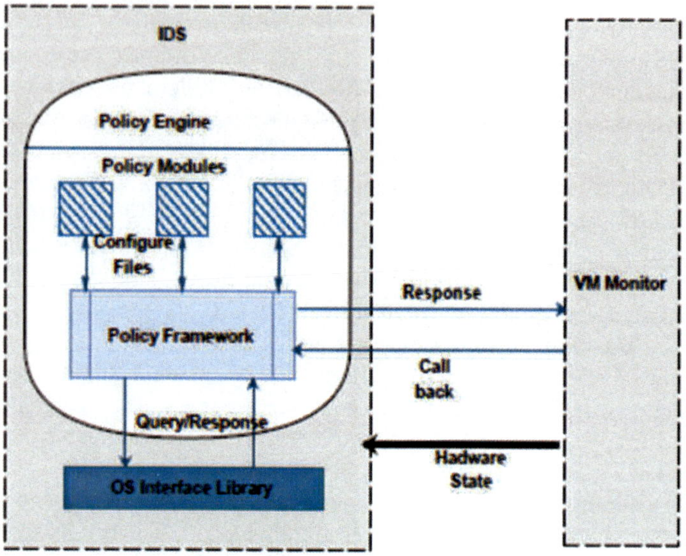

Fig. 3 IDS system for cloud environment [14]

The hardware states of cloud network and VM monitoring status are also defined in this network. There are number of IDS models proposed by the researchers in recent time to detect intrusion in cloud environment. Some of the most recent and promising IDS systems proposed by the researchers are listed and compared in Table 1. The algorithmic methods and analysis results claimed by these methods are discussed in this table.

The research work provided in Table 1 identified that various supervised and unsupervised learning methods were applied by the researchers for effective detection of intrusion within the cloud environment. Authors used clustering, Naive Bayes, decision tree, SVM, and ensemble learning methods for optimizing the performance of IDS. Some of the researchers have used feature selection methods to identify the effective communication attributes that can be used to predict intrusion accurately. The performance results identified that the ensemble learning and hybrid methods achieved the higher accuracy than the conventional machine learning methods. Krishnaveni et al. [18] method achieved 98.98% accuracy for honeypot and 99.935% accuracy for Kyoto dataset.

4 Swarm-Based Approaches for IDS

The optimization algorithms are the functional advancement to an existing model or algorithm to optimize the functionality against time, cost, prediction rate, or any other measure based on the application. In IDS system, the optimization algorithms

Table 1 IDS models for cloud environment

Author	Method/model	Description	Performance
Jaber et al. [15]	Fuzzy C means clustering (FCM) with support vector machine (SVM)	A membership function using cluster group based approach was applied on NSL-KDD dataset	97.37% accuracy U2L under different ML metrics
Singh et al. [16]	Ensemble learning	Combined Boosted tree, bagged tree, RUSBooted, and subspace discriminant machine learning models with voting scheme	97.24% accuracy
Srilatha et al. [17]	Kernel fuzzy CMeans clustering (KFCM) and optimal type-2 fuzzy neural network (OT2FNN)	Applied lion optimization with type-2 fuzzy neural network for feature selection and to detect the intruders	98.5% precision, 97.3% recall, and 96% Fmeasure
Krishnaveni et al. [18]	Ensemble classifier using SVM, Naive Bayes, logistic regression and decision tree	Used ensemble feature selection and classification method for identifying the selected and effected features and applied ensemble classifier on that	98.98% accuracy on honeypot dataset, 99.935% for kyoto 2006 dataset, 96.6 for NSL-KDD dataset
Seth et al. [19]	Light gradient boosting machine (LightGBM) with hybrid feature selection	A Hybrid feature processor is applied for feature filtration and followed by fast gradient boosting framework for effective Intrusion detection	97.73% accuracy, 97.57% FScore, 96% sensitivity
Punitha et al. [20]	Ensemble neural network	New centralized cloud information accountability integrity with imperialist competitive key generation algorithm (CCIAI-ICKGA). The cipher text policy attribute-based encryption (CP-ABE) was integrated for enhancing security in the system	6.2% security increase for 200 cloud users, 10.15% increase over CIA framework, 6.7% increase over other existing methods

(continued)

Table 1 (continued)

Author	Method/model	Description	Performance
Wen [21]	Artificial Bee colony based Back Propagation Neural Network (ABC-BPNN)	Back propagation neural network was combined with artificial bee colony optimization for effective intrusion detection	92.67% accuracy

are effective enough to improve the attack detection rate, prediction accuracy, and to reduce the error rate. The optimization algorithms can be rule based, evolutionary and nature or swarm-inspired techniques. The nature-inspired algorithms are comparatively light weighted and easy to implement. These are flexible, self-configured, and highly adaptable technique that can be applied in on different models, applications, and environments. In IDS, the scope and significance of nature-inspired algorithms is increasing extensively. These algorithms can be applied on different layers and algorithms of different networks to recognize the abnormal activities and behavior of network nodes and components. These are non-deterministic algorithm that uses the natural phenomenon to generate the solutions. The nature or social behavior of species, birds and animal is adapted by this algorithm to generate non-conventional solution. These algorithms act as global optimizer that can be applied on any domain or framework. The local and global search stages are included in these algorithms to generate optimum results [2]. In the recent years, the nature-inspired optimization ensures the competency to solve NP-Hard problems in different domains. These algorithms are adaptive to solve real-life problems. These algorithms use the nature of swarms such as birds, frogs, bees, wolf, cats, bat, bacteria, monkeys, fireflies, and whales. The biological and functional behavior of these particles or swarms can be used in different applications and environments to improve the performance, accuracy, and behavior of an existing application or algorithm. These swarm-based algorithms are integrated and have evolution, organization and learning stages to improve the competence of the algorithms. These algorithms can be applied for optimizing scheduling, attack mitigation, and improving the performance of pattern or object recognition systems [22, 23]. In this paper, the scope, use, and significance of swarm-based algorithm are discussed for intrusion detection system. These systems are applied on cloud and IoT-based network. In the recent years, various swarm-based applications and models are proposed by the researchers to optimize the significance of IDS in cloud environment. Table 2 shows some of the recent contributions of researchers in which different swarm-based methods are applied within intrusion detection system to improve their performance.

Table 2 shows that the integration of swarm-based optimization methods within conventional and hybrid models have improved the accuracy of intrusion detection systems. The researchers used PSO, ABC, GWO, and other optimization methods at feature selection and classification stages. Most of the researchers have worked

Table 2 Swarm-based IDS models

Author	Method/model/technique	Network/application /domain	Performance
Habib et al. [24]	Multi-objective particle swarm optimization with Levy flight randomization component (MOPSO-Levy)	IoT	93.7% on Baby monitor dataset, 98.9% on Dammini Doorbell, 95.6% on Security Camera PT737, 96.1% on Security Camera PT838
Kunhare et al. [25]	Random forest for feature selection and PSO is applied on identified selected features for intrusion detection	NSL-KDD Dataset	99.32% accuracy, 99.31 F1Score, 99.37 precision score
Agarwal et al. [26]	Feature-selection whale optimization algorithm-deep neural network (FS-WOA-DNN)	CICIDS2017 Dataset	95.35% accuracy, 96.9% Specificity, 90.71% Sensitivity, 0.0928 Error
Kala et al. [27]	Combined feed forward neural network (FFNN) with probabilistic neural network (PNN) and oppositional particle swarm optimization (OPSO)	NSL-KDD	99.44% accuracy, 78.88% sensitivity and 1 specificity
Dwivedi et al. [28]	Multi-parameter adaptive grasshopper optimization technique	NSL-KDD, AWID-ATK-R, NGIDS-DS	99.87% accuracy on NSL-KDD, 99.75% on AWID-ATK-R, 91.57% NGIDS-DS
Giri et al. [29]	Combined adaptive particle swarm optimization (APSO) with Adaptive artificial bee colony (AABC) algorithms	NSL-KDD	97.85% precision rate
Dwivedi et al. [30]	Ensemble feature selection and grasshopper optimization algorithm (EGSGOA)	NSL-KDD, KDD Cup 99	99.98% accuracy on NSL-KDD and 99.89% on KDD Cup 99 dataset
Keserwani et al. [31]	Gray wolf optimization (GWO) and particle swarm optimization (PSO)	KDDCup99, NSL-KDD, CICIDS-2017	99.66% Accuracy

(continued)

Table 2 (continued)

Author	Method/model/technique	Network/application /domain	Performance
Alamiedy et al. [32]	Multi-objective gray wolf optimization (GWO)	NSL-KDD	93.64% accuracy against DoS, 91.01% against Probe, 57.72% against R2L, 53.7% against U2R
Devi et al. [33]	Hybrid gray wolf optimizer cuckoo search optimization (HGWCSO)	NSL KDD Cup	91% for DoS, 93% for Probe, 78% for U2R, 83% for R2L
Singh et al. [34]	Cuckoo optimization	NIMS 2	71.69% average accuracy

on NSL-KDD datasets. The maximum accuracy of 99.87% was obtained for multi-parameter adaptive grasshopper optimization technique.

5 Conclusion

The intruders are the primary concern for any network. When the network is available in public domain and sensitive and personal information is communicated over the network, there are lots of chances of intrusion attack. The cloud and IoT-based networks are sensitive networks against such kind of attacks. The varied kind of data is passed through different layers. To ensure the reliability and security of the network, various intrusion detection systems were proposed by researchers. In this paper, a detailed study and result analysis are provided on various intrusion detection systems applied in cloud and IoT networks. Many of the researchers used various swarm-based methods for optimizing the performance of intrusion detection and classification systems. The paper also provided the analytical study on these swarm-based methods. The analysis results identified that the integration of swarm-based methods within an intrusion detection system optimized the performance and accuracy of intrusion detection system and achieved better results against conventional systems.

References

1. Thakkar A, Lohiya R (2021) A survey on intrusion detection system: feature selection, model, performance measures, application perspective, challenges, and future research directions. Artif Intell Rev. https://doi.org/10.1007/s10462-021-10037-9
2. Thakur K, Kumar G (2021) Nature Inspired Techniques and Applications in Intrusion Detection

Systems: Recent Progress and Updated Perspective. Arch Computat Methods Eng 28:2897–2919. https://doi.org/10.1007/s11831-020-09481-7
3. Thakkar A, Lohiya R (2021) A Review on Machine Learning and Deep Learning Perspectives of IDS for IoT: Recent Updates, Security Issues, and Challenges. Arch Computat Methods Eng 28:3211–3243. https://doi.org/10.1007/s11831-020-09496-0
4. Ayyagari MR, Kesswani N, Kumar M et al (2021) Intrusion detection techniques in network environment: a systematic review. Wireless Netw 27:1269–1285. https://doi.org/10.1007/s11276-020-02529-3
5. Keserwani PK, Govil MC, Pilli ES (2021) An effective NIDS framework based on a comprehensive survey of feature optimization and classification techniques. Neural Comput & Applic. https://doi.org/10.1007/s00521-021-06093-5
6. Kocher G, Kumar G (2021) Machine learning and deep learning methods for intrusion detection systems: recent developments and challenges. Soft Comput 25:9731–9763. https://doi.org/10.1007/s00500-021-05893-0
7. Mehanovic D, Keco D, Kevric J et al (2021) Feature selection using cloud-based parallel genetic algorithm for intrusion detection data classification. Neural Comput & Applic 33:11861–11873. https://doi.org/10.1007/s00521-021-05871-5
8. S. Kalaivani, A. Vikram and G. Gopinath, "An Effective Swarm Optimization Based Intrusion Detection Classifier System for Cloud Computing," 2019 5th International Conference on Advanced Computing & Communication Systems (ICACCS), 2019, pp. 185–188, doi: https://doi.org/10.1109/ICACCS.2019.8728450.
9. Hao Sun, "Improved BP Algorithm Intrusion Detection Model based on KVM," 2016 7th IEEE International Conference on Software Engineering and Service Science (ICSESS), 2016, pp. 442–445, doi: https://doi.org/10.1109/ICSESS.2016.7883104.
10. Zhou L, Liu Y, Chen G (2011) A Feature Selection Algorithm to Intrusion Detection Based on Cloud Model and Multi-Objective Particle Swarm Optimization. Fourth International Symposium on Computational Intelligence and Design 2011:182–185. https://doi.org/10.1109/ISCID.2011.147
11. Mayuranathan M, Murugan M, Dhanakoti V (2021) Best features based intrusion detection system by RBM model for detecting DDoS in cloud environment. J Ambient Intell Human Comput 12:3609–3619. https://doi.org/10.1007/s12652-019-01611-9
12. Santos L, Gonçalves R, Rabadão C et al (2021) A flow-based intrusion detection framework for internet of things networks. Cluster Comput. https://doi.org/10.1007/s10586-021-03238-y
13. T. S. Kala and A. Christy, "An Intrusion Detection System using Opposition based Particle Swarm Optimization Algorithm and PNN," 2019 International Conference on Machine Learning, Big Data, Cloud and Parallel Computing (COMITCon), 2019, pp. 184–188, doi: https://doi.org/10.1109/COMITCon.2019.8862237
14. Ravindranath V, Ramasamy S, Somula R, Sahoo KS, Gandomi AH (2020) Swarm Intelligence Based Feature Selection for Intrusion and Detection System in Cloud Infrastructure. IEEE Congress on Evolutionary Computation (CEC) 2020:1–6. https://doi.org/10.1109/CEC48606.2020.9185887
15. Jaber AN, Rehman SU (2020) FCM–SVM based intrusion detection system for cloud computing environment. Cluster Comput 23:3221–3231. https://doi.org/10.1007/s10586-020-03082-6
16. Singh P, Ranga V (2021) Attack and intrusion detection in cloud computing using an ensemble learning approach. Int. j. inf. tecnol. 13:565–571. https://doi.org/10.1007/s41870-020-00583-w
17. Srilatha D, Shyam GK (2021) Cloud-based intrusion detection using kernel fuzzy clustering and optimal type-2 fuzzy neural network. Cluster Comput 24:2657–2672. https://doi.org/10.1007/s10586-021-03281-9
18. Krishnaveni S, Sivamohan S, Sridhar SS et al (2021) Efficient feature selection and classification through ensemble method for network intrusion detection on cloud computing. Cluster Comput 24:1761–1779. https://doi.org/10.1007/s10586-020-03222-y
19. Seth S, Singh G, Kaur Chahal K (2021) A novel time efficient learning-based approach for smart intrusion detection system. J Big Data 8:111. https://doi.org/10.1186/s40537-021-00498-8

20. Punitha AAA, Indumathi G (2021) A novel centralized cloud information accountability integrity with ensemble neural network based attack detection approach for cloud data. J Ambient Intell Human Comput 12:4889–4900. https://doi.org/10.1007/s12652-020-01916-0
21. Wen L (2021) Cloud Computing Intrusion Detection Technology Based on BP-NN. Wireless Pers Commun. https://doi.org/10.1007/s11277-021-08569-y
22. Kaul S, Kumar Y, Ghosh U et al (2021) Nature-inspired optimization algorithms for different computing systems: novel perspective and systematic review. Multimed Tools Appl. https://doi.org/10.1007/s11042-021-11011-x
23. Ahsan MM, Gupta KD, Nag AK, Poudyal S, Kouzani AZ, Mahmud MAP (2020) Applications and Evaluations of Bio-Inspired Approaches in Cloud Security: A Review. IEEE Access 8:180799–180814. https://doi.org/10.1109/ACCESS.2020.3027841
24. Habib M, Aljarah I, Faris H (2020) A Modified Multi-objective Particle Swarm Optimizer-Based Lévy Flight: An Approach Toward Intrusion Detection in Internet of Things. Arab J Sci Eng 45:6081–6108. https://doi.org/10.1007/s13369-020-04476-9
25. Kunhare N, Tiwari R, Dhar J (2020) Particle swarm optimization and feature selection for intrusion detection system. Sadhana 45:109. https://doi.org/10.1007/s12046-020-1308-5
26. Agarwal A, Khari M, Singh R (2021) Detection of DDOS Attack using Deep Learning Model in Cloud Storage Application. Wireless Pers Commun. https://doi.org/10.1007/s11277-021-08271-z
27. Sree Kala T, Christy A (2021) HFFPNN classifier: a hybrid approach for intrusion detection based OPSO and hybridization of feed forward neural network (FFNN) and probabilistic neural network (PNN). Multimed Tools Appl 80:6457–6478. https://doi.org/10.1007/s11042-020-09804-7
28. Dwivedi S, Vardhan M, Tripathi S (2021) Multi-Parallel Adaptive Grasshopper Optimization Technique for Detecting Anonymous Attacks in Wireless Networks. Wireless Pers Commun 119:2787–2816. https://doi.org/10.1007/s11277-021-08368-5
29. Velliangiri S, Karthikeyan P (2020) Hybrid optimization scheme for intrusion detection using considerable feature selection. Neural Comput & Applic 32:7925–7939. https://doi.org/10.1007/s00521-019-04477-2
30. Dwivedi S, Vardhan M, Tripathi S (2021) Building an efficient intrusion detection system using grasshopper optimization algorithm for anomaly detection. Cluster Comput 24:1881–1900. https://doi.org/10.1007/s10586-020-03229-5
31. Keserwani PK, Govil MC, Pilli ES et al (2021) A smart anomaly-based intrusion detection system for the Internet of Things (IoT) network using GWO–PSO–RF model. J Reliable Intell Environ 7:3–21. https://doi.org/10.1007/s40860-020-00126-x
32. Alamiedy TA, Anbar M, Alqattan ZNM et al (2020) Anomaly-based intrusion detection system using multi-objective grey wolf optimisation algorithm. J Ambient Intell Human Comput 11:3735–3756. https://doi.org/10.1007/s12652-019-01569-8
33. Roopa Devi, E. M., and R. C. Suganthe. "Enhanced transductive support vector machine classification with grey wolf optimizer cuckoo search optimization for intrusion detection system." Concurrency and Computation: Practice and Experience 32, no. 4 (2020): e4999.
34. Singh, D. Asir Antony Gnana, R. Priyadharshini, and E. Jebamalar Leavline. "Cuckoo optimisation based intrusion detection system for cloud computing." International Journal of Computer Network and Information Security 11, no. 11 (2018): 42.

Smart Monitoring of Vital Sign Parameters in IoT-Based Fiber Bragg Grating Sensing Technology

Maitri Mohanty, Ambarish G. Mohapatra, and Premansu Sekhara Rath

Abstract The adaptation of Internet of Things (IoT) technologies in healthcare applications is growing rapidly over the globe. Keeping track of vital sign parameters act as an early diagnosis tool for various diseases. Further, the optical sensing technologies integrated with IoT-enabled systems can solve various healthcare diagnosis requirements. In this article, the intervention of Fiber Bragg Grating (FBG) sensors to effectively monitor vital cardiac parameters and present diagnosis reports using an innovative approach are discussed. The design and construction of a Fiber Bragg Grating sensor are also discussed in the methodology section of the article. Also, the recorded raw Ballistocardiogram (BCG) signal is filtered using advanced filtering algorithms. The heart rate (HR) of different subjects is evaluated in real time after successful filtering of the noisy cardiac signature.

Keywords Fiber Bragg grating · Optical sensors · IoT · Ballistocardiogram · Heart rate

1 Introduction

Cardiovascular disease is related to the irregular function of the heart and blood vessels, thus a major issue of death across the globe according to the survey of worldwide healthcare organizations and other sources [1]. An abnormal respiratory rate indicates the patient's disturbance of psychophysical state [2]. Researchers have proposed various sensors and electronic gadgets for monitoring human vital sign

M. Mohanty (✉) · P. S. Rath
Department of Computer Science & Engineering, GIET University, Gunupur, Odisha, India
e-mail: maitri.mohanty@giet.edu

P. S. Rath
e-mail: premansusekhararath@giet.edu

A. G. Mohapatra
Department of Electronics and Instrumentation Engineering, Silicon Institute of Technology, Bhubaneswar, Odisha, India
e-mail: ambarish.mahapatra@silicon.ac.in

S. Das et al. (eds.), *Advances in Data-driven Computing and Intelligent Systems*, Lecture Notes in Networks and Systems 653, https://doi.org/10.1007/978-981-99-0981-0_47

parameters. But existing electronic and electrical sensor has limited application in the healthcare sector. So, they have not taken place its position in the economic and commercial market till now. Monitoring of human respiratory and cardiac parameters is an essential assessment during medical treatment diagnoses like sleep apnea, Parkinson's, chronic bronchitis, emphysema, bradycardia, and heart attack [2, 3]. Unfortunately, monitoring of vital signs is not possible in all medical environments like causalities and is also a challenging task in an MRI environment. In the present scenario, monitoring of vital sign parameters is only done in a special ICU environment. But real-time information about a patient is not stored in any database and cannot use as feedback by doctors in the future. Also, the physical presence of doctors and nurses in ICU is not possible to round the clock and keeps track of every heartbeat of a patient. FBG sensors are suitable for real-time medical environments like MRI full-body scanning of old-age patients, children, disabled patients, anesthetized and coma patients, patients with artificial pacemakers, and insensible patients. Hence, there is a need to develop a smart medical application for monitoring various vital parameters round the clock to detect the patient's conditions and prevent the worst situation by giving treatment at right time. FBG sensors are attracted to the healthcare market due to their small size, low cost, high sensitivity, inert, nontoxic, resistive to a harsh environment, multiplexing capacity, and gives remotely sensing capacity. These sensors are replaced by electrical and electronic sensors in an adverse environment like an MRI environment. These sensors give vital sign information without compromising the quality and comfort level of the subject. The recorded raw BCG signal is corrupted due to various high frequencies, low-frequency components, and various movements of the subject body. So, in today's digital world, researchers have proposed various digital filters to improve signal quality. Therefore, passive FBG sensors are used in medical applications to measure HR, RR, chest expansion, and body temperature. Twelve FBG sensors are cascaded along with a single fiber and are used to monitor the respiration rate [4]. Similarly, an FBG sensor is placed inside a garment-like T-shirt to monitor HR and RR. In this, similar context, many researchers proposed FBG sensors for monitoring vital parameters with their limited application. This happens due to the design issue of sensors and grant of license to be used in real-life medical applications. Emerging Internet of Things technology progresses the various sectors like smart cities [5], smart homes [6], smart climate [7], smart agriculture [8], and smart healthcare application [9]. The advancement of the internet of things with the FBG sensing-based approach flies up the medical application to a new height. Here, the proposed sensor is implemented in distributed healthcare applications for the registration of vital sign parameters with the integration of the Internet of Things.

This proposed research work has presented the monitoring of cardiac parameters using FBG sensing technology which is focused on six different sections. The second part of the review section presents the application of recent sensing technology of FBG sensors in monitoring cardiac parameters. The third section describes the design, working principle of the Fiber Bragg grating sensor, and proposed methodology. Further, the cardiac parameters of a person are captured, filtered, and evaluated. The conclusion section of this article is the proposed methodology and the results of cardiac parameters of a subject followed by the reference section.

2 Literature Review

A Fiber Bragg Grating-based sensing technology with the internet of things is presented here for registration of cardiac and respiratory parameters in smart health-care applications. Nedoma et al. [10] present that FBG is encapsulated inside fiber-glass (composition of glass fiber, fabric, and cured synthetic resin) and placed inside a rectangular belt. The sensor signal is filtered using a third-order Butterworth band-pass filter with cut-off frequencies from 0 to 0.5 Hz. Furthermore, the processed signal is centered and normalized to detect the peaks. A median filter with window size 3 is used to detect a smooth RR curve. HR is detected with a cut-off frequency of 5–20 Hz. A median filter with a window size of 7 is used to smooth HR over time. The relative error for RR is 4.64% and HR is 4.87%.

Dziuda et al. [11] describe signals obtained from the FBG sensor made up of polymethyl placed inside a bed mattress. It is processed through a bandpass filter within the frequency range of 2–60 Hz to obtain RR and HR. The maximum relative error for RR = 7.67% and HR = 6.61%. So the total error of the system is less than 8%. Fajkus et al. [12] describe a wavelength division technique used on the spectral division of individual grating. The sensitivity of the sensor is increased four times using the probe. Two FBGs are encapsulated inside a PolydimethylSiloxane polymer in a probe to measure RR, HR, and body temperature. The blood pressure and heart rate measurements are done by using Butterworth second-order bandpass filter with frequencies ranging [0.75–5] Hz. After filtering the data with this range, Fourier series analysis is performed to calculate the dominant frequency from filtered data to calculate the heart rate. The relative error of RR is 3.9% and body temp is 0.36%. Furthermore, Zhu et al. [13] present three FBG sensor arrays (each array contains six sensors) placed inside the mat. The signals from six sensors of the same array are fused by applying a cepstrum. Recognizing peaks in a cepstrum of a fused signal is measured to detect heart rate. The best result is obtained from the chest portion of the heart.

Similarly, Gurkhan et al. [14] present a strain FBG sensor that monitors different normal and abnormal heartbeat measurements such as pulmonary hyperostosis, summation gallop sound, mitral stenosis, and muscular movements. Heartbeat sounds are translated into an electrical signal which is recorded by an FBG sensor that governs information about the change in reflection waveform. Time-domain analysis of signal using FFT gives distinct shapes than frequency analysis. Also, Dziuda et al. [15] present that FBG is put inside a pneumatic cushion. Breathing rate and heart rate are calculated by filtering the averaging of the sensor signal by using the Fabry–Perot filter and spectrally scanning filter. Two sensors are placed perpendicular to each other to get BR and HR. The maximum relative error is 14% for BR and 12% for HR.

J. De. Jonckheere et al. [16] present an FBG sensor placed inside a garment such as a t-shirt. Here spectroscopic technique with an optical spectral analyzer (OSA) is used to detect HR and RR. The sensing textile can monitor the elongation between 0.1 and 5%. Moreover, Wehrle et al. [17] present an FBG strain sensor placed inside a polymeric foil. Here, a fixed filter with OSA is used to detect RR by using an FBG strain sensor. During calibration of the signal, a lock-in amplifier is used to increase the signal-to-noise ratio.

Silva et al. [18] present that the FBG sensor is placed inside a polymeric foil. Here, bilinear technique in the digital domain is used to detect HR and RR by using the FBG sensor. One bandpass filter in the range 0.1–0.4 Hz is used to measure respiration rate, and another bandpass filter in the range of 0.5–1.3 Hz is used to measure HR. Elsarnagawy et al. [19] present that FBG is embedded and woven into a nylon textile that covers the torso portion of the body. Bandpass filters with FFT to measure HR and RR by using a single FBG sensor. The respiration peaks are measured in the range 0.1–0.4 Hz, and heartbeat peaks are measured in the range 0.9–2 Hz. In the future, they have also proposed to measure the temperature and volumes inhaled by a person in optimized form. Chethana [20] et al. present FBG is placed inside a Velcro strap. BCG signal is filtered using a low-pass filter of 0.5HZz to detect RR and a high-pass filter of cut-off frequency 0.5 Hz to detect HR.

Tosi et al. [21] describe an intensity a based technique using a fixed wavelength laser source and a photodetector to detect HR using a vibroacoustic sensor. Lau et al. [22] present a Micro-bend Fiber Optic Sensor (MFOS) that BR is calculated using bandpass filters which detect the local peaks in the time domain. Chen et al. [23] present that MFOS is placed inside a cushion to measure BR and HR. Respiratory and heart rates are separated in BCG signal using bandpass filters, then using a Savitsky-Golay method to smooth the signal. Zhu et al. [24] present MFOS placed inside a mat. The bandpass filter is applied to the BCG signal with fast fourier transform to measure BR and HR. A detailed analysis on various research works carried out related to the proposed cardiac sensing methodology is portrayed in Table 1.

3 Principle of FBG Sensor and Proposed Methodology

The FBG sensor is gained popularity in today's era market due to its simple principle of light reflection phenomenon. FBG is inscribed in an optical fiber due to the change in the refractive index of the core along the light propagation direction. When light incidents on grating region, some amount of same phase light is reflected to form one wavelength is called Bragg wavelength and the condition is called Bragg condition as shown in the Fig. 1. According to the Couple mode theory, Bragg wavelength is expressed as shown in Eq. 1 [11].

$$\lambda_B = 2\eta_{eff}\Lambda \tag{1}$$

Table 1 Vital sign monitoring using fiber optic sensor

S. No.	Author details	Sensor type and material	Measured parameters	Method used
1	Nedoma et al. [10]	FBG, fiberglass	RR, HR	Using Butterworth and median filter
2	Dziuda et al. [11]	FBG Strain, poly methyl	HR, RR	Using moving average filter
3	Fajkas et al. [12]	Two FBG, PDMS	HR, RR	Using Butterworth bandpass filter with FFT
4	Zhu et al. [13]	Three FBGs, inside mat	HR	Using Cepstrum analysis
5	J. De. Jonckheere et al. [16]	FBG, inside T-shirt	RR, HR	Optical Spectral analyzer
6	Elaarnagawy [19]	FBG, Nylon textile	HR, RR	Using the bandpass filters with FFT
7	Chethana et al. [20]	FBG, Velcro strap	HR, RR	Using lowpass filter

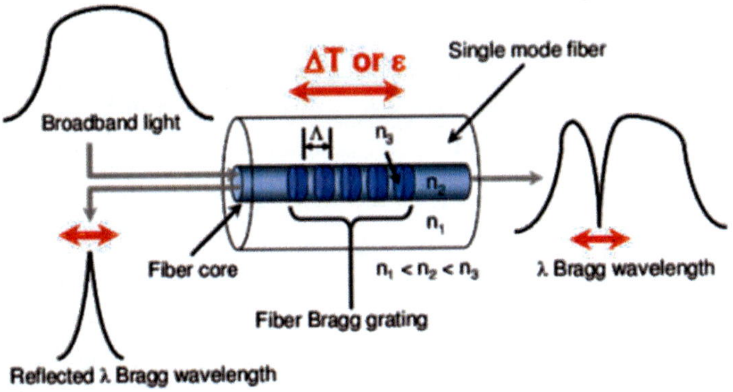

Fig. 1 Fiber Bragg Grating sensor working principle and optical phenomena

where η_eff is the effective refractive index of the optical fiber core and Λ is the periodicity of the grating period.

The change in Bragg wavelength to the principal Bragg wavelength can also be expressed as a differential Eq. 2 [11].

$$\frac{\Delta\lambda_B}{\lambda_B} = \left(\frac{1}{\Lambda}\frac{\partial\Lambda}{\partial\varepsilon} + \frac{1}{\eta_{\text{eff}}}\frac{\partial\eta_{\text{eff}}}{\partial\varepsilon}\right)\Delta\varepsilon + \left(\frac{1}{\Delta}\frac{\partial\Lambda}{\partial T} + \frac{1}{\eta_{\text{eff}}}\frac{\partial\eta_{\text{eff}}}{\partial T}\right)\Delta T$$

where $\partial\Lambda$ is the change in periodicity concerning the change in applied strain $\partial\varepsilon$ and change in temperature ∂T, $\partial\eta_{\text{eff}}$ is the change in refractive index concerning the change in applied strain $\partial\varepsilon$ and change in temperature ∂T.

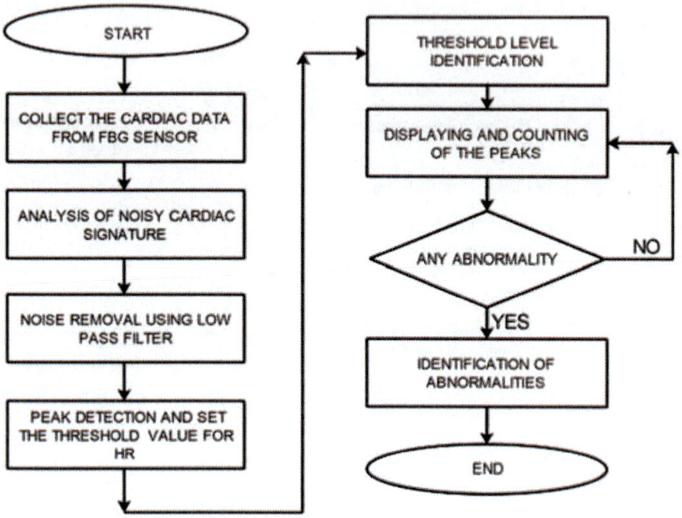

Fig. 2 Flow diagram of the proposed cardiac measurement scheme

The proposed cardiac sensing scheme consists of several measurement steps as shown in Fig. 2. The individual steps of the proposed sensing scheme are collection/acquisition of the cardiac signature, noise analysis, filtering/noise removal scheme, peak detection, thresholding, and condition-based abnormality detection.

4 Results and Analysis

Here, the FBG region is developed due to incident ultraviolet light on silica optical fiber in an experimental laboratory. The grating region is treated as a sensing point on fiber and multiplexed on various points. The designed sensor element simulation is performed using the COMSOL MultiPhysicsSoftware. Various static and dynamic analyzes are performed on the element. The simulation involved with COMSOL Multiphysics Software is

- Evaluation of various numerical methods.
- Application of meshing structure.
- Post-processing of sensor element.

The sensitivity of the sensor is increased due to the encapsulation of various materials on optical fiber, the size of the sensor element, and the periodicity of grating. The experimental research proposed that encapsulation of polydimethylsiloxane increases the sensitivity of the sensor four times more than the bare sensor [25]. Here, the sensor element is developed using a 3D model of COMSOL Multiphysics software, and the dimension of the element is given in Table 2.

Table 2 Dimension of the FBG sensing element

Parameters	Dimension in mm
Length	40
Width	40
Thickness	20

Further finite element analysis (FEA) method is carried out on the sensor element and meshing operation is applied to it. Principal stress analysis is performed to calculate the maximum sensitive region. Simulation of the sensor element is done using the parameters listed in Table 3.

FBG has been encapsulated inside 1 mm polydimethylsiloxane in a sandwich structure using standard adhesive (Anabond 202). FBG is woven into a chest wearable smart belt to capture human cardiac and respiratory parameters. Deposition of PDMS on FBG gives maximum sensitivity to the center region of bonded portion of the element. The final version of the PDMS deposited FBG sensing element is shown in Fig. 3.

The FBG interrogator with a SLED light source is used to capture the reflected wavelength. A software application is developed using the National Instrument LabVIEW platform to estimate the peak wavelength in reflected light. The complete experimental setup is shown in Fig. 4. The fabricated sensor element is tested in the laboratory environment by wearing it on the chest portion of a subject. In recorded raw signal R wave is visualized clearly, but P, Q, S, and T waves are not properly monitored. So, to improve the quality of the recorded raw signal, a digital filter

Table 3 Simulation parameters of the FBG sensing element

Simulation configuration	Parameters
Material	Polydimethylsiloxane (PDMS)
Applied frequency	0.33 Hz
Acts/min	20 acts per minute

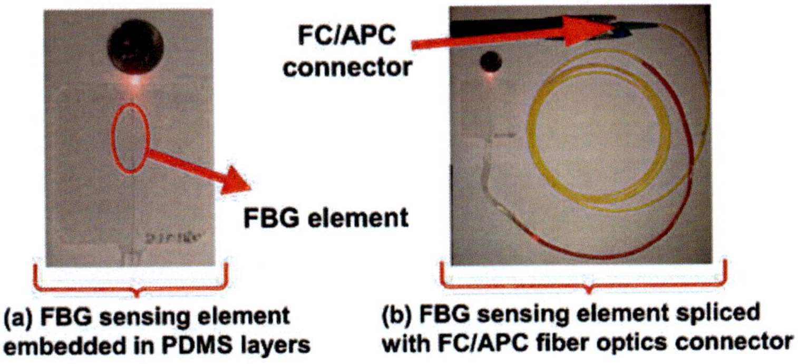

Fig. 3 FBG sensor element deposited inside PDMS layers for signal acquisition

is designed to remove various noise components from the signal. Thereafter these waves can be visualized clearly by filtering the signal with advanced filter signal processing algorithms. Here, a 30 Hz low-pass filter is designed to filter the noisy signal. The raw cardiac signature collected using the FBG sensing element is shown in Fig. 5. Further, the frequency spectrums present inside the noisy cardiac signature are estimated using fast Fourier transform (FFT) technique, and the frequencies present in the noisy signature are shown in Fig. 6.

Further, a low-pass filter is designed to remove the noisy frequency components present inside the recorded cardiac signature. It is visible that there are noisy frequencies above 30 Hz. The magnitude response of the designed low-pass filter is shown in Fig. 7. In a similar context, the noisy cardiac signature is filtered out using a low-pass

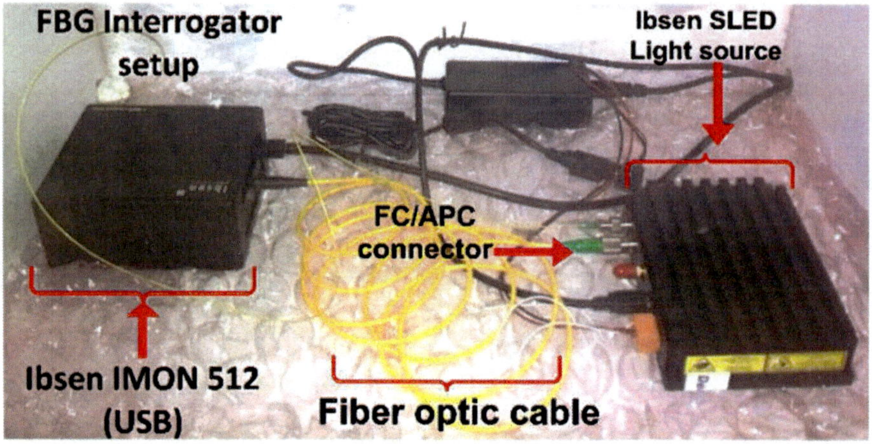

Fig. 4 Deposition of PDMS on Silica Fiber

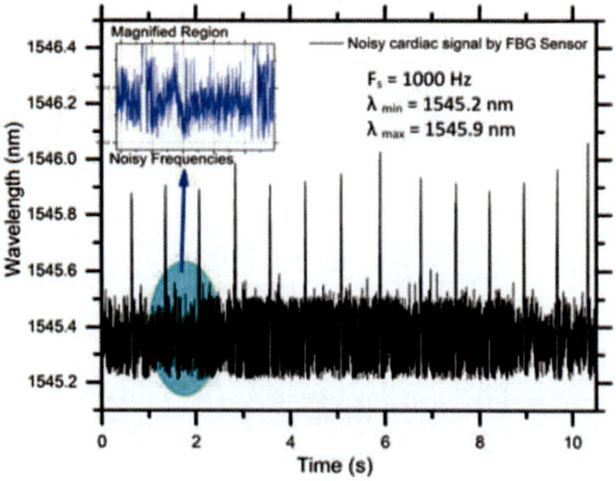

Fig. 5 Noisy cardiac signal

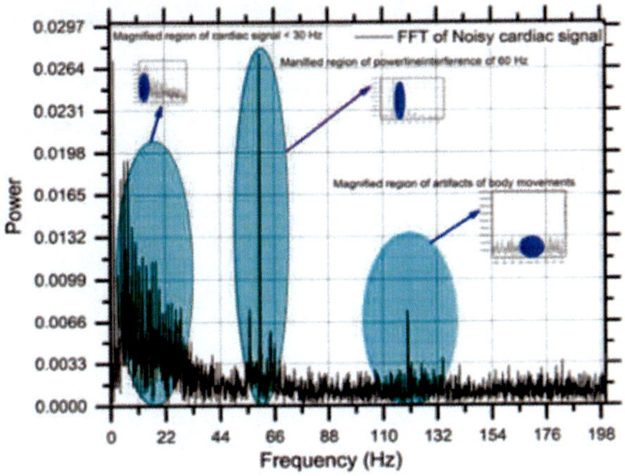

Fig. 6 Frequency spectrum present inside the noisy cardiac signal using FFT

filter, and the FFT of the filtered signature is shown in Fig. 8. It is observed that the filtered signature contains only the cardiac signature, and the filtered signature of the cardiac information is shown in Fig. 9.

The filtered cardiac signature collected from the laboratory-driven FBG sensing element is sufficient to extract useful cardiac information like HR, standard deviation of normal-to-normal (SDNN) interval, NN50, root mean square of successive differences (RMSSD) between normal heartbeats, and body temperature. The complete use case of the proposed non-invasive optical fiber cardiac measurement technique is presented as an architectural diagram in Fig. 10.

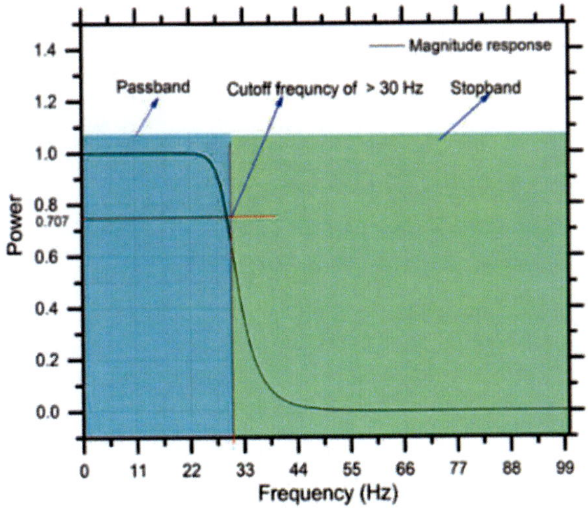

Fig. 7 Magnitude response of the designed low-pass filter

Fig. 8 FFT of filtered cardiac signature

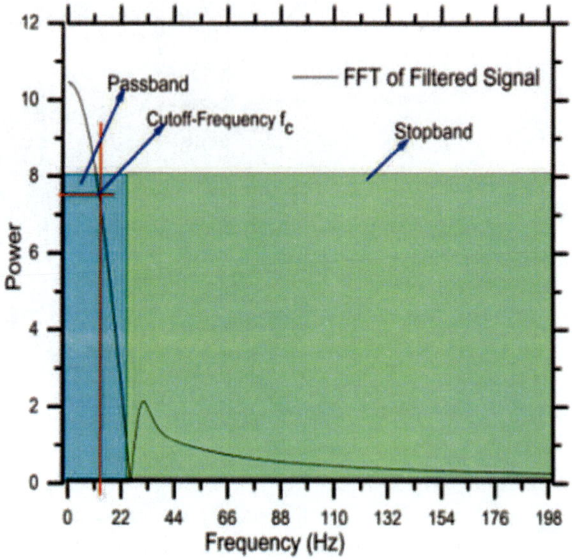

Fig. 9 Filtered signal using 30 Hz low-pass filter

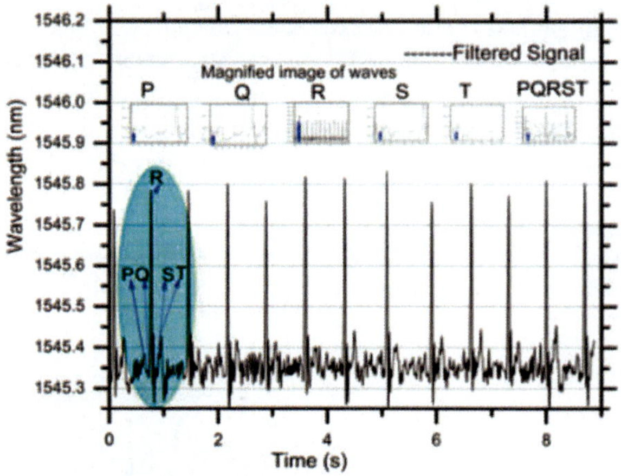

The introduced new non-invasive passive optical element is used for the acquisition of vital sign parameters in a real-life distributed environment. The acquisition of real-time signals is processed to give information about various cardiac and respiratory parameters constantly.

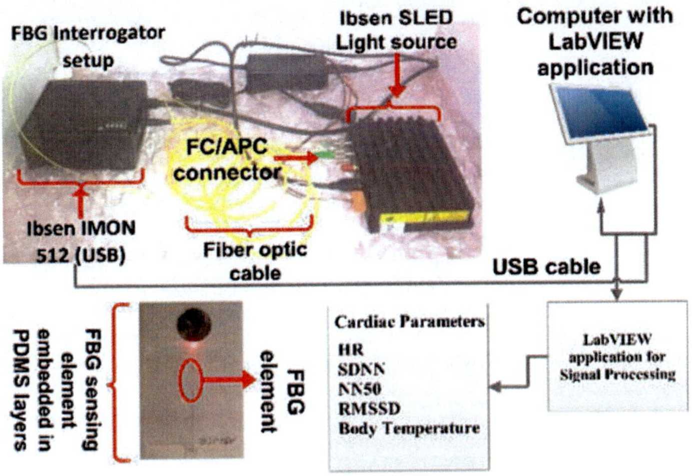

Fig. 10 Architectural diagram of the non-invasive optical fiber cardiac measurement technique

5 Conclusion

FBG sensors are the rugged optical elements used for high-speed applications and are mostly preferred in critical sensing applications. The proposed non-invasive optical fiber cardiac measurement technique is successfully developed in the laboratory, and several experimental tests have been carried out using a PDMS-embedded FBG sensing element. The design considerations of the sensing element are also validated using finite element analysis (FEA), and the fabricated sensor is tested for the acquisition of cardiac signature. Complete structural analysis is performed on the sensor element and deposition of polydimethylsiloxane on the FBG sensor element increases sensitivity. The heart rate of the normal subject is calculated, and a remote monitoring scheme is also proposed in this article.

Acknowledgements The authors thank the Silicon Institute of Technology, Bhubaneswar, to provide continuous support in fabricating the FBG sensor during the research work. The authors also thank the Silicon Institute of Technology, Bhubaneswar to provide license software like the LabVIEW development platform and FBG interrogator to conduct this experiment successfully.

References

1. Jadhav, U. M. Cardio-metabolic disease in India—the upcoming tsunami. Annals of translational medicine, 6(15), 2018
2. Smith I, Mackay J, Fahrid N, Krucheck D (2011) Respiratory rate measurement: a comparison of methods. British Journal of Healthcare Assistants 5(1):18–23

3. Patel S, Park H, Bonato P, Chan L, Rodgers M (2012) A review of wearable sensors and systems with application in rehabilitation. J Neuroeng Rehabil 9(1):1–17
4. Hao J, Jayachandran M, Kng PL, Foo SF, Aung PWA, Cai Z (2010) FBG-based smart bed system for healthcare applications. Frontiers of Optoelectronics in China 3(1):78–83
5. Zanella A, Bui N, Castellani A, Vangelista L, Zorzi M (2014) Internet of things for smart cities. IEEE Internet Things J 1(1):22–32
6. Al-Ali AR, Zualkernan IA, Rashid M, Gupta R, AliKarar M (2017) A smart home energy management system using IoT and big data analytics approach. IEEE Trans Consum Electron 63(4):426–434
7. Mois G, Folea S, Sanislav T (2017) Analysis of three IoT-based wireless sensors for environmental monitoring. IEEE Trans Instrum Meas 66(8):2056–2064
8. Ayaz M, Ammad-Uddin M, Sharif Z, Mansour A, Aggoune EHM (2019) Internet-of-Things (IoT)-based smart agriculture: Toward making the fields talk. IEEE Access 7:129551–129583
9. Islam MM, Rahaman A, Islam MR (2020) Development of smart healthcare monitoring system in IoT environment. SN computer science 1:1–11
10. Nedoma, J., Fajkus, M., Martinek, R., & Nazeran, H., Vital sign monitoring and cardiac triggering at 1.5 teslas: a practical solution by an MR-ballistocardiography fiber-optic sensor. Sensors, 19(3), 470, 2019.
11. Dziuda Ł, Krej M, Skibniewski FW (2013) Fiber Bragg grating strain sensor incorporated to monitor patient vital signs during MRI. IEEE Sens J 13(12):4986–4991
12. Fajkus M, Nedoma J, Martinek R, Vasinek V, Nazeran H, Siska P (2017) (2017), A non-invasive multichannel hybrid fiber-optic sensor system for vital sign monitoring. Sensors 17(1):111
13. Zhu, Y., Fook, V. F. S., Jianzhong, E. H., Maniyeri, J., Guan, C., Zhang, H., ... & Biswas, J., Heart rate estimation from FBG sensors using cepstrum analysis and sensor fusion, In 2014 36th annual international conference of the IEEE engineering in medicine and biology society (pp. 5365–5368). IEEE, 2014.
14. Gurkan, D., Starodubov, D., & Yuan, X., Monitoring of the heartbeat sounds using an optical fiber Bragg grating sensor, In SENSORS, 2005 IEEE (pp. 4-pp). IEEE, 2005.
15. Dziuda L, Skibniewski FW, Krej M, Lewandowski J (2012) Monitoring respiration and cardiac activity using fiber Bragg grating-based sensor. IEEE Trans Biomed Eng 59(7):1934–1942
16. Narbonneau, F., D'angelo, L. T., Witt, J., Paquet, B., Kinet, D., Kreber, K., & Logier, R., FBG-based smart textiles for continuous monitoring of respiratory movements for healthcare applications, In The 12th IEEE International Conference on e-Health Networking, Applications and Services (pp. 277–282). IEEE, (2010, July).
17. Wehrle G, Nohama P, Kalinowski HJ, Torres PI, Valente LCG (2001) A fiber optic Bragg grating strain sensor for monitoring ventilatory movements. Meas Sci Technol 12(7):805
18. Silva AF, Carmo JP, Mendes PM, Correia JH (2011) Simultaneous cardiac and respiratory frequency measurement based on a single fiber Bragg grating sensor. Meas Sci Technol 22(7):075801
19. Elsarnagawy T (2015) A simultaneous and validated wearable FBG heartbeat and respiration rate monitoring system. Sens Lett 13(1):48–51
20. Fresvig T, Ludvigsen P, Steen H, Reikerås O (2008) Fiber optic Bragg grating sensors: an alternative method to strain gauges for measuring deformation in bone. Med Eng Phys 30(1):104–108
21. Tosi D, Olivero M, Perrone G (2008) Low-cost fiber Bragg grating vibroacoustic sensor for voice and heartbeat detection. Appl Opt 47(28):5123–5129
22. Lau, D., Chen, Z., Teo, J. T., Ng, S. H., Rumpel, H., Lian, Y., ... & Kei, P. L., Intensity-modulated microbend fiber optic sensor for respiratory monitoring and gating during MRI, IEEE Transactions on Biomedical Engineering, 60(9), 2655 -2662, 2013.
23. Chen, Z., Teo, J. T., Ng, S. H., Yang, X., Zhou, B., Zhang, Y., ... & Thong, M., Monitoring respiration and cardiac activity during sleep using microbend fiber sensor: A clinical study and new algorithm, In 2014 36th annual international conference of the IEEE engineering in medicine and biology society (pp. 5377–5380). IEEE, 2014.

24. Zhu, Y., Zhang, H., Jayachandran, M., Ng, A. K., Biswas, J., & Chen, Z. ,Ballistocardiography with fiber optic sensor in headrest position: A feasibility study and a new processing algorithm, In 2013 35th annual international conference of the IEEE engineering in medicine and biology society (EMBC) (pp. 5203–5206). IEEE, 2013.
25. Nedoma J, Fajkus M, Bednarek L, Frnda J, Zavadil J, Vasinek V (2016) Encapsulation of FBG sensor into the PDMS and its effect on spectral and temperature characteristics. Advances in Electrical and Electronic Engineering 14(4):460–466

Analysis of Stock Price-Prediction Models

Yash Mehta, Parth Singh, Dipak Ramoliya, Parth Goel, and Amit Ganatra

Abstract Everything has two sides in the world just like coins where one is good and the other is not in our favor where stock market also has two sides, on one hand, being the lucrative source of income and the largest investment option with the highest return on investment while the second side reveals drawbacks and losses. In the current expanding market, there are a lot of people facing failure. Here, major reasons for market failure are lack of financial understanding and influenced trading. Scholars have developed a number of models in past few years for time series analysis; every model predicts at a particular level according to their algorithms. Here, we will analyze available models for time series analysis and provide a thorough analysis for each model allowing investors to select the best model for their next trade analysis.

Keywords LSTM · CNN · ARIMA · SMA · EMA · RNN · GRU · ANN first keyword · Second keyword · Third keyword

1 Introduction

In the financially growing world where people from different domains are taking interest in financial awareness and have started investing in different domains such

Y. Mehta (✉) · P. Singh · D. Ramoliya · P. Goel · A. Ganatra
Department of Computer Science and Engineering, Faculty of Technology and Engineering (FTE), Devang Patel Institute of Advance Technology and Research (DEPSTAR), Charotar University of Science and Technology (CHARUSAT), Anand, India
e-mail: developer.0208@gmail.com

P. Singh
e-mail: parthsingh123.ps@gmail.com

D. Ramoliya
e-mail: dipakramoliya.ce@charusat.ac.in

P. Goel
e-mail: parthgoel.ce@charusat.ac.in

A. Ganatra
e-mail: amitganatra.ce@charusat.ac.in

S. Das et al. (eds.), *Advances in Data-driven Computing and Intelligent Systems*, Lecture Notes in Networks and Systems 653, https://doi.org/10.1007/978-981-99-0981-0_48

as stock markets, real estate, as well as different government schemes; the stock market plays a very important role in the investment sector where if someone wants to examine the performance of stock market as well as any securities, one needs to assess indices first. They are calculated by adding up stock price variations in the exchange market for all firms or a specific class of existing companies. As a result, investors will need to study stock price index prediction models in order to turn the securities market into a successful region, due to the high competition among players in financial markets. On the one hand, this attracts financiers. Knowledge being updated in order to make effective portfolio selection judgments, and on the other side, it leads to the provision of foreign investment options for investors. Financial market prediction, however, is a difficult undertaking due to the noisy, nonstationary, and irregular nature of financial time series. In this literature, an effort has been made to analyze different models of time series analysis for stocks and investments.

Here analysis will flow in a way where first basic models where averages of all values will decide prediction for the future close in which basic to advance starting with simple moving average (SMA) then exponential moving average (EMA) [1] and at the end complex ARIMA model were analyzed. After completion of basic model analysis, analysis based on neural network models will start where initially basic feed-forward network artificial neural network (ANN) will be analyzed then for recurrent neural network, three levels of RNN will be analyzed where first we have basic RNN then gated recurrent unit and at last long short-term memory (LSTM) will be analyzed, and at last most complex model convolutional neural network (CNN) will be analyzed. In the end when all models will be analyzed a detailed output will be there to help users to decide which model will be useful in which condition to trade better.

2 Related Work

For the stock price prediction model analysis, following research papers' results are taken into consideration where the research paper about EMA VERSUS SMA by Svetlana Šaranda represents the results of using EMA and SMA on the S&P 500 and OMX BALTIC BENCHMARK [2]. In the case of ARIMA model, the paper by Jung-Hua Wang where the Taiwan stock exchange weighted stock index, abbreviated as TSEWSI [2] was used which represented pretty good accuracy by ARIMA. Moving towards the neural network paper on ANN by Mruga Gurjar "STOCK MARKET PREDICTION USING ANN" where the dataset used was a basic NSE dataset for price prediction [3]. At last, model motivations and analysis for advanced neural networks (RNN, LSTM, and CNN) came from the paper "STOCK PRICE PREDICTION USING LSTM, RNN AND CNN-SLIDING WINDOW MODEL" by Sreelekshmy Selvin where datasets for Infosys, TCS, and Cipla stocks were used, and for every dataset CNN showed best performance expect for TCS in which LSTM worked better [3].

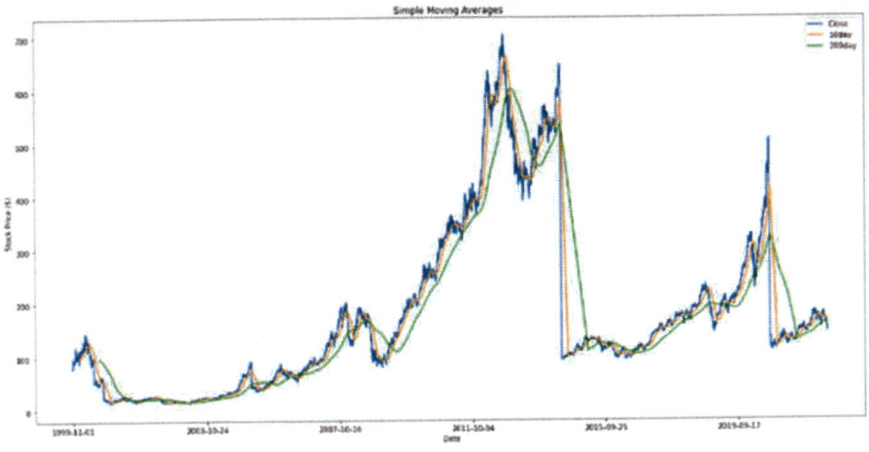

Fig. 1 Simple moving average forecasting on AAPL data shows the prediction for AAPL time series while predicting with SMA model

3 Models and Algorithms

Starting with basic models for the analysis paper will move forward for advanced neural network models.

3.1 Simple Moving Average (SMA)

An SMA is a simple computation of a stock's average price over a defined number of days (typically using the close price).

Algorithm

(1) Summation of all data points starting with $i = 1$ to $i = n$ and divide that by n to find the average value for a certain time period.
(2) SMA = $(P1 + P2 + P3 \cdots + Pn/N)$, Here Pn—Value at a certain time & N—data point range (Fig. 1).

3.2 Exponential Moving Average (EMA)

As an upgradation for the simple moving average, exponential moving average EMA doesn't allocate equal weight to every time-series data point. EMA assigns much weight to recent data points, like the last 50 days' prices data points. The weighting factor's magnitude is determined by the number of time periods. EMA has an edge

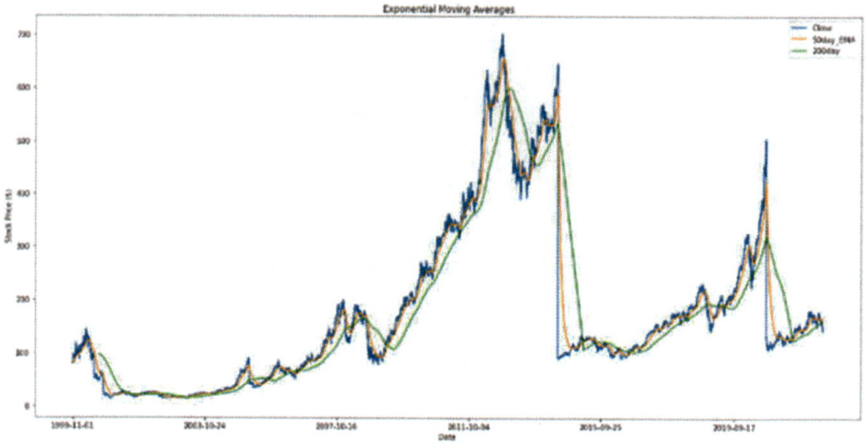

Fig. 2 Exponential moving average forecasting on AAPL data shows price prediction model output for AAPL data using EMA model

over SMA as EMA is more approachable to price change, which helps in short-term trading.

Algorithm

(1) Equation: EMA $= Pt * k +$ EMA $t - 1 * (1 - K)$.

$Pt =$ data value at timestamp t.
EMA $t - 1 =$ last EMA value (at time t − 1).
$N =$ data range.
Here weighting average factor $k = 2(\text{N} + 1)$ (Fig. 2).

3.3 AutoRegressive Integrated Moving Average (ARIMA)

Moving to the final upgradation for moving average-based algorithms autoregressive integrated moving average is the most used prediction algorithm for historical stock data analysis. Here, ARIMA algorithm can capture a suite of various standard data structures. To deep dive into the algorithm, moving toward the basics of the algorithm in which Autoregressive (AR) suggest that the model gets information using the connection between normal observations and lagged observations which may be called time lag. Where Integrated (I) suggest that the model differentiate basic results to develop the steady time series and moving average (MA) expresses the algorithm exploits the relationship between the residual error and the observation.

Model parameters:

Input parameters are:

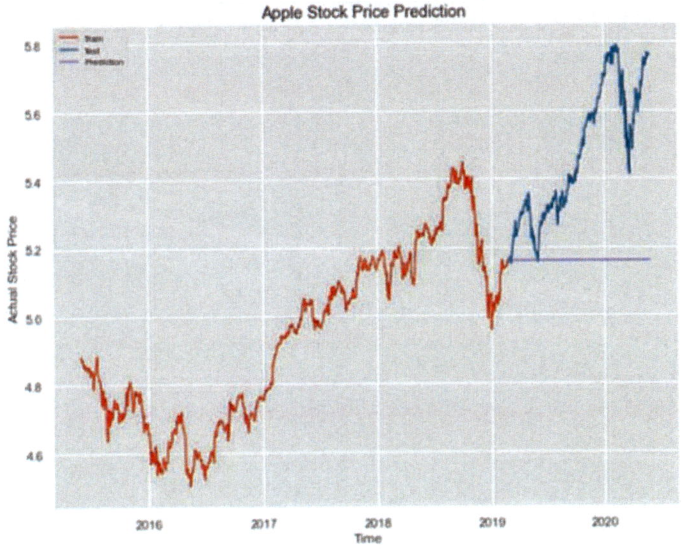

Fig. 3 Using ARIMA model AAPL price prediction provide output in the following manner

p–lag rate
d–differentiation value
q–moving average window size.

Using ARIMA model AAPL price prediction provide output in the following manner shown in Fig. 3.

3.4 Artificial Neural Network (ANN)

ANN is a basic feed-forward network. In the stock price prediction model, backpropagation is used, whereas the training phase model uses forward propagation and due to the forward pass model will get output values at the output node [4]. As ANN is a basic feed-forward network like other FFN, output is generated using the activation function.

The calculation for the activation function works in the way where

$$\text{Total Input} = n1 * w1 + n2 * w2 + \cdots + nm * wm + b$$

In which, $n1$, $n2$, $n3$, … nm reflects Input neurons & $w1$, $w2$, $w3$, …wm reflects input weights associated with respective input neurons at last b reflects weight bias.

For output activation function $= 1/(1 + e^{(-\text{Total Input})})$, total input is total input to the neuron.

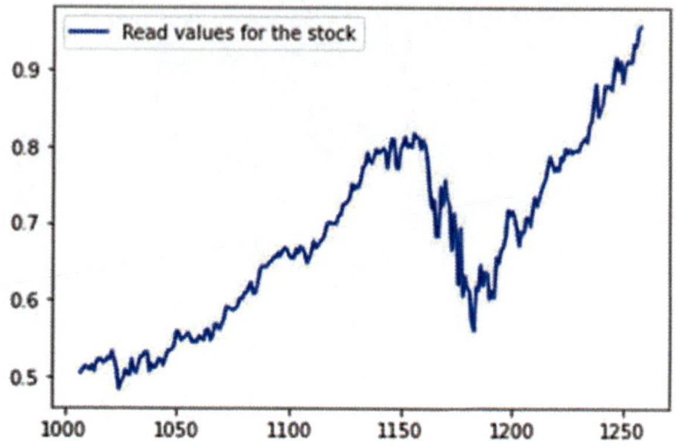

Fig. 4 Represents original time series value of AAPL stock

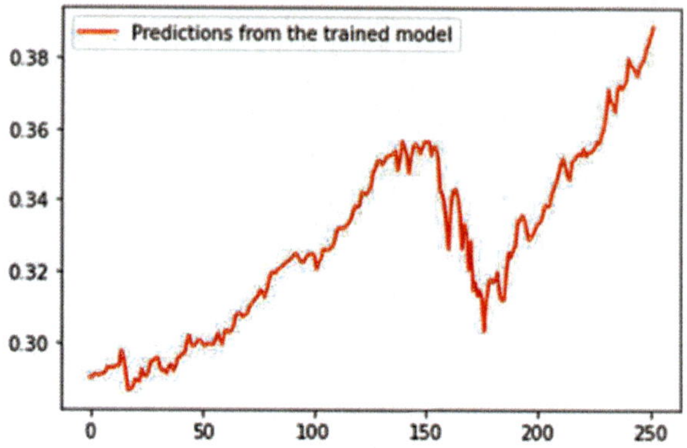

Fig. 5 Output graph for AAPL time series analysis using the ANN model

A regularly used method of training an ANN is the backpropagation of errors in which optimization is obtained using gradient descent. Two phases of cycle propagation and weight updation work repeatedly in the algorithm. Image below shows original time series values for AAPL data (Figs. 4 and 5).

3.5 Recurrent Neural Network (RNN)

RNN are classes of neural network which follows methodologies where previous outputs are used as input in hidden layers which represents the concept of memory in RNN. Analysis for the recurrent neural network has three levels.

(1) Basic Recurrent Neural Network

The basic RNN has at least one feedback relationship which causes a loop flow round in the activation function, so the network process mortal processing and learn sequences. Considering *Xt* as the input and *Yt* as output at timestamp *t* afterward

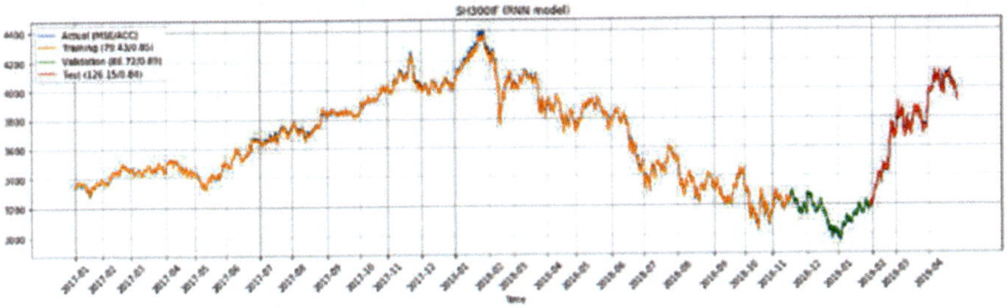

Fig. 6 Training and testing output for time series data using RNN

all model does is develop a feedback connection starting with a hidden layer to self to use that retrieved information at timestamp $t - 1$ which implies a delay of the single time unit. At last, it can be stated that the feedback loop helps information to be carried from the first phase to the next which represents memory in the network.

$$ht = f(\mathrm{Wh}^T * ht - 1 + wx^T Xt + bh)$$

$$Yt = \mathrm{softmax}(W0^T ht + b0)$$

$$F = \mathrm{Sigmoid}, \tan h, \mathrm{ReLu}$$

where,

ht = current state
$ht - 1$ = last state
Xt = Input at timestamp t
Wh = Weight at the recurrent neuron
Wx = Weight at input neuron
Yt = Output at timestamp t (Fig. 6).

(2) Gated Recurrent Unit

The GRU was developed to resolve the problem of vanishing gradient which was one of the drawbacks of a basic RNN. Now, as a solution, GRU uses the update and reset gate method, in which output is dependent on the two vectors. The specialty of the GRU is that they can be developed to store information for long period of time without forgetting information that is not relevant to the prediction [5].

- Update Gate

Update gate for time stamp t will be calculated by:

When $x\,t$ is plugged into the network unit, its weight W is multiplied (z). The same is true for h ($t - 1$), which contains data for the past $t - 1$ units and is multiplied by its own weight U. (z). Both results are combined together, and the result is squashed between 0 and 1 using a sigmoid activation function [2].

- Reset Gate

To decide the value of information to be forgotten reset gate is used. We plug in the blue line *h* (*t* − 1) and the purple line *x t*, multiply them with their weights, add the results, and apply the sigmoid function.

- Current memory value

Multiplication of input *x_t* with a weight *W* and *h_*(*t* − 1) with a weight *U*.

Hadamard (element-by-element) product between *r t* and *Uh_*(*t* − 1) will determine which time steps should be removed. Let's imagine we have a sentiment analysis situation where we need to figure out how someone feels about a book based on a review he published. "This is a fantasy novel that demonstrates…" begins the text, which finishes with "I didn't completely appreciate the book since I believe it captures too much information." [6] Only the last section of the review is required to evaluate the overall level of satisfaction with the book. As the neural network gets closer to the end of the text, it will learn to assign r t vectors near to 0, washing out the previous data and focusing exclusively on the present [7].

- Add results of steps 1 & 2

Use nonlinear activation function tan*h*.

- Final memory at timestamp *t*

For the final value, network will calculate *h*(*t*−*a*) vector containing a value for the last unit [8] and provide information to the network as the last step for that update gate is required which will determine what to get from the current memory content—*h*'*t* and what model got from last state—*h*()*d* what (Fig. 7).

(3) Long short-term memory (LSTM)

An upgradation in the available recurrent neural network (RNN) long short-term memory algorithm a sequential network is used for time series analysis. In the algorithm, initial cell states are referred to as short-term memory whereas accumulated eight are called long-term memory [5]. The main problem in RNN (vanishing gradient

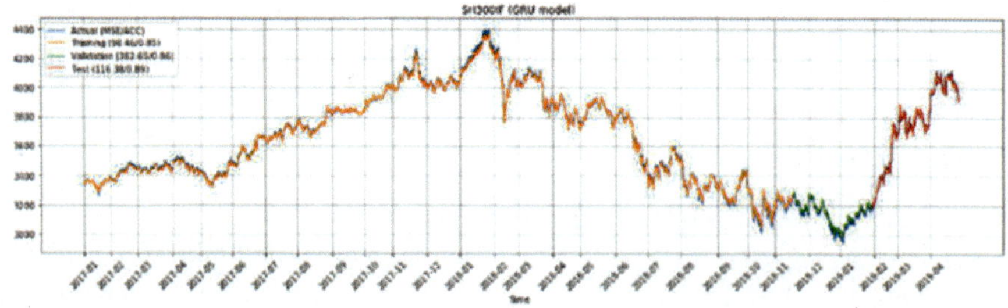

Fig. 7 Training and testing for time series analysis using GRU show the following output

problem) can be solved by changing the middle layers with LSTM blocks this is the most important distinction in LSTM as per Hochreiter and Schmidhuber, 1997 [1]. As previously, RNNs were not capable to learn long-term memory; LSTM has the capability to do so which is a plus point over RNN. The network weight must be changed to determine the next close price. Which demands keeping the initial time-series data. As RNNs have the limitation that they can learn only a limited number of short-term relations, as well as they, fail to learn long-term time series like 2000 close data. On the other hand, as per Schmidhuber LSTM can go for these long-term affiliations. In the LSTM formation, memory blocks are a set of cyclic subnetworks where each block consists of one or more autoregressive memory cells that also have three multiple units of 'output, input, and forgetting' [8]. which works similarly for uninterrupted writing, reading, as well as cell regulation. LSTM blocks also contain encoder-decoder LSTMs, stacked LSTMs, bidirectional LSTMs, generative LSTM as per Brownlee and convolutional neural network LSTMs.

When a computational problem can be solved using a certain algorithm, the time complexity as well space complexity for that solution is the next priority factor to be examined after solvability [add]. When applied to large circumstances, the algorithm's behavior is extremely remarkable [5].

Required memory for an LSTM cell (when f d is the length of the input vector whereas the number of neurons in the hidden layer is $d\ h$) is 0. Because the result of the next cell $(t + 1)$ will be replaced in the same memory as the old values, the required memory in long short-term memory is $O(d\ h)$ [8]. Because computational models have an infinite amount of memory. Space complexity is usually not an issue. The time complexity of an algorithm is what is most interesting applications, so when we speak about an algorithm's difficulty, we always refer to its time complexity.

There are three vector and eight matrix multiplications in an LSTM unit. The entire time complexity of an LSTM unit is $O(d\ h)$ [8]. As each vector multiplication is $O(d)$ or $O(h)$ and matrix multiplication is $O(d)$ or $O(h)$.

In the time series analysis model learning, time is not important as the model will finish learning before the system runs, so there will be a lot of time to learn. The accuracy of learning and the reduction of errors are the most essential factors (Fig. 8).

3.6 *Convolutional Neural Network (CNN)*

The primary purpose of the convolutional neural network is image recognition, but the advantages of CNN over the classic neural networks can be used to predict future stock prices much better. The working structure of CNN:

CNN consists of four layers

1. Convolution layer
2. ReLu layer
3. Pooling
4. Fully Connected.

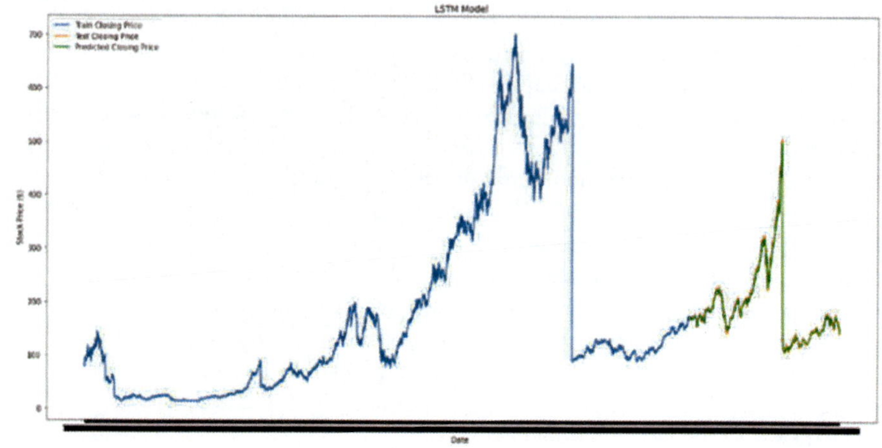

Fig. 8 Prediction graph for AAPL time series analysis using LSTM price prediction model

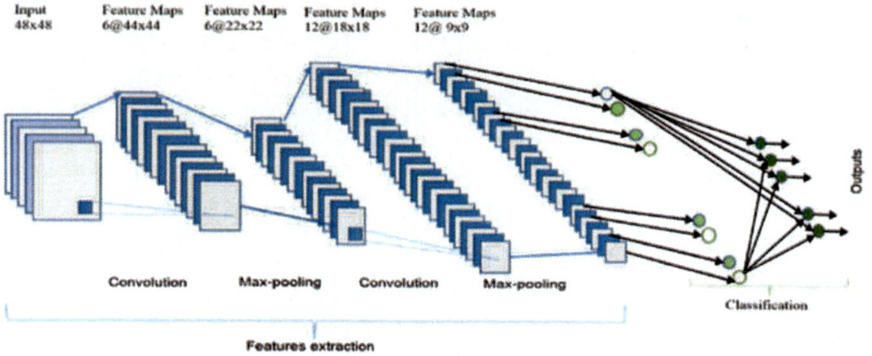

Fig. 9 CNN working methodology consisting four layers of cnn

In LSTM, the basic concept is timestamps used as basic data, whereas in the case of CNN, data is represented as a matrix [9]. The two-dimensional CNN algorithm accepts an input of $N * m * 1$, where.

N–number of time stamps

m–the number of features in each time (Fig. 9).

4 Analysis

During the thorough analysis of all the eight models' results, the execution time and root mean square error (RMSE) for each model are taken into consideration, and for that, a similar dataset is used for analysis. Here are dataset details.

For initial prediction model generation, daily data of Nifty 50(^NSEI) from January 2012 to March 2021 were used in this analysis. NIFTY 50 is a diverse stock market index containing the top 50 leading diversified companies in India, representing different market sectors such as pharma, IT, engineering, and financial services. The weighting of these 50 equities is determined by their free-float market capitalization [2]. It is used to benchmark fund portfolios, index-based contracts, and index funds, among other things. National stock exchange Indices Limited formally known as Indian Index Services & Product Limited manages Nifty 50 [3]. NSE Indices is a specialist company in India that focuses on indexes as a key product [10–12].

The data for this study was gathered from the Yahoo Finance website. Python 3.10 was used to code time series analysis LSTM model and other machine learning approaches, as well as to compare the results and choose the best algorithm. Mean square error (MSE), MAE, and root mean square error (RMSE) are some of the error evaluation criteria. Images given below show data references and graph for AAPL data. Figures 10 and 11 show input data and its graph.

Analysis details.

This Table 1 represents execution time and RMSE for all models where observations tell that CNN shows the best accuracy having minimum RMSE.

	High	Low	Open	Close	Volume
Date					
2019-01-02	10895.349609	10735.049805	10868.849609	10792.500000	309700.0
2019-01-03	10814.049805	10661.250000	10796.799805	10672.250000	286200.0
2019-01-04	10741.049805	10628.650391	10699.700195	10727.349609	296600.0
2019-01-07	10835.950195	10750.150391	10804.849609	10771.799805	269400.0
2019-01-08	10818.450195	10733.250000	10786.250000	10802.150391	277700.0
...	...	...	...	...	...
2022-03-08	16028.750000	15671.450195	15747.750000	16013.450195	543600.0
2022-03-09	16418.050781	15990.000000	16078.000000	16345.349609	462200.0
2022-03-10	16757.300781	16447.900391	16757.099609	16594.900391	486400.0
2022-03-11	16694.400391	16470.900391	16528.800781	16630.449219	343700.0
2022-03-14	16868.599609	16606.500000	16633.699219	16862.699219	0.0

Fig. 10 Dataset values and attributes are as per shown in the figure

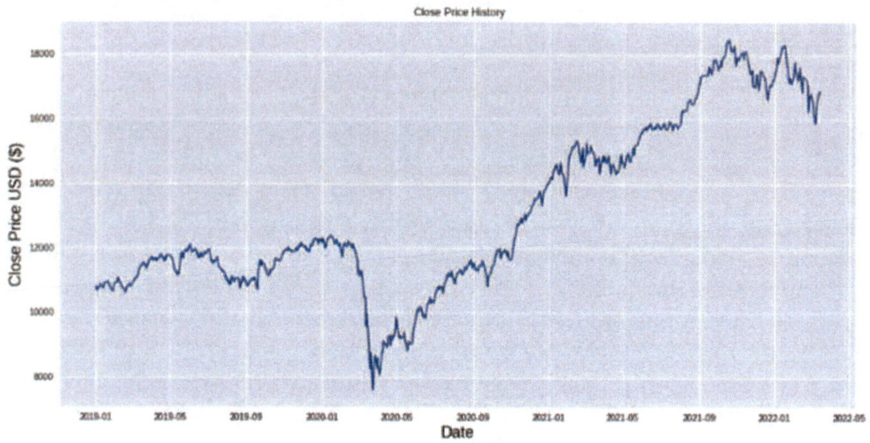

Fig. 11 Close data chart for AAPL stock starting from 2008

Table 1 Analysis table containing execution time and RMSE

Model	Execution time	RMSE
LSTM	4 M 42 s	1.274
EMA	7 s	36.0523
SMA	6 s	43.036
CNN	22 min	0.7056
ARIMA	1 min 22 s	1.4727
RNN	3 min	11.23
GRU	2 min	10.79
ANN	3 min 40 s	14.36

5 Conclusion

Stock price prediction models developed over decades are all best in their own way, whereas basic models EMA, SMA, and ARIMA in which complex sequential networks of algorithms are not there executes in minimum time, but these models show big RMSE which can cause big losses when talking about the task which contains huge financial risk. So moving toward a neural network, basic ANN, RNN, and GRU have similar execution times and about equal RMSE. LSTM and CNN drags toward the conclusion that when the user's basic requirement is accuracy, CNN is a better algorithm than all others, but when taken into consideration that if the user wants to create a real-time application for stock price prediction, LSTM is better to take in consideration as it shows less execution time with less RMSE which helps in real-time forecasting.

References

1. Aydin AD, Cavdar SC (2015) Comparison of prediction performances of artificial neural network (ANN) and vector autoregressive (VAR) models by using the macroeconomic variables of gold prices
2. Pawar CS et al (2021) Use of machine learning services in cloud. In: Computer networks, Big Data and IoT. Springer, Singapore, pp 43–52.
3. Hiteshkumar G, Gour P, Steeve N, Parmar T, Singh P (2021) Cleanliness automation: YOLOv3. In: 2021 6th International conference for convergence in technology (I2CT), 2021, pp 1–6. https://doi.org/10.1109/I2CT51068.2021.9418056
4. Atsalakis GS, Valavanis KP (2009) Surveying stock market forecasting techniques—part II: soft computing methods. Expert Syst Appl 36(3):5932–5941
5. Intelligent Systems and Applications. Springer Science and Business Media LLC, 2020
6. Anon. (1998) Glossary of terms. Mach Learn 30(2–3):271–274
7. Bao W, Yue J, Rao Y (2017) A deep learning framework for financial time series using stacked autoencoders and long-short term memory. PLoS ONE 12(7):e0180944
8. Nikou M, Mansourfar G, Bagherzadeh J (2019) Stock price prediction using DEEP learning algorithm and its comparison with machine learning algorithms. Intell Syst Account, Finance, Manag
9. Borsa Istanbul (BIST) 100 index and US dollar–Turkish lira (USD/TRY) exchange rates. Procedia Econ Finance 30:3–14
10. Chan M-C, Wong C-C, Lam C-C (2000) Financial time series forecasting by neural network using conjugate gradient learning algorithm and multiple linear regression weight initialization. Comput Econ Finance 61
11. Jia H (2016) Investigation into the effectiveness of long short term memory networks for stock price prediction. arXiv preprint arXiv:1603.07893
12. Batres-Estrada G (2015) Deep learning for multivariate financial time series. ser. Technical Report, Stockholm

Multimodal Recommendation Engine for Advertising Using Object Detection and Natural Language Processing

S. Rajarajeswari, Manas P. Shankar, D. S. Kaustubha, Kaushik Kampli, and Manish Manohar

Abstract In today's world, there is an explosion in online advertising due to high levels of activity of users online. With this comes the intrinsic issue of promoting a certain product or service to the right set of users without breaching their privacy, or by limiting invasion of privacy. We plan to address this issue by employing a multimodal recommender system composed of two fundamental components. The first component detects objects and their frequencies from the video, while the second component extracts text from the audio of the same video being watched by the user, using speech recognition and Natural Language Processing. The findings of these two components are combined via a recommender engine to determine the most appropriate advertisement to recommend to the user. By design, the user's personal data are not used during the entire process. In this way, the proposed model recommends advertisements in a novel way, aiming to improve the relevancy of the advertisement relative to the video being viewed. It further aims to address the important issue of netizen privacy while browsing their favorite video streaming sites.

Keywords Object detection · Deep learning · Recommendation engine · Natural language processing

1 Introduction

As we enter the twenty-first century, we find that younger and middle-aged demographics are increasingly shifting away from traditional forms of entertainment consumption. From the rise of YouTube as a popular site for content creators and users to on-demand streaming services like Netflix, Amazon Prime, and Hotstar, it is safe to conclude that traditional advertising strategies have lagged. Consumers are no longer compelled to endure 5–10-min-long regular ad interruptions between TV

S. Rajarajeswari (✉) · M. P. Shankar · D. S. Kaustubha · K. Kampli · M. Manohar
Department of Computer Science and Engineering, Ramaiah Institute of Technology MSR Nagar, Bengaluru, India
e-mail: raji@msrit.edu

S. Das et al. (eds.), *Advances in Data-driven Computing and Intelligent Systems*, Lecture Notes in Networks and Systems 653, https://doi.org/10.1007/978-981-99-0981-0_49

shows and movies. Advertising on YouTube, for example, lasts no more than 20 s for unskippable ads. Ads longer than a minute can be avoided after 5 s. Small- and medium-sized businesses have doubled the number of times they have advertised on YouTube in the past two years. Streaming services deliver recommender-based advertisements using the watch history on their platforms. Recommender systems have proven to be quite effective as a means of targeted advertising of the most recent TV episodes and movies on such platforms. As a result, targeted advertising is the name of the game, and the effectiveness of product sales is frequently tied to product marketing.

A multimodal recommendation engine for advertising is proposed in this study, employing object detection and Natural Language Processing. The central concept is that of a recommender system that analyzes the objects and audio in a video. It then recommends the most relevant ad to the customer based on the frequencies of the objects and words detected in the video, as well as other parameters in the context-defined business model. The paper is structured as follows. Section 2 discusses current advances in the fields of object detection, speech recognition, and recommender systems. The proposed methodology is discussed in Sect. 3, followed by the experimental setup and findings in Sect. 4. In Sect. 5, we summarize the paper's main idea, uniqueness, and limits, followed by the references.

2 Related Work

The majority of object detection research is in the disciplines of the face and pedestrian identification and also used in fields like image retrieval and video surveillance. Speech detection is a subfield of NLP that has seen substantial study in areas such as development of end-to-end speech recognition pipelines, speech augmentation. Meanwhile, much of the research into recommender systems has focused on collaborative filtering and content-based filtering.

Kang et al. [1] propose a Tubelet Proposal Network (TPN), as well as a Long Short-Term Memory (LSTM) sub-network for assessing object confidences using temporal information. To improve object recognition speed in surveillance video cameras, the video quality can be lowered such that meaningful sections in frames are maintained. Galteri et al. [2] suggest a method for video encoding for the same. Tung et al. [3] offer a fresh effort to assess consistency of YOLO for large-scale applications, using photos from a network of public cameras. Zhu et al. [4] attempt to improve video object detection performance with two novel methods: 'Sparse Feature Propagation' and 'Dense Feature Aggregation'. Ramesh et al. [5] propose an object detection and event-based clothing recommendation system. The type of event is first identified, and using the nearest neighbor's strategy, the most frequently used clothing is later recommended. Wang et al. [6] discuss Salient Object Detection (SOD), classifying existing deep SOD models based on network architecture, level of supervision, learning paradigms, etc., as well as evaluate their performance. Bahar et al. [7] specify the use of an auxiliary connectionist temporal classification

(CTC) loss for improved convergence. Jiang et al. [8] propose developing an app with an intelligent search and recommendation system utilizing speech recognition. The task of normalizing Vietnamese transcribed texts in Speech to Text systems is presented by Tran et al. [9]. Carlini et al. [10] discuss the use of targeted adversarial instances on automatic speech recognition that corresponds to another audio waveform. Bansal et al. [11] provide a solution to the problem of speech-to-text translation in low-resource circumstances where neither automated speech recognition (ASR) nor machine translation (MT) is available. FAIRSEQ S2T, developed by Tang et al. [12], is a modeling extension for ST applications such as end-to-end voice recognition and translation. Dey et al. [13] conduct a survey to detect how user's ad preferences are altered when considering five personality traits. Li et al. [14] created a context-aware advertising system that mixes the user's static personal interests with dynamic news from friends. Zhou et al. [15] propose a similarity network based on ad distribution, as well as a deep learning model framework for predicting ad click through rates. Choi et al. [16] seek machine learning techniques applicable in improving targeted web advertising. Liao et al. [17] discuss the exploitation of social media users' personal information for targeted advertising and create an optimal privacy policy which decides how much data can be shared. Smith et al. [18] propose 'EarSketch,' a sound recommendation system for an online learning environment that engages learners in writing code to generate music.

3 Proposed Methodology

In this section, we describe the recommender engine's approach as well as the overall system design. Figure 1 displays the overall system design, with data flow between various components. The fundamental motivation for developing the methodology was to make the recommendation engine's decision-making genuinely multimodal. We intended to extensively analyze any given video for objects and audio and extract those components as completely as possible. This would provide all the information required to calculate the recommendation scores and apply a weighted average technique for combining those scores.

To discuss the system's methodology, as seen in Fig. 1, the first phase entails uploading a video to the website, similar to uploading a video to any major video sharing network. The uploaded video is then doubly transferred in the following stage, to the YOLO model (for object detection) and the wav2vec model (for speech recognition). Sects. 3.2 and 3.3 describe the YOLO and wav2vec methodologies, respectively. We next obtain dictionaries of detected words and their frequencies (from wav2vec) and of detected objects and their frequencies (from YOLO). This is fed into the recommender engine in the following stage. Three recommendation scores are calculated by the recommender engine. The first score r1 is the score for each detected object, based on their detected frequencies. The second score r2 is the score for each detected word, again based on their detected frequencies. In the following stage, r1 and r2 are combined using a weighted average technique to get final_r,

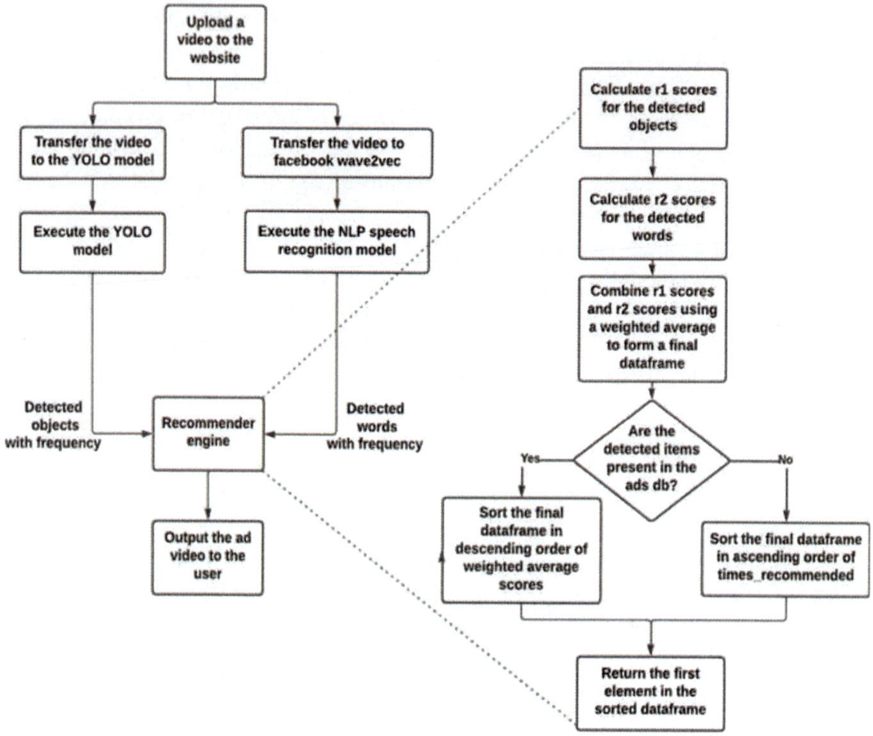

Fig. 1 Overall flow of data for the multimodal method of recommendation

forming the final output data frame. A decision block is used in this stage to compare the detected items in the result to the actual items in the ads database. If the detected items are found in the ads database, the output data frame is sorted in descending order of final_r scores. If the detected items are not found in the ads database, the output data frame is sorted in ascending order of parameter "times_recommended", placing the least recommended ad at the top of the data frame. In the final stage, the first element in the sorted data frame is returned, containing the ad. Section 3.4 describes the algorithms used to calculate r1, r2, and final_r scores. The implementation details of the methodology is described in Fig. 4, in Sect. 4.

3.1 Data Collection

Due to the lack of a dedicated dataset of detected objects and detected words in ordinary videos found on popular social media platforms, we were motivated to collect a varying set of videos. Google Colab, a free cloud service providing GPU access, was used to train the YOLO and wav2vec models. A sample of 50 videos were obtained from different domains, sourced primarily from YouTube.

3.2 *Object Detection*

For object detection, the YOLOv3 model is used. Only one forward propagation run with the Convolutional Neural Network (CNN) is needed for predictions, "looking only once" at the input image. A single CNN is applied to the whole frame of the video and repeated for each subsequent frame. The image in the frame is divided into 13×13 cells, each cell responsible for predicting and applying probabilities to five bounding boxes. The model uses binary cross-entropy to determine loss for each label and logistic regression for class predictions and probabilities. Figure 2 describes the DarkNet-53 architecture, used as the backbone of YOLOv3. It uses 53 convolutional layers for feature extraction, beginning with a convolution layer, followed by batch normalization layers and a leaky ReLU activation for each operation. To downsample feature maps, a convolutional layer with stride 2 is used, preventing the loss of low-level features frequently attributed to pooling. Images of dimensions 448×448 are used as inputs.

Fig. 2 DarkNet-53 architecture [19]

	Type	Filters	Size	Output
	Convolutional	32	3 × 3	256 × 256
	Convolutional	64	3 × 3 / 2	128 × 128
1x	Convolutional	32	1 × 1	
	Convolutional	64	3 × 3	
	Residual			128 × 128
	Convolutional	128	3 × 3 / 2	64 × 64
2x	Convolutional	64	1 × 1	
	Convolutional	128	3 × 3	
	Residual			64 × 64
	Convolutional	256	3 × 3 / 2	32 × 32
8x	Convolutional	128	1 × 1	
	Convolutional	256	3 × 3	
	Residual			32 × 32
	Convolutional	512	3 × 3 / 2	16 × 16
8x	Convolutional	256	1 × 1	
	Convolutional	512	3 × 3	
	Residual			16 × 16
	Convolutional	1024	3 × 3 / 2	8 × 8
4x	Convolutional	512	1 × 1	
	Convolutional	1024	3 × 3	
	Residual			8 × 8
	Avgpool		Global	
	Connected		1000	
	Softmax			

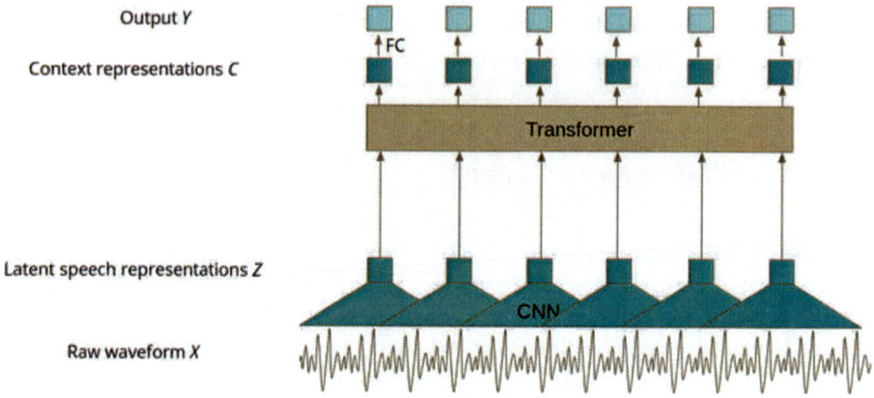

Fig. 3 wav2vec 2.0 model architecture for self-supervised training [20]

3.3 Speech-to-Text Analysis

Wav2vec 2.0 is a model for automatic voice recognition using a self-supervised training strategy. The model's architecture has three major components: convolutional layers to process raw waveform X (generating latent representations Z), transformer layers that provide context representations C, and the linear projection that produces Y. Its goal is a weighted sum of the contrastive loss and the diversity loss. Figure 3 depicts the wave2vec 2.0 model architecture. To preprocess detected words, remove stop words, perform lemmatization, and use embeddings, Natural Language Toolkit (NLTK) was used.

3.4 Recommendation

This section describes the algorithms used for recommendations. In Algorithm 1, the video uploaded to the website is taken as input. Firstly, the uploaded video is sent to YOLOv3. The model processes the video and generates object labels, with counts of each detected object in each frame of the video. These labels and counts are saved in the database for R1 to process. The r1 score for each object is calculated by dividing the object frequency by the frequency of all detected objects. These results are added to a column in the output R1_DF data frame and are sent to recommender R3.

Algorithm 1: Recommender R1
INPUT: video uploaded to website, having certain resolution and dimension
OUTPUT: List of scores of each ad based on frequency of detected objects

(continued)

(continued)

Algorithm 1: Recommender R1
BEGIN Recommender R1 : Transfer uploaded video to YOLO model : Execute YOLO model on video, extract labels and count_of_labels in each frame of video : Store labels,count_of_labels to Database : Calculate the r1_score for each of the ads by dividing the frequency of the detected object by the total frequency of all the detected objects : Store the r1_score of all ads in the ads data frame R1_DF : Output R1_DF to Recommender R3 **END** Recommender R1

Algorithm 2 has the same inputs as Algorithm 1. The video is sent to wav2vec and NLTK models. The model processes the video and generates word labels with counts of each word in each frame of the video. The labels and their counts are saved in the database for R2 to process. The r2 score for a word is calculated by dividing the word frequency by the frequency of all detected words. These results are added to a column in the output R2_DF data frame and sent to recommender R3.

Algorithm 2: Recommender R2
INPUT: video uploaded to website, having certain resolution and dimension
OUTPUT: List of scores of each ad based on frequency of detected words
BEGIN Recommender R2 : Transfer uploaded video to wav2vec model : Execute facebook's wav2vec model and the natural language processing on video, extract labels and count_of_labels in each frame of video : Store labels,count_of_labels to Database : Calculate the r2_score for each of the ads by dividing the frequency of the detected words by the total frequency of all the detected words : Store the r2_score of all ads in the ads data frame R2_DF : Output R2_DF to Recommender R3 **END** Recommender R2

Algorithm 3 takes the results of R1 and R2 as input. The first step is calculating a weighted average of r1 and r2 scores and merging the data frames into FINAL_DF. If none of the detected items are in the ads database, the least recommended ad is chosen. If the detected items are in the ads database, FINAL_DF is sorted in descending order of final_r. The times_recommended attribute is updated, and the first object in FINAL_DF is returned to the user, containing the recommended ad.

Algorithm 3: Recommender R3
INPUT: R1_DF and R2_DF from Recommender R1 and Recommender R2 respectively
OUTPUT: Recommended advertisement to be displayed to the user

(continued)

(continued)

Algorithm 3: Recommender R3

BEGIN Recommender R3
: Calculate a weighted average of r1_score and r2_score called final_r from R1_DF and R2_DF respectively after merging them into FINAL_DF
If no detected item is in ads database:
: Sort FINAL_DF in ascending order of final_r based on the number of times the ad was recommended i.e. times_recommended
Else:
: Sort FINAL_DF in descending order of final_r scores
: Update times_recommended
: Output the top most in FINAL_DF as the recommended ad to the user
END Recommender R3

When a video is played on the website, a dictionary containing labels and frequencies for that video is sent to the recommender engine. It selects the most appropriate ad based on the algorithms above and returns the ad to the user, played after the video.

4 Experiments and Results

In this section, we discuss the experimental setting and statistics of video samples. We also describe observations, as well as deliberate reasons for ad mismatches.

4.1 Experimental Setup

Figure 4 depicts the recommender's many processing elements, with the relevant inputs and outputs.

Video recommendation. The website runs on a browser at the client's end. Before the video begins, we obtain an object from database1, containing the video URL and the dictionary of detected objects. This is sent to server hosting the recommendation engine. Here, a list of ads containing the ad URLs and other characteristics are fetched from database2. The AD_VIDEO is displayed to the user once the video has finished.

Generating the dictionary. We obtain the dictionaries of detected objects and words with their frequencies from the uploaded video. This is sent to server, hosting the YOLO and wav2vec model. The models produce labels for detected items and words and track their frequencies, generating dictionaries and saving it in database1.

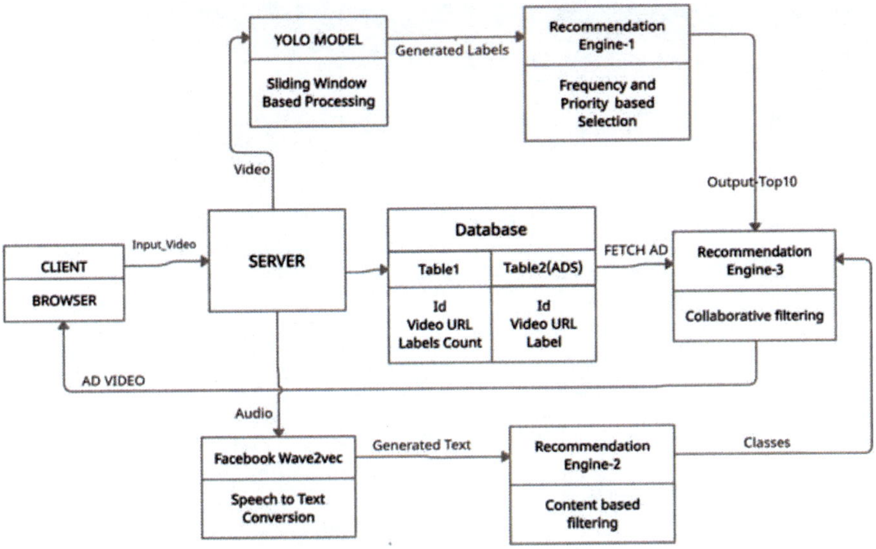

Fig. 4 Overall view of the multimodal recommender representing the various processing elements along with the inputs and outputs

4.2 *Sampling, Detection, and Recommendation Results*

Figure 5 shows snapshots of bounding boxes generated by YOLO on frames of several videos. A Google Form was used to anonymously collect user relevancy scores for the ads recommended. Table 1 summarizes the average relevance scores for various domains between our model and YouTube's model. It is observed that the relevance scores of ads recommended by our model are higher compared to YouTube. The main reason for this is that our model offers advertising relevant to the current video being watched, while YouTube promotes ads based on the user's past history, which may or may not be relevant to the current video being watched.

The recommendation model provided here is currently a proof-of-concept model, with the goal of demonstrating that appropriate recommendations may be made using the given methodology. Hence, it must be highlighted that the model has certain limitations.

Some explanations for the mismatches include the model's present ability to give recommendations only across 81 types of objects. While this is a large number of objects, it is still rather insufficient in this day and age of wide variance in product availability. Another potential source of mismatches is that the model is currently completely content-based, which means that the context of the video is not taken into account appropriately. Most production-level recommendation algorithms experience this issue as well, but they take steps to address it, such as by utilizing video metadata. Furthermore, the recommender engine can presently only recommend ads

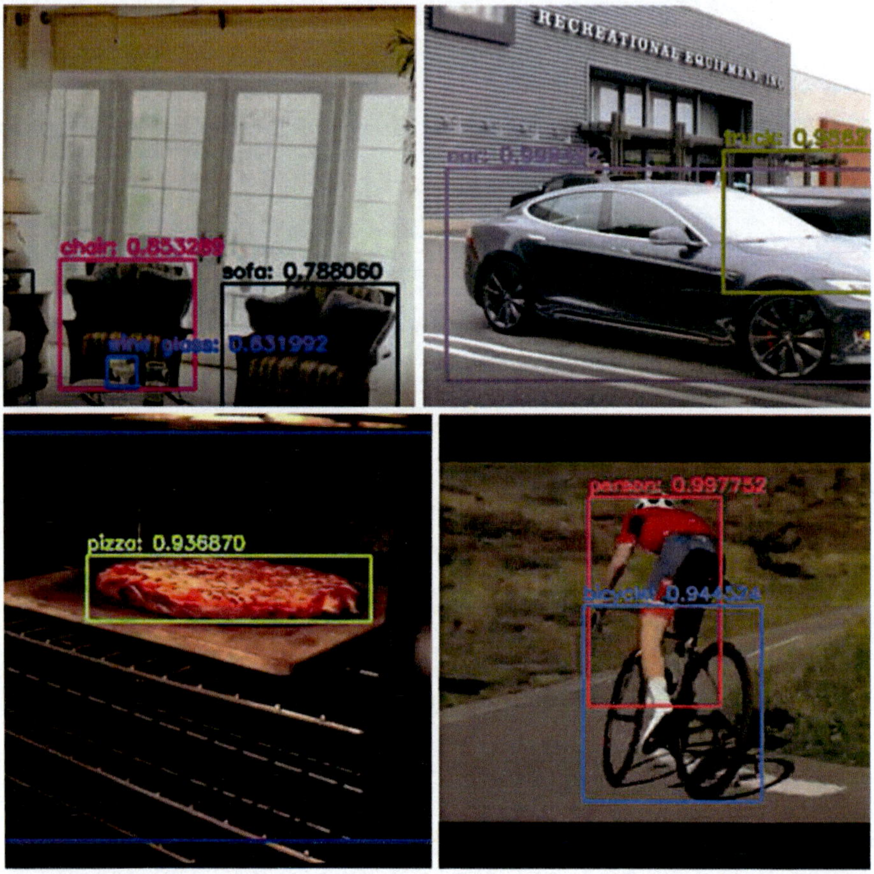

Fig. 5 Object detection bounding boxes in sample videos

Table 1 Relevance scores through a survey for the various videos grouped by their domains for our recommendation engine's recommended ad and YouTube's recommended ad

Domain	Average relevance score for our recommendation engine	Average relevance score for YouTube's recommendations
Electronics	8.700	5.800
Sports	9.400	5.900
Food	6.600	5.900
Kitchenware	9.000	6.667
Automobiles	6.734	4.600
Miscellaneous	10.000	7.000

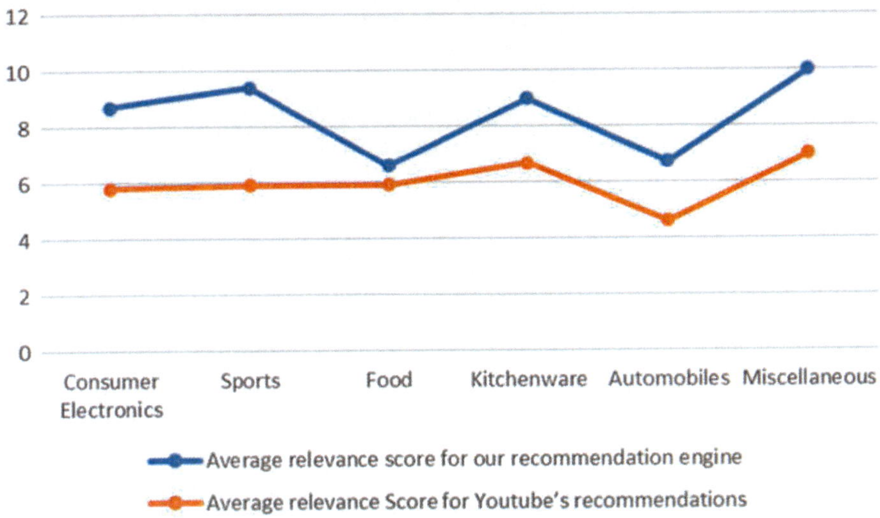

Fig. 6 Comparison of the average relevance scores given by users grouped by the various domains for our recommendation engine and YouTube

for videos spoken in English. We also remark that the ad's relevance to the content being shown is subjective and varies from person to person.

4.3 Performance Analysis

To better put across the idea of ad relevance, we highlight a poll that was done as part of a research project into the communication effectiveness of YouTube advertising. It discovered a substantial positive link between the informativeness of the commercial and the advertiser's brand recognition [21]. The graph in Fig. 6 demonstrates that our recommendation engine's average relevance scores are greater than YouTube's for all domains. The average relevance score of YouTube's ad is extremely close to the average relevance score in the "Food" domain. One thing to keep in mind is that YouTube's recommended ad will differ for various users and may also differ for the same person who views the same video but at a different time.

5 Conclusion

In this research, a unique advertising method involving a multimodal approach to recommend ads is proposed, based on word frequencies and detected objects. Besides improving ad relevance, privacy is also improved since there are no data collected on the user's end. The proposed model is a proof-of-concept, to demonstrate that appropriate recommendations can be made using the described methodology, and our experiments and observations confirm this. The model's drawbacks include the restricted number of classes it can detect, content-centric approach, support for only English, and the time required to run the model on long videos. In the future, we hope to address all the above limitations, as well as carry out research regarding context-defined business models. Overall, the proposed model has the advantages of innovation, demonstrated functionality, and relative simplicity. Given proper investment, both to address the model's limitations as well as improve on its abilities, it is capable of performing as an optimal ad recommendation system.

References

1. Kang K, Li H, Xiao T, Ouyang W, Yan J, Liu X, Wang X (2017) Object detection in videos with tubelet proposal networks. Published in: Proceedings of the IEEE conference on computer vision and pattern recognition, pp 727–735
2. Galteri L, Bertini M, Seidenari L, Del Bimbo A (2018) Video compression for object detection algorithms. In: 2018 24th International conference on pattern recognition (ICPR). IEEE, pp 3007–3012
3. Tung C, Kelleher MR, Schlueter RJ, Xu B, Lu YH, Thiruvathukal GK, Lu Y (2019) Large-scale object detection of images from network cameras in variable ambient lighting conditions. In: 2019 IEEE conference on multimedia information processing and retrieval (MIPR). IEEE, pp 393–398
4. Zhu X, Dai J, Yuan L, Wei Y (2018) Towards high performance video object detection. In: Proceedings of the IEEE conference on computer vision and pattern recognition, pp 7210–7218
5. Ramesh N, Moh TS (2018) Outfit recommender system. In: 2018 IEEE/ACM international conference on advances in social networks analysis and mining (ASONAM). IEEE, pp 903–910
6. Wang W, Lai Q, Fu H, Shen J, Ling H, Yang R (2021) Salient object detection in the deep learning era: an in-depth survey. IEEE Trans Pattern Anal Mach Intell
7. Bahar P, Bieschke T, Ney H (2019) A comparative study on end-to-end speech to text translation. In: 2019 IEEE automatic speech recognition and understanding workshop (ASRU). IEEE, pp 792–799
8. Jiang J, Wang HH (2021) Application intelligent search and recommendation system based on speech recognition technology. Int J Speech Technol 24(1):23–30
9. Tran OT, Bui VT (2021) Neural text normalization in speech-to-text systems with rich features. Appl Artif Intell 35(3):193–205
10. Carlini N, Wagner D (2018) Audio adversarial examples: targeted attacks on speech-to-text. In 2018 IEEE security and privacy workshops (SPW). IEEE, pp 1–7
11. Bansal S, Kamper H, Lopez A, Goldwater S (2017) Towards speech-to-text translation without speech recognition. arXiv preprint arXiv:1702.03856
12. Wang C, Tang Y, Ma X, Wu A, Okhonko D, Pino J (2020) fairseq s2t: fast speech-to-text modeling with fairseq. arXiv preprint arXiv:2010.05171
13. Dey S, Duff BR, Chhaya N, Fu W, Swaminathan V, Karahalios K (2020) Recommendation for video advertisements based on personality traits and companion content. Published in: Proceedings of the 25th international conference on intelligent user interfaces, pp 144–154
14. Li Y, Zhang D, Lan Z, Tan K-L (2016) Context-aware advertisement recommendation for high-speed social news feeding. Published in: The 2016 IEEE 32nd international conference on data engineering (ICDE), held in Helsinki, Finland

15. Zhou L (2020) Product advertising recommendation in e-commerce based on deep learning and distributed expression. Published in: Electronic commerce research, Springer, pp 321–342
16. Choi JA, Lim K (2020) Identifying machine learning techniques for classification of target advertising. ICT Express
17. Liao G, Chen X, Huang J (2020) Privacy policy in online social network with targeted advertising business. In IEEE INFOCOM 2020-IEEE conference on computer communications. IEEE, pp 934–943
18. Smith J, Weeks D, Jacob M, Freeman J, Magerko B (2019) Towards a hybrid recommendation system for a sound library. In IUI workshops
19. Baevski A, Zhou Y, Mohamed A, Auli M (2020) wav2vec 2.0: a framework for self-supervised learning of speech representations. Adv Neural Inf Process Syst 33:12449–12460
20. Farhadi A, Redmon J (2018) Yolov3: an incremental improvement. In Computer vision and pattern recognition, 1804-02767
21. Anthony SJ, Liu V, Cheng C, Fan F (2020) Evaluating communication effectiveness of Youtube advertisements. Int J Inf Res Rev 7(4):6896–6901

Hybrid Integration of Transforms-Based Fusion Techniques for Anaplastic Astrocytoma Disease Affected Medical Images

Bharati Narute and Prashant Bartakke

Abstract As a diagnostic tool with several clinical and healthcare applications, Hybrid Multimodal Medical Image Fusion preserves crucial elements of images and details the various images from the source to create a visually robust enough single fused image. In this paper, the Non-Subsampled Shearlet Transform and Non-Subsampled Contourlet Transform-based hybrid algorithms for Multimodal Medical Image Fusion are proposed. Initial images are broken into low and high frequencies using NSST in the suggested method. The average fusion rule is applied to the components with a lower frequency. For high-frequency components, the "maximum fusion rule" is applied. Each frequency band's coefficients is inversely converted to produce merged images. There are no distortions or misleading artefacts in the proposed algorithms' merged images. The proposed strategy is compared to conventional methods. The images produced by combining the content of both sources using the aforesaid algorithm provide the best visualization and diagnosis of the disease.

Keywords First keyword · Second keyword · Third keyword

1 Introduction

The advancement of medical imaging and information processing technology has made it possible to obtain a wide range of medical pictures for use in clinical diagnosis. These are frequently used in a variety of applications, including disease diagnosis, surgery, and radiotherapy. Every sensor obtained through a variety of imaging

B. Narute (✉) · P. Bartakke
Department of Electronics and Telecommunication, College of Engineering, Pune 411005, India
e-mail: bharati.narute@mescoepune.org

P. Bartakke
e-mail: ppb.extc@coep.ac.in

B. Narute
Department of Electronics and Telecommunication, M.E.S. College of Engineering, Pune 411001, India

S. Das et al. (eds.), *Advances in Data-driven Computing and Intelligent Systems*, Lecture Notes in Networks and Systems 653, https://doi.org/10.1007/978-981-99-0981-0_50

modalities has a unique set of advantages and provides unique information about the human body. Since no one image can provide a complete diagnosis, doctors must employ a variety of imaging modalities to get more precise information about the tissue or organ. A single medical image modality is unable to provide full and precise information because of its limitations. Therefore, not every modality may be capable of displaying all pertinent information regarding a particular illness. Therefore, clinicians always recommend that patients undergo a variety of imaging modalities before arriving at a final diagnosis. Almost all health centres lack the ability to get combined information on numerous modalities from a single system, which is a significant limitation. A hybrid modality imaging system is not available in any Indian hospital due to the prohibitive cost of the technology. Even if scanners become more widely available, the cost of imaging will remain prohibitively expensive for most individuals. These services may not be available in the near future for patients from nations like India that are economically developing. With a PET–CT scanner, you may get the best of both worlds in one convenient machine. The development of a new PET–MRI scanner is ongoing. An urgent need for software that can combine data from multiple imaging techniques in one image at a reasonable cost has emerged as a result of this trend. One such software solution is multimodality medical image fusion (MMIF). To build a new and improved single frame, all relevant and complementary information from two or more modality photos is combined. Radiologists should be able to better see abnormalities using this technology, which should make it easier for them to collect information from all available imaging modalities.

2 Literature Review

Wang et al. [1] proposed a shift-invariant shearlet transform to develop a new dependency model, called as explicit generalized Gaussian density dependency model to solve the concern. Fusion techniques in three-dimensional shearlet space were developed by Wang et al. [2]. For global data, you will have to use the average–maximum fusion rule. A local window region is where it has been used. Yang et al. [3] developed a new image fusion technique based on NSST and compressive sensing (CS) theory for computed tomography and MRI. According to studies, the information in these fused images was enhanced and the complexity of calculation reduced. The shearlet transform and the CS model were used to demonstrate a new way for combining medical images Ling et al. [4]. This method outperforms current methods like non-negative matrix factorization when compared with simulated experiments. Using the singular value decomposition and the shearlet transform, Narasimha et al. [5] came up with a new method for enhancing clinical pictures. Here, two different PET and MRI scans have been combined. With the help of the moving frame-based decomposition framework and NSST, Liu et al. [6] came up with a new MMIF method. Studies show that the method provides superior visual effects and objective criteria compared to more traditional methods. Wan et al. [7] proposed remote sensing picture fusion in the

NSST area. The lack of spatial information in procedures after the intensity–hue–saturation colour space transforms that are depending on multiresolution analysis (MRA) is addressed using a modelled enhancement strategy. Fusion of multimodal sensor medical pictures utilizing local differences in non-subsampled domain was proposed by Weiwei Kong and colleagues [8]. When compared to other methods, the presented fusion technique has shown itself to be simple and effective, both in terms of subject perception and an objective evaluation of the data. It has been used in variety of tests. Yin et al. [9] showed a revolutionary MMIF technique in the NSST area. In this case, the Non-Subsampled Shearlet Transformation decomposition yields representations of the source pictures that are multiscale and multidirectional. Tang [10] proposed Bayesian prediction model implemented and adaptive PCNN applied for image fusion. The method's effectiveness and ability to produce better outcomes in the algorithm are proven in the studies and comparisons are demonstrated. CT and MRI image fusion implemented using NSST and PET images transformed into HSV space and S-PCNN employed by Jin et al. [11]. Multimodal medical images, Xia et al. [12] proposed MMIF method that incorporates Sparse Representation and PCNN in NSCT domain. Multiscale transformation and deep convolutional network proposed by Kia-Jia-Xia et al. [13] and results of CNN-based method are improved. Rajlingam et al. [14] developed a method for merging medical images using discrete fractional wavelet transform and dual tree complex wavelet transform. Rajlingam et al. [15] suggested an improved hybrid method based on NSCT and DTCWT. He implemented techniques to analyse quantitatively the multimodality data of images for diagnosing and understanding the prognostic value of Alzheimer's disease (AD) at the MCI or preclinical stages. Hybrid fusion transform-based implementation proposed to improve result of image fusion which can be utilized for clinical diagnosis.

3 Proposed Methodology

Non-Subsampled Shearlet Transform (NSST). Because of its deep mathematical structure, the shearlet transform is an extremely useful tool for multiscale geometric analysis (MGA). In this regard to the spatial domain, it is well localized and decays quickly. Shearlets are in accordance with the law of parabolic scaling. It has a high degree of directional sensitivity. The number of possible directions increases by a factor of two with each next finer scale. However, because it is not shift invariant, the pseudo-Gibbs phenomenon and other inefficiencies emerge in the fusion findings, as well as other inefficiencies. As a result, a shift-invariant version of the shearlet transform, known as the non-subsampled shearlet transform (NSST), has been created to overcome these shortcomings. NSST is capable of producing optimal sparse approximation of pictures. An affine system with composite dilations is used to construct the system. Instead of performing a series of operations on a single function, shearlets can be created by applying a family of operations to a single function and are therefore associated with multiresolution analysis. As a result, the discrete shearlet transform decomposition can be broken down into two steps: multiscale subdivision and directed localization. The non-subsampled pyramid transform is used to accomplish the multiscale subdivision of the shearlet transform. By substituting the convolution operation for the subsampling process in the shearlet transform, it is possible to achieve shift invariance in the transform. A corollary of this is that the pseudo-Gibbs phenomenon, which causes smoothness around the singularities, is no longer present. In the second phase, which is directed localization, shift-invariant shearing filters are used to accomplish this. It divides the frequency plane into a low-frequency subband and multiple trapezoidal high-frequency subbands, which are then divided further. The non-subsampled pyramid transform is used to decompose the low-frequency subband even further. After then, the process is repeated until the appropriate level of breakdown is reached. As previously stated, NSST has the potential to be more effective than the fundamental reforms (e. g. wavelet, contourlet). However, while the Non-Subsampled Contourlet Transform (NSCT) can overcome some of the primary shortcomings of the earlier transforms (wavelet, contourlet, and so on), it has a limited directionality and high computational complexity.

NSLP and a large number of shearing filters are used to create NSST, which is capable of handling tiny changes in opposite directions (within an image) more efficiently than other methods. To perform image analysis using the multi-scalability and multi-directionality qualities, the NSST first deconstructs a picture into low- and high-frequency components, and then employs direction filtering to produce distinct subbands in different directions, as illustrated in Fig. 1. In regard to direction filtering, this type of transformation is distinguished from others by the usage of a shear matrix (ShF). The image is subdivided into $m + 1 = 4$ subbands at the decomposition level $m = 3$, yielding a total of $m + 1 = 4$ subbands. The image is subdivided at the decomposition level $m = 3$. (one LFS and 3 HFSs). Here are the properties that distinguish the shearlet transform from other multiscale geometric analysis (MGA) tools in a nutshell:

Non-Subsampled Contourlet Transform (NSCT). To overcome the limitations of the wavelet, the contourlet uses square brush strokes and a large number of "dots" to generate a sparse expansion for smooth outlines. It uses the Directional Filter Bank (DFB) for directional decomposition and the Laplacian Pyramid (LP) for multiscale decomposition (DFB). The number of directions in a decomposition level can vary, but there are only three in a wavelet. Unfortunately, the original contourlet includes both LP and DFB examples of down and up samplers. Since it is not shift-invariant, it generates pseudo- Gibbs events at singularities. NSCT eliminates the need for down and up samplers during the decomposition and reconstruction of images. Pyramid filter banks NSDFB and NSPFB are utilized in NSCT without being sampled. Non-subsampled two-dimensional filter banks are used to achieve this. Up sampling filters are employed in the DFB tree structure for each two-channel filter bank that has both down and up samplers disabled. The NSCT is shift-invariant in addition to sharing many of the same characteristics as contourlets. Using this technique, the images fused together have identical subband dimensions, making it simple to establish the link between separate subbands. The NSCT decomposition framework is depicted in Fig. 2.

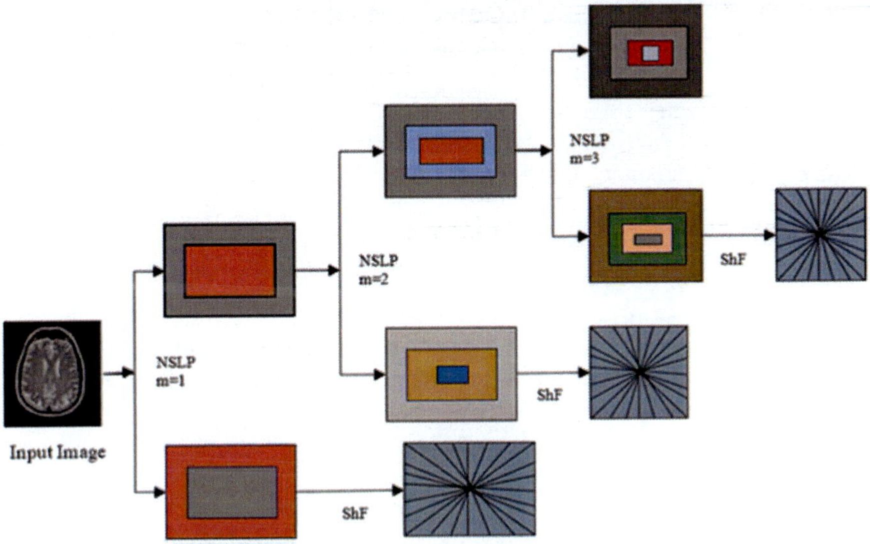

Fig. 1 Image decomposition framework of shearlet transform

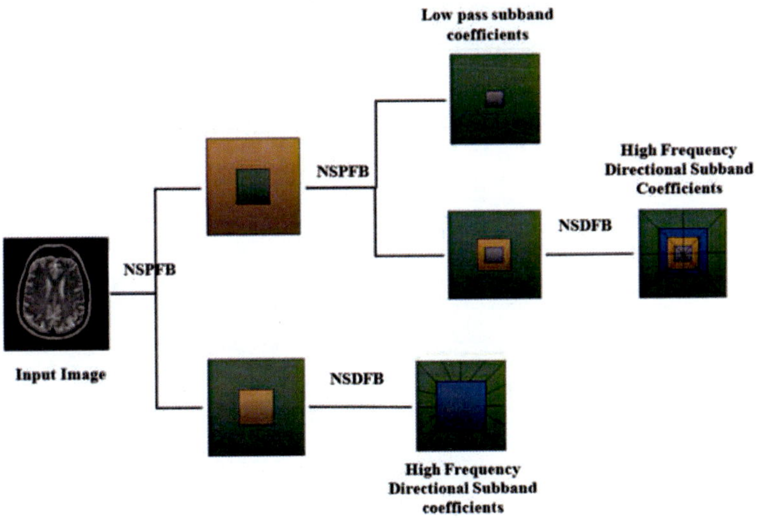

Fig. 2 Decomposition framework of the NSCT

4 Hybrid Fusion Algorithm

These hybrid image fusion techniques are proposed in this research work to overcome the limitations of existing traditional image fusion techniques, such as spatial domain, transform domain, fuzzy logic, and neural network techniques, as well as the limitations of existing traditional image fusion technique. Spatial domain techniques, transform domain techniques, fuzzy logic techniques, and neural network techniques are examples of such techniques. One of the earliest techniques that has been offered is a combination of the NSCT and the NSST techniques. Image 3—A block diagram of the proposed fusion technology (NSCT–NSST), which is composed of multiple subsystems, is depicted in Fig. 3.

Proposed Hybrid Fusion Algorithm Procedural Steps (NSCT–NSST). When applying the suggested method to both input multimodal medical images, the techniques of NSCT and NSST are used.

- Read the two photographs that were provided in the first step of the process.
- 2nd Step: The input images are downsized to 256 by 256 pixels in the second step.
- To breakdown the input images into complex coefficient sets, NSCT is used in Step 3.
- For both coefficient sets, thresholds are generated for each decomposition level for each source picture separately for each of the two coefficient sets' coefficient sets.

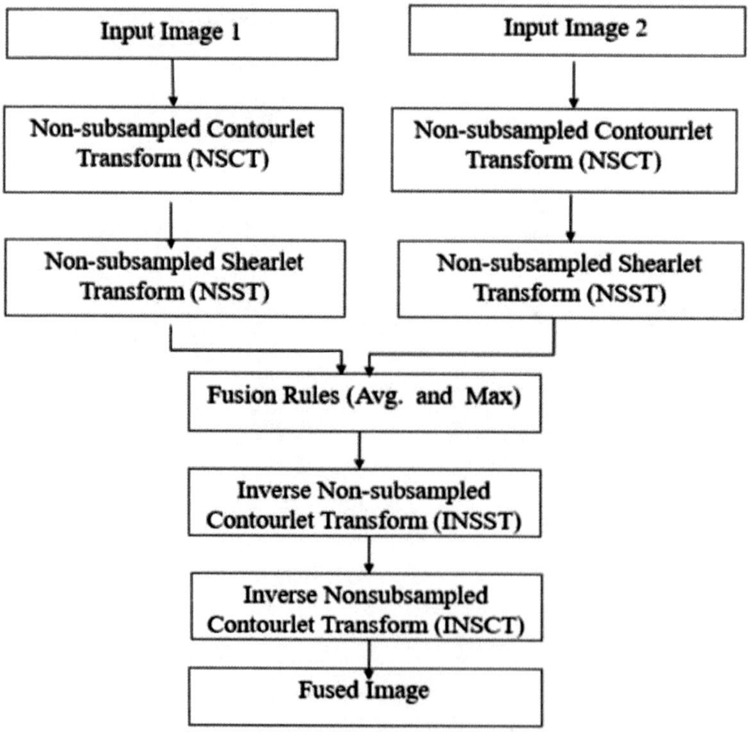

Fig. 3 Proposed hybrid fusion algorithm

- Then, for each of the above-mentioned deconstructed images, we do the NSST.
- Using the fusion rules, adjust the low-frequency coefficients so that they follow the average fusion rule and the high-frequency coefficients follow the maximum fusion rule, as shown in Step 5.
- INSCT and INSST are used to obtain the final proposed fused picture from the inverse transform of the fused image, which is INSCT and INSST, respectively.

Because the human visual system is not sensitive to individual pixels, but rather to edge, direction, and texture information in the image, image fusion based on the transform domain often yields satisfactory fusion results when the fusion rules of maximum and average rules are employed to pick coefficients. In this situation, the visual system can be well satisfied by fusion principles based on regional energy. Fusion rules for higher bands are quite complicated, and this has an impact on the computation speed as well. If you look at the structure and texture of a medical image, you will see that it's simple to apply fusion rules to both the low-frequency and the high-frequency subbands coefficients to produce the desired fusion effect while keeping the computation speed in check. When two medical images are fused together, the resulting image is called the fused image. The picture is first decomposed into decomposition coefficients, which are then utilized to reconstruct the image with the help of the NSCT–NSST algorithm. A factor of the maximum number of the coefficients is applied to the absolute values of the coefficients to produce fusion images. To fuse low-frequency coefficients, the average fusion rule is used; to fuse high-frequency coefficients, the maximum fusion rule is applied. The Xin Jin demonstrates [11] maximum and average fusion rules in more detail:

$$C_F = \begin{cases} C_i^1, \text{ if } C_i^1 > C_i^2 \\ C_i^2, \text{ if } C_i^1 < C_i^2 \end{cases} \tag{1}$$

$$C_F = \frac{1}{2}(C_i^1 + C_i^2) \tag{2}$$

where C_F is the combined coefficient, C^1, C^2 are the coefficients of input images X, Y. Finally, the fused coefficients C_F are utilized to reconstruct the fused image **through** inverse NSST and inverse NSCT.

5 Results and Discussion

The suggested HMMIF technique is being evaluated on pilot study sets that include MRI and SPECT scans of the brain from patients with anaplastic disease, as well as healthy volunteers. Anatomical and functional similarities between the input pictures for each pair of both MRI and SPECT slices guide the selection of input images from the same patient for each pair of MRI and SPECT slices. The results of the

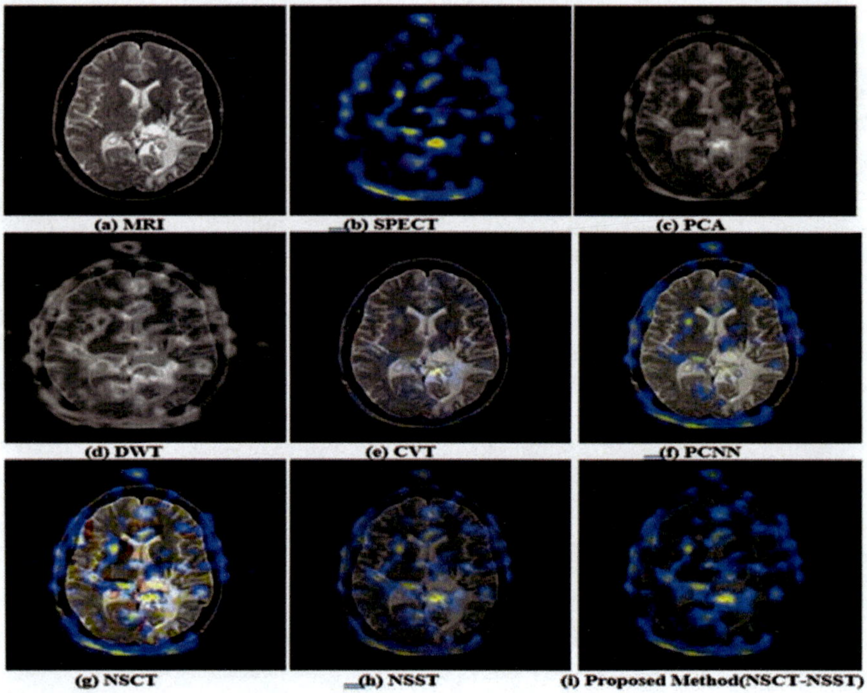

Fig. 4 Anaplastic astrocytoma disease experimental outcomes (Set 1)

suggested hybrid image fusion approach, as well as the results of other current techniques, are depicted in Figs. 4, 5, 6 and 7. It is necessary to take four independent sets of MRI/SPECT brain pictures to create the image fusion for patients with anaplastic astrocytoma illness. Results of the identical input pictures were fused using PCA, DWT, Curvelet, PCNN, NSCT, and NSST, with the proposed hybrid technique employing the combination of NSCT and NSST as the fusion algorithm. Out of all classical fusion procedures, the simulation result shown of the suggested hybrid technique delivers the best performance in both qualitative, and Table 1 shows quantitative analysis, compared to the other techniques.

Table 2 shows proposed methods evaluation metrics performance better than current methods.

6 Conclusion

As per the study, it is possible to combine several images from different medical imaging techniques using an innovative hybrid method built on the NSCT and NSST. To evaluate the efficacy of various image fusion algorithms, four separate sets of

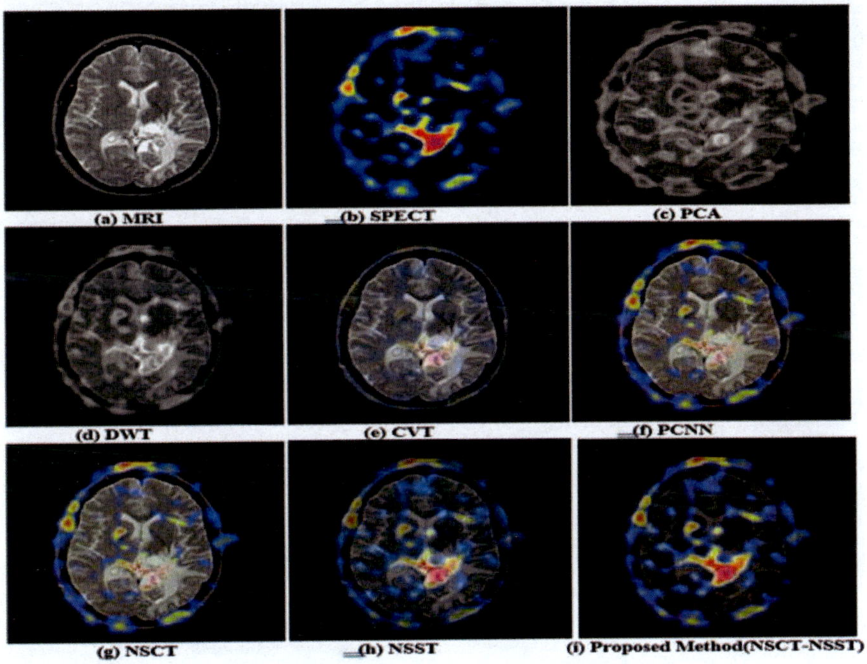

Fig. 5 Anaplastic astrocytoma disease experimental outcomes (Set 2)

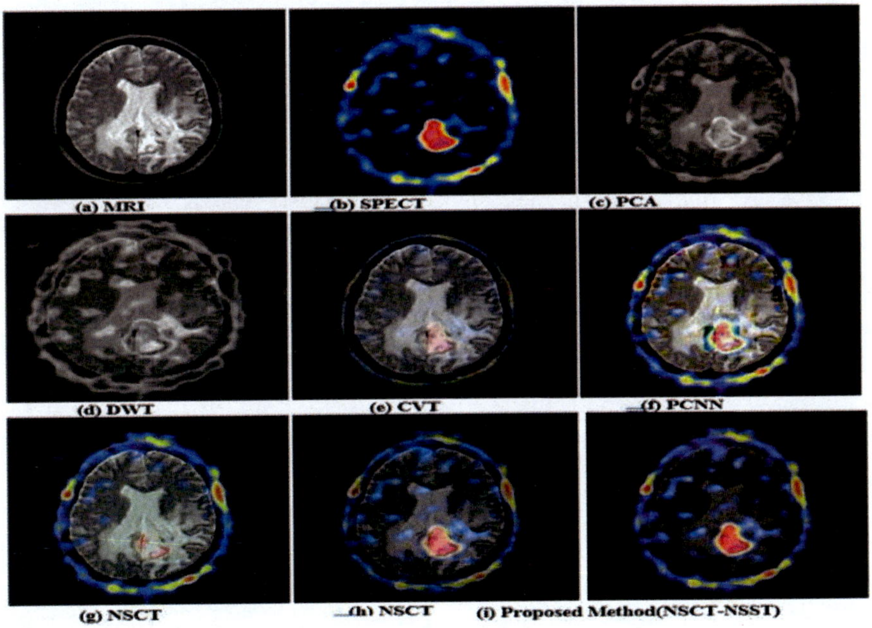

Fig. 6 Anaplastic astrocytoma disease experimental outcomes (Set 3)

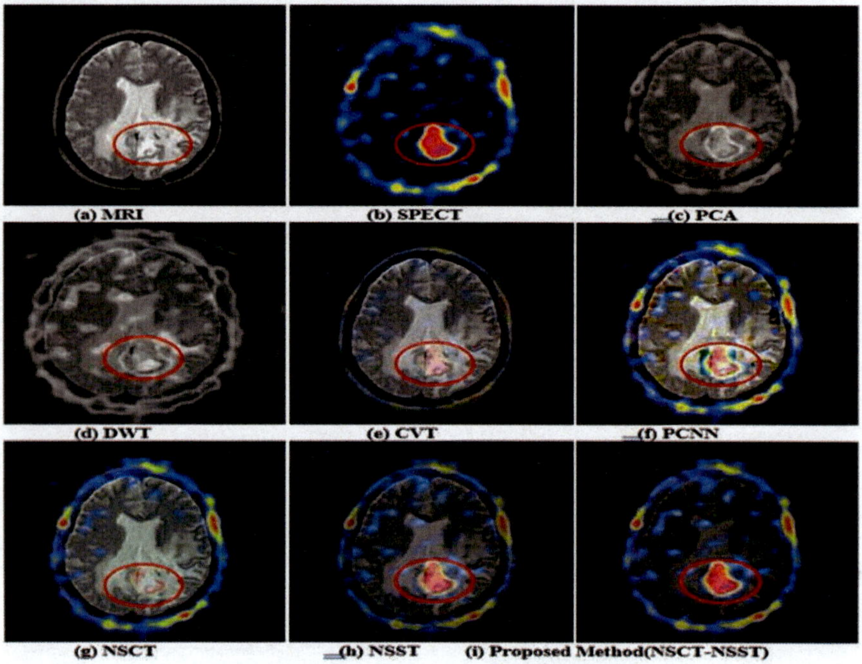

Fig. 7 Anaplastic astrocytoma disease experimental outcomes (Set 4)

MRI/SPECT medical images must be collected. We applied to fuse low-frequency coefficients, use the average fusion rule; otherwise, we use the maximum fusion rule. The quality of the fusion has been evaluated using different performance matrices. Cross-entropy values are found to be at their lowest in the proposed hybrid image fusion strategy, while the values of four parameters such as fusion factor, IQI, mSSIM, and EQM are found to be at their highest. Image fusion based on NSCT–NSST outperforms over all other conventional methods.

Table 1 Comparison of different fusion metrics (Set 1, Set 2, Set 3 and Set 4)

Study set	Algorithm	Fusion factor	Image quality index	mSSIM	Cross entropy	Edge quality measure
Set 1	PCA	1.5201	0.5061	0.4393	2.7102	0.4521
	DWT	1.8513	0.5281	0.4917	2.3122	0.4919
	CVT	1.9023	0.5512	0.5216	2.2511	0.5173
	PCNN	1.9976	0.5991	0.5568	2.0043	0.5517
	NSCT	2.0141	0.6031	0.5902	1.8981	0.5712
	NSST	3.8282	0.8712	0.8191	0.9612	0.8714

(continued)

Table 1 (continued)

Study set	Algorithm	Fusion factor	Image quality index	mSSIM	Cross entropy	Edge quality measure
	Proposed (NSCT–NSST)	4.0191	0.9012	0.9012	0.7581	0.9412
Set 2	PCA	1.2161	0.4331	0.4367	2.5279	0.4768
	DWT	1.519	0.5292	0.4729	2.0752	0.5616
	CVT	1.7416	0.6112	0.5482	1.8225	0.6096
	PCNN	1.6219	0.6528	0.5899	1.7319	0.6438
	NSCT	1.7998	0.8026	0.6455	1.2031	0.6999
	NSST	2.9515	0.9884	0.9295	0.951	0.9048
	Proposed (NSCT–NSST)	3.0163	0.9721	0.9545	0.9463	0.9153
Set 3	PCA	2.3511	0.4981	0.4016	2.012	0.6812
	DWT	2.4125	0.5218	0.5289	1.9920	0.7128
	CVT	2.5612	0.5592	0.5618	1.8623	0.7518
	PCNN	2.7189	0.5717	0.5812	1.2517	0.7518
	NSCT	3.1812	0.6218	0.6912	1.1620	0.7819
	NSST	5.1782	0.8818	0.9062	0.9743	0.8012
	Proposed (NSCT–NSST)	5.5192	0.9402	0.9409	0.9502	0.9257
Set 4	PCA	1.8871	0.5062	0.4152	2.0182	0.4805
	DWT	2.1832	0.5574	0.5491	2.1823	0.5821
	CVT	2.3361	0.5908	0.6021	1.9562	0.6721
	PCNN	3.0763	0.6362	0.7627	1.7923	0.7231
	NSCT	3.2123	0.8054	0.8361	1.6732	0.8943
	NSST	3.4985	0.8329	0.9275	1.5824	0.361
	Proposed (NSCT–NSST)	5.2318	0.9286	0.9480	1.1743	0.9573

Table 2 Comparison of the proposed methods' performance metrics with current methods

Methods		Evaluation metrics	
	Mutual information	Standard deviation	$Q^{AB/F}$
Xia et al. [12]	2.242	0.5144	0.5887
Yang et al. [3]	3.663	0.4329	0.6197
Proposed method 1(NSCT-NSST)	4.278	0.5745	0.878

References

1. Wang L, Li B, Tian L (2014) Multi-modal medical image fusion using the inter-scale and intra-scale dependencies between image shift-invariant shearlet coefficients. Inf Fusion 19:20–28
2. Wang, Li B, Tian L (2014) EGGDD: an explicit dependency model for multi-modal medical image fusion in shift-invariant shearlet transform domain. Inf Fusion 19:29–37
3. Yang J, Wu J,Wang Y, Xiong Y (2016) A novel fusion technique for CT and MRI medical image based on NSST. In: IEEE 28th Chinese control and decision conference (CCDC), vol 28, pp 4367–4372
4. Ling N, Mei-Xia D (2016) Fusion for medical images based on shearlet transform and compressive sensing model. Int J Signal Process, Image Process Pattern Recogn 9(4):1–10
5. Narasimha KN, Murthy, Kusuma J (2017) Fusion of medical image using STSVD. In: Proceedings of the 5th international conference on frontiers in intelligent computing: theory and applications. Springer 516(2):69–79
6. Liu X, Mei W, Du H (2018) Multi-modality medical image fusion based on image decomposition framework and non-subsampled shearlet transform. Biomed Signal Process Control 40:343–350
7. Wan W, Yang Y, Lee HJ (2018) Practical remote sensing image fusion method based on guided filter and improved SML in the NSST domain. Signal, Image Video Process 12(5):959–966
8. Kong W, Ma J (2018) Medical image fusion using non-subsampled shearlet transform and improved PCNN. Intell Sci Big Data Eng 11266:635–645
9. Yin M, Liu X, Liu Y, Chen X (2018) Medical image fusion with parameter-adaptive pulse coupled-neural network in nonsubsampled shearlet transform domain. IEEE Trans Instrum Meas 68(1):49–64
10. Tang L, Tian C, Xu K (2017) IGM-based perceptual multimodal medical image fusion using free energy motivated adaptive PCNN. Int J Imaging Syst Technol 28(2):1–7
11. Jin X, Chen G, Hou J, Jiang Q, Zhou D, Yao S (2018) Multimodal sensor medical image fusion based on non-sub sampled shearlet transform and S-PCNNs in HSV space. Signal Process 153:379–395
12. Xia J, Chen Y, Chen A, Chen Y (2018) Medical image fusion based on sparse representation and PCNN in NSCT domain. Comput Math Methods Med, Hindawi 2018:1–12
13. Xia KJ, Yin H, Wiang J (2019) A novel improved deep convolutional neural network model for medical image fusion. Cluster Comput 22(1):1515–1527
14. Rajalingam B, Priya R, Bhavani R (2019) Hybrid multimodal medical image fusion algorithms for astrocytoma disease analysis. In: Emerging technologies in computer engineering: microservices in Big Data analytics, vol 985. Springer, pp 336–348
15. Rajalingam B, Priya R, Bhavani R (2019) Hybrid multimodal medical image fusion using combination of transform techniques for disease analysis. Procedia Comput Sci 152:150–157

GeoAI-Based Covid-19 Prediction Model

Jyoti Kumari and Dipti P. Rana

Abstract The outbreak and diffusion of Severe Acute Respiratory Syndrome Coronavirus 2 (SARS-CoV-2), responsible for the Covid-19, has caused an emergency in the health system worldwide. The topic of whether a woman over the age of 50 can be save from the Covid-19 in the next summer requires scientific analysis in relation to the propagation and transmission of the virus. In this study new methodology has been discussed that is use to predict the risk for a particular gender, specific age group, and a specific weather condition. The association between numerous characteristics and the risk of Covid-19 is identified in this work using a variety of supervised machine learning models, including Logistic Regression, Random Forest, Naive Bayes, K-Nearest Neighbors, Support Vector Machine, and Artificial Neural Networks. We have gathered the DS4C Covid-19 dataset, to produce desirable outcome with a significant level of prediction accuracy.

Keywords GeoAI · Covid-19 · Gender · Age · Weather · Prediction

1 Introduction

A virus known as SARS-CoV-2 has been rapidly affecting Wuhan city, China, from December 2019. It later expanded to numerous other nations around the globe via international travel from China [1]. As a result, Covid-19 soon rose to prominence as the key case studies across all scientific disciplines. The pandemic has compelled the entire world to respond. Instead that worrying about how bad the issue is, many individuals are more concerned about how it will play out in the near future as well as which areas, genders, and age groups are more at risk [2]. Our goal is to detail the epidemiological, clinical, and laboratory, outcomes for patients with Covid-19 infection [3].

An emerging scientific field called geospatial artificial intelligence (GeoAI) that combines innovations in spatial science, artificial intelligence methods in machine

J. Kumari · D. P. Rana (✉)
Sardar Vallabhbhai National Institute of Technology, Surat, India
e-mail: dpr@coed.svnit.ac.in

S. Das et al. (eds.), *Advances in Data-driven Computing and Intelligent Systems*, Lecture Notes in Networks and Systems 653, https://doi.org/10.1007/978-981-99-0981-0_51

learning to extract knowledge from spatial big data. This concept is quite effective since it enables people to be aware of events before anyone else has even heard of them [4]. To process geospatial large data at scale and analyze it intelligently, GeoAI makes use of current advances in machine learning and high-end computing [5]. Additionally, a number of AI techniques are used in healthcare to analyze data and make decisions [6]. Location or place is a key factor in health intelligence since it can have a big impact on health, regardless of the intended scale of the population or individual [7].

It is crucial to pay special attention to the relationship during the dispersion of Covid-19 and weather factors in the impacted areas. Since the most infected areas where outbreaks occur, including parts of South Korea, China's Hubei province in the center, Iran, the North Eastern United state, Spain, Japan, Italy, Germany, and England, all have roughly similar weather information, such as average temperatures of 8–12 °C and 50–80% moisture in January and February 2020. In South Korea, areas with temperatures above 25 °C and a humidity of 75% have a lower chance of Covid-19 cases [8].

1.1 Motivation

The globe has been negatively impacted by Covid-19 in every aspect, not just economically but also in terms of human's physical health. This study and research aim to build a framework which will predict the risk of Covid-19 for a specific gender of specific age-group Covid-19 affected people in a specific weather condition. It helps in predicting risk of Covid-19 in a certain region. From this we can save people from Covid-19 and it also reduces the cost of treatment. A faster and better prediction model will help in the early discovery and forecast the risk of Covid-19.

1.2 Objective

The objective of the work is to predict Covid-19 related findings, patterns, understanding, and visualization of the Covid-19 data which will help in deriving informative Covid-19 data insights. This study will help in understanding the impact and predicting Covid-19 risk for a specific gender of specific age-group Covid-19 affected people in a specific weather condition. For this we need information of patient's gender, age, and weather attribute based on location. A dataset with all required information has not yet been discovered for the Covid-19 statistics. Therefore, we are going to combine several datasets and train a model to forecast the probability of Covid-19 risk of covid affected people using this dataset.

1.3 Organization

The organization of the remaining section is as follows. This chapter provides an introduction to the problem statement. Section 2 discusses related work and the problem with previous approaches. Section 3 discusses the proposed framework and steps for a prediction model. Section 4 cover performance results and analysis of the implemented study followed by a conclusion and future work in Sect. 5 on the Covid-19 data analysis with the predictive model.

2 Related Work

A large amount of research has already been done on prediction and diagnosing of Covid-19 using various data analysis techniques and machine learning models. This chapter is discussing the literature review of different research papers useful for this research.

2.1 Literature Survey

Muhammad et al. [9] used epidemiological data from Covid-19 patients in South Korea to build data mining algorithms for prediction. The suggested model predicts the smallest and largest amount of periods needed for a patient to heal from Covid-19, as well as the age range of patients who are more vulnerable to a pandemic of Covid-19, as well as those who are most likely to recover.

Khanam et al. [10] concentrated on the data visualization and analysis of publically available Covid-19 datasets. They presented data analytics addresses several different features of Covid-19, including indications, differences between Covid-19 and other diseases caused by Severe Acute Respiratory Syndrome Coronavirus 2 (SARS-CoV-2), Swine Flu, and Middle East Respiratory Syndrome Coronavirus 2 (MERS-CoV-2).

Demongeot et al. [11] demonstrated that the virulence of coronavirus infections caused by viruses such as SARS-CoV-2 and MERS-CoV-2 lowers under humid and warmer weather and rise in cold, dry weather environmental conditions.

Sajadi et al. [12] created a basic model that shows a region with an increased risk of the transmission of Covid-19 using atmospheric prediction. To predict which locations are often to be more at higher risk of Covid-19 communal propagation in the coming weeks, this model analyzed environmental data, climatologic data, and persistence forecasting data.

Iwendi et al. [13] proposed boosting the AdaBoost algorithm to fine-tune a Random Forest classification model. The Covid-19 patient's location data, health information, and demographic data were utilized to predict the patient's Covid-19 severity and the likelihood of recovery, or death. The study found a link between patient fatalities and gender, and discover that the majority of patients who test positive for Covid-19 are between the ages of 20 and 70.

Muhammad et al. [14] used an epidemiology dataset of Covid-19 cases across Mexico, which included both positive and negative corona patients data. According to the outcomes, the Naive Bayes Model has providing a detailed information, the Support Vector Machine Model had the better performance when compared, and the decision tree model functioned the best.

According to Narin et al. [15], coronavirus is spreading rapidly over the world. Due to limited number of test kits available and the growing number of cases, an automatic detection system for Covid-19 diagnosis is required. On chest X-ray radiographs, CNN-based models such as InceptionV3, ResNet50, and Inception-ResNetV2 were utilized with fivefold cross validation.

Xu et al. [16] proposed a statistical research that utilized extensive information of the Covid-19 epidemic since April 2020. These researchers developed and tested a technique for estimating the number of affected people. They believe that the results of their research may be used to help restrict the transmission of the infection by taking into account factors including detection latency, density of population.

Cartocci et al. [17] proposed a model to reassemble and examine sexual identity and age-grouped Covid-19 data. It was suggested to use an age- and sex-distributed Susceptible-Infected-Recovered-Deceased (SIRD) model that changes over period. The vaccination of elderly and youngest team with a high probability of interaction was advised method in this model.

Her et al. [18] demonstrated the effect of gender with in fatality of persons with Covid-19 infection in South Korea. During January 20, 2020, and April 30, 2020, Korea Centers for Disease Control and Prevention (KCDC) collected data on 5628 probable patients diagnosed who were hospitalized in South Korea with an Covid-19 disease. The group made up of 3308 women (59%) and 2320 men (41%). Women in South Korea who were hospitalized with Covid-19 infection had a considerably lower probability of dying in the hospital.

Several research, including laboratory, epidemiological analysis and mathematical modeling, have suggested that ambient temperature has a role in viral survival and transmission [19]. Since epidemics exhibit distinct seasonal and regional patterns, observers have long suspected that environmental factors are among the major determinants of influenza transmission [20]. Absolute humidity levels, degrees of vulnerability, changes in population mixing, and contact rates all influence when pandemic influenza outbreaks occur [21]. One of the leading causes of sickness and mortality in the globe is a virus [22]. Some findings have given rise to direct experimental proof that the dynamics of influenza transmission are significantly influenced by weather [23]. Cold temperatures and low relative humidity brought on by interior heating are characteristics of winter that encourage the propagation of the influenza virus [24].

2.2 *Problem with Previous Approaches*

The related literature surveys that have been considered, studied, and reviewed so far shows that Data Analysis, Data Mining, Machine Learning techniques, and various Data Science fields play important roles in Covid-19 detection, prediction, treatment, and diagnosis, which benefits limited healthcare systems in the fight against the pandemic. In terms of what degree of temperature limits the Covid-19 spread, some weather-related research has been done that shows a connection between climate conditions and the breakout of Covid-19. And some study has been done that takes gender and age into account and demonstrates a link between older age groups and male having a higher risk of Covid-19. Till here our understanding can tell that, no one has investigated and shown proof for the predicting of Covid-19 risk by taking into account several factors such as weather, gender, and age. Therefore, this research will focus on these characteristics.

3 Proposed Framework

To carry out the Covid-19 risk forecasting for covid infected people, we have examined the relationship of climate variables, age, gender with and the state of patient. The suggested model's crucial step-by-step development process is depicted in Fig. 1. This model is divided into five sections namely Data Collection, Data Preprocessing, Feature Engineering, Machine Learning Model, and Result Analysis. This study is able to estimate the risk of Covid-19 for a particular gender of Covid-19 affected individuals in a particular age group and particular weather condition. Once the model has been built a separate test set is used to verify the model's performance.

1. **Data Collection**
   To create the model for the desired result, data collecting is an essential first step. In order to find the ideal and practical data for the study, several data repositories, Internet sites, and digital tool are investigated in this process. In this study, the proposed model is used to utilize Data Science for COVID-19 (DS4C) Dataset which obtained from KCDC and made available on Kaggle Website [8]. It consist of 11 tables in .CSV format but required dataset are PatientInfo, Region, and Weather.
2. **Data Preprocessing**
   Any data project or research must go through the step of data preparation prior to moving on to utilize the dataset. Because the data we need is not available in a single .CSV file, there is need of integrate multiple datasets in order to gather the relevant features for our research. Data that is not adequately evaluated might produce false results, hence data preprocessing is an important element of information evaluation.

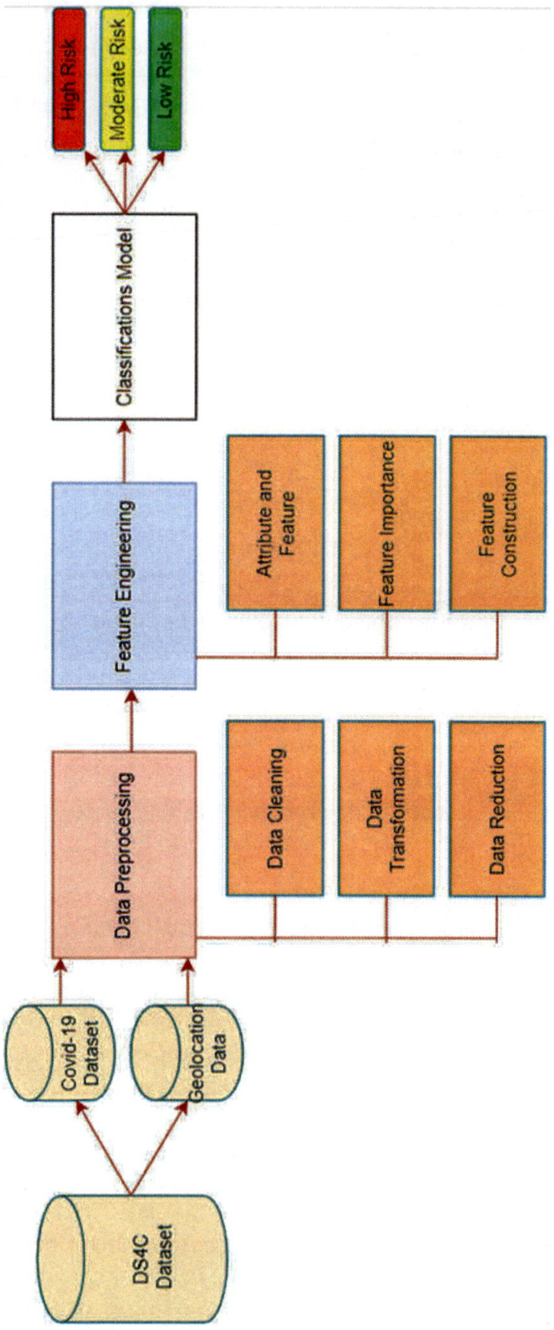

Fig. 1 The general steps for the proposed models

This comprises Data Cleaning, Data Transformation, and Data Reduction.
Data Cleaning It is the process of updating, replacing, or eliminating inaccurate or unnecessary data from a data set of creating a new data.
Age feature: Null values are replaced with various age group, and those with the best accuracy are chosen to replace the missing value in the corresponding model which is shown in Fig. 4. Gender feature: Missing value updated with most common attribute.
Data Transformation The process of changing and translating datasets into an adequate and suitable form of data processing is known as data transformation.
Data Reduction By reducing the amount of data, this approach makes assessment simpler while still producing the same or very similar results.

(a) Country column has been removed because we are considering only one country.
(b) In terms of enhancing the effectiveness of the algorithm, longitude and latitude columns have been eliminated in favor of region code columns.
(c) ID column has been removed because it's of no use.
(d) Released date column has been removed because it's of no use.

3. **Feature Engineering**
Machine learning feature engineering seeks to enhance prediction results.
This process includes recognizing the features that helps with Feature Creation and importance analysis, Feature Importance, and Feature Construction.
Attribute and Feature In the context of an issue, a feature is an attribute that is relevant and meaningful.
Feature Construction Feature construction means creation of new features from raw data features for the model usability. Risk feature is created by taking different scenarios into consideration and by using 'Age', 'Gender', and 'Weather attributes'.
If the state of a patient is deceased, then risk is 'VERY HIGH'.
If the state of a patient is isolated, then risk is 'HIGH/MODERATE'.
If the state of a patient is released, then risk is 'LOW'.
Feature Importance Features are calculated with a score in Feature Importance, and then they are ranked according to their scores.
Scores for each feature have been calculated with different methods.
4. **Classification Model**
A classification model is a supervised machine learning model that makes an effort to derive a result from data that are seen. A classification model makes an effort to predict the value using one or more outputs from one or more inputs. The following classification models are used to categorize data: LR, NB, RF, SVM, KNN, and NN. These algorithms were chosen because they are the most often utilized in Covid-19 prediction and other prediction-related analysis.
5. **Result Analysis**
This study focuses on estimating the risk for individuals with Covid-19 for a certain gender, age group, and weather condition. Results show that focusing

exclusively on patient demographics, such as age and gender features, or exclusively on the weather, does not always result in more accurate form. In order to assess the correctness of the results from data mining algorithms, evaluation approaches are used [14]. Using either a machine learning algorithm or a data mining algorithm, the procedures evaluate the model's quality and effectiveness. Accuracy, Precision Score, Recall Score, and $F1$ Score are the key performance evaluation approaches for the data mining model.

4 Experimental Results and Analysis

To evaluate the proposed method and baselines this chapter presents the details of dataset used, evaluation metrics, and implementation results. All implemented code is available at [25].

4.1 Experimental Setup

Machine Configuration

Operation System: Windows 10
RAM: 8.00 GB
Processor: Intel(R) Core(TM) i3-1005G1 CPU @ 1.20 GHz 1.19 GHz
System Type: 64-bit operating system
Programming Language and Tool: Python, Virtual Studio Code.

4.2 Dataset Analysis

Exploratory data analysis of Dataset is illustrated in Figs. 2, 3, 5 and 6.

4.3 Prediction Result

This section is showing different experimental results obtained from the study and analysis.

Missing Value of Age: Each age group fills in the dataset's missing value for the age property in order to determine the best fit. Figure 4 shows the accuracy for each age group to every model, and for the age group with the highest accuracy, respective training model is used (Figs. 5 and 6).

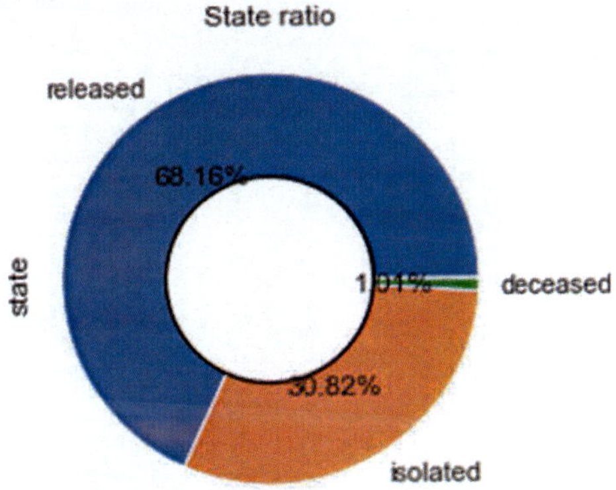

Fig. 2 State of patient distribution

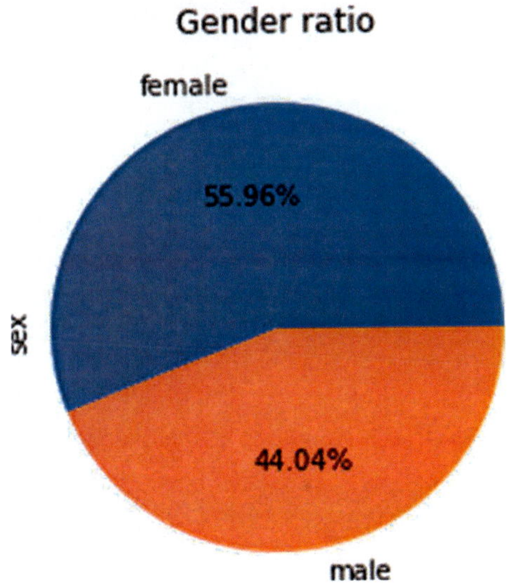

Fig. 3 Gender of patient distribution

Train and Test Size: The provided algorithms need to be trained on a large portion of the dataset in order to predict the risk of Covid-19. The size of the dataset is crucial to the training process and has an impact on how well the suggested algorithms work. The ratio of training and testing dataset which have more accuracy is taken for the train and test the respective model. Table 1 also shows that we are taking into account the train and test dataset sizes that are providing the best accuracy for each model.

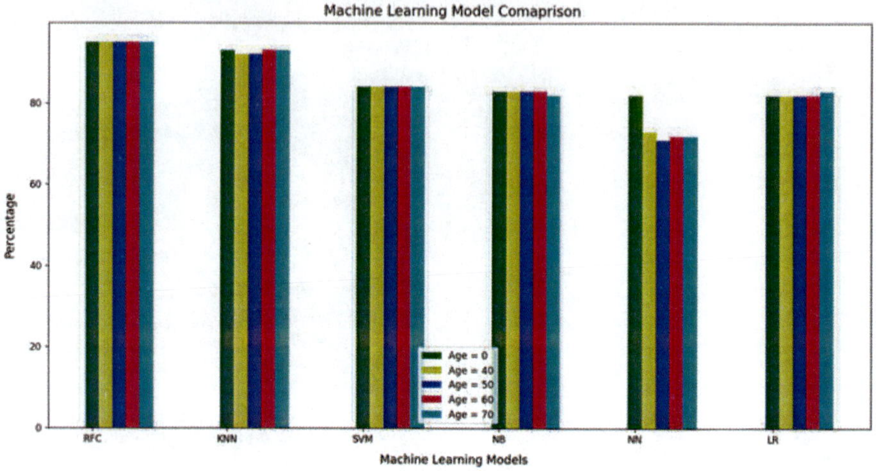

Fig. 4 Handling the missing value

Fig. 5 Age wise distribution of gender of patient

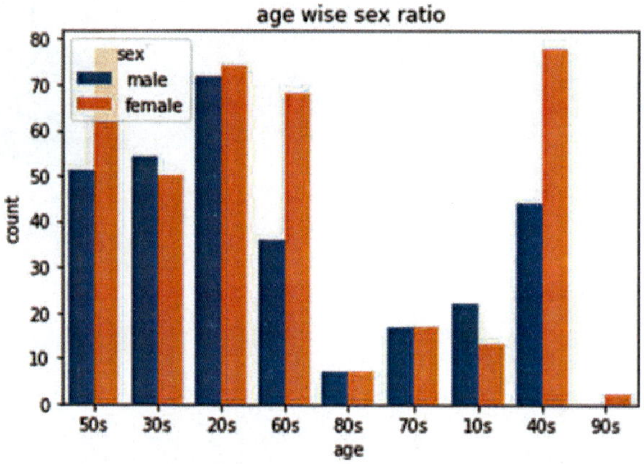

Performance Analysis: We have demonstrated that environmental factors like temperature and humidity play a role in the propagation of the virus based on the experimental results depicted in Fig. 7. Merged dataset contains all feature such as age, gender, and weather information. And patient dataset contains all feature except weather information. Table 1 and Fig. 8 display the performance evaluation for each model's risk prediction using merged dataset.

4.4 Summary of Results

The results of proposed network model can be summarized as:

Table 1 Evaluated result

Model	Train split (%)	Test split (%)	Accuracy (%)	$F1$ score (%)	Precision (%)	Recall (%)
Random Forest Classifier	80	20	95	94	95	94
KNN Classifier	75	25	93	91	91	91
Support Vector Machine	80	20	84	81	81	82
Neural Network	55	45	83	79	79	80
Naive Bayes Classifier	75	25	82	78	79	77
Logistic Regression	60	40	82	78	77	70

1. It has been shown through experimentation that focusing solely on patient demographics such as age and gender characteristics or just the weather information alone does not guarantee more accurate outcomes.
2. We developed a more generic machine learning model that takes into account both of the features, such as patient demography and environmental condition, in order to overcome this limitation.
3. The model also identified that more females than males are infected, despite the fact that females also recover more quickly on average.
4. For both males and females, the age categories with the highest proportion of infected people were 20–29 years old and 50–59 years old. In contrast, the greatest number of deaths occurred in the older groups and in particular in males and females aged 80–89 years.
5. This research covers an awareness of criteria that can minimize or maximize the risk of the disease, assisting people in better planning and preparing for everyday activities based on age, gender of person, and weather conditions of region.

5 Conclusion and Future Work

In this study, we are inspired to investigate the impact of a gender, age, and climate condition for the Covid-19 risk on certain region. In this research, LR, NB, RF, SVM, KNN, and NN classification algorithms are used on the DS4C dataset. The dataset is created utilizing several data preparation procedures, many data visualizations, and feature engineering techniques. This research then uses data analysis techniques to improve the performance of a supervised machine learning model. This study focused on predicting the risk for a specific gender of specific age-group in a specific weather condition for covid affected people. Results indicate that concentrating only on patient demographics, such as age and gender characteristics or only the weather, does not guarantee more accurate outcomes. This study shown that weather information is equally essential for forecasting the risk of Covid-19 by focusing on all these three features. Implementation result shows that Random Forest algorithm

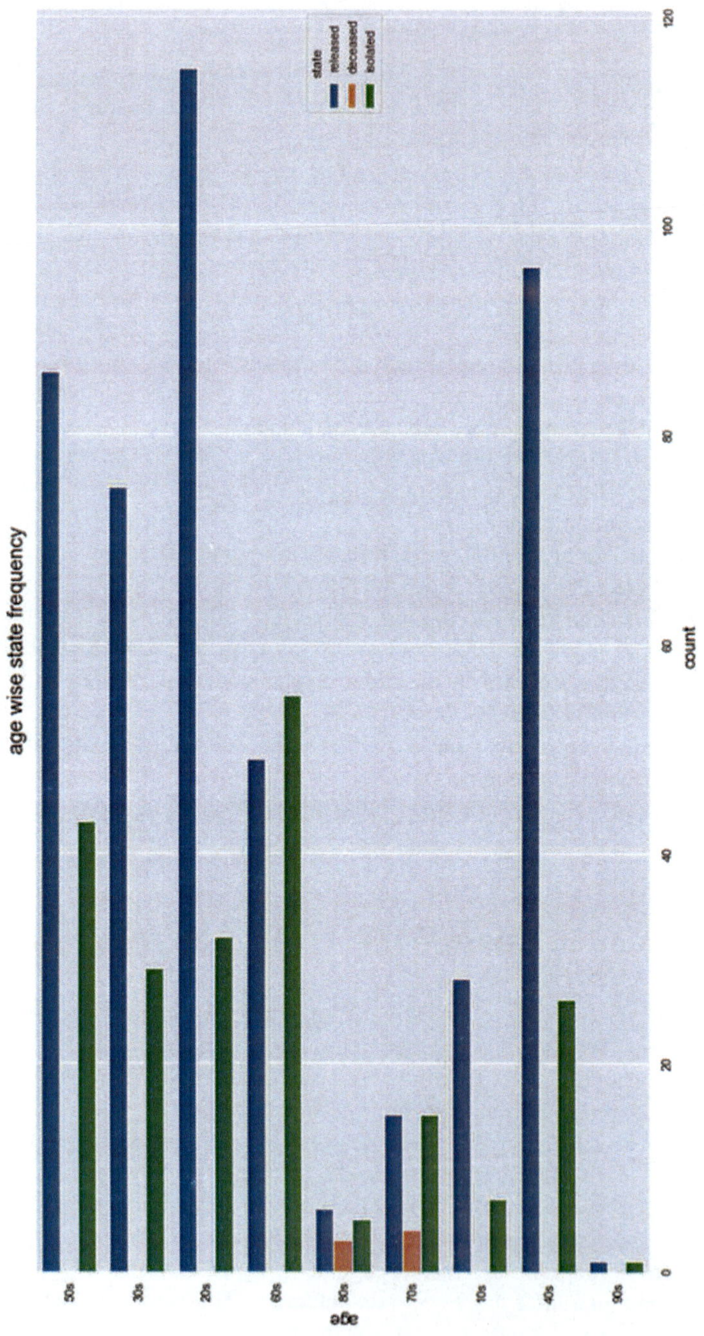

Fig. 6 Age wise distribution of state of patient

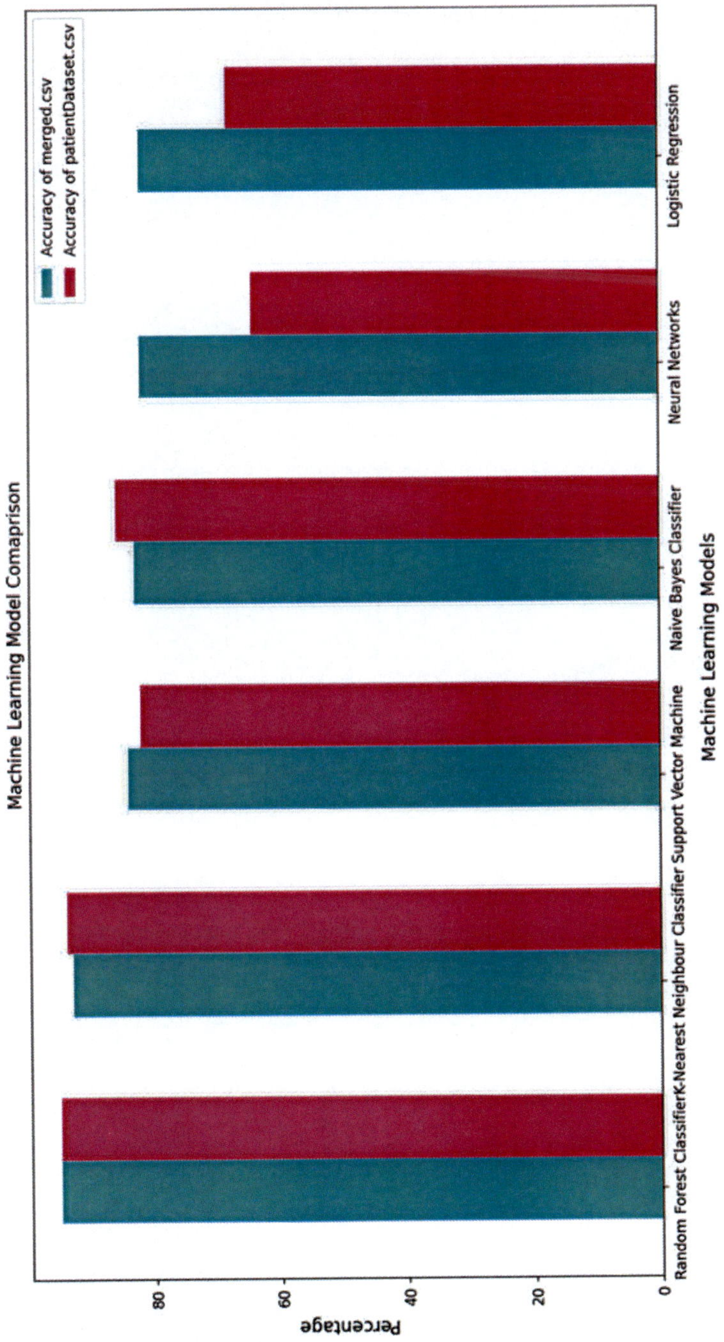

Fig. 7 Bar plot of model with two set of dataset

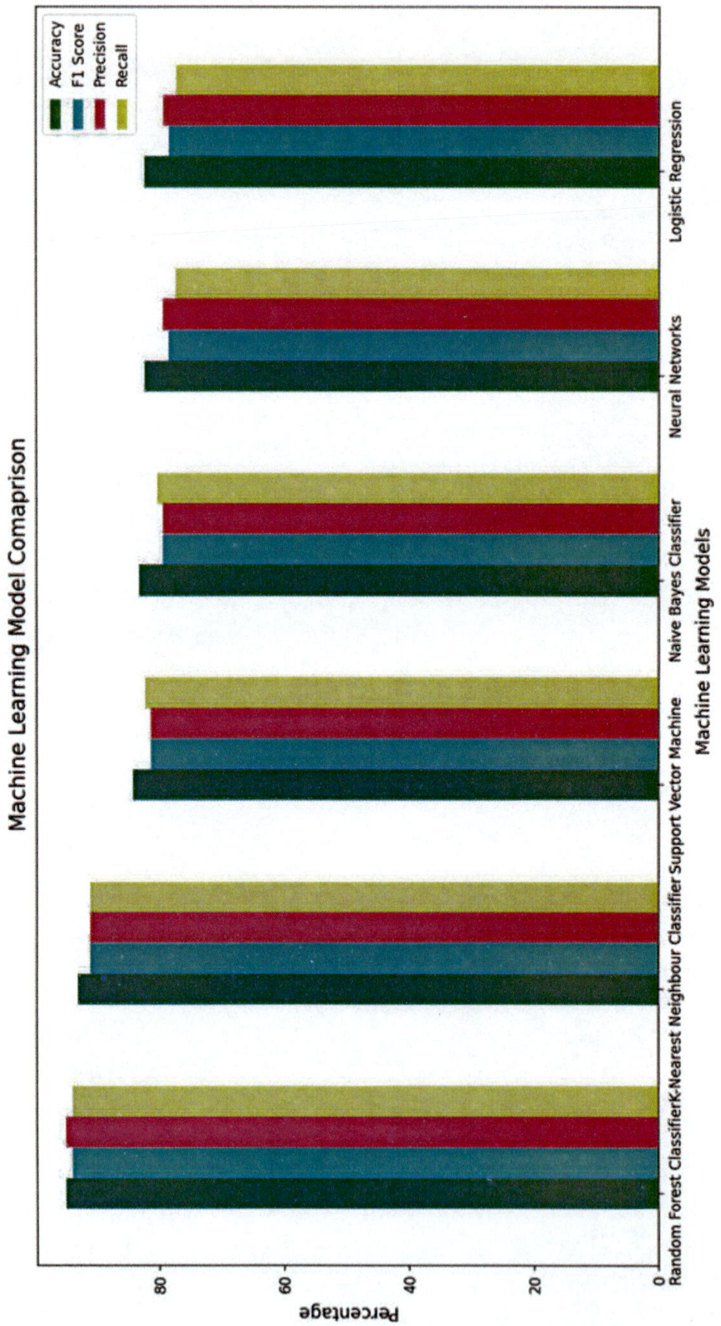

Fig. 8 Bar plot of model with merged dataset

gives best result with data preprocessing and feature engineering with accuracy as 95.40% and precision value as 94.52% followed by KNN algorithm with accuracy as 94.90% and precision value as 93.74%.

To keep upgrading this work with new analysis and scenarios, need large dataset with minimum one year of weather and patient information, risk of patient by different variation of Covid-19, population density, migration patterns that may directly influence what will be the risk of Covid-19 of a normal person at particular location. As a result, the creation of classification model related to environmental circumstances and medical regulations is a valuable piece of information for humanity at this critical time.

References

1. Malki Z, Atlam ES, Hassanien AE, Dagnew G, Elhosseini MA, Gad I (2020) Association between weather data and COVID-19 pandemic predicting mortality rate: machine learning approaches. Chaos Solitons Fract 138:110137
2. Dangi RR, George M (2020) Temperature, population and longitudinal analysis to predict potential spread for COVID-19. In: Population and longitudinal analysis to predict potential spread for COVID-19, 24 Mar 2020
3. Huang C, Wang Y, Li X, Ren L, Zhao J, Hu Y, Cao B (2020) Clinical features of patients infected with 2019 novel coronavirus in Wuhan, China. Lancet 395(10223):497–506
4. VoPham T, Hart JE, Laden F, Chiang YY (2018) Emerging trends in geospatial artificial intelligence (geoAI): potential applications for environmental epidemiology. Environ Health 17(1):1–6
5. Li W (2020) GeoAI: where machine learning and big data converge in GIScience. J Spat Inf Sci 20:71–77
6. Melin P, Monica JC, Sanchez D, Castillo O (2020) Analysis of spatial spread relationships of coronavirus (COVID-19) pandemic in the world using self organizing maps. Chaos Solitons Fract 138:109917
7. Kamel Boulos MN, Peng G, VoPham T (2019) An overview of GeoAI applications in health and healthcare. Int J Health Geogr 18(1):1–9
8. Dataset collection. https://www.kaggle.com/datasets/kimjihoo/coronavirusdataset. Accessed 12 June 2022
9. Muhammad LJ, Algehyne EA, Usman SS, Ahmad A, Chakraborty C, Mohammed IA (2021) Supervised machine learning models for prediction of COVID-19 infection using epidemiology dataset. SN Comput Sci 2(1):1–13
10. Khanam F, Nowrin I, Mondal MRH (2020) Data visualization and analyzation of COVID-19. J Sci Res Rep 26(3):42–52
11. Demongeot J, Flet-Berliac Y, Seligmann H (2020) Temperature decreases spread parameters of the new Covid-19 case dynamics. Biology 9(5):94
12. Sajadi MM, Habibzadeh P, Vintzileos A, Shokouhi S, Miralles-Wilhelm F, Amoroso A (2020) Temperature, humidity, and latitude analysis to estimate potential spread and seasonality of coronavirus disease 2019 (COVID-19). JAMA Netw Open 3(6):e2011834
13. Iwendi C, Bashir AK, Peshkar A, Sujatha R, Chatterjee JM, Pasupuleti S, Jo O (2020) COVID-19 patient health prediction using boosted random forest algorithm. Front Public Health 8:357
14. Muhammad LJ, Islam M, Usman SS, Ayon SI (2020) Predictive data mining models for novel coronavirus (COVID-19) infected patients' recovery. SN Comput Sci 1(4):1–7
15. Narin A, Kaya C, Pamuk Z (2021) Automatic detection of coronavirus disease (COVID-19) using X-ray images and deep convolutional neural networks. Pattern Anal Appl 24(3):1207–1220

16. Xu R, Rahmandad H, Gupta M, DiGennaro C, Ghaffarzadegan N, Amini H, Jalali MS (2020) The modest impact of weather and air pollution on COVID-19 transmission. MedRXiv
17. Cartocci A, Cevenini G, Barbini P (2021) A compartment modeling approach to reconstruct and analyze gender and age-grouped CoViD-19 Italian data for decision-making strategies. J Biomed Inform 118:103793
18. Her AY, Bhak Y, Jun EJ, Yuan SL, Garg S, Lee S, Shin ES (2022) Sex-specific difference of in-hospital mortality from COVID-19 in South Korea. PLoS One 17(1):e0262861
19. Tamerius J, Nelson MI, Zhou SZ, Viboud C, Miller MA, Alonso WJ (2011) Global influenza seasonality: reconciling patterns across temperate and tropical regions. Environ Health Perspect 119(4):439–445
20. Barreca AI, Shimshack JP (2012) Absolute humidity, temperature, and influenza mortality: 30 years of county-level evidence from the United States. Am J Epidemiol 176(Suppl 7):S114–S122
21. Shaman J, Goldstein E, Lipsitch M (2011) Absolute humidity and pandemic versus epidemic influenza. Am J Epidemiol 173(2):127–135
22. Żuk T, Rakowski F, Radomski JP (2009) Probabilistic model of influenza virus transmissibility at various temperature and humidity conditions. Comput Biol Chem 33(4):339–343
23. Żuk T, Rakowski F, Radomski JP (2009) A model of influenza virus spread as a function of temperature and humidity. Comput Biol Chem 33(2):176–180
24. Lowen AC, Mubareka S, Steel J, Palese P (2007) Influenza virus transmission is dependent on relative humidity and temperature. PLoS Pathog 3(10):e151
25. Jyoti K (2022) Code that you may find on my GitHub page. https://github.com/Jyoti-ops/Mtech-Dissertation

Neuro-Fuzzy-Based Supervised Feature Selection: An Embedded Approach

Madhusudan Rao Veldanda and V. N. Sastry

Abstract In problems of machine learning, selection of important features is an important pre-processing step. Many large datasets contain redundant information in the form of non-informative features. Sometimes the features can be irrelevant and noisy. Feature selection is the process of obtaining a minimal feature subset that sufficiently represents the original set of features. This subset of features can subsequently be used for tasks such as classification, clustering and inference. Most of the work on feature selection found in literature is either supervised or unsupervised. In this paper, an Adaptive Neuro-Fuzzy Inference System (ANFIS)-based feature ranking and selection approach which is supervised in nature is presented. For feature ranking, keep one-leave the rest and leave one-keep the rest strategies are used. A greedy forward feature selection approach is also presented, which aims to select a feature subset. The results of the proposed approaches are compared with the results obtained by a well-known unsupervised neuro-fuzzy feature selection algorithm and the supervised Relief-F algorithm. The findings are quite encouraging.

Keywords Feature selection · Feature ranking · Supervised · Unsupervised · ANFIS · Neuro-fuzzy · Fuzzy inference

1 Introduction

In many real world datasets some degree of redundancy is present in the form of identical object instances, irrelevant features and features that are not independent. This redundancy can mislead a classifier into the extraction of false rules and can also make learning a very time taking process. The aim of feature selection is to filter out the irrelevant and dependent features and is regarded as a pre-processing

M. R. Veldanda (✉)
Geethanjali College of Engineering and Technology, Hyderabad, India
e-mail: madhuveldanda.cse@gcet.edu.in

V. N. Sastry
IDRBT, Masab Tank, Hyderabad, India
e-mail: vnsastry@idrbt.ac.in

S. Das et al. (eds.), *Advances in Data-driven Computing and Intelligent Systems*, Lecture Notes in Networks and Systems 653, https://doi.org/10.1007/978-981-99-0981-0_52

step in knowledge extraction or classifier training. Application of feature selection algorithms has been found on datasets from different domains such as image recognition as reported by Jelonek and Stefanowski [10], bioinformatics as in Cedeño and Agrafiotis [2] and Xiong et al. [24]. Forman [6] and Jensen and Shen [11] have used feature selection in text classification. Most of the feature selection algorithms are supervised in nature and can be applied to labeled datasets. Whereas unsupervised feature selection algorithms are relatively small in number.

In feature selection algorithms, the original meaning and semantics of the data is preserved. Whereas in feature extraction methods, the original set of attributes are transformed, loosing the original semantics thereby. Feature ranking methods reported by Varshavsky [22] and He et al. [7] rank the features according to their importance, calculated according to some criteria such as consistency, entropy. Given 'n' number of features, the problem of feature selection can be viewed as a search for the minimal feature subset. In an exhaustive search, the search space is made up of the 2^n candidate subsets. As we can readily see, the search space increases exponentially with the number of features. This makes an exhaustive search prohibitively expensive in the context of time. Many feature selection methods adopt either heuristic or random search strategies, sacrificing optimality in favor of time.

Whereas the feature selection algorithms select a subset of the full feature set, ranking methods rank the features according to some evaluation criteria. After ranking the features, the user sometimes specifies the number of most important features to be selected. The ranking criterion is useful as a guiding principle for the purpose of feature selection either in the forward selection approach or in the backward elimination approach. Feature selection approaches can be categorized as (a) filter, (b) wrapper or (c) embedded.

Fuzzy systems which are built on fuzzy rules have been successfully used to carry out various classification tasks as reported by Bezdek et al. [1], Nozaki et al. [16]. Fuzzy systems depend on rules involving linguistic variables which are either provided by domain experts or extracted from training data by a variety of methods as reported by Chiu [4] and Pal et al. [17]. The work reported by Jang [9], Jang and Sun [8], Kasabov and Song [12] show that fuzzy systems can be efficiently implemented by neural networks. Such systems are broadly named as neural fuzzy systems and one such system is used in this current work. Neuro-fuzzy approaches can also be used for feature selection as shown by Chakraborty and Pal [3] and Pal et al. [18].

Features having good predictive power for classifying instances are useful and therefore most feature selection methods attempt to find them. However, as there may exist correlation or non-linear dependence between such features, we are likely to obtain a better performance by removing such redundant features.

Chung et al. [5] propose a novel machine learning method that imposes a penalty on the use of dependent or correlated features during the system identification phase along with feature selection. Feature subset selection is important in the areas of data mining and information processing. However, most methods only consider the relevance factor while ignoring diversity. This makes it difficult to exploit the hidden information. To address this, Tan et al. [20] propose a hybrid model named intuitionistic fuzzy (IF) rough set. The proposed model combines the technical advantages of rough sets and IF set and can effectively consider both relevance and diversity. To enable the fuzzy rough set for big data analysis, Kong et al. [13] propose a novel distributed fuzzy rough set (DFRS)-based feature selection, which separates and assigns the tasks to multiple nodes for parallel computing. Tavakoli et al. [21] propose a new online approach to adapt some systematic parameters of a typical fuzzy inference system. They select some training and systematic parameters during the training procedure to be constant or to be adaptive. Neural networks have the ability to learn input/output mappings through minimization of a suitable error function by means of a back propagation algorithm. Unfortunately, neural networks cannot represent knowledge explicitly as they function as black boxes. On the other hand, a fuzzy system can represent knowledge through fuzzy if-then rules. Thus, an integration of neural networks and fuzzy systems can yield systems, which are capable of learning as well as decision making.

This paper is organized as follows. In the next section, we present a brief introduction to the concept of fuzzy inference and ANFIS. Two new supervised feature ranking algorithms and one feature subset selection algorithm based on ANFIS, are presented in Sect. 3. Section 4 contains the details of the experimental setup used in the current work and the results obtained. Section 5 contains discussion and conclusions.

2 Fuzzy Inference and ANFIS

A fuzzy inference system is a model that maps input attributes to input membership functions and then input membership function to fuzzy rules. After that, rules are mapped to a set of output characteristics, output characteristics to output membership functions, and then the output membership function to a decision associated with the output. Two of the most popularly used fuzzy inference systems are proposed by [15] and an another proposed by [19].

ANFIS, the acronym, stands for adaptive neuro-fuzzy inference system. From a given labeled data set, ANFIS constructs a fuzzy inference system (FIS) whose membership function parameters are adjusted using either a back propagation algorithm or a combination with a least squares-based method. This tuning makes the fuzzy systems to learn from the data they are supposed to model.

Inputs are mapped via a network kind of a structure, by input membership functions and associated parameters, and then by output membership functions and associated parameters to outputs. The parameters corresponding to the input and output membership functions change throughout the learning process. The computation of these parameters is done with the help of a gradient vector. The gradient vector indicates how well the FIS is modeling the data for a given set of parameters. When the gradient vector is obtained, any of different optimization techniques can be applied in order to tune the parameters to reduce error. In many cases, the error measure is usually based on the sum of the squared difference between actual and expected outputs. In ANFIS, either back propagation or a combination thereof with least squares based methods are used for estimating membership function parameters. ANFIS uses a similar modeling approach to many system identification methods. To begin with, a parameterized model is chosen and then, input-output data in an appropriate form that is usable by ANFIS for training is collected. The FIS model is then evolved to represent the training data by continuously modifying the membership function parameters.

3 The Proposed Feature Selection Algorithms

Three feature selection algorithms which are embedded in nature have been proposed here. In the first strategy, an ANFIS system is created initially with all the features and then optimized. From the resulting model, only the rule that contains the feature that is being evaluated is removed keeping the rest of the rules. The reduced model is then used for inference on the test data and the Root Mean Squared Error (RMSE) is noted. The feature whose removal results in a larger RMS is ranked higher. This is referred to as Leave-one-out strategy.

Algorithm 1 Leave one-keep the rest

1: A training data set and a testing dataset are created from a given dataset, $D_{(M \times N)}$
2: Using the training dataset, an initial Fuzzy Inference System (FIS) is created.
3: Then, using the testing dataset, the ANFIS system is optimized for the appropriate membership function
4: The optimized ANFIS is then used to test the test data.
5: The average RMS error in prediction is obtained
6: **for** feature $= 1, 2, \ldots, N$ **do**
7:     From each rule, remove this feature from the antecedent.
8:     Using rules with reduced antecedents, test the data again.
9:     Obtain the average RMS error in prediction.
10: **end for**
11: Rank the features according to the amount of error introduced in their absence. That is, if the error obtained by removing the feature x is more than the error obtained by removing the feature y, x is ranked higher than y

In the second approach, all the rules are removed from the original ANFIS model except the rule containing the feature being evaluated. The resulting model, which contains only one rule is used in fuzzy inference and the RMS is noted. The feature responsible for the least RMS is ranked highest and so on.

The third approach is a greedy forward selection method. In this approach the top ranked feature is selected from the second approach. This feature is combined separately with each one of the remaining features, resulting in several feature subsets of size two. Each one of these subsets containing two features is evaluated by keeping the corresponding rules in the ANFIS model and removing the rest. The subset with the least RMS is selected in this step. This process is repeated until the resulting RMS is either less or equal to the RMS obtained with the full feature set.

Algorithm 2 Keep one-leave the rest

1: A training dataset and a testing dataset are created from a given dataset, $D_{(M \times N)}$
2: Using the training dataset, an initial Fuzzy Inference System (FIS) is created.
3: Then, using the testing dataset, the ANFIS system is optimized for the appropriate membership function
4: The optimized ANFIS is then used to test the test data.
5: The average RMS error in prediction is obtained
6: **for** feature $= 1, 2, \ldots, N$ **do**
7:     In each rule, keep only this feature
8:     Using rules with reduced antecedents, test the data again.
9:     Obtain the average RMS error in prediction.
10: **end for**
11: Rank the features according to the amount of error introduced in their absence. That is, if the error obtained by removing the feature x is more than the error obtained by removing the feature y, x is ranked higher than y

Algorithm 3 Nuero-fuzzy greedy forward selection (NFGFS)

1: A training data set and a testing dataset are created from a given dataset, $D_{(M \times N)}$
2: Using the training dataset, an initial FIS (Fuzzy Inference System) is created.
3: Then, using the testing dataset, the ANFIS system is optimized for the appropriate membership function
4: The optimized ANFIS is then used to test the test data.
5: The average RMS error in prediction, $E(C)$, where C is the full feature set, is obtained
6: Let R be an empty set
7: **while** $E(R) > E(C)$ **do**
8:     Assign R to T
9:     **for** $\forall x \in C - R$ **do**
10:         Calculate $E(R \cup x)$
11:         If $E(R \cup x) < E(T)$ then assign $(R \cup x)$ to T
12:     **end for**
13:     Assign T to R
14: **end while**
15: Output the selected feature subset R

4 Experimental Setup and Result Analysis

The proposed ANFIS-based feature evaluation approaches are applied to three benchmark datasets from the UCIML repository. In order to the put them in perspective, the results are compared with those obtained with Relief-F which is a well-known supervised feature ranking algorithm, proposed by Kononenko [14]. A comparison is also carried out with another neuro-fuzzy feature ranking algorithm, proposed by Pal et al. [18], referred to in this work as NFA. The characteristics of the datasets used are presented in Table 1. The feature rankings obtained from the existing methods when applied to the datasets are presented in Table 2. There are four features in the Iris dataset. If the rankings are given as 4, 3, 1, 2 for example, it means that the 4th feature is ranked first, 3rd feature is ranked second and so on. The rankings

Table 1 UCIML datasets

S. No.	Data sets	Instances	Features	Classes
1	Iris	150	04	3
2	Glass	214	09	7
3	Pima diabetes	768	08	2

Table 2 Feature ranking by well-known algorithms

Data set	NFA	Relief-F
Iris	3, 4, 2, 1	4, 3, 1, 2
Pima diabetes	2, 3, 4, 8, 5, 6, 1, 7	2, 6, 4, 1, 8, 7, 3, 5
Glass	8, 3, 1, 7, 6, 9, 4, 5, 2	3, 4, 8, 7, 2, 1, 5, 6, 9

Table 3 Feature ranking by proposed algorithms

Data set	Alg1	Alg2	NFGFS
Iris	4, 3, 2, 1	4, 3, 1, 2, 7	4, 3, 2
Pima diabetes	2, 8, 1, 6, 5, 7, 3, 4	8, 2, 6, 5, 4, 1, 7, 3	8, 2, 1, 7
Glass	3, 9, 4, 8, 5, 1, 7, 6, 2	3, 7, 1, 8, 6, 5, 4, 9, 2	3, 9, 1, 4, 8

obtained from the first two algorithms and the feature subsets as obtained by the NFGFS algorithm are presented in Table 3.

4.1 Performance Evaluation of the Algorithms

For a given dataset and algorithm, the following experimental procedure is adopted:

1. Features are ranked by applying the algorithm to the dataset.
2. To begin with, the dataset is represented by keeping only the first most important feature and removing the rest of the features.
3. To this dataset of reduced dimensionality, the J48 classifier from the Weka machine learning tool box [23] is applied and the percentage error in classification is obtained.
4. Then, two most important features are kept and the others are removed. Error in classification is again noted with this dataset of reduced dimensionality.
5. Like this, the same procedure is repeated with increasing number of most important features (according to the ranking order).
6. Graphs are plotted with the number of important features used to represent the dataset on the X-axis and the corresponding percentage of classification error on the Y-axis.

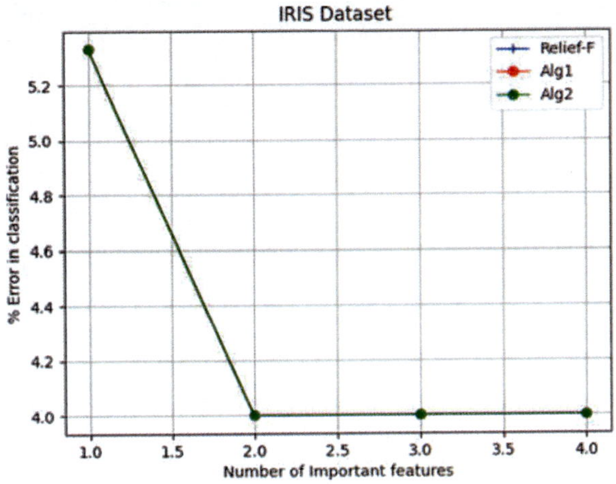

Fig. 1 Performance evaluation against Relief-F (IRIS Dataset)

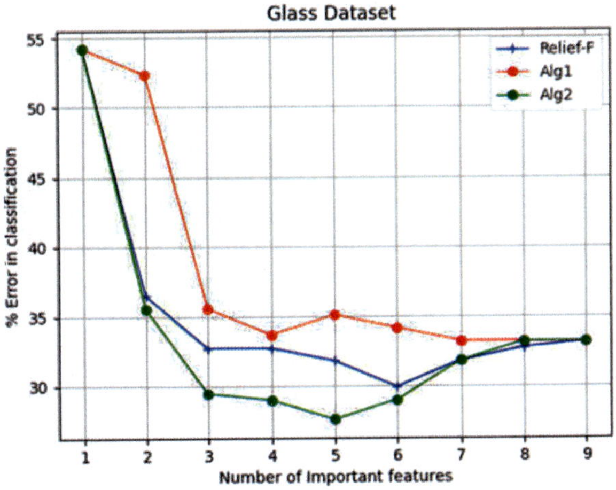

Fig. 2 Performance evaluation against Relief-F (Glass Dataset)

4.2 Comparison with a Supervised Algorithm (Relief-F)

The classification error rates obtained for the Relief-F algorithm and Alg1 and Alg2 are presented in Figs. 1, 2, and 3 for the Iris, glass and pima diabetes datasets, respectively. It can be observed that for the Iris dataset the performance of all the algorithms is equal. For the glass dataset, Alg2 has clearly outperformed the other algorithms. For the Pima diabetes dataset, the algorithms performed more or less equally except for the first feature, where algorithm 2 seems to err a bit more.

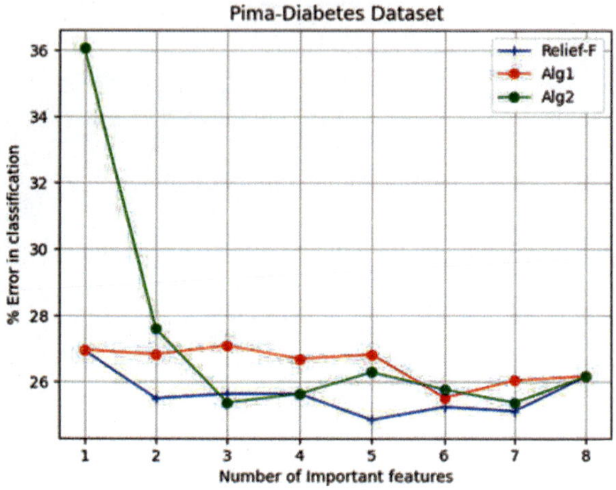

Fig. 3 Performance evaluation against Relief-F (Pima Diabetes Dataset)

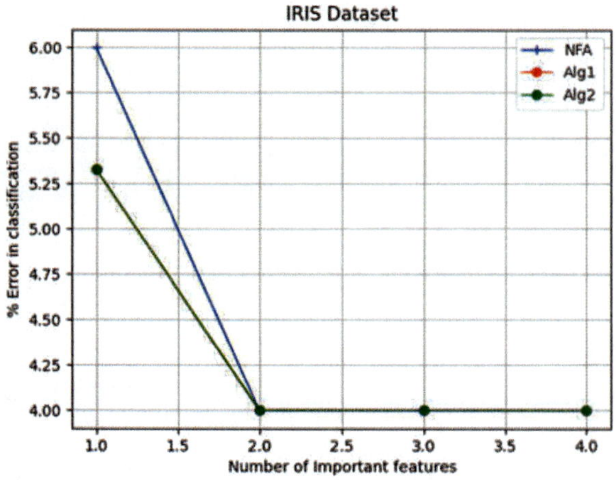

Fig. 4 Performance evaluation against NFA (IRIS Dataset)

4.3 Comparison with an Unsupervised Neuro-Fuzzy Algorithm

The proposed algorithms are further compared with the unsupervised neuro-fuzzy approach for feature ranking. The results are presented in Figs. 4, 5, and 6 for the Iris, glass and pima diabetes datasets, respectively. It can be observed that for the Iris dataset the performance of all the algorithms is more or less equal. For the glass dataset, Alg2 has clearly outperformed the other algorithms. For the pima diabetes dataset also, Alg2 has performed well in comparison with the other two giving the least error for a lesser number of features (3 top features).

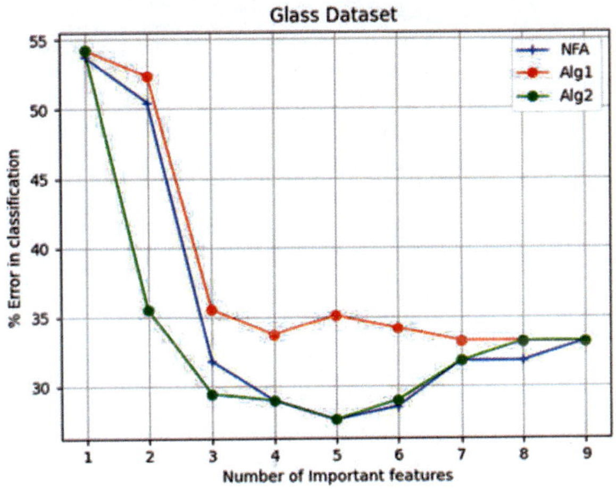

Fig. 5 Performance evaluation against NFA (Glass Dataset)

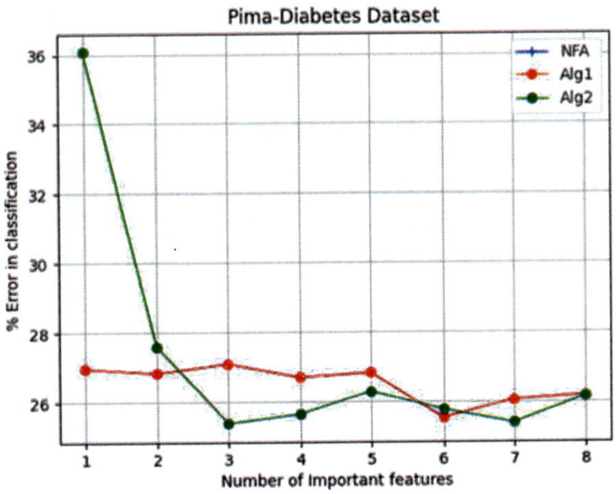

Fig. 6 Performance evaluation against NFA (Pima Diabetes Dataset)

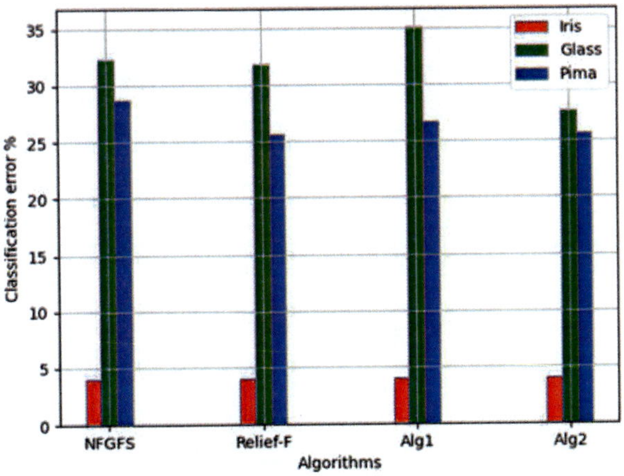

Fig. 7 Overall Performance evaluation against NFA

4.4 Evaluation of NFGFS

NFGFS has selected a feature subset of size 3 for Iris, feature subset of size 5 for glass and a feature subset of size 4 for pima diabetes dataset as shown in Table 3. In order to compare its performance with other algorithms, the classification error rates with feature subsets of the same size are used. It can be observed from the results shown in Fig. 7 that Alg2 again either outperforms or equals all other algorithms.

5 Conclusion

This paper empirically compares the performance of three new feature selection approaches between themselves and with the well-known Relief-F algorithm which is supervised in nature and with an existing neuro-fuzzy approach which is unsupervised. Experimental results suggest that the performance of Alg2, which employs the keep one-leave the rest approach, has consistently equaled and sometimes even bettered the performance of other approaches. In this paper, the criterion for feature importance or saliency is the root mean squared error as obtained either by the absence or the presence of the feature in the antecedents of fuzzy rules of the ANFIS model.

Future work may include the use of other criteria such as the unbiased criterion and the regularity criterion and different search strategies such as Monte Carlo methods for generating feature subsets.

References

1. Bezdek JC, Keller J, Krishnapuram R, Pal NR (1999) Fuzzy models and algorithms for pattern recognition and image processing. Kluwer, Watham, MA
2. Cedeño W, Agrafiotis DK (2003) Using particle swarms for the development of QSAR models based on k-nearest neighbor and kernel regression. J Comput-Aided Mol Des 17:255–263
3. Chakraborty D, Pal NR (2004) A neuro-fuzzy scheme for simultaneous feature selection and fuzzy rule-based classification. IEEE Trans Neural Netw 15(1):110–123
4. Chiu SL (1994) Fuzzy model identification based on cluster estimation. J Int Fuzzy Syst 2:267–278
5. Chung IF et al (2018) Feature selection with controlled redundancy in a fuzzy rule based framework. IEEE Trans Fuzzy Syst 26(2)
6. Forman G (2003) An extensive empirical study of feature selection metrics for text classification. J Mach Learn Res 3:1289–1305
7. He X, Cai D, Niyogi P (2005) Laplacian score for feature selection. In: NIPS: advances in neural information processing systems, Vancouver, Canada, 8–10 Dec 2005
8. Jang J-SR, Sun C-T (1995) Neuro-fuzzy modeling and control. Proc IEEE 83:378–406
9. Jang J-SR (1993) ANFIS: adaptive-network-based fuzzy inference system. IEEE Trans Syst Man Cybern 23:665–685
10. Jelonek J, Stefanowski J (1997) Feature subset selection for classification of histological images. Artif Intell Med 9(3):227–239

11. Jensen R, Shen Q (2004) Fuzzy-rough attribute reduction with application to web categorization. Fuzzy Sets Syst 141(3):469–485
12. Kasabov N, Song Q (2002) DENFIS: dynamic evolving neural-fuzzy inference system and its application for time series prediction. IEEE Trans Fuzzy Syst 10
13. Kong L et al (2020) Distributed feature selection for big data using fuzzy rough sets. IEEE Trans Fuzzy Syst 28(5):846–857. https://doi.org/10.1109/TFUZZ.2019.2955894
14. Kononenko I (1994) Estimating attributes: analysis and extensions of RELIEF. In: European conference on machine learning, pp 171–182
15. Mamdani EH (1977) Applications of fuzzy logic to approximate reasoning using linguistic synthesis. IEEE Trans Comput 26(12):1182–1191
16. Nozaki K, Ishibuchi H, Tanaka H (1996) Adaptive fuzzy rule-based classification systems. IEEE Trans Fuzzy Syst 4:238–250
17. Pal NR, Pal K, Bezdek JC, Runkler TA (1997) Some issues in system identification using clustering. In: International joint conference on neural networks, ICNN 1997, Piscataway, NJ. IEEE, pp 2524–2529
18. Pal SK, De RK, Basak J (2000) Unsupervised feature evaluation: a neuro-fuzzy approach. IEEE Trans Neural Netw 11(2):366–376
19. Sugeno M (1985) Industrial applications of fuzzy control. Elsevier Science Pub. Co.
20. Tan A et al (2019) Intuitionistic fuzzy rough set-based granular structures and attribute subset selection. IEEE Trans Fuzzy Syst 27(3)
21. Tavakoli P, Vaezi N, Fahim P, Karimpour A (2021) A new approach based on fuzzy-adaptive structure and parameter learning applied in meta-cognitive algorithm for ANFIS. In: 2021 7th international conference on control, instrumentation and automation (ICCIA), pp 1–5. https://doi.org/10.1109/ICCIA52082.2021.9403539
22. Varshavsky R (2006) Novel unsupervised feature filtering of biological data. Bioinformatics 22(14):e507–e513
23. Weka 3: data mining software in Java. http://www.cs.waikato.ac.nz/ml/weka/
24. Xiong M, Li W, Zhao J, Jin L, Boerwinkle E (2001) Feature (gene) selection in gene expression-based tumor classification. Mol Genet Metab 73(3):239–247

An Overview of Hybridization of Differential Evolution with Opposition-Based Learning

Shweta Sharma, Vaishali Yadav, Ashwani Kumar Yadav, and Anuj Arora

Abstract The paper is focused on comprehensive survey on the concept of hybridizing opposition-based learning with Differential Evolution Algorithm. In this paper, we focused on all phases such as population initialization, mutation, crossover and generation jumping, on which the methods applied, application part and motivation.

Keywords Opposition-based learning · Differential evolution algorithm · Optimization technique · Hybridization

Abbreviations

PI	Population Initialization
NP	Newly generated Population
MT	Mutation
SL	Selection
σ	Standard deviation
D	Dimension size
Jr	Jumping Rate
TVJR1	Higher jumping rate in exploration
TVJR2	Lower jumping rate in exploitation
NA	Not Applicable
DE	Differential Evolution
ODE	Opposition-based Differential Evolution
SDE	Shuffled Differential Evolution
CDOR	Coordination of directional overcurrent relays

S. Sharma · V. Yadav (✉)
Manipal University Jaipur, Jaipur, Rajasthan, India
e-mail: vaishaliyadav26@gmail.com

A. K. Yadav · A. Arora
Amity University Rajasthan, Jaipur, Rajasthan, India

S. Das et al. (eds.), *Advances in Data-driven Computing and Intelligent Systems*, Lecture Notes in Networks and Systems 653, https://doi.org/10.1007/978-981-99-0981-0_53

ADE Adaptive Differential Evolution Algorithms

1 Introduction

Optimization is a process of determining the best or it is an act of creating the best or making the optimal/finest use of any situation or resources. It may also be termed as 'Finding the perfect or as effective as possible alternative or result by maximizing the desired and minimizing the undesired factors under the given constrictions. With respect to the field of Mathematics, Computer Science and Operations Research, optimization can be defined in terms of the selection of a best facet (as per certain norms) from a set of available alternatives. In this process, input values are systematically selected from an allowed set to maximize or minimize a real function and computes its value.

2 Overview

Researchers are always keen to find new and innovative ideas for revolution that are useful for humans in their day today life. That is why huge number of technologies are taken place on human efforts. In this context, machine learning [1] helped a lot to solve real world problems through different nature inspired metaheuristic algorithms [2, 3]. Evolutionary computation (EC), artificial neural networks (NN), artificial immune systems (AIS), fuzzy systems and swarm intelligence (SI) are some of the important areas of nature inspired algorithm (NIA) where they contributed like never before for problem solving system. For the same, Evolutionary Computation paradigm is chosen to explore the opportunities of research.

Population initialization techniques are classified on the bases of randomness, compositionality and generality. One of the most prevalent techniques for population generation based on compositionality is opposition-based learning (OBL).

2.1 *Opposition Based Learning*

In the present era, 'opposition' is very powerful word that can see in various areas. Opposition exists almost everywhere in existing world, irrespective of what quantity, quality, strength and shape its presence is always come across, the behaviour of entities and its reverse entities can be understandable in various ways [4]. The variety and intricacy if opposition may be understood by these set of words: antipodal, hostile, disparate antipodeans, contradictory, antagonistic, contrary, diametrical, polar, antipathetic, unalike, counter, adverse, dissimilar, reverse, inverse, converse, divergent antithetical, etc. These all the words that explain some concept of opposition in addition it can easily be used at various contexts to represent various relationships.

2.2 *Opposite Point Concept*

Numerous machine intelligence algorithms are encouraged by various natural systems. Neural networks, bee colonies, genetic algorithms and reinforcement agents, etc., to name a few, well recognized practices stimulated by human nervous system, evolution and animal intelligence psychology. As a new scheme in machine intelligence, opposition-based learning was introduced in [5]. The main idea of this optimization is to find an improved result. The technique that simultaneously consider the estimate and its equivalent opposite estimate is nearer to the global optimum. To find improved results, both actual and opposite directions will help to resolve the problem quickly and efficiently. The major foundation of the approach is weights and opposite weights, estimates and counter-estimates and actions versus counteractions.

Opposition-based learning was developed by Tizhoosh [5]. The core foundation of this unique approach was estimates and counter-estimates, weights and opposite weights and actions versus counter-actions. Nowadays, it is incorporated on many nature inspired algorithms for optimization. OBL is a popular initialization technique which is widely used on different steps of evolutionary algorithms. The expansion of this method initially started on the exploration of the search space so that each small area of search can be exploited. As the time grows, researcher shows their interest in this particular area and the concept utilized drastically in various steps of algorithms like DE, PSO, ABC, BBO, neural networks, reinforcement learning, fuzzy systems, harmony search, cuckoo search, etc.

3 Differential Evolution

Differential Evolutionary algorithms [6] are stochastic algorithms. The term 'Differential Evolution' is derived from word 'Evolution', i.e. the process of development of any living thing from its former forms. 'The survival of fittest', by Herbert Spencer, arises from evolutionary theory, a process of explaining the concept of natural selection. This means that whosoever is best adapted to the current environment refers to 'fit' as opposed to 'unfit'. Every individual struggle to reproduce its offspring and the fittest has the best chance to reproduce. DE is population-based random probability distribution process in which best must survive for its upcoming generations. The process goes through different stages that are population initialization, mutation, crossover and selection.

The concept of OBL is used to overcome the problems faced by the various researchers. To determine the utilization of OBL in Differential Evolutional (DE) Algorithm, Table 1 is given below to describe the work done on different stages of DE on the bases of opposition-based learning. For the same, various papers are reviewed which are relevant to the title. The papers which are relevant to the title are reviewed, and according to the results, the papers are listed below. These papers were published between 2005 and 2022. The table includes the detailed explanation of each paper and is differentiated with respect to its authors, work they have done, the phase on which OBL applied, application where the algorithm used and finally the motivation why that algorithm is designed (Fig. 1).

Table 1 Comprehensive survey of opposition-based learning in differential evolution algorithm

Author	Work Done	Phase	Application	Motivation
Rahnamayan et al. [7]	In this study, the concept of OBL is incorporated in Differential Evolution Algorithm first time for initializing the population in noise free environment and in generation jumping process	PI	Basic DE	Fast convergence
Rahnamayan et al. [8]	A versatile nature of OBL is integrated in initial population that can discover and develop the search space in noisy problems	PI	Noisy function of DE ($\sigma = 0, 0.25, 0.50, 0.75$)	Fast convergence
Rahnamayan et al. [9]	Due to the no limitation of dimensions to generate opposite vector, OBL easily accommodated with basic DE. Acceleration rate of algorithm is compared with different dimension size. Also, outcomes are compared with respect to random initialization technique and opposing based technique	PI	Basic DE (10 > D > 10)	Acceleration of EA
Rahnamayan et al. [10]	Three types of generation jumping trend is established named as deterministic, adaptive and self-adaptive is introduced that are changeable according to time for each entity to jump to the opposite direction	PI, NP	ODE ($J_r = 0.3$, 0.6, TVJR1, TVJR2)	Fast convergence

(continued)

Table 1 (continued)

Author	Work Done	Phase	Application	Motivation
Rahnamayan et al. [11]	A new extension of ODE is proposed where middle point (quasi) is founded between the original point and its opposite point of new solution candidate. It narrows down the search space by updating the boundaries from quasi and opposite points	NP	ODE	ODE acceleration
Rahnamayan et al. [12]	An experimental study is done using ODE for exploring and exploiting large scale problems	PI	Large scale functions ($D =$ 500, 1000)	Comparative analysis
Boskovic et. al [13]	An adaptive mutation is incorporated with OBL for tuning the chess program based on Differential Evolution	PI, MT	Chess Programme	NA
Ventresca et al. [4]	Opposition concept is used as a framework for population-based incremental learning that control the diversity boundaries for a specified population	PI	Probability update rule and Sampling procedure	Controlled exploration
Rahnamayan et al. [14]	The binary and normalized grey-level images are compared to find the dissimilarities between original image of grey scale and the threshold image generated from various methods through utilizing Micro DE and Micro-ODE (low dimension size) via pixel-to-pixel method	PI	Image processing	Exploitation
Rahnamayan et al. [15]	It is an extension of [15] with respect to dimensionality, population size, mutation operators, jumping rates and finally the comparison with various variants of DE	PI	Basic DE	Accuracy of ODE

(continued)

Table 1 (continued)

Author	Work Done	Phase	Application	Motivation
Wang et al. [16]	Scalability test is done on different dimension size by incorporating Generalized OBL	PI	Basic DE	Acceleration of DE
Mahamed et al. [17]	In mutation, free mind searching sense of animals is incorporated to enhance the optimization process. For that, initial population chosen randomly on the bases of worst in initial and opposite marked location. It has a limitation of performance in the presence of noise and also fixed number of iterations	PI	Basic DE	NA
Omrana et al. [18]	A perfect combination of chaotic search, opposition-based learning, Differential Evolution Algorithm and quantum mechanics is applied constrained functions	PI	ODE	NA
Tang et al. [19]	An improved ODE uses randomly adaptive value of scaling factor (F) combine with multiple opposite point as well as opposition calculated at the end of the process	PI	DE and ODE	Fast convergence
Xu et al. [20]	It is suggested that opposite points are to keep away from global optimum. It has to be used for current optimum that leads to better function optimization	NP	ODE	Optimization

(continued)

Table 1 (continued)

Author	Work Done	Phase	Application	Motivation
Wang et al. [21]	Generalize OBL is incorporated with ODE. Depend upon the opponent generation, if the values are satisfactory then GODE executes with random (GOBL-R) and if not, then algorithm automatically adapt basic DE	PI	ODE	NA
Thangaraj et al. [22]	Another variant of ODE is proposed where the value of scaling factor adapts dynamically via chaotic sequence	PI	ODE	Global optimization
Rahnamayan et al. [23]	Author prove the theory in a simpler way as compared with existing ones by generalizing higher dimensions, also analyse the intuition of opposition	PI	DE	Distance optimization
Jianghua et al. [24]	Permutation of three different strategy of mutation with three setting of control parameters are used as an adaptive approach	MT	DE	Adaptive DE
Alinia et al. [25]	Subdivision and shuffling of SDE and OBL are merged to accelerate DE	PI, MT	SDE	Accelerate DE
Karthikeyan et al. [26]	Virtual Mapping Procedure (VMP) is introduced in Generation Expansion Planning Problem (GEP) to convert the combination of candidate units in dummy decision variables by incorporating Oppositional Based Differential Evolution (ODE)	PI	Generation expansion planning problem	NA

(continued)

Table 1 (continued)

Author	Work Done	Phase	Application	Motivation
Wanga et al. [27]	A new approach known as GOjDE, is introduced where generalized opposition-based learning (GOBL) and self-adapting control parameters are combined to give better candidate solutions	PI	Graphics processing unit and central processing unit	High-dimensional optimization problems
Salehinejad et al [28]	Inverse-meta-modelling techniques is used for mapping of objective value with respect to variables values	PI	ODE	Enhance ODE
Chelliah et al. [29]	Coordination of directional over current relays is optimized by the use of chaotic approach with OBL	PI	CDOR	NA
Hu et al. [30]	Partial OBL is proposed where the vector generate partial opposite point rather than whole opposites in high dimensions	PI	ADE	Evaluation of DE on real parameters
Liu et al. [31]	An adaptive GODE algorithm is proposed known as AGODE that present the concept of adaptation with opposition mechanism for probability tuning of opposite points automatically throughout the search process	NP	GODE	Accelerate DE
Padma et al. [32]	Opposition-based modified differential evolutionary (OMDE) algorithm is presented in the direction of determining the optimal location of SSVR for refining distribution scheme performance	PI	SSVR device	NA

(continued)

Table 1 (continued)

Author	Work Done	Phase	Application	Motivation
Park and Lee [33]	It uses selection switching scheme ((μ, λ) selection and ($\mu + \lambda$) selection) and a partial dimensional changing scheme combine with diverse beta distribution like convex and concave. Also, OBL is utilized with beta distribution for selection switching and partial dimensional change and it combine with DE to improve the searchability and convergence speed	PI	DE	NA
Wang et. al [34]	A detailed comparative analysis is done on nine variants including basic DE. The mean value of each variant is compared on the bases of different runs and shows the outperformance in all the cases	PI	DE	Performance analysis
Salehinejad et al. [35]	Ensemble mutation scheme is used to select the population and opposition-based learning used to increase the diversity in MDE at the phase of mutation and selection	MT, SL	MDE	Fast convergence
Yanan et al. [36]	The DE and OBL is used as a tool for MFEA. To deal with diversity and convergence issue, they applied on single objective multi-tasking evolutionary optimization. Both the operators are combined with SBX for the exploration of search space	PI	MFEA	Better diversity and convergence

(continued)

Table 1 (continued)

Author	Work Done	Phase	Application	Motivation
Sharma et al. [37]	The concept of OBL is emerged with crossover phase of DE and applied on 20 benchmark functions	CR	20 benchmark functions	Better convergence
Wood et al. [38]	OBL is partially used to generate reference point. Then average of actuals and references' taken into consideration	PI	CEC-2014 benchmark functions on $D =$ 30, 50, and 100	Composite of best candidates
Mousavirad et al. [39]	A novel MLNN is proposed for training algorithm. Three techniques are merged together where Cen-S is for centroid of the best individuals then for the improvement of exploitation and last is DOBL strategy for generation of opposite leads to enhanced exploration	PI	26 conventional and population-based algorithms	Exploitation
Mousavirad [40]	The paper based on quasi-opposition-based differential evolution approach and applied on FFNN learning for the improvement	PI	Feedforward neural network	Find the proper weights for connections and biases

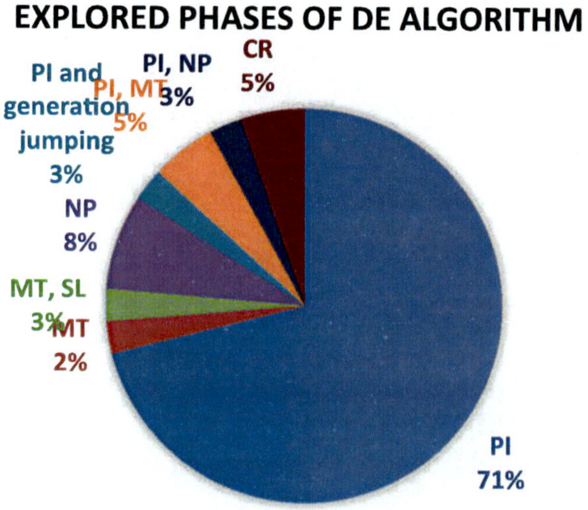

Fig. 1 Work done by various researchers on different phases of differential evolution algorithm

4 Research Gaps

The algorithmic assembly of DE provides its ability to create a tremendous fit singular that provides clues/leads to a search space till another individual of superior performance is born. Here, researchers found a prospect as a package of breach where DE stuck in stagnation problem. In this tricky stage, offspring of individuals is not being able to perform well as compared to parent–offspring. The ability to mutant itself is lost. Contrariwise, the perseverance of a tremendous fit singular after some time of associates does not ensure the deprived working of the algorithm. It is an accepted leg of evolution that one of the offsprings somehow restructured itself throughout the run [21]. The circumstance described above possibly related to the fact that the model of DE has number of nonexistence important features which can be witnessed in organic evolution [22].

5 Conclusion

This chapter gives a survey of Opposition Based Learning in terms of papers projects; articles, etc., more than a hundred papers are reviewed according to OBL in various algorithms such as opposition-based learning in DE algorithms.

References

1. Shapiro J (2001) Genetic algorithms in machine learning. In: Paliouras G, Karkaletsis V, Spyropoulos CD (eds) Machine learning and its applications. ACAI 1999. Lecture Notes in Computer Science, vol 2049. Springer, Berlin, Heidelberg
2. Yang XS (2010) Nature-inspired metaheuristic algorithms. Luniver press
3. Fister I, Yang XS, Brest J, Fister D (2013) A brief review of nature-inspired algorithms for optimization. Elektroteh. Vestnik/Electrotechnical Rev 80(3):116–122
4. Tizhoosh HR, Ventresca M (2008) A diversity maintaining population-based incremental learning algorithm. Inf Sci (Ny) 178(21):4038–4056
5. Tizhoosh HR (2005) Opposition-Based learning : a new scheme for machine intelligence. In: International conference on computational intelligence for modelling control and automation international conference intelligence agents, web technology internet commerce, vol 1, pp 695–701
6. Fleetwood K (2009) An introduction to differential evolution. New ideas Optim, pp 79–108
7. Rahnamayan S, Tizhoosh HR, Salama MMA (2006) Opposition-based differential evolution algorithms. In: 2006 IEEE congress evolutionary computation CEC 2006, pp 2010–2017
8. Rahnamayan S, Tizhoosh HR, Salama MMA (2006) Opposition-Based differential evolution for optimization of noisy problems. In: 2006 IEEE international conference evolutionary computation, no 519, pp 1865–1872
9. Rahnamayan S, Tizhoosh HR, Salama MMA (2007) A novel population initialization method for accelerating evolutionary algorithms. Comput Math Appl 53:1605–1614
10. Rahnamayan S, Tizhooshl HR, Salama MMA (2007) Opposition-Based differential evolution (ODE) with variable jumping rate. In: IEEE symposium on foundations of computational intelligence, 2007, pp 81–88

11. Rahnamayan S, Tizhoosh HR, Salama MMA (2007) Quasi-Oppositional differential evolution. In: 2007 IEEE congress evolutionary computation, pp 2229–2236
12. Wang GG, Wangsfuca G (2008) Solving large scale optimization problems by opposition-based differential solving large scale optimization problems by opposition-based differential evolution (ODE). WSEAS Trans Comput 7(10):1792–1804
13. Žumer BBGBZ (2008) An adaptive differential evolution algorithm with opposition-based mechanisms, applied to the tuning of a chess program. Springer, Berlin, Heidelberg
14. Rahnamayan S, Tizhoosh HR (2008) Image thresholding using micro opposition-based differential evolution (Micro-ODE). In: 2008 IEEE congress evolutionary computation (IEEE World Congress Computation Intelligence), pp 1409–1416
15. Rahnamayan S, Tizhoosh HR, Salama MM (2008) Opposition-based differential evolution. Stud Comput Intell 143(1):155–171
16. Wang H, Wu Z, Rahnamayan S, Kang L (2009) A Scalability test for accelerated DE using generalized opposition-based learning. In: 2009 Ninth international conference intelligence system design applications, no 2, pp 1090–1095
17. Omran MGH, Engelbrecht AP (2009) Free search differential evolution. In: 2009 IEEE congress evolutionary computational CEC 2009, pp 110–117
18. Omran GH, Salaman A (2009) Constrained optimization using CODEQ. Chaos Solitons Fractals 42(2):662–668
19. Tang J, Zhao X (2010) On the improvement of opposition-based differential evolution. In: 2010 Sixth international conference on natural computation, vol 5, no Icnc, pp 2407–2411
20. Xu Q, Wang L, He B, Wang N (2011) Modified opposition-based differential evolution for function optimization. J Comput Inf Syst 7(5):1582–1591
21. Wang H, Wu Z, Rahnamayan S (2011) Enhanced opposition-based differential evolution for solving high-dimensional continuous optimization problems. Soft Comput A Fusion Found Methodol Appl 15(11):2127–2140
22. Thangaraj R, Pant M, Chelliah TR, Abraham A (2012) Opposition based chaotic differential evolution algorithm for solving global optimization problems. In: 2012 Fourth world congress nature biologically inspired computing, pp 1–7
23. Rahnamayan S, Wang GG, Ventresca M (2012) An intuitive distance-based explanation of opposition-based sampling. Appl Soft Comput 12(9):2828–2839
24. Li J (2012) A hybrid differential evolution algorithm with opposition-based learning. In: Proceedings of 2012 4th international conference intelligent human-machine systems cybernetics IHMSC 2012, vol 1, no 1, pp 85–89
25. Ahandani MA (2016) Opposition-based learning in the shuffled bidirectional differential evolution algorithm. Swarm Evol Comput 26:64–85
26. Karthikeyan K, Kannan S, Baskar S, Thangaraj C (2013) Application of opposition-based differential evolution algorithm to generation expansion planning problem. J Electr Eng Technol 8(4):686–693
27. Wang H, Rahnamayan S, Wu Z (2013) Parallel differential evolution with self-adapting control parameters and generalized opposition-based learning for solving high-dimensional optimization problems. J Parallel Distrib Comput 73(1):62–73
28. Salehinejad H, Member S, Rahnamayan S, Member S, Tizhoosh HR (2014) Type-II opposition-based differential evolution. In: 2014 IEEE congress evolutionary computation, pp 1768–1775
29. Chelliah TR, Thangaraj R, Allamsetty S, Pant M () Coordination of directional overcurrent relays using opposition based chaotic differential evolution algorithm. Int J Electr Power Energy Syst 55:341–350
30. Hu Z, Bao Y, Xiong T (2014) Partial opposition-based adaptive differential evolution algorithms: Evaluation on the CEC 2014 benchmark set for real-parameter optimization. In: Proc. 2014 IEEE congress evolutionary computation CEC 2014, pp 2259–2265
31. Liu H, Wu Z, Wang H, Rahnamayan S, Deng C (2014) Improved differential evolution with adaptive opposition strategy. In: Proceedings of 2014 IEEE congress evolutionary computation CEC 2014, pp 1776–1783

32. Padma K (2015) Oposition-based modified differential evolution algorithm with SSVR device under different load conditions. IN: 2015 IEEE power, communication information technology conference, pp 935–940
33. Park S, Lee J (2016) Stochastic opposition-based learning using a beta distribution in differential evolution. IEEE Trans Cybern. 46(10):2184–2194
34. Wang W, Wang H, Sun H (2016) Using opposition-based learning to enhance differential evolution : a comparative study. In: 2016 IEEE congress evolutionary computation, pp 71–77
35. Salehinejad H, Rahnamayan S, Tizhoosh HR (2017) Opposition-based ensemble micro-differential evolution, pp 0–7
36. Optimization M (2019) Multifactorial differential evolution with opposition-based learning for multi-tasking optimization. In: 2019 IEEE congress evolutionary computation, pp 1898–1905
37. Sharma S, Yadav A, Sinwar D, Naruka B, Yadav V (2019) Cross-Opposition based differential evolution optimization. Int. J. Recent Technol. Eng. (IJRTE), 8(2):2887–2893. ISSN: 2277-3878
38. Wood B, Patel V, Rahnamayan S (2020) A smart scheme for variable selection in partial opposition-based differential evolution. In: 2020 IEEE symposium series on computational intelligence (SSCI), 2020, pp 329–336. https://doi.org/10.1109/SSCI47803.2020.9308504
39. Mousavirad SJ, Oliva D, Hinojosa S, Schaefer G (2021) Differential evolution-based neural network training incorporating a centroid-based strategy and dynamic opposition-based learning. In: 2021 IEEE congress on evolutionary computation (CEC), 2021, pp 1233–1240. https://doi.org/10.1109/CEC45853.2021.9504801
40. Mousavirad SJ, Rahnamayan S (2020) Evolving feedforward neural networks using a quasi-opposition-based differential evolution for data classification. In; 2020 IEEE symposium series on computational intelligence (SSCI), 2020, pp 2320–2326. https://doi.org/10.1109/SSCI47803.2020.9308591

Prediction of Nitrogen Deficiency in Paddy Leaves Using Convolutional Neural Network Model

Swami Nisha Bhagirath, Vaibhav Bhatnagar, and Linesh Raja

Abstract At various stages of development, including flowering and fruit production, plants need a variety of minerals and nutrients to grow. The phenotype, quality, and yield of crops are all influenced by the nitrogen level in precision agriculture. In the future, it won't be possible to achieve without the application of nitrogen fertilizer. A useful and cutting edge technique for diagnosing the nitrogen nutrition of crops is needed for an effective and appropriate nitrogen fertilizer management system. Plant diseases that have a substantial impact on agricultural output are brought on by nutrient deficiency. Rice producers can reduce output loss significantly by taking essential measures with the help of early disease identification. Deep learning, an effective machine learning method, has recently demonstrated considerable potential in the task of classifying images. The convolutional neural network has been trained after classification of the affected leaf regions. In this study, the lack of nitrogen in the paddy crop was identified and predicted using leaf data. The model achieved 95% of accuracy for identifying nitrogen deficiency in leaves from available dataset.

Keywords Convolutional neural network · Rice diseases · DenseNet · Augmentation

1 Introduction

To keep plants healthier and less vulnerable to pests, the nutrients in the soil should be in the right proportions. Population growth is accompanied by an increase in food demand. Roughly, half of the world's population eats rice every day. Farmers must increase output and balance the economy to minimize losses in order to meet the world's growing demand for food. Rice is one of the three main crops grown worldwide [1]. 78% of the entire rice crop is for human use. More than half of the world's population more than 3.5 billion people rely on rice as their main food source for more than 20% of their daily calories. 90% of the world's rice is consumed by Asia,

S. N. Bhagirath · V. Bhatnagar (✉) · L. Raja
Manipal University Jaipur, Jaipur, Rajasthan, India
e-mail: Vaibhav.bhatnagar15@gmail.com

S. Das et al. (eds.), *Advances in Data-driven Computing and Intelligent Systems*, Lecture Notes in Networks and Systems 653, https://doi.org/10.1007/978-981-99-0981-0_54

and its demand for rice as a whole is rising. The growth of global rice consumption is still being driven by population and economic growth, particularly in many Asian and African nations [2]. In the current digital era, it is crucial for farmers to utilize cutting edge technology for effective crop management. Due to fungi, bacteria, and viruses, most farmers face a variety of difficulties when growing rice. Farmers typically fail to recognize illnesses properly and in a timely manner, which results in a loss of expected productivity [3]. Agriculture experts are called in to assess the situation and provide farmers with the best advice when a rice disease strikes. It could occasionally be difficult for a small team of scientists to cover the entire area impacted by the illness in order to do a visual assessment of the disease. Farmers are unable to identify the illness and take appropriate action in a timely manner as a result. Cellphones can now be used by farmers to take pictures of the diseased rice leaves. Therefore, if images of these disease affected leaves can be used to detect the diseases, farmers do not need to wait for the professionals to visit their fields in order to treat the diseases; instead, they act fast [4]. The level of nitrogen in the soil is a crucial indicator of the health of the rice crop. The growth and productivity of the rice plant are greatly influenced by nitrogen. In order to anticipate the lack of nitrogen in rice, this research suggests a convolutional neural network-based method [5]. Numerous researches have already been done to identify the lack of nitrogen in various crops. The majority of these research classified various diseases using attributes that were derived using machine learning methods like support vector machines and genetic algorithms. Researchers are using convolutional neural network technology to detect nitrogen deficiency because its recent performance in image categorization has been so promising [6]. The adaptable CNN model was developed specifically to handle image data and has the ability to autonomously identify critical functions without operator intervention. To minimize losses, it is necessary to change the weights and learning rate in particular neural networks. The optimizers are algorithms that are used to change these attributes in neural networks and are in charge of decreasing losses and boosting accuracy [7].

The role of this paper is a convolutional neural network model which identifies the nitrogen deficiency from the available dataset for rice leaves images. The order of the paper remaining sections is as follows. Section 2 provides a summary of the literature. In Sect. 3, the experimental work is explained using dataset and convolutional neural network. Discussion of the experimental work is explained in Sect. 4. The paper is concluded in Sect. 5.

2 Related Work

Various researchers have published several techniques to enhance the CNN model and achieve high accuracy. For the purpose of optimizing CNN hyperparameters, it was suggested to use a powerful genetic algorithm with variable length chromosomes. The author [8] suggested a model under various lighting and leaf shelter conditions, MobileNetv2-YOLOv3 models were utilized to identify the grey leaf spot disease on tomato leaves. The results are then contrasted. The GIoU bounding box is used to improve the regression box's accuracy when spotting gray leaf spots on tomato plants. MobileNetv2-YOLOv3 outperformed other architectures in terms of detection speed, the $F1$ score, and the AP value. The author [9] suggested using transfer learning and many deep neural networks to detect nutritional deficiencies in a black gram leaf image. A collection of 4088 photographs of black gram leaves produced less than seven distinct treatment complete nutrition, iron deficiency, magnesium deficiency, nitrogen deficiency, potassium deficiency, phosphorus deficiency, and calcium deficiency was used in the experiments. The deep CNN model ResNet50, with an F-measure of 66.15% and a test accuracy of 65.44%, emerged as the top model. The classification of rice illnesses and the estimation of their severity have both been studied by the author [10]. Their approach can be used to diagnose three common rice diseases, including Brown Spot, Leaf Blast, and Bacterial Leaf Blight. K-means clustering is utilized to segregate the portion of the body affected by the disease. The parts of the photos that were damaged by the disease were used to extract color, texture, and shape attributes. Based on the stage of the disease, recommendations for pesticides and fertilizers are made. Rice diseases are identified using an SVMclassifier. This method has a 90.95% accuracy rate for Brown Spot, a 94.11% accuracy rate for Leaf Blast and an 85.71% accuracy rate for Bacterial Leaf Blight. Although accurate, this work only distinguished three disorders. A system for recognizing and classifying diseases that damage rice has been proposed by the author [11]. The data in the UCIMachine Learning Repository is used with the background elimination method they developed. K-means clustering was then used to distinguish the ill areas from the photos of the leaves. They were able to achieve 88 features from the segmented images and 73.33% test accuracy and 100% training accuracy. The author proposed [12] that healthy leaves and three different types of rice diseases—Leaf Blight, Rice Blast, and Brown Spot—had been classified using CNN and transfer learning. There are 1509 entries in the dataset, and 92.46% transfer learning accuracy was found. This work might be more effective if brightness enhancement were used during the pre-processing. The author [13] uses 900 images to classify four different disease types. The backdrop was divided in half for each shot, blur was eliminated, and gray scale was added. The accuracy was 97.4% when they used CNN. The dataset used for this analysis included a good quantity of photos for each of the four illnesses, augmentation, which would have increased effectiveness of the method. A convolution neural network (CNN) based on Inception-ResNet v2 was recommended by the author [14] as a way to classify and predict the calcium and potassium status of tomato plants. The effectiveness of the Inception-ResNet v2 is verified using actual fruit photographs. In this work, additive merging is used to increase the efficiency and precision of picture recognition training. The writer [15] recommended two training approaches like fine tuning and transfers learning on utilizing Inception-ResNet technology to identify nitrogen Okra plants are deficient in potassium, phosphorus, and magnesium. Adam optimizer is employed in the transfer process for better results. Freezing of some early data is used for learning while fine tuning layers. The outcomes of the experiment demonstrate that fine-tuning better efficiency of the method over transfer learning accuracy. The author [16] suggested a methodology for employing image processing to identify a nitrogen deficit in cotton plants. Two preparatory steps one for histogram analysis and another for calculating leaf area are used in the suggested methodology. The area of the nitrogen deficient leaf is lower than the area of the normal leaf, and the histogram for the nitrogen deficient leaf has smaller peak amplitude. In order to determine nitrogen deficit in the cotton plant, scientists compared deficient leaves with normal leaf. The author [17] described a smartphone app that helped rice farmers identify nitrogen insufficiency based on the color of the plant. The device can be used instead of or along with the usual nitrogen use. It was suggested that farmers adopt an intuitive technology without any prior instruction. In order to produce results with a high degree of accuracy, this paper introduced automated image processing techniques. The z-score statistical method was used to calculate the desired outcomes.

3 Experimental Model

In this paper, CNN is used to approach the image based nitrogen deficiency detection for rice leaves. Any application based on machine learning must begin with the gathering of data and creation of a dataset. There are around 1000 images in dataset of 300 dpi. The dataset consist of different folders which are training dataset folder and testing dataset folder. The training folder contains two sub-folders that have images of rice leaves with nitrogen and other folder which contains images of rice leaves with nitrogen deficiency. The dataset of rice leaves images is used from Kaggle [18] and modified that data according to the requirement. Some of the images of rice leave of different categories from dataset is shown in Fig. 1.

Computer vision issues like object identification and image categorization are successfully handled by deep neural network models utilizing CNN [19]. The community of researchers has been encouraged to explore and suggest novel CNN architectures as a result of the widespread success of CNN in several domains. Additionally, pre-trained CNN models are increasingly being used to extract information from images. A unique kind of neural network called a CNN uses an image or video frame as its input, employs layers to extract the useful features, and weights various image parameters to help the model execute labeling or object detection tasks. Some of the various parts or layers that make up a CNN model include convolution layers, pooling layers, fully linked layers, and Softmax or sigmoid functions [20]. For the convolution layer, various hyperparameters are accessible, including the quantity of filters, filter size, strides, activation, etc. The filters are convolved with the input image to produce a reduced image with the data required for prediction. Low level information such as edge structure, gradient, color is retrieved as activation maps in the first convolution layer. In order for the model to learn high level features, the following layers build the relationships between various low level features [21].

The spatial and temporal correlations in an input image are captured by CNN using filters. The image matrix size is reduced by pooling layers. The pooling layer achieves the goal of dimensionality reduction by reducing the input spatial dimensions in

Fig. 1 Rice leaves images from dataset [18]

accordance with the predominant features. Pooling layer settings include pool size, strides, padding, and pooling type. The final convolutional or pooling layer's flattened feature map is used as input to a fully linked layer [20]. Each input in the dense layer or completely connected layer is weighed and connected to each output. When building the CNN model, the quantity of covert layers and the number of neurons in each layer are important factors. The Softmax function is utilized for multi-class categorization.

4 Discussion

In this study, images of both healthy and nitrogen affected rice leaves were gathered from various sources and included to a unique dataset. The consistency and effectiveness of the strategy might be defended by analyzing how well it performed on three separate datasets. The convolutional neural network is implemented using the Google Colab which is cloud based deep learning tool. First all the required libraries are imported in Colab file. The dataset is splinted in three parts, train data, test data and validation data. On these images, pre-processing techniques including background removal and scaling were used. After that, picture segmentation was carried out to isolate the disease affected areas. The convolutional neural network-based classifier algorithm used these photos as its input [21]. An image augmentation technique has been used to increase the number of photographs during the training phase. The pre-processing of the test and validation sets' photos was also identical. These pre-treatment methods helped to make the images similar in a number of ways, making them appropriate for CNN processing. Additionally, two other freely accessible datasets were employed [22]. CNN is practically capable of solving the majority of image recognition problems efficiently, provided that the architecture and hyperparameters of the model such as batch size, optimizer type, and number of epochs, layers, number of filters in the different layers, and number of neurons in the fully connected layer are properly chosen.

The confusion matrix for the prediction of validation data which gives the true positives, true negatives, false positives, and false negatives is displayed in Fig. 2. The confusion matrix is created for binary classification. In binary classification, each input sample is assigned to one of two classes. These two classes are assigned labels like 1 and 0, or *positive* and *negative*. The true positive predicted images are 63, true negative predicted images are 84, the false positive predicted images are 4 and the false negative predicted 3 images. Data for training and validation are chosen at random. The input image measures are (224, 224, and 3). Augmentation is used to resize and convert images from gray to RGB. By contrasting the measured labels with the expected labels, the confusion matrix is created. The performance measure parameters, such as accuracy, sensitivity, specificity, FPR, and $F1$ score, are calculated from the confusion matrix. The provision of more data for network training is the second advantage of augmentation. The confusion matrix for the prediction of testing data is shown in Fig. 3.

The true positive predicted images are 17, true negative predicted images are 22, the false positive predicted images are 0, and the false negative predicted 1 images. To determine classification error and testing accuracy, the CNN model is trained and tested. The test accuracy is used to obtain the fitness function. The Adam optimizer is used in this model. Convolutional neural network model is trained using 15 epochs, 3 steps per epochs and batch size of 32. For the first epoch, the accuracy obtained is 62%. When the epoch reached at 15, the accuracy obtained is 95%. The CNN model that is being considered for use by the fitness function is output that shows

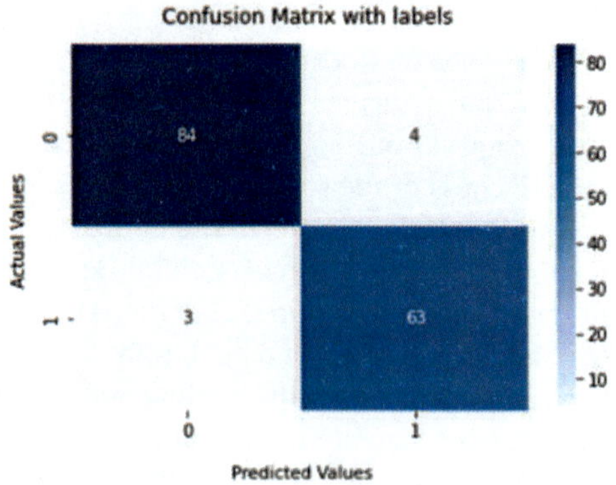

Fig. 2 Confusion matrix for the prediction of validation data

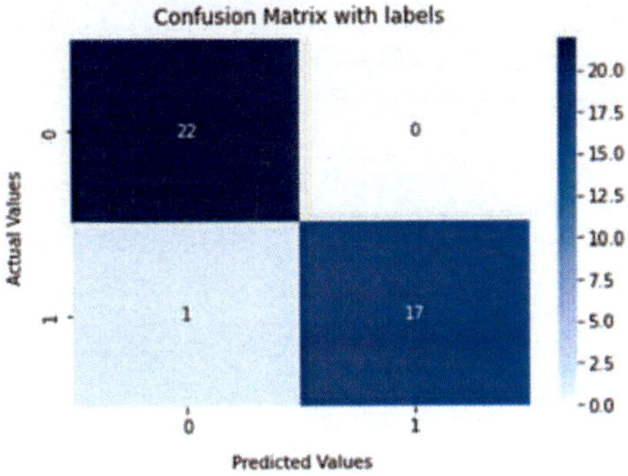

Fig. 3 Confusion matrix for the prediction of test data

how effective the CNN model is at classifying images of rice. Python is used to implement and assess the suggested strategy against other modern machine learning image classifiers using Google Colab and GPU Tesla P100-PCIE-16 GB for nitrogen prediction [23].

5 Conclusion

This paper proposes a convolutional neural network model to classify the rice leaves image dataset for identifying the nitrogen deficiency. Using leaf data, it was possible to identify and anticipate the paddy crop's nitrogen deficiency. There are 1000 photos of leaves in the dataset. For recognizing nitrogen deficit leaves from the given dataset, the model had an accuracy of 95%. In the future work, the convolutional neural network results will be optimized through genetic algorithm optimization using the hyperparameter tuning. Also, the convolutional neural network model will be compared using other optimizers.

References

1. Zhang Y et al (2015) TOND1 confers tolerance to nitrogen deficiency in rice. Plant J 81(3):367–376
2. Rautaray SK et al (2020) Energy efficiency, productivity and profitability of rice farming are using Sesbania as green manure-cum-cover crop. Nutrient Cycl Agroecosyst 116(1):83–101
3. Islam A et al (2021) Rice leaf disease recognition using local threshold based segmentation and deep CNN. Int J Intell Syst Appl 13(5):35–45
4. Upadhyay SK, Kumar A (2022) A novel approach for rice plant diseases classification with deep convolutional neural network. Int J Inf Technol 14(1):185–199
5. Khaki S, Wang L, Archontoulis SV (2020) A cnn-rnn framework for crop yield prediction. Front Plant Sci 10:1750
6. Murugan D (2022) Paddy Doctor: a visual image dataset for paddy disease classification. arXiv preprint arXiv:2205.11108

7. Liu J, Wang X (2020) Early recognition of tomato gray leaf spot disease based on MobileNetv2-YOLOv3 model. Plant Methods 16(1):1–16
8. Han KAM, Watchareeruetai U (2019) Classification of nutrient deficiency in black gram using deep convolutional neural networks. In: 2019 16th International joint conference on computer science and software engineering (JCSSE). IEEE
9. Han KAM, Watchareeruetai U (2020) Black gram plant nutrient deficiency classification in combined images using convolutional neural network. In: 2020 8th International electrical engineering congress (iEECON). IEEE
10. Pinki FT, Khatun N, Mohidul Islam SM (2017) Content based paddy leaf disease recognition and remedy prediction using support vector machine. In: 2017 20th International conference of computer and information technology (ICCIT). IEEE
11. Prajapati HB, Shah JP, Dabhi VK (2017) Detection and classification of rice plant diseases. Intell Dec Technol 11(3):357–373
12. Ghosal S, Sarkar K (2020) Rice leaf diseases classification using CNN with transfer learning. In: 2020 IEEE Calcutta conference (CALCON). IEEE
13. Al-Amin M, Karim DZ, Bushra TA (2019) Prediction of rice disease from leaves using deep convolution neural network towards a digital agricultural system. In: 2019 22nd International conference on computer and information technology (ICCIT). IEEE
14. Choi JW et al (2018) A nutrient deficiency prediction method using deep learning on development of tomato fruits. In: 2018 International conference on fuzzy theory and its applications (iFUZZY). IEEE
15. Wulandhari LA et al (2019) Plant nutrient deficiency detection using deep convolutional neural network. ICIC Express Lett 13(10):971–977
16. Ayane, SS, Khan MA, Agrawal SM (2013) Identification of nitrogen deficiency in cotton plant by using image processing. IJPRET 1(8):112–118
17. Dela Cruz GB (2019) Nitrogen deficiency mobile application for rice plant through image processing techniques. Int J Eng Adv Technol 8(6):2950–2955
18. Nutrient_Deficiency_Symptoms_in_Rice. https://kaggle.com/code/naufalauliaadam/nutrient-deficiency-symptoms-in-rice. Accessed 10 Aug 2022
19. Li L, Zhang S, Wang B (2021) Plant disease detection and classification by deep learning—a review. IEEE Access 9:56683–56698
20. Saxena S, Shukla S, Gyanchandani M (2020) Pre-trained convolutional neural networks as feature extractors for diagnosis of breast cancer using histopathology. Int J Imaging Syst Technol 30(3):577–591
21. Wani JA et al (2021) Machine learning and deep learning based computational techniques in automatic agricultural diseases detection: methodologies, applications, and challenges. Arch Comput Methods Eng 1–37
22. Daniya T, Vigneshwari S (2019) A review on machine learning techniques for rice plant disease detection in agricultural research. System 28(13):49–62
23. Bari BS et al (2021) A real-time approach of diagnosing rice leaf disease using deep learning-based faster R-CNN framework. PeerJ Comput Sci 7:e432

SegCon: A Novel Deep Neural Network for *Seg*mentation of *Con*junctiva Region

Junaid Maqbool, Tanvir Singh Mann, Navdeep Kaur, Aastha Gupta, Ajay Mittal, Preeti Aggarwal, Krishan Kumar, Munish Kumar, and Shiv Sajan Saini

Abstract Anemia is a major health problem that primarily affects women and children worldwide. Early anemia detection in children is necessary, and non-invasive diagnostic methods are advised. The patterns of the conjunctiva's pallor and the blood's hemoglobin concentration is used to diagnose anemia. Computer-aided diagnosis systems facilitate the physicians by automatic detection and classification of anemia from eye images. A key component of any anemia diagnosis system is precise segmentation of the conjunctiva region. In this study, a deep neural network, *SegCon,* is proposed to extract the conjunctiva region from eye images. The eye image dataset used in this study is collected from a tertiary healthcare facility in Chandigarh. It has been demonstrated through analysis of numerous assessment indicators that the region of interest is accurately determined and the methodology employed is efficient. The proposed model, SegCon has achieved the accuracy of 96.43% that outperforms the state-of-the-art methods.

Keywords Conjunctiva region · Convolution neural networks · Medical image segmentation · Anemia detection · Non-invasive method for anemia detection

---

J. Maqbool · T. S. Mann · A. Mittal (✉) · P. Aggarwal · K. Kumar
UIET, Panjab University, Chandigarh, India
e-mail: ajaymittal@pu.ac.in

N. Kaur
MCM DAV College for Women, Sector 36-A, Chandigarh, India

A. Gupta
Department of Mathematics, Punjab Engineering College, Chandigarh, India

M. Kumar
Maharaja Ranjit Singh Punjab Technological University, Bathinda, Punjab, India

S. S. Saini
Post Graduate Institute of Medical Education and Research, Chandigarh, India

S. Das et al. (eds.), *Advances in Data-driven Computing and Intelligent Systems*, Lecture Notes in Networks and Systems 653, https://doi.org/10.1007/978-981-99-0981-0_55

1 Introduction

Anemia and malnutrition are significant problems in children below the age of five years. Anemia is the condition of a shortage of Red Blood Cells (RBCs) or abnormal RBCs. As per a survey, performed by Ghosal et al. [1], 23% of the world's population is anemic, amongst those 93% belong to low-economic and middle-economic countries. Specifically, in India, 21.7% of men and 53.02% of women suffer from anemia [2]. Anemia is one of the leading contributors to the global burden of diseases, thereby directly impacting work productivity and quality of life [3]. National Family Health Survey (NFHS-4) revealed that the prevalence of anemia among children under five years is 58% in India. The primary causes of anemia include nutritional deficiencies, oncological diseases and treatments, and infections like malaria, hookworm, and HIV. Some patients can inherit it from parents as RBC disorders like sickle cell disease, thalassemia, and myelodysplastic syndrome [4].

Anemia is generally curable if timely detected and treated. Thus, early detection and accurate identification of anemia is of utmost importance. Hemoglobin (Hb) concentration in blood is the most reliable indicator of anemia and is generally measured using invasive and painful venous or capillary blood extraction [5]. These methods are time-consuming, uncomfortable, and need trained professionals for sample collection. Until urgent, these methods are not recommended for infants and elderly people. Traditionally, anemia gets detected by analyzing physical features of pallor of conjunctiva, palms, fingernails, and tongue by clinicians, which are susceptible to human errors and bias. Developing a non-invasive procedure is of great need to overcome these shortcomings. In low-income and middle-income countries, these methods also have an operational advantage. Several non-invasive medical devices and mobile applications have been proposed to predict the Hb values using either image of the fundus, conjunctiva, fingernails, tongue, or palms [6, 7]. Diagnosis of malnutrition and anemia using these techniques will make screening in densely populated areas inexpensive and efficient [1].

Disease detection and classification is considered the major research problem in computer vision. For instance, in ophthalmology, some methods can detect diabetic retinopathy and glaucoma from eye images with accuracy at par with clinicians [8]. Several machine learning algorithms [3, 9] detect anemia by utilizing digital photographs to analyze various features of region of interest ROI and predict the corresponding Hb values. The drawback of prevalent techniques is the need to explicitly define the ROI [10]. A key component of automatic anemia diagnosis system is precise segmentation of the conjunctiva region. The major contribution of this study is the development of a deep neural network to automatically segment the region of interest, i.e., conjuctiva region, from eye images. The images used in this study are captured using special biomedical devices, which can be further used to detect anemia by predicting Hb values.

The rest of the paper is organized as follows. The related work is described in detail in Sect. 2. Material and methods used are summarized in Sect. 3. Discussion about the experimentation results is presented in Sect. 4. Future scope and conclusion of this research work is discussed in Sect. 5.

2 Related Work

This section presents the related work in two aspects: general medical image segmentation and conjunctiva segmentation in particular.

2.1 *Medical Image Segmentation*

Several methods and algorithms are used to segment the ROI from the medical images. Ramesh et al. [11] show that there is no single technique which works best for all varied segmentation needs. Image segmentation can be grouped based on techniques used like thresholding, region-based techniques, clustering, edge detection, and various model-based methods [12]. Thresholding involves binarization of an image into ROI and background based on a threshold value of the pixels [13] which can be determined manually through hit and trial or using some strategies like Ostu's methods [14], K-means clustering [14], or maximum entropy method [15]. Region-based methods like region growing select the region starting from a seed point and allow it to expand based on morphological similarities [16]. Furthermore, other region-based approaches are: region split and merge [17], and watershed approach [18]. For segmentation K-means clustering is one of fundamental clustering technique employed, however other techniques such as Fuzzy C-Means (FCM) [19], Kernelized FCM [20], Type-II FCM [11] are also utilized for successful segmentation. In contrast, edge detection is used in segmentation by analyzing pixels on region boundary and then isolating a closed region as the ROI [21]. Model-based algorithms like deep convolutional neural networks, Markov random fields [11], UNet [22], UNet++ [23], LinkNet [24] have also been utilized for medical image segmentation.

2.2 *Conjunctiva Region Segmentation*

Several methods are employed to develop non-invasive techniques for the automatic detection and classification of anemia. The first crucial step in the process is to segment the ROI from the images, which, in most cases, are various portions of eye conjunctiva. Brea et al. [25] performed segmentation of bulbar conjunctiva and extracted features of the region based on the color and presence of blood vessels. The bulbar conjunctiva has been extracted from eye pictures using various thresholding techniques. Iroshan et al. [26] used the Chan–Vese algorithm to segment ROI using the initial contour selected by user and number of regions. In the case of more than one region, the largest region was selected as ROI. Similarly, Brea et al. [27] used spline, ellipse, and contour location-based approaches for bulbar conjunctiva segmentation. Curti et al. [28] used a variety of CNN architectures from DenseNet to lighter UNet variants and finally, Mobile UNet with skip connections was used for conjunctiva segmentation.

Jain et al. [9] performed manual segmentation of the ROI on a dataset of 99 images which were further augmented to increase the generalization and size of the dataset. Sevani et al. [29] and Dimauro et al. [7] performed manual segmentation. Thresholding and other basic image processing technique were used by Jha et al. [30] and derived the threshold value of 210 for segmentation empirically. Sedki et al. [31] proposed using a manual threshold of 225 red channels in addition with a median filter gives better results. Image processing techniques depend on the condition and circumstances in which images are acquired and are hence less robust. Machine learning-based techniques have been used for this task to decrease sensitivity to picture acquisition problems.

Deep learning has performed excellently in computer vision. CNN is proficient at processing spatial context and performs well in image processing and segmenting medical images. A limited amount of research has been done on conjunctiva segmentation based on deep learning. Saldivar-Espinoza et al. [32] used 7-layer CNN to segment conjunctiva from eye images acquired using a smartphone. Kasiviswanathan et al. [6] employed transfer learning using UNet [22] for conjunctiva segmentation. Normalized cuts approach along with heat map as a bias was used to effectively segment the conjunctiva portion from the eye image by Dimauro and Simone [33] with an accuracy of 93.79% (Fig. 1).

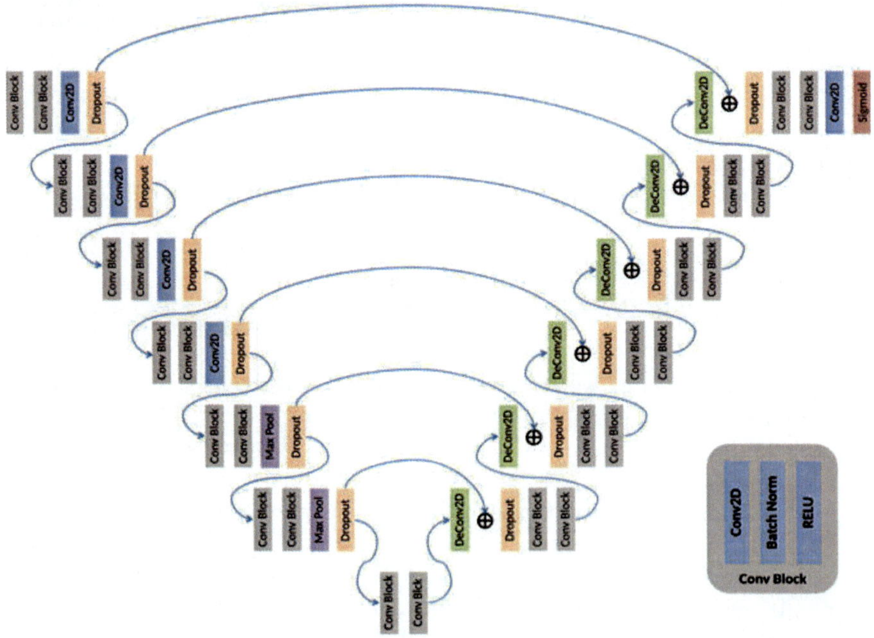

Fig. 1 The architecture of the proposed model

Deep learning methods have witnessed high accuracy rates in other image segmentation techniques. Thus, there is scope to improve the accuracy, i.e., 93.79% of state-of-the-art methods used for conjuctiva segmentation.

3 Material and Methods

3.1 Dataset

In this study, the dataset used for training and testing is collected by medical professional at a tertiary healthcare facility in North India from March-2022 till May-2022. The dataset contains 324 eye images of anemic and non-anemic pediatric population. The images are captured using a smartphone mobile application with four primary cameras in an uncontrolled environment. The resolution of the cameras being 64 MP, 8 MP, 2 MP, and 2 MP. The RGB color images are captured with auto-focus setting from a distance of 5–14 cm and then uploaded to a remote server with reference to each patient in .png format. Informed consent of guardian/parent was obtained before capturing the images of these 162 subjects. The ground truth polygon masks for conjunctiva are manually created using CVAT tools. These ground truth masks along with eye images are used for training, testing, and validation of the proposed model. These 324 images are randomly split into training, and testing dataset in the ratio of 80%, 20%.

Data augmentation and pre-processing Since convolution procedures are translation invariant, the training data is augmented using angular rotation at angle increments of 90°. Furthermore, horizontal and vertical flips are also performed on the dataset. These augmentation techniques are applied to all images of training set and also to the corresponding masks of each image. The images are downscaled and resolution is reduced to 512×512 to improve the training time and reduce the memory. To serve as model input, the data had to be normalized.

3.2 Proposed Model

In this paper, we proposed a deep neural network, namely SegCon, for segmentation of conjuctiva region from eye images. SegCon is inspired from the two path approach of UNet architecture. UNet (a CNN-based deep learning architecture) has demonstrated quick and superior segmentation results on minimal/modest data in the biomedical image segmentation and other semantic segmentation disciplines. UNet architectures employ two path approach to retrieve and localize complete context of an image. For detailed understanding of the UNet architecture, reader are referred to read. A down path is employed to extract the context and an up path layer is used

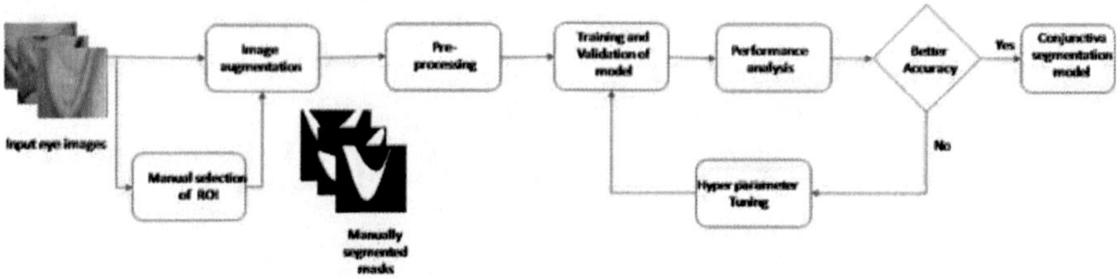

Fig. 2 The flow graph of training of the proposed model, SegCon

to enable precise location. In Fig. 2, we outline the steps involved in developing the proposed model. The proposed model differs from traditional UNet model in following ways:

- The traditional UNet model comprises four contracting and four expanding layers whereas to improve the specificity SegCon comprises six contracting and six expanding layers.
- In UNet model max pooling layers appear at each stage, whereas in SegCon, max pooling is only attempted at the last two stages and replaced with convolutional layers with stride 2 in preceding levels.

The architecture of the proposed model, depicted in Fig. 1, shows shrinking (left) and growing (right) phases. All the layer blocks in shrinking phase include two convolutions, followed by ReLU activation and convolution with stride 2. Some of the information is discarded by the pooling layers. Thus, max pool layers in UNet are replaced with convolution with stride 2 layer. In shrinking phase the feature channels are halved. In growing phase, these features are concatenated with features of corresponding layer of growing phase, followed by two convolutional layers and ReLU. The final output of the growing phase is fed to a convolutional layer followed by sigmoid at the end to generate mask of conjuctiva region. To prevent the overfitting of the model, dropout is applied at end of all layers in both the phases.

The proposed model, SegNet takes RBG image of size 512×512 and generated the binary mask of conjuctiva region. The qualitative analysis of proposed model indicate high specificity (98.49%) in case of conjunctiva segmentation.

3.3 Evaluation Metrics

The model may fall into one of the categories listed below depending on whether the predictions it makes are accurate or inaccurate.

1. True Negative (TN): The model properly infers the lack of the label when it is absent in the ground truth.
2. True Positive (TP): The model successfully infers a label when it matches the ground truth.

3. False Positive (FP): This occurs when the model infers a label that does not actually exist in the ground truth.
4. False Negative (FN): This occurs when the model fails to recognize a label that is actually present in the ground truth.

We utilized the following common measures to assess how well the proposed deep neural network, SegCon performed:

1. **Accuracy**: The model's accuracy is determined by dividing the number of correct predictions by the total number of predictions the model made. We can compute it mathematically as:

$$\text{Accuracy} = \frac{\text{TP} + \text{TN}}{\text{FP} + \text{FN} + \text{TP} + \text{TN}}.$$

2. **Sensitivity**: Sensitivity (true positive rate) is the likelihood that a test will result in a true positive outcome, and is mathematically determined as:

$$\text{Sensitivity} = \frac{\text{TP}}{\text{FN} + \text{TP}}.$$

3. **Specificity**: Specificity (true negative rate) refers to the probability of a negative test, conditioned on truly being negative, i.e.,

$$\text{Specificity} = \frac{\text{TN}}{\text{TN} + \text{FP}}.$$

4. **Dice or F1**: Dice score is measured as harmonic mean of precision and recall, i.e.,

$$\text{Dice} = \frac{2\text{TP}}{2\text{TP} + \text{FN} + \text{FP}}.$$

3.4 *Experimental Setup*

The experimental model has been trained and tested on hardware machine equipped with Nvidia Quadro P5000 Graphics Processing Unit (GPU), 3.30 GHz Intel(R) Xeon(R) W-2155 Central Processing Unit (CPU) and GB RAM. Training has been done on a variety of hyperparameter combinations, including learning rate and optimizer, to determine the optimal configuration for the proposed model to perform well for segmentation tasks. To reduce the vanishing gradient problem Adam optimizer [34] was used as it is invariant to diagonal rescaling of gradients to train the model. A ratio of 80:20 is utilized to divide 324 eye images, and the latter portion is used to verify and test the model. The training part of dataset underwent augmentation to increase the size of dataset.

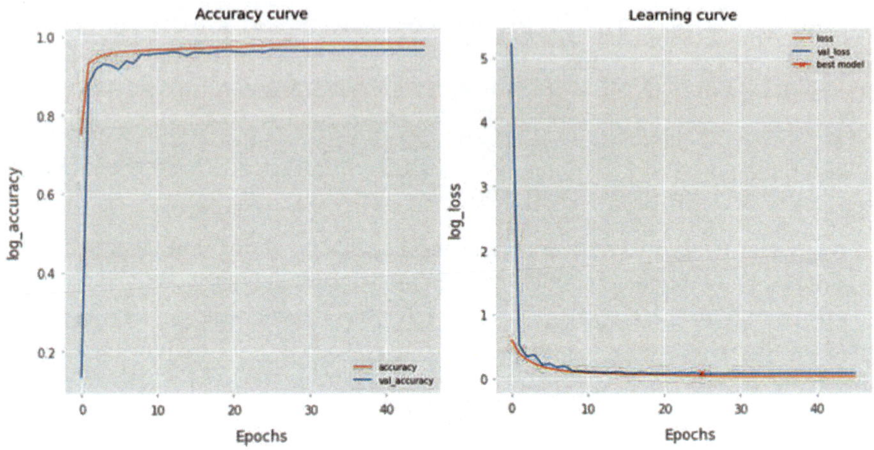

Fig. 3 Accuracy and learning curve

Table 1 Performance comparison of proposed model with state-of-the-art methods

Model used	Dataset	F1-score	Accuracy	Sensitivity	Specificity	IoU
Normalized cuts [33]	Custom dataset (94 patients)	73.63	93.79	86.73	94.63	NA
Viola-Jones algorithm [35]	Custom dataset (115 images)	NA	92.2	NA	NA	NA
UNet-based [6] conjunctiva segmentation model	Custom dataset (135 images)	NA	NA	NA	NA	85.7
Proposed model (SegCon)	Custom dataset (324 images)	84.68	96.43	84.45	98.49	NA

4 Results

The results of proposed model, SegCon are evaluated on our own dataset. In total 324 images captured from 162 subjects and after are randomly split into training, and testing dataset in the ratio of 80%, 20%. The accuracy and training loss curves depicted in Fig. 3, shows that the proposed model, SegCon, is well-trained. The averaged results of the comparison between manually segmented images (ground truth masks) and automatically segmented images of conjunctiva (by SegCon) are shown in Table 1. All the metrics depict that in most of cases a fine meaningful part of conjunctiva has been extracted.

The proposed model, SegCon, outperforms state-of-the-art models in most of the evaluation metrics. The model was able to segment the correct conjunctiva portion with an accuracy of 96.43% and sensitivity of 98.49%. Table 1 presents the comparison of various accuracy metrics with the normalized cuts approach model

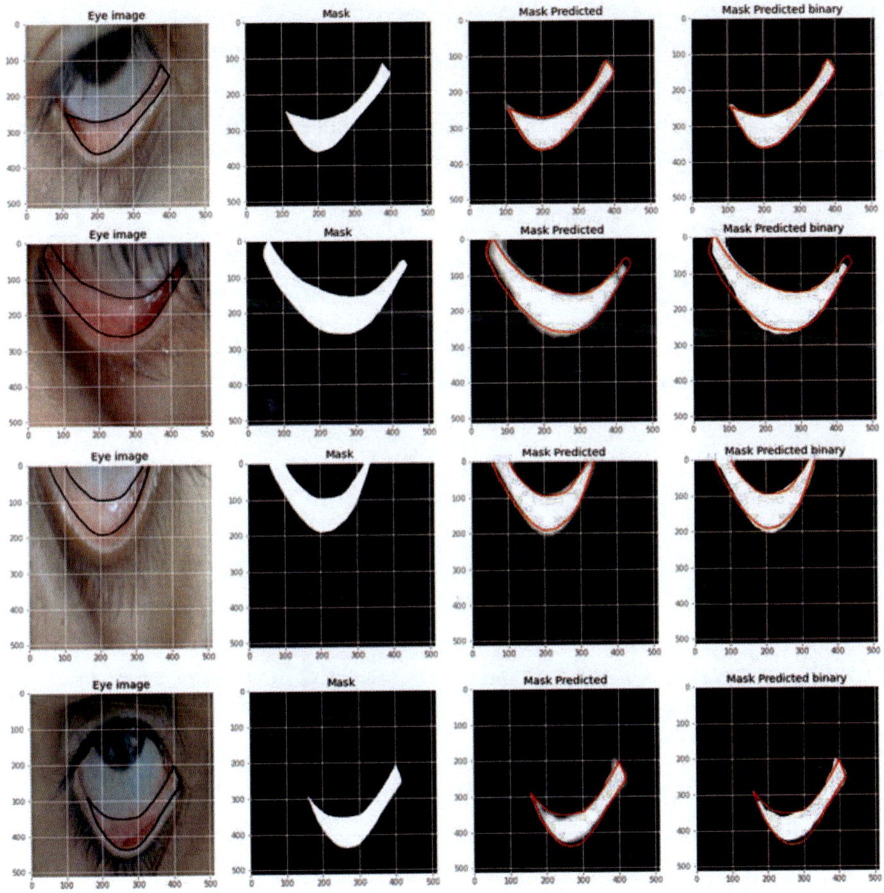

Fig. 4 Qualitative analysis of state-of-art comparison with proposed model

proposed by Dimauro and Simone [33]. These evaluation metric scores confirm the effectiveness and viability of the proposed model.

With the intention of estimating the performance of the proposed model and visualizing the automatically segmented portions of conjunctiva and then contrasting it with the manually chosen ROI in accordance with established accuracy metrics. It can been seen from Fig. 4, the first column of images show the original images with manual segmentation masks, second column displays the mask, then the subsequent image in same row displays the mask predicted by proposed model furthermore a binarized mask is shown in last image in each row when thresholding at 0.5 was done.

5 Conclusion and Future Scope

In this work, a fully automated segmentation model, SegCon, is proposed specifically for conjunctiva segmentation. The dataset is collected from the tertiary medical facility. A total of 324 eye images of children are captured and then conjuctiva region is manually segmented to create ground truth. The automated segmented ROI and the manual mask are quantitatively compared on a number of accuracy metrics. From the results it is evident that the model is able to segment the conjunctiva portion of the eye successfully with the accuracy of 96.43%. The proposed model paves way to explore more deep learning architectures for conjunctiva and other medical image segmentation tasks. The extracted ROI can be utilized for various diagnosis like anemia detection by checking correlation with the Hb values.

Acknowledgements The work done in this study is supported by funds received from Technology Development and Transfer Division, Department of Science and Technology, Govt. of India, New Delhi vide Grant No. TDP/BDTD/45/2021(G).

References

1. Ghosal S, Das D, Udutalapally V, Talukder AK, Misra S (2020) sHEMO: smartphone spectroscopy for blood hemoglobin level monitoring in smart anemia-care. IEEE Sens J 21(6):8520–8529
2. Didzun O, De Neve J-W, Awasthi A, Dubey M, Theilmann M, Bärnighausen T, Vollmer S, Geldsetzer P (2019) Anaemia among men in India: a nationally representative cross-sectional study. Lancet Glob Health 7(12):e1685–e1694
3. Mitani A, Huang A, Venugopalan S, Corrado GS, Peng L, Hammel NW, Liu Y, Varadarajan AV (2020) Detection of anaemia from retinal fundus images via deep learning. Nat Biomed Eng 4(1):18–27
4. An R, Huang Y, Man Y, Valentine RW, Kucukal E, Goreke U, Sekyonda Z, Piccone C, Owusu-Ansah A, Ahuja S et al (2021) Emerging point-of-care technologies for anemia detection. Lab Chip 21(10):1843–1865
5. McLean E, Cogswell M, Egli I, Wojdyla D, De Benoist B (2009) Worldwide prevalence of anaemia, who vitamin and mineral nutrition information system, 1993–2005. Public Health Nutr 12(4):444–454
6. Kasiviswanathan S, Bai Vijayan T, Simone L, Dimauro G (2020) Semantic segmentation of conjunctiva region for non-invasive anemia detection applications. Electronics 9(8):1309
7. Dimauro G, Caivano D, Girardi F (2018) A new method and a non-invasive device to estimate anemia based on digital images of the conjunctiva. IEEE Access 6:46968–46975
8. Ting DSW, Cheung CY-L, Lim G, Tan GSW, Quang ND, Gan A, Hamzah H, Garcia-Franco R, San Yeo IY, Lee SY et al (2017) Development and validation of a deep learning system for diabetic retinopathy and related eye diseases using retinal images from multiethnic populations with diabetes. JAMA 318(22):2211–2223
9. Jain P, Bauskar S, Gyanchandani M (2020) Neural network based non-invasive method to detect anemia from images of eye conjunctiva. Int J Imaging Syst Technol 30(1):112–125
10. Tamir A, Jahan CS, Saif MS, Zaman SU, Islam MM, Khan AI, Fattah SA, Shahnaz C (2017) Detection of anemia from image of the anterior conjunctiva of the eye by image processing and thresholding. In: 2017 IEEE region 10 humanitarian technology conference (R10-HTC). IEEE, pp 697–701

11. Ramesh KKD, Kumar GK, Swapna K, Datta D, Rajest SS (2021) A review of medical image segmentation algorithms. EAI Endorsed Trans Pervasive Health Technol 7(27):e6
12. Egmentation S, Pham DL, Xu C, Prince JL (2000) Current methods in medical image segmentation. Annu Rev Biomed Eng 2(1):315–337
13. Kang W-X, Yang Q-Q, Liang R-P (2009) The comparative research on image segmentation algorithms. In: 2009 first international workshop on education technology and computer science, vol 2. IEEE, pp 703–707
14. Haralick RM, Shapiro LG (1985) Image segmentation techniques. Comput Vis Graph Image Process 29(1):100–132
15. Gonzalez RC (2009) Digital image processing. Pearson Education India
16. Wu J, Poehlman S, Noseworthy MD, Kamath MV (2008) Texture feature based automated seeded region growing in abdominal MRI segmentation. In: 2008 international conference on biomedical engineering and informatics, vol 2. IEEE, pp 263–267
17. Thakur A, Anand RS (2004) A local statistics based region growing segmentation method for ultrasound medical images. Statistics 11:12
18. Belaid LJ, Mourou W (2009) Image segmentation: a watershed transformation algorithm. Image Anal Stereol 28(2):93–102
19. Ahmed MN, Yamany SM, Mohamed N, Farag AA, Moriarty T (2002) A modified fuzzy c-means algorithm for bias field estimation and segmentation of MRI data. IEEE Trans Med Imaging 21(3):193–199
20. Zhang D-Q, Chen S-C (2004) A novel kernelized fuzzy c-means algorithm with application in medical image segmentation. Artif Intell Med 32(1):37–50
21. Kaganami HG, Beiji Z (2009) Region-based segmentation versus edge detection. In: 2009 fifth international conference on intelligent information hiding and multimedia signal processing. IEEE, pp 1217–1221
22. Ronneberger O, Fischer P, Brox T (2015) U-net: convolutional networks for biomedical image segmentation. In: International conference on medical image computing and computer-assisted intervention. Springer, pp 234–241
23. Jonmohamadi Y, Takeda Y, Liu F, Sasazawa F, Maicas G, Crawford R, Roberts J, Pandey AK, Carneiro G (2020) Automatic segmentation of multiple structures in knee arthroscopy using deep learning. IEEE Access 8:51853–51861
24. Chaurasia A, Culurciello E (2017) Linknet: exploiting encoder representations for efficient semantic segmentation. In: 2017 IEEE visual communications and image processing (VCIP). IEEE, pp 1–4
25. Brea MLS, Barreira Rodríguez N, Mosquera González A, Evans K, Pena-Verdeal H (2016) Defining the optimal region of interest for hyperemia grading in the bulbar conjunctiva. Comput Math Methods Med 2016
26. Iroshan KA, De Zoysa ADN, Warnapura CL, Wijesuriya MA, Jayasinghe S, Nanayakkara ND, De Silva AC (2018) Detection of diabetes by macrovascular tortuosity of superior bulbar conjunctiva. In: 2018 40th annual international conference of the IEEE engineering in medicine and biology society (EMBC). IEEE, pp 1–4
27. Brea LS, Barreira Rodríguez N, Mosquera González A, Pena-Verdeal H, Yebra-Pimentel Vilar E (2018) Precise segmentation of the bulbar conjunctiva for hyperaemia images. Pattern Anal Appl 21(2):563–577
28. Curti N, Giampieri E, Guaraldi F, Bernabei F, Cercenelli L, Castellani G, Versura P, Marcelli E (2021) A fully automated pipeline for a robust conjunctival hyperemia estimation. Appl Sci 11(7):2978
29. Sevani N, Persulessy GBV et al (2018) Detection anemia based on conjunctiva pallor level using k-means algorithm. IOP Conf Ser Mater Sci Eng 420:012101. IOP Publishing
30. Jha P, Das M, Mishra A (2018) Image segmentation of eye for non-invasive detection of anemia. Available at SSRN 3282850
31. Sedki AG, Shaban SA, Elsheweikh DL (2020) A proposed image processing framework for diagnosis of anemia with providing proper nutrition. Int J Comput Sci Inf Secur (IJCSIS) 18(7)

32. Saldivar-Espinoza B, Núñez-Fernández D, Porras-Barrientos F, Alva-Mantari A, Leslie LS, Zimic M (2019) Portable system for the prediction of anemia based on the ocular conjunctiva using artificial intelligence. arXiv preprint arXiv:1910.12399
33. Dimauro G, Simone L (2020) Novel biased normalized cuts approach for the automatic segmentation of the conjunctiva. Electronics 9(6):997
34. Kingma DP, Ba J (2014) Adam: a method for stochastic optimization. arXiv preprint arXiv:1412.6980
35. Delgado-Rivera G, Roman-Gonzalez A, Alva-Mantari A, Saldivar-Espinoza B, Zimic M, Barrientos-Porras F, Salguedo-Bohorquez M (2018) Method for the automatic segmentation of the palpebral conjunctiva using image processing. In: 2018 IEEE international conference on automation/XXIII congress of the Chilean association of automatic control (ICA-ACCA). IEEE, pp 1–4

Data Consumption Behaviour and Packet Delivery Delay Analysis in OTT Services Using Machine Learning Techniques

Rohit Kumar Thakur and Raj Kumari

Abstract Network operators must monitor their networks and analyse their customer's consumption patterns. The data gathered on consumption patterns enables the creation of new data plans targeted at specific consumers, as well as a proper network perspective. The high amount of network resources consumed by over-the-top (OTT) apps are well-known. This data is transferred through protocols that can give us valuable information about consumption and time delay. With this in mind, using data mining and traditional machine learning, a concept for tailoring consumption behaviour and time delay for users, has been considered. Data set with 1581 instances and 87 attributes are there which is sufficient to find the behaviour of the user. Two protocols are extracted from the data set on which various machine learning classifiers are implemented to find the best accuracy. Only four attributes are used for the analysis of consumption patterns: Flow ID, Source IP, Destination IP, and timestamp. After reviewing the results, the best suitable algorithm from the traditional approach was Random Forest with an accuracy of 99.67% while the best classifier from the incremental approach is the decision tree with accuracy 99.38%.

Keywords Over-the-top · Machine learning classification · User consumption behaviour · Data set

1 Introduction

The technological tool market is developing quickly. Internet Service Providers' (ISPs') traditional business model is beginning to undergo a significant change as a result of the ongoing generation of service and application provider that use an over-the-top (OTT) business model as a platform for their new services. Skype, YouTube,

R. K. Thakur (✉) · R. Kumari
University Institute of Engineering and Technology, Panjab University, Chandigarh, India
e-mail: thakur.rohit1997@gmail.com

R. Kumari
e-mail: rajkumari@puchd.ac.in

S. Das et al. (eds.), *Advances in Data-driven Computing and Intelligent Systems*, Lecture Notes in Networks and Systems 653, https://doi.org/10.1007/978-981-99-0981-0_56

Telegram, Netflix, WhatsApp and a slew of other companies and apps have sprung up to satisfy the demands of consumers for new connections and functionality [1].

For network operators, network monitoring and consumption behaviour analysis are necessary. Data on consumption patterns can be used to create data plans targeted at particular clients and to better understand the network. It's common knowledge that OTT applications use up a lot of network resources. A relatively typical strategy for limiting the amount of data that can be sent is service degradation. It is frequently utilised broadly, affecting user-use of programmes while ignoring personal preferences and behaviour. In the light of this, an idea for tailoring service degradation rules enforced to users has been explored using data mining and conventional machine learning.

Network operators must track and examine the data usage trends of their consumers in order to offer new data plans that are customised for certain users. This enables them to, among other things, identify possible risks, preserve service quality, and avoid network collapse [2].

As consumption of the Internet can be monitored through the application layer with great accuracy. The consumption can be monitored through the protocols used in the application layer. Some of the examples of the application layer protocols are Hypertext Transfer Protocol (HTTP), HTTP proxy, Secure Socket layer (SSL), WebSocket, etc. These all protocols work with some attributes which are important for their working like source id, destination id, timestamp and flow id. In this paper by taking these objects, finding consumption behaviour and time delay in the delivery of packet is to find out which protocol consumes more data in different OTT applications. These all findings are only done for finding the user behaviour for the OTT like on which application users spend more time so that finding the relevant application for the user.

2 Related Work

The approach of Artificial Intelligence technique to govern and run networks as a result of ongoing technologies is backed by the authors in [4] about Knowledge Defined Networking (KDN). Another foundational notion is incremental learning, which refers to machine learning approaches capable of continual model adaptation based on a continuously incoming data stream, which is often present anytime systems must behave independently (e.g., autonomous robotics or driving) [5, 6]. In the light of this, the current state of the art will be provided following the implementation of a systematic mapping of academic texts, as described in [7], to offer an overview of the study area.

The monitoring and analysis of network traffic flow is the major focus of this research. The information gathered is utilised to construct smart user consumption profiles and service degradation policies. As a result, we looked at works published in the realm of information and telecommunication technology, which encompasses this topic. In addition, we included other publications linked to social networks and mobile services in our list since we felt some of the results to be useful. Following that, we'll go over these works in more detail.

In the light of this, the current state of the art will be given after adopting an appropriate mapping of academic texts based on the state of the art to provide an overview of the methodologies offered in [3] to identify the amount and kind of data in the study area works that are relevant. The following topics caught our attention: service degradation, with a focus on identifying the features that are directly associated with service deterioration; Quality of Service (QoS), with a purpose of identifying how Internet Service Providers (ISPs) handle this resource control mechanism in their networks. OTT services, highlighting the fact that it is a relatively new research subject, to better understand how this issue has developed in the study field; user categorization in a mobile network to comprehend how network operators handle users; classification of network traffic, with a focus on identifying the approaches used and, in particular, how incremental learning has been applied within that framework.

The extent to which phone actions can be correlated with users was investigated by authors in [8]. A multilayer perceptron designed as an auto-encoder was used to do this. To uncover unique characterizations of user behaviour, they used a self-organising map (SOM). User behaviour was predicted using data from the application, mobile towers, services accessed by him and websites.

In [9], authors looked at the possibilities for determining a user's gender based on application snapshots on mobile devices. Their major objective was to make smartphone systems easier to use and give more tailored services and goods. They discovered that gender may be deduced from usage patterns. Through an empirical analysis, they discovered considerable disparities across application kinds, functionality and icon designs. Their prediction model was accurate to the tune of 76.62%. The technique used by author in [9] in traditional learning approach which is not so efficient in many cases when there more than 50 protocols for analysis.

Network traffic analysis was used by authors [10–12] in as a promising tool for user identification, but they fail to minimise the delay in packet delivery. To describe and identify users, they created the idea of flow-bundle-level characteristics. A collection of flow records containing 65 users was used to assess four distinct machine learning models. Using the Random Forest technique, the authors were able to identify people with an accuracy of 83% which is due to large number of null values in various attributes.

3 Architecture for Collection of Data

This section describes the architecture used to collect and preprocess data collected on the Universidad del Cauca (Unicauca) campus in order to study the network's users' OTT consumption habits. It's worth noting that the design seeks to include a knowledge plane to the network by gathering information apropos to users OTT consumption behaviour and assisting network managers in the creation of individualised service mortification rules by adopting the KDN idea [4]. Working architecture

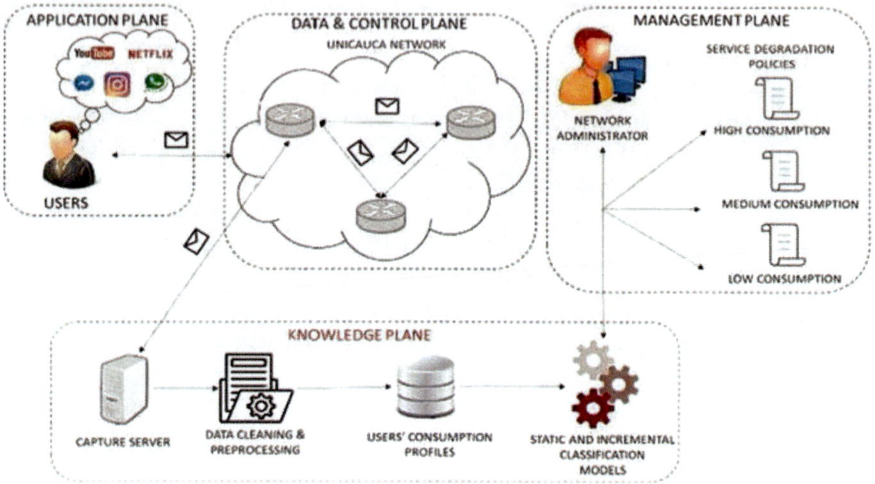

Fig. 1 Architecture of protocol working [13]

of protocols is shown in Fig. 1 where application plane is platform for various application then in second layer data and control plane is there. Management plane helps in the network administration, whereas knowledge plane is used for data fetching.

3.1 Actors

Anyone with access to a gadget that can connect to the Internet is considered a network user. Through the use of over-the-top (OTT) apps and other similar activities that can be performed online, their main activity is the generation of network traffic. A person or group of persons who have technical expertise in network architecture and device operations is known as a network expert. This actor is capable of establishing network devices, collecting and summarising network data, and creating testing grounds for networks. All data analysis and machine learning models must be built by the data analyst. These procedures include cleansing and preparing data, data transformation, grouping and pattern identification, as well as training, choosing, and deploying machine learning models. Last but not least, the network analyst is the individual or group in charge of network upkeep, network policy execution, resource allocation, and business strategy decision-making. The model and the helpful knowledge of the data analyst can be used by this actor to make important decisions. Occasionally, a person or group may perform the roles of both the network analyst and the network expert depending on the quantity of labour available. The individualised service degradation criteria that are suggested are relevant to the decision-making process in our case study.

3.2 *Components*

All the tasks that the network expert must complete are grouped under the heading of data gathering and flow generation. The knowledge of network architecture and hardware is frequently all in the hands of the network professional. These procedures cover all steps involved in configuring network devices, gathering and archiving IP packets, aggregating packets to create flows, and labelling applications (packet persistence and flow generation). Data pretreatment and model selection encompass all tasks that fall under the purview of the data analyst's competencies. Data cleaning, clustering, user consumption estimation, flow purification, and pattern recognition are all covered. Last but not least, decision-making refers to the actions that the network analyst must do in the light of their information and the feedback from model users. As a result, the network analyst will be better equipped to evaluate how network resources are allocated, QoS standards, service pricing, and service delivery. In this case study, decision-making is used to advise service degradation regulations based on user classification in line with their OTT consumption behaviour, which helps businesses better understand their consumers' demands.

3.3 *Workflow*

Workflow is done in three steps. In the first step, the data is transferred in the IP packets hence the packet persistence in the first step in our workflow. The packets are saved in a format known as packet capture (PCAP) files once they have been captured. PCAPs can be hundreds of gigabytes in size, depending on the amount of traffic. Flow generation is the second phase. The communications generated by user devices are represented statistically by network flows [14]. They store information such as the quantity of data sent and the length of the conversation, among other things. The flow cleansing procedure is the subject of the third stage. Unnecessary information is deleted at this stage. Superfluous data includes flows that don't correlate to user device communications. It is usually necessary for this operation to combine the skills of the network professional and data analyst.

4 Proposed Scenario

Three main sections have been established with the goal of proposing a set of personalised service mortification policies taking into account the user's consumption behaviour around OTT and other applications such as food ordering application and online cab booking applications.

4.1 Data Set Contents

The data given in this research was gathered over six days (April 26, 27, 28, and May 9, 11, and 15) in 2017 by a network segment from Universidad Del Cauca, which performed packet captures at different times during the morning and afternoon. A total of 16,545,768 instances were gathered and kept in CSV file. The data set presented in this paper is available at [15]. As can be seen in the provided link, the available data sets are The 16,545,768 instances with 87 attributes that were recorded over the duration of six days are stored in the data set. The data set was split into six CSV files as the data size is big for each day (one per day). Each instance contains information about an IP flow created by a network device, such as source and destination IP addresses, ports, interarrival periods, and the layer 7 protocol (application) used on that flow as the class, among other things. The majority of the attributes are numerical because of the timestamp, but there are also nominal and date types of data that are turned into numerical while working with machine learning to obtain precise results. This data set also includes each and every one of the 79 application labels that were found during the Deep Packet Inspection (DPI) procedure. These all attributes groups, attributes, and their description are given in Table 1.

4.2 Data Preprocessing

Preprocessing data before feeding it into our model is important because the quality of data and the valuable information that can be gathered from it has a direct impact on our model's ability to learn as a result, preprocessing data before feeding it into our model is critical step. The data preprocessing includes various steps in its operational steps that is Handling Null Values, Standardisation, Handling Categorical Variables, One-Hot Encoding, and Multicollinearity. During the implementation phase, we employ a label encoder to turn the labels into a numeric form that is machine-readable [16]. In this paper, we convert various attributes into numeric form like protocol name, label, flow id, source id, timestamp, and destination id. From data set we abstract only two protocols which have maximum number of values. Many protocols have very little values which is not sufficient for the implementation of machine learning so here only two protocols are used: "HTTP_PROXY" and "SSL" protocols.

In this implementation, we did data cleaning in the data preprocessing by filtering various protocols which contain null values. Normalisation is a data preparation process that involves changing the values of numeric columns in a data set to a similar scale. This is especially important if the features in your machine learning model have a wide range of values.

In the actual world, a scenario like this is rather frequent; one attribute may be fractional and vary between zero and one, while another may range between zero and a thousand. If we're trying to forecast something using regression, this characteristic will have a bigger impact on the outcome because of its higher value, even if it's not the most important predictor. Here, we took a normalizer for the normalisation of the data. The data set used by classification model is in two phases 30% of data for testing and 70% of data for training of model [17]. Result of the various classification models of machine learning are discussed in next section.

5 Result Analysis

This section gives the results generated after the implementation of classification models. After the implementation of classification model, we left with confusion matrix and accuracy of various classifiers. Some classifiers uses incremental learning like Random Forest while SVM uses traditional learning which makes differences in their result. Similarly in case of Logistic Regression, it gives best results in case of complex algorithm while for complex network decision tress classifiers is used

Table 1 Data set description [1]

Attribute Group	Attributes	Description
Network identifiers (7 attributes)	Flow ID; Source IP; Source Port; Destination IP; Destination Port; Protocol; Timestamp	The IP addresses, transport layer protocol, and ports over which the packet is to be transferred are all contained in these properties, together with other data pertaining to the source and destination of an Internet flow
Interarrival times (15 attributes)	Flow Duration; Flow IAT Mean; Flow IAT std; Flow IAT Max; Flow IAT Min; Fwd IAT Total; Fwd IAT Mean; Fwd IAT Std; Fwd IAT Max; Fwd IAT Min; Bwd IAT Total; Bwd IAT Mean; Bwd IAT Std; Bwd IAT Max; Bwd IAT Min	These properties contain all the data pertaining to the forward and reverse interarrival times
Flag features (12 attributes)	Fwd PSH flags; Bwd PSH flags; Fwd URG flags; Bwd URG flags; FIN Flag Count; SYN Flag Count; RST Flag Count; PSH Flag Count; ACK Flag Count; URG Flag Count; CWE Flag Count; ECE Flag Count	These properties display details about each flag that is present in a packet's header, including Push, Urgent, Finish, and other flags
Header descriptors (5 attributes)	Fwd Header Length; Bwd Header Length; Average Packet Size; Fwd Header Length 1	The header-related information is saved

(continued)

Table 1 (continued)

Attribute Group	Attributes	Description
Flow descriptors (36 attributes)	Total Fwd Packets; Total Bwd Packets; Total Length of Fwd Packets; Total Length of Bwd Packets; Fwd Packet Length Max; Fwd Packet Length Max; Fwd Packet Length Min; Fwd Packet Length Mean; Fwd Packet Length Std; Bwd Packet Length Max; Bwd Packet Length Min; Bwd Packet Length Mean; Bwd Packet Length Std; Flow Bytes S; Flow Packets S; Min Packet Length; Max Packet Length; Packet Length Mean; Packet Length Std; Packet Length Variance; Down Up Ratio; Avg Fwd Segment Size; Avg Bwd Segment Size; Fwd Avg Bytes Bulk; Fwd Avg Packets Bulk; Fwd Avg Bulk Rate; Bwd Avg Bytes Bulk; Bwd Avg Packets Bulk; Bwd Avg Bulk Rate; Init Win bytes forward; Init Win bytes backward; act data pkt fwd; min seg size forward; Label; L7Protocol; ProtocolName	These properties contain all the data pertaining to the Internet flow, including the volume, standard deviation, and number of packets in both the forward and backward directions
Subflow descriptors (4 attributes)	Subflow Fwd Packets; Subflow Fwd Bytes; Subflow Bwd Packets; Subflow Bwd Bytes	If there were subflows, these properties include all relevant data regarding their volume and forward and reverse packet count in both directions
Flow timers (8 attributes)	Active Mean; Active Std; Active max; Active min; Idle Mean; Idle std; Idle max; Idle min	Information about the active and inactive periods for each flow

so they work according to different data sets. From this we can say that results of various classifiers varies according to the data set.

5.1 Confusion Matrix

Machine learning classification models performance is measured using a confusion matrix. It's a type of table that lets you see how well a classification model performs on a set of test data for which the true values are known. For above confusion matrix is given Fig. 2 (Table 2).

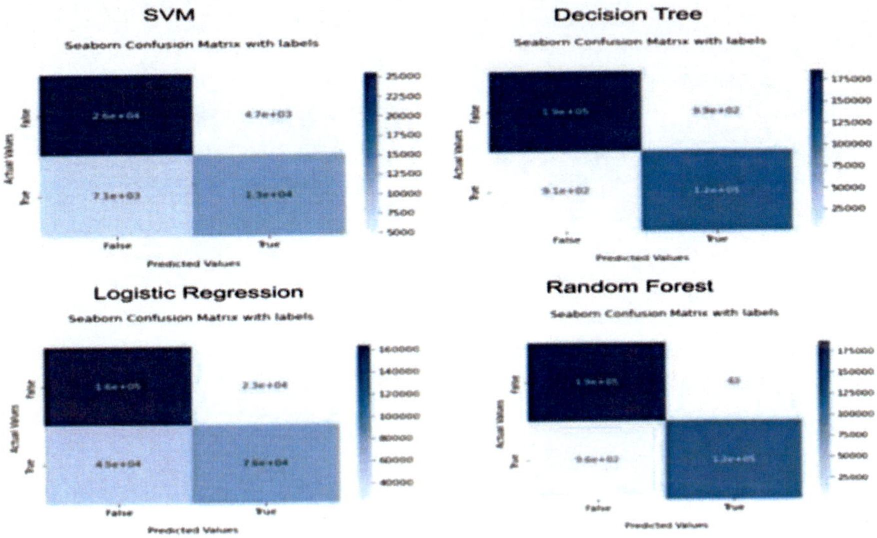

Fig. 2 Confusion matrix for different machine learning classifier

Table 2 Accuracy table for precision, recall, and $F1$ score

Classifier Name	Precision	Recall	F1 Score	Accuracy (%)
SVM	0.76	0.76	0.76	76
Random forest	1.00	1.00	1.00	99.67
Decision tree	0.78	0.78	0.77	99.38
Logistic regression	0.78	0.78	0.77	77.91

5.2 *Accuracy of Different Classifier*

Implementation of various classification models [18] on the data and their accuracy level is discussed here in this section. For evaluation of accuracy, we use feature extraction which is a crucial step in machine learning. Two most popular protocols SSL and HTTP Proxy which are used frequently and contain more data fields in the data set are extracted from the data set. Confusion matrix is made for the data set which helps in prediction of best protocol. Various classification models are implemented to predict the protocol which is best for over-the-top (OTT) or other applications which run over the Internet. The precision, recall, and F1 score is shown in Fig. 3. The results of various classifiers in machine learning with test and train split of 30% and 70%, respectively, are shown in Fig. 4.

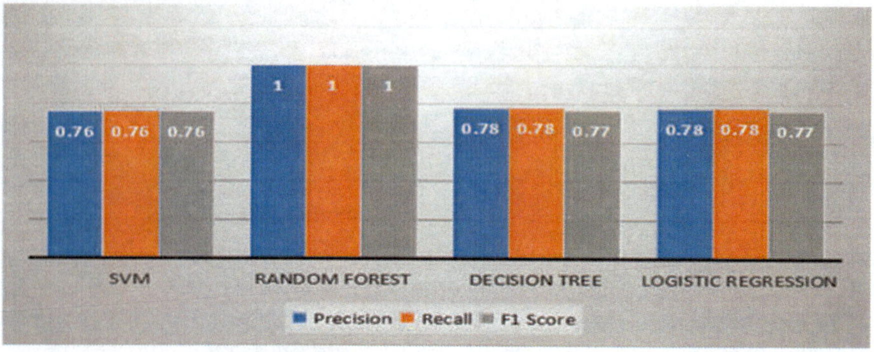

Fig. 3 Chart for precision, recall, and $F1$ score

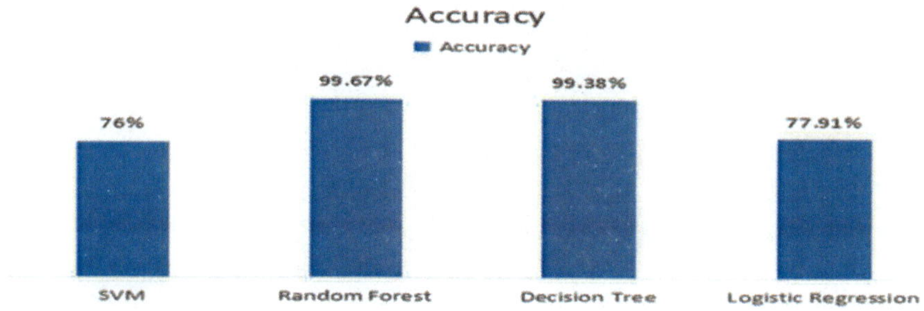

Fig. 4 Chart for accuracy of various classifiers

6 Conclusion and Future Scope

With the unpredictable market and rapid change that the Internet, particularly OTT applications, exhibit, this paper describes a performance comparison of traditional and incremental machine learning algorithms with the goal of dynamically personalising the service degradation policies that Internet service provider companies apply to users once their data plan consumption limit is exceeded. On using this approach, ISP can also recommend plans according to the user's needs. This comparison was done since a typical categorization model can't take into consideration the Internet's quick changes and, consequently, the changes that a user's OTT consumption behaviour may show over time. The development of an application that helps in finding out the machine learning classifiers model to be used to classify users based on their consumption behaviour over the Internet, as well as the study and implementation of a mechanism that allows the enforcement of personalised service degradation policies within the architecture of a real network. In order to get data that is more reminiscent of a real-world network scenario, an inquiry into developing a synthetic data generator that does not assume statistical independence across characteristics is being conducted. Previous studies used traditional learning approaches for the training of models but we have used incremental learning models. Only four attributes are taken for the analysis of behaviour and packet delivery delay and only these four attributes are important as they contain all information about the data like type of data, source IP, destination IP, and timestamp.

In future, there is also scope of implementation of deep learning when we have sufficient data in the data set for the implementation of deep learning. Deep learning is a powerful technique when we have to test and train a big data set. Various other machine learning classifiers also can be used for better performance analysis. Similarly with the increase in size of data set, the accuracy increases as we are working with an incremental learning approach.

References

1. Rojas JS, Rendón A, Corrales JC (2019) Consumption behavior analysis of over the top services: incremental learning or traditional methods? IEEE Access 7
2. Carela-Español V (2014) Network traffic classification? From theory to practice. Ph.D. dissertation, Department d'Arquitectura Computadors, University Politècnica Catalunya, Barcelona, Spain
3. Petersen K, Feldt R, Mujtaba S, Mattsson M (2008) Systematic mapping studies in software engineering. In: Proceedings of 12th International Conference on Evaluation and Assessment in Software Engineering Swindon, U.K., pp 68–77
4. Mestres A, Rodriguez-Natal A, Carner J, Barlet-Ros P, Alarcón E, Solé M et al (2017) Knowledge-defined networking. ACM SIGCOMM Comput Commun Rev 47:2–10
5. Gepperth A, Hammer B (2016) Incremental learning algorithms and applications. Eur Symp Artif Neural Netw (ESANN)
6. Silver DL (2011) Machine lifelong learning: challenges and benefits for artificial general intelligence. In: Artificial general intelligence. Springer, Berlin, Germany, pp 370–375
7. Petersen K, Feldt R, Mujtaba S, Mattsson M (2008) Systematic mapping studies in software engineering. In: Proceedings of 12th international conference evaluation and assessment software engineering, pp 68–77
8. Swartz C, Joshi A (2014) Identification in encrypted wireless networks using supervised learning. In: Proceedings of IEEE military communications conference, pp 210–215
9. Zhao S, Xu Y, Ma X, Jiang Z, Luo Z, Li S et al (Feb.2020) Gender profiling from a single snapshot of apps installed on a smartphone: an empirical study. IEEE Trans Ind Informat 16(2):1330–1342
10. Paul I, Khattar A, Kumaraguru P, Gupta M, Chopra S (2019) Elites tweet? Characterizing the Twitter verified user network. arXiv:1812.09710. (online) Available: http://arxiv.org/abs/1812.09710
11. Chen S, Zeng K, Mohapatra P (Dec.2014) Efficient data capturing for network forensics in cognitive radio networks. IEEE/ACM Trans Netw 22(6):1988–2000
12. Lal S, Kulkarni P, Singh U, Singh A (2013) An efficient approach for network traffic classification. In: Proceeding of IEEE international conference on computational intelligence and computing research, pp 1–5
13. Rojas JS, Rendón Á, Corrales JC (2019) Consumption behavior analysis of over the top services: incremental learning or traditional methods? IEEE Access 7:136581–136591. https://doi.org/10.1109/ACCESS.2019.2942782
14. Hofstede R, Celeda P, Trammell B, Drago I, Sadre R, Sperotto A et al (2014) Flow monitoring explained: From packet capture to data analysis with NetFlow and IPFIX. IEEE Commun Surv Tuts 16(4):2037–2064

15. Dataset Unicauca—2018—Google Drive. https://drive.google.com/drive/folders/1FcnKUlSqRb4q5PkGfAGHz-g7bVKL8jmu?usp=sharing
16. Jackson E, Agrawal R (2019) Performance evaluation of different feature encoding schemes on cybersecurity logs. SoutheastCon 2019:1–9. https://doi.org/10.1109/SoutheastCon42311.2019.9020560
17. Data Splitting. Z. Reitermanov. Introduction. Cross-Validation Techniques (2019). (online) Available: https://docplayer.net/26609777-Data-splitting-z-reitermanova-introduction-cross-validation-techniques.html
18. Williams N, Zander S, Armitage G (2006) A preliminary performance comparison of five machine learning algorithms for practical IP traffic flow classification. SIGCOMM Comput Commun Rev 36(5):5–16

Multilevel Deep Learning Model for Fabric Classification and Defect Detection

Pranshu Goyal, Abhiroop Agarwal, Kriti Singhal, Basavraj Chinagundi, and Prashant Singh Rana

Abstract Detecting defects in fabrics is a difficult task as there are a lot of variations in the type of fabric and the defect itself. Many methods have been proposed to solve this problem, but their detection rate and accuracy were very low depending on the model tested. To eliminate the variations and to improve the performance, we implemented multilevel modeling in our approach. This article proposes an enhanced and more accurate approach to detecting tissue defects. Here, we compare the performance of various advanced deep learning models such as MobileNetV2, Xception, VGG19, and InceptionV3 and how their performance changes with the type of fabric. First, a Convolutional Neural Network model is used to classify the fabric into different types with an accuracy of 97.6%, and then on the basis of the type of fabric, the best model is used to detect defects in the fabric. This has a significant advantage in improving the overall performance of fabric defect detection. In addition, K-fold cross-validation has been carried out to verify the coherence of the proposed model.

Keywords Fabric defect detection · Deep learning · MaxPool · Multilevel modeling · TensorFlow · ImageNet

1 Introduction

Texture assessment is exceptionally critical in the textile industry [1], including the detection of defects in fabrics. The nature of texture depends on indispensable processes of texture review to distinguish the deformities of fabric. The benefits of industrialism have been diminished because of texture imperfections causing repulsive losses. Conventional deformity discovery strategies are directed in numerous ventures by proficient human auditors who physically curate the fabric. In any case, such recognition strategies have a few weaknesses, such as fatigue, dreariness, carelessness, mistake, confusion, and tediousness which cause to diminish the finding of shortcomings. To address these shortcomings, various image processing techniques

P. Goyal (✉) · A. Agarwal · K. Singhal · B. Chinagundi · P. Singh Rana
Thapar Institute of Engineering and Technology, Patiala, Punjab, India
e-mail: goyalpranshu267@gmail.com

S. Das et al. (eds.), *Advances in Data-driven Computing and Intelligent Systems*, Lecture Notes in Networks and Systems 653, https://doi.org/10.1007/978-981-99-0981-0_57

have been executed to naturally and productively distinguish and recognize texture defects.

To make increments in the textile sector, there is a requirement of huge investments to bring improvements to the output product [2]. Just like various other industries, textile industry faces various issues. This includes insurance to reduce the impact of domestic accidents, customer disappointments, wasted time, and more. Fabric flaws are undoubtedly the most significant test facing the textile industry. Fabric is made daily using fibers and a commonly used material. Most tissues are delivered to different stages of manufacture. Various tools and processes are used at different phases of production.

Throughout the production process, tissues are subjected to various pressures and constraints that can lead to defects. Faults have several names, as stated in their structures and directions. The textile industry has identified over 70 defects, including holes, scales, tips, and petroleum marks. Unplanned work may cause various defects to the tissue surface during tissue production. Tissue manufacturing is one of the more critical traditional industries where tissue inspection systems can be crucial in increasing production rates. The inspection process is essential in any manufacturing process, particularly from an industry perspective. The inspection process aims to recognize errors or defects and prevent the inspector from checking the manufacturing process and removing faulty products.

There are two main types of inspection models are used to detect fabric defects. the main model is the human inspection system [2]. The second framework is an automated inspection system. Detecting human error by experts can therefore quickly prove to be a perplexing and delicate task. As such, having a competent and automated framework nearby is essential for improving consistent quality and speeding up quality control, which can increase profitability. The topic of automatic error detection has been explored in many studies over the past decades. Despite the fact that there is no widely used methodology to address this problem, several strategies have been proposed in recent years that rely on image processing methods. These techniques have been used to detect defects at the image level, resulting in poor accuracy and difficulty in accurately locating defects. This method cannot be stretched into different fabrics. Recently, various techniques according to the local image level have been proposed [3], which use the base unit as the basic activity object to extract the features of the image.

Recognizing defects in fabrics is especially difficult due to a large number of variables. As discussed in the paragraph above, there are many different types of defects, but these defects are different for different types of fabrics. During the course of our research, we tried to find the correlation between the fabric type and its defect detection accuracy for various state-of-the-art deep learning models. We were able to see that certain models performed better for certain types of fabric. Hence, we proceeded further to first determine the type of fabric (Type A, Type B, Type C, Type D) using a CNN model, and then we identify whether it has defects or not.

2 Types of Defects in the Fabric

Different types of fabric materials are used to prepare diverse categories and variations of fabric articles in the industry. Each fabric has other qualities. Consequently, the quality of the yarn or commercial defects influences the quality of the fabric. It was estimated that tissue defects would reduce the price of tissues from 45 to 65% [4, 5]. It is due to defects such as dye/stain marks, soft chains, faulty pattern cards, and holes. In a cloth, defects can occur due to machine defects, color bleeding, thread problems, excessive stretching, etc. The textile company has identified more than 70 types of default [3]. These defects are shown in illustration in Fig. 1.

Hence, it is very difficult to identify the exact defect and classify it into types. We focused on classifying whether the fabric has defects or not for the course of this research. To do this, we used various state-of-the-art deep learning models like VGG 19, Xcpetion, Inception V3, and MobileNet V2.

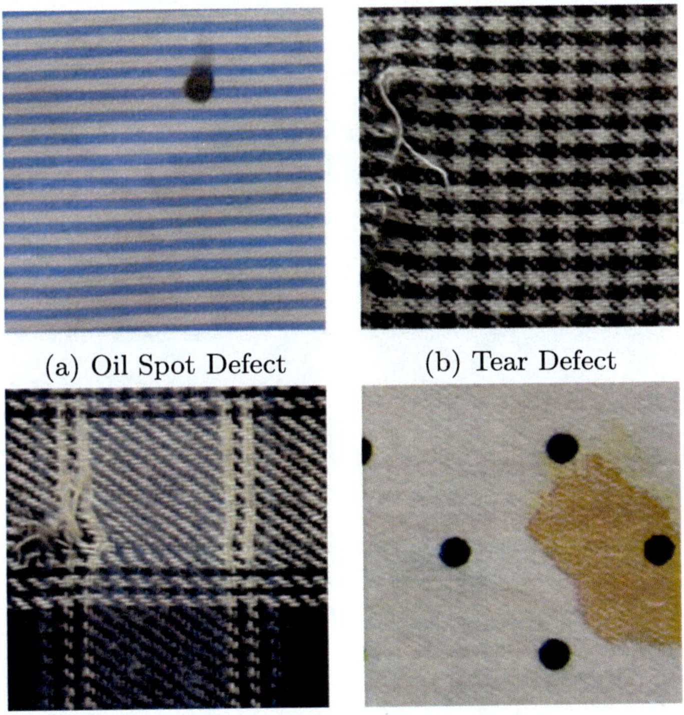

(a) Oil Spot Defect (b) Tear Defect

(c) Shear Defect (d) Paint Spot Defect

Fig. 1 Types of defects in fabrics

3 Fabric Classification

There are many different types of defects, but there are also different types of fabrics that one comes across in the textile industry. Fabric can be distinguished on the basis of pattern, design, and the weaving method. Some of the different types of fabric are shown in Fig. 2. As we can see, the fabric is broadly classified into four types on the basis of the pattern. These are: (1) Dot Pattern, (2) Thin Stripes, (3) Twill Plaid, and (4) Houndstooth. Each fabric has different sewing patterns, due to which the defects found in each of them be very different. Hence, if we classify the fabric type first and then detect the defect, the prediction will be distinct and achieve better accuracy.

These four types of fabrics will be classified using a CNN model, which we developed. These fabrics are taken from the ZJU-Leaper dataset [6] which is a benchmark dataset for fabric defect detection and comparative study. This dataset plans to set a benchmark suite for vision-based texture deformity recognition, including picture dataset, assessment convention, and pattern tests. Once we are able to identify the type of fabric, we can further analyze which deep learning model will give the user the best accuracy for detecting the defect.

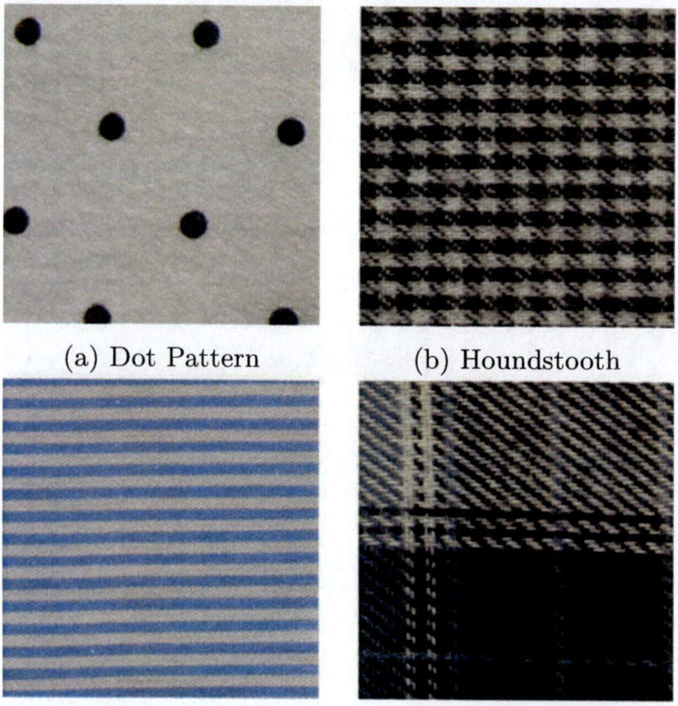

(a) Dot Pattern (b) Houndstooth (c) Thin Stripes (d) Twill Plaid

Fig. 2 Different types of fabrics

4 Proposed Methodology for Defect Detection

Three major groups can be used to organize traditional AI-based [7] defect detection methodologies: statistical, structural and model-based methodology. Let us analyze these methodologies and discuss how our approach differs from them. The statistical approach uses grayscale properties to describe the texture properties of texture images. This is referred to as first-order statistics and higher-order statistics. First-order statistics can measure the variance of gray-level intensities among various features between defect regions and the background. Higher-order statistics rely on the joint probability distributions of the pixel sets. In any case, the drawback of this strategy is that the defect size must be large enough to allow constrained estimation of texture properties. Therefore, this methodology is efficient in handling large sized defects. However, this methodology lacks behind in handling small local defects. Moreover, computing higher-order statistics is cumbersome. Needless to say, this approach requires a high-quality image of the texture rather than the entire fabric image. The structural approach generally applies to the properties of defect-free fabric texture primitives near defective areas and their associated placement rules. Apparently, the utility of this methodology is to overload these textures with uniform macro-textures.

In the class model-based methodology below, the policies commonly used are Markov Random Fields and Gaussian Markov Random Fields. This approach relies on focusing on the textures considered, which means that more precise spatial relationships between greyscale in a texture can be found. However, similar to secondary statistical methodologies, it is more challenging for model-based methods to identify small gaps. Therefore, this methodology also requires a relatively large texture area for estimating model parameters.

The above-mentioned methodologies have their own advantages and disadvantages. We came up with a multilevel model-based approach consisting of a CNN model for categorizing the type of fabric, and then we selected the best deep learning model for detecting whether it has defects or not. As an intermediate step, we try to find a correlation between the type of fabric and the deep learning model, as some models are better at detecting defects in a particular type of fabric.

Step 1—The dataset is prepared from the ZJU-Leaper dataset [6]. The dataset consists of four types of fabrics—Type A, Type B, Type C, and Type D. Each type of fabric has distinct patterns and characteristics. The dataset has a total of 2000 images, where each type of fabric has 500 images, out of which 250 are defective, and 250 are non-defective.

Step 2—A convolutional neural network model is developed for the classification of the fabric. Figure 3 depicts the architecture of the CNN model.

Step 3—The proposed CNN model is trained on the dataset to predict the type of fabric. Hyperparameter tuning is done to achieve an accuracy of 97.6%.

Step 4—Various state-of-the-art deep learning models—InceptionV3, Xception, VGG19, and MobileNetV2—were selected and then trained to predict if the fabric

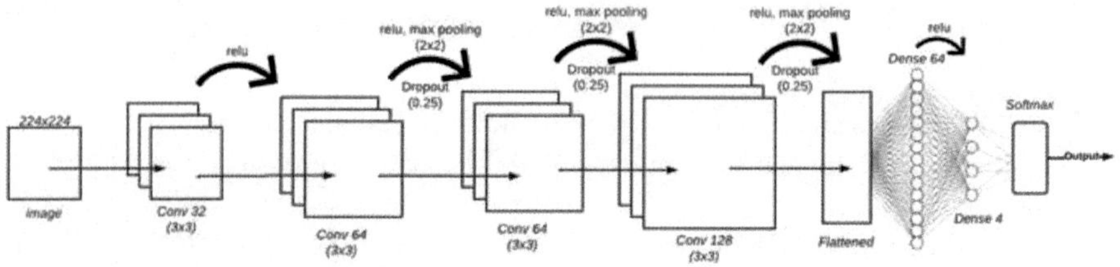

Fig. 3 CNN architecture diagram

Table 1 Number of parameters for deep learning models

Deep learning model	Number of parameters
MobileNetV2	3,608,678
Xception	22,998,606
VGG19	143,667,240
InceptionV3	23,885,392

Table 2 Comparison of accuracy of the various deep learning models for different type of fabrics

	Deep learning models			
	MobileNetV2	Xception	VGG19	InceptionV3
Type A	**0.96**	0.95	0.92	0.95
Type B	0.90	0.94	0.90	**0.95**
Type C	0.74	**0.76**	0.68	0.72
Type D	0.83	0.83	0.72	**0.91**

The bold signifies the accuracy for the best model for the given type of fabric

was defective or non-defective. Table 1 gives the number of parameters of each model.

Step 5—The models are compared, and it is concluded that different models give more accurate predictions for different types of fabric. Table 2 gives the comparison between models.

Step 6—Two dense layers and one GlobalAveragePooling layer are added to increase the accuracy of the models. Table 3 gives the improved accuracies of the models. It is observed that different models give more accurate predictions for different types of fabric.

Step 7—A multilevel machine learning model is proposed, where a CNN model is used for fabric classification, and the best-suited state-of-the-art deep learning model is used to detect defects in the fabric.

Step 8—K-fold cross-validation is performed to verify the consistency of the CNN model and the best model selected for each type of fabric.

The flowchart of our proposed methodology is shown in Fig. 4.

Table 3 Comparison of improved accuracy after adding 2 dense layers and global average pooling

	Deep learning models			
	MobileNetV2	Xception	VGG19	InceptionV3
Type A	0.96	**0.97**	0.93	0.96
Type B	0.92	0.93	0.93	**0.95**
Type C	**0.80**	0.76	0.71	0.76
Type D	0.86	0.89	0.72	**0.94**

The bold signifies the accuracy for the best model for the given type of fabric

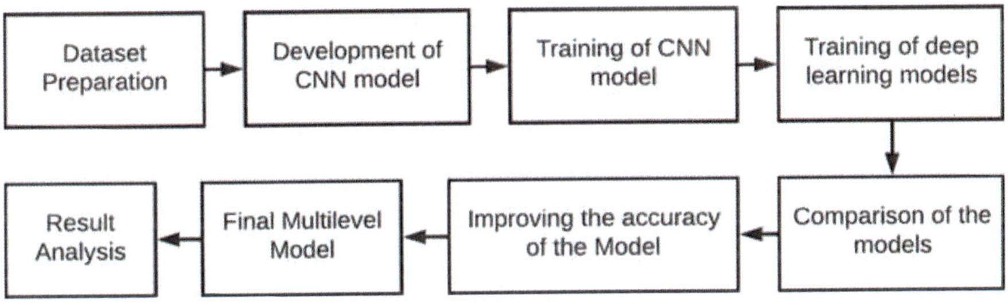

Fig. 4 Flowchart of the proposed methodology

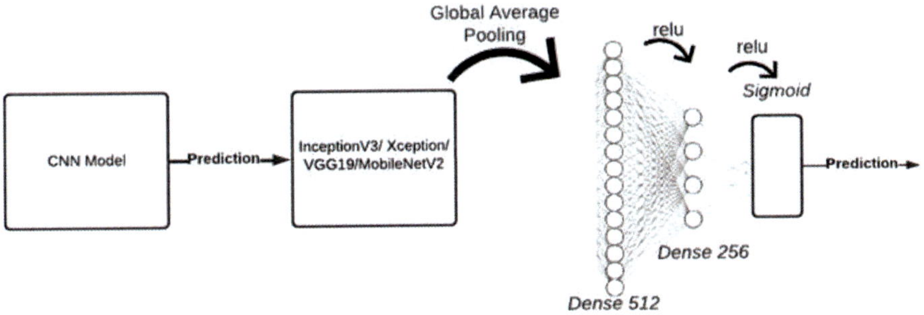

Fig. 5 Proposed multilevel model using deep learning models

5 Deep Learning Architectures

In the proposed approach, we have utilized profound convolutional brain networks in light of VGG (VGG19), GoogLeNet (Inception V3 and Xception) designs, and MobileNet (MobileNet V2) engineering, pre-prepared for fabric defect detection task on the given dataset (Fig. 5).

1. **VGG Architecture**: 16 layers (VGG16) and 19 layers (VGG19) VGG networks [8] were the preconditions for the participation of the Visual Geometry Group (VGG) in the ImageNet Challenge 2014. Where the VGG group obtained first place in the restrictive route and second place in the control route. VGG Engi-

neering begins with five blocks of convoluted layers, followed by three fully connected layers. The convolutive layer uses a 3×3 portion of step 1 and a cushion of 1 to ensure that each starting map contains spatial aspects similar to the previous layer. The upright vertical unit (ReLU) starts immediately after each convolution and uses the maximum pooling activity toward the end of each block to minimize the spatial appearance. The max pooling layer uses 2×2 bits with the stride of 2 along with no cushioning to ensure that all spatial components of the ratifying map from the previous layer are shared. Then, two fully allocated layers with 4096 ReLU units are used before the final 1000 fully allocated softmax layers. The drawback of the VGG19 model is that it is costly to evaluate and has extensive storage and usage limits. The VGG19 is about 143 million borders. Most of these limits (around 100 million) are in the first fully connected layer, and these fully connected layers can be removed without scaling up. Finally, the number of critical limits is known to go down to essential limits.

2. **GoogLeNet Architecture**: The design of GoogLeNet [9] was introduced as GoogLeNet (Inception V1), then improved as Inception V2, and more recently as Inception V3. While the Inception modules are artfully convolutive, they have all the attributes needed to learn more extravagant representations with fewer limitations. The traditional convolutive layer attempts to teach channels in a 3D space, with two spatial aspects (width and level) and a channel aspect. This way, a portion of the solitary convolution is entrusted while planning the relationships between the channels and the spatial connections. The Inception module aims to make this cycle easier and more competent by clearly integrating it into a progression of tasks. It would be open to looking at inter-channel relationships and spatial connections. The Xception design [10] is an extension of Inception engineering, which replaces standard Inception modules with detailed detachable convolutions. Rather than segmenting input information into a few blocks, it maps spatial relations for each output channel independently. After that plays a 1×1 depth-wise convolution to grab a cross-channel connection, this is essentially the same as a common activity known as "separate deep convolution", which includes deep convolution. This is a spatial convolution performed freely for each channel followed by a one-time convolution (a convolution of 1×1 for all channels). Xception slightly beats InceptionV3 on the ImageNet dataset and incomprehensibly submerges a larger image aggregation dataset with 17,000 classes. In particular, it has a similar number of borders as Inception V3, suggesting greater computational capability. Xception has 22,855,952 education frontiers, while Inception V3 has 23,626,728 education frontiers.
3. **MobileNet Architecture**: MobileNetV2 [11] is a fundamental improvement over MobileNetV1 and pushes the boundaries of versatile visual assertion, including the collection, revelation of objects, and division of semantics. Mobile NetV2 is part of the TensorFlow-Slim image classification library, where you can quickly start exploring MobileNetV2 within Collaboratory. Then, you can download the notebook and look at it locally using Jupyter. MobileNetV2 is also available as a plugin for TF-Hub. MobileNetV2 develops MobileNetV1 considerations, including profoundly separable convolution as feasible design blocks.

In all cases, V2 familiarizes two new functionalities with the design: (1) the direct bottlenecks between layers and (2) safeguarding the action relationship between the bottlenecks.

6 Multilevel Modeling

The proposed framework consists of three components. (1) Data collection, (2) Fabric type classification using a convolutional neural network, and (3) Defect detection using the most suitable deep learning model for that type of fabric. The block diagram of the proposed framework is shown in Fig. 6. The framework takes 224×224 resolution of images, classifies the type of fabric and then uses the best state-of-the-art deep learning model to detect whether the fabric is defective or non-defective.

Phase I: The first phase includes identifying four different types of fabric with distinct features and characteristics from the ZJU-Leaper dataset [6]. The dataset of 2000 images is then prepared, where each type of fabric has 500 images, out of which 250 are defective and 250 are non-defective. Training and testing data are generated in the ratio of 80:20, respectively.
Phase II: The CNN model [12] is used for fabric classification and predicts the type of fabric with an accuracy of 97.6%.
Phase III: On the basis of the predictions from Phase II, the best-suited model among MobileNetV2, Xception, VGG 19, and InceptionV3 is used to predict if the fabric is defective or non-defective. Figure 7 shows the entire workflow of the proposed methodology constituting the phases involved.

7 Model Evaluation

Various parameters such as accuracy [13] and K-fold cross-validation [14] are calculated to evaluate the performance of the proposed multilevel model. The results are compiled in the form of tables and have been shown below. To verify the model's robustness, K-fold cross-validation has been done repeatedly.

7.1 *Model Evaluation Parameters*

The accuracy achieved is used to calculate the parameters for the model evaluation. The classifier's accuracy serves as a barometer of its correctness. Accuracy is computed as:

$$\text{Accuracy} = (TP + TN)/\text{Total Data} \quad (1)$$

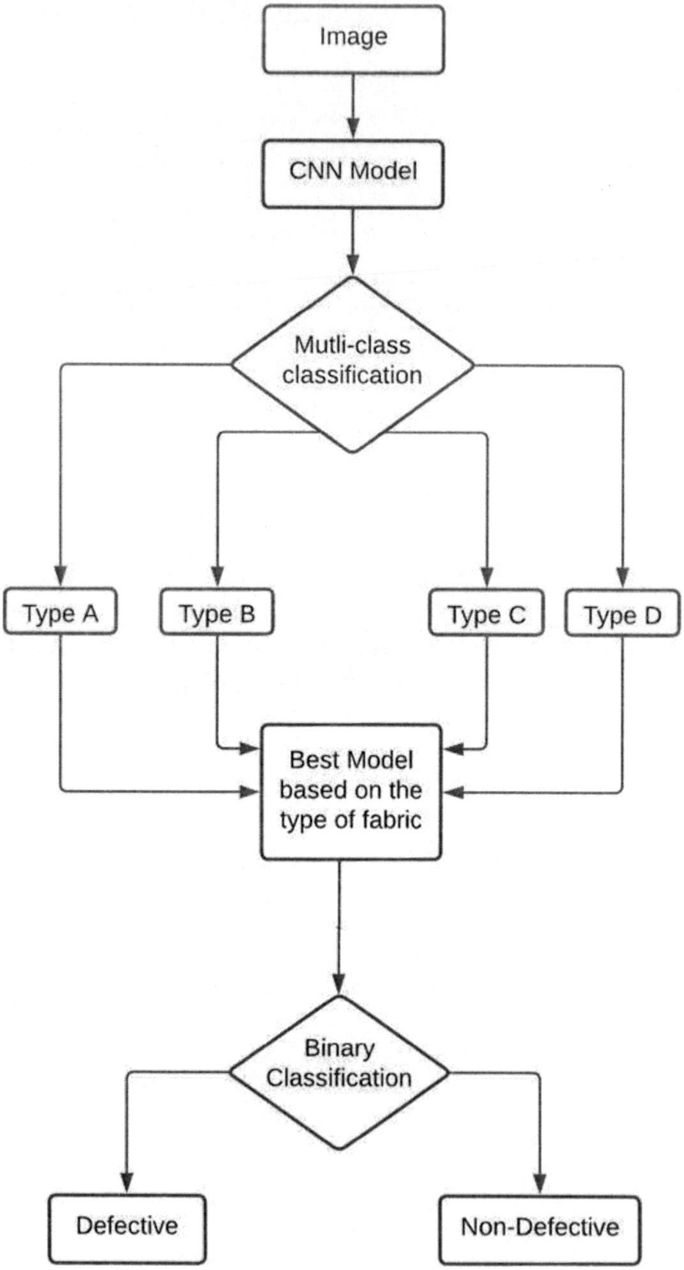

Fig. 6 Block diagram of the multilevel model

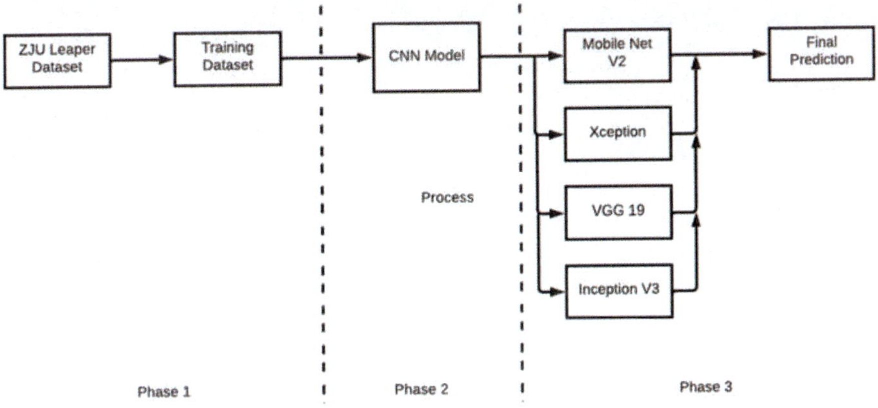

Fig. 7 Workflow of the proposed methodology

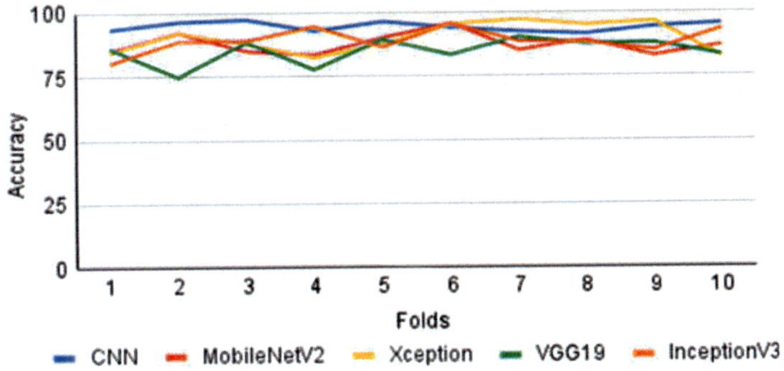

Fig. 8 Graph of K-fold cross-validation

7.2 *K-Fold Cross-Validation*

Estimating a grader's accuracy is essential in predicting its future projections' accuracy and selecting a grader from a given set (model selection). Repeated cross-validation is performed to ensure that the proposed multi-step model is compatible with low bias and low variance. In this work, cross-validate five times is repeated ten times. This cross-validation process applies to the four deep learning models and the CNN model. Figure 8 shows the K-fold cross-validation relative to the accuracy of each model, and the overlapping lines indicate that the proposed multi-step model is robust.

8 Result Analysis

The models are trained on the training dataset and are further tested on the testing dataset. The multilevel ensemble model is a combination of a CNN model and four deep learning models. The models are evaluated on various parameters, as mentioned above. The proposed CNN model has achieved an accuracy of 97.6% for fabric classification. Table 2 gives a comparison between the accuracies of different state-of-the-art models such as MobileNetv2, Xception, VGG19, and InceptionV3 for predicting defects in different types of fabrics. Table 3 gives that the performance of the models has been further improved by the proposed models. These tables help conclude that there is a correlation between the type of fabric and deep learning models as different models perform better for defect detection in different types of fabrics.

A problem that may occur while training is overfitting. To deal with the issues of overfitting, the model should be cross-validated, and if the resultant accuracy after various runs is consistent in all the runs, then the trained models are not overfitted. The accuracy is validated by applying tenfold cross-validation five times, as shown in Fig. 8. After the above analysis, we can conclude that the proposed model is not overfitted. Hence, the multilevel model approach of first classifying the type of fabric and then using the best-suited state-of-the-art model for defect detection provides a significant advantage in improving the overall performance of fabric defect detection.

9 Conclusion and Future Work

In today's world, Fabric Defect Detection can find numerous applications in eliminating wasteful production and improving cost-effectiveness. A multilevel model for fabric detection is proposed, which combines a CNN model and four deep learning models (InceptionV3, Xception, VGG19, and MobileNetV2), which achieves an average accuracy of 97.6%. In the future, we intend to study different deep learning architectures like artificial neural networks and stacked autoencoders for fabric defect detection on a larger dataset and with higher computational power. We also intend to further classify the defects detected as a future scope for this research.

References

1. Siew LH, Hodgson RM, Wood EJ (1988) Texture measures for carpet wear assessment. IEEE Trans Pattern Anal Mach Intell 10(1):92–105
2. Roland TC, Harlow CA (1982) Automated visual inspection: a survey ieee transactions on pattern analysis and machine intelligence
3. Rao Ananthavaram RK, Srinivasa Rao O, Prasad MHMK (2012) Automatic defect detection of patterned fabric by using rb method and independent component analysis. Int J Comput Appl 39(18):52–56

4. Sengottuvelan P, Wahi A, Shanmugam A (2008) Automatic fault analysis of textile fabric using imaging systems. Res J Appl Sci 3(1):26–31
5. Dastoor PH, Radhakrishnaiah P, Srinivasan K, Jayaraman S (1994) SDAS: a knowledge-based framework for analyzing defects in apparel manufacturing. J Text Inst 85(4):542–560
6. Zhang C, Feng S, Wang X, Wang Y (2020) Zju-leaper: a benchmark dataset for fabric defect detection and a comparative study. IEEE Trans Artif Intell 1(3):219–232
7. Bertram N, Karl-Heinz S, Schmalfuß H (1993) Automatic textile inspection. Automatische warenschau
8. Simonyan K, Zisserman A (2014) Very deep convolutional networks for large-scale image recognition. arXiv:1409.1556
9. Szegedy C, Vanhoucke V, Ioffe S, Shlens J, Wojna Z (2016) Rethinking the inception architecture for computer vision. In: Proceedings of the IEEE conference on computer vision and pattern recognition, pp 2818–2826
10. Chollet F (2017) Xception: deep learning with depthwise separable convolutions. In: Proceedings of the IEEE conference on computer vision and pattern recognition, pp 1251–1258
11. Sandler M, Howard A, Zhu M, Zhmoginov A, Chen L-C (2018) Mobilenetv2: inverted residuals and linear bottlenecks. In: Proceedings of the IEEE conference on computer vision and pattern recognition, pp 4510–4520
12. Zhao X, Zhang M, Zhang J (2021) Ensemble learning-based cnn for textile fabric defects classification. Int J Cloth Sci Technol
13. Wong T-T, Yeh P-Y (2019) Reliable accuracy estimates from k-fold cross validation. IEEE Trans Knowl Data Eng 32(8):1586–1594
14. Ron K et al (1995) A study of cross-validation and bootstrap for accuracy estimation and model selection. Ijcai, Montreal 14:1137–1145

Resource Allocation for Device-to-Device Networks Using WOA and PSO Algorithms

A. Vijaya Lakshmi, Banothu Pavankalyan, and Avanghapuram Surya Prakash Reddy

Abstract Future cellular networks focus on efficiency, scalability, and device-to-device (D2D) communication. D2D allows mobile phone users to engage directly with one another without the need for a base station. The growing use of smart devices and mobile apps has resulted in a massive surge in mobile traffic demand. In optimum device selection, power, and resource allocation approaches, resource management solutions for network-assisted D2D communication across the cellular spectrum are being investigated. Due to the fundamental disadvantages of standard methodologies, tackling such a problem is usually tricky and requires unique solutions. A particle swarm optimization (PSO)-based mode selection and resource allocation technique is developed. Simulation findings demonstrate that it outperforms alternative throughput and maximum power allocation approaches. Our major work is to investigate the applicability of WOA to address resource allocation problems in D2D networks as an alternative to existing approaches and to provide applications of WOA in D2D network resource allocation.

Keywords D2D communication · Resource allocation · Particle swarm optimization · Meta-heuristic optimization · Edge computing · WOA

1 Introduction

D2D is a fundamental method for 5G mobile communication networks that are still developing. The unauthorized user equipment (UEs) can access unused approved cellular UE bands via the D2D approach, which improves bandwidth consumption efficiency. Because data is transferred without passing via the base station, the D2D can maximize communications services in nearby locations. For example, when regional transmission and Wi-Fi are combined, more nearby users can access multimedia services, identify a nearby buddy, have real-time communications chats, and view product advertisements.

A. Vijaya Lakshmi (✉) · B. Pavankalyan · A. S. P. Reddy
Department of ECE, Vardhaman College of Engineering, Hyderabad, India
e-mail: vijayalakshmivardhan@gmail.com

S. Das et al. (eds.), *Advances in Data-driven Computing and Intelligent Systems*, Lecture Notes in Networks and Systems 653, https://doi.org/10.1007/978-981-99-0981-0_58

Based on the final goal/objective of the resource allocation process, the various resource allocation algorithms are divided into four broad classes: (1) throughput optimization, (2) fairness, (3) user QoS criteria fulfillment, and (4) combined power and subcarrier allocation. Furthermore, cellular user equipment that uses multi-sharing D2D communication can share its radio resources with many D2D devices. This study allocates resource blocks (RBs) to UEs based on evolutional methods for LTE systems. Increased transmission energy efficiency, energy savings, and improved coverage rate are all advantages of D2D approaches. However, most UEs can meet the minimum transmission rate requirements by installing a relay station for transmission. As a result, the system's performance beats direct transmission systems following the use of a relay station for communication in a severe interference environment. In the previous decade, the mobile communications sector saw a significant increase in mobile users and the volume of data traffic generated by those subscribers.

For a long time, voice traffic reigned supreme on mobile networks. On the other hand, data traffic exploded as intelligent gadgets and mobile apps became more widely available. Global monthly data traffic currently outnumbers voice traffic by over seventeen times, according to [1]. Because of better web content, multimedia file sharing, audio, and, most crucially, HD video streaming, wireless networks will witness a surge in traffic in the future. According to Cisco's newest Visual Networking Index research, a smartphone generates about 4.0 GB of monthly data traffic, five times more than the average monthly data traffic of 819 MB in 2014 [2]. The increase in global mobile traffic will be aided by the development of wireless devices that use mobile networks for purposes other than personal communication (e.g., machine-to-machine communication) [3]. The next generation of mobile devices, commonly referred to as the fifth generation, will surely confront a tremendous increase in traffic as its principal issue (5G). 5G networks are expected to produce 1000 times more cellular data volume per region, 10–100 times faster user data rates, and 10–100 times more connected devices than current cellular systems [4–6]. Designing wireless networks that may satisfy these lofty ambitions while balancing cost, energy, and radio spectrum restrictions is challenging.

MIMO and other multi-antenna transmission technologies are required to densify the network and increase the number of base stations (BSs) [4, 7–9]. It is common to boost network capacity in highly populated areas by deploying smaller cells in heterogeneous networks (i.e., business districts, universities, malls, etc.) [10–12]. In cellular networks, D2D communication helps both users and network operators. As a result, it is receiving more attention from the 3GPP LTE2 standard. Mobile users can enjoy high data speeds and little latency while conserving power and energy. Cell coverage and capacity per region may be enhanced for the second time without increasing infrastructure expenses. Users at the cell's edge often have low performance and can engage directly or through a relay. D2D communication and a link to the cell network are formed in this case. Spectrum efficiency can be enhanced by allowing direct D2D communications to share airwaves with standard cellular networks, allowing for more concurrent transmissions [13–15]. By permitting local control of short-distance data flows, D2D communication offers the extra benefit of alleviating network congestion and traffic management effort at core network nodes [16, 17]. As a result of incorporating D2D communication into future wireless networks, new challenges will be described in the next section. There are a variety of resource management challenges in D2D networks. Some of these concerns are addressed in this thesis, which also verifies D2D's promise.

1.1 Device-to-Device Communication Possibilities

D2D communication scenarios are expected to improve the efficacy of current proximity-based solutions while also introducing new ones.

1.1.1 Social and Commercial Services

The growing popularity of proximity-based services [18], for which standard uplink/downlink transmissions may be inefficient, encourages the adoption of the direct connection between adjacent devices. Local information sharing occurs in crowded circumstances (e.g., a baseball field or a performance) when many users seek the same popular material or when groups of individuals in the exact location (e.g., a shopping mall or a college) want to engage with one another. Mobile multiplayer gaming requires high performance, latency, and battery life. D2D communication is a possible new channel for local marketing and promotion from retailers and restaurants to nearby consumers, as well as local broadcasts of public transportation information, such as train timetables in subway stations or flight updates in airports [19–22]. While current technologies (such as Bluetooth) may provide all of these proximity services, they cannot give the same level of flexibility, security, or assured quality of service as mobile networks.

1.1.2 Public Safety

D2D communication is a compelling alternative for public safety organizations, frequently called upon to react during natural disasters or crowded events [23, 24]. The cellular network may fail in these scenarios due to infrastructure damage or excessive congestion and overload due to intense communication [25].

1.1.3 Traffic Safety and D2D Relay

Vehicle-to-vehicle (V2V) communication allows cars to communicate close to minimize collisions and enhance traffic flow. D2D communication appears to be a perfect fit for the purpose [26] because of its stringent requirements regarding communication dependability and latency. Machine-to-Machine (M2M) communication is another new technology for wireless cellular networks, allowing many devices to

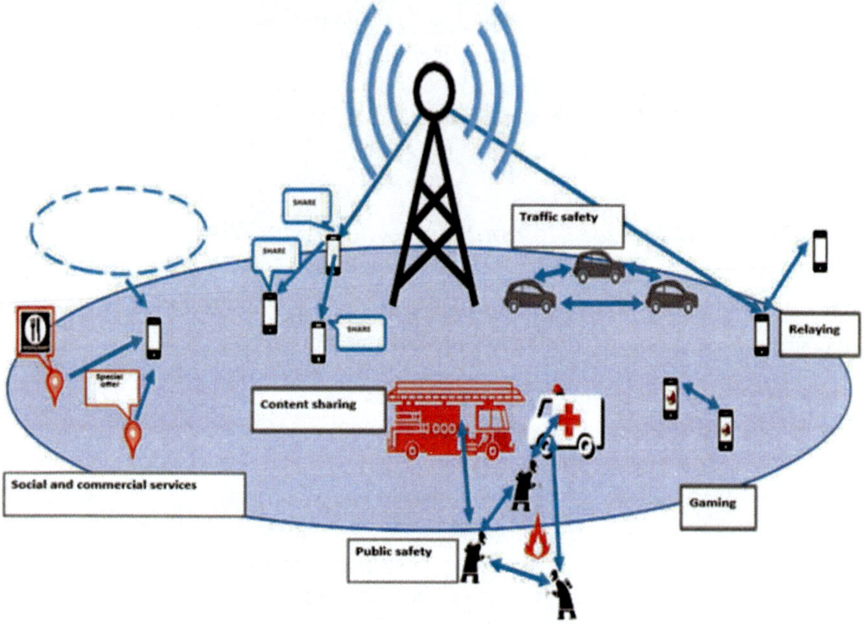

Fig. 1 D2D communication in cellular networks

connect to the cellular network for applications such as wide-scale environmental sensing, health monitoring, and so on [27]. Because such devices are often low-powered, a dependable D2D link between them and a smart device can serve as a cellular infrastructure relay. This example demonstrates how the D2D idea can be extended to mobile relaying, which can enable the communication of devices in places where cellular coverage is weak [28]. The primary conceptual use cases for D2D communications are depicted in Fig. 1. [29, 30] It has a long list of use-case descriptions.

2 Literature Survey

D2D communication, which provides ultra-low latency for user communication, is expected to play a large part in the future cellular networks. This new model can operate on both legal and unlicensed frequencies. It's a unique approach to mobile communication.

However, it has many financial and technological challenges that must be overcome before it can completely integrate into the cellular ecosystem. The basic elements of D2D communication are discussed in this article, including usage scenarios, architecture, technical features, and active research topics.

Paval et al. [31] According to this study, D2D communication, or direct communication between two or more devices without a base station, is a potential technology for improving cellular network spectrum and energy efficiency. Despite its widespread usage in non-cellular technologies such as Bluetooth and Wi-Fi, the D2D communication architecture has yet to be incorporated entirely into current cellular networks. As detailed in this paper, the 3GPP standardization community has accepted a new proposal to integrate D2D communication into LTE-A. D2D communication in cellular networks raises several significant difficulties, including interference management and deciding whether or not devices should interact directly. This survey presents a complete review of the state of the art in D2D communication, particularly within 3GPP LTE/LTE-A. For starters, they offer an in-depth classification of publications that examine D2D from various angles. Then, papers addressing all significant D2D challenges and topics are given, and the methodologies provided in the documents are compared using predetermined criteria. They identify areas not properly addressed so far and outline essential issues for future studies on the efficient integration of D2D in cellular networks based on the surveyed studies.

Arash Asadi et al. [32] According to this research, device-to-device (D2D) interactions were originally offered as a new paradigm for boosting network performance in cellular networks. As a result of the introduction of new applications such as content sharing and location-aware advertising, new use cases for D2D communications over cellular networks have evolved. In preliminary studies, D2D communications were shown to have enhanced spectrum efficiency and decreased communication time. However, this communication mode is still being investigated regarding interference management overhead and protocols. The feasibility of D2D communications in long-term evolution advanced is being investigated by academia, business, and standards authorities. They give a taxonomy based on the D2D communication spectrum and a comprehensive review of the literature under the proposed classification. Furthermore, they provide new insights into under and over-explored regions, identifying open research subjects in cellular network D2D communications.

Pimmy Gandotra et al. [33] According to this study, mobile communication networks have developed from the (1G) to the (3G) as a consequence of a continuing need to boost network capacity to satisfy the rising demand of users (5G). In the near future, there will be billions of linked gadgets. Because of the variety of such a vast number of connections, faster data rates, reduced latency, higher scalability, and higher throughput are predicted. Because spectrum resources are limited, mobile network operators (MNOs) must be flexible in their utilization to satisfy rising demand. D2D communication is a new facilitator with high data rates (NGNs). This paper provides a comprehensive overview of D2D communication, including its benefits, key open challenges such as peer invention and resource allocation, and some of its integrant innovations, such as millimeter-wave D2D (mm-Wave), ultra-dense networks (UDNs), cognitive D2D, D2D handover, and its numerous use cases. The design is proposed to satisfy all subscriber requirements efficiently. D2D communication resource allocation architecture is suggested. D2D communication is being investigated as technological innovation of next-generation networks to meet rising subscriber expectations and supply them with relevant services. The base station employs a sectored antenna to divide the coverage area into three 120° sectors for optimal resource sharing between D2D and cellular users. As a result, less interference between both two groups of users will occur, resulting in improved system performance.

A probabilistic integrated resource allocation approach and a quasi-convex optimization algorithm based on statistical channel characteristics are provided for device-to-device (D2D) communication mode selection and resource optimization in the 5G communication network [34]. A D2D resource allocation method for joint power control successfully handled the outliers in the clustering problem by determining its initial cluster number and cluster center based on an upgraded FCM algorithm [35].

3 Proposed System

D2D multicast communication is a novel (5G) network technology that can effectively handle the rising demand for user-to-user content sharing. It allows for direct communication between nearby devices with increasing spectral efficiency by reusing licensed cellular airwaves. The proposed system block diagram is shown in Fig. 2.

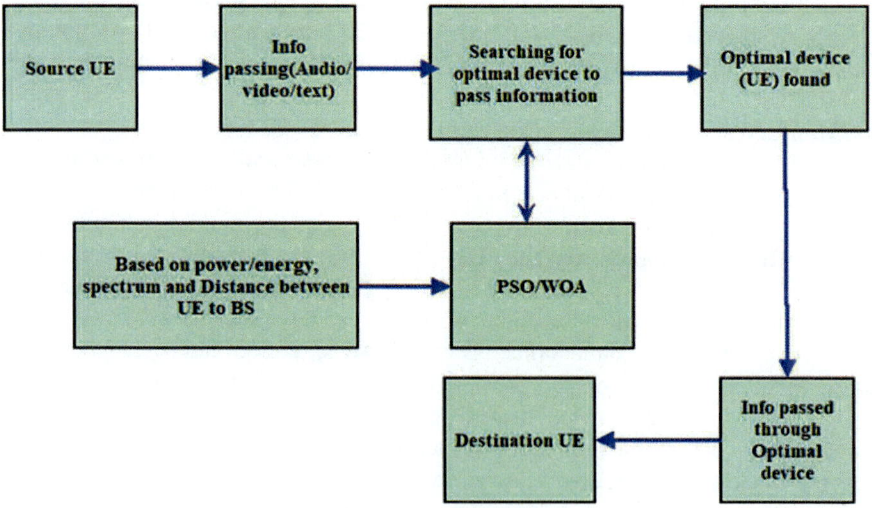

Fig. 2 Proposed system model

Source UE

The source UE is the device from where the data is transmitted, for example, mobile phone 1 (UE1). The information is passed through a wireless channel. The information may be in the form of audio, videos, texts, etc.

Searching for the optimal device to pass the information

This search uses optimization algorithms such as Particle Swarm Optimization (PSO) and Whale Optimization (WOA) Algorithms. The search is based on parameters like power, energy, and distance between UEs and BS.

3.1 PSO Algorithm

Optimization strategies have been particularly beneficial in tackling the difficulties mentioned above. An optimization algorithm compares many solutions iteratively until the best or most pleasing one is found. Since the development of computers, optimization has been a part of computer-aided design processes [36–39].

Particle

Juneja and Nagar [37] developed PSO in 1995 based on the social behavior of birds or schools of fish. In a large valley, a flock of birds is searching for food. Only one global minimum exists throughout this search area. Although none of the particles is aware of the location of the global minimum, they all have fitness values that must be optimized using the fitness function.

The flowchart of PSO is shown in Fig. 3. Particle defined by its coordinates.

$$p_i^t = [x_{0,i}^t, x_{1,i}^t, x_{2,i}^t, x_{3,i}^t \ldots x_{n,i}^t] \quad (1)$$

Each of these particles has a velocity that allows them to update their position over time to discover the global minimum.

The velocity in each direction defines the particle's velocity.

$$V_i^t = [v_{0,i}^t, v_{1,i}^t, v_{2,i}^t, v_{3,i}^t \ldots v_{n,i}^t] \quad (2)$$

Swarm

PSO resembles evolutionary computing technologies like Genetic Algorithms in many ways (GA). The system will start with a random population of solutions and then upgrades generations to find the best ones. Each particle's speed is stochastically increased in the search space as it approaches its (personal best) and the group's best solution (global best).

$$P_i^{t+1} = P_i^t + V_i^{t+1}$$

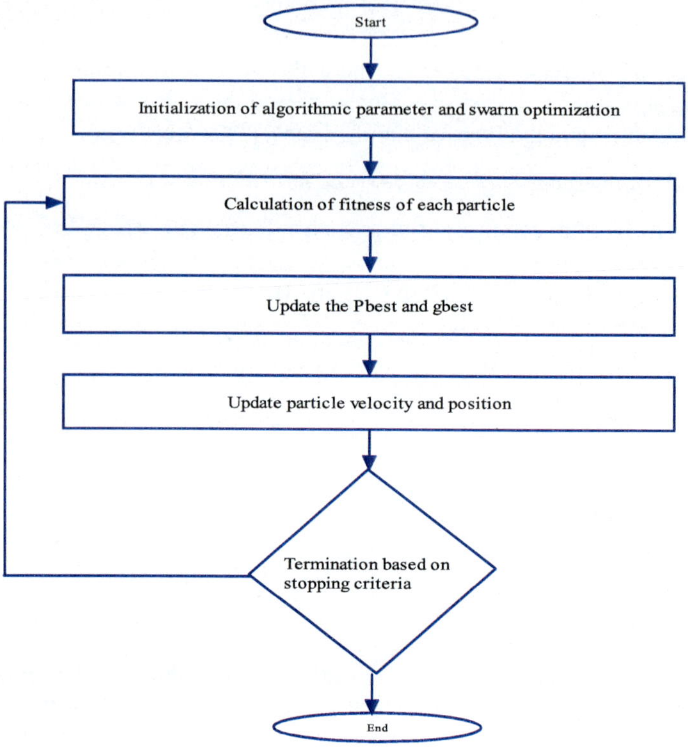

Fig. 3 Flowchart of the PSO algorithm

$$V_i^{t+1} = wV_i^t + c_1 r_1 (P_{best(i)}^t - P_i^t) + c_2 r_2 (P_{bestglobal}^t - P_i^t) \tag{3}$$

To put it another way, each iteration changes the velocity of each particle. The two best values determine this velocity found so far and is subject to inertia. As a result, each particle has its personal best answer and the global best solution stored in its memory.

Optimization

The inertia, cognition, and sociality coefficients were calculated. These factors have an impact on the levels of exploration and exploitation. Particles' ability to target the effective answer found so far is referred to as "exploitation." Investigation refers to a particle's ability to evaluate the research space. Our emotionally driven birds tend to wake up on the wrong side of the bed daily. Then, they'll want to follow their instincts and the group, more or less. At each cycle, the acceleration is weighted by random terms. The weights r_1 and r_2 are used to alter cognitive and social acceleration stochastically.

$$w \in R^+ \text{ Inertia, } r_2 \in [0, 2] \quad c_2 \in R^+ \text{ Social(global),}$$

$$r_1 \in [0, 2] \quad c_1 \in R^+ \quad \text{cognitive(personal)} \tag{4}$$

Different bird species can be found in the wild. These many species have the propensity to shift their course throughout time. The hyperparameter w defines the swarm's capacity to shift direction. The inertia of the particles is related to the coefficient w. To further appreciate the influence of this coefficient w, envision the three swarms of particles above (also known as inertia weight). We can observe that the stronger the convergence is, the smaller the coefficient w. It's important to note that $w > 1$ must be avoided because it might cause our particles to split. Each species is generally inclined to follow its instincts (personal) while also focusing on the communal experience (social). The c_1 hyperparameter can define the group's capacity to be impacted by the best personal solutions discovered across iterations. The c_2 hyperparameter can be used to describe the group's ability to be affected by the global best solution discovered across iterations. Figure 3 depicts the flowchart of PSO. On the GIF, we can see that the solution exploration is not optimum and that exploitation of the best global solution is critical (as seen in iteration 40). As a result, the coefficients c_1 and c_2 are complimentary. Exploration and exploitation are both boosted when the two are combined.

3.2 *Whale Optimization Algorithm*

The WOA algorithm is based on meta-heuristics. Mirjalili and Lewis, 2016 [40] presented this approach for solving numerical issues, as the name suggests.

WOA was motivated by the social behavior of humpback whales in the waters and bubble-net hunting. The world's most giant animals are humpback whales. They're also one of seven whale species. The bubble-net feeding method is a one-of-a-kind hunting mechanism. Because their brains include spindle cells, they are clever. The development of unique bubbles in the shape of a spiral or route allows for this type of foraging activity. As a result, consider the following hierarchy of whale leadership:

When the leader whale spots the prey, he dives 12 m below the surface, forms a spiral-shaped bubble around this one, and swims up to the surface, following the bubbles. A seasoned whale makes a call in time with the leader. The rest of the team forms a line behind the leader and lunges in the same direction.

Mathematical Model of the Whale Optimization Algorithm

The WOA algorithm mimics humpback whales' social behavior and hunting style in the waters. So, whale hunting contains three steps: 1. Exploration phase: searching for the prey 2. Encircling the prey 3. Exploitation phase: attacking the prey using a bubble-net method.

1. *Exploration Phase: Searching Model*

The search agent (humpback whale) randomly looks for the optimal solution (prey), based on each agent's position. If $|A| > 1$, the search agent should be compelled to stray from a whale that acts as a model, as shown in Eq. (7).

The mathematical model behind this phase is

$$\vec{D} = \left|\vec{C} * \vec{X}'(t) - \vec{X}(t)\right| \tag{5}$$

$$\vec{X}(t+1) = \vec{X}'(t) - \vec{A} * \vec{D} \tag{6}$$

$\vec{X}$ rand is a random position vector selected based on the population presently, and $\{\vec{A}, \vec{C}\}$ are coefficient vectors. Besides, here are the equations of $\{\vec{A}$ and $\vec{C}\}$ that can be used to find the best search agents:

$$\vec{A} = 2 * \vec{a} * \vec{r} - \vec{a} \tag{7}$$

$$\vec{C} = 2 * \vec{r} \tag{8}$$

where $\vec{r}$ is a random vector in the range [0, 1], and during the iterations, the value $\vec{r}$ drops steadily from 2 to 0.

2. *Encircling Model*

During hunting, humpback whales encircle the prey. As a result, the current best feasible solution is regarded as the best and most close to the perfect solution. In a nutshell, the following is the encircling behavior model that is used to keep track of the other whales' were about to the best search agent:

$$\vec{D} = \left|\vec{C} * \vec{X}'(t) - \vec{X}(t)\right| \tag{9}$$

$$\vec{X}(t+1) = \vec{X}'(t) - \vec{A} * \vec{D} \tag{10}$$

where t is the current iteration, $\vec{X}'$ is the position of the best solution, $\vec{X}$ refers to the position vector of a solution, and $\vec{A}, \vec{C}$ are coefficient vectors as given in Eqs. (7) and (8).

3. *Bubble-Net Attacking Model in the Exploitation Phase*

This phase concentrates on attacking the prey using a bubble net. Two techniques are used in this bubble-net system.

- *Shrinking encircling mechanism*

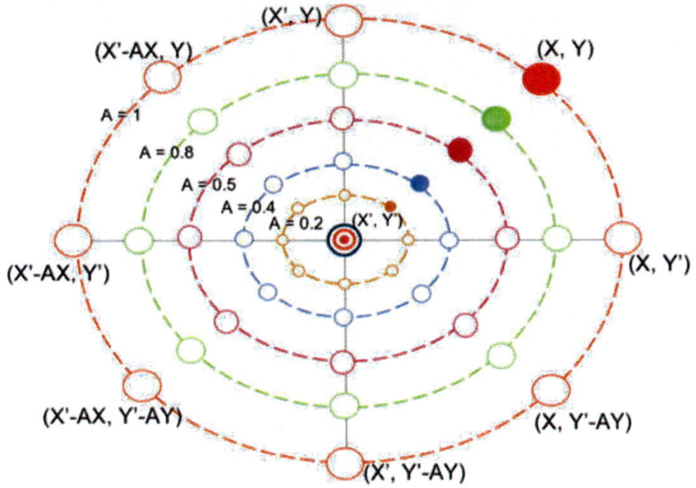

Fig. 4 Shrinking encircling mechanism

The value of A is chosen randomly from the range [− a, a], as indicated in Eq. (7), and the value of a decreases from 2 to 0 throughout repetitions. Specify some random values for A in the range [− 1, 1]. A new position for a search agent can be defined between the best agent's current position and the agent's initial position. The graph Fig. 4 shows the various places from (X, Y) to (X', Y').

This method determines the distance between the whale (X, Y) and the food (X', Y'). To establish the location between whale and prey, here's the spiral Eq. (11) underpinning humpback whales' helix-shaped movement:

- *Mechanism for updating location in a spiral*

$$\vec{X}(t+1) = \vec{D}'' * e^{bl} * \cos(2\pi l) + \vec{X}' \quad (11)$$

where $\vec{D}'' \left| \vec{X}'(t) - \vec{X}(t) \right|$ represents the distance between both the whale and the food (best solution obtained so far), 1 is a random number between [− 1, 1], and b is a constant defining the shape of the logarithmic spiral.

The mathematical model shows how a humpback whale circles its prey in a decreasing circle while following a spiral-shaped path as follows.

$$\vec{X}(t+1) = \begin{cases} X'(t) - \vec{A} * \vec{D} & \text{if} \quad p < 0.5 \\ \vec{D}'' * e^{bl} * \cos(2\pi l) + \vec{X}' & \text{if } p \geq 0.5 \end{cases} \quad (12)$$

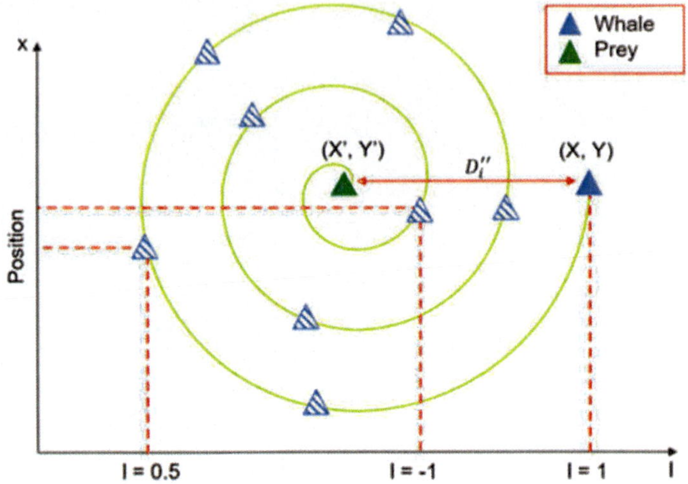

Fig. 5 Spiral updating position mechanism

where p is the likelihood of using one of these two strategies to update whale positions, suppose the possibility of selecting one of the two techniques is 50%. Then, p is an integer chosen randomly between [0, 1]. Figure 5 shows the spiral updating position.

Flowchart of a Whale Optimization Algorithm

The whale population n and values must be initialized first, as shown in Fig. 6.

The search agent's fitness solution value is then evaluated. The number of iterations is then set to one. After that, we seek prey using Eq. (5). After discovering the prey, we use Eq. (9) to surround it. Then, using the bubble-net technique in Eq. (12), we change the position of search agents to attack the prey.

Then, we update the values of $\{a, A, C\}$ with the new position of the search agent. After that, we verify the equality and inequality constraints of A and p for each search agent's new location before increasing the iteration number.

Finally, if we hit *max_iter*, we'll stop and record the fitness value as the best solution; otherwise, we'll restart the process until we find a solution.

3.3 *Optimal Device (UE) Found*

The optimal device was found based on the fitness function, which is fit to pass the information to the next UE. In this way, based on the signal strength, the data is passed through different machines which are strong enough and, at last, reach the destination (UE). Note: There may be at least one UE that may be connected to the base station.

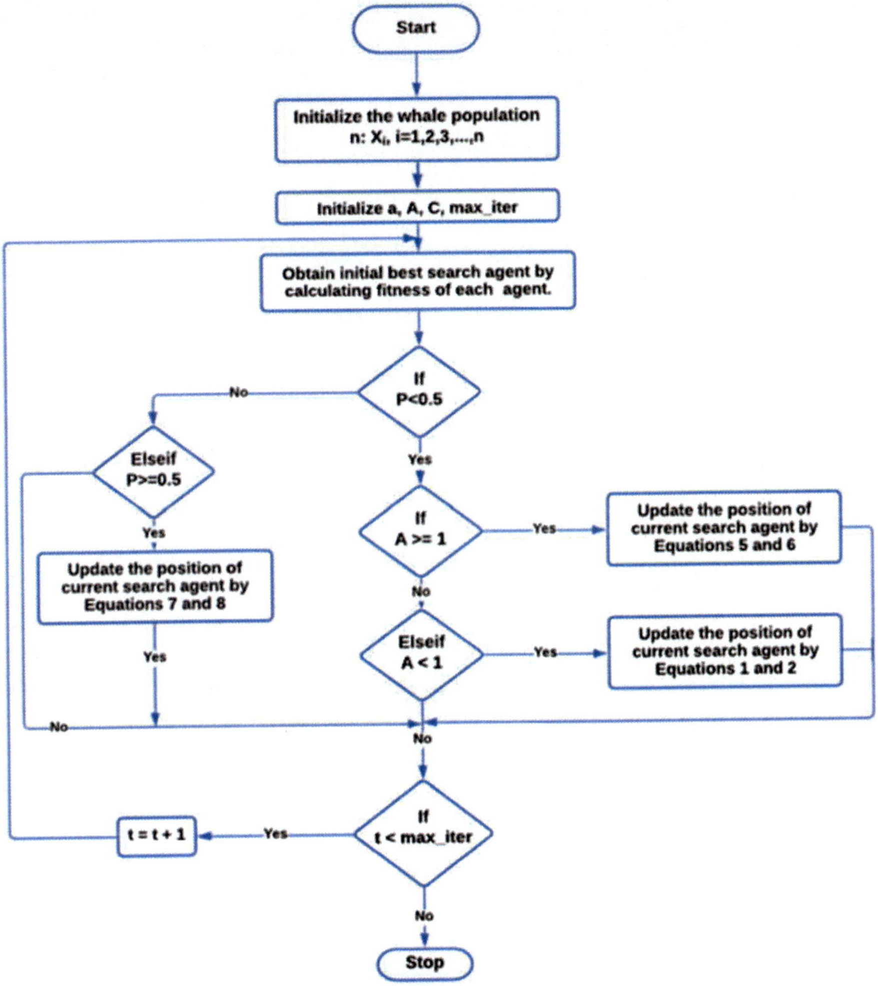

Fig.6 Flowchart of a whale optimization algorithm

3.4 D2D Using PSO Algorithm

The PSO approach simulates the bird's swarm foraging issue, which includes particles searching for food space. Each particle, in addition to its natural movement, imitates a bird, which seeks the finest possible moving experience. s^{pbest} It stands for "personal best experience." Following that, the global experience is referred to as the best globally, symbolized and based on the group's best moving experience. The ultimate convergence finds the best response based on the types mentioned above of data after repeated development. The target function's value defines the efficiency of the iteration for the particle in this research, which has M particles. The best reaction comes from the particle with the highest target function value, whereas all relay stations serve the UEs of a particular RB, the PSO algorithm optimizes resource block allocation as the particle *si*. Particles are equally scattered in the solution space. There are 13 RBs accessible, as well as three relay stations. In the $(g+1)$-generation, the movement velocity of the initial particle is given as

$$V_i^{g+1} = c_1 \times \text{rand}() \times (s^{\text{pbest}} - s_i^g) + c_2 \times \text{rand}() \times (s^{\text{pbest}} - s_i^g) \quad (13)$$

where s_i^g is the ith particle's location vector in the g-generation; $\mathbf{s}_i^g = [\mathbf{s}_{i_1}^g, \mathbf{s}_{i_2}^g, \mathbf{s}_{i_3}^g]$; and $\mathbf{s}_{i_l}^g$ is the location of the lth relay station's ith particle. All particles' local optimum locations are represented by $s^{\text{pbest}} = [s^{\text{pbest1}}, s^{\text{pbest2}}, s^{\text{pbest3}}]$ where s^{pbestl} is the lth relay station's local optimum location. s^{gbest} is the global best location for all particles in the next generation. Both c_1 and c_2, also known as the individual and societal factors, are acceleration coefficients. $c_1 = 2$ and $c_2 = 2$ are commonly used to set the factors. Rand () is a random function with a uniform distribution in the range [0, 1]. Furthermore, at $(g+1)$-generation, position updating may be stated as

$$s_i^{g+1} = s_i^g + v_i^{g+1} \quad (14)$$

PSO assigns RBs to all K UEs in the relay station's coverage region and composites them into a particle I s with a $1 \times K$ array. There are 13 RBs distributed across the UEs at three relay stations, and the M particles are evenly dispersed throughout the solution space. The object function in PSO is described as follows:

$$fc = \sum_{ul=1}^{k} y_{ul} \quad \text{where } y_{ul} = \begin{cases} 1, \sum_{n=1}^{N} x_{ul}^{(n)} R_{ul}^{(n)} \geq R_{\text{th}} \\ 0, \text{ otherwise} \end{cases} \quad (15)$$

A constraint-handling approach, namely the penalty function, is used to create a fitness function to solve the restricted optimization issue. The method's fundamental idea is to use a penalty function in the objective function to change a constrained issue into an unconstrained one. The penalty function and the fitness function are built as follows:

$$\text{penalty} = \sum_k \left[\min\left(o, \sum_n \sum_q x_{k,n}^{(q)} r_{k,n}^{(q)} \right] \right.,$$

$$\text{Fitness} = \sum_k \sum_n \sum_q x_{k,n}^{(q)} r_{k,n}^{(q)} - P \sum_k \left[\min\left(o, \sum_n \sum_q x_{k,n}^{(q)} r_{k,n}^{(q)} - R_{\min} \right]^2 \right. \quad (16)$$

Here, $P \in \mathbf{R}^{+}$ is a penalty factor. For a feasible solution, Penalty $= 0$ for a viable solution, and the maximum of Fitness is the best answer to the PSO. The penalty function is crucial in encouraging particles to fly as quickly as possible out of the non-feasible zone or as near the feasible region. Furthermore, the fitness function provides consistency, allowing us to fully exploit the typical PSO's advantage in looking for optimal solutions.

3.5 *D2D Using WOA Algorithm*

This paper aims to thoroughly examine how WOA may be used to address three key D2D network optimization issues: Maximizing the maximum–minimum secrecy rate, optimizing the EE-SE trade-off, and offloading MEC computation. The first example shows how WOA may be used to address an unconstrained large-scale optimization issue, while the second example demonstrates how to solve a limited continuous optimization problem, and the third demonstrates how to solve an MINLP problem. While the WOA can handle the first problem directly, the second example necessitates a constraint-handling strategy, and BWOA is in charge of the third issue, which employs both decomposition and constraint-handling approaches [38, 39].

A. *Power Distribution to Achieve the Highest Rate of Secrecy*

Consider an M-user, interference-limited D2D network. Each user may be thought of as a single-antenna transceiver, forming a communication link. The problem of the max–min secrecy rate (MMSR) can be described as follows [41]:

$$\begin{aligned} \max_{p} \quad & \phi(p) = \min_{i=1,\dots M}[R_i(p) - \Gamma_i(p)] \\ s.t \quad & 0 \leq p_i \leq p_i^{\max}, \quad i = 1, \dots M \end{aligned} \tag{17}$$

where $p_i^{\max}$ signifies the user's maximum transmit power *i*. p_i is the user's *i* transmit power, $R_i(\boldsymbol{p})$ is the user's *i* data rate $\Gamma_i(\boldsymbol{p})$ is the user *i*'s wiretapped rate at the EV.

Sheng et al. developed a path-following technique for addressing the MMSR issue instead of employing the d.c. (difference of two concave functions) formulation of the secrecy rate [41]. Because the MMSR issue is a continuous optimization problem with no constraints, the original WOA technique may be used to solve it directly.

B. *Power Allocation for a Tradeoff Between Energy and Spectrum Efficiency*

With the same notations used in Section 4.3-A, consider an interference-limited D2D network. The total power consumption (P_{tot}) is broken down into two parts: p_i^C the power consumption of the circuit and p_i the power consumption of the transmitting device. The difficulty of maximizing global EE (GEE) while adhering to minimum rate constraints and transmit power limits may be characterized as a tradeoff between EE and spectral efficiency (SE). As a result, for the EE-SE trade-off [42], the following optimization issue is taken into account:

$$\begin{aligned} \max_{p} \quad & q = \sum_{i=1}^{M} R_i(p)/P_{\text{tot}}(p) \\ s.t \quad & C1 : R_i(p) \geq R_i^{\text{req}} \quad , \quad i = 1, \ldots M \\ & C2 : 0 \leq p_i \leq p_i^{\max} \quad i = 1, \ldots M \end{aligned} \tag{18}$$

Bits/Joule/Hz is the unit of q, while R_i^{req} representing the user's minimum needed rate. Because the aforementioned optimization issue is non-convex and NP-Hard, finding a polynomial-time solution is challenging [42, 43]. Bisection and sequential convex approximation (SCA) are used to solve the issue in (14). The issue in (14) may be rewritten as follows in the case of a certain value of q:

$$\max_{p} \sum_{i=1}^{M} R_i(p) - q P_{\text{tot}}(p) \; s.t. C1 \text{ and } c2 \tag{19}$$

At each iteration, the issue in (15) can be addressed by resolving a standard multi-objective optimization problem with the new objective function's d.c. representation and reorganizing the constraint $C1$. The criterion is used to halt the algorithm in the same way that it is used to terminate the path-following technique above. In this scenario, users may be thought of as search agents (humpback whales), and the transmit power p indicates the location of search agents' X. The transmission power $p(t)$ (equivalent to $X(t)$) could be updated at iteration t via one of three major mechanisms: reducing the surrounds, spiral updating the location, or hunting for food. The optimization process used to select the optimal search agent is as follows:

$$\text{Fitness}(p) = -\frac{\sum_{i=1}^{M} R_i(p)}{P_{\text{tot}}(p)} + \mu \sum_{i=1}^{M} F_i(f_i(p)) f_i^2(p) \tag{20}$$

where $f_i(\boldsymbol{p}) = R_i^{\text{req}} - R_i(\boldsymbol{p})$ and $\mu = 10^{14}$. A maximizing problem becomes a minimizing problem when the objective function is given a negative sign.

C. *Mobile Edge Computation Offloading*

Consider a MEC situation including M users and a single MEC server, which is generally referred to as the eNB. To maximize the benefits of compute offloading in terms of completion time and energy consumption, the overall utility may be described as follows:

$$v_i(a_i, p_i, f_i) = a_i(\beta_i^t \frac{T_i^l - T_i^r}{T_i^l} + \beta e \frac{E_i^l - E_i^r}{E_i^l}) \tag{21}$$

where β_i^t and β_i^e are the user's finishing time and energy usage choices, respectively. The challenge of maximizing total utility by optimizing the computation offloading choice and resource allocation may be stated as follows [44]:

$$\max_{a,p,f} \sum_{i=1}^{M} v_i(a_i, p_i, f_i)$$
$$C1 : a_i = \{0, 1, \ \forall i = 1 \ldots M, \ C2 : 0 \leq p_i \leq p_i^{\max},$$
$$\forall i = 1 \ldots M, \ C3 : f_i > 0, \forall i \in s,$$
$$C4 : \sum_{i \in s} f_i \leq f_0, \quad C5 : \sum_{i=1}^{M} a_i \leq N \tag{22}$$

$S = \{i = 1, \ldots, M | a_i = 1\}$ is the set of offloading users set, the maximum computing resource of the MEC server is f_0, and the number of subcarriers is N, implying that N users can have their computing work offloaded. The MEC server only assigns computational resources to unload subscribers. The total number of resources offered should be fewer than the highest computing capability $f0$, according to the limitations $C3$ and $C4$ in the problem (18). The authors of [44] recommended breaking the original issue into two subproblems since problem (18) is NP-hard. The first constraint (JCCR) aims to optimize communication and computation resources for a specific offloading option $\hat{a}$ in the following way:

$$\max_{p,f} \sum_{i \in s} v_i(a_i, p_i, f_i) \ s.t. \ C2, C3 \text{ and } C4 \tag{23}$$

The second constraint (COD) is to optimize unbiased decisions for a specified $\left(\hat{p}, \hat{f}\right)$ and is expressed as

$$\max_{a} \sum_{i \in s} v_i(a_i, \hat{p}_i, \hat{f}_i) \ s.t. \ C1 \text{ and } C5 \tag{24}$$

Because the transmission power p and compute resources f are decoupled for a given $\hat{a}$, the JCCR constraint may be further disintegrating into two distinct p and f problems. The convex optimization and bisection methods may then be used to solve the solutions of p and f, respectively. The COD subproblem's goal is demonstrated to be a submodular optimization issue; hence, the offloading choice a is addressed using a heuristic technique based on submodular optimization. We suggest utilizing the decomposition approach to get two subproblems of a and compare the BWOA-based solution's performance to that of the existing one (p, f). The JCCR constraint is addressed using the same method as in [42], but the COD subproblem is solved using the BWOA algorithm to yield the offloading decision answer. This part focuses on addressing the binary optimization problem (20), as contrast to Sections 4.3-A and 4.3-B, which employ the WOA method to handle ongoing optimization issues.

The truth is that the original problem can be solved using a mix of WOA and BWOA (18). However, because such methods are not the topic of this study, they are omitted. For more information on the algorithm of channel access and power management in D2D communications, see [45] and the references.

A basic way of dealing with the inequality $C5$ is to check to see if constraint $C5$ is fulfilled. The solution is viable if the constraint is met and the fitness value can be evaluated. When a constraint is mistreated, the answer must be abandoned, and a new one must be developed. This strategy is plainly inefficient and slow. The fitness value might be stated as follows to increase performance when using the penalty strategy:

$$\text{Fitness}(a) = -\sum_{i \in s} v_i(a_i, \hat{p}_i, \hat{f}_i) + \mu F(f(a)) f^2(a) \tag{25}$$

4 Simulation Results

Simulation results in this part to demonstrate the benefits of the suggested PSO and WOA techniques to maximize the ideal network solution. Simulation results for the PSO and the WOA. As a result, the simulation results are also given as two separate examples. Utilized the PSO to pick the optimal device in the first scenario and used the PSO to distribute transmit power to D2D users among all of the simulation's schemes. Evaluated the proposed WOA scheme against the HODA in the second scenario.

4.1 Simulation Results of D2D Using PSO

In the simulation system, all users are randomly placed in a typical hexagonal cell with a radius of 250 m, with the BS in the middle. To evaluate the suggested algorithm's performance benefits, a comparison is made between transmitter power (dB) and distance to the base station (BS), with various epsilon values. A D2D user can reuse a single CU resource at a time, and different parameters from the literature are changed to fit the system's needs. Figure 7 shows the performance comparison between transmitter power (dB) and distance to the base station (BS) with varied epsilon values. When compared to references, it can be shown that the suggested technique effectively improves the transmitter power of D2D communication. When the number of D2D users is small, the no. of shared sub-channels is large, allowing system resources to be fully used. C is inversely proportional to R.

$$C \; \alpha \; R^{-\varepsilon\alpha} \tag{26}$$

where $C =$ transmitter carrier power, $R =$ distance measured from the transmitter to the receiver, $(\alpha) =$ weighting factor and $(\epsilon) =$ path loss parameter.

As we can see at the path loss exponent ($\epsilon = 0$), there is no loss, so as the distance to the BS increases, there will be no change in the transmitted power. There is some loss at path loss exponent ($\epsilon = 0.5$), so as the distance to the base station increases, the required transmit power also increases. At path loss exponent ($\epsilon = 1$), there is some more loss, so as the distance to the base station increases, the transmit power also increases. In Fig. 7, it can be seen how the path loss component (ϵ) affects the power transmission of the proposed algorithm. The area under coverage (distance from CC UE to BS) decreases as the path loss (ϵ) value increases.

The fundamental reason for this is that the route loss between D2D users rises as the distance between them grows. The distance between D2D users dramatically influences how well the system works. At $\alpha = 2.5$, the distance coverage is less, and at $\alpha = 4$, the distance coverage is more. Here, α is the weighting factor. By observing Fig. 8, we can say that with the increase in α value, we can transfer the signal to the base station.

Figure 9 compares the performance of the maximum allowable transmitter power (dBm) to the distance from the transmitter to the base station (BS), with varying values (path loss), and the CC distance, which is simply the distance between any one device (UE) linked to the base station (BS). It can be observed how well the distances between D2D users impact the suggested method's power allocation. The power allocation increases as the distance between D2D users grows with different values of path losses and cc distances. The main reason behind this is that when the distance between D2D users grows, the route loss between them increases. The distance between D2D users has a big influence on how well the system works. Figure 9 depicts the power allocation performance of the proposed D2D with PSO

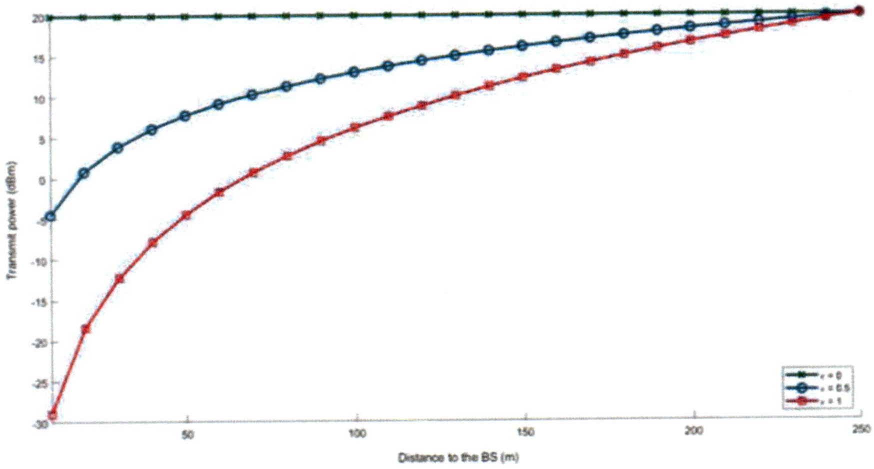

Fig. 7 Transmitted power versus distance to the base station

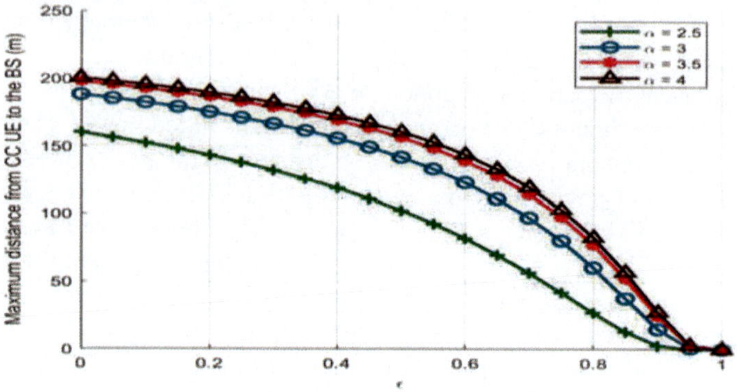

Fig. 8 Maximum distance from CC UE to the BS versus ϵ

for various cellular user locations. Moreover, the cellular user's (CU) position is changed such that the radius between both the cellular BS and the CU can range from 0 to 250 m, the D2D source and D2D destination are constant in their positions, and the D2D relays are placed at random inside the D2D source's communication range.

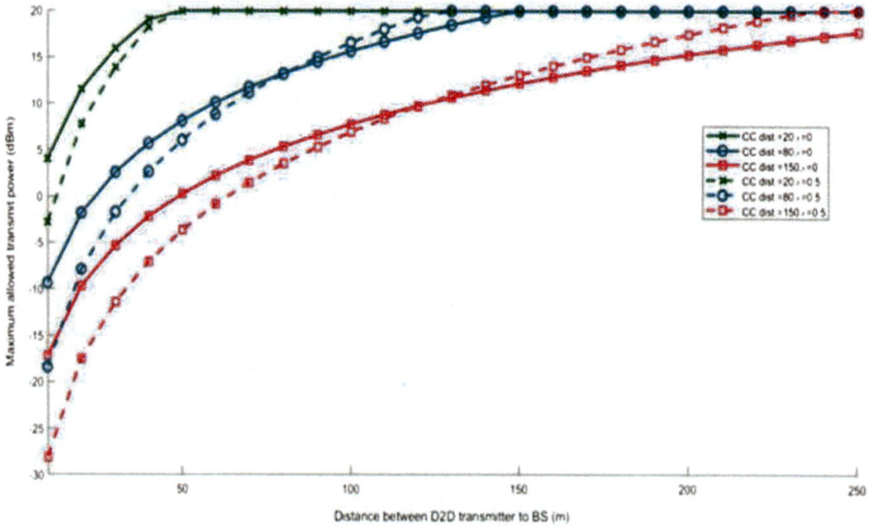

Fig. 9 Maximum allowed transmit power versus distance

4.2 Simulation Results of D2D Using WOA

The suggested WOA-based method and the path-following procedure in [41] are shown against the iteration index in Fig. 10. The path-following procedure takes 14 iterations, whereas the WOA-based approach takes 28 iterations, as illustrated in Fig. 10. The path-following approach does, however, necessitate solving a convex problem at each iteration. For the network scenario examined in this paper, the convex approximation problem develops M optimization variables and M linear constraints with $T = 14$. In contrast to [41], users represent the standard search agent at every iteration.

As a result, the WOA-based approach is computationally inexpensive, which is critical in real-world systems with many connections. Furthermore, the path-following method and the WOA-based methodology provide virtually equal results: The path-following technique uses 1.348 bps/Hz, while the WOA-based process uses 1.36 bps/Hz. Compared to the present technique [39], the WOA-based approach has a lower computational complexity.

The simulation settings are the same as in [42]. According to Fig. 11, the proposed technique might achieve convergence in just a few tens of repetitions, roughly three times the number of essential loops found in the GAP [42]. The proposed WOA-based technique is a one-loop system, while the GAP is a two-loop system, with the outer loop upgrading energy efficiency q and the inner loop optimizing transmit power for a given q. In addition, each inner loop must solve a set of approximation convex problems, the same as the path-following method presented in [41]. The WOA's computational complexity is $O(TN(m + M))$ $N = 30$, $m =$ M, and $T = 48$ (for

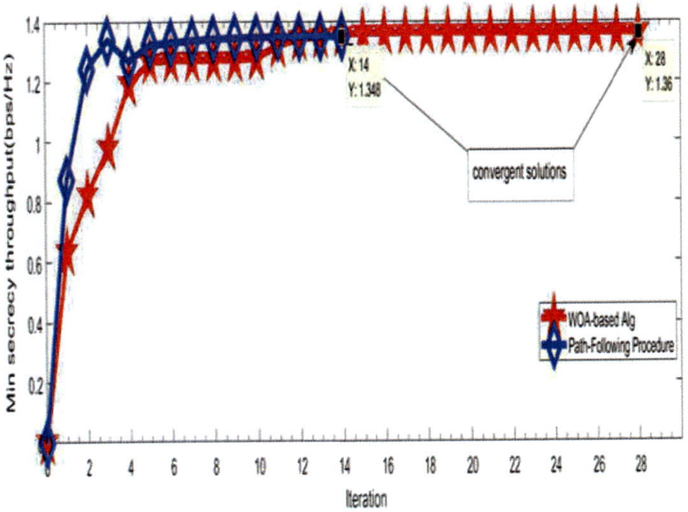

Fig. 10 WOA and the path-following procedure's between D2D transmitter to BS convergence progression (with $M = 4$ users)

$R_i^{req} = 1.0$ bit/s/Hz). On the other hand, the GAP has a high level of computational complexity. $T1 = 32$ and $T2 = 28$ iterations are necessary for the outer and interior loops, respectively. The WOA has a reduced computing complexity as compared to the GAP, which is important for practical applications.

Compare the proposed WOA-based approach to the HODA in [42] in terms of system utility, system-wide (i.e., total) processing overhead, and percentage of offloading. At the convergent solution, all of these metrics are evaluated:

1. The goal of the optimization issue is to maximize system utility,
2. The term "system-wide computation overhead" refers to the amount of time it takes for a system to perform the task, and it is represented as

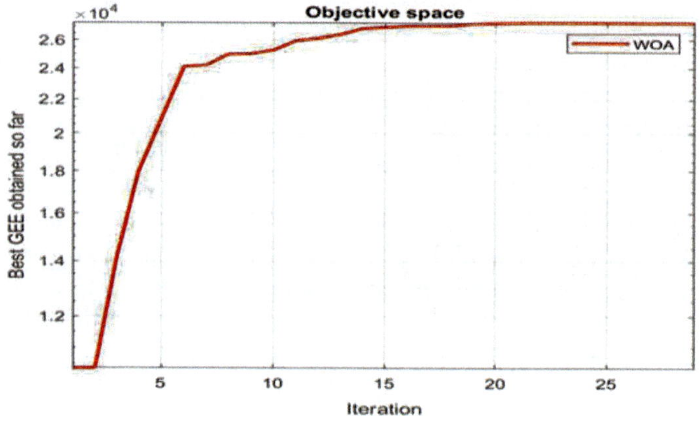

(a). Best GEE obtained so far vs iteration

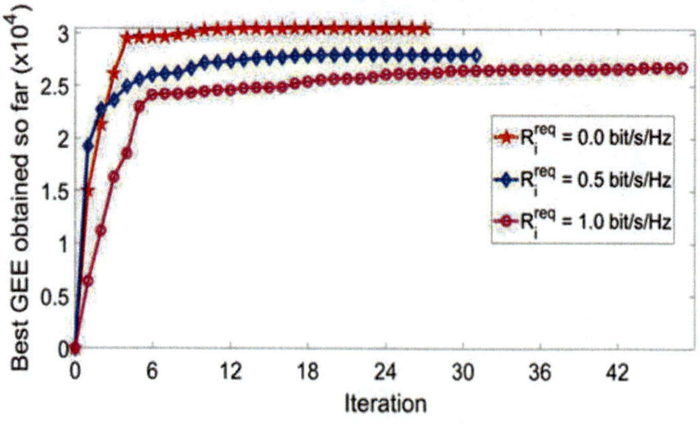

(b). Convergence evolution

Fig. 11 Evaluation of the WOA-based algorithm's performance **a** and **b**

$$Z_{\text{tot}} = \sum_{i=1}^{M} (1 - a_i)(\beta_i^t T_i^l + \beta_i^e E_i^l) + a_i(\beta_i^t (T_i^t + T_i^e) + \beta_i^e E_i^r) \tag{27}$$

1. The ratio of offloading users to total users is known as the percentage of offloading. i.e., $P_{\text{off}} = \boldsymbol{a}^T 1/N$. Our studies used the identical set of simulation settings as in HODA [42].

In addition, each plot is created by averaging the outcomes of 200 different realizations, with users allocated to each one at random. Both algorithms, as shown in Fig. 10, have a broader diversity of users with varying channel circumstances, local computing capabilities, and calculation duties as the number of users rises, allowing them to produce continual improvements in system usefulness. Because each user competes for compute offloading with other users, and due to a lack of subcarriers, certain users are still unable to offload, even though they would benefit from remote execution. As a result, as the number of users grows, so does the fraction of offloading and processing overhead throughout the whole system. The value of both increases as the no. of users increases. The WOA-based technique has lower complexity than the HODA challenges and best solutions.

Figure 12 shows closely how the WOA-based solution approaches the heuristic centralized HODA [44] answer. With $M = 8$, the WOA-based technique has a system utility of 1.62, while the HODA has a system utility of 1.6211. Furthermore, by performing an exhaustive search, [44] demonstrated that the HODA could retain its performance within 5 percent of the optimum response. Consequently, the WOA competes with the least amount of computing complexity possible. The three examples above prove that the WOA might use D2D networks to handle various resource allocation challenges. The WOA-based method obtains almost the same results as previous algorithms, with assured convergence in all tests. Exploration and exploitation are developed throughout the optimization process, so it's reasonable to conceive of the WOA as a global optimizer.

5 Conclusion

This work presents hybrid power management and resource allocation strategy to maximize energy efficiency and resource utilization by expanding the particle swarm algorithm. The strategy described in this work improves power efficiency and resource utilization compared to a scenario in which only one CU resource can be reused by a D2D user, resulting in a swarm intelligence-based optimization solution. We simulate optimal device power allocation. D2D communication enables high-data-rate services across short distances and improves 5G network performance. Two users in a D2D pair can choose between two means of communication: direct communication through CU sub-carriers and indirect communication via eNB. The selection of which spectrum (licensed or unlicensed frequency) D2D couples to use for direct communication may be regarded as mode selection in LTE-unlicensed. MIP

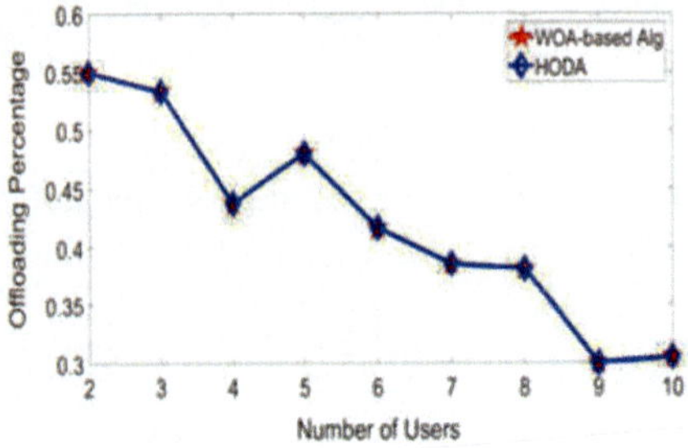

(a) Offloading users as a percentage

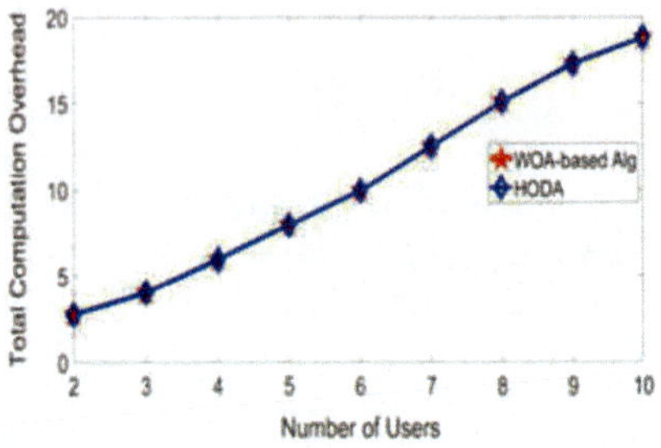

(b) Overhead in computation at the system level

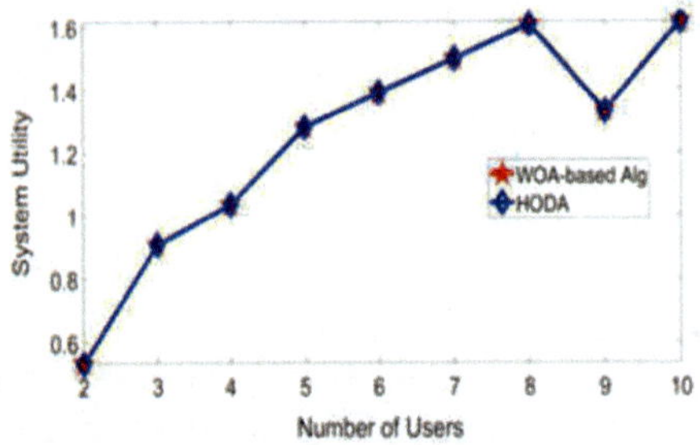

(c) System utility

Fig. 12 Comparison of performance of WOA and HODA when the number of users ranges between 2 and 10. **a**, **b**, **c**

issues commonly express D2D model selection and subcarrier assignment considerations. Matching theory, coalition games, and graph theory are potential techniques to find a high-quality binary answer faster and cheaper than an exhaustive search. WOA's approach to these problems is unique. Despite no convergence proof, numerical results can verify the binary WOA approach. We demonstrated that the WOA algorithm could match the centralized heuristic system.

References

1. Rima QureshiI ON, The PO, Networked THE (2013) Erricson mobility report. (February):4–7
2. Cisco (2015) White paper: Cisco visual networking index: global mobile data traffic forecast update, 2014–2019. Tech. rep., Cisco
3. Gonçalves V, Dobbelaere P (2010) Business scenarios for machine-to-machine mobile applications. In: International conference on mobile business and global mobility roundtable, pp 394–401
4. Andrews J, Buzzi S, Choi W, Hanly S, Lozano A, Soong A, Zhang J (2014) What will 5G be? IEEE J Sel Areas Commun 32:1065–1082

5. Osseiran A, Boccardi F, Braun V, Kusume K, Marsch P, Maternia M, Queseth O, Schellmann M, Schotten H, Taoka H, Tullberg H, Uusitalo M, Timus B, Fallgren M (2014) Scenarios for 5G mobile and wireless communications: the vision of the METIS project. IEEE Commun Mag 52:26–35
6. Ericsson (2015) 5G radio access. Tech. rep.
7. FP7 European Project 317669 METIS (Mobile and wireless communications enablers for the twenty-twenty information society), 2012
8. White paper: Strategies for mobile network capacity expansion. Tech. rep., November 2010
9. Bhushan N, Li J, Malladi D, Gilmore R, Brenner D, Damnjanovic A, Sukhavasi R, Patel C, Geirhofer S (2014) Network densification: the dominant theme for wireless evolution into 5G. IEEE Commun Mag 52:82–89
10. Chandrasekhar V, Andrews J, Gatherer A (2008) Femtocell networks: a survey. IEEE Commun Mag 46:59–67
11. Bendlin R, Ekpenyong T, Greenstreet D (2012) Paving the path for wireless capacity expansion. Tech. rep., Texas Instruments
12. Sesia S, Toufik I, Baker M (2011) LTE—the UMTS long term evolution: from theory to practice. Wiley
13. Fodor G, Dahlman E, Mildh G, Parkvall S, Reider N, Miklos G, Turanyi Z (2012) Design aspects of network assisted device-to-device communications. IEEE Commun Mag 50:170–177
14. Doppler K, Rinne M, Wijting C, Ribeiro C, Hugl K (2009) Device-to-device communication as an underlay to LTE-advanced networks. IEEE Commun Mag 47:42–49
15. Li Z, Moisio M, Uusitalo M, Lunden P, Wijting C, Sanchez Moya F, Yaver A, Venkatasubramanian V (2014) Overview on initial METIS D2D concept. In: 1st International conference on 5G for ubiquitous connectivity, pp 203–208
16. ICT-317669 METIS Project DD (2014) Mobile and wireless communications enablers for the twenty-twenty information society
17. Andreev S, Pyattaev A, Johnsson K, Galinina O, Koucheryavy Y (2014) Cellular traffic offloading onto network-assisted device-to-device connections. IEEE Commun Mag 52:20–31
18. Corson M, Laroia R, Li J, Park V, Richardson T, Tsirtsis G (2010) Toward proximity-aware internetworking. IEEE Wirel Commun 17:26–33
19. Mumtaz S, Rodriguez J (2014) Smart device to smart device communication. SpringerLink: Bücher, Springer International Publishing
20. Feng J (2013) Device-to-device communications in LTE-advanced network. Télécom Bretagne, Université de Bretagne-Sud, Theses
21. Lin X, Andrews J, Ghosh A, Ratasuk R (2014) An overview of 3GPP device-to-device proximity services. IEEE Commun Mag 52:40–48
22. 3GPP (2015) Delivering public safety communications with LTE. Tech. rep
23. Doumi T, Dolan M, Tatesh S, Casati A, Tsirtsis G, Anchan K, Flore D (2013) LTE for public safety networks. IEEE Commun Mag 51:106–112
24. Goratti L, Gomez K, Fedrizzi R, Rasheed T (2013) A novel device-to-device communication protocol for public safety applications. In: IEEE Globecom Workshops, pp 629–634
25. Townsend A (2005) Telecommunications infrastructure in disasters: preparing cities for crisis communications. University of New York
26. Sun W, Strom E, Brannstrom F, Sui Y, Sou KC (2014) D2D-based V2V com- munications with latency and reliability constraints. In: Globecom Workshops, pp 1414–1419
27. Pratas N, Popovski P (2014) Underlay of low-rate machine-type D2D links on downlink cellular links. In: IEEE international conference on communications workshops, pp 423–428
28. Abrardo A, Fodor G, Tola B (2015) Network coding schemes for device-to-device communications based relaying for cellular coverage extension. In: IEEE international workshop on signal processing advances in wireless communications, pp 670–674
29. 3GPP (2012) 3rd generation partnership project; technical specification group SA; feasibility study for proximity services (ProSe) (Release 12), tech. rep., TR 22.803 V1.0.0
30. Mach P, Becvar Z, Vanek T (2015) In-band device-to-device communication in OFDMA cellular networks: a survey and challenges. IEEE Commun Surv Tutorials 17(4):1885–1922

31. Mach P, Becvar Z, Vanek T (2015) In-band device-to-device communication in OFDMA cellular networks: a survey and challenges. IEEE Commun Surv Tutorials 17(4):1885–1922. https://doi.org/10.1109/COMST.2015.2447036
32. Asadi A, Wang Q, Mancuso V (2014) A survey on device-to-device communication in cellular networks. IEEE Commun Surv Tutorials 16(4):1801–1819. https://doi.org/10.1109/COMST.2014.2319555
33. Gandotra P, Jha RK (2016) Device-to-Device communication in cellular networks: a survey. J Netw Comput Appl 71:99–117. ISSN 1084-8045. https://doi.org/10.1016/j.jnca.2016.06.004
34. Li J, Lei G, Manogaran G, Mastorakis G, Mavromoustakis CX (2019) D2D communication mode selection and resource optimization algorithm with optimal throughput in 5G network. IEEE Access 7:25263–25273. https://doi.org/10.1109/ACCESS.2019.2900422
35. Zafar MH, Khan I, Alassafi MO (2022) An efficient resource optimization scheme for D2D communication. Digital Commun Netw
36. Sun S, Yoan S (2014) Resource allocation for D2D communication using particle swarm optimization in LTE networks. In: International conference on ICT convergence, pp 371–376. https://doi.org/10.1109/ICTC.2014.6983158
37. Juneja M, Nagar SK (2016) Particle swarm optimization algorithm and its parameters: a review. In: 2016 International conference on control, computing, communication and materials (ICCCCM), pp 1–5. https://doi.org/10.1109/ICCCCM.2016.7918233
38. Seyedali M, Andrew L (2016) The whale optimization algorithm. Adv Eng Softw 95:5167. ISSN09659978
39. Tan T-H, Chen B-A, Huang Y-F (2018) Performance of resource allocation in device-to-device communication systems based on evolutionally optimization algorithms †. Appl Sci 8:1271. https://doi.org/10.3390/app8081271
40. Pham Q, Mirjalili S, Kumar N, Alazab M, Hwang W (2020) Whale optimization algorithm with applications to resource allocation in wireless networks. IEEE Trans Veh Technol 69(4):4285–4297. https://doi.org/10.1109/TVT.2020.2973294
41. Sheng Z, Tuan HD, Nasir AA, Duong TQ, Poor HV (2018) Power allocation for energy efficiency and secrecy of wireless interference networks. IEEE Trans Wireless Commun 17(6):3737–3751
42. Li Y, Sheng M, Yang C, Wang X (2013) Energy efficiency and spectral efficiency tradeoff in interference-limited wireless networks. IEEE Commun Lett 17(10):1924–1927
43. Pham Q-V, Hwang W-J (2017) Fairness-aware spectral and energy efficiency in spectrum-sharing wireless networks. IEEE Trans Veh Technol 66(11):10 207–10 219
44. Lyu X, Tian H, Sengul C, Zhang P (2017) Multiuser joint task offloading and resource optimization in proximate clouds. IEEE Trans Veh Technol 66(4):3435–3447
45. Girmay GG, Pham Q, Hwang W (2019) Joint channel and power allocation for device-to-device communication on licensed and unlicensed band. IEEE Access 7(22):196–22 205
46. Yu G, Xu L, Feng D, Yin R, Li GY, Jiang Y (2014) Joint mode selection and resource allocation for device-to-device communications. IEEE Trans Commun 62(11):3814–3824

A 2 Stage Pipeline for Segmentation and Classification of Rooftop from Aerial Images Using MultiRes UNet Model

P. Uma Maheswari, Shruthi Muthukumar, Gayathri Murugesan, and M. Jayapriya

Abstract Meeting energy demands are one of the primary goals of sustainable ecosystem, but civilization has already wreaked enough damage. And a lot of valuable resources is spent in understanding urban dynamics to create a sustainable environment. Segmentation of buildings from aerial images plays a pivotal role in many applications including terrain mapping, simulation for PV panel placements, urban planning, and so on. In accordance, this paper focuses on creating a 2-staged automated pipeline to address the problem of identifying rooftops from satellite images and classifying the rooftops into different types which can be used further for the above applications. A deep learning MultiRes UNet model is used in the first stage for segmentation of buildings in the city of Christchurch, New Zealand. The experimental results show an average IoU of 95.25% for building segmentation. The extracted rooftops are then manually classified to create a dataset for roof type classification. Transfer learning models such as ResNet50, EfficientNetB4, VGG16, and shallow CNN models are used for classification, and a comparative analysis is made between the models. After analyzing the results from different models, it was decided to go with majority voting as an ensembling method as different models provided better results in learning different parts, and the classification accuracy was further improved by 5.67%.

Keywords Computer vision · Image segmentation · Image procesing · Transfer learning models · AIRS · MultiRes UNet

1 Introduction

Sustainable environment is required for the advancement of economic development, to improve energy security, access to energy, and mitigate climate change. And these days, a lot of significant energy is spent in remote sensing ad urban planning. Urban dynamic mapping is a tedious and time-consuming process which involves a lot of

P. U. Maheswari · S. Muthukumar (✉) · G. Murugesan · M. Jayapriya
College of Engineering, Anna University, Guindy, Chennai 600025, India
e-mail: shruthi.muthukumar@gmail.com

S. Das et al. (eds.), *Advances in Data-driven Computing and Intelligent Systems*, Lecture Notes in Networks and Systems 653, https://doi.org/10.1007/978-981-99-0981-0_59

manual work. Accurate mapping of urban buildings poses a lot of challenges, and one of the major concern is inadequate and poor resolution of images. Along side that, there is no proper delineation of buildings from vegetation.

We must note that so far no automated approaches have been used to detect buildings from a wide coverage satellite image and classify roof types. This is where our one-shot pipeline comes into play, which can segment and detect buildings from aerial/satellite images [1], extract each of the rooftops, and classify them based on their type which can be used in applications like urban mapping and smart city plans. This can easily automate the entire process of assessing hundreds of roof tops. The following contributions are made in order to ease the process of PV panels placement on rooftop of buildings from aerial images:

- To segment and detect buildings in a given satellite image using MultiRes UNet model and perform background subtraction to extract the rooftops.
- To label the extracted building rooftops into different classes—flat, gable, and hip and create a dataset.
- Train different deep learning models for roof type classification and provide a comparative analysis.

2 Related Work

Image databases and computer vision have got a lot of traction. Many breakthroughs in image segmentation have been made as a result of this. Bio-medical and remote sensing are the two industries where segmentation is critical in understanding the images.

A multi-scale convolutional neural network that adopts an encoder–decoder UNet architecture is utilized by Li et al. [16] for building segmentation. In this approach, a UNet is constructed as the main network, and the bottom convolution layer of UNet is replaced by a set of cascaded dilated convolution with different dilation rates and an auxiliary loss function added after the cascaded dilated convolution which has aided in network convergence. However, the segmentation of middle parts in buildings is misaligned, and the bulges on the boundaries are lost. Yet, another major issue is that the algorithm performs well only in one subset (countryside and forest) but not in another. Alamri et al. [2] have proposed a GAN architecture for footprint extraction.

A mask R-CNN with three steps is proposed to extract buildings in the city of Christchurch from aerial images by Chen and Li [9] to recognize small detached residences to understand the havoc caused by earthquakes. The regional proposal network (RPN) model is utilized to locate RoI and filter out the irrelevant bounding boxes using object and background classification, as well as bounding box regression. The RoIAlign method used instead of the RoI Pool gives better feature extraction. However due to a small training dataset, the model was unable to successfully demarcate building edges, resulting in low accuracy and precision when compared to other SOTA models.

A combination of image processing techniques, including adaptive edge detection and contours, is used by Kumar and Sreedevi [13] to segment out rooftop boundaries and obstacles present inside them along with polygon shape approximation. It provides a comparative analysis of the solar potential of buildings. In research done by Reinartz et al. [21], several types of the rooftop are considered to learn the intra-class variations. Because Google Maps India's satellite resolution is so low, the edges are not fully identified, and there are outliers plotting solar panels outside of the building's rooftop area. An integrated approach for edge detection is used for analyzing terrains of agricultural terrace by Li et al. [10].

The problem of semantic segmentation of buildings from remote sensor imagery is addressed by a novel framework called the ICTNet proposed by Chatterjee and Poullis [7]. ICTNet: a novel network with the underlying architecture of a fully convolutional network infused with feature re-calibrated dense blocks at each layer is combined with dense blocks and squeeze-and-excitation (SE) blocks. Dense blocks connect every layer to every other layer in a feed-forward fashion. Along with good gradient propagation, they also encourage feature reuse and reduce the number of parameters substantially as there is no need to relearn the redundant feature maps which allows the processing of large patch sizes.

Reconstruction is done by extruding the extracted boundaries of the buildings, and comparative analysis is made between the two in Stilla et al. [19]. With no 3D information on the buildings, the authors have used the building boundaries as a proxy for the reconstruction process and have got better overall IoU compared to other methods. The main limitation here is that there is no loss function for the reconstruction accuracy. Furthermore, due to the fact that ground truth photos used for training contain mistakes and are manually generated, there is a large variance in per-building IoU. In addition, the reconstruction accuracy is consistently lower than classification accuracy by an average of $4\% \pm 1.65\%$. A similar convolution network approach is executed by Yuille et al. [8] where the authors have incorporated fully connected CRFs for building semantic segmentation.

Rooftop classification is the important step in identifying the type of PV panels that can be fitted. Buyukdemircioglu et al. [5] have undertaken research to generate a roof type dataset from very high-resolution (10 cm) orthophotos of Cesme, Turkey and to classify the roof types using a shallow CNN architecture. UltraCam Falcon large-format digital camera is used to capture orthophotos with 10cm spatial resolution, and roofs are manually classified into 6 different labels. The prediction is investigated by comparing with three different pre-trained CNN models, i.e., VGG16, EfficientNetB4, and ResNet50. Roof type classification by Lithen et al. [3] is done via deep CNN on low resolution images.

Simple CNN models are hence easier to implement and require nominal hardware specifications. The shallow CNN model has achieved 80% accuracy. As the roof images were clipped automatically from orthophotos, there are few buildings with overlap. Half-hip roofs are not classified properly, and the F1-score obtained for them is very low. The authors have not experimented with alternate hyperparameter tweaking for the shallow CNN architecture, which is a serious flaw. The rooftops detected using semantic segmentation and Hough transform in Kirsten et al. [11] are used to determine the no. of solar arrays in the detected region. Other applications of roof type segmentation include solar photovolaic panel detection which [12, 15, 17, 18, 23] have researched.

Although the above models address the issue of building segmentation to a great extent, there are certain major flaws. In situations where shadows, vegetation and parking lots, and other obstacles enclose buildings, the above frameworks showed poor results in building detection as mentioned in Malof et al. [6]. In addition, there was no clear distinction to demarcate adjacent buildings, and boundaries of buildings were not properly detected. Furthermore, when a satellite or aerial image spanning, a large area with several buildings, was provided, only manual cropping was done to extract roofs for classification, and there were no automated mechanisms in the approaches outlined above.

To address the above problems, we propose a 2-staged pipeline for building segmentation and roof type classification with a simulation of PV panels fitted on it. In this work, we adapt MultiRes UNet, with additional convolution layers to extract spatial features at various scales. The skip connections in UNet are replaced with a chain of convolution operations to reduce the semantic gap between encoder and decoder features, and hence, building segmentation results can be improved drastically. This is then followed by a technique called background subtraction to extract the rooftops automatically which resolves the problem of manual cropping and selection of roofs. Roof type classification is performed with deep learning models.

3 System Design

The proposed model's overall system architecture is depicted in Fig. 1.

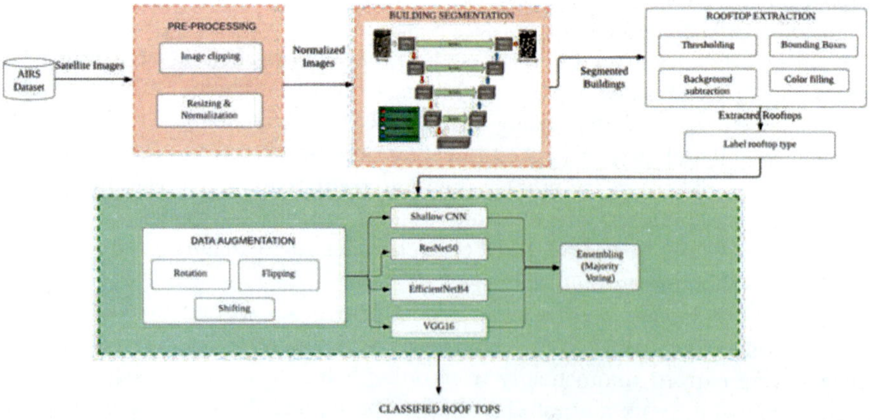

Fig. 1 Overall system architecture of building extraction and roof type classification for maximal PV panel installation

The AIRS dataset contains satellite images of Christchurch in New Zealand with 7.5 cm resolution. For high-resolution satellite imagery, building extraction is done using deep learning methods by Chen et al. [25]. To begin, the system includes two pre-processing steps. The pre-processing stages include clipping satellite photos and the accompanying ground truth masks, as well as scaling and normalization. After that, the model is trained using the MultiRes UNet architecture. Following this, background subtraction is applied by drawing bounding boxes to extract roof tops. The first module mainly emphasizes building detection from satellite images.

The extracted rooftops from the previous module are manually labeled to create a dataset of three classes—flat, gable, and hip which are then fed to three different models for roof type classification, and a comparative analysis is made. An ensemble method (majority voting) is used here to improve the efficiency of classification models.

4 Experimental Setup

Training and testing the MultiRes UNet network for building detection and different transfer learning models for rooftop classification are executed under TensorFlow backend and Keras framework in Colab Pro with a memory of 16 GB, T4 GPU. Flask framework is used to develop a Web application of the system.

5 Methodology

5.1 Dataset

The dataset used here is the aerial imagery for roof segmentation (AIRS) dataset, which covers the area of Christchurch in New Zealand with a spatial resolution of 7.5 cm and spatial dimensions of 10000×10000. The training set has 857 satellite images, and the validation and testing set each contain 90 images along with the corresponding ground truth masks. The roof type classification dataset has 1115 images manually labeled into three subcategories: flat, gable, and hip, with 1020 images in the training set, 50 images in the validation set, and 45 images in the testing set.

5.2 Detailed Module Design

The pipeline consists of two modules: building detection and extraction followed by roof type classification.

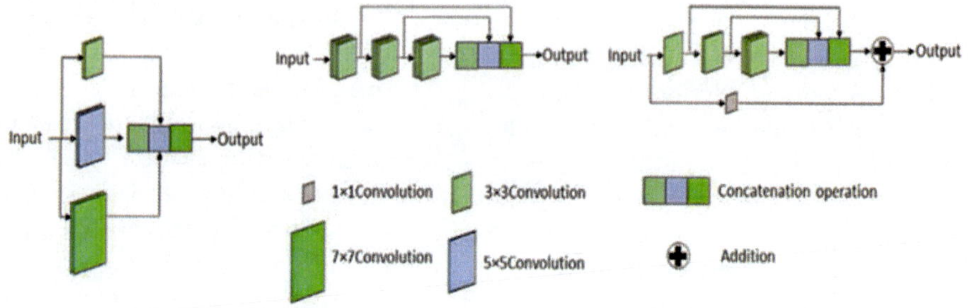

Fig. 2 Architecture of MultiRes block

5.3 *Building Detection*

Pre-processing: Satellite and aerial images [22] are typically too large to be segmented directly and hence need to be clipped into smaller patches or tiles. Aerial satellite images from AIRS dataset are of very large dimensions (10,000 $\times$ 10,000), and owing to the hardware constraints for training the model, as a first step in pre-processing, the images are clipped into (1536 $\times$ 1536) pixels using a sliding window technique, and consequently, we obtain 36 patches for a single satellite image. This is followed by resizing the images to 256 $\times$ 256 dimensions and further applying MinMax scaler as the normalization technique.

Building Segmentation: Once pre-processing is done, the normalized images are fed into the MultiRes UNet model. The MultiRes UNet model has two blocks: the ResPath and the MultiRes block. A sequence of 3 $\times$ 3, 5 $\times$ 5, and 7 $\times$ 7 convolution operations are utilized after each pooling layer and transposed convolutional layer in the MultiRes block as shown in Fig. 2. This is done to extract the spatial features learned from the image at multiple resolutions.

The skip connections in ResPath depicted in Fig. 3 allow the network to transfer spatial information from encoder to decoder [4] that is lost during the pooling procedure. To resolve this, a set of 3$\times$ 3, 1 $\times$ 1 convolutional layers are used along the shortcut links to reduce the discrepancy between the encoder and decoder characteristics [24]. The additional non-linear transformations applied to the features propagating from the encoder stage will account for the additional processing performed by the decoder.

Four MultiRes blocks are each used in the encoder and decoder stage. The number of filters in each of the MultiRes blocks is based on the formula: $W = \alpha \times U$ where α is the scalar coefficient whose value is set to 1.67, and U refers to the number of filters.

The parameter W preserves an analogous connection between the suggested MultiRes UNet network and the main UNet network. After every pooling or transposing of layers, the value of W became double, similar to the original UNet network. The values of U = [32, 64, 128, 256, 512] are set as follows. The hyperparameters used for the model are given in Table 1.

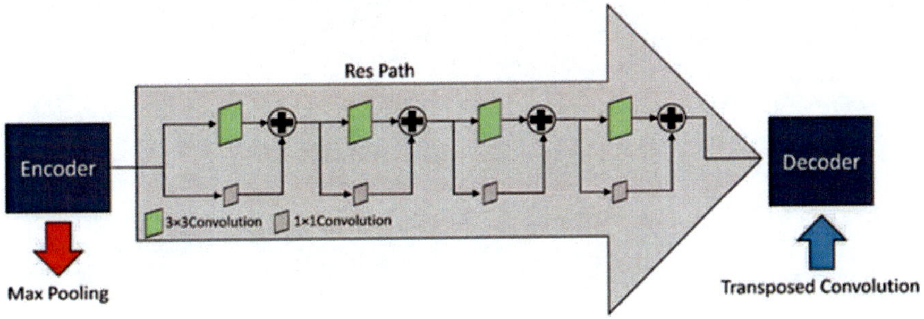

Fig. 3 Architecture of ResPath

Table 1 Hyperparameter values used in training our MultiRes UNet model

Hyperparameters	Values
Learning rate	0.0001
Epochs	100
Batch size	8
Optimizer	Adam
Loss	Binary cross entropy
Activation function	ReLU, sigmoid (for last layer)

After training the images with the MultiRes UNet model, some pixel values lie in the range between (0, 1) which do not clearly indicate if they belong to foreground or background. To resolve the ambiguity of delineating the building boundaries, simple thresholding is performed where values greater than 0.5 are assigned to the foreground, and rest to the background.

Figure 4 provides a comparison between the ground truth images with our prediction masks before and after applying threshold. After thresholding, our buildings are more effectively demarcated.

Rooftop Extraction: The segmented buildings are enclosed within a rectangle using bounding box which highlights the region of interest (RoI). This is followed by background subtraction of segmented mask with the satellite image. The extracted building rooftops as shown in Fig. 5 are stored in a database which is later used in the second stage for classification.

5.4 *Rooftype Classification and Boundary Detection*

The second phase in the pipeline is classifying the type of roofs and drawing boundaries on the rooftop.

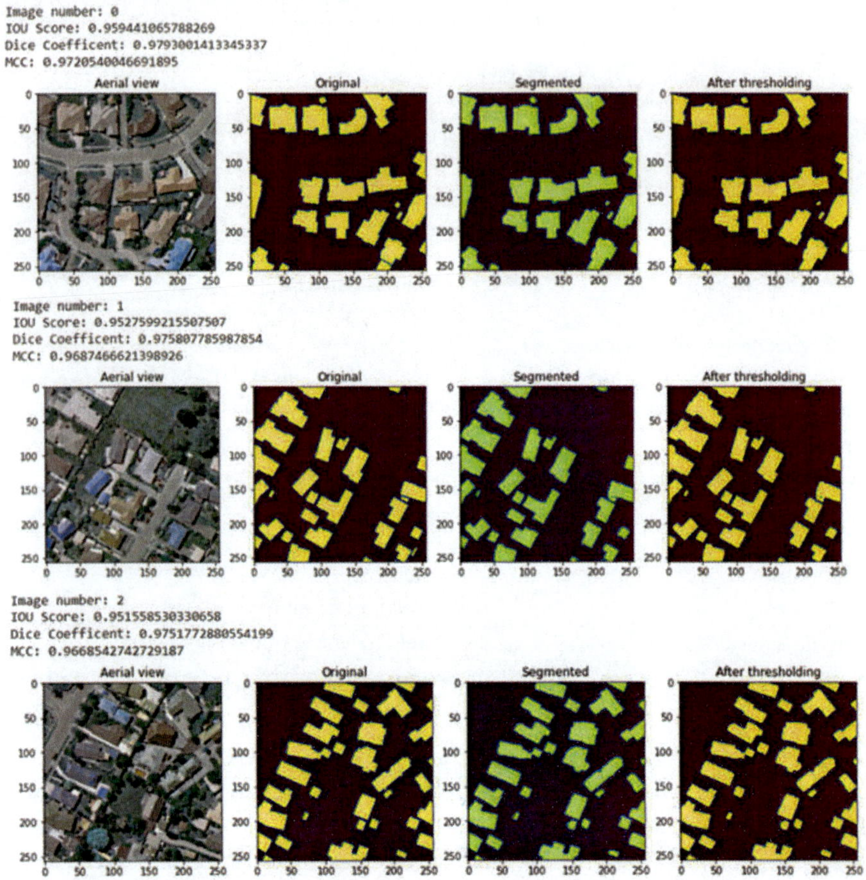

Fig. 4 Comparison of ground truth images with predictions from MultiRes before and after applying threshold

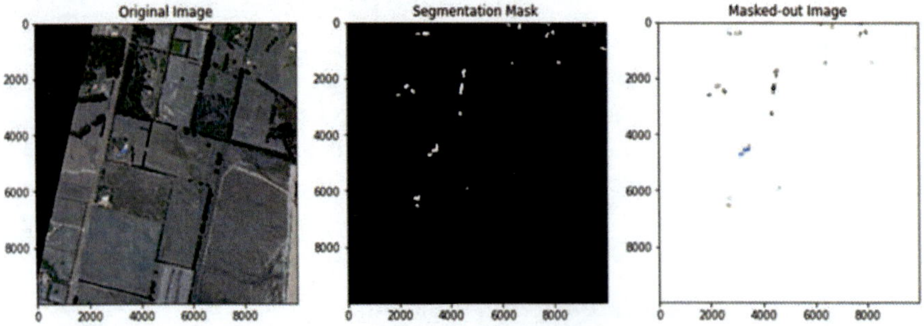

Fig. 5 Original satellite image (left), predicted masked image from MultiRes UNet model (center), extracted rooftops (right)

Table 2 Distribution of roof type dataset

Roof type	Training (91%)	Val (5%)	Testing (4%)	Total
Flat	303	14	18	335
Gable	410	23	18	451
Hip	307	13	9	329

Dataset Creation with Different Roof Types: After conducting rooftop extraction in the preceding step, a dataset including 1115 rooftop photos is populated into three different categories: flat, gable, and hip. As there was no supervised dataset for roof type classification available, it was decided to perform manual labeling. The dataset was cleaned up by removing roof types with poor resolution. A balance between the several roof type classifications was sought as much as possible when compiling the dataset. The distribution of different roof types is presented in Table 2. Table 3 gives a sample representation of the different roof tops belonging to each of the 3 classes.

Data Augmentation: For training neural network models as done by Scartezzini et al. [20], the size of the dataset needs to be large. However, owing to time constraints, only 1115 images were manually labeled. As a result, data augmentation approaches (rotation, shifting, and flipping) were employed to increase the dataset size for better model training.

Classification of Roof Type Images with Ensembling Approach: Four deep learning architectures are used here for roof type classification. Shallow CNN is used as a baseline model. Along with this, three pre-trained models such as ResNet50, EfficientNetB4, and VGG16 are fine-tuned, and a comparative analysis of the different models is made. Furthermore, majority voting is used as an ensembling approach to combine learnings from different models to provide better results to predict the roof type of unseen images.

6 Experimental Results

6.1 Performance Results on Building Segmentation

The first module involving building segmentation uses IoU/Jaccard coefficient and Dice coefficient as the primary metrics. Along with this, pixel accuracy and MCC are also used to evaluate the training and validation results.

The graphs in Fig. 6 allow us to easily interpret the training of the MultiRes UNet model. The blue line indicates the training details, while the red line shows the validation details. IoU, which penalizes heavily than dice coefficient, has achieved a remarkable value of 96.27% at the end of 100 epochs. It can be seen that the Dice coefficient starts with 58% at the beginning of training and gradually attains a value

Table 3 Samples of images of each class from Christchurch, New Zealand

Flat		
Gable		
Hip		

Table 4 Performance results of MultiRes model

Performance metrics	Training set	Validation set
IOU (%)	93.53	95.25
Dice coefficient (%)	96.25	97.56
MCC (%)	95.87	96.74
Accuracy (%)	98.51	97.83
Loss	0.1734	0.2033

of 98.5% with no major fluctuations. As can be seen from Fig. 7, the training and validation curve coincide with each other, and the difference in loss is very meager which shows the model does not overfit. The MCC value as inferred from Fig. 8 is close to 98% at the end of 100 epochs which indicates that buildings have been segmented near perfectly.

Furthermore, from Table 4, we infer that the model has performed remarkably well as the training loss hovers around 0.173, and validation loss is close to 0.2. This is a

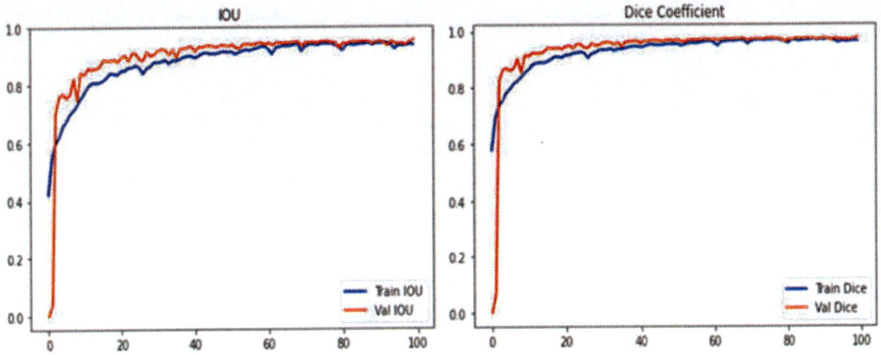

Fig. 6 IoU and dice coefficient graphs on training and validation set

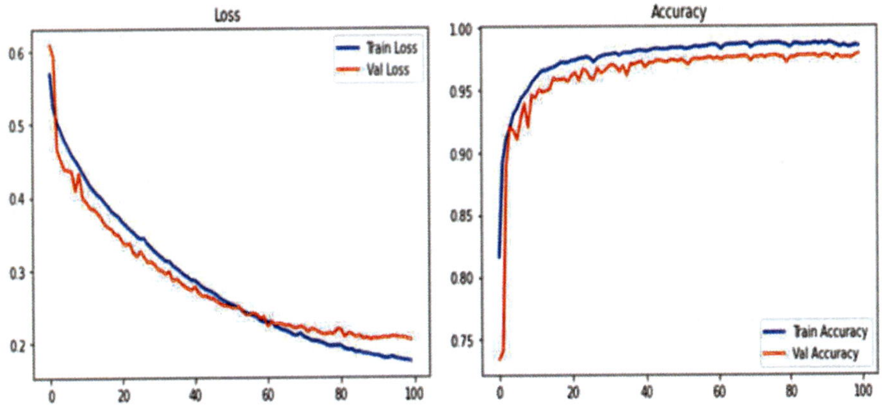

Fig. 7 Loss and accuracy graphs on training and validation set

Fig. 8 Matahew's correlation coefficient on training and validation set

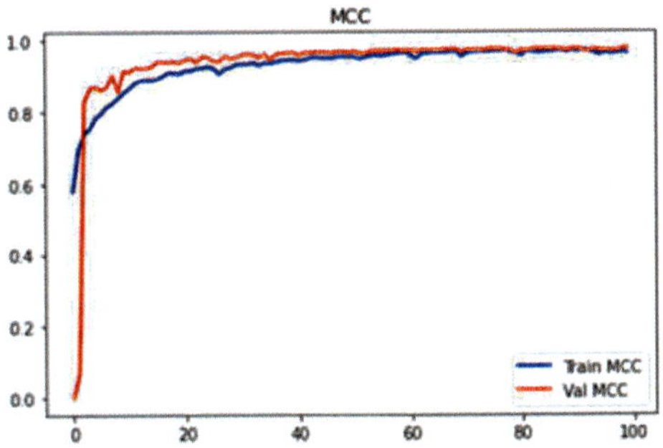

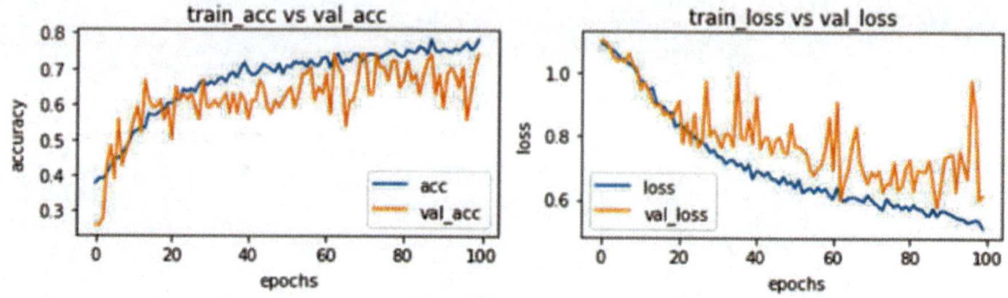

Fig. 9 Graphs showing training and validation accuracy, loss for CNN

reasonably good value which indicates that our model performs well in segmenting the buildings accurately even after the images are resized and normalized.

6.2 *Performance Results on Roof Type Classification*

The performance results for CNN model along with additional transfer learning models like VGG16, ResNet50, and EfficientNetB4 used for roof type classification are discussed below.

Performance of Shallow CNN Model: The CNN model achieved an accuracy of 78% on the training set and 74% on the validation set. From Fig. 9, it can be inferred that there is no generalization gap because of dropout and data augmentation. The model was likewise trained for 50 epochs at first, but because it did not produce satisfactory results, the model was further trained for another 50 epochs. The CNN model was able to classify gable and hip classes accurately to a larger extent than the flat class.

Performance of Fine-tuned ResNet50 Model: The ResNet50 has yielded a very good accuracy of 91.59% on the training set and 92.5% on the validation set. From Fig. 10, it can be witnessed that the loss is also very less at only 0.245 on the training set. All transfer learning models are only trained for 50 epochs as training for 100 epochs led to overfitting. The model was able to identify hip ad gable classes very precisely. Also, the classification of flat classes is better compared to the previous CNN model, and the misclassification of a few flat classes as gable could be due to the slight tilt angle of a few rooftop slopes

Performance of Fine-tuned EfficientNetB4 Model: The graphs in Fig. 11 exhibit that the accuracy of the model has steadily grown from 40 to 89%, and the model loss has significantly reduced from 1.25 to 0.33 at the end of 50 epochs. When compared with the ResNet50 model which had higher accuracy in correctly classifying hip class, EfficientNetB4 model has achieved higher F1-score and accuracy in identifying flat roof tops (Fig. 12). The average F1-score, precision, and recall hover around 91%.

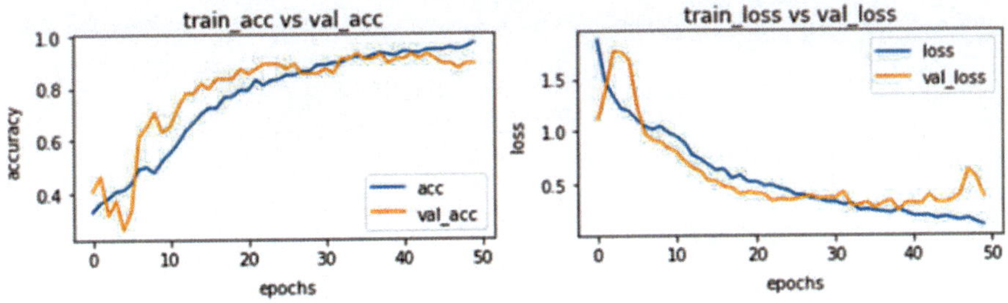

Fig. 10 Graphs showing training and validation accuracy, loss for ResNet50

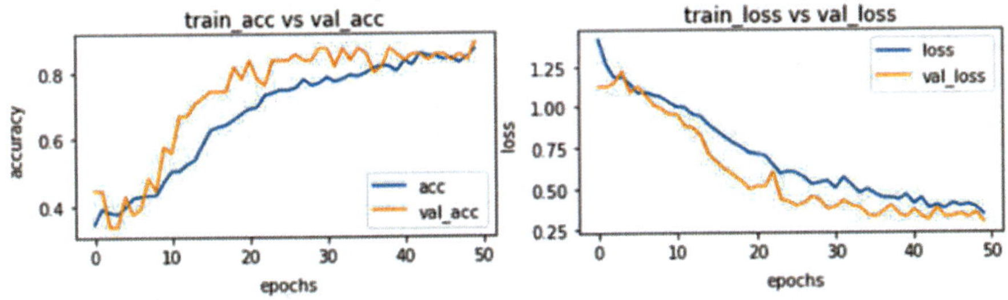

Fig. 11 Graphs showing training and validation accuracy, loss for EfficientNetB4

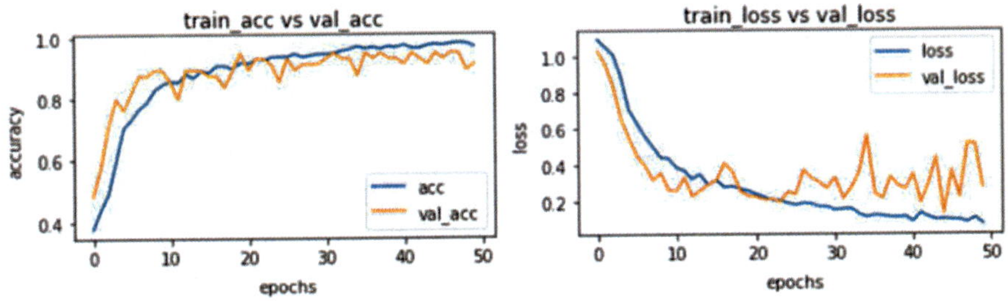

Fig. 12 Graphs showing training and validation accuracy, loss for VGG16

Performance of Fine-Tuned VGG16 Model: VGG16 model has achieved the highest accuracy compared to the other three models. The classification accuracy on the validation set is 94.45%, while the training accuracy is around 97%. In Fig. 12, it can be noticed that there is a gradual decline in training loss over 50 epochs, while there are some fluctuations with validation loss.

Majority Voting: It is inferred that transfer learning methods result in higher classification accuracy compared to shallow CNN model. However, these transfer learning models perform classification well on only some parts. Hence, an ensembling approach is used as it can make better predictions and achieve better performance than any single contributing model. Ensembling also reduces the spread or dispersion of the predictions and model performance. The ensembling method applied here is

Table 5 Comparative analysis for building segmentation using MultiRes UNet model

Performance metrics	MultiRes baseline	MultiRes with normalization
IOU (%)	93.14	95.25
MCC (%)	94.91	96.74
Dice coefficient (%)	96.45	97.56

majority voting. VGG16 was efficient in identifying hip roof types. Similarly, ResNet very well classified flat roofs from other rooftop types, and EfficientNet was able to classify a large chunk of gable roof types correctly. Majority voting is therefore incorporated to further boost the performance, and the accuracy has increased by 5.67%.

7 Conclusion and Future Work

A 2-step pipeline was proposed in this research to automate the process of PV panel placement on rooftops given a satellite image covering a wide range and different types of buildings. The first phase involved building segmentation on AIRS dataset with the MultiRes UNet model. Table 5 provides a comparative analysis of building segmentation with the MultiRes UNet baseline model and the model we have implemented after performing resizing and normalization with thresholding. Despite resizing the images to 256×256 dimensions, the findings show that the model works extremely well, with an average IoU of 95.25% and a Dice coefficient of 97.56%.

Proper segmentation of buildings led to proper extraction of different roof types which were manually labeled to create a dataset. Different deep learning models were trained and a comparative analysis proved that transfer learning models were much efficient than the shallow CNN. Majority voting was further used as an ensembling technique to improve the efficiency of classification. The study shows roof types which can be classified easily after performing building detection from a wide satellite image wihout manual assessment.

The classification of roof types is one of the areas of anticipated future development. We primarily seek to improve the outcomes of roof type classification by expanding the dataset size, either through labeling or by obtaining equivalent training data from other sources.

References

1. Abdollahi A, Pradhan B, Alamri AM (2020) An ensemble architecture of deep convolutional segnet and unet networks for building semantic segmentation from high-resolution aerial images

2. Abdollahi A, Pradhan B, Gite S, Alamri A (2020) Building footprint extraction from high resolution aerial images using generative adversarial network (GAN) architecture. IEEE Access Article 209517-209527. https://doi.org/10.1109/ACCESS.2020.3038225
3. Axelsson M, Soderman U, Berg A, Lithen T (2018) Roof type classification using deep convolutional neural networks on low resolution photogrammetric point clouds from aerial imagery. In: IEEE international conference on acoustics, speech and signal processing (ICASSP), pp 1293–1297. https://doi.org/10.1109/ICASSP.2018.846
4. Badrinarayanan V, Kendall A, Cipolla R (2017) Segnet: a deep convolutional encoder-decoder architecture for image segmentation. IEEE Trans Pattern Anal Machine Intell 39:2481–2495
5. Chatterjee B, Poullis C (2019) On building classification from remote sensor imagery using deep neural networks and the relation between classification and reconstruction accuracy using border localization as proxy. In: 16th conference on computer and robot vision (CRV), pp 41–48. https://doi.org/10.1109/CRV.2019.00014
6. Buyukdemircioglu M, Can R, Kocaman S (2021) Deep learning based roof type classification using VHR aerial imagery. The international archives of the photogrammetry. Remote Sens Spat Inform Sci XLIII-B3-2021:55–60. https://doi.org/10.5194/isprs-archives-XLIII-B3-2021-55-2021
7. Camilo J, Wang R, Collins LM, Bradbury K, Malof JM (2018) Application of a semantic segmentation convolutional neural network for accurate automatic detection and mapping of solar photovoltaic arrays in aerial imagery
8. Chen L-C, George P, Iasonas K, Kevin M, Yuille AL (2018) Deeplab: semantic image segmentation with deep convolutional nets, atrous convolution, and fully connected crfs. IEEE Trans Pattern Anal Mach Intell 40(4):834–848
9. Edun A, Harley J, Deline C, Perry K (2021) Unsupervised azimuth estimation of solar arrays in low-resolution satellite imagery through semantic segmentation and Hough transform. Appl Energy. https://doi.org/10.1016/j.apenergy.2021.117273
10. Dai W, Na J, Huang N, Hu G, Yang X, Tang G, Xiong L, Li F (2020) Integrated edge detection and terrain analysis for agricultural terrace delineation from remote sensing images. Int J Geogr Inform Sci 34(3):484–503. https://doi.org/10.1080/13658816.2019.1650363
11. Kumar A, Sreedevi I (2018) Solar potential analysis of rooftops using satellite imagery. ArXiv abs/1812.11606
12. Lee S, Iyengar S, Feng M, Shenoy P, Maji S (2019) Deeproof: a data-driven approach for solar potential estimation using rooop imagery. https://doi.org/10.1145/3292500.3330741
13. Chen L-C, George P, Iasonas K, Kevin M, Yuille AL (2018) Deeplab: semantic image segmentation with deep convolutional nets, atrous convolution, and fully connected crfs. IEEE Trans Pattern Anal Mach Intell 40(4):834–848
14. Marmanis D, Schindler K, Wegner JD, Galliani S, Datcu M, Stilla U (2018) Classification with an edge: improving semantic image segmentation with boundary detection. ISPRS J Photogr Remote Sens. https://doi.org/10.1016/j.isprsjprs.2017.11.009
15. Mohajeri N, Assouline D, Guiboud B, Bill A, Gudmundsson A, Scartezzini J-L (2018) A city-scale roof shape classification using machine learning for solar energy applications. Renew Energy 121:81–93. ISSN 0960-1481. https://doi.org/10.1016/j.renene.2017.12.096
16. Partovi T, Fraundorfer F, Azimi S, Marmanis D, Reinartz P (2017) Roof Type Selection based on patch-based classification using deep learning for high resolution satellite imagery. https://doi.org/10.5194/isprs-archives-XLII-1-W1-653-2017
17. Li P, Zhang H, Guo Z, Lyu S, Chen J, Li W et al (2021) Understanding rooftop PV panel semantic segmentation of satellite and aerial images for better using machine learning. Adv Appl Energy 4:100057. ISSN 2666-7924. https://doi.org/10.1016/j.adapen.2021.100057
18. Qi C, Wang L, Wu Y, Wu G, Guo Z, Steven W (2018) Aerial imagery for roof segmentation: a large-scale dataset towards automatic mapping of buildings. ISPRS J Photogram Remote Sens 147:42–55
19. Marmanis D, Schindler K, Wegner JD, Galliani S, Datcu M, Stilla U (2018) Classification with an edge: improving semantic image segmentation with boundary detection. ISPRS J Photogr Remote Sens. https://doi.org/10.1016/j.isprsjprs.2017.11.009

20. Qi L, Jiang M, Lv Y, Zhang Z, Yan J (2021) Techno-economic assessment of photo voltaic power generation mounted on cooling towers. https://doi.org/10.1016/j.enconman.2021.113907
21. Partovi T, Fraundorfer F, Azimi S, Marmanis D, Reinartz P (2017). Roof Type Selection based on patch-based classification using deep learning for high resolution satellite imagery. https://doi.org/10.5194/isprs-archives-XLII-1-W1-653-2017
22. Qi C, Wang L, Wu Y, Wu G, Guo Z, Steven W (2018) Aerial imagery for roof segmentation: a large-scale dataset towards automatic mapping of buildings. ISPRS J Photogram Remote Sens 147:42–55
23. Qi L, Jiang M, Lv Y, Zhang Z, Yan J (2021). Techno-economic assessment of photo voltaic power generation mounted on cooling towers. https://doi.org/10.1016/j.enconman.2021.113907
24. Wu G, Shao X, Guo Z, Chen Q, Yuan W, Shi X, Xu Y, Shibasaki R (2018). Automatic building segmentation of aerial imagery using multi-constraint fully convolutional networks. https://doi.org/10.3390/rs10030407
25. Xu Y, Wu L, Xie Z, Chen Z (2018) Building extraction in very high resolution remote sensing imagery using deep learning and guided filters

Land Cover Change Detection Using Multi-spectral Satellite Images

Galla Yagnesh, Mare Jagapathi, Kolasani Sai Sri Lekha, Duddugunta Bharath Reddy, and C. S. Pavan Kumar

Abstract Change detection in land use and land cover is critical for understanding the interplay of human activities with the environment; hence, the ability to mimic changes is required. Empirical observations revealed a change in land use and land cover surrounding the Krishna river basin, which is located in Andhra Pradesh's Krishna district. The purpose of this study is to examine how land use and land cover have evolved over the past eight years in the Krishna River Basin (2013–2021). As a remote sensing technique, the investigation was conducted utilizing USGS Landsat-8 collection 1 level 1 imaginaries from June 2013 to December 2021. GIS software is used to create thematic maps. The current study has revealed how the risk to biodiversity, industry, and change in cropping patterns and land use are all impacted by the limited availability of water resources. It frequently serves as a catalyst for conflict between states. Change has also occurred in wasteland, built-up areas, harvested areas, and agricultural land. The Krishna river basin's sustained development depends on sound land-use planning. The paper's conclusion states that present methods are helpful in gaining a general understanding of how land is used and 22% land change is observed which helps in planning development projects close to the Krishna River Basin.

Keywords Remote sensing · Principle component analysis · K-means · Change detection · Classification · Image processing · Multi-spectral images

1 Introduction

Changes in land use and land cover (LULC) are some of the most notable and obvious ways that humans have altered the surface of the planet. Land cover refers to the physical and biological components of land surfaces, such as vegetation, water bodies, bare plains, or man-made buildings. Alternately, land use describes a complex fusion

---

G. Yagnesh · M. Jagapathi · K. S. Sri Lekha · D. B. Reddy · C. S. Pavan Kumar (✉)
Department of Information Technology, Velagapudi Ramakrishna Siddhartha Engineering College, Vijaywada, Andhra Pradesh, India
e-mail: pavanchitturi@vrsiddhartha.ac.in

S. Das et al. (eds.), *Advances in Data-driven Computing and Intelligent Systems*, Lecture Notes in Networks and Systems 653, https://doi.org/10.1007/978-981-99-0981-0_60

of socioeconomic, management, and economic goals as well as the environments and frameworks in which lands are governed. Although we frequently group land use and land cover together, there is a significant distinction between the two [1]. While land use primarily focuses on human activities that are determined by the integration of natural and social scientific methods in various landscapes even with the same land cover, land cover specifically implies the spatial distribution of the various classes of land cover that can be assessed both qualitatively and quantitatively through remote sensing techniques.

The most prominent type of environmental change occurring on a global scale that is perceptible at various spatial and temporal dimensions and has a significant impact on our daily lives is LULC change. According to technical definitions, the terms "LULC change" and "land use and land cover change" refer to the mean quantitative changes in the spatial extent (growth or reduction) of a certain type of land cover and land use, respectively. The behavior of changes in land use and land cover is significantly influenced by both anthropogenic and environmental causes [2–6].

For the purpose of predicting potential future land-use change scenarios, it is essential to comprehend changes in LULC and subsequent modeling. There is a large amount of data generated by remote sensing, ranging from extremely high spatial resolution images to regional datasets produced on a regular to lower spatial resolution images that are now produced daily across the entire earth [1]. Now, we have valuable tool for researching changes in land use and land cover over time thanks to the temporal dynamics of the synoptic view of the earth's surface obtained with the help of satellites. The reflectance patterns of incident radiation are affected by changes in the vegetative cover, soil moisture, or numerous adjustments to the earth's surface [7]. These changes in land usage and land cover can be either natural or man-made. It is safe to extrapolate the changes in spatial extents and to assess the pace of changes because the changes in land use and land cover are largely unidirectional and do not oscillate considerably. The Geographic Information System (GIS) is a crucial tool in this regard. The Geographic Information System is a potent tool that allows for the user-friendly storage, organization, and retrieval of spatial data. A potential tool for environmental management is the combination of satellite remote sensing data with auxiliary data in a GIS environment along with GPS data.

2 Literature Review

T. Vignesh concluded that when using machine learning algorithms, a feature extraction phase is required before determining the change detection. The feature extraction phase is used to improve the accuracy of multi-spectral images. However, when using deep learning algorithms, a separate feature extraction phase is not required. The accuracy of classification is determined by the algorithms used to detect changes, as well as the image resolution [8–12].

In unsupervised change detection problems, Lorenzo Bruzzone proposed two techniques for analyzing the difference image. The main methodological innovation of this paper is the formulation of the unsupervised change detection problem in terms of Bayesian decision theory. They have proposed, in particular, an iterative technique that enables unsupervised estimations of the apriori probabilities and density functions associated with changed and unchanged pixels in the difference image. These estimates enable the use of supervised methods in the context of unsupervised change detection. Two automatic techniques for analyzing difference images have been presented within this framework. The first method assumes that the pixels in the difference image are independent of one another. The second technique analyses the difference image using an MRF approach that takes advantage of the interpixel class dependency context to improve the accuracy of the final change detection map [13–18].

Table 1 gives the details like methods used, pros and cons of the methods used. It is the collection of some of the research papers study.

3 Methodology

There are two proposed approaches to detect the changes in multi-temporal pictures. The first method is referred to as "Independent Data Transformation." Dimensional reduction is applied to all multi-temporal picture data, and components are obtained, according to it. After then, each image is classified separately from the others, and a post-classification change detection technique is carried out. The second method is referred to as "Unified Data Transformation" [7]. This concept involves registering multi-temporal satellite images to one another, then processing them into a single image data using methods like IR, ID, and others. PCA is then applied to the data.

The purpose of our study is to determine the change in Krishna District of Andhra Pradesh, India, using the second approach indicated above. In order to finish the registration procedure, the first part of the process involved using I1 and I2 Landsat satellite photographs taken between the same geographic area in 2013 and 2021.

In second step, I_1 image has been subtracted from I_2 image and image differencing (D) is obtained.

$$J = \sum_{j=1}^{k} \sum_{i=1}^{n} ||x_i^{(j)} - c_j||^2 \quad (1)$$

$$D = I_2 - I_1 \quad (2)$$

In the third step, ID has been divided into $h \times h$ non-overlapping blocks. Thus, a new vector set (V) is obtained from pixel values in blocks belonging to ID [19]. In order to shift the data to the center, with the average of the pixel values in $h \times h$ non-overlapping blocks, obtained in the previous step, average vector set are obtained.

Table 1 Literature review

Author(s)	Methods used	Advantages	Disadvantages
Kesikoglu et al. [19]	Fuzzy C-means and PCA	• Gives best result for overlapped dataset • Well-established mathematical technique for reducing the dimensionality of the data	• The performance of the FCM algorithm depends on the selection of the initial cluster center • PCA is sensitive to outliers
Ratha et al. [20]	K-means clustering	• Differencing method performs better than the single-channel intensity band ratio and the total power ratio	• Difficult to predict K-value
Sumaiya et al. [21]	Edge enhancement Algorithm	• Less computational cost, prefiltering is not required • Provides results directly wavelet domain	• Image may begin to look less natural, because the apparent sharpness of the overall image has increased, but the level of detail in flat, smooth areas has not
Marpu et al. [22]	IR-MAD	• Can converge to a better no change background even in the presence of huge changes by using the proposed initial change mask	• Fails to converge background when – there are large number of change pixels – no mask is used
Celik [23]	K-means clustering	• Less prone to errors caused	• Thicker boundaries of the resultant change map
de Bem et al. [24]	CNN, SharpMask, U-Net, and ResNet	• It provides fast and precise segmentation of image • Networks with large number of layers can be trained easily without increasing error percentage	• Computationally expensive • Deeper network usually requires weeks for training

The mathematical notation of the process is shown in the equation with number 3. (μ) is average vector set, and Q is the data shifted to the center.

$$\mu = \frac{1}{M} * \sum_{i=1}^{M} V. \tag{3}$$

The covariance matrix is computed in the fourth stage according to Eq. 4. You can obtain eigenvalues and eigenvectors with it. The term Q^T in the equation denotes the centering of the data through transposition.

$$Cov = \frac{1}{N} * \sum_{i=1}^{N} Q * Q^T. \tag{4}$$

In the fifth step, eigenvalues (Evals) are obtained by using covariance matrix. After that, Evals are sorted in decreasing order and three largest Evals are selected [19]. Then, Eigenvectors (Evecs) corresponding to three largest Evals are obtained. PCs including most of the information are some of the first components. Therefore, the three largest components are selected.

In the sixth step, PCs are obtained and feature vector space (K) is composed. This process is shown in the equation below:

$$K = Q^T * \text{EVec}. \tag{5}$$

In the seventh step, with the help of the technique of feature space formed by PCs has been divided into two classes and identifying the place where there is change and there is no change has been accomplished. Consequently, the areas which have change and do not have change are identified. According to the results in K-means technique, we will see that % part of our study area has changed or not. While the areas having change are white, the areas having no change is black in image.

4 Architect Diagram

By observing Fig. 1, proposed system follows the preprocessing of two timeline satellite compressed images. Then, create the difference images of two timeline satellite images. From the difference image, create $h \times h$ non-overlapping blocks. Then, PCA is implemented to construct eigenvector space. Then, using eigenvector space, feature vector space is generated. K-means clustering technique is used to generate two clusters, one cluster is for unchanged pixels and another one is for unchanged pixels. Then, using Euclidean distance, assign each feature vector to the cluster closest to it. Lastly, a change map is created.

5 Data Extraction and Preprocessing

There are numerous ways to retrieve data from Landsat satellite images. In order to extract satellite images, the US Geological Survey (USGS), a government-run geological survey provides assistance [2, 25]. It is necessary to stack all of the band

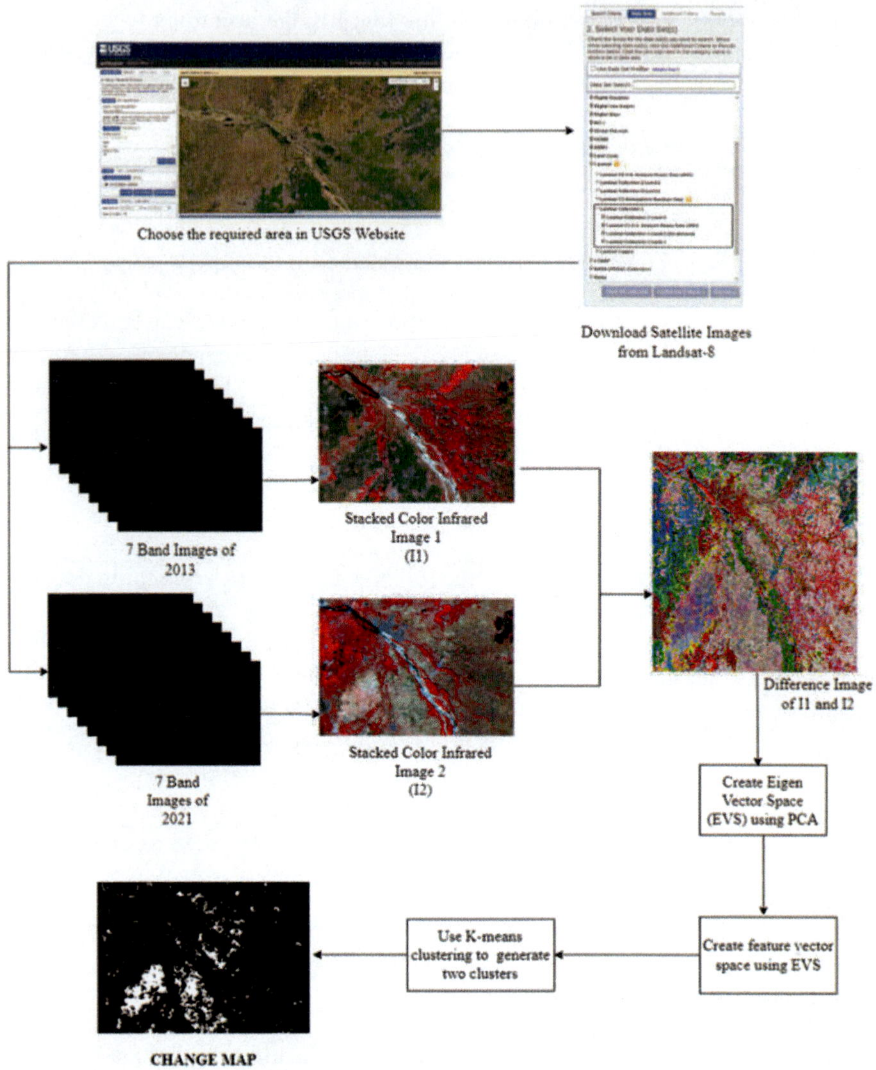

Fig. 1 Proposed architecture diagram

pictures from Krishna District's Landsat Collection 1 Level 1 in the years 2013 and 2021 after downloading them all. There are numerous Geographic Information Service (GIS) tools, such as ArcGIS, QGIS, for stacking and preprocessing. QGIS is a cross-platform desktop geographic information system application that allows you to view, edit, and analyze geographical data for free. In QGIS, there is plugin called Semi-Automatic Classification Plugin (SCP) which does the preprocessing process. Clip the satellite image to the required area and add reflectance to the image, it helps to sharpen the image.

6 Data Analysis

The sharpen images can be stacked according to the required band combinations of Landsat 8. The following Table 2 shows the band designations of Landsat 8 satellite [25]. Nine spectral bands are produced by the Landsat 8 Operational Land Imager (OLI) (Bands 1–9). The Landsat 8 bands from the OLI sensor are coastal, blue, green, red, NIR, SWIR-1, SWIR-2, and cirrus. On the other hand, the panchromatic band offers a finer resolution of 15 m [25]. Using two thermal bands, the Thermal Infrared Sensor (TIRS) gauges the thermal energy of the planet (bands 10 and 11). Both TIRS bands are long-wavelength infrared with a 100-m resolution.

Because vegetation may be easily seen in the image using this band combination, color infrared composite is utilized to display the color change because this paper primarily focuses on land cover. Urban areas are cyan blue or occasionally can appear yellow or gray, depending on their composition. Vegetation emerges in shades of red. Soils range from dark to light brown. Ice, snow, and clouds are a light cyan or white color. Coniferous trees will appear darker red than hardwood trees. Turbid water appears cyan, while clear water appears dark bluish.

Figures 2 and 3 are stacked images considered in this work.

Table 2 Band combinations and its composites for multi-spectral images

Composite name	Bands
Natural color	4-3-2
False color (urban)	7-6-4
Color infrared (vegetation)	5-4-3
Agriculture	6-5-2
Healthy vegetation	5-6-2
Land/water	5-6-4
Natural with atmospheric removal	7-5-3
Shortwave infrared	7-5-4
Vegetation analysis	6-5-4

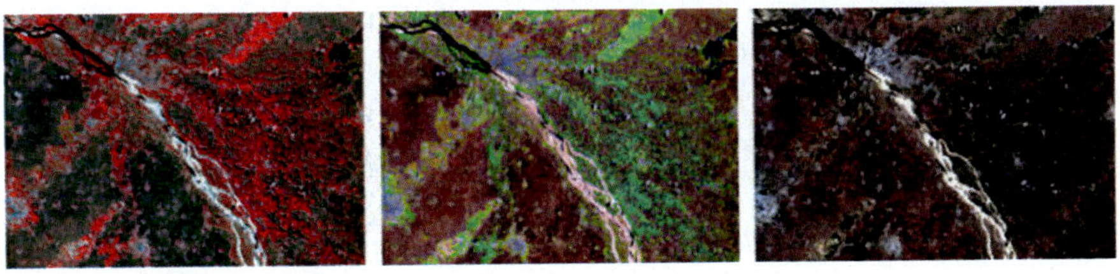

(a) (b) (c)

Fig. 2 These are the stacked images of Krishna District in the year 2013, when the composite band combinations are **a** color infrared, **b** agriculture, **c** natural color

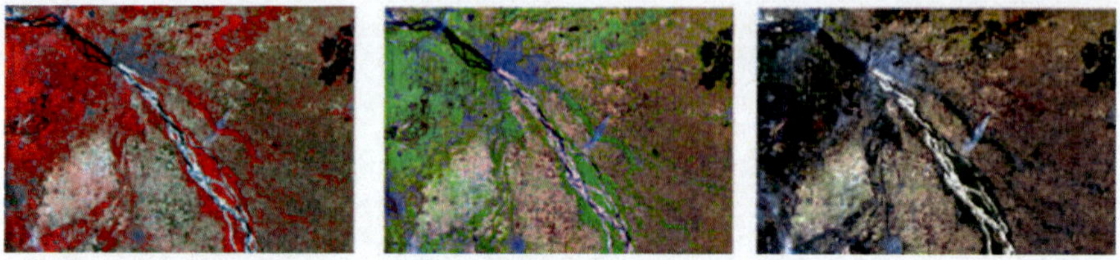

Fig. 3 These are the stacked images of Krishna District in the year 2021 when the composite band combinations are **a** color infrared, **b** agriculture, **c** natural color

7 Result Analysis

Before performing a classification, QGIS will need to know which specific portions of the image—and whose underlying data—belong to which class. The pixels of the image are divided into groups based on the type of ground cover using a remote sensing technique called classification. This is achieved by calculating the reflection values across different spectral bands. In contrast to unsupervised classification, which is the simple output of computer processing, supervised classification is based on sample classes that the user chooses. In this case, supervised classification is used. As a result, training inputs must be established. By selecting the "create training" under training input in the SCP Dock, you can create a new training input. The user chooses which classes to include and how many to have. The number of classes, on the other hand, should be maintained to a minimum for change detection. The change matrix and upcoming analysis will be increasingly difficult the more classes there are. The graphic below shows the classes that are used in this. Go to the Macro class list under training input in the SCP dock to create classes. Add classes, and then, click inside the MC info field to alter the name. Make sure that each class has a unique MC ID value like in the image below. You can adjust the map's visualization by using band rendering. Making land cover classes look more distinct from one another using this tool is helpful. For instance, the hue of the jungle and the water in the photograph below make it challenging to distinguish the two.

Visit the ROI signature list to begin adding ROIs after the classes have been created. There are two different kinds of ROIs that can be made. One method is to draw a polygon of an area that clearly belongs to a given class on your own. You can do this by selecting the "ROI polygon" symbol from the SCP toolbar. Now on the map, draw a polygon. If the ROIs are of sufficient quality, the classification can be applied to the entire image. There are now two levels of classification. Go to Postprocessing > Land Cover Change under SCP. Use the earliest classification layer as a reference and the most recent classification as a new classification when loading data. Ensure that the "report unaltered pixels" checkbox is selected as this offers crucial context for the interpretation. Next, select Run.

Land cover and land use can be detected by using land cover change matrix. In our paper, the land-cover change matrix cross-tabulates land cover at two different points in time. It shows how much of the opening stock of a land-cover category is still the same in the closing stock and the gross flows between the different categories of land cover. By using the above techniques and algorithms, we have seen that the 88% of our study area has changed. While the areas having change is white, the areas having no change are black in image. By observing Fig. 3, there are some land formation changes at the river Krishna basins, which are indicated as white visually. The unchanged land is observed in Table 3, so from unchanged, we have calculated the changed land area in m^2 and the accuracy is 89.25%.

According to Fig. 4e categorization index, the stacked images were categorized. As a result, we divided the stacked photographs into eight different classes for our paper.

Table 3 shows that land-cover change matrix cross-tabulates land cover at two different points in time. It shows how much of the opening stock of a land cover category is still the same in the closing stock and the gross flows between the different categories of land cover.

We can witness the change in land cover in the Krishna river basin at two different time stamps after performing change detection using PCA and the K-means algorithm. The final image is a grayscale image, in which the white area denotes change and the black area represents no change. The white area represents the shift in land cover along Krishna River.

The PCA and K-means algorithms were used to apply the metrics. The peak signal-to-noise ratio (PSNR) and the root mean square error (RMSE) were calculated. The difference between the source image and the segmented image is calculated using the RMSE. Between two integers, the PSNR block computes the peak signal-to-noise ratio in decibels. The original and compressed image qualities are compared using this ratio.

If RMSE is higher, the produced image will be of lesser quality. If PSNR is higher, the produced image will be of higher quality.

In Table 4, the obtained RMSE value is less and the PSNR value is more. Comparing our work with the Kesikoglu [19], the result of this paper is way better in accuracy which is increased from 78 to 89% by using GIS and because of PCA and *K*-means algorithm rather than using fuzzy C-means algorithm.

8 Conclusion

Using k-means clustering on feature vectors collected using $h \times h$ local data projection onto eigenvector space, an unsupervised change detection algorithm is constructed. PCA is used to build the eigenvector space from h × h non-overlapping difference picture blocks. The suggested technique extracts feature vectors for each pixel using $h \times h$ neighborhood, which automatically considers contextual information. Then, *K*-means clustering on feature vectors is obtained from $h \times h$ non-overlapping blocks on difference image with the use of PCA. The areas where there are change and no change are divided into two classes with *K*-means technique. While

Table 3 Land cover change matrix (m^2)

V_Reference class/new class	1	2	3	4	5	6	7	8	Total
1	15,616,800	5,434,200	2,881,800	1,768,500	261,900	1,006,200	102,600	680,400	27,752,400
2	511,200	2,169,000	871,200	7,030,800	8,748,900	13,106,700	14,210,100	3,020,400	49,668,300
3	2,594,700	6,205,500	3,041,100	13,742,100	22,103,100	63,836,100	13,497,300	6,204,600	131,224,500
4	1,151,100	2,911,500	2,374,200	40,581,900	30,042,900	276,609,600	31,707,900	67,713,300	453,092,400
5	1,161,900	4,045,500	2,237,400	57,780,900	224,537,400	703,653,300	99,948,600	60,106,500	1,153,471,500
6	274,500	270,900	235,800	2,993,400	4,208,400	35,373,600	2,274,300	1,737,900	47,368,800
7	979,200	2,013,300	1,127,700	30,609,000	71,688,600	208,940,400	167,011,200	41,271,300	523,640,700
8	3,061,800	5,828,400	2,133,900	8,839,800	28,646,100	63,859,500	28,475,100	66,003,300	206,847,900
Total	25,351,200	28,878,300	14,903,100	163,346,400	390,237,300	1,366,385,400	357,227,100	246,737,700	2,593,066,500

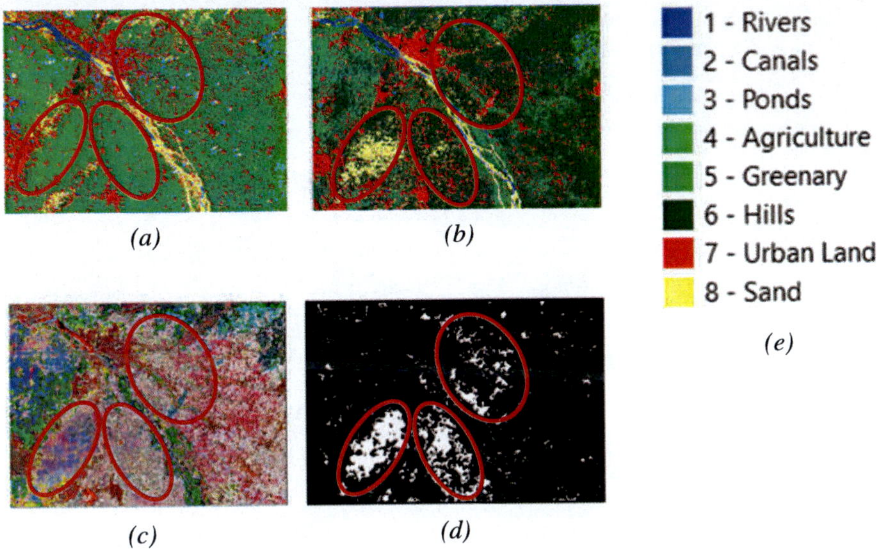

Fig. 4 **a** Corrected classified image of Krishna District in 2013. **b** Corrected classified image of Krishna District in 2021. **c** Difference image of Krishna District at different timelines 2013–2021. **d** Change map of Krishna District at different timelines 2013–2021. **e** Classification index

Table 4 Obtained metrics

Algorithm/metrics	RMSE	PSNR
PCA and *K*-means	0.040098414	27.937455212382844

identifying changing, Landsat 8 OLI images belonging to 2013 and 2021 years have been used. As a result, it is observed that 22% of the area has changed.

References

1. Mridha N, Chakraborty D, Roy A, Kharia SK, Role of remote sensing in land use and land cover modelling
2. Ayele GT, Tebeje AK, Demissie SS, Belete MA, Jemberrie MA, Teshome WM, Mengistu DT, Teshale EZ (2018) Time series land cover mapping and change detection analysis using geographic information system and remote sensing, Northern Ethiopia. Air Soil Water Res 11:1178622117751603
3. Abijith D, Saravanan S (2021) Assessment of land use and land cover change detection and prediction using remote sensing and CA Markov in the northern coastal districts of Tamil Nadu, India. Environ Sci Pollut Res 1–13
4. Aldhshan SRS, Shafri HZM (2019) Change detection on land use/land cover and land surface temperature using spatiotemporal data of Landsat: a case study of Gaza Strip. Arab J Geosci 12(14):1–14
5. Talukdar S, Singha P, Mahato S, Pal S, Liou Y-A, Rahman A (2020) Land-use land-cover classification by machine learning classifiers for satellite observations—a review. Remote Sens 12(7):1135

6. Vali A, Comai S, Matteucci M (2020) Deep learning for land use and land cover classification based on hyperspectral and multispectral earth observation data: a review. Remote Sens 12(15):2495
7. Munyati C (2004) Use of principal component analysis (PCA) of remote sensing images in wetland change detection on the Kafue Flats, Zambia
8. Vignesh T, Thyagharajan KK, Ramya K (2019) Change detection using deep learning and machine learning techniques for multispectral satellite images
9. Panuju DR, Paull DJ, Griffin AL (2020) Change detection techniques based on multispectral images for investigating land cover dynamics
10. Wang SW, Gebru BM, Lamchin M, Kayastha RB, Lee WK (2020) Land use and land cover change detection and prediction in the Kathmandu district of Nepal using remote sensing and GIS. Sustainability 12(9):3925
11. Karan SK, Samadder SR (2016) Accuracy of land use change detection using support vector machine and maximum likelihood techniques for open-cast coal mining areas. Environ Monit Assess 188(8):1–13
12. Chang N-B, Han M, Yao W, Chen L-C, Xu S (2010) Change detection of land use and land cover in an urban region with SPOT-5 images and partial Lanczos extreme learning machine. J Appl Remote Sens 4(1):043551
13. Bruzzone L, Prieto D (2000) Automatic analysis of the difference image for unsupervised change detection. IEEE Trans Geosci Remote Sens 38(3):1171–1182
14. Bidari I, Chickerur S, Talikoti RM, Kapali SS, Talawar S, Sangam S (2020) Performance analysis of change detection algorithms on multispectral imagery. In: 2020 12th International conference on computational intelligence and communication networks (CICN)
15. Katarki G, Ranmale H, Bidari I, Chickerur S (2019) Estimating change detection of forest area using satellite imagery. In: 2019 International conference on data science and communication (IconDSC), pp 1–8
16. Perez D, Lu Y, Kwan C, Shen Y, Koperski K, Li J (2018) Combining satellite images with feature indices for improved change detection
17. Bovolo F, Bruzzone L (2015) The time variable in data fusion: a change detection perspective. IEEE Geosci Remote Sens Mag
18. Chickerur S (2019) Introductory chapter: high performance parallel computing. High Performance Parallel Computing Satyadhyan Chickerur IntechOpen. (online) Available: https://www.intechopen.com/books/high-performance-parallel-computing/introductory-chapter-high-performance-parallel-computing
19. Kesikoglu MH, Atasever UH, Ozkan C (2013) Unsupervised change detection in satellite images using fuzzy c-means clustering and principal component analysis
20. Ratha D, De S, Celik T, Bhattacharya A (2017) Change detection in polarimetric SAR images using a geodesic distance between scattering mechanisms. IEEE Geosc Remote Sens Lett 14(7):1066–1070
21. Sumaiya MN, Shantha Selva Kumari R (2014) Unsupervised edge enhancement algorithm for SAR images using exploitation of wavelet transform coefficients. In: 2014 International conference on communication and network technologies (ICCNT)
22. Marpu PR, Gamba, P, Canty MJ (2011) Improving change detection results of IR-MAD by eliminating strong change. IEEE Geosci Remote Sens Lett 8
23. Celik T (2009) Unsupervised change detection in satellite images using principal component analysis and k-means clustering. IEEE Geosci Remote Sens Lett 6(4):772–776
24. de Bem PP, de Carvalho Junior OA, Fontes Guimarães R, Trancoso Gomes RA (2020) Change detection of deforestation in the Brazilian amazon using landsat data and convolutional neural networks. Remote Sens 12(6):901. https://doi.org/10.3390/rs12060901
25. GIS Geography Landsat 8 bands and band combinations article; website link: https://gisgeography.com/landsat-8-bands-combinations/#:~:text=Landsat%208%20bands%20from%20the,finer%20resolution%20of%2015%20meters

Hybrid Intermission—Cognitive Wireless Communication Network

M. Bindhu, S. Parasuraman, and S. Yogeeswran

Abstract We investigate a wirelessly powered cognitive wireless communication network (CWPCN) for Internet of Things (IoT) applications that consists of a main and secondary communication system. HTT and backscatter communication (BackCom) are two innovative methods that we propose for the secondary communications system. Cognitive users have two alternatives when a main channel is overcrowded and the information receiver is unable to receive information: backscattering or collecting energy for future information transfers. Both are possibilities. To transfer information when the main channel is inactive, CUs employ bistatic scatter (BS) and high-frequency transmission (HTT) modes. Time allocation between AB mode and energy harvesting, as well as BS mode and HTT, is studied in order to maximise secondary communication system throughput. To be more specific, we discover the optimum closed-form solution for a particular CU situation as well as the best combination of operating modes. According to numerical statistics, our proposed hybrid HTT and BackCom mode beats the benchmark mode.

Keywords BackCom · High-frequency transmission · Cognitive wireless communication

1 Introduction

In the Internet of Things (IoT) network, smart devices may communicate with one other and work together to improve our everyday lives [1–3]. Agriculture, smart

M. Bindhu (✉)
ECE, Sri Venkateshwara College of Engineering, Sriperumbudur, Chennai, Tamil Nadu 602117, India
e-mail: deepbind@gmail.com

S. Parasuraman
ECE, Karpaga Vinakaga College of Engineering and Technology, Chengalpattu, Tamil Nadu, India

S. Yogeeswran
ECE, P.T. Leechengalvarya Naicker College of Engineering and Technology, Veliyur, Tamil Nadu, India

S. Das et al. (eds.), *Advances in Data-driven Computing and Intelligent Systems*, Lecture Notes in Networks and Systems 653, https://doi.org/10.1007/978-981-99-0981-0_61

homes, health care, and public safety have all benefited from its usage in recent years. The limited network life cycle is one of several challenges that large-scale IoT adoption still confronts. A battery swap may be complicated when there are a large number of IoT devices, particularly if the batteries are small [2]. Now that energy harvesting has emerged as a practical energy source for IoT devices, it has the potential to significantly increase network life. A number of IoT devices may use EH techniques to capture wind and thermal energy and even radio frequency (RF) data. There has been a recent surge in interest in using RF signals to power IoT devices because of their reliability and ease of management. There has recently been a lot of research done on Wi-Fi-powered communication networks (WPCNs), which are often utilised in smart homes and agriculture, on the basis of RF-WPT. The authors in [4] suggest using a "harvest- then-send" (HTT) mode with the hybrid access point (HAP) to transmit data from IoT devices that do not need electricity.

For the HTT mode to work, there must be energy harvesting time, which may reduce the transmission time of information. By employing two antennas, a full-duplex (FD) WPCN was built in [5] by simultaneously transmitting and receiving information signals from IoT devices in order to boost system performance. There was a proposal in [6] for a user cooperative strategy to overcome the possible issue of WPCN IoT devices having a two-fold near-far disadvantage. By using the An energy-free IoT device and relay work together to first collect energy from the HAP before collaborating to send information in a mode called "harvest-then-cooperate" (HTC).

IoT network performance has been greatly boosted by the cooperative strategies presented in [6, 7]. Backscatter communication (BackCom) has, on the other hand, gotten a lot of interest since it allows IoT devices without inbuilt power supply to communicate [8–11]. Instead of actively radiating signals via variable antenna impedance, BackCom works by reflecting the incident signals. Carriers and digital modulation may be created without the need of ADCs and oscillators, which reduce the overall power consumption of the integrated circuit.

In the BackCom mode, IoT devices cannot communicate with each other if incident signals are absent. BackCom may be integrated into WPCNs to improve IoT applications like urgent data transmission and ubiquitous device connection, which is desired because of the advantages it provides. In the context of wirelessly powered heterogeneous networks, both HTT and BackCom IoT devices were examined in [12, 13]. Depending on whether a dislocated PB and an information receiver (IR) are used in BackCom-aided WPCNs or not, each IoT device operates in BackCom mode or HTT mode [14, 15]. According to [16], the hybrid BackCom mode defined by the authors for the WPCN may be activated while IoT devices are in BS or AB mode. Another concern for IoT is spectrum. Frequency band slicing is getting increasingly difficult due to the rapid growth of the Internet of Things (IoTs). IoT device cognitive radio capabilities are quickly rising to the top of research priorities [17]. Lee and Zhang [18] looked into how a secondary WPCN with a cognitive radio network (CRN), known as cognitive WPCN, may be used for the Internet of Things (IoT) (CWPCN). Devices in the secondary system, known as cognitive users (CUs), communicate data to the HAP while drawing energy from the secondary system's

HAP and the primary transmitter (PT). CWPCNs [19] have developed algorithms for managing online energy and spectrum. AB mode is used when the main channel is congested; HTT mode is used when it is not. In the CU, you may use both types of BackCom. AB and HTT modes were compared to see which one provided the best balance of throughput and power consumption. The paradigm in [20] can more practically be stretched to multiple CUs through IoT applications. During the "HTT-BackCom" phase of this project, a wireless sensor network (WSN) was used as an additional communication method. For example, in figures [20, 21], we may employ both the AB and BS modes to increase the transfer time by utilising the PT and an additional packet buffer. CUs must share spectrum with the principal communication pair when it comes to spectrum allotment [18]. We propose a hybrid HTT-BackCom mode if the PT is only active for a short period of time and the network can accurately identify its working status. Among them are [17, 18].

Here is a quick rundown of the paper's most significant contributions. HTT and BackCom modes may be combined for IoT applications in a hybrid HTT-BackCom mode that is dependent on the condition of the main channels (idle/busy). A good starting point is to compare how much throughput can be achieved using the AB and HTT modes in conjunction with energy harvesting vs. the BS and HTT modes. For a single CU instance, the KKT conditions provide the optimal closed-form solution, which exposes the optimum combination of the CU's operating modes. A significant improvement in secondary communication throughput may be achieved using the quantitatively proved hybrid mode with the optimal time allocation.

2 Model

Wireless charging, SU transmission, and various network and channel types, as well as relevant performance indicators, are all addressed in this section. Probability and transmission power are often denoted by the letters p and q, respectively, with the necessary subscripts. In addition, uppercase letters are used to denote random variables. In addition, the probability of occurrence is denoted by $P(\xi)$, and the expectation operation is shown by ξ and $E[\cdot]$.

2.1 Network and Channel Models

Wireless SUs use the licenced spectrum for PUs to their advantage, as seen in Fig. 1, in a cognitive WPCN. There are discrete intervals of time, each lasting a fixed amount of time. There are two important differences between our system model and that of [44]: (1) SUs are wirelessly powered by their receiver and (2) the SUs are not saturated. We use a two-dimensional HPPP model in which the SUs and PUs are distributed as separate HPPPs with spatial intensities equal to zero. To keep things simple, just a subset of PUs are really transmitting data at any given time; this subset

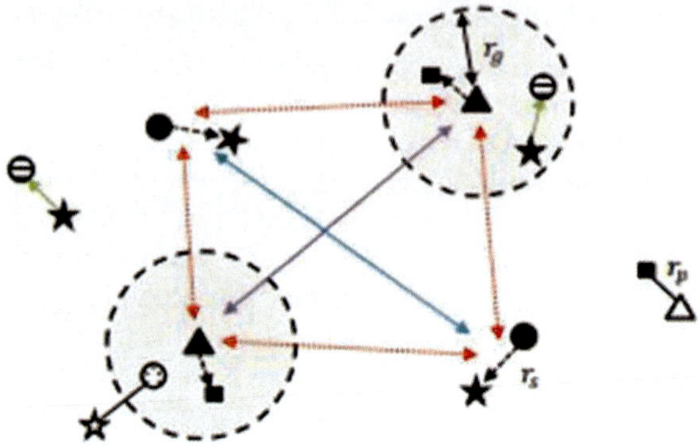

Fig. 1 Depicts a network of cognitive wireless communication devices

is selected by thinned HPPPs. Colour Theorem states that this subset creates another HPPP with spatial intensity $\Lambda p \leq \Lambda\ 0\ \mathrm{p}$ that is equal to or less than the original HPPP [21]. For the rest of this article, we will refer to the active PUs as PUs. If you want to know how much time it takes for an object to reach its full potential, you need to know how much time it takes for an object to reach its full potential (HPPP) $\Phi p = \{\times 1,\ \times 2, \ldots\} \subset R\,2$ and $\Phi s = \{y1, y2, \ldots\} \subset R\,2$. This article uses a bipolar network model to show that PUs send power *qp* to their intended receivers, which are randomly positioned at a distance *rp* from each other and at a fixed location. Thus, SUs send out power *qs* to their designated receivers, known as PBs, which are positioned at random distances *rs* distant from the particular SU's location. In the bipolar paradigm, the association between SUs and PBs is already established. As a convenience, we presume that a "guard zone" can be established around the PU receiver; however, this is not essential. A transmitter is thus more simpler to detect than a receiver (for example, by use of a simple energy detector) [10]. Let's presume that rgrp is utilised to protect the receiver of the PU from damage.

Because of the large-scale route loss, all channels are treated as quasi-static Rayleigh fading ones. The Euclidean distance between any two nodes in either the in-band or out-of-band WPT is the channel's effects. In-band and out-band loss are carried in SU, PU segments. Wireless charging on SUs and PUs occurs on a different frequency than the data transmission frequency utilised by these devices.

2.2 SU Transmission Model

For simplicity, we will assume that packets arrive at each source SU in a time-slot distribution with a probability of d 1, which is the pace at which new packets arrive. Its data buffer is large enough to hold the packets, and the FIFO queue discipline ensures that all packets are processed in the order they are received. To determine the likelihood that an SU's data queue has at least one data packet, we can use the formula [22].

$$\rho = \lambda d \cdot E[S],$$

For purposes of the next subsection, we shall use S to denote the SU's total service time. To make matters worse, even if a supercapacitor or other kind of energy storage is used, its capacity is restricted to the amount of power qs necessary to transmit a single slot. As previously mentioned, the charging procedure is started when an SU has a packet to broadcast. It begins transmitting data as soon as the battery is completely charged; otherwise, it remains silent until it exits a guard zone. Every transaction is limited to the transmission of a single data packet. The likelihood that a typical SU (y) is outside of any guard zone is equal to the probability that there is no PU in the disc centring at y with the radius rg. This disc is referred to as b y, rg. This Poisson random variable has an average value of r 2 Gp and is distributed according to the probability distribution of the number of pseudounits (PUs) in it.

If the following three requirements are met, the conventional SU is in transmission mode:

(1) It is equipped with a transmission package.
(2) When it has at least one packet to transmit, it has the energy it needs.
(3) It is not protected in any way.

3 Throughput Optimization Using Cooperative Spectrum Sensing

3.1 *Problem Formulation*

Cooperative spectrum sensing may benefit from the use of counting rules, as has been shown. Cooperative spectrum sensing throughput is maximised by the design of the fusion rule and the optimization difficulties across the sensing period in a frame. Because of the shadowing and fading that multipath signal transmission may cause, local spectrum sensing techniques lose some of their effectiveness. All local SS approaches, such as ED and MFD, have their own viewpoints and limits; therefore, there is no one scheme for all applications or circumstances. Because of this, CSS aims to take advantage of the wide variety of SUs throughout the world and finally leads to a worldwide choice [23–26]. According to the SU's method of disclosing detection data, centralised, distributed, and relay-assisted CSS technology may be divided into three main categories.

3.2 *Throughput Maximisation*

Problem (20) can be broken down into two suboptimal issues using our decoupling technique. Our data frame can only contain a certain number of symbols; therefore, we can apply a formula to get the optimal sensing time for each one of k. We can then compare the optimal sensing durations to determine which is best. The fusion rule-related parameter k is provided for the suboptimal problem A. The ideal sensing duration cannot be expressed in a closed form. However, under Assumption A, it is possible to show that 0 c T is a unimodal function of x and hence the continuous alternative to n. Within this sub optimal functions the time is calculated.

4 Proposed Method—Throughput Maximisation

There are "K" SUs and "C" clusters in the proposed system, depicted in Fig. 2. As a way of expressing this, "Ni" stands for "number of subclasses" in a cluster. There are two clusters in the system, each with seven participants. There are three parts to the suggested work: an SU-level decision, a CH-level decision, and a CH-level fusion decision. It is necessary to do the energy detection procedure in order to gather information about the channel availability in order to make a local choice.

As part of the global decision-making process, the NN fusion technique is used. Using a cluster-based strategy, all secondary users in the existing network are prioritised based on their needs.

Scheme: proposed clustering algorithm:

Stage one: local decision.

The PU channel signal is picked up by each SU.

At the same time, each SU computes an estimate of its SNR. The threshold should then be applied. PU is missing if the SU energy is lower.

- PU is present if the SU energy is high.

A choice and SNR value are sent to the appropriate cluster leader by each SU.
Stage two: cluster decision.

NN is employed at the CH level to make decisions.

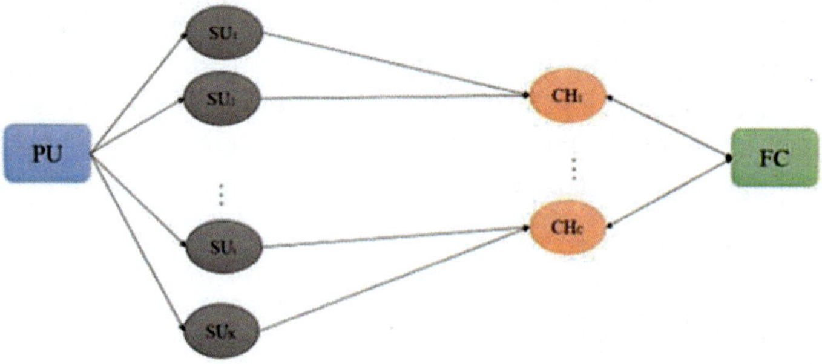

Fig. 2 Schematic of proposed diagram

- Make use of the local data to train the suggested NN-based model on the gathered characteristics and associated goals.
- Input features [SU-SNR and SU-local decision value]
- Targets [1—PU is present/0—PU is absent]

Each CH sends a message to the FC with its decision. Stage three: global decision

At FC, a worldwide decision is made using NN. NN is used to make a global decision at FC. Classifying data and determining the condition of a certain channel are done by FC.

- In order to inform the SUs of FC's final decision, CH is used.

5 Results

This section uses numerical simulations to assess the proposed hybrid HTT-BackCom mode. The benchmark's numerical results are also included for comparison. In our numerical analyses, the PT transmits signals at 915 MHz with a bandwidth of 6 MHz, whereas the PB transmits signals at 2.4 GHz with a bandwidth of 20 MHz. A total of 6 dBi of antenna gain is used across all CUs, whereas $0 = 1$ and $1 = 1$ are used to establish the BS mode reflectivity coefficients, respectively. About 1.1 dB of power is lost due to scattering efficiency. Otherwise, we set Pp = 17 kW, Ph = 20 dBm, the distance between Pp and Ph is 1 km, the distance between Ph and Ph is 1.5 m, and the distance between Ph and Ph is 2 m. In terms of system throughput, our suggested hybrid HTT-BackCom method is superior to the classic HTT mode.

The main channel's throughput is shown in Fig. 3 as a function of idle time. Figure 4 depicts the hybrid HTT-BackCom mode, which outperforms the HTT method in throughput. Clearly, this works in the way that it does. In the proposed hybrid HTT-BackCom mode, when the main channel is busy, AB mode is always utilised, while HTT mode is always used when the primary channel is idle. If you want to put it another way, the amount of time spent in AB and HTT modes reduces as the temperature increases. A graph of AB mode throughput vs. backscatter rate may be seen in Fig. 4.

Figure 5 shows the PB transmit power and secondary communication system throughput. When the transmit power of the PB is raised, the BS and HTT data rates improve. U1 transfers data in HTT mode while the main channel is not in use. In order to avoid information backscatter, U1 changes to BS mode when the PB's transmit power hits a certain threshold (e.g. 30 dBi). This results in a sudden increase in the data transmission rate of the BS mode.

According to the figures in Figs. 6 and 7, throughput may be shown in relation to distances between two points. Throughput begins to decline quickly as distance increases, and then steadily. The energy collected from the PB initially decreases dramatically and ultimately becomes insignificant, as seen in Fig. 6. When $d\ 1\ u =$ 0.5 m, the suggested and benchmark modes' throughputs are nearly identical (see

Fig. 3 Throughput versus primary channel idle time

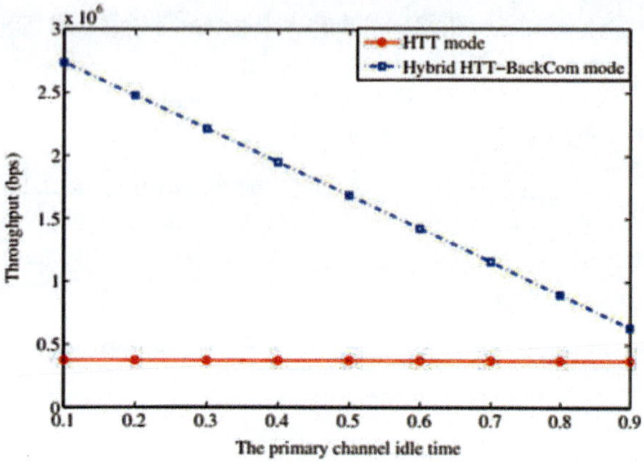

Fig. 4 Throughput versus AB mode's rate

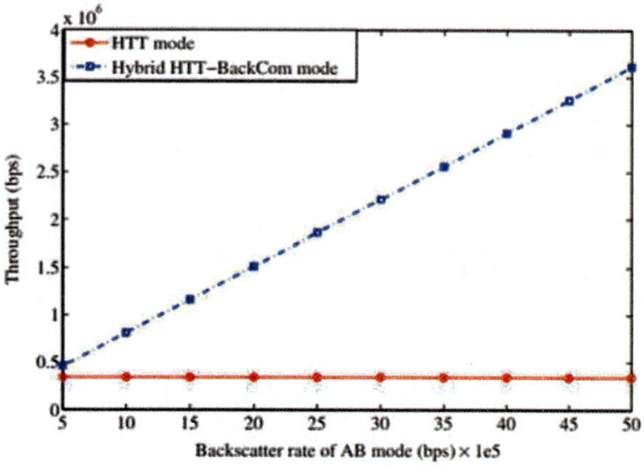

Fig. 5 Throughput versus PB's transmit power

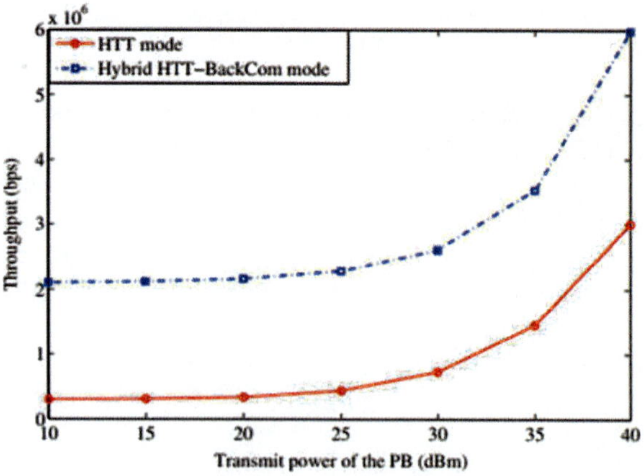

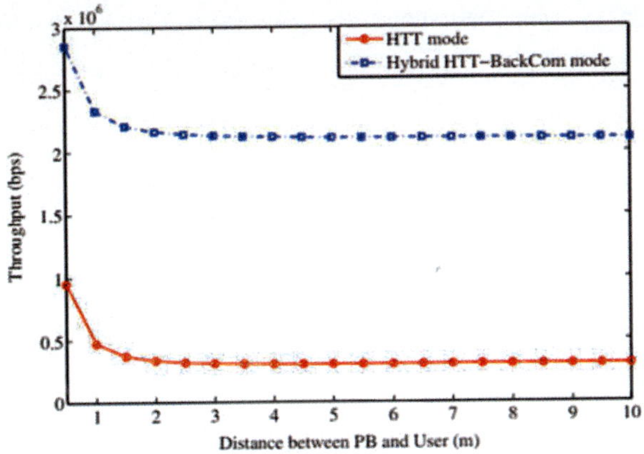

Fig. 6 Throughput versus distance between PB and user

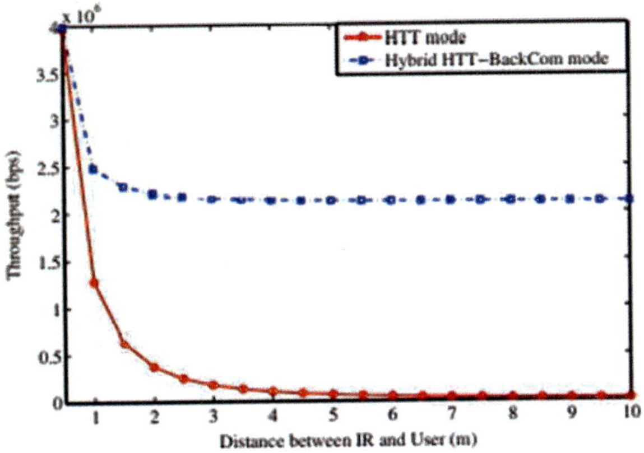

Fig. 7 Throughput versus distance between IR and user

Fig. 7). As a result, the suggested mode is dominated by the HTT mode's throughput here.

See Fig. 8 for a graph showing the relationship between throughput and the distance between the PT and the U1 points. As the distance grows, throughput decreases as seen in Figs. 6 and 7. The suggested hybrid HTT- BackCom and benchmark modes both achieve the same throughput at a reasonable distance since U1 operates entirely in the HTT mode for the proposed mode. To save energy, U1 switches from HTT mode to AB mode when it reaches a certain distance (e.g. 300 m). Graph depicting throughput and user count is shown in Fig. 9. For starters, in terms of speed and efficiency, the suggested setting exceeds the benchmark mode.

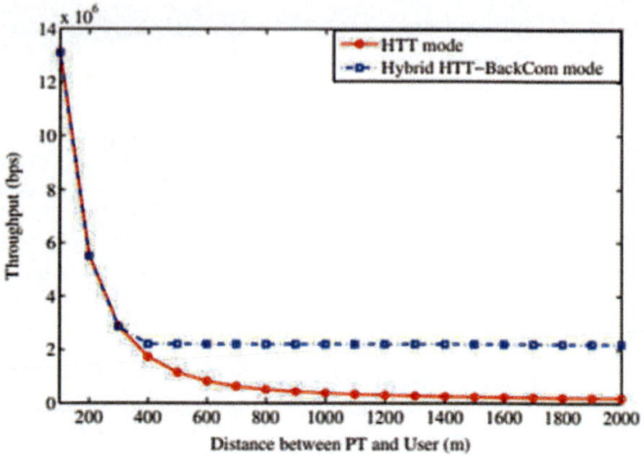

Fig. 8 Throughput versus distance between PT and user

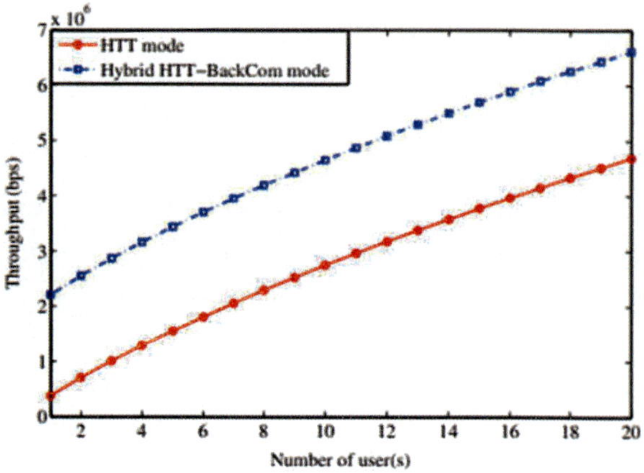

Fig. 9 Throughput versus number of user(s)

6 Conclusions

CWPCNs with IoT may now communicate with one another through HTT or AB thanks to the advent of the Internet of Things (IoT). Excitation signals for the AB and BS modes may be generated using the PT and PB in HTT mode. In order to maximise the throughput of the secondary communication system while the primary channel is idle, we investigated the optimal time allocation between AB mode and energy harvesting as well as between BS mode and HTT mode. For a signal CU condition, a closed-form solution and operating mode combination have been calculated and found. Our suggested combination of HTT and BackCom gives a greater throughput than each approach used alone, according to mathematical simulations. The PB's performance will be enhanced in the future using multi-antenna approaches.

References

1. Verma S et al (2017) A survey on network methodologies for real-time analytics of massive IoT data and open research issues. IEEE Commun Surv Tuts 19(3):1457–1477
2. Liu W et al (2017) Backscatter communication for Internet-of-Things: theory and applications. ArXiv e-prints. (Online). Available: https://arxiv.org/abs/1701.07588
3. Kawamoto Y et al (2017) A feedback control based crowd dynamics management in IoT system. IEEE Internet Things J 4(5):1466–1476
4. Ju H, Zhang R (2014) Throughput maximization in wireless powered communication networks. IEEE Trans Wireless Commun 13(1):418–428
5. Ju H, Zhang R (2014) Optimal resource allocation in full-duplex wireless powered communication network. IEEE Trans Commun 62(10):3528–3540
6. Ju H, Zhang R (2014) Uer cooperation in wireless powered communication networks. In: Proceeding of GLOBECOM, AUstin, TX, USA, pp 1430–1435
7. Chen H et al (2015) Harvest-then-cooperate: wireless-powered cooperative communications. IEEE Trans Signal Process 63(7):1700–1711
8. Kimionis J et al (2014) Increased range bistatic scatter radio. IEEE Trans Commun 62(3):1091–1104
9. Liu V et al (2013) Ambient backscatter: wireless communication out of thin air. In: Proceeding of SIGCOMM, Hong Kong, pp 39–50
10. Parks AN et al (2014) Turbocharging ambient backscatter communication. In: Proceeding of SIGCOMM, Chicago, USA, pp 619–630
11. Lu X et al (2017) Ambient backscatter networking: a novel paradigm to assist wireless powered communications. IEEE Wirel Commun 99:2–9. https://doi.org/10.1109/MWC.2017.160039
12. Lyu B et al (2017) Throughput maximization in backscatter assisted wireless powered communication networks. IEICE Trans Fundam E100-A(6)
13. Lyu B et al (2017) Throughput maximization in backscatter assisted wireless powered communication networks with battery constraint. In: Proceeding of WCSP, Nanjing, China, pp 1–5
14. Lyu B et al (2017) Wireless powered communication networks assisted by backscatter communication. IEEE Access 5:7254–7262. https://doi.org/10.1109/ACCESS.2017.2677521
15. Lyu B et al (2017) Optimal time allocation in backscatter assisted wireless powered communication networks. Sensors 17. https://doi.org/10.3390/s17061258
16. Kim SH, Kim DI (2017) Hybrid backscatter communication for wireless-powered heterogeneous networks. IEEE Trans Wireless Commun 16(10):6557–6570
17. Khan AA et al (2017) Cognitive-radio-based Internet of Things: applications, architectures, spectrum related functionalities, and future research directions. IEEE Wireless Commun 24(3):17–25
18. Lee S, Zhang R (2015) Cognitive wireless powered network: spectrum sharing models and throughput maximization. IEEE Trans Cogn Commun Netw 1(3):335–346
19. Zhang D et al (2016) Utility-optimal resource management and allocation algorithm for energy harvesting cognitive radio sensor networks. IEEE J Sel Areas Commun 34(12):3552–3565
20. Hoang DT et al (2016) The tradeoff analysis in RF-powered backscatter cognitive radio networks. In: IEEE GLOBECOM, Washington, USA, pp 1–6
21. Hoang DT et al (2017) Optimal time sharing in RF-powered backscatter cognitive radio networks. In: IEEE International conference on communications, Paris, France
22. Boyd S, Vandenberghe L (2004) Convex optimization. Cambridge University Pres
23. Ramezani P, Jamalipour A (2017) Toward the evolution of wireless powered communication networks for the future Internet of Things. IEEE Netw 31(6):62–69
24. Kawabata H et al (2017) Robust relay selection for large-scale energy- harvesting IoT networks. IEEE Internet Things J 4(2):384–392

25. Lu X et al (2015) Wireless networks with RF energy arvesting: a contemporary survey. IEEE Commun Surv Tuts 17(2):757–789
26. Lyu B et al (2018) The optimal control policy for RFpowered backscatter communication networks. IEEE Trans Vehi Tech 67(3):2804–2808

Social Media Fake Profile Classification: A New Machine Learning Approach

Nitika Kadam and Sanjeev Kumar Sharma

Abstract Social media fake profile serves various illegal social activities. Therefore, detection and prevention of these profiles are essential. The current approaches based on machine learning (ML) are just considering social media user profile attributes by providing a strict classification. This paper provides a view to utilize a scoring-based fake profile classification technique to monitor the activities of user by using profile attributes and published content. This paper first includes a review to know the dataset to be used and the technique to obtain data from social media platform. Then based on social media user's profile attribute-based ML model has been introduced to classify the fake and legitimate profiles. To train and validate the model, we have used five machine learning algorithms, namely artificial neural network (ANN), support vector machine (SVM), C4.5 decision tree, Bayes classifier, and k-nearest neighbor (k-NN). Here we have found ANN and SVM which is accurate classification technique as compared to others for this task. Finally, by updating the backpropagation neural network and a scoring method for profile a fake profile classification approach has been developed. The developed model is utilizing the content published by users and the basic profile information of public domain. The experiments have been carried out based on real twits and available profile attribute dataset in GitHub. The results are also compared with SVM and ANN algorithms. Based on the precision, recall, and F-score, the proposed technique outperforms as compared to other two implemented models and has been achieved up to 0.94 f-score.

Keywords Social media analysis · Security and privacy · Fake profile detection · Machine learning · Artificial intelligence

1 Introduction

Social networking or "online social networking (OSN)" [1] is a platform for sharing information and exploring new contacts. But there are many challenges, and it didn't

N. Kadam (✉) · S. K. Sharma
Computer Science Engineering, Oriental University, Indore, India
e-mail: kadamnitika01@gmail.com

S. Das et al. (eds.), *Advances in Data-driven Computing and Intelligent Systems*, Lecture Notes in Networks and Systems 653, https://doi.org/10.1007/978-981-99-0981-0_62

work legitimately due to some extreme issues. Among them, fake social media accounts are the key issue. At present, many social media sites include legitimate as well as malicious users. Mostly fake profiles are prepared by malicious users, when such user join social network, distribute unsocial content such as hate, porn, violence, he/she try to cheat someone and others. The fake profiles can harm an innocent user by socially and financially. The profiles can also involve in various kinds of cybercrimes. Therefore, to make the social media environment secure, we need an automated tool to analyze the social media profiles and able to distinguish between legitimate and fake user.

In this context, ML techniques are very useful which is delivering new things for humans. It is helpful in decision making, pattern recognition, predictions, and others. Due to its ability to automate data analysis, it is being accepted in different applications of engineering, medical, business, banking, and more [2]. In this paper, the ML techniques are employed in social media security by classifying social media fake profile. Based on our research, we divide user base into two categories: first who are technically sound and know the social media usages and limitations. On the other hand, some of them are naive. These "naïve users" have less experience with technology and its usages [3, 4]. So they become soft targets of attackers [5]. The goal of this paper is to classify the social media fake profiles with focused on the following objectives.

- Extraction of social media profile characteristics: in social media, profile characteristics express the nature of profile operations. Thus, social media profile attributes are identified for distinguishing fake and legitimate profile.
- The content posted by users to verify the social media user activity: The profile attributes are not enough for making conclusion of a user's profile; thus, we include the NLP-based features to analyze the structure of text. It can help to identify the context and intension of the posted contents.
- Performance analysis and selection of ML techniques to accurately classify profiles: ML techniques can be used to recognize fake profiles by using the identified features. Thus, a ML model is implemented for fake profile classification.

2 Dataset Collection

The initial training samples are required to train the ML algorithms. Once the learning is completed, we can apply the trained model for pattern reorganization. Thus, to prepare a dataset, first, we need a social media platform. That can disclose the data using some Application Programming Interface (API) for experimental use. So, some recent contributions are explored to find the dataset. Here we explored 8 recent papers based on different social media platforms as given in Table 1. The trend of social media data utilization is also provided in Fig. 1.

According to the considered research works, different social media platforms were used. Among them, Twitter dataset is frequently used. However, the proposed

Table 1 Various datasets we have used and its contributions

References	Dataset used	Contributions
[6]	Twitter	Works on the Twitter dataset for fake profile detection
[7]	Facebook	Works for fake profile detection using machine learning (ML) and natural language processing (NLP)
[8]	Spam user dataset	Publicly available dataset of 1337 fake users and 1481 genuine users were self-prepared
[9]	Twitter	Usages the Twitter dataset for fake profile analysis using profile characteristics and activity-based patterns
[6]	Twitter	The user's fake profiles are identified using Twitter dataset with two ML algorithms, namely random forest and deep convolutional neural network (CNN)
[10]	Academia	Works on social bots' identification therefore several bot datasets collected by academia have been included
[11]	Twitter	Detects the spam as well as the spammer or fake profile using the Twitter data
[12]	LinkedIn	Contributes data on LinkedIn with the clustering technique

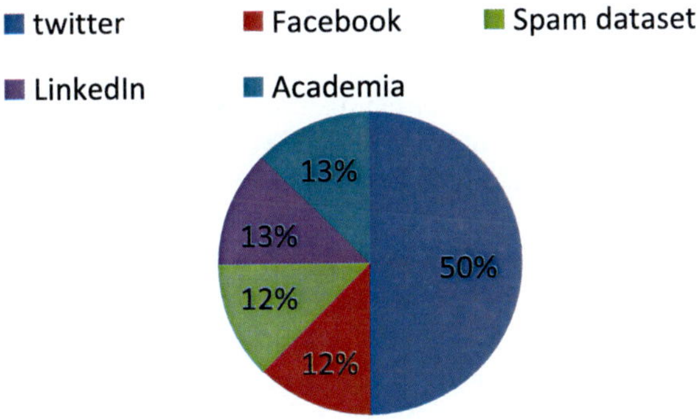

Fig. 1 Dataset used

work includes both kinds of data (i.e., profile attributes and the published contents). Thus, we need some recently published contents of users, and thus we need some API. Almost every OSN publishes its data and feeds using specific APIs. The APIs help to connect, query, and securely extract of social media data. We have decided to work with the Twitter. Therefore, the Twitter fake profile dataset is obtained from [13, 14]. We have found some tutorials also to extract the Twitter data feeds and profile information. The Twitter API describes the method for extracting the user-published data. Additionally, a dataset in Twitter API and [15] is also used. A fake profile dataset is also available for Twitter [15, 16]. This dataset is hosted on GitHub and contains 22 attributes. That dataset [16] is available in comma-separated values (CSV) format. A total of 33 attributes are available. The profile features are distributed into two separate files. The first file contains 1338 instances, and the file name is fake, and the second file is containing 1482 instances that contain legitimate profiles. After combining both the files, a total of 2820 instances and two class labels were found. The dataset contains a total of 34 attributes and two class labels. The refined dataset for experimental use is demonstrated in Table 2.

3 Proposed Work

Supervised methods are helpful for analyzing structured as well as unstructured data [2]. The proposed work is indented to utilize these methods for identifying or/and differentiating between fake and legitimate profiles and to make it suitable for every age group required a friendly and secure OSN ecosystem. According to the recent research contributions in social media fake profile detection, there are three major streams and we have found:

- Spam or unsolicited post-identification
- Detection of spam bots
- Detection of fake profile.

In the literature, most of the work focused on just profile attributes or by published contents. But there are fewer methods exist that utilize both (i.e., content and profile).

In this investigation, we are utilizing both (i.e., content and profile attributes). Additionally, some of the features are developed by own to estimate the risk level. The initial idea of the required system is demonstrated in Fig. 2. The social media experimental data is the first component of the proposed fake profile detection model. The data collection for utilizing the proposed model is described in the previous section. The data extracted from social media or from other sources may contain a significant amount of noise and unwanted data. Additionally, sometimes we found incomplete datasets that contain the missing or null values, special characters, and others. To minimize noise, data preprocessing is used. The preprocessing techniques are adopted to clean the data and improve quality. Data preprocessing from log files is time-consuming, which attempts to identify the necessity of preprocessing [17]. In the investigation, we have found several techniques available for data preprocessing. According to the findings, the technique used for preprocessing is depended on the application and the nature of data. Therefore, we include the techniques for handling the missing values and noisy attributes. During the data preprocessing, we are trying to reduce, refine, and engineer some new attributes. Therefore among 34 attributes, we consider only 17 attributes with some modification on their existing values. After consolidation and transformation, we have 17 attributes and two class labels. The list of refined attributes is given in Table 2. The preprocessed data is used for further decision-making process to identify the fake profile.

Table 2 Modified dataset for experimental use

S. No	Attribute name	Data type	Created by	Description
1	Id	Numeric	ID, Name, screen_name	Used for identifying the person or profile
2	statuses_count	Numeric	statuses_count	Number status posted
3	followers_count	Numeric	followers_count	Total number of followers
4	friends_count	Numeric	friends_count	Total number of friends
5	favourites_count	Numeric	favourites_count	Total number of favorites
6	created_at	Numeric (Days)	List_count, created_at	To know age of a profile
7	time_zone	Text	lang, time_zone and location	We consider time zone from other two
8	default_profile_image	Boolean	default_profile and default_profile_image	
9	geo_enabled	Boolean		Transformed into Boolean
10	profile_image_url	Boolean		Transformed into Boolean
11	profile_banner_url	Boolean		Transformed into Boolean
12	profile_use_background_image	Boolean	profile_use_background_image and profile_background_image_url_https	Transformed into Boolean
13	profile_background_tile	Boolean		Transformed into Boolean
14	Protected	Boolean		Transformed into Boolean
15	Verified	Boolean		Transformed into Boolean
16	Description	Boolean		Transformed into Boolean
17	Updated	Numeric (Days)		For freshness of profile

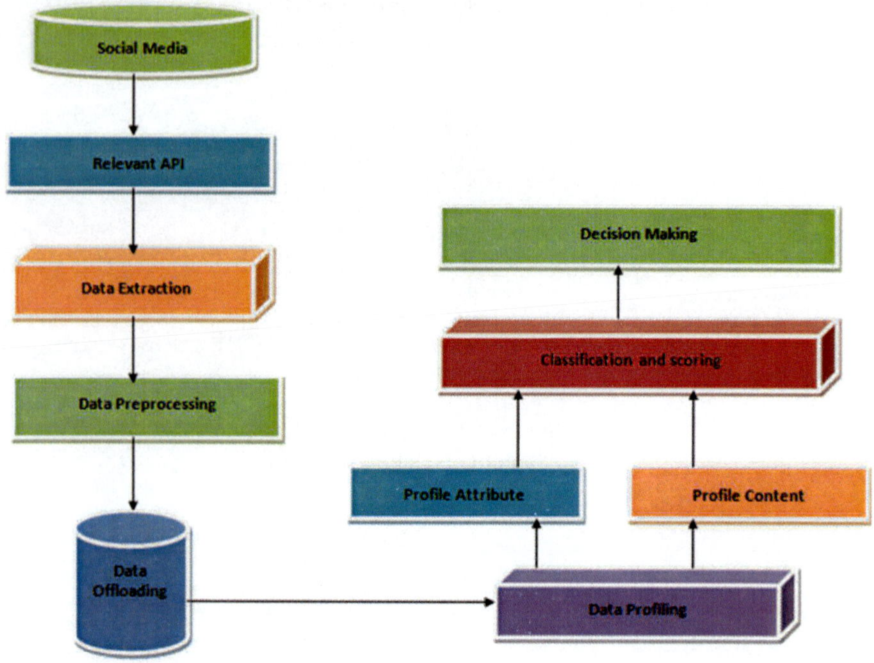

Fig. 2 Initial concept of fake profile detection system

4 Proposed Work

- **Conclusion 4.1**: Twitter is providing the APIs, and other sources for experimental datasets are also available. The preprocessing technique can impact the performance of classifier. Additionally, preprocessing of a dataset depends on the structure of data and the application. The preprocessed data is preserved in a local database. That is used to create training and testing datasets. Normally, the 70% of data is used for training and 30% is used for testing. Here we have used two different ratios for experimentation, i.e., 70–30 and 80–20. To learn on data and to classify fake profiles using only the profile attributes, an experimental model is given in Fig. 3. In order to train the system, five supervised learning algorithms are considered, namely C4.5 decision tree, support vector machine (SVM), artificial neural network (ANN), Bayes classifier, and k-nearest neighbor (KNN) algorithm. These models are utilizing the training datasets and produce a model. After learning these models, accepting the test dataset in four folds performs classification.

After employing these algorithms on the dataset, the performance is measured and a comparative performance study is reported. The aim is to find an efficient and accurate classifier for utilizing in further social media fake profile detection model

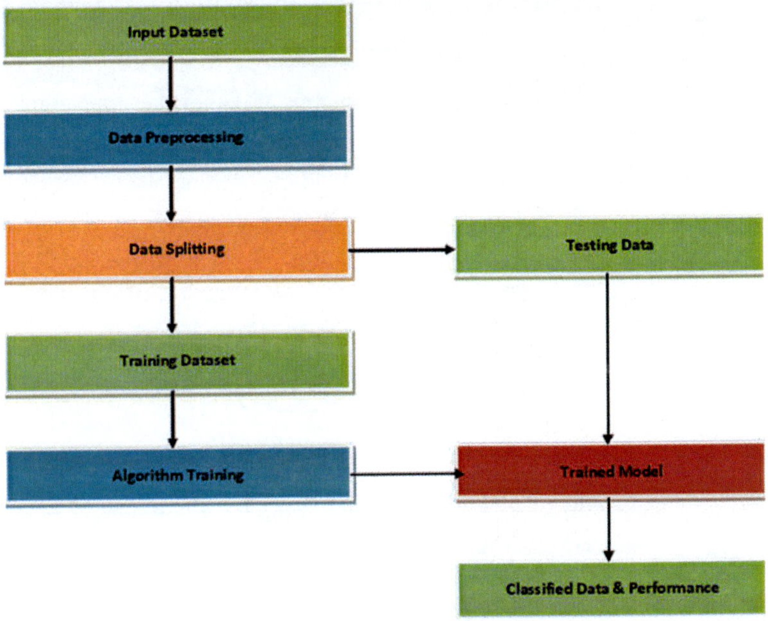

Fig. 3 Profile attribute classification model

development. Table 3 contains the algorithms for different performance parameters. Among the first parameter is accuracy, which can be explained as the measurement of correctness that can be measured using the following equation:

$$\text{accuracy} = \frac{\text{total correctly classified}}{\text{total patterns to classify}} \times 100$$

The parameter is error rate, which demonstrates the misclassification rate of the classification. That can be calculated using the following equation:

Table 3 Comparison among classification algorithms

Algorithms	70–30 Ratio				80–20 Ratio			
	Accuracy (%)	Error rate (%)	Memory (KB)	Time (MS)	Accuracy (%)	Error rate (%)	Memory (KB)	Time (MS)
C4.5	86.5	13.5	14,029	267	83.4	16.6	13,526	256
Bays	84.3	15.7	13,898	289	82.9	17.1	13,294	278
ANN	97.4	2.6	15,294	365	95.7	4.3	14,882	346
SVM	96.5	3.5	15,164	876	94.2	5.8	14,736	755
KNN	84.2	15.8	13,772	398	83.5	16.5	13,339	389

$$\text{error rate} = \frac{\text{total misclassified samples}}{\text{total samples to classify}} \times 100$$

Next the time consumption is measured which is also known as time complexity. The amount of time consumed for training is given as time consumption and measured using:

$$\text{time consumed} = \text{end time} - \text{start time}$$

Finally, the amount of memory utilized for the execution of an algorithm is measured as the memory usages. It is computed using:

$$\text{memory usage} = \text{total memory} - \text{free memory}$$

The accuracy of the algorithms is given in Fig. 4a and Table 3 for two different validation ratios (i.e., 80–20 and 70–30). The accuracy is notified on the Y-axis, and X-axis shows the validation ratios. The accuracy of the algorithm is given in terms of percentage (%). The accuracy is found effective for the 80–20 ratio as compared to the 70–30 ratio. Additionally, we found that the SVM and ANN show higher accuracy as compared to the other classifiers implemented. Thus, both the algorithms can be used for the implementation of the fake profile detection model. The error rate is given in Fig. 4b. The X-axis shows the validation ratios, and additionally Y-axis shows the error rate (%). The error rate demonstrates the ANN and SVM report fewer error rates. The less error rate is a good indicator of classification algorithm's correctness.

The performance in terms of training time is given in Fig. 4c. It is measured here in milliseconds (MS). That is the total amount of time to complete the training. The X-axis represents the validation ratios, and Y-axis shows the time in MS. According to the results, the ratio 70–30 consumes a higher amount of time as compared to 80–20 because this ratio contains fewer amounts of data as compared to the 70–30 ratio. Similarly, memory usage is also an essential parameter of performance. The memory usage is explained in Fig. 4d. The X-axis contains the validation ratios, and Y-axis shows the memory utilized in kilobytes (KB). The results show the validation ratio of 80–20 requires less amount of memory as compared to the 70–30 ratio because the ratio 80–20 requires less amount of data storage on the main memory as compared to 70–30 ratio.

- **Conclusion 4.2:** in this experiment we have refined the profile attributes. Additionally demonstrated a profile-based fake profile classification.

The classification is performed with various machine learning algorithms out of which SVM and ANN are most accurate classifiers. But the time consumption during training, SVM shows higher amount of training time; thus, for further implementation we have utilized the ANN with the modification for accurate classification.

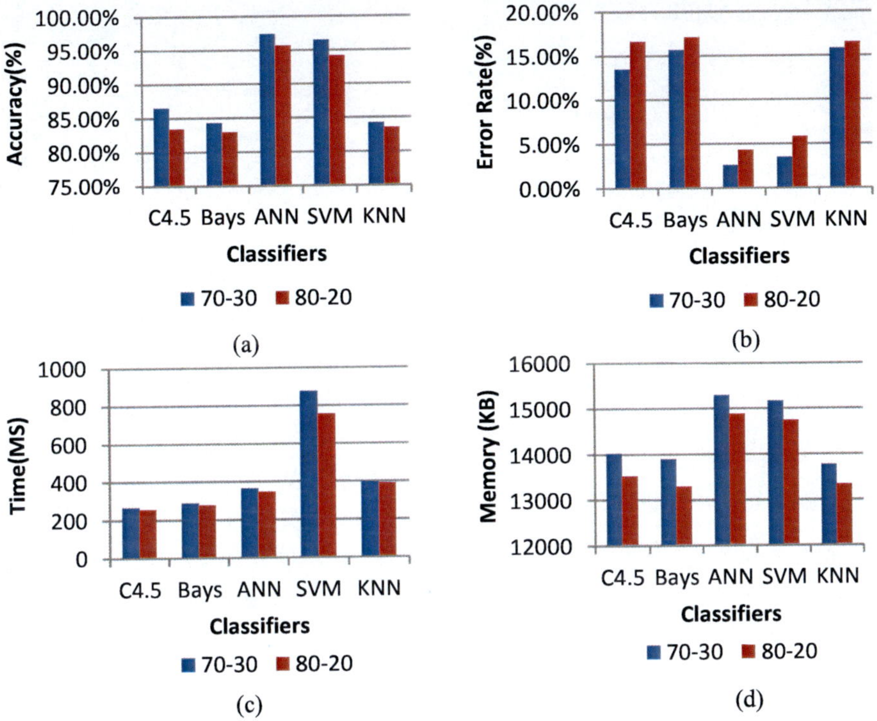

Fig. 4 Comparative performance of different ML algorithms for fake profile classification based on profile attributes in terms of **a** accuracy **b** error rate **c** time consumed and **d** memory usages

5 Proposed Classification Model

In this section, a new model for fake profile classification has been introduced as given in Fig. 5. The aim of this model is to utilize the user profile attributes as well as user-published contents to develop this model. The user profile-based dataset is described in Table 2, and the user-published contents are collected from Twitter social media using the Spark API. This API enables us to make query to the Twitter and extract the query keyword-relevant published contents. In order to utilize the extracted published contents, we first preprocess the data using a method reported in Table 4. This algorithm requires a list of stop words S_m and a list of special characters ⟦C⟧ _n and the extracted twits E_c.

Twitter data E_c is used with find and replaces function to remove all the stop words and special characters. After data cleaning, the outcome is stored in a variable P as preprocessed data. The preprocessed contents are further used with NLP parser to recover the part of speech tags (POS). The NLP parser library available for the JAVA is used to implement. It helps to parse the sentences in form of parse tree. In

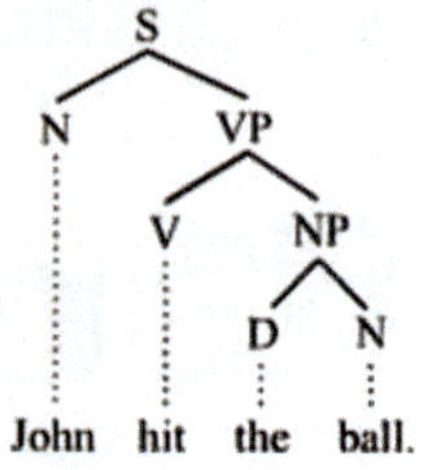

Fig. 5 NLP parser tree

Table 4 Text data preprocessing

Input: extracted contents E_c, List of Stop words S_m, List of Special characters C_n
Output: Preprocessed content P

Process:
1. $for(i = 1; i < m; i++)$
    a. $P = Find\ and\ Replace(E_c, S_i)$
2. $end\ for$
3 $for(j = 1; j < n; j++)$
    a. $P = Find\ and\ Replace(P, C_j)$
4. End for
5. Return P

this tree, the leaf nodes contain the features of the text. An example of the parser tree is given in Fig. 5.

For instance, for an input string "John hit the ball" after parsing the sentence the data is divided as noun →john, verb →hit, determiner →the, and the noun →ball. These features are computed and stored in the database as given in Table 5.

Using the NLP parser, the data is converted as vector. This vectored data is used with the profile attributes. The entire user's data (published content and profile attributes) are useful to represent the user's behavior and activity (Table 6).

The normal and fake both kinds of profiles were publishing the contents; thus for identifying the spammers and fake users, we need profile attributes and also published contents. Thus, an individual user profile attributes are paired with ten different social media posts (Fig. 6).

Thus, the profile attribute-based dataset is extended with the NLP feature set. This combined data can be used with ML classifiers to train and classify the profiles. But in place of utilizing the data with classifier, we have proposed a scoring system. That score is used to classify user profiles in terms of legitimate and fake profiles. The proposed scoring-based classification technique requires a ML classifier. Based on the recent experimental study, we have found that the ANN is accurate and efficient

Table 5 NLP features

Noun	Pronoun	Verb	Adverb	Adjective	……
1	2	1	1	1	

Table 6 Proposed scoring-based classifier

Input: Profile Attributes A and Profile Contents C Output: class labels L
Process: 1. $A_n = ReadProfileAttribute(A)$ 2. $C_m = ReadPublishedContent(C)$ 3. $PA_n = preProcess(A_n)$ 4. $PC_n = preProcess(C_n)$ 5. $TPA, TSPA = Split(PA_n, 70, 30)$ 6. $TCA, TSCA = Split(PC_n, 70, 30)$ 7. $TrainData = TPA + TCA$ 8. $TrainModel = MBPN.Train(TrainData)$ 9. $for(i = 1; i < TSPA; i++)$ a. $C_{temp} = TrainModel.predict(TSPA_i)$ b. $for(j = 1; j < TSCA; j++)$ i. $C^j_{temp} = TrainModel.predict(TSCA_j)$ c. $endfor$ d. $D = \frac{C_{temp} + C^j_{temp}}{2}$ e. $If(D \geq 0.75)$ i. $L = Legitimate$ f. $Elseif(D \leq 0.25)$ i. $L = fake$ g. End if 10. $endfor$ 11. Return L

algorithm. Thus, we modify the backpropagation neural network (BPN) train on the profile attributes and contents. The modification includes the data normalization and network initialization process. In order to normalize the data, we utilize the min–max normalization technique which is defined as:

$$\text{Norm} = \frac{\text{value} - \text{min}}{\text{max} - \text{min}}$$

Additionally, to initialize the neural network for fast convergence we utilize the normalized data with linear regression to find suitable weights to initialize the weights of BPN. After training of the modified BPN, we will use the trained model for final decision-making process. Thus, when a profile is appeared to test, we first use the profile attributes and the initial class labels are approximated and denoted as C_{temp}. After that, all the instances of published contents that belong to this profile are classified using the modified BPN. Let the target user posted N post and among them, X number of posts are recognized as legitimate, and Y number of contents as fake then

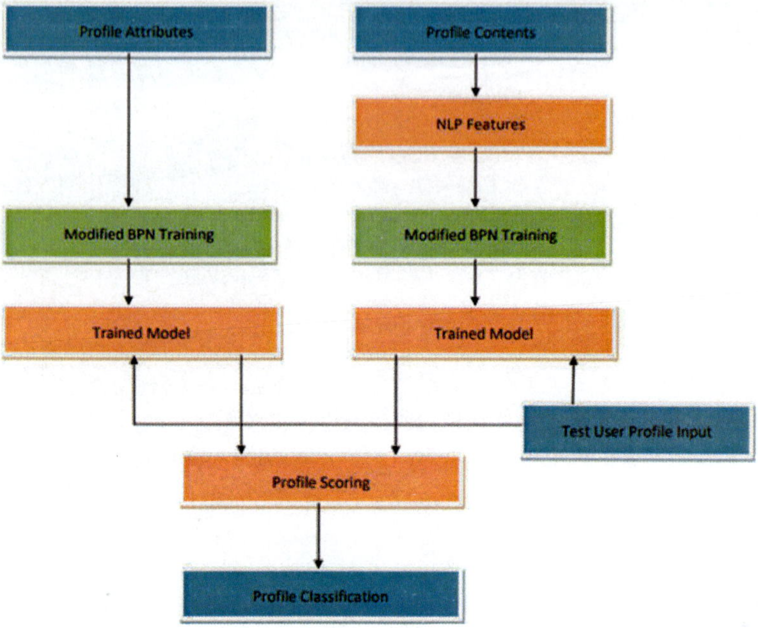

Fig. 6 Proposed fake profile classification model

$$C_{\text{temp}}^{j} = \frac{X}{N}$$

This provides a value between 0 and 1, and to estimate the final decision, we use the following equation.

$$D = \frac{C_{\text{temp}} + C_{\text{temp}}^{j}}{2}$$

Finally

$$D = \begin{cases} \text{Legitimate} & \text{if } D \geq 0.75 \\ \text{fake} & \text{if } D \leq 0.25 \end{cases}$$

That is the outcome of the fake profile detection system. Using the outcomes of analysis, the conclusion is made about the profile in terms of fake or legitimate.

This section provides details of the proposed methodology and the working progress. This model provides help in identifying the fake profiles more accurately. This model is also compared with the two popular classifiers, namely SVM and ANN in the next section.

6 Results Analysis

The experimental analysis of the proposed fake profile classification system in comparison with two popular classifiers is reported in this section. The method includes profile attributes and the content features of the OSN user for identifying them as fake or legitimate. There are two validation ratios used, i.e., 70–30% and 80–20%. Here we have proposed and implemented a new classification technique. This model includes:

- Utilization of profile attributes and published content of the user for the fake profile classification.
- Utilization of a modified BPN algorithm and scoring technique to achieve better accuracy.
- Compare the proposed technique with two popular classifier's performance, i.e., SVM and ANN.

The performance analysis of the extended fake profile classification technique using profile and content features is reported here and compared with the ANN and SVM algorithm. Thus, to measure the correctness of the models precision, recall, and f-score are measured. Precision is the fraction of relevant instances and total instances classified. That is computed using:

$$\text{precision} = \frac{\text{true positive}}{\text{true positive} + \text{false positive}}$$

The recall of a classification system demonstrates the impure outcome of the classifier. The recall is also called the sensitivity. It is a fraction of objects correctly classified, and data used for classification:

$$\text{recall} = \frac{\text{true positive}}{\text{true positive} + \text{false negative}}$$

The f-measures or F-score shows the fluctuations in the performance on the basis of precision and recall. That is also recognized as the harmonic mean of precision and recall. The f-measures can be approximated using:

$$F\text{ - measures} = 2.\frac{\text{precision} \times \text{recall}}{\text{Precision} + \text{recall}}$$

In addition, the training time and memory usage of the model has also been evaluated and compared. The obtained performance is demonstrated in Table 7. The precision shows the correctness of classification outcomes. Figure 7a demonstrates the precision of fake profile classification. The X-axis contains the algorithms used, and Y-axis demonstrates the classification precision of the implemented algorithms. Results show the proposed technique, improving the precision as compared to both traditional algorithms.

Table 7 Performance summary

S. No	Parameters	70–30			80–20		
		Proposed	ANN based	SVM based	Proposed	ANN based	SVM based
1	Precision	0.94	0.92	0.91	0.98	0.94	0.95
2	Recall	0.91	0.85	0.90	0.93	0.86	0.91
3	*f*-measures	0.92	0.89	0.91	0.95	0.90	0.94
4	Time	372	320	726	365	335	884
5	Memory	13,267	13,475	13,865	13,324	13,524	15,874

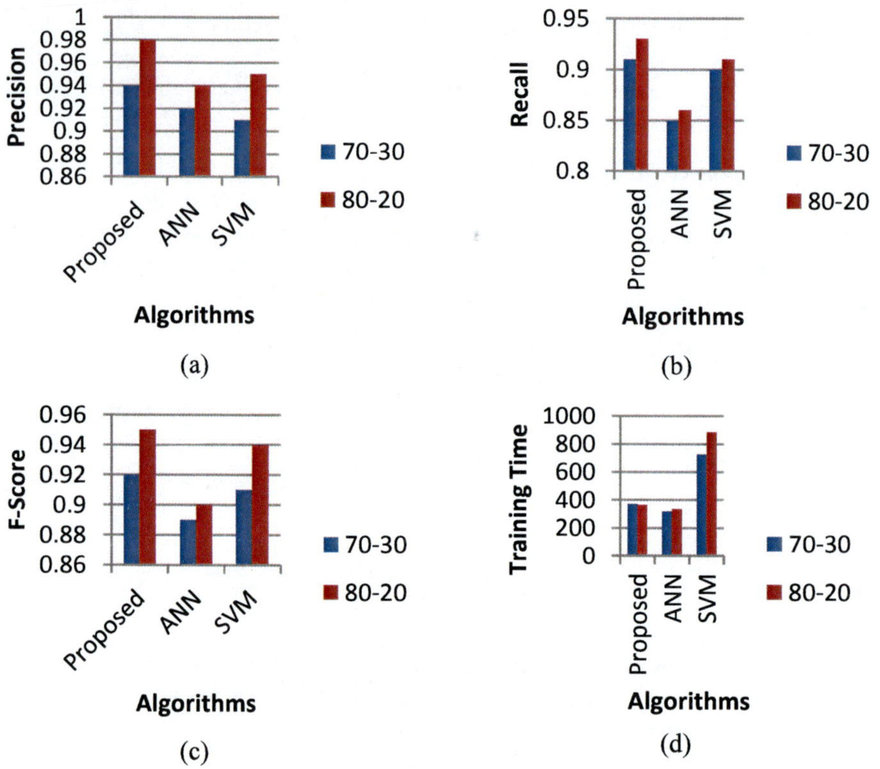

Fig. 7 Comparative performance of the proposed and traditional classification techniques for fake profile classification in terms of **a** precision **b** recall **c** *F*-score and **d** training time

The experimental results of the models in terms of recall are shown in Fig. 7b. In this figure, *X*-axis contains the employed algorithms and *Y*-axis shows the recall ratio. The proposed scoring-based technique shows improved performance as compared to the method based on SVM and ANN. The recall of the implemented algorithms is reported in Fig. 7c. Here the *X*-axis includes the algorithms used, and the *Y*-axis

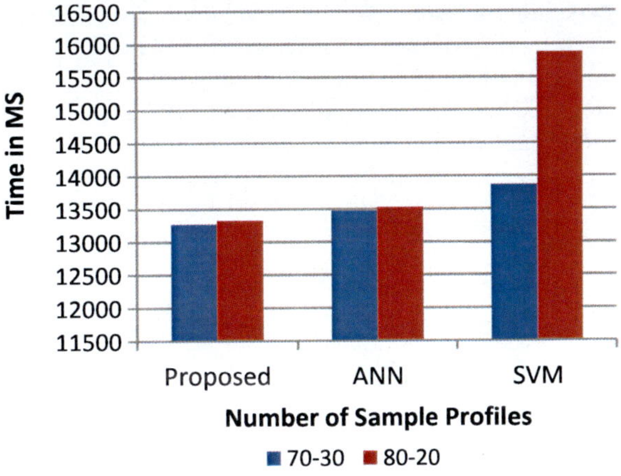

Fig. 8 Time usages

includes the *F*-measures. According to the results, the proposed technique outperforms as compared to SVM- and ANN-based techniques. Similarly, the performance of the system in terms of training time is measured and reported in Fig. 7d. In this figure, *X*-axis shows the implemented algorithm and *Y*-axis shows the training time utilized by algorithm; however, the proposed model utilizes the time higher than ANN but it is much better than the SVM-based classification approach. Here the training time is measured in terms of milliseconds (MS). Finally, the memory usages of the models are demonstrated in Fig. 8. In this figure, *X*-axis shows the algorithms used and *Y*-axis consists of the memory utilization. According to the results, the memory usages of the proposed technique are lower as compared to the SVM- and ANN-based models.

According to the results obtained based on different performance parameters, we found that the score-based fake profile classification model is working efficiently and accurately and additionally provides the better decision for managing social media surroundings healthy.

7 Conclusion and Future Work

The proposed work is aimed to deal with the OSN security and privacy issue by classifying the fake and legitimate profiles. Thus, a new classification model is designed and implemented. The proposed model provides a method to identify those social media profiles which are not utilize social media platforms legitimately. The proposed model is also promising to stop misuse of these platforms by preventing those profiles who are promoting malicious, hate, porno, phishing, and others. The basic concept is to utilize the ML techniques to analyze the OSN contents and profile attributes for discovering fake and legitimate profiles. To do this task, first a survey on existing fake profile detection techniques has been carried out and recovered frequently used platforms, datasets, data extraction techniques, and the useful classifiers. During this investigation, we have found that most of the researchers consumed the social media user profile attributes. But not effectively considering the published contents of profile. Therefore, in this paper a new technique is proposed which utilizes both profile attributes and the published contents. The prepared dataset using both the features is further used with modified BPN algorithms for accurate profile classification. Finally, the experiments have been carried out and compared with ANN and SVM classifiers. The performance is compared in terms of precision, recall, and f-score. Additionally, training time and memory usage are also measured. Based on the comparative results, the prepared fake profile classification models is efficient and accurate for identifying the profile legitimacy.

The proposed model is acceptable and provides a better solution among available techniques, but the extension on this solution may help to make more suitable models. Therefore, the following extension is proposed for the future:

- The proposed model can help to monitor the social media platforms to prevent misuse of the social media platforms.
- The proposed model is currently limited to work with the text and thus needs to include image and video analysis-based data processing techniques.
- The presented approach is not classifying profiles as hard threshold; thus, it can also be extended to notification and warning during violation of social media policies.

References

1. Guille A, Hacid H, Favre C, Zighed DA (2013) Information diffusion in online social networks: a survey. ACM 42(2)
2. Papalexakis EE, Faloutsos C, Sidiropoulos ND (2016) Tensors for data mining and data fusion: models, applications, and scalable algorithms. ACM Trans. Intell. Syst. Technol. 8(2)
3. Wüest C, The risks of social networking. Senior Software Engineer at Symantec Security Response
4. Ravichandran T, Enhancing soft skills and personality. Indian Institute of Technology Kanpur, National Programme on Technology Enhanced Learning
5. Romanov, A, Semenov A, Veijalainen, J, Revealing fake profiles in social networks by longitudinal data analysis. In: Proceeding of the 13th international conference on web information system and technology, pp 51–58. ISBN: 978-989-758-246-2
6. Shahane P, Gore D, Detection of fake profiles on twitter using random forest & deep convolutional neural network. Int J Manage Techn Eng IX(VI)
7. Dam JW, Velden M (2015) Online profiling and clustering of Facebook users. Decis Support Syst 70:60–72
8. Gayathri, A, Radhika S, Jayalakshmi SL, Detecting fake accounts in media application using machine learning. Spec Iss Pub Int J Adv Netw Appl
9. Sen, I, Aggarwal A, Mian S, Singh S, Kumaraguru P, Datta A (2018) Worth its weight in likes: towards detecting fake likes on Instagram. In: WebSci '18, May 27–30, 2018, Amsterdam, Netherlands ACM
10. Yang KC, Varol O, Davis CA, Ferrara E, Flammini A, Menczer F (2019) Arming the public with artificial intelligence to counter social bots. Hum Behav Emerg Tech 1:48–61

11. Masood F, Ammad G, Almogren A, Abbas A, Khattak HA, Din IU, Guizani M, Zuair M (2019) Spammer detection and fake user identification on social networks. IEEE 7:2169–3536
12. Xiao C, Freeman DM, Hwa T (2015) Detecting clusters of fake accounts in online social networks. In: AISec'15, Oct 16, 2015, Denver, Colorado, USA, © 2015 ACM
13. Liu L, Lu Y, Luo Y, Zhang R, Itti L, Lu J (2016) Detecting "Smart" spammers on social network: a topic model approach. In: Proceedings of NAACL-HLT
14. Feng S, Tan Z, Wan H, Wang N, Chen Z, Zhang B, Zheng Q, Zhang W, Lei Z, Yang S, Feng X, Zhang Q, Wang H, Liu Y, Bai Y, Wang H, Cai Z, Wang Y, Zheng L, Ma Z, Li J, Luo M (2022) TwiBot-22: towards graph-based twitter bot detection. arXiv:2206.04564
15. Cresci S, Di Pietro R, Petrocchi M, Spognardi A, Tesconi M (2015) Elsevier decision support systems. 80
16. Gupta HK, Detect fake profiles in online social networks using support vector machine. Neural Network and Random Forest
17. Munk M, Drlik M, Benko L, Reichel J (2017) Quantitative and qualitative evaluation of sequence patterns found by application of different educational data preprocessing techniques. IEEE 5:2169–3536

Design of a Metaphor-Less Multi-objective Rao Algorithms Using Non-dominated Sorting and Its Application in I-Beam Design Optimization

Jatinder Kaur and Pavitdeep Singh

Abstract Recently, Rao has proposed single-objective metaphor-less set of Rao algorithms to solve wide variety of optimization problems. The beauty of these algorithms lies in its simple implementation with no algorithm specific configuration of any parameters for the effective working of the algorithms. In this paper, we have attempted to extend these set of algorithms to multi-objective optimization using non-dominated sorting and crowding distance mechanisms. Subsequently, the proposed three variants, namely, Non-dominated Sorting Rao1 algorithm (abbreviated as NSRao1), Non-dominated Sorting Rao2 algorithm (abbreviated as NSRao2) and Non-dominated Sorting Rao3 algorithm (abbreviated as NSRao3) are successfully applied to solve multi-objective I-Beam optimization problem under various strength and geometric constraints. The three variants differ in the way the new solutions are formed from the best and worst solutions and their arbitrary interactions with other candidate solutions. NSRao1 technique was able to find better set of solutions towards minimizing the displacement area and where as it didn't perform well towards the other boundary values compared to few standard evolutionary techniques. However, NSRao2 and NSRao3 are able to yield superior results in finding the optimal set of solutions towards both the boundary values with better diversity compared to other evolutionary algorithms.

Keywords Multi-objective optimization problem (MOOP) · Teaching-learning-based optimization (TLBO) · Non-dominating sorting TLBO (NSTLBO) · Non-dominating sorting Rao1 algorithm (NSRao1) · Non-dominating sorting Rao2 algorithm (NSRao2) · Non-dominating sorting Rao3 algorithm (NSRao3)

J. Kaur (✉)
Chandigarh University, Mohali, India
e-mail: jatinder.kaur401@gmail.com

P. Singh
Natwest Group, Gurgaon, India

S. Das et al. (eds.), *Advances in Data-driven Computing and Intelligent Systems*, Lecture Notes in Networks and Systems 653, https://doi.org/10.1007/978-981-99-0981-0_63

1 Introduction

Recently, meta-heuristic algorithms have gained a lot of importance in solving complex real-life optimization techniques. These algorithms were able to find near optimal solutions for complex problems in finite time. However, there was a dire need to set up the right values for the algorithmic specific parameters to achieve the desired results which itself is a labour-intensive task. Moreover, in case the algorithmic parameters are not fine-tuned, this might lead to divergence and the algorithm can never find the optimal solutions. In view of these limitations, Rao in 2012 proposed a parameter less algorithm called Teaching-learning-based optimization (TLBO) which doesn't require any parameter to be setup during simulation of the algorithm. The technique mimics the teaching-learning process in a classroom. The student or learner can be considered a probable solution for an optimization problem and subject or variable refers to the dimensionality of the that problem. It is based upon the concept where the student enhances his skills through teacher-student interaction called the teacher phase and through student-student interaction called the learner phase of the TLBO algorithm. Till now, TLBO has been successfully applied to different benchmark functions (both constrained and unconstrained) and have been successfully applied to different industrial and engineering problems.

In 2018, Kaur and Chauhan [1] proposed an improved parameter less TLBO called GTLBOLE incorporating group learning and experiential learning concepts into the original TLBO algorithm. The proposed technique has been successfully applied to find the optimal solutions for various unimodal, multimodal and rotated global functions. In [2], Kaur and Chauhan proposed a multi-objective variant of the GTLBOLE called NSGLTLBOLE which incorporates key methods like non-dominated sorting, crowding distance, group learning and experiential learning. The proposed technique was able to generate higher number of non-dominated diversified set of optimal solutions for the bi-objective benchmark test functions when compared with some of the existing standard algorithms. In [3], Kaur and Chauhan successfully applied NSGLTLBOLE to solve multi-objective Two-Bar truss structure problem in structural and civil engineering.

In early 2020, Rao [4] proposed a set of new metaphor-less simple algorithms to find the optimal solution of a single-objective optimization problem. In the proposed technique, equations were designed for computing the objective function values based upon the best and worst solutions for an iteration within a population. The results were really encouraging for the unconstrained benchmark test functions.

There have been several attempts made in the past to solve the multi-objective I-Beam design optimization problem using different evolutionary algorithms. Yang and Deb [5] proposed a multi-objective cuckoo search algorithm using blend of vectorized mutation, crossover by permutation and selective elitism to find the optimal solutions and applied the proposed the technique to solve the I-beam and disk brake design optimizations in structural engineering. In [6], Miguel, proposed new methods for the construction of Pareto Front leveraging genetic algorithms (GAs) and multi-objective genetic algorithms (MOGAs) that produced good approximation of Pareto

Front and later on used these methods to successfully solve I-beam and gearbox design optimization problems.

In the subsequent sections, we will discuss the complete work flow of Non-dominated Sorting-based Rao algorithms. We will also look into the mathematical formulation of I-Beam design problem satisfying both strength and geometric constraints. Later on, simulated results will be compared with NSGA-II [7], GDE3 [8], PAES [9] and AbYSS [10] algorithms for solution accuracy.

2 Non-dominated Sorting of RAO Algorithms

In the recent times, multi-objective optimization has gained a lot of importance due to their applicability to solve non-trivial optimization problems comprising many objectives (some of them might be conflicting in nature). In the subsequent sections, we will discuss a multi-variate version of Rao algorithms (parameter less in nature) and its application in solving one of the significant optimization problems in structural engineering, namely I-Beam optimization problem.

2.1 Motivation

Generally, there are two standard ways to solve a multi-objective optimization problem, e.g. priori and posteriori-based approaches. In the priori-based approach, some sort of scalarization is used to reduce the multi-objective optimization problem to a single-objective optimization problem, e.g. aggerated weighted sum and ϵ-constraint approaches. In contrast, posteriori approach works on simultaneously optimizing all the objectives defined for the multi-objective optimization problem(MOOP) under consideration. However, in this approach, multiple set of solutions called Pareto optimal solutions are created and it is left to the decision maker to choose any of the solutions as per his/her requirements to solve the problem. In case of real-world multi-objective problems, Pareto-based approaches have gained substantial importance as they are able to yield better results. Non-dominated sorting is one of the most widely and commonly used posteriori approach to generate a set of Pareto optimal solutions in the world of multi-objective optimization problem. In this paper, we are proposing a non-dominated sorting and crowding distance mechanism to design the multi-objective version of Rao algorithms.

2.2 Flowchart of NSRAO Algorithms

In this section, we will briefly explain the multi-objective variant of three Rao algorithms. We will look into the details for generating the Pareto set of solutions by

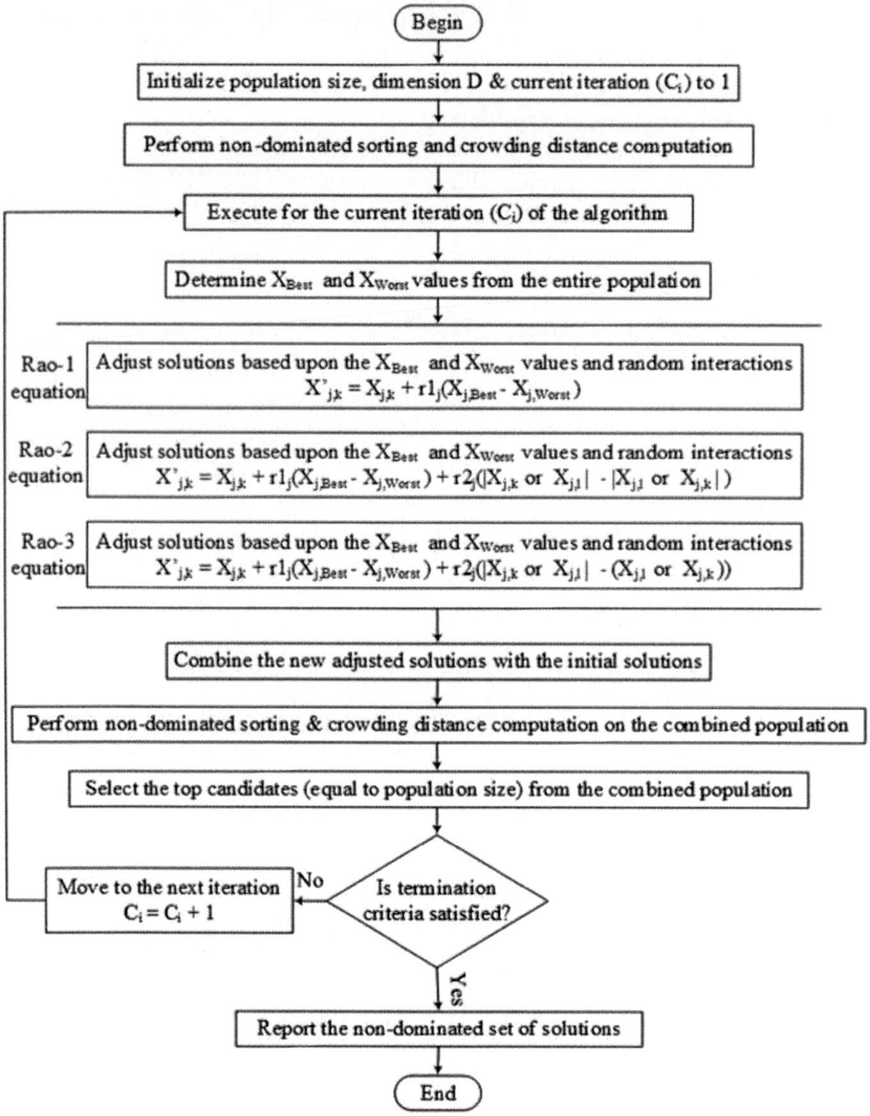

Fig. 1 Flowchart of the NSRao algorithms

the proposed algorithms. In contrast to traditional evolutionary algorithms, the proposed technique doesn't demand for setting-up and fine-tuning of algorithmic specific parameters for the efficient working of the algorithm.

In this paper, we propose a multi-variate of the single-objective Rao algorithms. Figure 1 depicts the flowchart of the proposed algorithms. At the heart of the algorithm is finding the best and worst solution among the population. As this is a multi-objective algorithm, there isn't a single-objective function value based upon which best and worst solution can be identified. Thus, we leverage the non-dominated sorting approach to rank all the individuals within the population. An individual having highest rank (rank = 1) is considered the best solution among the individuals. Similarly, an individual belonging to lowest rank is considered a worst solution. In case two or more individuals share the same rank, crowding distance computation decide the superiority of solutions. Crowding distance value of a particular individual refers to the average distance of its two neighbouring solutions. An individual having maximum crowding distance is adjudged as the best solution for the same rank individuals or solutions. Extreme or boundary solutions are assumed to have infinite crowding distance values. Thus, they are always preferred over other solutions. During each iteration of the simulation, a new value is updated based upon the different equations shown in the flowchart pertaining to three different Rao algorithms. In case a new value dominates the previous or current, it is updated in the population otherwise it is discarded. However, in case none of them dominates each other, new value is simply added to existing population. Subsequently, non-dominated sorting mechanism is again applied to get the best individuals equal to the original population size defined at the start of the algorithm. This entire process is repeated till the termination criteria is met. In the proposed algorithms, maximum number of function evaluations are used a termination criterion. The algorithm ends with the generation of non-dominated set of solutions. The complete process is depicted in Fig. 1.

Let current iteration is C_i, assuming there are m design variables and n candidate solutions within a population. If $X_{j,k}$ is the value of jth variable for kth candidate solution, then the new value of $X_{j,k}$ called $X'_{j,k}$ is calculated as per below equations. These three are the main equations in NSRao1, NSRao2 and NSRao3 algorithms, respectively, and are used to generate the updated postions based upon random interactions.

$$X'_{j,k} = X_{j,k} + r1_j(X_{j,\text{Best}} - X_{j,\text{Worst}}) \quad (1)$$

$$X'_{j,k} = X_{j,k} + r1_j(X_{j,\text{Best}} - X_{j,\text{Worst}}) + r2_j(|X_{j,k} \text{ or } X_{j,l}| - |X_{j,l} \text{ or } X_{j,k}|) \quad (2)$$

$$X'_{j,k} = X_{j,k} + r1_j(X_{j,\text{Best}} - X_{j,\text{Worst}}) + r2_j(|X_{j,k} \text{ or } X_{j,l}| - (X_{j,l} \text{ or } X_{j,k})) \quad (3)$$

where $r1_j$ and $r2_j$ are two random or arbitrary numbers in the range [0, 1] for the jth variable during C_ith iteration. Moreover, $X_{j,\text{Best}}$ and $X_{j,\text{Worst}}$ represent the best and worst values for jth variable in C_ith iteration.

3 Multi-objective Optimization of a I-Beam Design Problem

In this section, we will explore a multi-objective optimization problem called I-Beam design problem which is quite significant in civil engineering. Figure 2 presents the dimensions of a typical beam which aims to satisfy the strength and geometric constraints and to optimize the below conflicting objectives.

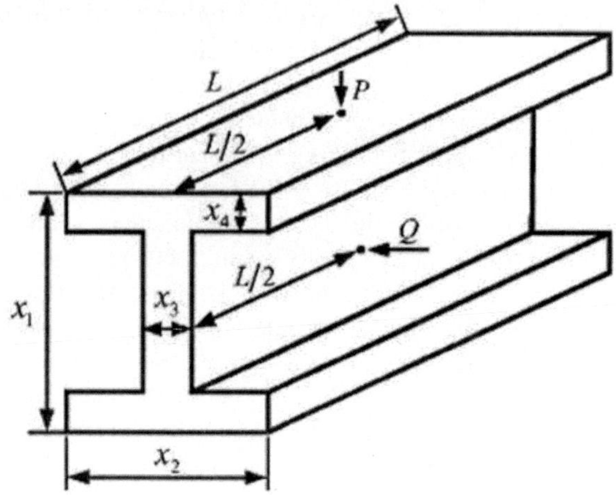

Fig. 2 I-beam design problem

- Minimizing the cross section area of the beam under consideration where the volume is reflected by its length.
- Minimizing the static deflection of the beam under the force P.

Subject to strength constraint

$$\frac{M_y}{W_y} + \frac{M_z}{W_z} \leq K_g$$

where M_y and M_z pertains to the maximum bending moments in Y and Z directions, respectively. W_y and W_z pertain to section moduli in Y and Z directions, respectively. K_g represents the allowed bending stress of the beam.

Subject to various geometric constraints

$$10 < x_1 \leq 80, 10 \leq x_2 \leq 50, 0.9 \leq x_3, x_4 \leq 5$$

where x_1, x_2, x_3 and x_4 are the design variables of the vector $\vec{x} = (x_1, x_2, x_3, x_4)$ which are measured in centimetres. These are truly conflicting objectives as attempt to minimize the cross section will yield in increase in displacement value of the beam and vice versa. Thus, the solution will involve finding non-dominated set of solutions for different combination of cross section area and displacement values.

Various parameters used in the design of the beam are mentioned below

1. Allowed bending stress of the beam Kg = 16 KN/cm^2
2. Young's modulus of elasticity E = 20,000 KN/cm^2
3. Bending force at P = 600 KN
4. Bending Force at Q = 50 KN
5. Maximum bending moments in Y direction M_y = 30,000 KN.cm
6. Maximum bending moments in Z direction M_z = 2500 KN.cm
7. Section Moduli in Y direction $W_y = \frac{x_3(x_1-2x_4)^3+2x_2x_4[4x_4^2+3x_1(x_1-2x_4)]}{6x_1}$

8. Section Moduli in Z direction $W_z = \frac{(x_1-2x_4)x_3^3+2x_4x_2^3}{6x_2}$
9. Length of beam = 200 cm
10. Inertia of moment $I = \frac{x_3(x_1-2x_4)^3+2x_2x_4[4x_4^2+3x_1(x_1-2x_4)]}{12}$

Mathematically, the two objectives can be formulated as

$$f(\overrightarrow{x_1}) = 2x_2x_4 + x_3(x_1 - 2x_4)\ \text{cm}^2 \tag{4}$$

$$f(\overrightarrow{x_2}) = \frac{Pl^3}{48EI}\text{cm} \tag{5}$$

4 Simulation Results and Discussion

jMetal Framework 4.3 [11, 12], which is a Java-based framework, is used to carry out the experiments for generating the set of optimal solutions for different metaheuristics techniques.

4.1 Experimental Settings

In this section, we will define the common parameters to be used during simulation of the results for various algorithms. As the algorithms may yield different results for different configuration of parameters, there is utmost a requirement to define and assign the same values to these different common parameters like population size, number of function evaluations before executing these algorithms. However, there are few algorithmic specific parameters which also needs to be set up and defined correctly for NSGA-II, GDE3, PAES and AbYSS algorithms to run properly. We will discuss these parameters and their values. During simulation runs, these parameters values will be used to generate and compare the results of different algorithms.

4.2 Common Parameters

There are few parameters which are common to every evolutionary algorithm like population size, termination criteria which can maximum number of generations or maximum function evaluations, number of independent runs and number of variables used during simulation of the algorithms.

Various common parameters and their values used are mentioned below.

- Population Size: 100
- Independent Runs: 10

- Maximum Function Evaluations: 25,000
- No of Variables: 4, (x_1, x_2, x_3 and x_4)

4.3 Algorithmic Specific Parameters

The empirically selected parameters for the NSGA-II, GDE3, PAES and AbYSS are defined in Tables 1, 2, 3 and 4 respectively.

The size of the design constraints defined for Vector x determines the search space of the solution. Standard evolutionary techniques like NSGA-II, GDE3, PAES and AbYSS are used to compare the results generated by three NSRao techniques. We have used the penalty method during calculation of the objectives function values for strength and geometric constraint violations. Same population size and function evaluations are used to generate the non-dominated set of solutions for the design problem. Following observations are made during simulations of results.

Figure 3 depicts the results obtained using multi-objective version of the simplest of RAO algorithms namely NSRao1 technique compared to NSGA-II, GDE3, PAES and AbYSS algorithms. It has been observed that it did perform well generating the set of Pareto solutions at one of the boundary values (having minimum displacement values against cross section area). However, when it comes to generating set of solutions for minimum cross section area (near the other end of boundary values), the results are not so encouraging and also performed slightly inferior compared to standard evolutionary algorithms.

Table 1 Different parameters set for NSGA-II algorithm

S. No.	Parameter name	Value
1	Probability (for crossover)	0.9
2	Distribution index (for crossover)	20
3	Probability (for mutation)	0.33
4	Distribution index (for mutation)	20
5	Mutation type	Polynomial mutation
6	Crossover type	SBX crossover
7	Selection type	Binary tournament2

Table 2 Different parameters set for GDE3 algorithm

S. No.	Parameter name	Value
1	CR	0.5
2	F	0.5
3	Crossover type	Differential evolution crossover
4	Selection type	Differential evolution selection

Table 3 Different parameters set for PAES algorithm

S. No.	Parameter name	Value
1	Archive size	100
2	Bi-sections	5
3	Probability	0.33
4	Distribution index	20
5	Mutation type	Polynomial mutation

Table 4 Different parameters set for AbYSS algorithm

S. No.	Parameter name	Value
1	Probability (for crossover)	0.9
2	Distribution index (for crossover)	20
3	Probability (for mutation)	0.33
4	Distribution index (for mutation)	20
5	Mutation type	Polynomial mutation
6	Crossover type	SBX crossover
7	Reference set 1	10
8	Reference set 2	10
9	Archive size	100

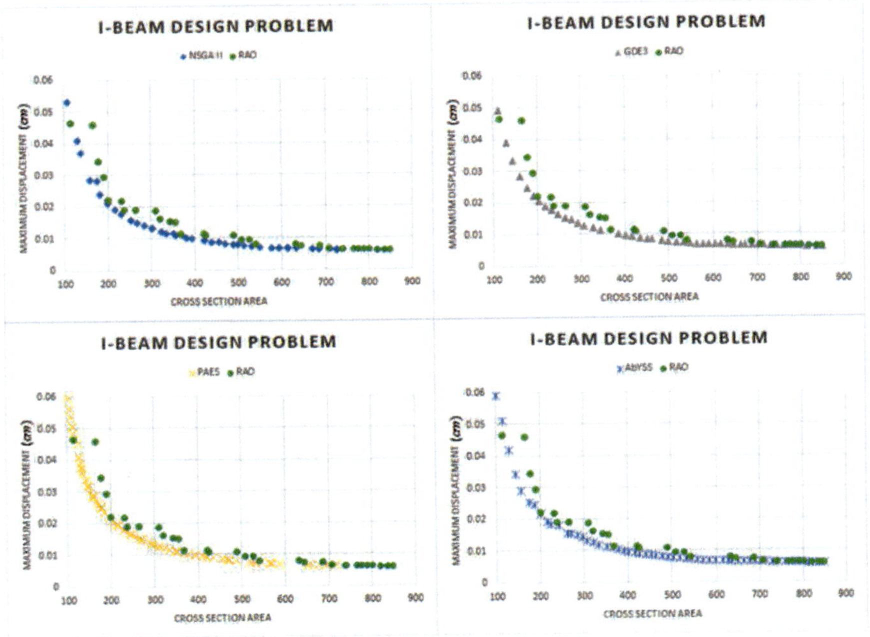

Fig. 3 Comparison of non-dominated solutions for the bi-objective I-beam design problem using NSRao1 with NSGA-II, GDE3, PAES and AbYSS algorithms

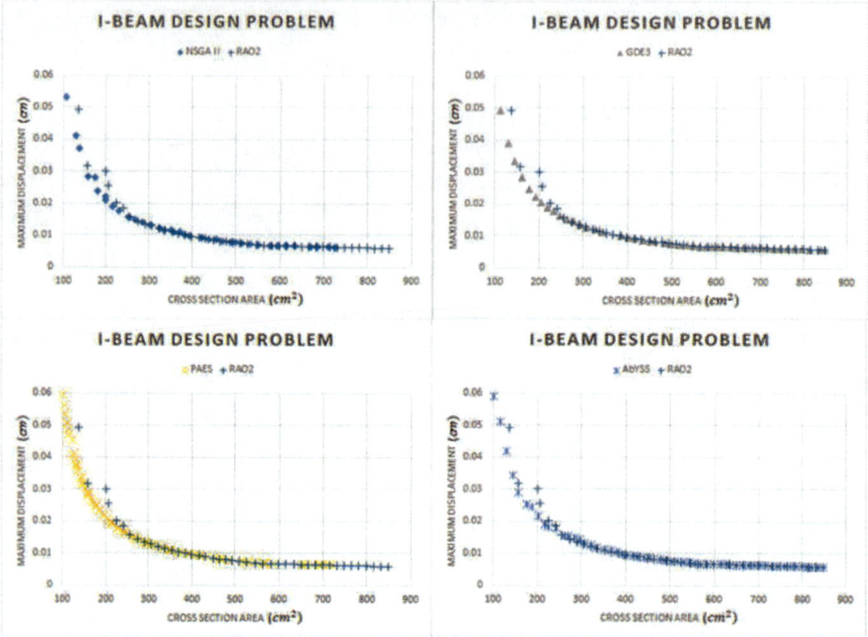

Fig. 4 Comparison of non-dominated solutions for the bi-objective I-beam design problem using NSRao2 with NSGA-II, GDE3, PAES and AbYSS algorithms

In Fig. 4, a good set of diverse Pareto optimal solutions are obtained for NSRao2 technique compared to most of the standard algorithms used for comparison. Subsequently, it is able to generate solutions at both the boundary values.

Lastly, in Fig. 5, NSRao3 algorithm is compared with the standard evolutionary algorithms for accuracy and effectiveness of the algorithm. Clearly, NSRao3 outperformed in terms of generating the set of non-dominated set of Pareto optimal solutions compared to other algorithms. It is able to generate diverse set of solutions for minimum displacement and cross section area as compared to most of the algorithms.

5 Conclusions and Future Scope

In this paper, we have extended the metaphor-less single-objective Rao algorithms to multi-objective versions using non-dominated sorting and crowding distance mechanisms. The proposed multi-objective techniques yield a set of pareto optimal solutions by ranking the individual solutions based upon their best and worst objective function values. These algorithms are simple to implement and can be easily applied to different optimization problems as it doesn't require any configuration in terms

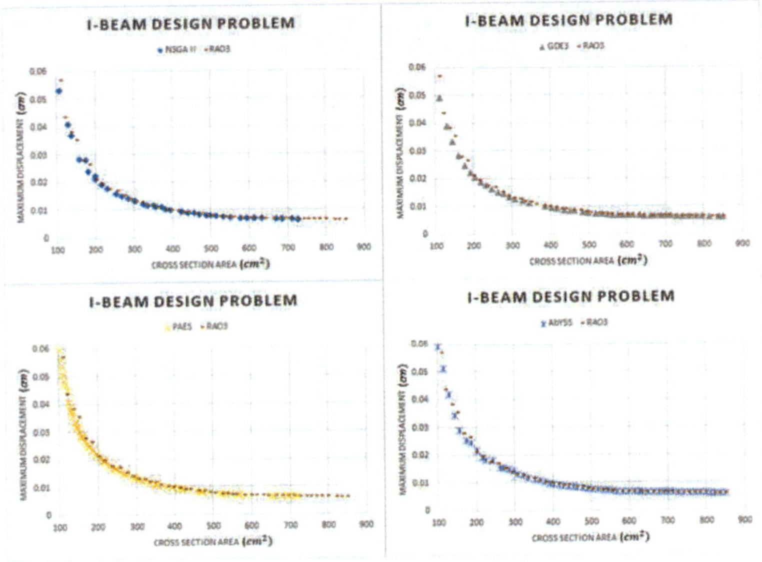

Fig. 5 Comparison of non-dominated solutions for the bi-objective I-beam design problem using NSRao3 with NSGA-II, GDE3, PAES and AbYSS algorithms

of algorithmic parameters to be setup or tuned. Later on, we have discussed and formulated the bi-objectives I-Beam design optimization problem subject to various strength and geometric constraints. The proposed algorithms, namely, NSRao1, NSRao2 and NSRao3 are then successfully applied to solve I-Beam design optimization problem in structural engineering. For solution accuracy and effectiveness of the proposed algorithms, these are simulated in comparison to few standard evolutionary algorithms, namely NSGA-II, GDE3, PAES and AbYSS techniques. Comparing the experimental results, few conclusions can be drawn. Firstly, NSRao1 is able to find good set of Pareto solutions at one of the extreme ends. However, the results are disappointing in terms of finding the near optimal set of solutions at the other extreme end of the Pareto front. In contrast, NSRao2 and NSRao3 have outperformed in yielding diverse set of solutions towards both the extreme of the Pareto front compared to NSGA-II, GDE3, PAES and AbYSS techniques.

Different variants of NSRao algorithms provides an excellent opportunity for the researchers to perform hybridization of these proposed algorithms with standard evolutionary techniques to further improve on the convergence rate and diversity factors. One of the prominent areas is to find ways to incorporate Elitism concept in these algorithms to replace the worst solutions with the best solutions found so far resulting in faster convergence. Nowadays, ensemble methods play a vital role in improving the diversity and convergence speed. Careful considerations need to be made while implementing ensemble methods with NSRao algorithms. Lastly, the researchers can apply the proposed algorithms to solve complex optimization problems in different field of engineering, social, economics, finance and banking to authenticate the accuracy and effectiveness of these algorithms.

References

1. Kaur J, Chauhan SS, Singh P (2019) An improved TLBO leveraging group and experience learning concepts for global functions. In: Advances in intelligent systems and computing, Springer Series, vol 2, pp 1221–1234. https://www.springer.com/in/book/9789811307607
2. Kaur J, Chauhan SS, Singh P (2019) NSGLTLBOLE: a modified non-dominated sorting TLBO technique using group learning and learning experience of others for multi-objective test problems. Adv Intell Syst Comput 900:243–251. https://doi.org/10.1007/978-981-13-3600-3_23
3. Kaur J, Chauhan SS, Singh P (2018) Multi-objective optimization of two-bar truss structure using non-dominated sorting TLBO leveraging group based learning and learning experience of others algorithm (NSGLTLBOLE). Int J Mech Eng Technol (IJMET)9:1016–1023
4. Rao RV (2020) Rao algorithms: three metaphor-less simple algorithms for solving optimization problems. Int J Ind Eng Comput 11:107–130
5. Yang X-S, Deb S (2011) Multi-objective cuckoo search for design optimization. Comput Operat Res 1:1–9. https://doi.org/10.1016/j.cor.2011.09.026
6. Martinez-Iranzo M, Herrero JM, Sanchis J, Blasco X, Garcia-Nieto S (2009) Applied Pareto multi-objective optimization by stochastic solvers. Eng Appl Artif Intell 22:455–465
7. Deb K, Pratap A, Agarwal S, Meyarivan T (2002) A fast and elitist multiobjective genetic algorithm: NSGA-II. IEEE Trans Evol Comput 6:182–197
8. Kukkonen S, Lampinen J (2005) GDE3: the third evolution step of generalized differential evolution. IEEE Congr Evol Comput 1:1–11
9. Knowles J, Corne D (1999) The Pareto archived evolution strategy: a new baseline algorithm for Pareto multiobjective optimisation. Proc Congr Evol Comput 1:98–105
10. Nebro AJ, Luna F, Alba E, Dorronsoro B, Durillo JJ, Beham A (2008) AbYSS: adapting scatter search to multiobjective optimization. IEEE Trans Evol Comput 12:439–457
11. Durillo JJ, Nebro AJ, Alba E (2010) The jmetal framework for multiobjective optimization: design and architecture. Proc IEEE Congr Evol Comput (CEC) 1:1–8
12. Durillo JJ, Nebro AJ (2011) The jmetal: a java framework for multi-objective optimization. Advances in engineering software. IEEE Trans Evol Comput 10:760–771

Computational Analysis for Candidate X-ray Images Using Generative Adversarial Network

Pradeep Kumar, Linesh Raja, and Ankit Kumar

Abstract Medical image analysis is becoming a critical deep learning application based on machine learning by contributing to developing a more sustainable health system, which probably overpowers doctors' workload drastically. With the advancement in deep learning technique, there is rise in the samples for training of diagnosis and treatment models. Generative Adversarial Networks (GANs) have sought attention in the field of medical image processing by their outstanding image generation capabilities and data generation without mapping the probability density function explicitly. GAN methods simulate the actual data distribution and reconstruct estimated accurate data. Medical images are available in less amounts, and the acquiring of medical image annotations is costly; therefore, generated data can solve the problem of data insufficiency or data imbalance. GANs are proven very useful in data augmentation and image translation. These qualities of GAN have fascinated researchers, and rapid adoption is noticed in the reconstruction, synthesis, segmentation, denoising detection, and classification of medical images. Finally, GAN models are extensively used for feature selection and extraction for medical image analysis and early diagnosis of diseases.

Keywords Images analysis · Generative adversarial networks · Machine learning · Neural network · Deep learning · Health care

1 Introduction

Image processing is a technique of manipulating the images acquired digitally using computers. This technique has several advantages: elasticity, adaptability,

P. Kumar (✉) · L. Raja
Department of Computer Applications, Manipal University Jaipur, Jaipur, India
e-mail: Pradeep.kumar@jaipur.manipal.edu

A. Kumar
Department of Computer Engineering and Applications, GLA University, Mathura, India

S. Das et al. (eds.), *Advances in Data-driven Computing and Intelligent Systems*, Lecture Notes in Networks and Systems 653, https://doi.org/10.1007/978-981-99-0981-0_64

data storing, and communication. Medical fields, such as medicine, diagnosis, and forensic, have numerous applications using digital image processing.

Medical imaging is getting images of body organs for diagnosing diseases [1]. Medical imaging is evolving speedily due to advances in image processing techniques, including image acquisition, classification, segmentation, reconstruction, denoising, and enhancement. Image processing improved accuracy in detecting the number of tissues in human body. Image acquisition is the process of retrieving an image from a device such as hardware systems like cameras, sensors, etc. Artificial intelligence-based image acquisition [2] can help for automatic scanning procedure and redesign the workflow with insignificant contact to patients and technicians. For example, CT and chest X-ray are broadly used in the diagnosis [3] and screening of COVID-19 infection. The procedure to tentatively formalize the image feature is known as candidate procedure.

Image synthesis is the procedure of synthetically generating images that comprise some required content. Image synthesis based on physics, e.g., MR imaging, has more than one image that are obtained from the identical structure with non-identical pulse chronological sequence and pulse sequence parameters so that underlying physical parameters can be calculated by reversing the so-called imaging calculations [4].

Medical image segmentation typically divides the spatial domain of a medical image into communally restricted subsets called regions, which are unvarying and homogeneous in respect of some properties like brightness, and whose property values are different in some substantial way from the characteristics value of each adjacent neighboring sections [5]. Recognition of edges which is separated from 'background' is very essential for medical image segmentation. Image reconstruction in health care is a mathematical procedure that can generate images from MRI and X-ray projection data attained at several different approaches around the patient. Accurate and effective image reconstruction techniques are essential to precisely generate images of the system features under investigation from the chronicled ultrasound data. Image reconstruction uses some filtration kernels in the Fourier sphere to process prior to prognosis image data to accomplish a firm image impression and contrast resolution [6].

All pragmatic machine learning techniques require supervised learning. Techniques iteratively do prediction on the labeled training data in supervised learning. Classification is a process which requires machine learning techniques and labeled training data [7] for achieving in title class label from the domain. Binary classification refers to predicting one of the binary classes: minority or majority. Image registration [8] is a method to map the given source images by using advertence image. The ultimate objective is to find the similarity with the resultant images on certain properties basis to assist for fusion. The first stage in image fusion process is to register the source images.

Features in medical image analysis are the characteristics of the objects of interest. Most of the features in medical imaging include color histogram features, shape features, texture features, and color moment features [9]. All past studies used global features for classification of medical images. Color and texture are the mostly used global features which are used in recognizing objects.

Generative adversarial network (GAN) is a deep learning technique which provides a path for learning deep representations with no comprehensively interpreted training data. This is achieved by using deriving backpropagation signals through a competitive process including couple of networks [10]. The representations which are realized by GAN are used in different applications which include synthesis, segmentation, registration, semantic editing, style transfer, reconstruction, and classification of images. GANs are a special network model based upon deep learning techniques where couple of different neural networks are trained simultaneously, with one emphasis on generating images and in parallel on the other side focusing on discrimination. Adversarial training gained huge interest in academics and industries because of its advantage in domain shift, and efficacy in image generation.

X-ray, CT, and MRI are the most frequently used imaging modalities for clinical examinations. Semantic knowledge of functional structures in these images is crucial to these applications, such as early diagnosis, evaluation, medication, and surgeries [11]. In medical imaging protection of patient, confidentiality is utmost; therefore, fetching these records is restricted to research units most of the times [12, 13]. Datasets in medical field are also extremely limited because they are suffering from imbalanced problems [14]. Such imbalanced datasets make the biased training of neural networks [15]. Various GAN models such as correlation-capturing Generative Adversarial Network (corGAN), are used for generating synthetic health records. This model utilizes CNN to identify the relationships between adjoining features in the data representation space by combination of convolutional autoencoders and CGAN [16].

2 Related Work

The diagnosis and analysis of medical images by using effective algorithm is an critical task for computer vision and the medical image analyst. The medical images require to be standardized [17] and labeled for early diagnosis. After preprocessing and the data entry into the deep architecture, the results of diagnosis may be achieved fast with accuracy. Convolution neural networks (CNNs), in recent past, have been gradually used to analysis medical image for increasing the efficiency.

Image candidate is the preprocessing activity for the image dataset prior to feeding the CNN or is vital for every imaging modality. Several preprocessing stages are taken into consideration for the medical images before feeding as source to the deep learning-based neural network model [18]. These activities include artificial noise removal, normalization, and registration. Normalization and artifact removal are the most preferred preprocessing steps across all the imaging modalities.

2.1 Imaging Modalities

CXR (chest X-ray) is frequently utilized for early finding of disease in the world of radiography. Chest X-ray can be divided into posteroanterior, anteroposterior (frontal), and lateral as per the orientation and position of patient and requirement of medical professionals [19]. Chest X-rays are used to generate images of your lungs, blood vessels, heart, and the skeletons of spine and chest.

Magnetic resonance imaging (MRI) is utilized to synthesize the structures of the human body organs due to its high-level spatial resolution and capability to differentiate soft tissue. Several clinical applications consist of MRI segmentation because it influences the outcomes of the whole analysis process [20]. Three-dimension cross-over-sectioned brain MRI is used to categorize patients with certain disease and to segment tissue types such as GM, CSF, and WM. For example, brain MRI reveals tissue as well as body organs of the head, spinal cord, eye, and skull [21]. Skull stripping to isolate non-brain tissue from the brain tissue requires to be achieved to distinguish the voxels as non-brain or brain.

Computed tomography (CT) is an essential diagnostic sensory system extensively used across globe of clinical indications for image-guided procedures and early diagnosis, and almost all CT images are digital nowadays, thus allowing more and more sophisticated image reconstruction techniques and analysis approaches within or as an addition to picture archiving and communication systems [22]. CT scanners are mostly fitted for imaging human bone structures due to superior hard and soft tissue and spatial resolution. CT and MR imaging are essential for early diagnosis of diseases, treatment, and clinical experiments, and it has become mandatory to use computers to support experts in clinical diagnosis. Ultrasound and PET scans are broad range of medical imaging modalities as the specialties of nuclear medicine in the spectrum.

2.2 Applications of GAN in Medical Imaging

Many datasets in medical field suffer from class imbalance issues because of sporadic nature in pathologies; hence, use of GAN models is increased to manage these constraints by producing images which looks like real image from a random distribution of noise. Several GAN models were proposed for analyzing medical image such as Cycle GAN, DCGAN, conditional GAN, Markovian GAN, and Wasserstein-GAN as the basic models in medical applications [23]. GAN-based models were applied on a variety of different modalities like CT, MRI, OCT, CXR, ultrasound, and microscopy. GAN models have been recommended to solve problems in medical images for application such as (a) image reconstruction, (b) image synthesis, (c) image classification, (d) image segmentation, and (e) image registration.

Deep learning-based GAN method has been proposed by Dang et al. [24] for reconstructing images through a given trained data and random vector. Firstly, a random vector was initiated and passed to GAN generator model. This latent tensor generates the reconstructed images using deconvolutional networks. Such images were similar in size of it to the input dataset which were used to train the discriminator model using convolutions and activation functions. The generated images and input images were used for training the GAN discriminator model for correct identification. CNN was used to find out whether images are fake or real. Model was examined on CXR dataset used in the NIH. Proposed model easily was classified to map it to one, and the fake score was less than 50% to map it to the zero.

U-net architecture is one of the most widely used fully convolutional network architectures for medical segmentation tasks. Lei et al. proposed an encoder-decoder module UNet-skip connection dense convolution [25] as the generator to generate segmented masks for lesion segmentation. Additionally, a dual discriminator was proposed to enhance the recognition which integrates the modules: skip connection dense convolution module and dual discrimination module. Generator module used convolution blocks which were densely dilated for generating a representation having deep information and the DD module in combination to decide whether the input image of discriminators is fake or real. Proposed GAN was assessed on the public ISIC Skin Lesion Challenge Datasets of 2017 and 2018.

GAN can be effectively used to augment data and increasing classification accuracy of disease in CXR (chest X-rays) for COVID-19 and pneumonia. Yuya et al. [26] used GAN-based models for image generation from computed tomography images of various cases with valid biopsy diagnosis. Proposed model used DCNN to examine the performance of classification among malignant and benign nodes, with factual and artificial nodes. The CNN-based DCNN was trained to make use of generated images through GAN and was tuned finely, by making ample use of the real node images to permit the DCNN for differentiation among malignant and benign nodes. The fine-tunned process and pre-training provided differentiation between 93.9% of malignant nodes and 66.7% of benign nodes. Thus, the proposed method enhanced the accuracy of classification by more than 21% compared to training utilizing the real images only.

Image registration in medical field is a process to find spatial or temporal correlations between image models and/or data. Heavy optimization load and parameter dependency restrict conventional registration methods from optimum registration mapping. Therefore, GAN models with its exceptional image makeover abilities have appeared as a candidate method to obtain more optimal registration mapping. Fan et al. [27] proposed a model based upon GAN for registering structural patterns which are defined in patches among different brain images. Proposed model consists of registration network like UNET to learn template image. Network is trained for registering the image precisely to assure the discriminator. A regularization is integrated in the training of R to preserve the efficiency in prediction. GANs improve the functioning of registration methods in medical settings.

Medical image synthesis is a process of modeling an association from the presented set of medical images to the unidentified object. GAN models have been applied with success for synthesis CXR images of retinal fundus, CT, and MRI. Zhang et al. proposed a SkrGAN [28] model introducing a sketch before restriction helping in generation of image in medical field. In proposed model, a module of sketch is developed for generating a structural sketch with superior quality from random distribution. After sketching, a colored mapping was utilized to implant the representations based on sketches resembling the backdrop presences. SkrGAN was applied on three public datasets of chest X-ray and achieved superior quality of medical images with accurate anatomical structures.

2.3 Feature Selection in Medical Image

Feature selection is the process to recognize the important features while developing a predictive model. Features of medical images include textural features, statistical features, and syntactical features. The size, color, hue, brightness, intensity value, and existing shapes in the medical image are vital in recognizing the correct class. Medical image datasets have another hundreds of features because the huge number of pixels, filters, embedded methods, and wrappers are considered as some feature selection methods [29]. Filters are not dependent on any learning method; hence, this method has a good capacity of generalization. Embedded methods need a learning method to perform feature selection while wrapper methods require an induction method to evaluate candidate subsets of features. Feature selection basis on correlation, INTERACT [30], and information gain are popular SOTA methods for feature selection. Some researchers also used nature-inspired algorithms and CNN methods for feature selection.

GAN models like WGAN and infoGAN can also be used for feature extraction by using discriminator as a feature extractor for classification. Hue et al. [31] proposed a method by combining WGAN and InfoGAN for representing feature in histopathology images. Proposed method worked well for extracting features through the discriminator by building a classifier at top level. Semi-supervised GANs are also used for chest abnormality classification, cardiac disease diagnosis, and patch-based retinal vessel classification. Furthermore, GAN has also been used to reduce adversarial loss in identifying cardiac deformities in chest X-ray.

3 Data Collection

Research on COVID and X-rays of the ordinary person's lungs was only possible because of the participation of several individuals and Internet groups who provided data. In addition to 3740 photographs of COVID chest X-rays, our collection includes 3740 images of typical patients' X-rays of their lungs. A combination of publicly available data from GitHub, Kaggle, and other places was used to create this dataset.

Figure 1 shows sample images of chest X-ray and X-ray images of COVID-19-infected patients. To train our convolutional neural network, we are required to extract

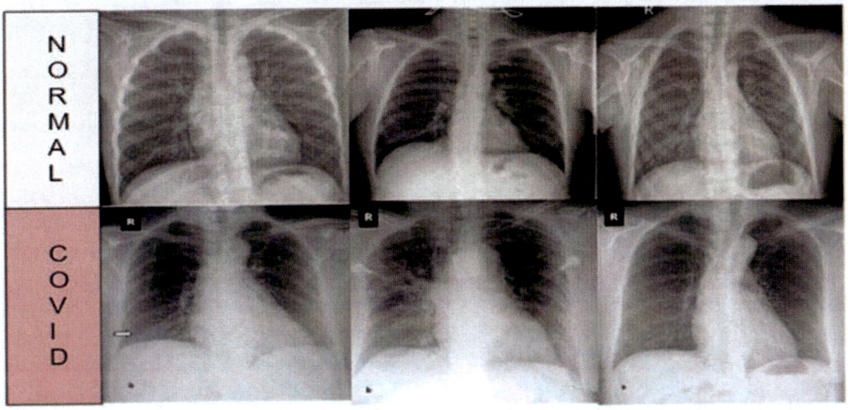

Fig. 1 COVID-19 sample image for identification of standard X-ray images of chest versus infected chest X-ray image of dataset which contains 7480 chest X-rays image

Table 1 My dataset source list

Data source names	Normal	COVID
COVID-19 radiography database	3740	3600
IEEE public GitHub repo	–	110
Actualmed COVID dataset from Taichung	–	30

the COVID sample pictures provided by a few dataset sources into a metadata.csv file.

Table 1 shows the number of photographs we have chosen from each of the datasets listed below.

COVID-19 radiography database (6).

COVID chest X-ray dataset from IEEE8023 public GitHub repository (7).

Accumulated COVID chest dataset from GitHub repo of agchung(8).

3.1 *Preprocessing the Data*

Image Processing

Since the image we will feed as the input should be exact and accurate, we must process the image. Image processing is a part in which we will enhance the quality of the selected images, and hence, the subsequent procedures must be followed.

Image Acquisition

This is the foundation step to the image processing here, and the images are retrieved soon after the attribute selection. The selected image should be of minimal noise. The essential part is to acquire images with appropriate quality and precision.

Image Preprocessing
The prime motive of image preprocessing is to improve the quality of the image so that we can scrutinize it better. We can defeat unsolicited distortions by preprocessing and intensifying some necessary features for our working model. This process includes resizing, orientation, and color correction.

Grayscale Conversion
In the grayscale conversion process, the color image that is the RGB image is converted into grayscale or shades of gray. Each pixel represents only the amount of light carried in the grayscale image. The grayscale intensity is set as an 8-bit integer of 256 practicable shades of gray from black to white. A grayscale image simplifies the algorithm and reduces the computational means.

Noise Removal
Noise removal removes the undesired pigment and unwanted hair from the given image. Following are the various filters used to remove noise.

Mean Filter: The arithmetic mean filter is one of the most uncomplicated known filters. This filter is used to blur the image or a part of the image. The function that is typically used is a Gaussian function. This filter is also utilized to restore every pixel value and the mean value of its adjacent pixels, including its pixel value in an image.

$$z(a, b) = \frac{1}{xy} \sum_{(i,j)\in S_{ab}} h(i, j) \tag{1}$$

where $z(a, b)$ depicts the output image and $h(I, j)$ portrays the image input with noise.

Median Filter: This is one of the order-statistics filters. It is mainly used to remove noises like 'salt-and-pepper' and 'speckle'. It is said to have high accuracy of 97.8%. Median filter is also identified as a non-linear digital filtering technique.

$$Z(s) = \text{median}\ ((M(s - S/2), M(s - S1 + 1), \ldots, M(s), \ldots, M(s + S/2)) \tag{2}$$

Here, s is the size of the window of the median filter.

Wiener Filter: Wiener filter is known as an optimum filter. It is availed to remove additional noise and upends blurring simultaneously. Wiener filter uses Discrete Fourier Transform (DFT) method.

$$F(x, y) = I(x, y).J(x, y) \tag{3}$$

where I is the Fourier transform of an 'ideal' version of a designated image, and J is the blurring function.

Alpha-Trimmed Mean Filter: This filter is a hybrid of mean and median filters. The main aim of this filter is to remove the most atypical element from the neighborhood and calculate the mean using the rest of them.

$$y(a,b) = \frac{1}{ij - m} \sum_{(c,d)\in k_{ab}} uv(c,d) \tag{4}$$

The value of m can be from the range 0 to $(ij - 1)$; when $m = 0$, the alpha-trimmed filter dwindles to the arithmetic mean filter $m = (ij - 1)/2$, and the filter will then become a median filter.

Adaptive Filter: We must consider the computational resource and the convergence speed while selecting this filter's algorithm. This is a linear filter.

In our model, the images that must be given as inputs consist of 'salt-and-pepper' noises. To remove these undesired noises, we must use median filter as we have already seen that the Median Filter effectively removes noises like 'salt-and-pepper' and 'speckle'. The caliber of the reconstructed image is estimated by two factors: (RMSE) Root Mean Square Error and the Peak.

Signal-to-Noise Ratio (PSNR)

Root Mean Square Error (RMSE) estimates the average error, weighted according to the square of the error.

$$\text{RMSE} = \frac{\sqrt{\sum (x(a,b) - y(a,b))^2}}{ij} \tag{5}$$

$x(a, b)$ is an indigenous aerial image with noise. $y(a, b)$ is the magnified image. i and j are the complete number of pixels in the image's vertical and horizontal dimensions.

PSNR is interpreted as the ratio of signal power to noise power that chiefly obtains the gray value difference between the resulting image and the original image

$$\text{PSNR} = 20 \log 10(225/\text{RMSE}) \tag{6}$$

3.2 *Proposed Work*

A conditional probability is the likelihood of conclusion C with given evidence E, whereas a dependence relationship exists between C and E.

This probability is signified as $P(C|E)$ where

$$P(C|E) = \frac{P(E|C)P(C)}{P(E)} \tag{7}$$

where

$P(C|E)$ is known as posterior probability.

P(*E*|*C*) is said to be a likelihood of classifier with a given attribute.

P(*C*) is called the class prior probability.

P(*E*) is known as predictor probability.

Bayesian networks perceptibly depict the combined probability distribution of a random variable set. An interpreted DAG which reveals a 'Joint probability distribution over a set of attributes *X*' is said to be a Bayesian network. A Bayesian network for *X* is a set of *B* = <DAG, *Θ*>, where nodes signify the attributes $X1, X2, X3 \ldots Xn$ and whose edges signify the direct dependencies between the attributes. It is also used to integrate the conditional probability of a node where the values are dedicated to the other nodes.

Presuming that $X1, X2, X3 \ldots Xn$ is the n attributes correlated to the nodes of the GAN network, and let us say an *n*-dimensional attribute vector *E* is represented as $x1, x2, x3 \ldots$on where $x1$ is the value of the attribute $X1$. Let x be the tuples and y be the training set of tuples. Let *C* signifies the class variable and *c* represents the class node in the GAN network, and then the class *c*(*E*) can be signified as a classifier by the Bayesian network as

$$c(E) = \text{argmaxp}(c)p(x1, x2, x3, \ldots, xn|c) \tag{8}$$

where $c \in C$.

Suppose there are n classes such as $C1$, $C2$, $C3$, and *Cn*. The naïve Bayesian classifier assumes the tuple *x*, which belongs to the class $C\alpha$ if and only if

$P(C\alpha|X) > P(C\beta|X)$ for $1 \leq \beta \leq n; \beta \neq \alpha$.

Now, we maximize $P(C\alpha|X)$. The class Cα for which $P(C\alpha|X)$ is maximized is known as the maximum posterior hypothesis. By Bayes theorem,

$$P(C\alpha|X) = \frac{P(X|C\alpha)P(C\alpha)}{P(X)} \tag{9}$$

Assume that the classes are equal, like $P(C1) = P(C2) = P(C3) = \ldots = P(Cn)$, and then maximize $P(X|C\alpha)\ P(C\alpha)$. Figure that the class prior probabilities may be approximated by $P(C\alpha) = |C\alpha, y|$, where $|C\alpha, y|$ is said to be the number of training tuples of class $C\alpha$ in *y*.

If there are no dependence relationships among the attributes, the class label of the tuple will be given. Thus,

$$P(X|C\alpha) = \prod_{\mu=1}^{n} P(x\mu|C\alpha) \tag{10}$$

$$= P(x1|C\alpha) \times P(x2|C\alpha) \times P(x3|C\alpha) \cdots \times P(xn|C\alpha) \tag{11}$$

Now, evaluate the probabilities of $P(\times 1|C\alpha)$, $P(\times 2|C\alpha)$, $P(\times 3|C\alpha) \ldots P(xn|C\alpha)$ from the training tuples. Here, $x\mu$ refers to the value of the attribute for tuple *x*. So, the naïve Bayes becomes more practical for learning restricted structures, and the naïve Bayesian classifier has a less complicated structure. The supposition made is that all attributes are not dependent on the given class and Eq. 2 takes the form

$$c(E) = \text{argmaxp}(c) \prod_{\alpha=1}^{n} p(x_\alpha | c) \tag{12}$$

where $c \in C$.Here, each attribute will have a class node as its predecessor only.

Algorithm 1: Innovative Method for Predicting COVID-19 Using GAN Network

BEGIN
    Input,
  build the training data set
    c = {(x1, C1), (x2, C2), (x3, C3),...... (xn,Cn)}
    X = (x1,x2,x3,...xn) new precedent to be classified
  for each labelled precedent(XαlCα) do
  if X has an unknown dataset, then
        X is average;
  else then
  for each proceeding in training data do
  calculate sum (X,Cα)
  if sum (X,Cα) = 1 then
      X is normal and exit;
  sum (X,Cα) from minimum to maximum (α = 1,2,3,...n)
  find a maximum score of sum (X, Cα);
  select the nb precedent to X: C
  authorize to X the most frequency class in C
  Ipsum.average = threshold then
      X is normal;
  else then
      X is abnormal;
  END.

Now, let us consider that an attribute X has many values, and the probability of the value $P(X = x\alpha|C = c)$ from Eq. 11 can be boundlessly small. So, the probability density estimation is used by presuming that X within class c is extracted from a Gaussian distribution.

$$\frac{1}{\sqrt{2\prod}\sigma_c} e^{\frac{(x\alpha - \mu_c)}{2\sigma_c^2}} \tag{13}$$

where σ_c is known as deviation and μc denotes the mean of attribute value from the training set.

Performance metrics of proposed work with exiting method are shown in Table 2. Performance of naïve Bayes, Decision tree, and proposed work are compared

Table 2 Performance metrics of Proposed Work with Exiting Method

Performance metrics	Naïve Bayes	Decision tree	Proposed work
Time	0.05	0.02	0.09
Kappa statistics	0.5156	0.5676	0.4353
Mean absolute error	0.064	0.3019	0.0633
Relative absolute error (%)	67.4033%	55.1686%	76.5609%
Root relative squared error (%)	88.9312%	98.9970%	95.9567%
Accuracy (%)	81.1176%	80.5130%	73.3528%
Sensitivity (%)	91.1011%	59.3509%	43.4304%
Specificity (%)	78.9048%	68.7589%	40.7306%

on various parameters such as time, Kappa statistics, accuracy, sensitivity, and specificity.

When there are no attribute and class label values, the frequency counts will be considered when the probability *P*(*x*|*c*) gets zero. To overcome this complication, classical access is to use the Laplace-m estimate.

$$P(C = c) = \frac{n_c + k}{N + n \times k} \tag{14}$$

where NC represents the number of satisfied instances $C = c$, N represents the number of training instances, *n* represents the number of classes, and $k = 1$.

$$P(X = x\alpha | C = c) = \frac{n_{c\alpha} + m \times P(X = x\alpha)}{n_c + m} \tag{15}$$

where $n_{c\alpha}$ represents the number of instances satisfying both $X = x\alpha$ and $C = c$, and here, *m* is said to be a constant and implies $m = 2$ and $P(X = x\alpha)$ valuated as *P* (*P*) $C = c$) as provided a top. We must consider a few unknown datasets among the attribute values.

Table 2 shows the comparative analysis of various exiting method, e.g., naïve Bayes and Decision tree with proposed work in terms of accuracy, sensitivity, specificity, etc.

Figure 2 shows the comparative analysis of proposed work with exiting method. The main disadvantage of this proposal is that if the attribute data does not follow a Gaussian distribution, then the valuation could not be dependent, suggesting the

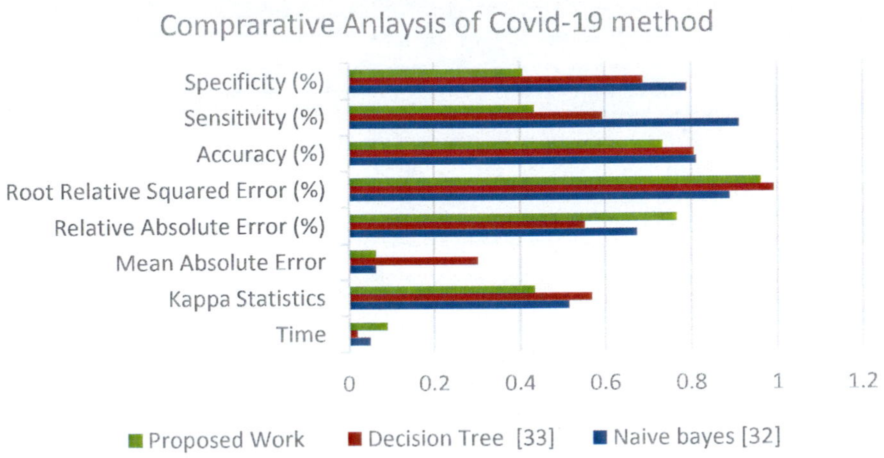

Fig. 2 Comparative analysis of proposed work with exiting method

other method, known as the kernel density estimation approach. It is the best way to evaluate the probability density function of a random variable.

4 Conclusion

This paper recaps frequently used GAN models for computational image analysis in medical science and the adversarial learning process in synthesizing another medical image. This paper analyzes the standard GAN models for image augmentation. It is concluded that GAN has greater possibility and development prospective in the field of medical image processing. The entire development of AI is in the direction of designing more sustainable health systems using unsupervised (deep) learning-based algorithms. Imminent trends of unsupervised prediction modeling, innovations in clinical requirement, and the necessity for GANs more fit for medical image analysis are also reviewed. Overall, the current image analysis methodologies in medical science are highly dependable, and the blend of GAN and other synthesis models also creates an excellent effect. The problems with deep learning techniques due to small amount of data and limited labels are also discussed by using techniques such as transfer learning and data augmentation. In this study, recent methods of feature selection for medical applications are reviewed. It is shown that feature selection is a very important tool for preprocessing which not only lessens the number of input features but also benefits the experts in understanding the causes of certain diseases in advance.

References

1. Bieniecki W, Grabowski S, Rozenberg W (2007) Image preprocessing for improving OCR accuracy. In: 2007 International conference on perspective technologies and methods in MEMS design. IEEE
2. Shi F, Wang J, Shi J, Wu Z, Wang Q, Tang Z, Shen D (2020) Review of artificial intelligence techniques in imaging data acquisition, segmentation, and diagnosis for COVID-19. IEEE Rev Biomed Eng 14:4–15
3. Ohata EF, Bezerra GM, das Chagas JVS, Neto AVL, Albuquerque AB, de Albuquerque VHC, Reboucas Filho PP (2020) Automatic detection of COVID-19 infection using chest X-ray images through transfer learning. IEEE/CAA J Automatica Sin 8(1):239–248
4. Giger ML (2018) Machine learning in medical imaging. J Am Coll Radiol 15(3):512–520
5. Balakrishnan R, Hernández MDCV, Farrall AJ (2021) Automatic segmentation of white matter hyperintensities from brain magnetic resonance images in the era of deep learning and big data–a systematic review. Comput Med Imaging Graph 88:101867
6. Schafer S, Siewerdsen JH (2020) Technology and applications in interventional imaging: 2D X-ray radiography/fluoroscopy and 3D cone-beam CT. In: Handbook of medical image computing and computer assisted intervention. Academic Press, pp 625–671
7. Johnson JM, Khoshgoftaar TM (2019) Survey on deep learning with class imbalance. J Big Data 6(1):1–54
8. Haskins G, Kruger U, Yan P (2020) Deep learning in medical image registration: a survey. Mach Vis Appl 31(1):1–18
9. Lai Z, Deng HF (2018) Medical image classification based on deep features extracted by deep model and statistic feature fusion with multilayer perceptron. In: Computational intelligence and neuroscience 2018
10. Creswell A, White T, Dumoulin V, Arulkumaran K, Sengupta B, Bharath AA (2018) Generative adversarial networks: an overview. IEEE Signal Process Mag 35(1):53–65
11. Zhang Y, Miao S, Mansi T, Liao R (2020) Unsupervised X-ray image segmentation with task driven generative adversarial networks. Med Image Anal 62:101664
12. Bertino E, Ooi BC, Yang Y, Deng RH (2005) Privacy and ownership preserving of outsourced medical data. In: 21st International conference on data engineering (ICDE'05). IEEE, pp 521–532
13. Salehinejad H, Valaee S, Dowdell T, Barfett J (2018) Image augmentation using radial transform for training deep neural networks. In: 2018 IEEE international conference on acoustics, speech and signal processing (ICASSP). IEEE, pp 3016–3020
14. Li D-C, Liu C-W, Hu SC (2010) A learning method for the class imbalance problem with medical data sets. Comput Biol Med 40(5):509–518
15. Salehinejad H, Valaee S, Dowdell T, Colak E, Barfett J (2018) Generalization of deep neural networks for chest pathology classification in x-rays using generative adversarial networks. In: 2018 IEEE international conference on acoustics, speech and signal processing (ICASSP). IEEE, pp 990–994
16. Torfi A, Fox EA (2020) CorGAN: correlation-capturing convolutional generative adversarial networks for generating synthetic healthcare records. In: The thirty-third international flairs conference, 2020
17. Yoon HJ, Jeong, Kang (2019) Medical image analysis using artificial intelligence. Prog Med Phys 30(2):49–58
18. Singh SP, Wang L, Gupta S, Goli H, Padmanabhan P, Gulyás B (2020) 3D deep learning on medical images: a review. Sensors 20(18):5097
19. Çallı E, Sogancioglu E, van Ginneken B, van Leeuwen KG, Murphy K (2021) Deep learning for chest X-ray analysis: a survey. Med Image Anal 72:102125
20. Debelee TG, Kebede SR, Schwenker F, Shewarega ZM (2020) Deep learning in selected cancers' image analysis—a survey. J Imaging 6(11):121

21. Mazurowski MA, Buda M, Saha A, Bashir MR (2019) Deep learning in radiology: an overview of the concepts and a survey of the state of the art with focus on MRI. J Magn Reson Imaging 49(4):939–954
22. Sharma N, Aggarwal LM (2010) Automated medical image segmentation techniques. J Med Phys/Assoc Med Phys India 35(1):3
23. Kazeminia S, Baur C, Kuijper A, van Ginneken B, Navab N, Albarqouni S, Mukhopadhyay A (2020) GANs for medical image analysis. Artif Intell Med
24. Dang N, Khurana M, Tiwari S (2020) MirGAN: medical image reconstruction using generative adversarial networks. In: 2020 5th International conference on computing, communication and security (ICCCS). IEEE
25. Lei B, Xia Z, Jiang F, Jiang X, Ge Z, Xu Y, Wang S (2020) Skin lesion segmentation via generative adversarial networks with dual discriminators. Med Image Anal 64:101716
26. Onishi Y, Teramoto A, Tsujimoto M, Tsukamoto T, Saito K, Toyama H, Fujita H (2019) Automated pulmonary nodule classification in computed tomography images using a deep convolutional neural network trained by generative adversarial networks. BioMed Res Int
27. Fan J, Cao X, Xue Z, Yap PT, Shen D (2018) Adversarial similarity network for evaluating image alignment in deep learning based registration. In: International conference on medical image computing and computer-assisted intervention. Springer, Cham, pp 739–746
28. Zhang T, Fu H, Zhao Y, Cheng J, Guo M, Gu Z, Liu J (2019) SkrGAN: sketching-rendering unconditional generative adversarial networks for medical image synthesis. In: International conference on medical image computing and computer-assisted intervention. Springer, Cham, pp 777–785
29. Guyon I, Gunn S, Nikravesh M, Zadeh LA (eds) (2008) Feature extraction: foundations and applications, vol 207. Springer
30. Zhao Z, Liu H (2009) Searching for interacting features in subset selection. Intell Data Anal 13(2):207–228
31. Hu B et al (2018) Unsupervised learning for cell-level visual representation in histopathology images with generative adversarial networks. IEEE J Biomed Health Inf 23(3):1316–1328

Sentiment Analysis Integrating with Machine Learning and Their Diverse Application

Bitthal Acharya, Sakshi Shringi, Nirmala Sharma, and Harish Sharma

Abstract Sentiment analysis has risen to prominence as a discipline that has attracted a significant amount of attention due to the vast range of applications that could benefit from its findings. Despite this, this region is still in the early stages of development, with urgent upgrades necessary on a variety of concerns, particularly in the display of opinion groups. Three main testing concerns influencing opinion arrangement are presented in this proposal, along with creative techniques to resolving these issues. First and foremost, message pre-handling has been identified as a vital component of the sensation grouping execution. As a result, for the opinion characterization procedure, a combination of a few known pre-processing techniques is presented. Second, the message qualities of monetary news are utilized to create models that predict how people will feel. Two distinct models are proposed, one that uses monetary events to predict financial news feelings, and another intriguing viewpoint that considers the evaluation per user view, rather than the example methodology that considers the opinion holder view. Another approach for capturing the individual user's mood is offered. Third, one of the characteristics of financial news is that it covers a wide range of topics; also, predicting public opinion across many disciplines is quite difficult. Different cross-space opinion examination methodologies have been suggested and fundamentally evaluated. And based on the evaluation and research results that show, the efficacy is about 95.6%.

Keywords Sentiment analysis · Machine learning · LSTM, Prediction

B. Acharya (✉) · N. Sharma · H. Sharma
Department of Computer Science and Engineering, Rajasthan Technical University, Kota, Rajasthan, India
e-mail: bitthalacharya97@gmail.com

N. Sharma
e-mail: nsharma@rtu.ac.in

H. Sharma
e-mail: hsharma@rtu.ac.in

S. Shringi
Department of Computer Science and Engineering, Manipal University Jaipur, Jaipur, Rajasthan, India
e-mail: sakshi.shringi@jaipur.manipal.edu

S. Das et al. (eds.), *Advances in Data-driven Computing and Intelligent Systems*, Lecture Notes in Networks and Systems 653, https://doi.org/10.1007/978-981-99-0981-0_65

1 Introduction

Sentiment analysis is a trending concept that has gained tremendous attention from people as well as from organizations previous ten years. This field encompasses a wide range of applications that have been addressed in many research studies [1].

Much effort has recently been expended to investigate the influence of various media methods on the financial world. The Internet has evolved into a vast repository of information for everyone. In the field of finance, there are a few types of data that are of interest. Financial backers, for example, are concerned about financial news related to their venture. Organizations are interested in the study of competitors, providers, materials, and client complaints. As a result, clients are interested in other client audits concerning products that they require to buy. To achieve this, several applications of feeling investigation have emerged in a variety of fields, including financial news opinion analysis, item audits, political judgments, and medical services [4]. In the current writing, Twitter, arguably the most well-known social media website, is gaining a lot of attention. The analysis of Twitter data or the alleged public mindset is used in a variety of sectors and applications, including social financial problems with applications to the securities exchange, public health, cataclysmic occurrences, and scourges.

Here, generally we have two types of learning that are used for training machines in the field of artificial intelligence that are named as supervised and unsupervised learning. These techniques are used in analyzing the sentiment. As in unsupervised learning, the data we use is not labeled that why clustering is necessary in this. On the other hand, if a model were assigned with some data which is labeled then it is called as supervised learning. These marked datasets are taught to create reasonable outcomes when they are met during decision-making [2].

This whole analysis relies on machine-based learning that helps us to acquire preferable understanding about the sentiment. The remaining portion of this paper is organized as follows. Sector second shows the work done for opinion research in various areas by various specialists in a nutshell. The sector third is the methodology we used for the feeling examination [4]. The fourth section discusses the complexities and consequences of the execution, followed by a discussion of the end and future work in the fifth region.

2 Background and Problem Definition

Opinion examination is considered a grouping issue in computational semantics. It involves normal language processing (NLP) on several levels and comes with its own set of challenges. There are numerous applications that could benefit from its outputs, including news analysis, advertising, question answering, and information gathering. The goal of this field is to improve the machine's ability to interpret texts in the same way as human readers can [9]. Taking use of the massive amount of emotions transmitted on the Internet, particularly on social networking websites, is critical for several organizations and foundations, whether it is for item criticism, public perception, or financial backer decisions.

The proposed proposal investigates different potential outcomes in order to improve feeling grouping execution. Three primary questions are investigated in order to resolve this issue. The first challenge is to use text pre-processing to better develop feeling characterization. The second step is to use text properties to further improve it. The third issue is to continue to grow it by constructing feelings, beginning with one space, and progressing to the next. In the accompanying, these issues are clarified [12].

The main goal of this hypothesis is to look into some crucial techniques for improving emotion grouping execution. There are three specific goals to achieve this. The primary purpose is to use text pre-handling to further develop sensation forecasts. A variety of pre-handling procedures are discussed, and an appropriate component selection strategy for the study is chosen. Report level opinion order is carried out in conjunction with a focus on item audits and the use of film audits, for example [2]. The following goal aims to improve opinion characterization by delving deeper into various text features.

Despite the fact that different strategies for the feeling examination have been presented for the various application mentioned in the preceding chapters, there are certain common characteristics that may be observed. The ability to see and accommodate these characteristics can aid in comprehending the difference between various application and relayed techniques, as well as allow an application to be designed in a systematic manner.

- At the document level, the analysis focuses on the total record and assigns an overall emotion to it, with the expectation that the report will transmit a single viewpoint.
- Sentence-level analysis examines sentences that express a single point of view and attempts to establish its direction.
- Phrase- or word-level analysis looks into the polarity of texts at a more granular level: the phrase level.
- In addition to the preceding layers of research, some studies look at user networks and estimate user sentiment based on the sentiment of nearby users.

2.1 Text Pre-processing

Online messages contain a slew of noise that might muddle the feeling organization process, such as HTML labels, advertising, hyperlinks, stop word, and words that have no bearing on the text direction. Every message is addressed by the vector in the examination, and each word is a passage [8]. As a result, each word represents a single aspect in the vector space. As a result, many text reports have high dimensionality, making the classifier's assignment more difficult. This is because increased dimensionally causes more scarcity making it difficult to find comparable properties among the arrangement objective.

2.2 *Text Property*

To create a model for feeling analysis, two aspects of financial news can be used. The key characteristic is that monetary news includes publicly reported events regarding publicly traded companies. These events have a direct impact on the stocks of the companies involved. This link could be used to build a model that predicts how people would feel based on the type of event described. Assessments can generally be seen from two perspectives [5]: assessment holder and assessment per user. The evaluation of clients who write online audits, all in all, the assessment holders, is the extent of feeling investigation in surveys.

2.3 *Cross Domain Sentiment*

Following a review of monetary news, it is clear that this category encompasses a wide range of topics. Some news outlets cover legal claims made by businesses, while others report yearly benefits or purchasing announcements made by businesses in the financial accounting realm. Consolidation and acquisitions might be justified or unjustifiable. The intriguing question is whether this space assortment has an impact on the opinion classifier's characterization execution. Before we go into this subject, it is important to remember that one of the most important aspects of opinion research is its area of application [11]. That means a single phrase or sentence might convey different points of view in two different contexts [10].

3 Purposed Approach

We took the Twitter dataset and analyzed it in our technique. A feature extraction that uses the uni-gram approach is taken under consideration for analyzing the labeling of data. A system has been taken having a processor that takes raw sentences as input and makes them more understandable. Additionally, different machine learning approaches utilize feature vectors to train the dataset, and semantic analysis gives a large number of equivalents and similarities, which is used to determine the polarity of the data.

3.1 Pre-processing of the Dataset

As in now, people express their emotions in various ways, which contains a lot of views about the data. The dataset which will be utilized in this analysis is already labeled. The dataset is labeled with two sets of polarity which is positive and negative, making the data set simple and easy to use. With polarized raw data, inconsistency and redundancy are usual issues [5]. The quality of the data plays an important role in the findings; hence, pre-processing the data has been taken under consideration which can be useful for quality increase. It is done with the data to eliminate the repeated words and mark of punctuation to enhance the data efficiency. "That painting is Beauuuutifull #", for example, becomes "painting Beautiful" after pre-processing. "@ Geet is Noww Hardworkingg" becomes "Geet is Noww Hardworkingg".

3.2 Extraction of Feature

The revised dataset has a number of distinctive characteristics after pre-processing. Using the approach of feature extraction, different aspects have been derived from the dataset. These aspects is used in determining the polarity of either positive and negative, which is required while using the uni-gram approaches for determining the thoughts of individual [5]. The adjectives are distinguished and extracted using the proposed model that is uni-gram. It simply ignores all the words from the sentences except adjectives. Using the approaches, for example, an adjective, "Beautifull" is extracted from the sentence, i.e., "Painting is beautiful".

3.3 Classification and Training

Supervised learning is a useful technique for tackling the categorization difficulties. That let us use various supervised learning algorithm that help to achieve the preferred outcome for analyzing the sentiment and thoughts [3]. As in this we categorize supervised learning techniques, Naive Bayes, support vector machine, and maximum entropy will be analyzed briefly in upcoming section, followed by an analysis named semantic which will get utilized in conjunction with the supervised learning to compute similarity.

3.4 Naive Bayes

Naive Bayes is one of the simple supervised learning algorithm used in both during training and classification stage [12]. It is said as classifier using probabilistic approach that can learn the sequence of analyzing the set of documents that has been categorized based on label. It matches the content with the list of words to classify the document for their right category.

$$C^* = \text{argmac}_c P_{\text{NB}}(c|d) \tag{1}$$

$$P_{N|B}(c|d) = (p(c) \sum_{i=1}^{m} P(f|c)^{n1|d} / p(d) \tag{2}$$

The Bayesian classification approach offers a helpful framework for comprehending and assessing a variety of learning algorithms. It generates precise probability for hypotheses and is resistant to input data noise.

In the Eq. (2), f represents a feature and (d) represents the count of feature fi found in tweet d. There are a total of m features. Parameters $P(c)$ and $p(f|c)$ are obtained through maximum likelihood estimates, and add-1 smoothing is utilized for unseen features.

3.5 Support Vector Machine

The SVM classification technique analyzes data, determines decision boundaries, and uses kernels to conduct calculations in input space. Two sets of m-dimensional vectors make up the input data. Then, for each item of data represented as a vector, a class is assigned. The goal now is to find a non-documented boundary between two classes. The distance determines the classifier's margin; increasing the margin reduces indecisive decisions [12]. SVM also aids in discovering the aspects that must be examined in order to correctly interpret it, and it supports regression and classification tasks, which become useful in statistical learning theory.

3.6 Semantic Analysis

The support vector machine evaluates the present data, establishes decision boundaries, and uses kernels to compute in input space. Two sets of m-dimensional vectors make up the input data. Then, for each item of data represented as a vector, a class is assigned. The goal now is to find a non-documented boundary between two classes. The distance determines the classifier's margin; increasing the margin reduces indecisive decisions. SVM also aids in discovering the aspects that must be studied in order to fully appreciate it [7], and it facilitates classification and regression, which are useful in statistical learning theory.

3.7 *Long Short-Term Memory, LSTMs*

Different arithmetic operation used like multiplication and addition are used is Long Short-Term Memory (LSTM) for making slight changes in data. LSTMs used a mechanism termed as cell states through which information travel for further procedure. In this way, LSTMs may selectively remember or forget things. The information at a particular cell state is dependent on three main things. In many areas, traditional neural network and RNNs (Recurrent Neural Network) are outperform by the LSTMs [2]. The reason behind this is due to their ability to memorize sequence for longer period of time. The goal of this post is to explain LSTM and show you how to apply it to real-world problems.

4 Result Based on Implementation

The supervised learning algorithm that we have included was classified and trained based on the programming language Python and a Tool kit used by Natural Language. We have considered a large amount of data set for training and validation analysis.

To determine sentiments from text, many symbolic and machine learning approaches are used. Symbolic approaches are more difficult and the time consumption of this technique is more than machine learning techniques. These strategies can be used to analyze Twitter sentiment. When it comes to finding emotional keywords from tweets with numerous keywords, there are a few issues to consider. Misspellings and slang terms are often tough to deal with. To address these challenges, an efficient feature vector is produced by extracting features in two steps following suitable pre-processing [6] (Fig. 1).

The following is a pseudo-code description of the process.

Algorithm 1. Pseudo code of Proposed work

4.1 *Result Analysis*

The first step is to extract and add the Twitter-specific feature to the feature vector. These features are then deleted from the tweets, and the feature extraction is performed as if it were on normal text. The characteristics are also incorporated into the feature vector. Different classifieds, such as Naive Bayes, SVMs (support vector machines), maximum entropy, and ensemble classifiers are used to test the feature vector's classification accuracy. For the new feature vector, all of these classifiers have nearly identical accuracy. This feature vector works well in the sector of electronic products [7]. As for the future scope, we see two possible other outcome, using new technologies like Tachyon to improve the suggested algorithm and implementation by lowering communication costs. Constructing and applying the technique of

```
Input:Labeled Dataset
Output:positive and negative polarity with Synonym of words and similarity between words.

Step-1 Pre-processing the tweets:
        pre-processing()
        Remove URL(Uniform Resource Locator):
        Remove special symbols
        convert to lower:

Step-2 Get the Feature Vector List:
        For w in words:
        Replace two or more words strips:
        if(w in stopwords)
            continue
        Else:
            Append the file
        Return feature vector

Step-3 Extract Feature from Feature vector List:
        For word in feature list
            Features=word in tweets_words
        Return features

Step-4 Combine Pre-processing Dataset and Feature Vector List:
        Pre-processed file = path name of the file
        Stopwords = file path name
        Feature Vector List = file path of feature vector list

Step-5 Training the step 4
        Apply classifier classes

Step-6 Find Synonym and similarity of the feature Vector
        For every sentences in feature list
        Extract feature vector in the tweets()
        For each Feature Vector:x
          For each Feature Vector:y
            Find the similarity(x,y)
            if(similarity>threshold)
             Match Found
             Feature Vector: x = Feature Vector: y
             Classify(x,y)
        Print: Sentiment Polarity with similar feature words.
```

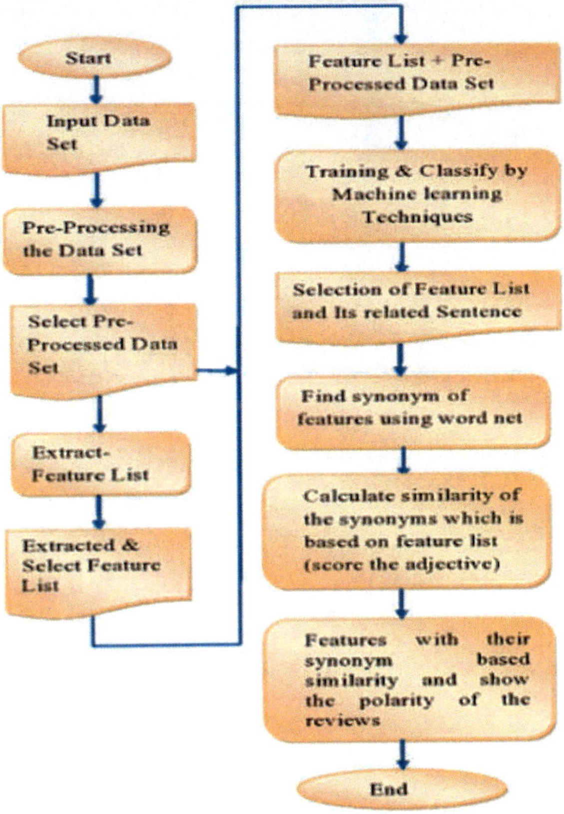

Fig. 1 Work flow of designed model

new linear algebra software libraries which are based on methods implemented for larger-scale matrices, the computational task includes decomposition using singular value, eigenvalue, and eigenvector.

Figure 2 is a graphical representation shows about the word count on the individual comments and frequency of comment based on their word count. It shows people thought and how they express their reaction either in negative or positive view.

Figure 3 shows that a portion of the features are set to zero during training (in this instance, 50% are using Dropout (0.5)). Testing involves utilizing all functionalities (and are scaled appropriately). Therefore, the model is more reliable during test time, which might result in improved testing accuracy.

Figure 4 shows that the dropout, which, as expected, and decrease into the validation loss with respect to the number of epoch.

Table 1 show the number of epoch = 50 with validation loss, accuracy along with loss, and accuracy in training.

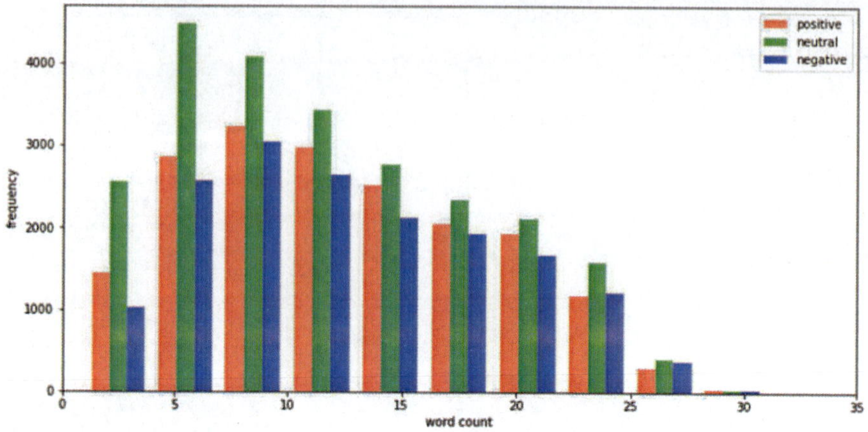

Fig. 2 A graph show the frequency versus word count in sentiment analysis

Fig. 3 A graphical representation of training and validation accuracy

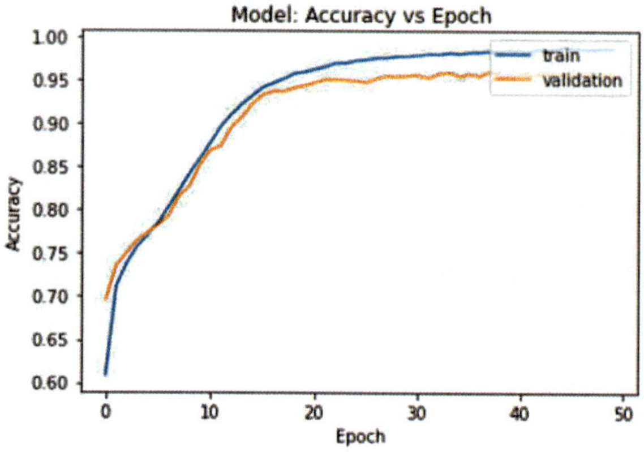

Fig. 4 A graphical representation of loss in training and validation

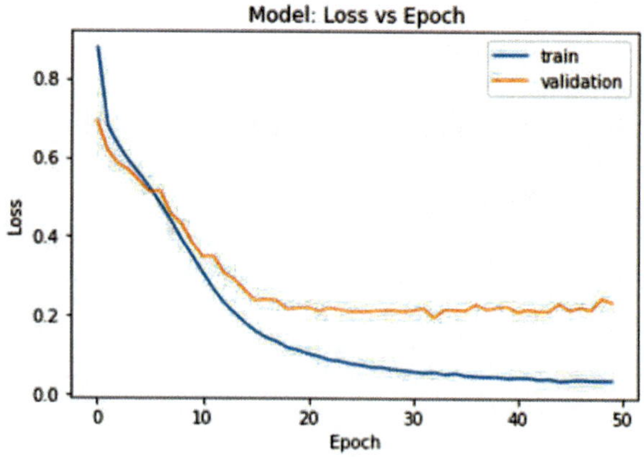

Table 1 Validation loss and accuracy

Epoch	Loss	Validation loss	Validation accuracy	Accuracy
5	0.5590	0.5329	0.7756	0.7582
10	0.516	0.3836	0.8565	0.8601
15	0.1800	0.2569	0.9219	0.9321
20	0.1087	0.2290	0.9456	0.9602
25	0.796	0.2151	0.9511	0.9712
30	0.603	0.2062	0.9521	0.9785
35	0.0495	0.2210	0.9548	0.9835
40	0.0407	0.2200	0.9536	0.9864
45	0.0373	0.2373	0.9550	0.9870
50	0.0363	0.2291	0.9558	0.9881

5 Conclusion

This article introduced a number of machine learning approaches for grouping sentences and semantic analysis in this work. In light of information from Twitter, item surveys are being conducted. The important thing to remember is to use the Twitter dataset, which may split down a large number of surveys which have yet to be named. The procedure for gullible byes, which provides the beneficial outcome over the maximum entropy used, and the uni-gram proposed model uses SVM to produce superior results rather than relying on it solely. The precision is advanced once more. The inquiry followed by semantics related to the WordNet strategy takes longer to get to 89.9% from 88.2%. In making preparations, the informational index can be used to further build the element vector associated sentence recognizable proof, and it can also be used to broaden WordNet for the audits' outline. It may provide a more accurate picture of the chemical, which will be beneficial to clients.

Due to the differences in how people express themselves in these domains, sentiment analysis systems trained on review data are frequently much less accurate when applied to data from other domains, such as news or social media. For example, journalists typically do not express sentiment in the same way that a reviewer does, which is different from how a poster on social media expresses sentiment. As a result, a machine learning system that has been trained on review data will typically not be able to predict sentiment in other domains using the patterns that it has learned to recognize.

6 Future Scope

Sentiment analysis is a particularly useful technique for businesses looking to understand the opinions, attitudes, and feelings of consumers regarding their brand. Businesses and brands have mostly been responsible for the majority of sentiment analysis projects so far, collecting data from social media, survey responses, and other user-generated content hubs. By examining and evaluating customer attitudes, these brands may get a behind-the-scenes look into consumer behavior and better serve their target markets with the goods, services, and experiences they offer.

The goal of sentiment analysis in the future is to go far beyond counting likes, comments, and shares to truly understand the significance of social media interactions and what they reveal about the people using the platforms. Brands will continue to use sentiment analysis, but so will individuals in the public eye, governments, nonprofits, educational institutions, and many other organizations, according to this estimate.

References

1. Agarwal B, Sharma VK, Mittal N (2013) Sentiment classification of review documents using phrase patterns. In: 2013 International conference on advances in computing, communications and informatics (ICACCI). IEEE, pp 1577–1580
2. Ahmed K, El Tazi N, Hossny AH (2015) Sentiment analysis over social networks: an overview. In: 2015 IEEE international conference on systems, man, and cybernetics. IEEE, pp 2174–2179
3. Eliacik AB, Erdoğan N (2015) User-weighted sentiment analysis for financial community on twitter. In: 2015 11th international conference on innovations in information technology (IIT). IEEE, pp 46–51
4. Fang X, Zhan J (2015) Sentiment analysis using product review data. J Big Data 2(1):1–14
5. Gautam G, Yadav D (2014) Sentiment analysis of twitter data using machine learning approaches and semantic analysis. In: 2014 seventh international conference on contemporary computing (IC3). IEEE, pp 437–442
6. Gupta B, Negi M, Vishwakarma K, Rawat G, Badhani P (2017) Study of twitter sentiment analysis using machine learning algorithms on python. Int J Comput Appl 165(9):29–34
7. Jagdale RS, Shirsat VS, Deshmukh SN (2016) Sentiment analysis of events from twitter using open source tool. Int J Comput Sci Mobile Comput 5(4):475–485
8. Joshi R, Tekchandani R (2016) Comparative analysis of twitter data using supervised classifiers. In: 2016 international conference on inventive computation technologies (ICICT), vol 3. IEEE, pp 1–6
9. Kharche SR, Bijole L (2015) Review on sentiment analysis of twitter data. Int J Comput Sci Appl 8
10. Li J, Luong M-T, Jurafsky D, Hovy E (2015) When are tree structures necessary for deep learning of representations? arXiv preprint arXiv:1503.00185
11. Neethu MS, Rajasree R (2013) Sentiment analysis in twitter using machine learning techniques. In: 2013 fourth international conference on computing, communications and networking technologies (ICCCNT). IEEE, pp 1–5
12. Varma R, Jat MG (2022) Analysis of fake news in a twitter data set for a food review

Underwater Image Enhancement and Large Composite Image Stitching of Poompuhar Site

B. Sridevi, S. Akash, A. Prawin, and K. A. Rohith Kumar

Abstract This paper proposes an efficient image stitching method for fusing the underwater images of the Poompuhar site. Underwater images suffer from poor visibility conditions because of medium scattering, light distortion, and inhomogeneous illumination. In our proposed model, the image initially goes through a fusion algorithm based on white balancing algorithm, color balance algorithm, and histogram stretching which greatly improves the visibility and clarity of those images that are to be stitched. In the second phase, the images are first oriented using (SOM) Self-organizing Maps and the features of those images are identified using SURF (Speeded Up Robust Feature) Registration Technique. With the assistance of the RANSAC (Random Sample Consensus) algorithm, finally the enhanced images of those sites are stitched together into a single image using wavelet transform.

Keywords Image enhancement · Fusion algorithm · PSF · Histogram stretching · Underwater image · CLAHE Algorithm

1 Introduction

Photographic stitching of enhanced images of the Poompuhar site is proposed to solve the mystery of the exact location of initial establishment of Poompuhar, its age, later shifts, along with periods, time-series spatial evolution in the present location at the mouth of river Cauvery, and the reasons and periods of its extinction. About 1000 years ago, this port city was swallowed by the sea due to sediment erosion and regular tsunamis. Recently, the Ministry of Technology's ICPS division has launched a project to digitally reconstruct this site, located 30 km from the currently submerged city of Poompuhar. Image stitching is another technology that allows you to combine multiple shots to create a larger image that exceeds the normal aspect ratio and resolution of each camera shot. Photo stitching is the process

B. Sridevi · S. Akash (✉) · A. Prawin · K. A. Rohith Kumar
Electronics and Communication Engineering, Velammal Institute of Technology, Thiruvallur, India
e-mail: akamcoc2000@gmail.com

S. Das et al. (eds.), *Advances in Data-driven Computing and Intelligent Systems*, Lecture Notes in Networks and Systems 653, https://doi.org/10.1007/978-981-99-0981-0_66

of matching the colors of multiple photos and merging them into one. Our role in this project is to restore and enhance the underwater images on this site and stitch them all into one image. Deep Underwater Image Restoration refers to a class of advanced image restoration techniques that employ deep learning models, such as Convolutional Neural Networks(CNNs), to restore the visual quality of degraded underwater images by learning to map low-quality input images to their corresponding high-quality counterparts [1]. A probabilistic representation is used for all aspects of the object: shape, appearance, occlusion, and relative scale. An entropy-based feature detector is used to select regions and their scale within the image [2]. We analyze the contributions and limitations of existing methods to facilitate the comprehensive understanding of underwater image restoration and enhancement [3].

2 Image Stitching

Image stitching is the process of removing overlapping areas and combining multiple images to clearly show higher resolution locations. Building panoramas from photographs were taken with an Uncalibrated Hand-Held Camera [4]. Our robust method corrects exposure disparity even if the two overlapping images are initially misaligned [5]. We present a novel method that utilizes synthesis of multiple homography to warp the images [6].

We can use this algorithm to stitch images of various sizes that are not otherwise supported. Aligning the images in the correct order and angle before merging the images will improve the efficiency of the mosaicked images. Our first contribution is a method for dealing with objects that move between different views of a dynamic scene. If such moving objects are left in, they will appear blurry and "ghosted." Treating such regions as nodes in a graph, we use a vertex cover algorithm to selectively remove all but one instance of each object. Our second contribution is a method for continuously adjusting exposure across multiple images in order to eliminate visible shifts in brightness or hue [7].

The aim is to maximally preserve the visual content while eliminating inconsistencies in the overlap region [8]. It builds on the blending of two images that are directly derived from a color-compensated and white-balanced version of the original degraded image [9].

The SURF and wavelet algorithms use RANSAC (Random Sample Consensus) to remove overlapping areas in the image. A novel method for large-scale image stitching is robust against repetitive patterns and featureless regions in the imagery [10].

3 Proposed Method

Images captured from underwater environment always suffer from color distortion, detail loss, and contrast reduction due to the medium scattering and absorption [11].

- Image acquisition and preprocessing.

The images are captured using underwater cameras. The images are then de-noised and de-hazed orienting the images into proper order and angle increases the efficiency of the mosaicked image. This preprocessing is performed by Self-organizing Map (SOM) algorithm.

A fusion algorithm is proposed for the restoration and enhancement of underwater images. Color balance, contrast optimization, and histogram stretching are carried out. To alleviate the effect of color shift in an underwater image, the scalar values of R, G, and B channels are renewed so that the distributions of the three channels in histogram are similar [12].

- Orient the images using Self-organizing Map.

Employing Self-organizing Maps used in Artificial Neural Networks (ANN) increases the probability of orienting the image in perfect angle. Artificial Neural Network (ANN) is defined as an information processing system that has characteristics resembling human neural tissue. The weights of the BMU and the images close to it in the SOM grid are adjusted toward the input vector. The update formula for an image x with weight vector $Wx(s)$ is,

$$Wv(s+1) = Wv(s) + \theta(u, v, s).\alpha(s).(D(t) - Wv(s))$$

where s is the step index, t is the index of training sample and u is the index of BMU of the input vector.

- Obtain the feature points using SURF registration.

The Speeded Up Robust Features (SURF) is a local feature detector and descriptor which is used for object recognition, image registration, classification, 3D reconstruction, etc.

In SURF method, the image is converted into coordinates using multi-resolution pyramid technique to obtain the same image with reduced bandwidth using Gaussian or Laplacian Pyramid shape. The algorithm has three parts: interest point detection, neighborhood description, and matching. The Euclidean distance between points $p = (p1, p2, \ldots, pn)$ and $q = (q1, q2, \ldots, qn)$ is Euclidean length of this displacement vector:

$$||Q - P|| = \surd(Q1 - P1)2 + (Q2 - P2)2 + \cdots + (QN - PN)2$$

We introduce a new class of distinguished regions based on detecting the most salient convex local arrangements of contours in the image. The regions are used in a similar way to the local interest points extracted from gray-level images, but they capture shape rather than texture [13–15].

- Eliminate the overlapping region RANSAC elimination.

Random Sample Consensus (RANSAC) is technique to estimate parameters of mathematical model from a set of observed data which contains outliers. Moving DLT that is able to tweak or fine-tune the projective warp to accommodate the deviations of the input data from the idealized conditions. This produces as-projective-as-possible image alignment that significantly reduces ghosting without compromising the geometric realism of perspective image stitching [16].

- Perform Wavelet Transform-based image fusion.

First, the images are blended using wavelet transform. Then, the images are decomposed, and then, the images are fused. Fusing the images is the next step. The original image is passed through high pass and low pass filters so as to get the detail an approximate component. Again, down sampling operation takes place followed by the next filtering stage to generate the low-low (LL), low–high (LH), high-low (HL), and high-high (HH) image sub-band components.

4 Deblurring

4.1 *Blind Deconvolution Technique*

Blind deconvolution is a deconvolution used in image processing to reconstruct a target scene from a collection of single or "blurred" images when the point spread function is inadequate or unknown (PSF). The PSF is calculated from the image or image collection of the blind deconvolution and can perform the deconvolution. A novel blind deconvolution scheme is proposed to recover a single handshake blurry image [17].

4.2 *PSF (Point Spread Function)*

The degree to which an optical system blurs (scatters) points of light is measured by the point spread function (PSF). The inverse Fourier transform of the optical transfer function is PSF (OTF). OTF describes the response of a linear position-invariant system to impulses in the frequency domain. The Fourier transform of points is known as OTF (PSF).

5 Histogram Equalization

This technique often improves the overall contrast of a large number of images, especially if the images are represented by a small number of intensity values. This change allows the intensity to be better distributed on the histogram. As a result, contrast may improve where local contrast is low. This is achieved by histogram equalization. We design a piecewise linear function for histogram transform, which is adaptive to the whole RGB value [18]

6 CLAHE Algorithm

In this article, a multicolor space based on Contrast Limited Adaptive Histogram Equalization (CLAHE) improves contrast and restores color to underwater images captured by camera sensors without suffering from inadequate detail or color casts.

The color space conversion from RGB to YIQ is linear, but the color space conversion from RGB to HSI is non-linear. The program then runs CLAHE in both the YIQ and HSI color spaces to generate two different extended images. The luminance component (Y) of the YIQ color space and the intensity component (I) of the HSI color space are emphasized. The number of tiles and clip limits are two parameters used by CLAHE. Adjusting the CLAHE parameter can change the result. A novel method is proposed to improve the performance of the SIFT (Scale–Invariant Feature Transformation) algorithm in adverse illumination conditions (in an outdoor environment at night); for this research, it is proposed to work with CLAHE (Contrast Limited Adaptive Histogram Equalization), adding a preprocessing stage to the traditional methodology of the SIFT algorithm [19].

6.1 CLAHE Steps

i. Scan the indexed color image and put it in the workspace.
ii. Convert an indexed image to a TrueColor (RGB) image, then an RGB image to an L * a * b * color space.
iii. Scale the value to the expected range with the Adapthisteq function [01].
iv. Run CLAHE on the L channel. Scale the result back to the range used in the L * a * b * color space.
v. Convert the resulting image back to RGB color space.
vi. Show the original and processed images.

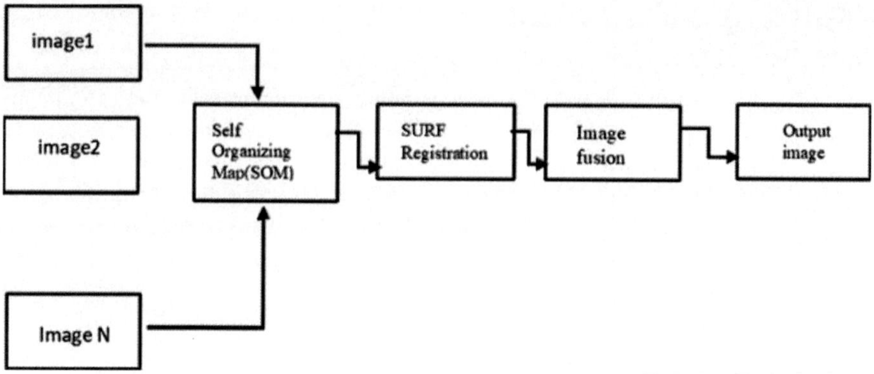

Fig. 1 Block diagram of image stitching

7 Steps in Proposed Method of Image Stitching

Here are the basic steps for merging underwater images to get a natural mosaic-like image as shown in Fig. 1.

1. SOM training
2. Image acquisition and preprocessing
3. Orientation with trained SOM
4. Recognition of feature points and points of interest
5. Elimination of matching feature points
6. Image fusion.

8 RANSAC Algorithm

Random Sample Consensus (RANSAC) is an iterative method for estimating the parameters of a mathematical model from a series of observational data containing outliers. The proposed method builds on random sampling consensus to exploit measurement redundancy and enables the network to determine outlier observations with local communications [20]. This local structure can be described and characterized by a vector of local features measured by local operators such as Gaussian derivatives or Gabor filters [21]. The basic RANSAC algorithm is given as follows:

i. Randomly choose the minimum number of points needed to determine the model parameters.
ii. Solve the parameters of the model.
iii. Determine the number of points that fit the predefined tolerances from all sets of points.

iv. If the ratio of the number of inliers to the total number of points in the set exceeds the predefined threshold τ, use all identified inliers to re-estimate the model parameters and exit.

v. Otherwise, repeat steps 1 to 4 (up to N times).

9 Image Fusion Based on Wavelet Transform

First, blend the images using the wavelet transform. Then, the images are disassembled, and then, the images are merged. In general, wavelet-based schemes perform better than standard schemes, particularly in terms of minimizing color distortion. Schemes that combine standard methods with wavelet transforms produce superior results than either standard methods or simple wavelet-based methods alone [22]. A number of pixel-based image fusion algorithms (using averaging, contrast pyramids, the discrete wavelet transform, and the dual-tree complex wavelet transform (DT-CWT) to perform fusion) are reviewed and compared with a novel region-based image fusion method which facilitates increased flexibility with the definition of a variety of fusion rules [23]. The result of image fusion is a new image which is more suitable for human and machine perception or further image-processing tasks such as segmentation, feature extraction, and object recognition [24].

The basic operations performed by wavelet transform are illustrated in Fig. 2.

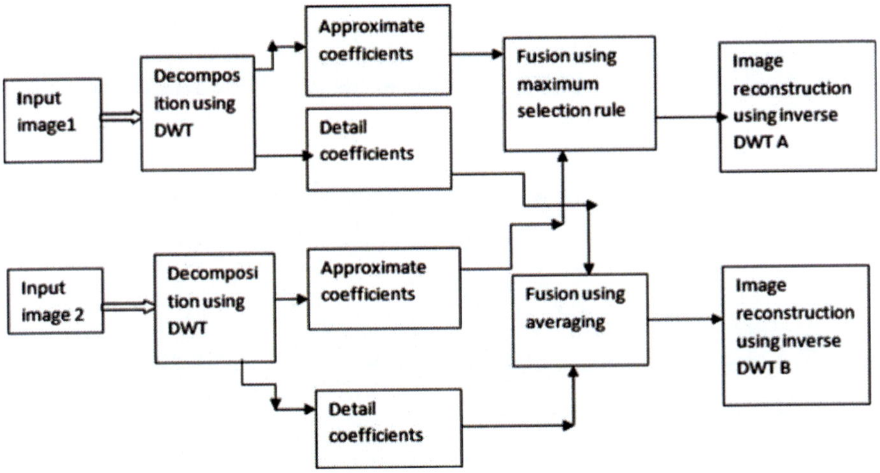

Fig. 2 Wavelet decomposition and reconstruction

10 Maximum Selection Fusion Rule

The Maximum selection Fusion Scheme just selects the largest absolute wavelet coefficient at each location from the input images as the coefficient at that location in the fused image.

The objective of image fusion is to combine relevant information from two or more images of the same scene into a single composite image which is suitable for human and machine perception [25]

11 Result and Conclusion

This paper proposes a fusion algorithm for underwater image restoration and enhancement, in which the deblurring of the underwater images is carried out and foggy images have been treated based on the mentioned right deblurring techniques for the right type of images.

11.1 Deblurring

In our technique, the blurred images taken underwater are deblurred using the blind deconvolution technique where the psf is taken four units above and deblurred and then taking four units below the psf value and deblurred and then the third step is for an initial PSF that is exactly of the same size as the true PSF, the right value of psf is taken and deblurred (refer Fig. 3).

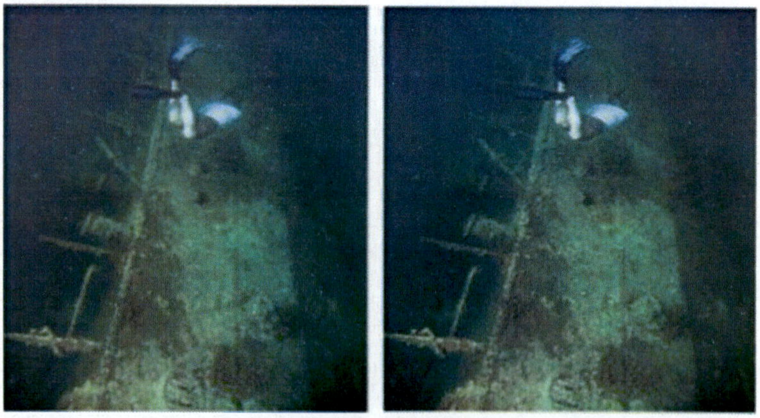

Fig. 3 Input blurred image and output of blind deconvolution image

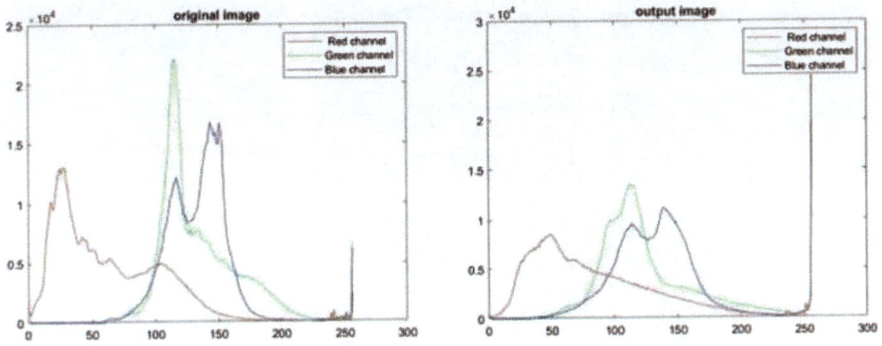

Fig. 4 Histogram of all color formats of the original and enhanced image

The above steps are taken for accuracy and precision purposes. Then, the edge of the image is found using the weight array by the edge function. The algorithm weights each pixel according to the WEIGHT array while restoring the image and the PSF.

11.2 CLAHE

After the deblurring process, a modified histogram approach is proposed to improve the contrast and clarity of underwater photos after dehazing, based on the features of underwater light attenuation.

High contrast and entropy underwater image contrast and entropy can be improved by at least 131.25% and 2.36%, respectively. The contrast and entropy of low contrast, low entropy images can be improved by at least 2495.52% and 24.66%, respectively, same as illustrated in Fig. 4.

11.3 Image Stitching

Figures 5 and 6 show the unstitched images acquired by unmanned vehicles which have overlapping regions.

11.4 Results and Discussion

The histogram of the output image verifies that the input images are stitched seamlessly, and thus, the complete information of a particular region is obtained with more quality.

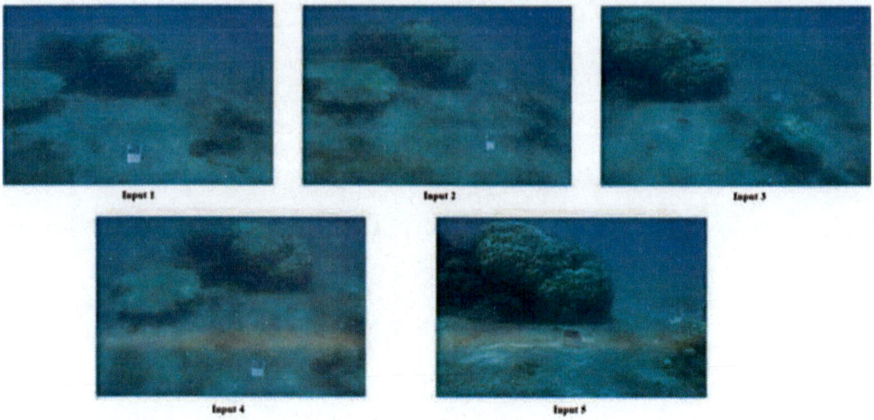

Fig. 5 Input images for stitching

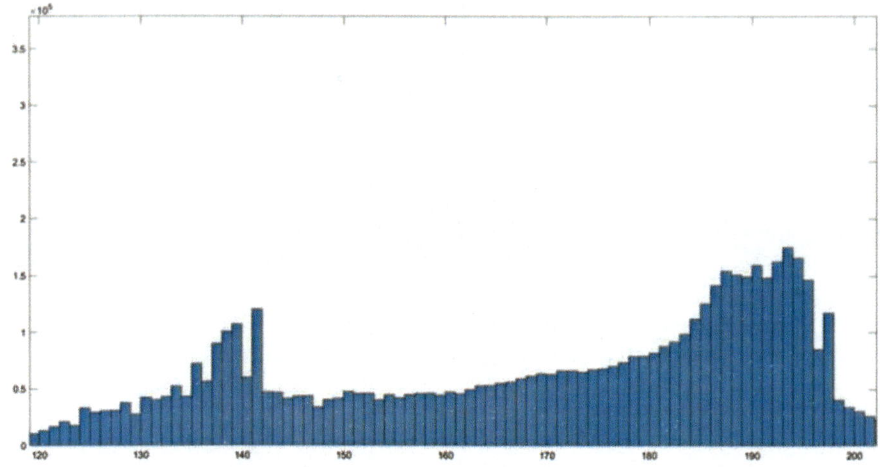

Fig. 6 Histogram for stitched output image

The hsv image has a psnr of 17.1 and a snr of 10.59. It is seen that the output image has a psnr of 15.9 and a snr of 9.43. When comparing the psnr and snr of both the hsv and output images, there is a significant rise in the values of the psnr and snr of the output image. Higher psnr values indicate good output resulting in a better image. According to this study, all algorithms work the same and give similar results in terms of PSNR and execution time, but performance is improved when the type of ambiguity is specified. If you do not know the type of blur, it is difficult to get better results.

As a result, you can use the blind convolution, but the disadvantage is that it takes longer to calculate the PSF of the convolution depending on the blur. After reviewing the literature on numerous new image blur algorithms provided by various scientists, restoring or blurring the average blur in a photo turns out to be a difficult task to solve. As you can see from the analysis above, ASDSAR's method of removing image blur is more accurate and less complicated than other methods. The Wiener filter concludes that performance is poor in the non-blind category compared to other methods with low PSNR (peak signal-to-noise ratio) values, while the LR method with high PSNR values is with other wells. Compare im-execution of approach.

Compared to the non-blind approach, the blind deconvolution method gives the best results. The algorithms described in this study have more powerful improvements than the other algorithms, but it is worth noting that they are slower to process. Real-time performance is not guaranteed with a large number of photos. By comparison, it is suitable for pre-treatment steps.

12 Conclusion

In this project, the problem of obtaining clear view of underwater region is solved in a more efficient way. The underwater image is deblurred using blind deconvolution algorithm. Then, the image is subjected to fusion algorithm of underwater white balancing method, color balancing method, and histogram equalization which enhances the visual appearance of the image.

For stitching multiple images, the images obtained from unmanned vehicles are perfectly aligned; overlapping regions are removed and then fused to obtain a complete view. This SURF and wavelet-based stitching method is more effective than the existing ORB and SIFT algorithms. The result of this image stitching system is found to be highly accurate and reliable.

Acknowledgements This work has been funded by ICPS Division of the Department of Science and Technology of India as a part of the Digital Poompuhar Project.

References

1. Dudhane A, Hambarde P, Patil P, Murala S (2020) Deep underwater image restoration and beyond. 27
2. Fergus R, Perona P, Zisserman A (2003) Object class recognition by unsupervised scale invariant learning. In: IEEE Conference on computer vision and pattern recognition, Madison, Wisconsin, pp 264–271
3. Yang M, Hu J, Li C, Rohde G, Du Y, Hu K (2019) An in-depth survey of underwater image enhancement and restoration
4. Brown M, Lowe DG (2003) Recognizing panoramas. In: Proceedings IEEE international conference on computer vision (ICCV 03), vol 2, pp 1218–1225
5. Jia J, Tang CK (2003) Image registration with global and local luminance alignment. In: Proceedings In: IEEE international conference on computer vision (ICCV03), vol 1, pp 156–163

6. Kim S, Uh Y, Byun H (2012) Generating panorama image by synthesizing multiple homography. In: Proceedings 19th IEEE international conference on image processing (ICIP), pp 2981–2984
7. Uyttendaele M, Eden A, Szeliski R (2001) Eliminating ghosting and exposure artifacts in image mosaics. In: Computer vision and pattern recognition (CVPR'01). IEEE Computer Society, pp 509–516
8. Huang C-M, Lin S-W, Chen J-H (2014) Efficient image stitching of continuous image sequence with image and seam selections, vol 15
9. Ancuti CO, Ancuti C, De Vleeschouwer C, Bekaert P (2018) Color balance and fusion for underwater image enhancement, vol 27
10. Pellikka M, Lahtinen V (2021) A robust method for image stitching. Pattern Anal Appl 24
11. Tao Y, Dong L, Xu W (2020) A novel two-step strategy based on white-balancing and fusion for underwater image enhancement, vol 8
12. Luo W, Duan S, Zheng J (2021) Underwater image restoration and enhancement based on a fusion algorithm with color balance, contrast optimization, and histogram stretching, vol 9
13. Motwani MC, Gadiya MC, Motwani RC, Harris Jr FC (2004) Survey of image denoising techniques. Citeseer
14. Wen L, Li X, Gao L, Zhang Y (2007) A new convolutional neural network based data-driven fault diagnosis method. IEEE Trans Ind Electron, (99):1
15. Jurie F, Schmid C (2015) Scale-invariant shape features for recognition of object categories
16. Zaragoza J, Chin TJ, Tran QH, Brown MS, Suter D (2014) As-projective-as-possible image stitching with moving DLT. IEEE Trans Pattern Anal Mach Intell 36(7):1285–1298
17. Wang W, Zheng J-J, Chen S, Xu S-H, Zhou H-J (2014) Two-stage blind deconvo deconvolution scheme using useful priors. Optik—Int J Light Electron Opt 125:1503–1506
18. Li C, Tang S, Kwan HK, Yan J, Zhou T (2020) Color correction based on CFA and enhancement based on Retinex with dense pixels for underwater images
19. Olvera R, Zeron E, Pedraza Ortega JC, Ramos Arreguin JM, Gorrostieta Hurtado E (2015) A feature extraction using SIFT with a preprocessing by adding CLAHE algorithm to enhance image histograms, 20–25
20. Montijano E, Martinez S, Sagues C (2015) Distributed robustconsensus using RANSAC and dynamic opinions. IEEE Trans Control Syst Technol 23(1):150–163
21. Schiele B, Crowley JL (2000) Recognition without correspondence using multidimensional receptive field histograms. Int J Comput Vision 36(1):31–50
22. Amolins K, Zhang Y, Dare P (2007) Wavelet based image fusion techniques—an introduction, review and comparison. ISPRS J Photogram Remote Sens 62:249–263
23. Lewis JJ, O'Callaghan RJ, Nikolov SG, Bull DR, Canagarajah N (2007) Pixel- and region-based image fusion with complex wavelets. Inf Fusion 8:119–130
24. G. Pajares, J.M. de la Cruz, A wavelet-based image fusion tutorial Pattern Recogn., 37 (September (9)) (2004), pp. 1855–1872.
25. Prakash O, Kumar A, Khare A (2014) Pixel-level image fusion scheme based on steerable pyramid wavelet transform using absolute maximum selection fusion rule, 765–770

Author Index

S. Das et al. (eds.), *Advances in Data-driven Computing and Intelligent Systems*, Lecture Notes in Networks and Systems 653, https://doi.org/10.1007/978-981-99-0981-0